Code of Federal Regulations

CODE OF FEDERAL REGULATIONS

Title 7
Agriculture

Parts 700 to 899

Revised as of January 1, 2022

Containing a codification of documents
of general applicability and future effect

As of January 1, 2022

Published by the Office of the Federal Register
National Archives and Records Administration
as a Special Edition of the Federal Register

T0175192

Table of Contents

Cite this Code: **CFR**

To cite the regulations in this volume use title, part and section number. Thus, 7 CFR 701.1 refers to title 7, part 701, section 1.

Explanation

The Code of Federal Regulations is a codification of the general and permanent rules published in the Federal Register by the Executive departments and agencies of the Federal Government. The Code is divided into 50 titles which represent broad areas subject to Federal regulation. Each title is divided into chapters which usually bear the name of the issuing agency. Each chapter is further subdivided into parts covering specific regulatory areas.

Each volume of the Code is revised at least once each calendar year and issued on a quarterly basis approximately as follows:

Title 1 through Title 16...as of January 1
Title 17 through Title 27 ....as of April 1
Title 28 through Title 41 ....as of July 1
Title 42 through Title 50...as of October 1

The appropriate revision date is printed on the cover of each volume.

LEGAL STATUS

The contents of the Federal Register are required to be judicially noticed (44 U.S.C. 1507). The Code of Federal Regulations is prima facie evidence of the text of the original documents (44 U.S.C. 1510).

HOW TO USE THE CODE OF FEDERAL REGULATIONS

The Code of Federal Regulations is kept up to date by the individual issues of the Federal Register. These two publications must be used together to determine the latest version of any given rule.

To determine whether a Code volume has been amended since its revision date (in this case, January 1, 2022), consult the "List of CFR Sections Affected (LSA)," which is issued monthly, and the "Cumulative List of Parts Affected," which appears in the Reader Aids section of the daily Federal Register. These two lists will identify the Federal Register page number of the latest amendment of any given rule.

EFFECTIVE AND EXPIRATION DATES

Each volume of the Code contains amendments published in the Federal Register since the last revision of that volume of the Code. Source citations for the regulations are referred to by volume number and page number of the Federal Register and date of publication. Publication dates and effective dates are usually not the same and care must be exercised by the user in determining the actual effective date. In instances where the effective date is beyond the cut-off date for the Code a note has been inserted to reflect the future effective date. In those instances where a regulation published in the Federal Register states a date certain for expiration, an appropriate note will be inserted following the text.

OMB CONTROL NUMBERS

The Paperwork Reduction Act of 1980 (Pub. L. 96–511) requires Federal agencies to display an OMB control number with their information collection request.

Many agencies have begun publishing numerous OMB control numbers as amendments to existing regulations in the CFR. These OMB numbers are placed as close as possible to the applicable recordkeeping or reporting requirements.

PAST PROVISIONS OF THE CODE

Provisions of the Code that are no longer in force and effect as of the revision date stated on the cover of each volume are not carried. Code users may find the text of provisions in effect on any given date in the past by using the appropriate List of CFR Sections Affected (LSA). For the convenience of the reader, a "List of CFR Sections Affected" is published at the end of each CFR volume. For changes to the Code prior to the LSA listings at the end of the volume, consult previous annual editions of the LSA. For changes to the Code prior to 2001, consult the List of CFR Sections Affected compilations, published for 1949-1963, 1964-1972, 1973-1985, and 1986-2000.

"[RESERVED]" TERMINOLOGY

The term "[Reserved]" is used as a place holder within the Code of Federal Regulations. An agency may add regulatory information at a "[Reserved]" location at any time. Occasionally "[Reserved]" is used editorially to indicate that a portion of the CFR was left vacant and not dropped in error.

INCORPORATION BY REFERENCE

What is incorporation by reference? Incorporation by reference was established by statute and allows Federal agencies to meet the requirement to publish regulations in the Federal Register by referring to materials already published elsewhere. For an incorporation to be valid, the Director of the Federal Register must approve it. The legal effect of incorporation by reference is that the material is treated as if it were published in full in the Federal Register (5 U.S.C. 552(a)). This material, like any other properly issued regulation, has the force of law.

What is a proper incorporation by reference? The Director of the Federal Register will approve an incorporation by reference only when the requirements of 1 CFR part 51 are met. Some of the elements on which approval is based are:

(a) The incorporation will substantially reduce the volume of material published in the Federal Register.

(b) The matter incorporated is in fact available to the extent necessary to afford fairness and uniformity in the administrative process.

(c) The incorporating document is drafted and submitted for publication in accordance with 1 CFR part 51.

What if the material incorporated by reference cannot be found? If you have any problem locating or obtaining a copy of material listed as an approved incorporation by reference, please contact the agency that issued the regulation containing that incorporation. If, after contacting the agency, you find the material is not available, please notify the Director of the Federal Register, National Archives and Records Administration, 8601 Adelphi Road, College Park, MD 20740-6001, or call 202-741-6010.

CFR INDEXES AND TABULAR GUIDES

A subject index to the Code of Federal Regulations is contained in a separate volume, revised annually as of January 1, entitled CFR INDEX AND FINDING AIDS. This volume contains the Parallel Table of Authorities and Rules. A list of CFR titles, chapters, subchapters, and parts and an alphabetical list of agencies publishing in the CFR are also included in this volume.

An index to the text of "Title 3—The President" is carried within that volume.

The Federal Register Index is issued monthly in cumulative form. This index is based on a consolidation of the "Contents" entries in the daily Federal Register.

A List of CFR Sections Affected (LSA) is published monthly, keyed to the revision dates of the 50 CFR titles.

REPUBLICATION OF MATERIAL

There are no restrictions on the republication of material appearing in the Code of Federal Regulations.

INQUIRIES

For a legal interpretation or explanation of any regulation in this volume, contact the issuing agency. The issuing agency's name appears at the top of odd-numbered pages.

For inquiries concerning CFR reference assistance, call 202–741–6000 or write to the Director, Office of the Federal Register, National Archives and Records Administration, 8601 Adelphi Road, College Park, MD 20740-6001 or e-mail *fedreg.info@nara.gov*.

THIS TITLE

Title 7—AGRICULTURE is composed of fifteen volumes. The parts in these volumes are arranged in the following order: Parts 1-26, 27-52, 53-209, 210-299, 300-399, 400-699, 700-899, 900-999, 1000-1199, 1200-1599, 1600-1759, 1760-1939, 1940-1949, 1950-1999, and part 2000 to end. The contents of these volumes represent all current regulations codified under this title of the CFR as of January 1, 2022.

The Food and Nutrition Service current regulations in the volume containing parts 210-299, include the Child Nutrition Programs and the Food Stamp Program. The regulations of the Federal Crop Insurance Corporation are found in the volume containing parts 400-699.

All marketing agreements and orders for fruits, vegetables and nuts appear in the one volume containing parts 900-999. All marketing agreements and orders for milk appear in the volume containing parts 1000-1199.

For this volume, Robert J. Sheehan, III was Chief Editor. The Code of Federal Regulations publication program is under the direction of John Hyrum Martinez, assisted by Stephen J. Frattini.

Title 7—Agriculture

(This book contains parts 700 to 899)

1

Subtitle B—Regulations of the Department of Agriculture (Continued)

CHAPTER VII—FARM SERVICE AGENCY, DEPARTMENT OF AGRICULTURE

EDITORIAL NOTE: 1. Nomenclature changes to chapter VII appear at 59 FR 60299, Nov. 23, 1994, as corrected at 59 FR 66438, Dec. 27, 1994, and at 60 FR 64297, Dec. 15, 1995.

SUBCHAPTER A—AGRICULTURAL CONSERVATION PROGRAM

SUBCHAPTER A—AGRICULTURAL CONSERVATION PROGRAM

PART 700 [RESERVED]

PART 701—EMERGENCY CONSERVATION PROGRAM, EMERGENCY FOREST RESTORATION PROGRAM, AND CERTAIN RELATED PROGRAMS PREVIOUSLY ADMINISTERED UNDER THIS PART

AUTHORITY: 16 U.S.C. 2201–2206; Sec. 101, Pub. L. 109–148, 119 Stat. 2747; and Pub. L. 111–212, 124 Stat. 2302

SOURCE: 69 FR 10302, Mar. 4, 2004, unless otherwise noted.

Subpart A—General

§ 701.1   Administration.

(a) Subject to the availability of funds, this part provides the terms, conditions and requirements of the Emergency Conservation Program (ECP) and the Emergency Forest Restoration Program (EFRP) administered by the Farm Service Agency (FSA).

Neither program is an entitlement program and payments will only be made to the extent that the Deputy Administrator announces the eligibility of benefits for certain natural disasters, the areas in which such benefits will be available, the time period in which the disaster and the rehabilitation must occur, and only so long as all the conditions for eligibility specified in this part and elsewhere in law are met. However, the Deputy Administrator will not apply any non-statutory limitation on payments provided for in this part in such a way that it would necessarily result in the non-expenditure of program funds required to otherwise be made by law.

(b) ECP and EFRP are administered by the Administrator, FSA through the Deputy Administrator, FSA, and shall be carried out in the field by State and county FSA committees (State and county committees), subject to the availability of funds. Except as otherwise provided in this rule, discretionary determinations to be made under this rule will be made by the Deputy Administrator. Matters committed to the discretion of the Deputy Administrator shall be considered in all cases to be permissive powers and no person or legal entity shall, under any circumstances, be considered to be entitled to an exercise of such power in their favor.

(c) State and county committees, and representatives and employees, do not have authority to modify or waive any regulations in this part.

(d) The State committee may take any action authorized or required of the county committee by this part, but which the county committee has not taken, such as:

(1) Correct or require a county committee to correct any action taken by such county committee that is not in accordance with this part; or

(2) Require a county committee to withhold taking any action that is not in accordance with this part.

(e) No provision or delegation herein to a State or county committee shall preclude the Administrator, FSA, or a designee, from determining any question arising under the program or from reversing or modifying any determination made by a State or county committee.

(f) The Deputy Administrator may authorize State and county committees to waive or modify deadlines and other requirements in cases where lateness or failure to meet such other requirements does not adversely affect the operation of the program.

(g) The Deputy Administrator may limit the authority of state and county committees to approve cost share in excess of specified amounts.

(h) Data furnished by the applicants will be used to determine eligibility for program benefits. Furnishing the data is voluntary; however, the failure to provide data could result in program benefits being withheld or denied.

(i) FSA may consult with any other Federal agency, State agency, or other provider of technical assistance for such assistance as is determined by FSA to be necessary to implement ECP or EFRP. FSA is responsible for the technical aspects of ECP and EFRP but may enter into a Memorandum of Agreement with another party to provide technical assistance. If the requirement for technical assistance results in undue delay or significant hardship to producers in a county, the State committee may request in writing that FSA waive this requirement for that county. However, nothing in this paragraph or in this part creates a right of appeal or action for an applicant with respect to provisions relating to internal procedures of FSA.

(j) The provisions in this part shall not create an entitlement in any person or legal entity to any ECP or EFRP cost share or claim or any particular notice or form or procedure.

(k) Additional terms and conditions may be set forth in the application or the forms participants will be required to sign for participation in the ECP or EFRP.

[69 FR 10302, Mar. 4, 2004, as amended at 75 FR 70087, Nov. 17, 2010]

§ 701.2 Definitions.

(a) The terms defined in part 718 of this chapter shall be applicable to this part and all documents issued in accordance with this part, except as otherwise provided in this section.

(b) The following definitions shall apply to this part:

Agricultural producer means an owner, operator, or tenant of a farm or ranch used to produce for food or fiber, crops (including but not limited to, grain or row crops; seed crops; vegetables or fruits; hay forage or pasture; orchards or vineyards; flowers or bulbs; or field grown ornamentals) or livestock (including but not limited to, dairy or beef cattle; poultry; swine; sheep or goats; fish or other animals raised by aquaculture; other livestock or fowl) for commercial production. Producers of animals raised for recreational uses only are not considered agricultural producers.

Annual agricultural production means production of crops for food or fiber in a commercial operation that occurs on an annual basis under normal conditions.

Applicant means a person or legal entity who has submitted to FSA a request to participate in the ECP or EFRP.

Commercial forest land means forest land with trees intended to be harvested for commercial purposes that has a productivity potential greater than or equal to 20 cubic feet per year of merchantable timber.

Cost-share payment means the payment made by FSA to assist a program participant under this part to establish practices required to address qualifying damage suffered in connection with a qualifying disaster.

Deputy Administrator means the Deputy Administrator for Farm Programs, FSA, the ECP Program Manager, or designee.

Farmland means land devoted to agricultural production, including land used for aquaculture, or other land as may be determined by the Deputy Administrator.

Natural disaster means wildfires, hurricanes or excessive winds, drought, ice storms or blizzards, floods, or other naturally-occurring resource impacting events as determined by FSA. For EFRP, a natural disaster also includes insect or disease infestations as determined by FSA in consultation with other Federal and State agencies as appropriate.

Nonindustrial private forest land means rural commercial forest lands with existing tree cover, or which are suitable for growing trees, that are owned by a private non-industrial forest landowner as defined in this section.

Owners of nonindustrial private forest means, for purposes of the EFRP, an individual, group, association, corporation, Indian Tribe, or other legal private entity owning nonindustrial private forest land or who receives concurrence from the landowner for making the claim in lieu of the owner; and, for practice implementation, the one who holds a lease on the land for a minimum of 10 years. Owners or lessees principally engaged in the primary processing of raw wood products are excluded from this definition. Owners of land leased to lessees who would be excluded under the previous sentence are also excluded.

[69 FR 10302, Mar. 4, 2004, as amended at 75 FR 70087, Nov. 17, 2010; 84 FR 32841, July 10, 2019]

§§ 701.3–701.12 [Reserved]

§ 701.13 Submitting requests.

(a) Subject to the availability of funds, the Deputy Administrator shall provide for an enrollment period for submitting ECP or EFRP cost-share requests.

(b) Requests may be accepted after the announced enrollment period, if such acceptance is approved by the Deputy Administrator and is in accordance with the purposes of the program.

[69 FR 10302, Mar. 4, 2004, as amended at 75 FR 70088, Nov. 17, 2010]

§ 701.14 Onsite inspections.

(a) An onsite inspection must be made before approval of any request for ECP or EFRP assistance.

(b) Notwithstanding paragraph (a) of this section, onsite inspections may be waived by FSA, in its discretion only, where damage is so severe that an onsite inspection is unnecessary, as determined by FSA.

[69 FR 10302, Mar. 4, 2004, as amended at 75 FR 70088, Nov. 17, 2010]

§ 701.15 Starting practices before cost-share request is submitted; non-entitlement to payment; payment subject to the availability of funds.

(a) Subject to paragraphs (b) and (c) of this section, costs will not be shared for practices or components of practices that are started before a request for cost share under this part is submitted with the applicable county FSA office.

(b) Costs may be shared for drought and non-drought practices or components of practices that are started before a request is submitted with the county FSA office, only if:

(1) Considered and approved on a case-by-case basis in accordance with instructions of the Deputy Administrator;

(2) The disaster that is the basis of a claim for cost-share assistance created a situation that required the producer to take immediate action to prevent further losses;

(3) The Deputy Administrator determines that the request for assistance was filed within a reasonable amount of time after the start of the enrollment period; and

(4) The practice was started no more than 60 days before the ECP or EFRP designation was approved for the applicable county office.

(c) Any action taken prior to approval of a claim is taken at the producer's own risk.

(d) An application for relief may be denied for any reason.

(e) All payments under this part are subject to the availability of funds.

[69 FR 10302, Mar. 4, 2004, as amended at 75 FR 70088, Nov. 17, 2010]

§ 701.16 Practice approval.

(a) Requests shall be prioritized before approval based on factors deemed appropriate by FSA, which include, but are not limited to:

(1) Type and degree of damage;

(2) Type of practices needed to address the problem;

(3) Availability of funds;

(4) Availability of technical assistance;

(5) Environmental concerns;

(6) Safety factors; or

(7) In the case of ECP, welfare of eligible livestock.

(b) Requests for cost-share assistance may be approved if:

(1) Funds are available; and

(2) The requested practice is determined eligible.

[69 FR 10302, Mar. 4, 2004, and amended at 75 FR 70088, Nov. 17, 2010]

§§ 701.17–701.20 [Reserved]

§ 701.21 Filing payment application.

Cost-share assistance is conditioned upon the availability of funds and the performance of the practice in compliance with all applicable specifications and program regulations.

(a) *Completion of practice.* After completion of the approved practice, the participant must certify completion and request payment by the payment request deadline. FSA will provide the participant with a form or another manner to be used to request payment.

(b) *Proof of completion.* Participants shall submit to FSA, at the local county office, the information needed to establish the extent of the performance of approved practices and compliance with applicable program provisions.

(c) *Payment request deadline.* The time limits for submission of information shall be determined by the Deputy Administrator. The payment request deadline for each ECP practice will be provided in the agreement after the application is approved. Time limits may be extended where failure to submit required information within the applicable time limits is due to reasons beyond the control of the participant.

§ 701.22 Eligibility to file for cost-share assistance.

Any eligible participant, as defined in this part, who paid part of the cost of an approved practice may file an application for cost-share payment.

§ 701.23 Eligible costs.

(a) Cost-share assistance may be authorized for all reasonable costs incurred in the completion of the practice, up to the maximums provided in §§ 701.126, 701.127, and 701.226.

(b) Eligible costs shall be limited as follows:

(1) Costs for use of personal equipment shall be limited to those incurred

beyond the normal operation of the eligible land.

(2) Costs for personal labor shall be limited to personal labor not normally required in the operation of the eligible land.

(3) Costs for the use of personal equipment and labor must be less than that charged for such equipment and labor by commercial contractors regularly employed in such areas.

(4) Costs shall not exceed those needed to achieve the minimum performance necessary to resolve the problem being corrected by the practice. Any costs above those levels shall not be considered to be eligible costs for purposes of calculations made under this part.

(c) Costs shall not exceed the practice specifications in §701.112(d) or §701.212(d) for cost-share calculations.

(d) The gross amount on which the cost-share eligibility may be computed will not include any costs that were reimbursed by a third party including, but not limited to, an insurance indemnity payment.

(e) Total cost-share payments from all sources shall not exceed the total of eligible costs of the practice to the applicant.

[69 FR 10302, Mar. 4, 2004, and amended at 75 FR 70088, Nov. 17, 2010]

§701.24 Dividing cost-share among more than one participant.

(a) For qualifying cost-share assistance under this part, the cost shall be credited to the participant who personally performed the practice or who paid to have it performed by a third party. If a payment or credit was made by one participant to another potential participant, paragraph (c) of this section shall apply.

(b) If more than one participant contributed to the performance of the practice, the cost-share assistance for the practice shall be divided among those eligible participants in the proportion they contributed to the performance of the practice. FSA may determine what proportion was contributed by each participant by considering the value of the labor, equipment, or material contributed by each participant and any other factors deemed relevant toward performance.

(c) Allowance by a participant of a credit to another participant through adjustment in rent, cash or other consideration, may be considered as a cost of a practice to the paying party only if FSA determines that such credit is directly related to the practice. An applicant who was fully reimbursed shall be considered as not having contributed to the practice performance.

§701.25 Practices carried out with aid from ineligible persons or ineligible legal entities.

Any assistance provided by someone other than the eligible participant, including assistance from a State or Federal agency, shall be deducted from the participant's total costs incurred for the practice for the purpose of computing ECP or EFRP cost shares. If unusual conditions exist, the Deputy Administrator may waive deduction of such contributions upon a request from the State committee and demonstration of the need for such a waiver.

[69 FR 10302, Mar. 4, 2004, as amended at 75 FR 70088, Nov. 17, 2010]

§§701.26–701.30 [Reserved]

§701.31 Maintenance and proper use of practices.

(a) Each participant receiving cost-share assistance is responsible for the required maintenance and proper use of the practice. Some practices have an established life span or minimum period of time during which they are expected to function as a conservation practice with proper maintenance. Cost-share assistance shall not be authorized for normal upkeep or maintenance of any practice.

(b) If a practice is not properly maintained for the established life span, the participant may be required to refund all or part of cost-share assistance received. The Deputy Administrator will determine what constitutes failure to maintain a practice and the amount that must be refunded.

§ 701.32 Failure to comply with program provisions.

Costs may be shared for performance actually rendered even though the minimum requirements otherwise established for a practice have not been satisfied if a reasonable effort was made to satisfy the minimum requirements and if the practice, as performed, will adequately address the need for the practice.

§ 701.33 Death, incompetency, or disappearance.

In case of death, incompetency, or disappearance of any participant, any cost-share payment due shall be paid to the successor, as determined in accordance with part 707 of this chapter.

§ 701.34 Appeals.

Part 11 of this title and parts 614 and 780 of this chapter apply to determinations made under this part.

[69 FR 10302, Mar. 4, 2004, as amended at 75 FR 70088, Nov. 17, 2010]

§ 701.35 Compliance with regulatory measures.

Participants who perform practices shall be responsible for obtaining the authorities, permits, rights, easements, or other approvals necessary to the performance and maintenance of the practices according to applicable laws and regulations. The ECP or EFRP participant shall be wholly responsible for any actions taken with respect to the project and shall, in addition, be responsible for returning and refunding any ECP or EFRP cost shares made, where the purpose of the project cannot be accomplished because of the applicants' lack of clearances or other problems.

[69 FR 10302, Mar. 4, 2004, as amended at 75 FR 70088, Nov. 17, 2010]

§ 701.36 Schemes and devices and claims avoidances.

(a) If FSA determines that a participant has taken any action designed to defeat, or has the effect of defeating, the purposes of this program, the participant shall be required to refund all or part of any of the program payments otherwise due or paid that participant or related person or legal entity for that particular disaster. These actions include, but are not limited to, failure to properly maintain or deliberately destroying a practice and providing false or misleading information related to practices, costs, or arrangements between entities or individuals that would have an effect on any eligibility determination, including, but not limited to, a payment limit eligibility.

(b) All or any part of cost-share assistance that otherwise would be due any participant may be withheld, or required to be refunded, if the participant has adopted, or participated in, any scheme or device designed to evade the maximum cost-share limitation that applies to the program or to evade any other requirement or provision of the program or this part.

(c) If FSA determines that a participant has employed any scheme or device to deprive any other person or legal entity of cost-share assistance, or engaged in any actions to receive payments under this part that also were designed to avoid claims of the United States or its instrumentalities or agents against that party, related parties, or third parties, the participant shall refund all or part of any of those program payments paid to that participant for the project.

(d) For purposes of this section, a scheme or device can include, but is not limited to, instances of coercion, fraud, or misrepresentation regarding the claim for ECP or EFRP assistance and the facts and circumstances surrounding such claim.

(e) A participant who has knowingly supplied false information or filed a false claim shall be ineligible for cost-share assistance related to the disaster for which the false information was filed, or for any period of time FSA deems appropriate. False information or a false claim includes, but is not limited to, a request for payment for a practice not carried out, a false billing, or a billing for practices that do not meet required specifications.

[69 FR 10302, Mar. 4, 2004, and amended at 75 FR 70088, Nov. 17, 2010]

§ 701.37 Loss of control of the property during the practice life span.

In the event of voluntary or involuntary loss of control of the land by the

ECP or EFRP cost-share recipient during the practice life-span, if the person or legal entity acquiring control elects not to become a successor to the ECP or EFRP agreement and the practice is not maintained, each participant who received cost-share assistance for the practice may be jointly and severally liable for refunding any ECP or EFRP .cost-share assistance related to that practice. The practice life span, for purposes of this section, includes any maintenance period that is essential to its success.

[69 FR 10302, Mar. 4, 2004, as amended at 75 FR 70088, Nov. 17, 2010]

§§ 701.38–701.40 [Reserved]

§ 701.41 Cost-share assistance not subject to claims.

Any cost-share assistance or portion thereof due any participant under this part shall be allowed without regard to questions of title under State law, and without regard to any claim or lien against any crop or property, or proceeds thereof, except liens and other claims of the United States or its instrumentalities. The regulations governing offsets and withholdings at parts 792 and 1403 of this title shall be applicable to this program and the provisions most favorable to a collection of the debt shall control.

§ 701.42 Assignments.

Participants may assign ECP cost-share assistance payments, in whole or in part, according to part 1404 of this title.

§ 701.43 Information collection requirements.

Information collection requirements contained in this part have been approved by the Office of Management and Budget under the provisions at 44 U.S.C. Chapter 35 and have been assigned OMB Number 0560–0082.

§ 701.44 Agricultural Conservation Program (ACP) contracts.

Contracts for ACP that are, or were, administered under this part or similar contracts executed in connection with the Interim Environmental Quality Incentives Program, shall, unless the Deputy Administrator determines otherwise, be administered under, and be subject to, the regulations for ACP contracts and the ACP program that were contained in the 7 CFR parts 700 to 899, edition revised as of January 1, 1998, and under the terms of the agreements that were entered into with participants.

[69 FR 10302, Mar. 4, 2004, as amended at 75 FR 70088, Nov. 17, 2010]

§ 701.45 Forestry Incentives Program (FIP) contracts.

The regulations governing the FIP as of July 31, 2002, and contained in the 7 CFR parts 700 to 899, edition revised as of January 1, 2002, shall continue to apply to FIP contracts in effect as of that date, except as provided in accord with a delegation of the administration of that program and such delegation and actions taken thereunder shall apply to any other FIP matters as may be at issue or in dispute.

[69 FR 10302, Mar. 4, 2004, as amended at 75 FR 70088, Nov. 17, 2010]

Subpart B—Emergency Conservation Program

§§ 701.100–701.102 [Reserved]

§ 701.103 Eligible losses, objective, and payments.

(a) FSA will provide cost-share assistance to farmers and ranchers to rehabilitate farmland damaged by wind erosion, floods, hurricanes, wildfire, or other natural disasters as determined by the Deputy Administrator, and to carry out emergency water conservation measures during periods of severe drought, subject to the availability of funds and only for areas, natural disasters, and time periods approved by the Deputy Administrator.

(b) The objective of the ECP is to make cost-share assistance available to eligible participants on eligible land for certain practices, to rehabilitate farmland damaged by floods, hurricanes, wildfire, wind erosion, or other natural disasters, and for the installation of water conservation measures during periods of severe drought.

(c) Payments may also be made under this subpart for:

(1) Emergency water conservation or water enhancement measures (including measures to assist confined livestock) during periods of severe drought; and

(2) Floodplain easements for runoff and other emergency measures that the Deputy Administrator determines is necessary to safeguard life and property from floods, drought, and the products of erosion on any watershed whenever fire, flood, or other natural occurrence is causing or has caused, a sudden impairment of the watershed.

(d) Payments under this part are subject to the availability of appropriated funds and any limitations that may otherwise be provided for by Congress.

[69 FR 10302, Mar. 4, 2004. Redesignated and amended at 75 FR 70088, Nov. 17, 2010; 84 FR 32841, July 10, 2019]

§ 701.104   Producer eligibility.

(a) To be eligible to participate in the ECP the Deputy Administrator must determine that a person or legal entity is an agricultural producer with an interest in the land affected by the natural disaster, and that person or legal entity must be liable for or have paid the expense that is the subject of the cost share. The applicant must be a landowner or user in the area where the qualifying event has occurred, and must be a party who will incur the expense that is the subject of the cost share.

(b) Federal agencies and States, including all agencies and political subdivisions of a State, are ineligible to participate in the ECP.

(c) All producer eligibility is subject to the availability of funds and an application may be denied for any reason.

[69 FR 10302, Mar. 4, 2004. Redesignated and amended at 75 FR 70088, Nov. 17, 2010]

§ 701.105   Land eligibility.

(a) For land to be eligible, the Deputy Administrator must determine that land that is the subject of the cost share:

(1) Will have new conservation problems caused as a result of a natural disaster that, if not treated, would:

(i) Impair or endanger the land;

(ii) Materially affect the productive capacity of the land;

(iii) Represent unusual damage that, except for wind erosion, is not of the type likely to recur frequently in the same area; and

(iv) Be so costly to repair that Federal assistance is or will be required to return the land to productive agricultural use. Conservation problems existing prior to the disaster are not eligible for cost-share assistance.

(2) Be physically located in a county in which the ECP has been implemented; and

(3) Be one of the following:

(i) Land expected to have annual agricultural production,

(ii) A field windbreak or a farmstead shelterbelt on which the ECP practice to be implemented involves removing debris that interferes with normal farming operations on the farm and correcting damage caused by the disaster; or

(iii) A farm access road on which debris interfering with the normal farming operation needs to be removed.

(b) Land is ineligible for cost share if the Deputy Administrator determines that it is, as applicable:

(1) Owned or controlled by the United States;

(2) Owned or controlled by States, including State agencies or other political subdivisions of a State;

(3) Protected by a levee or dike that was not effectively and properly functioning prior to the disaster, or is protected, or intended to be protected, by a levee or dike not built to U.S. Army Corps of Engineers, NRCS, or comparable standards;

(4) Adjacent to water impoundment reservoirs that are subject to inundation when the reservoir is filled to capacity;

(5) Land on which levees or dikes are located;

(6) Subject to frequent damage or susceptible to severe damage according to paragraph (c) of this section;

(7) Subject to flowage or flood easements and inundation when water is released in normal operations;

(8) Between any levee or dike and a stream, river, or body of water, including land between two or more levees or dikes;

(9) Located in an old or new channel of a stream, creek, river or other similar body of water, except that land located within or on the banks of an irrigation canal may be eligible if the Deputy Administrator determines that the canal is not a channel subject to flooding;

(10) In greenhouses or other confined areas, including but not limited to, land in corrals, milking parlors, barn lots, or feeding areas;

(11) Land on which poor farming practices, such as failure to farm on the contour, have materially contributed to damaging the land;

(12) Unless otherwise provided for, not considered to be in annual agricultural production, such as land devoted to stream banks, channels, levees, dikes, native woodland areas, roads, and recreational uses; or

(13) Devoted to trees including, but not limited to, timber production.

(c) To determine the likely frequency of damage and of the susceptibility of the land to severe damage under paragraph (b)(6) of this section, FSA will consider all relevant factors, including, but not limited to, the location of the land, the history of damage to the land, and whether the land was or could have been protected by a functioning levee or dike built to U. S. Army Corps of Engineers, NRCS, or comparable standards. Further, in making such determinations, information may be obtained and used from the Federal Emergency Management Agency or any other Federal, State (including State agencies or political subdivisions), or other entity or individual providing information regarding, for example, flood susceptibility for the land, soil surveys, aerial photographs, or flood plain data or other relevant information.

[69 FR 10302, Mar. 4, 2004. Redesignated at 75 FR 70088, Nov. 17, 2010]

§§ 701.106–701.109 [Reserved]

§ 701.110 Qualifying minimum cost of restoration.

(a) To qualify for assistance under § 701.103(a), the eligible damage must be so costly that Federal assistance is or will be required to return the land to productive agricultural use or to provide emergency water for livestock.

(b) The Deputy Administrator shall establish the minimum qualifying cost of restoration. Each affected State may be allowed to establish a higher minimum qualifying cost of restoration.

(c) A producer may request a waiver of the qualifying minimum cost of restoration. The waiver request shall document how failure to grant the waiver will result in environmental damage or hardship to the producer and how the waiver will accomplish the goals of the program.

[69 FR 10302, Mar. 4, 2004; 69 FR 22377, Apr. 26, 2004. Redesignated and amended at 75 FR 70088, Nov. 17, 2010]

§ 701.111 Prohibition on duplicate payments.

(a) *Duplicate payments.* Participants are not eligible to receive funding under the ECP for land on which the participant has or will receive funding under:

(1) The Wetland Reserve Program (WRP) provided for in 7 CFR part 1467;

(2) The Emergency Wetland Reserve Program (EWRP) provided for in 7 CFR part 623;

(3) The Emergency Watershed Protection Program (EWP), provided for in 7 CFR part 624, for the same or similar expenses.

(4) Any other program that covers the same or similar expenses so as to create duplicate payments, or, in effect, a higher rate of cost share than is allowed under this part.

(b) *Refund.* Participants who receive any duplicate funds, payments, or benefits shall refund any ECP payments received.

[69 FR 10302, Mar. 4, 2004, as amended at 71 FR 30265, May 26, 2006. Redesignated at 75 FR 70088, Nov. 17, 2010]

§ 701.112 Eligible ECP practices.

(a) Cost-share assistance may be offered for ECP practices to replace or restore farmland, fences, or conservation structures to a condition similar to that existing before the natural disaster. No relief under this subpart shall be allowed to address conservation problems existing before the disaster.

(b) The practice or practices made available when the ECP is implemented shall be only those practices authorized by FSA for which cost-share assistance is essential to permit accomplishment of the program goals.

(c) Cost-share assistance may be provided for permanent vegetative cover, including establishment of the cover where needed, only in conjunction with eligible structures or installations where cover is needed to prevent erosion and/or siltation or to accomplish some other ECP purpose.

(d) Practice specifications shall represent the minimum levels of performance needed to address the ECP need.

[69 FR 10302, Mar. 4, 2004. Redesignated and amended at 75 FR 70088, Nov. 17, 2010]

§§ 701.113–701.116 [Reserved]

§ 701.117 Average adjusted gross income limitation.

To be eligible for payments issued from the $16 million provided under the U.S. Troop Readiness, Veterans' Care, Katrina Recovery, and Iraq Accountability Appropriations Act, 2007 (Pub. L. 110–28, section 9003), each applicant must meet the provisions of the Adjusted Gross Income Limitations at 7 CFR part 1400 subpart G.

[72 FR 45880, Aug. 16, 2007. Redesignated at 75 FR 70088, Nov. 17, 2010]

§§ 701.118–701.125 [Reserved]

§ 701.126 Maximum cost-share percentage.

(a) In addition to other restrictions that may be applied by FSA, an ECP participant shall not receive more than 75 percent of the total allowable costs, as determined by this part, to perform the practice.

(b) However, notwithstanding paragraph (a) of this section, a producer who is a limited resource, socially disadvantaged, or beginning farmer or rancher that participates in ECP may receive up to 90 percent of the total allowable costs expended to perform the practice as determined under this part.

(c) In addition to other limitations that apply, in no case will the ECP payment exceed 50 percent of what the Deputy Administrator has determined

is the agricultural value of the affected land.

[69 FR 10302, Mar. 4, 2004. Redesignated and amended at 75 FR 70088, Nov. 17, 2010; 84 FR 32841, July 10, 2019]

§ 701.127 Maximum ECP payments per person or legal entity.

A person or legal entity, as defined in part 1400 of this title, is limited to a maximum ECP cost-share of $500,000 per person or legal entity, per natural disaster.

[75 FR 7088, Nov. 17, 2010, as amended at 84 FR 32841, July 10, 2019]

§ 718.128 Repair or replacement of fencing.

(a) With respect to a payment to an agricultural producer for the repair or replacement of fencing, the agricultural producer has the option of receiving up to 25 percent of the projected payment, determined based on the applicable percentage of the fair market value of the cost of the repair or replacement, as determined by FSA before the agricultural producer carries out the repair or replacement.

(b) If the funds provided under paragraph (a) of this section are not spent by the agricultural producer within 60 calendar days of the date on which the agricultural producer receives those funds, the funds must be returned to FSA by a date determined by FSA.

(c) Payments made under this section are subject to the availability of funds.

[84 FR 32841, July 10, 2019]

§§ 701.129–701.149 [Reserved]

§ 701.150 2005 hurricanes.

In addition benefits elsewhere allowed by this part, claims related to calendar year 2005 hurricane losses may be allowed to the extent provided for in §§ 701.150 through 701.157. Such claims under those sections will be limited to losses in counties that were declared disaster counties by the President or the Secretary because of 2005 hurricanes and to losses to oyster reefs. Claims under §§ 701.151 through 701.157 shall be subject to all normal

ECP limitations and provisions except as explicitly provided in those sections.

[71 FR 30265, May 26, 2006. Redesignated and amended at 75 FR 70088, 70089, Nov. 17, 2010]

§ 701.151 Definitions.

The following definitions apply to §§ 701.152 through 701.157:

Above-ground irrigation facilities means irrigation pipes, sprinklers, pumps, emitters, and any other integral part of the above ground irrigation system.

Barn means a structure used for the housing of animals or farm equipment.

Commercial forest land means forest land with trees intended to be harvested for commercial purposes that has a productivity potential greater than or equal to 20 cubic feet per year of merchantable timber.

Date of loss means the date the hurricane damage occurred in calendar year 2005.

Eligible county means any county that was declared a disaster county by the President or the Secretary because of a calendar year 2005 hurricane, that otherwise meets the eligibility requirements of this part.

Forest management plan means a plan of action and direction on forest lands to achieve a set of results usually specified as goals or objectives consistent with program policies prepared or approved by a natural resource professional, such as a State forestry agency representative.

Poultry house means a building used to house live poultry for the purpose of commercial food production.

Private non-industrial forest land means rural commercial forest lands with existing tree cover, or which are suitable for growing trees, that are owned by a private non-industrial forest landowner as defined in this section.

Private non-industrial forest landowner means, for purposes of the ECP for forestry, an individual, group, association, corporation, Indian tribe, or other legal private entity owning non-industrial private forest land or who receives concurrence from the landowner for making the claim in lieu of the owner, and for practice implementation and who holds a lease on the land for a minimum of 10 years. Owners or lessees principally engaged in the primary processing of raw wood products are excluded from this definition. Owners of land leased to lessees who would be excluded under the previous sentence are also excluded.

Shade house means a metal or wood structure covered by a material used for shade purposes.

[71 FR 30265, May 26, 2006. Redesignated and amended at 75 FR 70088, 70089, Nov. 17, 2010]

§ 701.152 Availability of funding.

Payments under §§ 1701.53 through 701.157 are subject to the availability of funds under Public Law 109–148.

[71 FR 30265, May 26, 2006. Redesignated and amended at 75 FR 70088, Nov. 17, 2010]

§ 701.153 Debris removal and water for livestock.

Subject to the other eligibility provisions of this part, an ECP participant addressing damage in an eligible county from hurricanes during calendar year 2005 may be allowed up to 90 percent of the participant's actual cost or of the total allowable cost for cleaning up structures such as barns, shade houses and above-ground irrigation facilities, for removing poultry house debris, including carcasses, and for providing water for livestock.

[71 FR 30265, May 26, 2006. Redesignated and amended at 75 FR 70088, 70089, Nov. 17, 2010]

§ 701.154 [Reserved]

§ 701.155 Nursery.

(a) Subject to the other eligibility provisions of this part except as provided explicitly in this section, assistance may be made available in an eligible county under this section for the cost of removing nursery debris such as nursery structures, shade houses, and above ground irrigation facilities, where such debris was created in calendar year 2005 by a 2005 hurricane.

(b) Notwithstanding § 701.126, an otherwise eligible ECP participant may be allowed up to 90 percent of the participant's actual cost or of the total allowable cost for losses described in paragraph (a) of this section.

[71 FR 30265, May 26, 2006. Redesignated and amended at 75 FR 70088, 70089, Nov. 17, 2010]

§ 701.156 Poultry.

(a) Subject to the other eligibility provisions of this part except as provided explicitly in this section, assistance may be allowed under this section for uninsured losses in calendar year 2005 to a poultry house in an eligible county due to a 2005 hurricane.

(b) Claimants under this section may be allowed an amount up to the lesser of:

(1) The lesser of 50 percent of the participant's actual or the total allowable cost of the reconstruction or repair of a poultry house, or

(2) $50,000 per poultry house.

(c) The total amount of assistance provided under this section and any indemnities for losses to a poultry house paid to a poultry grower, may not exceed 90 percent of the total costs associated with the reconstruction or repair of a poultry house.

(d) Poultry growers must provide information on insurance payments on their poultry houses. Copies of contracts between growers and poultry integrators may be required.

(e) Assistance under this section is limited to amounts necessary for reconstruction and/or repair of a poultry house to the same size as before the hurricane.

(f) Assistance is limited to poultry houses used to house poultry for commercial enterprises. A commercial poultry enterprise is one with a dedicated structure for poultry and a number of poultry that exceeds actual noncommercial uses of poultry and their products at all times, and from which poultry or related products are actually, and routinely, sold in commercial quantities for food, fiber, or eggs. Unless otherwise approved by FSA, a commercial quantity is a quantity per week that would normally exceed $100 in sales.

(g) Poultry houses with respect to which claims are made under this section must be reconstructed or repaired to meet current building standards.

[71 FR 30265, May 26, 2006. Redesignated at 75 FR 70088, Nov. 17, 2010]

§ 701.157 Private non-industrial forest land.

(a) Subject to the other eligibility provisions of this part except as provided explicitly in this section, assistance made available under this section with respect to private, non-industrial forest land in an eligible county for costs related to reforestations, rehabilitation, and related measures undertaken because of losses in calendar year 2005 caused by a 2005 hurricane. To be eligible, a non-industrial private forest landowner must have suffered a loss of, or damage to, at least 35 percent of forest acres on commercial forest land of the forest landowner in a designated disaster county due to a 2005 hurricane or related condition. The 35 percent loss shall be determined based on the value of the land before and after the hurricane event.

(b) During the 5-year period beginning on the date of the loss, the eligible private non-industrial forest landowner must:

(1) Reforest the eligible damaged forest acres in accordance with a forest management plan approved by FSA that is appropriate for the forest type where the forest management plan is developed by a person or legal entity with appropriate forestry credentials, as determined by the Deputy Administrator;

(2) Use the best management practices included in the forest management plan; and

(3) Exercise good stewardship on the forest land of the landowner while maintaining the land in a forested state.

(c) Notwithstanding § 701.126, an ECP participant shall not receive under this section more than 75 percent of the participant's actual cost or of the total allowable cost of reforestation, rehabilitation, and related measures.

(d) Payments under this section shall not exceed a maximum of $150 per acre for any acre.

(e) Requests will be prioritized based upon planting tree species best suited to the site as stated in the forest management plan.

[71 FR 30265, May 26, 2006. Redesignated and amended at 75 FR 70088, 70089, Nov. 17, 2010]

Subpart C—Emergency Forest Restoration Program

Source: 75 FR 70889, Nov. 17, 2010, unless otherwise noted.

§§ 701.200–701.202 [Reserved]

§ 701.203 Eligible measures, objectives, and assistance.

(a) Subject to the availability of funds and only for areas, natural disasters, and time periods for the natural disaster and rehabilitation approved by the Deputy Administrator, FSA will provide financial assistance to owners of nonindustrial private forest land who carry out emergency measures to restore land damaged by a natural disaster as determined by FSA.

(b) The objective of EFRP is to make financial assistance available to eligible participants on eligible land for certain practices to restore nonindustrial private forest land that has been damaged by a natural disaster.

[75 FR 70889, Nov. 17, 2010, as amended at 84 FR 32841, July 10, 2019]

§ 701.204 Participant eligibility.

(a) To be eligible to participate in EFRP, a person or legal entity must be an owner of nonindustrial private forest land affected by a natural disaster, and must be liable for or have the expense that is the subject of the financial assistance. The owner must be a person or legal entity (including an Indian tribe) with full decision-making authority over the land, as determined by FSA, or with such waivers as may be needed from lenders or others as may be required, to undertake program commitments.

(b) Federal agencies and States, including all agencies and political subdivisions of a State, are ineligible for EFRP.

(c) An application may be denied for any reason.

§ 701.205 Land eligibility.

(a) For land to be eligible, it must be nonindustrial private forest land and must, as determined by FSA:

(1) Have existing tree cover or have had tree cover immediately before the natural disaster and be suitable for growing trees;

(2) Have damage to natural resources caused by a natural disaster that, if not treated, would impair or endanger the natural resources on the land and would materially affect future use of the land; and

(3) Be physically located in a county in which EFRP has been implemented.

(b) Land is ineligible for EFRP if FSA determines that the land is any of the following:

(1) Owned or controlled by the United States; or

(2) Owned or controlled by States, including State agencies or political subdivisions of a State.

[75 FR 70889, Nov. 17, 2010, as amended at 84 FR 32841, July 10, 2019]

§§ 701.206–701.209 [Reserved]

§ 701.210 Qualifying minimum cost of restoration.

(a) FSA will establish the minimum qualifying cost of restoration, which may vary by State or region.

(b) An applicant may request a waiver of the qualifying minimum cost of restoration. The waiver request must document how failure to grant the waiver will result in environmental damage or hardship to the person or legal entity, and how the waiver will accomplish the goals of the program.

§ 701.211 Prohibition on duplicate payments.

(a) Participants are not eligible to receive funding under EFRP for land on which FSA determines that the participant has or will receive funding for the same or similar expenses under:

(1) The Emergency Conservation Program provided for in subpart B of this part;

(2) The Wetland Reserve Program (WRP) provided for in part 1467 of this title;

(3) The Emergency Wetland Reserve Program (EWRP) provided for in part 623 of this chapter;

(4) The Emergency Watershed Protection Program (EWP), provided for in part 624 of this chapter; or

(5) Any other program that covers the same or similar expenses so as to create duplicate payments, or, have the

effect of creating in total, otherwise, a higher rate of financial assistance than is allowed on its own under this part.

(b) Participants who receive any duplicate funds, payments, or benefits must refund any EFRP payments received, except the Deputy Administrator may reduce the refund amount to the amount determined appropriate by the Deputy Administrator to ensure that the total amount of assistance received by the owner of the land under all programs does not exceed an amount otherwise allowed in this part.

§ 701.212 Eligible EFRP practices.

(a) Financial assistance may be offered to eligible persons or legal entities for EFRP practices to restore forest health and forest-related resources on eligible land.

(b) Practice specifications must represent the minimum level of performance needed to restore the land to the applicable FSA, NRCS, Forest Service, or State forestry standard.

§§ 701.213–701.225 [Reserved]

§ 701.226 Maximum financial assistance.

(a) In addition to other restrictions that may be applied by FSA, an EFRP participant will not receive more than 75 percent of the lesser of the participant's total actual cost or of the total allowable costs, as determined by this subpart, to perform the practice.

(b) A person, or legal entity, as defined in part 1400 of this title, is limited to a maximum cost-share of $500,000 per person or legal entity, per natural disaster.

[75 FR 70889, Nov. 17, 2010, as amended at 84 FR 32841, July 10, 2019]

PART 707—PAYMENTS DUE PERSONS WHO HAVE DIED, DISAPPEARED, OR HAVE BEEN DECLARED INCOMPETENT

ment pursuant to the regulations in this part.

AUTHORITY: 7 U.S.C. 1385 and 8786.

SOURCE: 30 FR 6246, May 5, 1965, unless otherwise noted.

§ 707.1 Applicability.

This part applies to all programs in title 7 of the Code of Federal Regulations which are administered by the Farm Service Agency under which payments are made to eligible program participants. This part also applies to all other programs to which this part is applicable by the individual program regulations.

§ 707.2 Definitions.

"Person" when relating to one who dies, disappears, or becomes incompetent, prior to receiving payment, means a person who has earned a payment in whole or in part pursuant to any of the programs to which this part is applicable. "Children" shall include legally adopted children who shall be entitled to share in any payment in the same manner and to the same extent as legitimate children of natural parents. "Brother" or "sister", when relating to one who, pursuant to the regulations in this part, is eligible to apply for the payment which is due a person who dies, disappears, or becomes incompetent prior to the receipt of such payment, shall include brothers and sisters of the half blood who shall be considered the same as brothers and sisters of the whole blood. "Payment" means a payment by draft, check or certificate pursuant to any of the Programs to which this part is applicable. Payments shall not be considered received for the purposes of this part until such draft, check or certificate has been negotiated or used.

§ 707.3 Death.

(a) Where any person who would otherwise be eligible to receive a payment dies before the payment is received, payment may be released in accordance with this section so long as, and only if, a timely program application has been filed by the deceased before the death or filed in a timely way before or after the death by a person legally authorized to act for the deceased. Timeliness will be determined under the relevant program regulations. All program conditions for payment under the relevant program regulations must have been met for the deceased to be considered otherwise eligible for the payment. However, the payment will not be made under this section unless, in addition, a separate release application is filed in accordance with § 707.7. If these conditions are met, payment may be released without regard to the claims of creditors other than the United States, in accordance with the following order of precedence:

(1) To the administrator or executor of the deceased person's estate.

(2) To the surviving spouse, if there is no administrator or executor and none is expected to be appointed, or if an administrator or executor was appointed but the administration of the estate is closed (i) prior to application by the administrator or executor for such payment or (ii) prior to the time when a check, draft, or certificate issued for such payment to the administrator or executor is negotiated or used.

(3) If there is no surviving spouse, to the sons and daughters in equal shares. Children of a deceased son or daughter of a deceased person shall be entitled to their parent's share of the payment, share and share alike. If there are no surviving direct descendants of a deceased son or daughter of such deceased person, the share of the payment which otherwise would have been made to such son or daughter shall be divided equally among the surviving sons and daughters of such deceased person and the estates of any deceased sons or daughters where there are surviving direct descendants.

(4) If there is no surviving spouse and no direct descendant, payment shall be made to the father and mother of the deceased person in equal shares, or the whole thereof to the surviving father or mother.

(5) If there is no surviving spouse, no direct descendant, and no surviving parent, payment shall be made to the brothers and sisters of the deceased person in equal shares. Children of a deceased brother or sister shall be enti-

tled to their parent's share of the payment, share and share alike. If there are no surviving direct descendants of the deceased brother or sister of such deceased person, the share of the payment which otherwise would have been made to such brother or sister shall be divided equally among the surviving brothers and sisters of such deceased person and the estates of any deceased brothers or sisters where there are surviving direct descendants.

(6) If there is no surviving spouse, direct descendant, parent, or brothers or sisters or their descendants, the payment shall be made to the heirs-at-law in accordance with the law of the State of domicile of the deceased person.

(b) If any person who is entitled to payment under the above order of precedence is a minor, payment of his share shall be made to his legal guardian, but if no legal guardian has been appointed payment shall be made to his natural guardian or custodian for his benefit, unless the minor's share of the payment exceeds $1,000, in which event payment shall be made only to his legal guardian.

(c) Any payment which the deceased person could have received may be made jointly to the persons found to be entitled to such payment or shares thereof under this section or, pursuant to instructions issued by the Farm Service Agency, a separate payment may be issued to each person entitled to share in such payment.

[30 FR 6246, May 5, 1965, as amended at 75 FR 81835, Dec. 29, 2010]

§ 707.4 Disappearance.

(a) Where any person who would otherwise be eligible to receive a payment disappears before the payment is received, payment may be released in accordance with this section so long as, and only if, a timely program application has been filed by that person before the disappearance or filed timely before or after the disappearance by someone legally authorized to act for the person involved. Timeliness will be determined under the relevant program regulations. All program conditions for payment under the relevant program regulations must have been met for the person involved to be considered otherwise eligible for the payment. However, the payment will not be made unless, in addition, a separate release application is filed in accordance with § 707.7. If these conditions are met, payment may be released without regard to the claims of creditors other than the United States, in accordance with the following order of precedence:

(1) The conservator or liquidator of his estate, if one be duly appointed.

(2) The spouse.

(3) An adult son or daughter or grandchild for the benefit of his estate.

(4) The mother or father for the benefit of his estate.

(5) An adult brother or sister for the benefit of his estate.

(6) Such person as may be authorized under State law to receive payment for the benefit of his estate.

(b) A person shall be deemed to have disappeared if (1) he has been missing for a period of more than 3 months, (2) a diligent search has failed to reveal his whereabouts, and (3) such person has not communicated during such period with other persons who would be expected to have heard from him. Evidence of such disappearance must be presented to the county committee in the form of a statement executed by the person making the application for payment, setting forth the above facts, and must be substantiated by a statement from a disinterested person who was well acquainted with the person who has disappeared.

[30 FR 6246, May 5, 1965, as amended at 75 FR 81835, Dec. 29, 2010]

§ 707.5 Incompetency.

(a) Where any person who would otherwise be eligible to receive a payment is adjudged incompetent by a court of competent jurisdiction before the payment is received, payment may be released in accordance with this section so long as, and only if, a timely and binding program application has been filed by the person involved while capable or by someone legally authorized to file an application for the person involved. Timeliness is determined under the relevant program regulations. In all cases, the payment application must have been timely under the relevant program regulations and all program conditions for payment must have been met by or on behalf of the person involved. However, the payment will not be made unless, in addition, a separate release application is filed in accordance with § 707.7. If these conditions are met, payment may be released without regard to the claims of creditors other than the United States, to the guardian or committee legally appointed for the person involved. In case no guardian or committee had been appointed, payment, if for not more than $1,000, may be released without regard to claims of creditors other than the United States, to one of the following in the following order for the benefit of the person who was the subject of the adjudication:

(1) The spouse.

(2) An adult son, daughter, or grandchild.

(3) The mother or father.

(4) An adult brother or sister.

(5) Such person as may be authorized under State law to receive payment for the person (see standard procedure prescribed for the respective region).

(b) In case payment is more than $1,000, payment may be released only to such person as may be authorized under State law to receive payment for the incompetent, so long as all conditions for other payments specified in paragraph (a) of this section and elsewhere in the applicable regulations have been met. Those requirements include the filing of a proper and timely and legally authorized program application by or for the person adjudged incompetent. The release of funds under this paragraph will be made without regard to claims of creditors other than the United States unless the agency determines otherwise.

[75 FR 81836, Dec. 30, 2010]

§ 707.6 Death, disappearance, or incompetency of one eligible to apply for payment pursuant to the regulations in this part.

In case any person entitled to apply for a release of a payment pursuant to the provisions of § 707.3, § 707.4, § 707.5, or this section, dies, disappears, or is adjudged incompetent, as the case may be, after he has applied for such payment but before the payment is received, payment may be made upon proper application therefor, without regard to claims of creditors other than the United States, to the person next entitled thereto in accordance with the order of precedence set forth in §707.3, §707.4, or §707.5, as the case may be.

[30 FR 6246, May 5, 1965, as amended at 75 FR 81836, Dec. 29, 2010]

§707.7 Release application.

No payment may be made under this part unless a proper program application was filed in accordance with the rules for the program that generated the payment. That application must have been timely and filed by someone legally authorized to act for the deceased, disappeared, or declared-incompetent person. The filer can be the party that earned the payment themselves—such as the case of a person who filed a program application before they died—or someone legally authorized to act for the party that earned the payment. All program conditions for payment must have been met before the death, disappearance, or incompetency except for the timely filing of the application for payment by the person legally authorized to act for the party earning the payment. But, further, for the payment to be released under the rules of this part, a second application must be filed. That second application is a release application filed under this section. In particular, as to the latter, where all other conditions have been met, persons desiring to claim payment for themselves or an estate in accordance with this part 707 must do so by filing a release application on Form FSA-325, "Application for Payment of Amounts Due Persons Who Have Died, Disappeared or Have Been Declared Incompetent.If the person who died, disappeared, or was declared incompetent did not apply for payment by filing the applicable program application for payment form, such program application for payment must also be filed in accordance with applicable regulations. If the payment is made under the Naval Stores Conservation Program, Part II of the Form FSA-325 shall be executed by the local District Supervisor of the U.S. Forest Service. In connection with applications for payment under all other programs itemized in §707.1, Form FSA-325, and program applications for payments where required, shall be filed with the FSA county office where the person who earned the payment would have been required to file his application.

[30 FR 6246, May 5, 1965, as amended at 75 FR 81836, Dec. 29, 2010]

PART 708—RECORD RETENTION REQUIREMENTS—ALL PROGRAMS

AUTHORITY: Sec. 4, 49 Stat. 164, secs. 7–17, 49 Stat. 1148, as amended; 16 U.S.C. 590d, 590g–590q.

§708.1 Record retention period.

For the purposes of the programs in this chapter, no receipt, invoice, or other record required to be retained by any agricultural producer as evidence tending to show performance of a practice under any such program needs to be retained by such producer more than two years following the close of the program year of the program.

[25 FR 105, Jan. 7, 1960. Redesignated at 26 FR 5788, June 29, 1961]

SUBCHAPTER B—FARM MARKETING QUOTAS, ACREAGE ALLOTMENTS, AND PRODUCTION ADJUSTMENT

PART 714—REFUNDS OF PENALTIES ERRONEOUSLY, ILLEGALLY, OR WRONGFULLY COLLECTED

AUTHORITY: Secs. 372, 375, 52 Stat. 65, as amended, 66, as amended; 7 U.S.C. 1372, 1375.

SOURCE: 35 FR 12098, July 29, 1970, unless otherwise noted.

§ 714.35 Basis, purpose, and applicability.

(a) *Basis and purpose.* The regulations set forth in this part are issued pursuant to the Agricultural Adjustment Act of 1938, as amended, for the purpose of prescribing the provisions governing refunds of marketing quota penalties erroneously, illegally, or wrongfully collected with respect to all commodities subject to marketing quotas under the Act.

(b) *Applicability.* This part shall apply to claims submitted for refunds of marketing quota penalties erroneously, illegally, or wrongfully collected on all commodities subject to marketing quotas under the Act. It shall not apply to the refund of penalties which are deposited in a special deposit account pursuant to sections 314(b), 346(b), 356(b), or 359 of the Agricultural Adjustment Act of 1938, as amended, or paragraph (3) of Pub. L. 74, 77th Congress, available for the refund of penalties initially collected which are subsequently adjusted downward by action of the county committee, review committee, or appropriate court, until such penalties have been deposited in the general fund of the Treasury of the United States after determination that no downward adjustment in the amount of penalty is warranted. All prior regulations dealing with refunds of penalties which were contained in this part are superseded upon the effective date of the regulations in this part.

§ 714.36 Definitions.

(a) *General terms.* In determining the meaning of the provisions of this part, unless the context indicates otherwise, words imparting the singular include and apply to several persons or things, words imparting the plural include the singular, words imparting the masculine gender include the feminine as well, and words used in the present tense include the future as well as the present. The definitions in part 719 of this chapter shall apply to this part. The provisions of part 720 of this chapter concerning the expiration of time limitations shall apply to this part.

(b) *Other terms applicable to this part.* The following terms shall have the following meanings:

(1) "Act" means the Agricultural Adjustment Act of 1938, and any amendments or supplements thereto.

(2) "Claim" means a written request for refund of penalty.

(3) "Claimant" means a person who makes a claim for refund of penalty as provided in this part.

(4) "County Office" means the office of the Agricultural Stabilization and Conservation County Committee.

(5) "Penalty" means an amount of money collected, including setoff, from or on account of any person with respect to any commodity to which this part is applicable, which has been covered into the general fund of the Treasury of the United States, as provided in section 372(b) of the Act.

(6) "State office" means the office of the Agricultural Stabilization and Conservation State Committee.

§ 714.37 Instructions and forms.

The Deputy Administrator shall cause to be prepared and issued such instructions and forms as are necessary for carrying out the regulations in the part.

§714.38 Who may claim refund.

Claim for refund may be made by:

(a) Any person who was entitled to share in the price or consideration received by the producer with respect to the marketing of a commodity from which a deduction was made for the penalty and bore the burden of such deduction in whole or in part.

(b) Any person who was entitled to share in the commodity or the proceeds thereof, paid the penalty thereon in whole or in part and has not been reimbursed therefor.

(c) Any person who was entitled to share in the commodity or the proceeds thereof and bore the burden of the penalty because he has reimbursed the person who paid such penalty.

(d) Any person who, as buyer, paid the penalty in whole or in part in connection with the purchase of a commodity, was not required to collect or pay such penalty, did not deduct the amount of such penalty from the price paid the producer, and has not been reimbursed therefor.

(e) Any person who paid the penalty in whole or in part as a surety on a bond given to secure the payment of penalties and has not been reimbursed therefor.

(f) Any person who paid the whole or any part of the sum paid as a penalty with respect to a commodity included in a transaction which in fact was not a marketing of such commodity and has not been reimbursed therefor.

§714.39 Manner of filing.

Claim for refund shall be filed in the county office on a form prescribed by the Deputy Administrator. If more than one person is entitled to file a claim, a joint claim may be filed by all such persons. If a separate claim is filed by a person who is a party to a joint claim, such separate claim shall not be approved until the interest of each person involved in the joint claim has been determined.

§714.40 Time of filing.

Claim shall be filed within 2 years after the date payment was made to the Secretary. The date payment was made shall be deemed to be the date such payment was deposited in the general fund of the Treasury as shown on the certificate of deposit on which such payment was scheduled.

§714.41 Statement of claim.

The claim shall show fully the facts constituting the basis of the claim; the name and address of and the amount claimed by every person who bore or bears any part or all of the burden of such penalty; and the reasons why such penalty is claimed to have been erroneously, illegally, or wrongfully collected. It shall be the responsibility of the county committee to determine that any person who executes a claim as agent or fiduciary is properly authorized to act in such capacity. There should be attached to the claim all pertinent documents with respect to the claim or duly authenticated copies thereof.

§714.42 Designation of trustee.

Where there is more than one claimant and all the claimants desire to appoint a trustee to receive and disburse any payment to be made to them with respect to the claim, they shall be permitted to appoint a trustee. The person designated as trustee shall execute the declaration of trust.

§714.43 Recommendation by county committee.

Immediately upon receipt of a claim, the date of receipt shall be recorded on the face thereof. The county committee shall determine, on the basis of all available information, if the data and representations on the claim are correct. The county committee shall recommend approval or disapproval of the claim, and attach a statement to the claim, signed by a member of the committee, giving the reasons for their action. After the recommendation of approval or disapproval is made by the county committee, the claim shall be promptly sent to the State committee.

§714.44 Recommendation by State committee.

A representative of the State committee shall review each claim referred by the county committee. If a claim is sent initially to the State committee, it shall be referred to the appropriate county committee for recommendation as provided in §714.43 prior to action being taken by the State committee. Any necessary investigation shall be made. The State committee shall recommend approval or disapproval of the claim, attaching a statement giving the reasons for their action, which shall be signed by a representative of the State committee. After recommending approval or disapproval, the claim shall be promptly sent to the Deputy Administrator.

§714.45   Approval by Deputy Administrator.

The Deputy Administrator shall review each claim forwarded to him by the State committee to determine whether, (a) the penalty was erroneously, illegally, or wrongfully collected, (b) the claimant bore the burden of the payment of the penalty, (c) the claim was timely filed, and (d) under the applicable law and regulations the claimant is entitled to a refund. If a claim is filed initially with the Deputy Administrator, he shall obtain the recommendations of the county committee and the State committee if he deems such action necessary in arriving at a proper determination of the claim. The claimant shall be advised in writing of the action taken by the Deputy Administrator. If disapproved, the claimant shall be notified with an explanation of the reasons for such disapproval.

§714.46   Certification for payment.

An officer or employee of the Department of Agriculture authorized to certify public vouchers for payment shall, for and on behalf of the Secretary of Agriculture, certify to the Secretary of the Treasury of the United States for payment all claims for refund which have been approved.

PART 718—PROVISIONS APPLICABLE TO MULTIPLE PROGRAMS

Subpart A—General Provisions

Subpart B—Determination of Acreage and Compliance

Subpart C—Reconstitution of Farms, Allotments, Quotas, and Base Acres

Subpart D—Equitable Relief From Ineligibility

AUTHORITY: 7 U.S.C. 1501–1531, 1921–2008v, 7201–7334, and 15 U.S.C. 714b.

SOURCE: 61 FR 37552, July 18, 1996, unless otherwise noted.

Subpart A—General Provisions

SOURCE: 68 FR 16172, Apr. 3, 2003, unless otherwise noted.

§718.1 Applicability.

(a) This part is applicable to all programs specified in chapters VII and XIV of this title that are administered by the Farm Service Agency (FSA) and to any other programs that adopt this part by reference. This part governs how FSA administers marketing quotas, allotments, base acres, and acreage reports for those programs to which this part applies. The regulations to which this part applies are those that establish procedures for measuring allotments and program eligible acreage, for determining program compliance, farm reconstitutions, application of finality, and equitable relief from compliance or ineligibility.

(b) For all programs, except for those administered under parts 761 through 774 of this chapter:

(1) The provisions of this part will be administered under the general supervision of the Administrator, FSA, and carried out in the field by State and county FSA committees (State and county committees);

(2) State and county committees, and representatives and employees thereof, do not have authority to modify or waive any regulations in this part;

(3) No provisions or delegation herein to a State or county committee will preclude the Administrator, FSA, or a designee, from determining any question arising under the program or from reversing or modifying any determination made by a State or county committee;

(4) The Deputy Administrator, FSA, may authorize State and county committees to waive or modify deadlines and other requirements in cases where lateness or failure to meet such other requirements does not adversely affect the operation of the program.

(c) The programs under parts 761 through 774 will be administered according to the part, or parts, applicable to the specific program.

[72 FR 63284, Nov. 8, 2007, as amended at 80 FR 41994, July 16, 2015]

§718.2 Definitions.

Except as provided in individual parts of chapters VII and XIV of this title, the following terms shall be as defined herein:

Administrative variance (AV) means the amount by which the determined acreage of tobacco may exceed the effective allotment and be considered in compliance with program regulations.

Allotment means an acreage for a commodity allocated to a farm in accordance with the Agricultural Adjustment Act of 1938, as amended.

Allotment crop means any tobacco crop for which acreage allotments are established pursuant to part 723 of this chapter.

Barley means barley that follows the standard planting and harvesting practice of barley for the area in which the barley is grown.

Base acres means, with respect to a covered commodity on a farm, the number of acres in effect on September 30, 2013, as defined in the regulations in part 1412, subpart B, of this title that were in effect on that date, subject to any reallocation, adjustment, or reduction. The term "base acres" includes any generic base acres as specified in part 1412 planted to a covered commodity as specified in part 1412.

Beginning farmer or rancher means a person or legal entity (for legal entities to be considered a beginning farmer or rancher, all members must be related by blood or marriage and all members must be beginning farmers or ranchers) for which both of the following are true for the farmer or rancher:

(1) Has not operated a farm or ranch for more than 10 years; and

(2) Materially and substantially participates in the operation.

CCC means the Commodity Credit Corporation.

Combination means consolidation of two or more farms or parts of farms, having the same operator, into one farm.

Common land unit means the smallest unit of land that has an identifiable border located in one physical location (county), as defined in this part, and all of the following in common:

(1) Owner;

(2) Management;

(3) Cover; and

(4) Where applicable, producer association.

Common ownership unit means a distinguishable parcel of land consisting of one or more tracts of land with the same owners, as determined by FSA.

Constitution means the make-up of the farm before any change is made because of change in ownership or operation.

Contiguous means sharing any part of a boundary but not overlapping.

Contiguous county means a county contiguous to the reference county or counties.

Contiguous county office means the FSA county office that is in a contiguous county.

Controlled environment means, with respect to those crops for which a controlled environment is required or expected to be provided, including but not limited to ornamental nursery, aquaculture (including ornamental fish), and floriculture, as applicable under the particular program, an environment in which everything that can practicably be controlled with structures, facilities, growing media (including but not limited to water, soil, or nutrients) by the producer, is in fact controlled by the producer.

Controlled substances means the term set forth in 21 CFR part 1308.

Corn means field corn or sterile high-sugar corn that follows the standard planting and harvesting practices for corn for the area in which the corn is grown. Popcorn, corn nuts, blue corn, sweet corn, and corn varieties grown for decoration uses are not corn.

County means the county or parish of a state. For Alaska, Puerto Rico and the Virgin Islands, a county shall be an area designated by the State committee with the concurrence of the Deputy Administrator.

County committee means the FSA county committee.

Crop reporting date means the latest date upon which the Administrator, FSA will allow the farm operator, owner, or their agent to submit a crop acreage report in order for the report to be considered timely.

Cropland. (a) Means land which the county committee determines meets any of the following conditions:

(1) Is currently being tilled for the production of a crop for harvest. Land which is seeded by drilling, broadcast or other no-till planting practices shall be considered tilled for cropland definition purposes;

(2) Is not currently tilled, but it can be established that such land has been tilled in a prior year and is suitable for crop production;

(3) Is currently devoted to a one-row or two-row shelter belt planting, orchard, or vineyard;

(4) Is in terraces that, were cropped in the past, even though they are no longer capable of being cropped;

(5) Is in sod waterways or filter strips planted to a perennial cover;

(6) Is preserved as cropland in accordance with part 1410 of this title; or

(7) Is land that has newly been broken out for purposes of being planted to a crop that the producer intends to, and is capable of, carrying through to harvest, using tillage and cultural practices that are consistent with normal practices in the area; provided further that, in the event that such practices are not utilized other than for reasons beyond the producer's control, the cropland determination shall be void retroactive to the time at which the land was broken out.

(b) Land classified as cropland shall be removed from such classification upon a determination by the county committee that the land is:

(1) No longer used for agricultural production;

(2) No longer suitable for production of crops;

(3) Subject to a restrictive easement or contract that prohibits its use for the production of crops unless otherwise authorized by the regulation of this chapter;

(4) No longer preserved as cropland in accordance with the provisions of part 1410 of this title and does not meet the conditions in paragraphs (a)(1) through (a)(6) of this definition; or

(5) Converted to ponds, tanks or trees other than those trees planted in compliance with a Conservation Reserve Program contract executed pursuant to part 1410 of this title, or trees that are used in one-or two-row shelterbelt plantings, or are part of an orchard or vineyard.

Current year means the year for which allotments, quotas, acreages, and bases, or other program determinations are established for that program. For controlled substance violations, the current year is the year of the actual conviction.

Deputy Administrator means Deputy Administrator for Farm Programs, Farm Service Agency, U.S. Department of Agriculture or their designee.

Determination means a decision issued by a State, county or area FSA committee or its employees that affects a participant's status in a program administered by FSA.

Determined acreage means that acreage established by a representative of the Farm Service Agency by use of official acreage, digitizing or planimetering areas on the photograph or other photographic image, or computations from scaled dimensions or ground measurements.

Direct and counter-cyclical program (DCP) cropland means land that currently meets the definition of cropland, land that was devoted to cropland at the time it was enrolled in a production flexibility contract in accordance with part 1413 of this title and continues to be used for agricultural purposes, or land that met the definition of cropland on or after April, 4, 1996, and continues to be used for agricultural purposes and not for non-agricultural commercial or industrial use.

Division means the division of a farm into two or more farms or parts of farms.

Double cropping means, as determined by the Deputy Administrator on a regional basis, consecutive planting of two specific crops that have the capability to be planted and carried to maturity for the intended uses, as reported by the producer, on the same acreage within a 12-month period. To be considered double cropping, the planting of two specific crops must be in an area where such double cropping is considered normal, or could be considered normal, for all growers under normal growing conditions and growers are typically able to repeat the same cycle successfully in a subsequent 12-month period.

Entity means a corporation, joint stock company, association, limited partnership, limited liability partnership, limited liability company, irrevocable trust, estate, charitable organization, or other similar organization, including any such organization participating in the farming operation as a partner in a general partnership, a participant in a joint venture, or a participant in a similar organization.

Extra Long Staple (ELS) Cotton means cotton that follows the standard planting and harvesting practices of the area in which the cotton is grown, and meets all of the following conditions:

(1) American-Pima, Sea Island, Sealand, all other varieties of the Barbandense species of cotton and any hybrid thereof, and any other variety of cotton in which 1 or more of these varieties is predominant; and,

(2) The acreage is grown in a county designated as an ELS county by the Secretary; and,

(3) The production from the acreage is ginned on a roller-type gin.

Family member means an individual to whom a person is related as spouse, lineal ancestor, lineal descendant, or sibling, including:

(1) Great grandparent;

(2) Grandparent;

(3) Parent;

(4) Child, including a legally adopted child;

(5) Grandchild;

(6) Great grandchildren;

(7) Sibling of the family member in the farming operation; and

(8) Spouse of a person listed in paragraphs (1) through (7) of this definition.

Farm means a tract, or tracts, of land that are considered to be a separate operation under the terms of this part provided further that where multiple tracts are to be treated as one farm, the tracts must have the same operator and must also have the same owner except that tracts of land having different owners may be combined if all owners agree to the treatment of the multiple tracts as one farm for these purposes.

Farm inspection means an inspection by an authorized FSA representative using aerial or ground compliance to determine the extent of producer adherence to program requirements.

Farm number means a number assigned to a farm by the county committee for the purpose of identification.

Farmland means the sum of the DCP cropland, forest, acreage planted to an eligible crop acreage as specified in 1437.3 of this title and other land on the farm.

Field means a part of a farm which is separated from the balance of the farm by permanent boundaries such as fences, permanent waterways, woodlands, and croplines in cases where farming practices make it probable that such cropline is not subject to change, or other similar features.

GIS means Geographic Information System or a system that stores, analyzes, and manipulates spatial or geographically referenced data. GIS computes distances and acres using stored data and calculations.

GPS means Global Positioning System or a positioning system using satellites that continuously transmit coded information. The information transmitted from the satellites is interpreted by GPS receivers to precisely identify locations on earth by measuring distance from the satellites.

Grain sorghum means grain sorghum of a feed grain or dual purpose variety (including any cross that, at all stages of growth, having characteristics of a feed grain or dual purpose variety) that follows the standard planting and harvesting practice for grain sorghum for the area in which the grain sorghum was planted. Sweet sorghum is not considered a grain sorghum.

Ground measurement means the distance between 2 points on the ground, obtained by actual use of a chain tape, GPS with a minimum accuracy level as determined by the Deputy Administrator, or other measuring device.

Intended use means for a crop or a commodity, the end use for which it is grown and produced.

Joint operation means a general partnership, joint venture, or other similar business organization.

Landlord means one who rents or leases farmland to another.

Limited resource farmer or rancher means a farmer or rancher who is both of the following:

(1) A person whose direct or indirect gross farm sales do not exceed $176,800 (2014 program year) in each of the 2 calendar years that precede the most immediately preceding complete taxable year before the relevant program year that corresponds to the relevant program year (for example, for the 2014 program year, the two years would be 2011 and 2012), adjusted upwards in later years for any general inflation; and

(2) A person whose total household income was at or below the national poverty level for a family of four in each of the same two previous years referenced in paragraph (1) of this definition. (Limited resource farmer or rancher status can be determined using a Web site available through the Limited Resource Farmer and Rancher Online Self Determination Tool through National Resource and Conservation Service at *http://www.lrftool.sc.egov.usda.gov.*)

(3) For legal entities, the sum of gross sales and household income must be considered for all members.

Measurement service means a measurement of acreage or farm-stored commodities performed by a representative of FSA and paid for by the producer requesting the measurement.

Measurement service after planting means determining a crop or designated acreage after planting but before the farm operator files a report of acreage for the crop.

Measurement service guarantee means a guarantee provided when a producer requests and pays for an authorized FSA representative to measure acreage for FSA and CCC program participation unless the producer takes action to adjust the measured acreage. If the producer has taken no such action, and the measured acreage is later discovered to be incorrect, the acreage determined pursuant to the measurement service will be used for program purposes for that program year.

Minor child means an individual who is under 18 years of age. For the purpose of programs under chapters VII and XIV of this title, State court proceedings conferring majority on an individual under 18 years of age will not change such an individual's status as a minor.

Nonagricultural commercial or industrial use means land that is no longer suitable for producing annual or perennial crops, including conserving uses, or forestry products.

Normal planting period means that period during which the crop is normally planted in the county, or area within the county, with the expectation of producing a normal crop.

Normal row width means the normal distance between rows of the crop in the field, but not less than 30 inches for all crops.

Oats means oats that follows the standard planting and harvesting practice of oats for the area in which the oats are grown.

Operator means an individual, entity, or joint operation who is determined by the FSA county committee to be in control of the farming operations on the farm.

Owner means one who has legal ownership of farmland, including:

(1) Any agency of the Federal Government; however, such agency is not eligible to receive any program payment;

(2) One who is buying farmland under a contract for deed; or

(3) One who has a life-estate in the property.

Partial reconstitution means a reconstitution that is made effective in the current year for some crops, but is not made effective in the current year for other crops. This results in the same farm having two or more farm numbers in one crop year.

Participant means one who participates in, or receives payments or benefits in accordance with any of the programs administered by FSA.

Pasture means land that is used to, or has the potential to, produce food for grazing animals.

Person means an individual, or an individual participating as a member of a joint operation or similar operation, a corporation, joint stock company, association, limited stock company, limited partnership, irrevocable trust, revocable trust together with the grantor of the trust, estate, or charitable organization including any entity participating in the farming operation as a partner in a general partnership, a participant in a joint venture, a grantor of a revocable trust, or a participant in a similar entity, or a State, political subdivision or agency thereof. To be considered a separate person for the purpose of this part, the individual or other legal entity must:

(1) Have a separate and distinct interest in the land or the crop involved;

(2) Exercise separate responsibility for such interest; and

(3) Be responsible for the cost of farming related to such interest from a fund or account separate from that of any other individual or entity.

Physical location means the political county and State determined by FSA for identifying a tract or common land unit, as applicable, under this part. FSA will consider all the DCP cropland within an original tract to be in one single physical location county and State based upon 95 percent or more of the tract's DCP cropland. For DCP cropland that FSA determines lies outside the physical location (county) of the original tract that is 10 acres or more and more than 5 percent of the original tract, FSA will divide that land from the original tract and establish a new tract for that area.

Planted and considered planted (P&CP) means with respect to an acreage amount, the sum of the planted and prevented planted acres on the farm approved by the FSA county committee for a crop. P&CP is limited to initially planted or prevented planted crop acreage, except for crops planted in an FSA approved double-cropping sequence. Subsequently planted crop acreage and replacement crop acreage are not included as P&CP.

Producer means an owner, operator, landlord, tenant, or sharecropper, who shares in the risk of producing a crop and who is entitled to share in the crop available for marketing from the farm, or would have shared had the crop been produced. A producer includes a grower of hybrid seed.

Quota means the pounds allocated to a farm for a commodity in accordance with the Agricultural Adjustment Act of 1938, as amended.

Random inspection means an examination of a farm by an authorized representative of FSA selected as a part of an impartial sample to determine the adherence to program requirements.

Reconstitution means a change in the land constituting a farm as a result of combination or division.

Reported acreage means the acreage reported by the farm operator, farm owner, farm producer, or their agent on a Form prescribed by the FSA.

Required inspection means an examination by an authorized representative of FSA of a farm specifically selected by application of prescribed rules to determine adherence to program requirements or to verify the farm operator's, farm owner's, farm producer, or agent's report.

Rice means rice that follows the standard planting and harvesting practices of the area excluding sweet, glutinous, or candy rice such as Mochi Gomi.

Secretary means the Secretary of Agriculture of the United States, or a designee.

Sharecropper means one who performs work in connection with the production of a crop under the supervision of the operator and who receives a share of such crop for its labor.

Skip-row or strip-crop planting means a cultural practice in which strips or rows of the crop are alternated with strips of idle land or another crop.

Socially disadvantaged farmer or rancher means a farmer or rancher who is a member of a socially disadvantaged group whose members have been subjected to racial, ethnic, or gender prejudice because of their identity as members of a group without regard to their individual qualities. Socially disadvantaged groups include the following and no others unless approved in writing by the Deputy Administrator:

(1) American Indians or Alaskan Natives,

(2) Asians or Asian-Americans,

(3) Blacks or African-Americans,

(4) Hispanics or Hispanic-Americans,

(5) Native Hawaiians or other Pacific Islanders, and

(6) Women.

Staking and referencing means determining an acreage before planting by:

(1) Measuring or computing a delineated area from ground measurements and documenting the area measured; and, (2) Staking and referencing the area on the ground.

Standard deduction means an acreage that is excluded from the gross acreage in a field because such acreage is considered as being used for farm equipment turn-areas. Such acreage is established by application of a prescribed percentage of the area planted to the crop in lieu of measuring the turn area.

State committee means the FSA State committee.

Subdivision means a part of a field that is separated from the balance of the field by temporary boundary, such as a cropline which could be easily moved or will likely disappear.

Subsequent crop means a crop following an initial crop that is not in an approved double cropping combination.

Tenant means:

(1) One who rents land from another in consideration of the payment of a specified amount of cash or amount of a commodity; or

(2) One (other than a sharecropper) who rents land from another person in consideration of the payment of a share of the crops or proceeds therefrom.

Tolerance means a prescribed amount within which the reported acreage and/or production may differ from the determined acreage and/or production and still be considered as correctly reported.

Tract means a unit of contiguous land under one ownership located in one physical location (county), as defined in this part, which is operated as a farm, or part of a farm.

Tract combination means the combining of two or more tracts if the tracts have common ownership and are contiguous.

Tract division means the dividing of a tract into two or more tracts because of a change in ownership or operation.

Turn-area means the area across the ends of crop rows which is used for operating equipment necessary to the production of a row crop (also called turn row, headland, or end row).

United States means all 50 States of the United States, the District of Columbia, the Commonwealth of Puerto Rico and any other territory or possession of the United States.

Upland cotton means planted and stub cotton that is not considered extra long staple cotton, and that follows the standard planting and harvesting practices of the area and is produced from other than pure strain varieties of the Barbadense species, any hybrid thereof, or any other variety of cotton in which one or more of these varieties predominate. For program purposes, brown lint cotton is considered upland cotton.

Veteran farmer or rancher means a farmer or rancher who has served in the United States Army, Navy, Marine Corps, Air Force, and Coast Guard, including the reserve components and who:

(1) Has not operated a farm or ranch;

(2) Has operated a farm or ranch for not more than 10 years; or

(3) Is a veteran (as defined as a person who served in the active duty or either active duty for training or inactive duty during which the individual was disabled, and who was discharged or released therefrom under conditions other than dishonorable) who has first obtained status as a veteran during the most recent 10-year period.

Wheat means wheat for feed or dual purpose variety that follows the standard planting and harvesting practice of wheat for the area in which the wheat is grown.

[68 FR 16172, Apr. 3, 2003; 69 FR 250, Jan. 5, 2004, as amended at 79 FR 74571, Dec. 15, 2014; 80 FR 41994, July 16, 2015; 84 FR 45886, Sept. 3, 2019]

§718.3 State committee responsibilities.

(a) The State committee shall, with respect to county committees:

(1) Take any action required of the county committee, which the county committee fails to take in accordance with this part;

(2) Correct or require the county committee to correct any action taken by such committee, which is not in accordance with this part; or

(3) Require the county committee to withhold taking any action which is not in accordance with this part.

(b) The State committee shall submit to the Deputy Administrator requests to deviate from deductions prescribed in §718.109, or the error amount or percentage for refunds of redetermination costs as prescribed in §718.112.

[61 FR 37552, July 18, 1996, as amended at 80 FR 41994, July 16, 2015]

§718.4 Authority for farm entry and providing information.

(a) This section applies to all farms that have a tobacco allotment or quota under part 723 of this chapter and all farms that are currently participating in programs administered by FSA.

(b) A representative of FSA may enter any farm that participates in an FSA or CCC program in order to conduct a farm inspection as defined in this part. A program participant may request that the FSA representative present written authorization for the farm inspection before granting access to the farm. If a farm inspection is not allowed within 30 days of written authorization:

(1) All FSA and CCC program benefits for that farm shall be denied;

(2) The person preventing the farm inspection shall pay all costs associated with the farm inspection;

(3) The entire crop production on the farm will be considered to be in excess of the quota established for the farm; and

(4) For tobacco, the farm operator must furnish proof of disposition of:

(i) All tobacco which is in addition to the production shown on the marketing card issued with respect to such farm; and

(ii) No credit will be given for disposing of excess tobacco other than that identified by a marketing card unless disposed of in the presence of FSA in accordance with §718.109 of this part.

(c) If a program participant refuses to furnish reports or data necessary to determine benefits in accordance with paragraph (a) of this section, or FSA determines that the report or data was erroneously provided through the lack of good faith, all program benefits relating to the report or data requested will be denied.

(d) Program participants requesting program benefits as a beginning farmer or rancher, limited resource farmer or rancher, socially disadvantaged farmer or rancher, or veteran farmer or rancher must provide a certification of their status as a member of one of those groups as required by the applicable program provisions.

[68 FR 16172, Apr. 3, 2003, as amended at 84 FR 45886, Sept. 3, 2019]

§ 718.5   Rule of fractions.

(a) Fractions shall be rounded after completion of the entire associated computation. All mathematical calculations shall be carried to two decimal places beyond the number of decimal places required by the regulations governing each program. In rounding, fractional digits of 49 or less beyond the required number of decimal places shall be dropped; if the fractional digits beyond the required number of decimal places are 50 or more, the figure at the last required decimal place shall be increased by "1" as follows:

Required decimal	Computation	Result
Whole numbers	6.49 (or less)	6
	6.50 (or more)	7
Tenths	7.649 (or less)	7.6
	7.650 (or more)	7.7
Hundredths	8.8449 (or less)	8.84
	8.8450 (or more)	8.85
Thousandths	9.63449 (or less)	9.634
	9.63450 (or more)	9.635
0 thousandths	10.993149 (or less)	10.9931
	10.993150 (or more)	10.9932

(b) The acreage of each field or subdivision computed for tobacco and CCC disaster assistance programs shall be recorded in acres and hundredths of an acre, dropping all thousandths of an acre. The acreage of each field or subdivision computed for crops, except tobacco, shall be recorded in acres and tenths of an acre, rounding all hundredths of an acre to the nearest tenth.

§ 718.6   Controlled substance.

(a) The following terms apply to this section:

(1) *USDA benefit* means the issuance of any grant, contract, loan, or payment by appropriated funds of the United States.

(2) *Person* means an individual.

(b) Notwithstanding any other provision of law, any person convicted under Federal or State law of:

(1) Planting, cultivating, growing, producing, harvesting, or storing a controlled substance in any crop year is ineligible during the crop year of conviction and the four succeeding crop years, for any of the following USDA benefits:

(i) Any payments or benefits under part 1412 of this title;

(ii) Any payments or benefits for losses to crops or livestock covered under disaster programs administered by FSA;

(iii) Any price support loan available in accordance with part 1421 of this title;

(iv) Any price support made under the Commodity Credit Corporation Charter Act;

(v) A farm storage facility loan made under section 4(h) of the Commodity Credit Corporation Charter Act or any other Act;

(vi) Crop Insurance under the Federal Crop Insurance Act;

(vii) A loan made or guaranteed under the Consolidated Farm and Rural Development Act or any other law administered by FSA's Farm Loan Programs.

(2) Possession or trafficking of a controlled substance, is ineligible for any or all USDA benefits:

(i) At the discretion of the court,

(ii) To the extent and for a period of time the court determines.

(c) If a person denied benefits under this section is a shareholder, beneficiary, or member of an entity or joint operation, benefits for which the entity or joint operation is eligible will be reduced, for the appropriate period, by a percentage equal to the total interest of the shareholder, beneficiary, or member.

[72 FR 63284, Nov. 8, 2007, as amended at 84 FR 45886, Sept. 3, 2019]

§ 718.7   Furnishing maps.

(a) A reasonable number, as determined by FSA, of reproductions of photographs, mosaic maps, and other maps will be made available to the owner of a farm, an insurance company reinsured by the Federal Crop Insurance Corporation (FCIC), or a private party contractor performing official duties on behalf of FSA, CCC, and other USDA agencies.

(b) For all others, reproductions will be made available at the rate FSA determines will cover the cost of making such items available.

[80 FR 41994, July 16, 2015]

§718.8 Administrative county and servicing FSA county office.

(a) FSA farm records are maintained in an administrative county determined by FSA. Generally, a farm's administrative county is based on the physical location county of the farm. If all land on the farm is physically located in one physical location county, the farm's records will be administratively located in that physical location county.

(b) In cases where there is no FSA office in the county in which the farm is physically located or where a servicing FSA county office is responsible for more than one administrative county, the farm records will be administratively located as specified in paragraph (a) of this section and with a servicing FSA county office that FSA as designated as responsible for that administrative county.

(c) Farm operators and owners can conduct their farm's business in any FSA county office. FSA's designation of a farm's administrative county is based on where land of the farm is located as specified in paragraph (a) of this section or as might be required under paragraph (b) of this section.

(d) Farm operators and owners can request a change to their servicing FSA county office and that request may necessitate a change to the farm's administrative county as specified in paragraph (a) or (b) of this section. If the requested servicing FSA county office is not responsible for and does not have an administrative county for the physical location of the farm according to paragraphs (a) or (b) of this section and FSA approves the request for change of servicing FSA county office, FSA will designate the administrative county for the farm from those available in the requested servicing FSA county office.

(e) If a county contiguous to the county in which the farm is physically located in the same State does not have a servicing FSA county office, the farm will be administratively located by FSA in a contiguous county in another contiguous State that is convenient to the farm operator and owner. Requests for changes to a farm's servicing FSA county office, which may or may not result in a change to a farm's administrative county under this section, must be submitted to FSA by August 1 of each year for the change to take effect that calendar year.

(f) When land on the farm is physically located in more than one county, the farm will be administered by a servicing FSA county office determined by FSA to be the administrative county responsibility for administration of programs for one or more of the physical counties involved in the farm's constitution. Paragraph (b), (c), or (d) of this section applies if changes occur to the servicing FSA county office and administrative county.

(g) Farm operators and owners cannot request a change to a farm's administrative county. The operator and owner of a farm serviced by an FSA county office responsible for a farm's administrative county can request a change of servicing FSA county office to another FSA servicing county office in the same State by August 1 for the change to take effect that calendar year. Review and approval of any change to the servicing FSA county office is solely at the discretion of FSA. Requests for change in servicing FSA county office, which may or may not result in a change to a farm's administrative county, will be reviewed and approved by county committee if all the following can be determined to apply:

(1) The requested change does not impact the constitution of a farm;

(2) The requested change will not result in increased program eligibility or additional benefits for the farm's producers that would not be earned absent the change in servicing FSA county office and, if applicable, administrative county being made; and

(3) The change is not to circumvent any of the provisions of other program regulations to which this part applies.

(h) The State committee will submit all requests for exceptions from regulations specified in this section to the Deputy Administrator.

[84 FR 45886, Sept. 3, 2019]

§718.9 Signature requirements.

(a) When a program authorized by this chapter or chapter XIV of this title requires the signature of a producer, landowner, landlord, or tenant, then a spouse may sign all such FSA or CCC documents on behalf of the other spouse, except as otherwise specified in this section, unless such other spouse has provided written notification to FSA and CCC that such action is not authorized. The notification must be provided to FSA for each farm.

(b) A spouse may not sign a document on behalf of the other spouse with respect to:

(1) Program document required to be executed in accordance with part 3 of this title;

(2) Easements entered into under part 1410 of this title;

(3) Power of attorney;

(4) Such other program documents as determined by FSA or CCC.

(c) An individual; duly authorized officer of a corporation; duly authorized partner of a partnership; executor or administrator of an estate; trustee of a trust; guardian; or conservator may delegate to another the authority to act on their behalf with respect to FSA and CCC programs administered by USDA service center agencies by execution of a Power of Attorney, or such other form as approved by the Deputy Administrator. FSA and CCC may, at their discretion, allow the delegations of authority by other individuals through use of the Power of Attorney or such other form as approved by the Deputy Administrator.

(d) Notwithstanding another provision of this regulation or any other FSA or CCC regulation in this title, a parent may execute documents on behalf of a minor child unless prohibited by a statute or court order.

(e) Notwithstanding any other provision in this title, an authorized agent of the Bureau of Indian Affairs (BIA) of the United States Department of Interior may sign as agent for landowners with properties affiliated with or under the management or trust of the BIA. For collection purposes, such payments will be considered as being made to the persons who are the beneficiaries of the payment or may, alternatively, be considered as an obligation of all persons on the farm in general. In the event of a need for a refund or other claim may be collected, among other means, by other monies due such persons or the farm.

(f) Documents that were previously acted on and approved by the FSA county office or county committee will not subsequently be determined inadequate or invalid because of the lack of signature authority of any person signing the document on behalf of the applicant or any other individual, entity, general partnership, or joint venture, unless the person signing the program document knowingly and willfully falsified the evidence of signature authority or a signature. However, FSA may require affirmation of the document by those parties deemed appropriate for an affirmation, as determined by the Deputy Administrator. Nothing in this paragraph relieves participants of any other program requirements.

[68 FR 16172, Apr. 3, 2003; 69 FR 250, Jan. 5, 2004, as amended at 80 FR 41995, July 16, 2015]

§718.10   Time limitations.

Whenever the final date prescribed in any of the regulations in this title for the performance of any act falls on a Saturday, Sunday, national holiday, State holiday on which the office of the county or State Farm Service Agency committee having primary cognizance of the action required to be taken is closed, or any other day on which the cognizant office is not open for the transaction of business during normal working hours, the time for taking required action shall be extended to the close of business on the next working day. Or in case the action required to be taken may be performed by mailing, the action shall be considered to be taken within the prescribed period if the mailing is postmarked by midnight of such next working day. Where the action required to be taken is with a prescribed number of days after the mailing of notice, the day of mailing shall be excluded in computing such period of time.

§718.11   Disqualification due to Federal crop insurance violation.

(a) Section 515(h) of the Federal Crop Insurance Act (FCIA) provides that a person who willfully and intentionally provides false or inaccurate information to the Federal Crop Insurance Corporation (FCIC) or to an approved insurance provider with respect to a policy or plan of FCIC insurance, after notice and an opportunity for a hearing on the record, will be subject to one or more of the sanctions described in section 515(h)(3). In section 515(h)(3), the FCIA specifies that in the case of a violation committed by a producer, the producer may be disqualified for a period of up to 5 years from receiving any monetary or non-monetary benefit under a number of programs. The list includes, but is not limited to, benefits under:

(1) The FCIA.

(2) The Agricultural Market Transition Act (7 U.S.C. 7201 *et seq.*), including the Noninsured Crop Disaster Assistance Program under section 196 of that Act (7 U.S.C. 7333).

(3) The Agricultural Act of 1949 (7 U.S.C. 1421 *et seq.*).

(4) The Commodity Credit Corporation Charter Act (15 U.S.C. 714 *et seq.*).

(5) The Agricultural Adjustment Act of 1938 (7 U.S.C. 1281 *et seq.*).

(6) Title XII of the Food Security Act of 1985 (16 U.S.C. 3801 *et seq.*).

(7) The Consolidated Farm and Rural Development Act (7 U.S.C. 1921 *et seq.*).

(8) Any law that provides assistance to a producer of an agricultural commodity affected by a crop loss or a decline in prices of agricultural commodities.

(b) Violation determinations are made by FCIC. However, upon notice from FCIC to FSA that a producer has been found to have committed a violation to which paragraph (a) of this section applies, that person will be ineligible for payments under the programs specified in paragraph (a) of this section that are funded by FSA for the same period of time for which, as determined by FCIC, the producer will be ineligible for crop insurance benefits of the kind referred to in paragraph (a)(1) of this section. Appeals of the determination of ineligibility will be administered under the rules set by FCIC.

(c) Other sanctions may also apply.

[72 FR 63284, Nov. 8, 2007]

Subpart B—Determination of Acreage and Compliance

SOURCE: 68 FR 16176, Apr. 3, 2003, unless otherwise noted.

§718.101 Measurements.

(a) Measurement services include, but are not limited to, measuring land and crop areas, measuring quantities of farm-stored commodities, and appraising the yields of crops in the field when required for program administration purposes. The county committee will provide measurement service if the producer requests such service and pays the cost, except that measurement service is not available and will not be provided to determine total acreage or production of a crop when the request is made:

(1) For acreage, after the established final reporting date for the applicable crop, unless a late filed report is accepted as provided in §718.104; or

(2) After the farm operator has furnished production evidence when required for program administration purposes except as provided in this subpart.

(b) Except for measurements and determinations performed by FSA in accordance with late-filed acreage reports filed in accordance with §718.104, when a producer requests, pays for, and receives written notice that measurement services have been furnished, the measured acreage is guaranteed to be correct and used for all program purposes for the current year even though an error is later discovered in the measurement.

[84 FR 45887, Sept. 3, 2019]

§718.102 Acreage reports.

(a) In order to be eligible for benefits, participants in the programs specified in paragraphs (b)(1) through (b)(6) of this section must submit accurate information annually as required by these provisions.

(b)(1) Participants in programs for which eligibility for benefits is tied to base acres must report the acreage of fruits and vegetables planted for harvest on a farm enrolled in such program;

(2) Participants in the programs governed by parts 1421 and 1427 of this title must report the acreage planted to a commodity for harvest for which a marketing assistance loan or loan deficiency payment is requested;

(3) Participants in the programs governed by part 1410 of this title must report the intended use of land enrolled in such programs;

(4) All participants in the programs governed by part 1437 of this title must report all acreage and intended use of the eligible crop in the country in which the producer has a share;

(5) Participants in the programs governed by part 723 of this chapter and part 1464 of this title must report the acreage planted to tobacco by kind on all farms that have an effective allotment or quota greater than zero;

(6) All participants in the programs governed by parts 1412, 1421, and 1427 of this title must report the use of all cropland on the farm.

(7) All producers reporting acreage as prevented planted acreage or failed acreage must provide documentation that meets the provisions of § 718.103 to the FSA county office where the farm is administered.

(c) The annual acreage reports required in paragraph (a) of this section must be filed with the county committee by the farm operator, farm owner, producer of the crop on the farm, or duly authorized representative by the final reporting date applicable to the crop as established by the Deputy Administrator.

(d) Participants in programs to which this part is applicable must report all crops, in all counties, in which they have an interest. This includes crops on cropland and noncropland, including native or improved grass that will be hayed or grazed.

[68 FR 16176, Apr. 3, 2003, as amended at 71 FR 13741, Mar. 17, 2006; 79 FR 74571, Dec. 15, 2014; 80 FR 41995, July 16, 2015]

§ 718.103 Prevented planted and failed acreage.

(a) Prevented planting is the inability to plant an eligible crop with proper equipment during the planting period as a result of an eligible cause of loss, as determined by CCC. The eligible cause of loss that prevented the planting must have:

(1) Occurred after a previous planting period for the crop;

(2) Occurred before the final planting date for the crop in the applicable crop year or, in the case of multiple plantings, the harvest date of the first planting in the applicable planting period, and

(3) Similarly affected other producers in the area, as determined by CCC.

(b) FSA may approve acreage as "prevented planted acreage" if all other conditions for such approval are met and provided the conditions in paragraphs (b)(1) through (6) of this section are met.

(1) Except as specified in paragraph (b)(2) of this section, producers must report the acreage, on forms specified by FSA, within 15 calendar days after the final planting date determined for the crop by FSA.

(2) If the acreage is reported after the period identified in paragraph (b)(1) of this section, the application must be filed in time to permit:

(i) The county committee or its authorized representative to make a farm visit to verify eligible disaster conditions that prevented the specified acreage or crop from being planted; or

(ii) The county committee or its authorized representative the opportunity to determine, based on visual inspection, that the acreage or crop in question was affected by eligible disaster conditions such as damaging weather or other adverse natural occurrences that prevented the acreage or crop from being planted.

(3) A farm visit to inspect the acreage or crop is required for all late-filed acreage reports where prevented planting credit is sought. Under no circumstance may acreage reported after the 15-day period referenced in paragraph (b)(1) of this section be deemed acceptable unless the criteria in paragraph (b)(2) of this section are met. State and county committees do not have the authority to waive the field inspection and verification provisions for late-filed reports.

(4) All determinations made during field inspections must be documented on each late-filed acreage report, with

results also recorded in county committee minutes to support the documentation.

(5) The acreage must have been prevented from being planted as the result of a natural disaster and not a management decision.

(6) The prevented planted acreage report was approved by the county committee. The county committee may disapprove prevented planted acreage credit if it is not satisfied with the documentation provided.

(c) To receive prevented planted credit for acreage, the producer must show to the satisfaction of FSA that the producer intended to plant the acreage. Documentation supporting such intent includes documents related to field preparation, seed purchase, and any other information that shows the acreage could and would have been planted and harvested absent the natural disaster or eligible cause of loss that prevented the planting.

(d) Prevented planted acreage credit will not be given to crops where the prevented-planted acreage was affected by drought, unless:

(1) On the final planting date for non-irrigated acreage, the area that is prevented from being planted has insufficient soil moisture for germination of seed and progress toward crop maturity because of a prolonged period of dry weather, as determined by CCC; and

(2) Prolonged precipitation deficiencies exceeded the D2 level as determined using the U.S. Drought Monitor; and

(3) Verifiable information is collected from sources whose business or purpose it is to record weather conditions, as determined by CCC, and including but not limited to the local weather reporting stations of the U.S. National Weather Service.

(e) Prevented planted acreage credit under this part applies to irrigated crops where the acreage was prevented from being planted due to a lack of water resulting from drought conditions or contamination by saltwater intrusion of an irrigation supply resulting from drought conditions if there was not a reasonable probability of having adequate water to carry out an irrigation practice.

(f) Acreage ineligible for prevented planting coverage includes, but is not limited to, acreage:

(1) With respect to which the planting history or conservation plans indicate it would remain fallow for crop rotation purposes;

(2) Used for conservation purposes or intended to be or considered to have been left unplanted under any program administered by USDA, including the Conservation Reserve and Wetland Reserve Programs;

(3) Not planted because of a management decision;

(4) Affected by the containment or release of water by any governmental, public, or private dam or reservoir project, if an easement exists on the acreage affected for the containment or release of water;

(5) Where any other person receives a prevented planted payment for any crop for the same crop year, unless the acreage meets all the requirements for double cropping under this part;

(6) Where pasture or other forage crop is in place on the acreage during the time that planting of the crop generally occurs in the area;

(7) Where another crop is planted (previous or subsequent) that does not meet the double cropping definition;

(8) Where any volunteer or cover crop is hayed, grazed, or otherwise harvested on the acreage for the same crop year;

(9) Where there is an inadequate supply of irrigation water beginning on the Federal crop insurance sale closing date for the previous crop year or the Noninsured Crop Disaster Assistance Program (NAP) application closing date for the crop as specified in part 1437 of this title through the final planting date of the current year;

(10) On which a failure or breakdown of irrigation equipment or facilities, unless the failure or breakdown is due to a natural disaster;

(11) That is under quarantine imposed by a county, State, or Federal government agency;

(12) That is affected by chemical or herbicide residue, unless the residue is due to a natural disaster;

(13) That is affected by drifting herbicide;

(14) On which a crop was produced, but the producer was unable to obtain a market for the crop;

(15) Involving a planned planting of a "value loss crop" as that term is defined for NAP as specified in part 1437 of this title, including, but not limited to, Christmas trees, aquaculture, or ornamental nursery, for which NAP assistance is provided under value loss procedure;

(16) For which the claim for prevented planted credit relates to trees or other perennials unless the producer can prove resources were available to plant, grow, and harvest the crop, as applicable;

(17) That is affected by wildlife damage;

(18) Upon which, the reduction in the water supply for irrigation is due to participation in an electricity buy-back program, or the sale of water under a water buy-back or legislative changes regarding water usage, or any other cause which is not a natural disaster; or

(19) That is devoted to non-cropland.

(g) CCC may allow exceptions to acreage ineligible for prevented planting coverage when surface water or ground water is reduced because of a natural disaster (as determined by CCC).

(h) Failed acreage is acreage that was planted with the proper equipment during the planting period but failed as a result of an eligible cause of loss, as determined by CCC.

(i) To be approved by CCC as failed acreage the acreage must have been reported as failed acreage before disposition of the crop, and the acreage must have been planted under normal conditions but failed as the result of a natural disaster and not a management decision. Producers who file a failed acreage report must have the request acted on by the county committee. The county committee will deny the acreage report if it is not satisfied with the documentation provided.

(j) To receive failed acreage credit the producer must show all of the following:

(1) That the acreage was planted under normal conditions using the proper equipment with the intent to harvest the acreage.

(2) Provide documentation that the crop was planted using farming practices consistent for the crop and area, but could not be brought to harvest because of disaster-related conditions.

(k) The eligible cause for failed acreage must have:

(1) Occurred after the crop was planted, and

(2) Before the normal harvest date for the crop in the applicable crop year or in the case of multiple plantings, the harvest date of the first planting in the applicable planting period, and

(3) Other producers in the area were similarly affected as determined by CCC.

(l) Eligible failed acreage will be determined on the basis of the producer planting the crop under normal conditions with the expectation to take the crop to harvest.

(m) Acreage ineligible for failed acreage credit includes, but is not limited to acreage:

(1) Which was planted using methods that could not be considered normal for the area and without the expectation of harvest;

(2) Used for conservation purposes or intended to be or considered to have been un-harvested under any program administered by USDA, including the Conservation Reserve and Wetland Reserve Programs; or

(3) That failed because of a management decision.

[71 FR 13741, Mar. 17, 2006, as amended at 80 FR 41995, July 16, 2015; 84 FR 45887, Sept. 3, 2019]

§ 718.104 Late-filed and revised acreage reports.

(a) Late-filed acreage reports may be accepted after the final reporting date through the crop's immediately subsequent crop year's final reporting date and processed by FSA if both of the following apply:

(1) The crop or identifiable crop residue remains in the field, permitting FSA to verify and determine the acreage and

(2) The crop acreage and common land unit for which the reported crop acreage report is being filed has not already been determined by FSA.

(b) Acreage reports submitted later than the date specified in paragraph (a) of this section will not be processed by FSA and will not be used for program purposes.

(c) The person or legal entity filing a report late must pay the cost of a farm inspection and measurement unless FSA determines that failure to report in a timely manner was beyond the producer's control. The cost of the inspection and measurement is equal to the amount FSA would charge for measurement service; however, FSA's determination of acreage as a result of the inspection and measurement is not considered a paid for measurement service under § 718.101. The acreage measured will be entered as determined acres.

(d) When an acceptable late-filed acreage report is filed in accordance with this section, the reported crop acreage will be entered for the amount that was actually reported to FSA before FSA determined acres, and the determined crop acreage will be entered as it was determined and established by FSA.

(e) Revised acreage reports may be filed to change the acreage reported if:

(1) The acreage has not already been determined by FSA; and

(2) Actual crop or residue is present in the field.

(f) Revised reports will be filed and accepted:

(1) At any time for all crops if the crop or residue still exists in the field for inspection to verify the existence and use made of the crop, the lack of the crop, or a disaster condition affecting the crop; and

(2) If the producer was in compliance with all other program requirements at the reporting date.

[71 FR 13742, Mar. 17, 2006, as amended as 80 FR 41996, July 16, 2015; 84 FR 45887, Sept. 3, 2019]

§ 718.105 Tolerances and adjustments.

(a) Tolerance is the amount by which the determined acreage for a crop may differ from the reported acreage or allotment for the crop and still be considered in compliance with program requirements under §§ 718.102(b)(1), (b)(3) and (b)(5).

(b) Tolerance rules apply to those fields for which a staking and referencing was performed but such acreage was not planted according to those measurements or when a measurement service is not requested for acreage destroyed to meet program requirements.

(c) Tolerance rules do not apply to:

(1) Program requirements of §§ 718.102(b)(2), (b)(4) and (b)(6);

(2) Official fields upon which the entire field is devoted to one crop;

(3) Those fields for which staking and referencing was performed and such acreage was planted according to those measurements; or

(4) The adjusted acreage for farms using measurement after planting which have a determined acreage greater than the marketing quota crop allotment.

(d) If the acreage report for a crop is outside the tolerance for that crop:

(1) FSA may consider the requirements of §§ 718.102 (b)(1), (b)(3) and (b)(5) not to have been met;

(2) Participants may be ineligible for all or a portion of payments or benefits subject to the requirements of §§ 718.102 (b)(1), (b)(3) and (b)(5); and

(3) Participants may be ineligible for all or a portion of payments or benefits under a program that requires accurate crop acreage reports under rules governing the program.

[68 FR 16176, Apr. 3, 2003, as amended at 80 FR 41996, July 16, 2015; 84 FR 45887, Sept. 3, 2019]

§ 718.106 Non-compliance and false acreage reports.

(a) Participants who provide false or inaccurate acreage reports may be ineligible for some or all payments or benefits, subject to the requirements of § 718.102(b)(1) and (3).

(b) [Reserved]

[80 FR 41996, July 16, 2015]

§ 718.107 Acreages.

(a) If an acreage has been established by FSA for an area delineated on an aerial photograph or within a GIS, such acreage will be recognized by the county committee as the acreage for the area until such time as the boundaries of such area are changed. When boundaries not visible on the aerial photograph are established from data furnished by the producer, such acreage shall not be recognized as official acreage until an authorized representative of FSA verifies the boundaries.

(b) Measurements of any row crop shall extend beyond the planted area by the larger of 15 inches or one-half the distance between the rows.

(c) The entire acreage of a field or subdivision of a field devoted to a crop shall be considered as devoted to the crop subject to a deduction or adjustment except as otherwise provided in this part.

§ 718.108 Measuring acreage including skip row acreage.

(a) When one crop is alternating with another crop, whether or not both crops have the same growing season, only the acreage that is actually planted to the crop being measured will be considered to be acreage devoted to the measured crop.

(b) Subject to the provisions of this paragraph and section, whether planted in a skip row pattern or without a pattern of skipped rows, the entire acreage of the field or subdivision may be considered as devoted to the crop only where the distance between the rows, for all rows, is 40 inches or less. If there is a skip that creates idle land wider than 40 inches, or if the distance between any rows is more than 40 inches, then the area planted to the crop shall be considered to be that area which would represent the smaller of; a 40 inch width between rows, or the normal row spacing in the field for all other rows in the field—those that are not more than 40 inches apart. The allowance for individual rows would be made based on the smaller of actual spacing between those rows or the normal spacing in the field. For example, if the crop is planted in single, wide rows that are 48 inches apart, only 20 inches to either side of each row (for a total of 40 inches between the two rows) could, at a maximum, be considered as devoted as the crop and normal spacing in the field would control. Half the normal distance between rows will also be allowed beyond the outside planted rows not to exceed 20 inches and will reflect normal spacing in the field.

(c) In making calculations under this section, further reductions may be made in the acreage considered planted

if it is determined that the acreage is more sparsely planted than normal using reasonable and customary full production planting techniques.

(d) The Deputy Administrator has the discretionary authority to allow row allowances other than those specified in this section in those instances in which crops are normally planted with spacings greater or less than 40 inches, such as in case of tobacco, or where other circumstances are present which the Deputy Administrator finds justifies that allowance.

(e) Paragraphs (a) through (d) of this section shall apply with respect to the 2003 and subsequent crops. For preceding crops, the rules in effect on January 1, 2002, shall apply.

§ 718.109 Deductions.

(a) Any contiguous area which is not devoted to the crop being measured and which is not part of a skip-row pattern under § 718.108 shall be deducted from the acreage of the crop if such area meets the following minimum national standards or requirements:

(1) A minimum width of 30 inches;

(2) For tobacco—three-hundredths (.03) acre. Turn areas, terraces, permanent irrigation and drainage ditches, sod waterways, non-cropland, and subdivision boundaries each of which is at least 30 inches in width may be combined to meet the 0.03-acre minimum requirement; or

(3) For all other crops and land uses—one-tenth (.10) acre. Turn areas, terraces, permanent irrigation and drainage ditches, sod waterways, non-cropland, and subdivision boundaries each of which is at least 30 inches in width and each of which contain 0.1 acre or more may be combined to meet any larger minimum prescribed for a State in accordance with this subpart.

(b) If the area not devoted to the crop is located within the planted area, the part of any perimeter area that is more than 217.8 feet (33 links) in width will be considered to be an internal deduction if the standard deduction is used.

(c) A standard deduction of 3 percent of the area devoted to a row crop and zero percent of the area devoted to a close-sown crop may be used in lieu of measuring the acreage of turn areas.

§718.110 Adjustments.

(a) The farm operator or other interested producer having excess tobacco acreage (other than flue-cured or burley) may adjust an acreage of the crop in order to avoid a marketing quota penalty if such person:

(1) Notifies the county committee of such election within 15 calendar days after the date of mailing of notice of excess acreage by the county committee; and

(2) Pays the cost of a farm inspection to determine the adjusted acreage prior to the date the farm visit is made.

(b) The farm operator may adjust an acreage of tobacco (except flue-cured and burley) by disposing of such excess tobacco prior to the marketing of any of the same kind of tobacco from the farm. The disposition shall be witnessed by a representative of FSA and may take place before, during, or after the harvesting of the same kind of tobacco grown on the farm. However, no credit will be allowed toward the disposition of excess acreage after the tobacco is harvested but prior to marketing, unless the county committee determines that such tobacco is representative of the entire crop from the farm of the kind of tobacco involved.

§718.111 Notice of measured acreage.

(a) FSA will provide notice of measured acreage and mail it to the farm operator. This notice constitutes notice to all parties who have ownership, leasehold interest, or other interest in such farm.

(b) [Reserved]

[80 FR 41996, July 16, 2015]

§718.112 Redetermination.

(a) A redetermination of crop acreage, appraised yield, or farm-stored production for a farm may be initiated by the county committee, State committee, or Deputy Administrator at any time. Redetermination may be requested by a producer with an interest in the farm if the producer pays the cost of the redetermination. The request must be submitted to FSA within 5 calendar days after the initial appraisal of the yield of a crop, or before the farm-stored production is removed from storage. A redetermination will be undertaken in the manner prescribed by the Deputy Administrator. A redetermination will be used in lieu of any prior determination unless it is determined by the representative of the Deputy Administrator that there is good cause not to do so.

(b) FSA will refund the payment of the cost for a redetermination when, because of an error in the initial determination:

(1) The appraised yield is changed by at least the larger of:

(i) Five percent or 5 pounds for cotton;

(ii) Five percent or 1 bushel for wheat, barley, oats, and rye; or

(iii) Five percent or 2 bushels for corn and grain sorghum; or

(2) The farm stored production is changed by at least the smaller of 3 percent or 600 bushels; or

(3) The acreage of the crop is:

(i) Changed by at least the larger of 3 percent or 0.5 acre; or

(ii) Considered to be within program requirements.

[68 FR 16176, Apr. 3, 2003, as amended at 80 FR 41996, July 16, 2015]

Subpart C—Reconstitution of Farms, Allotments, Quotas, and Base Acres

Source: 68 FR 16178, Apr. 3, 2003, unless otherwise noted.

§718.201 Farm constitution.

(a) In order to implement FSA programs and monitor compliance with regulations, FSA must have records on what land is being farmed by a particular producer. This is accomplished by a determination of what land or group of lands "constitute" an individual unit or farm. Land that was properly constituted under prior regulations will remain so constituted until a reconstitution is required by paragraph (c) of this section. The constitution and identification of land as a "farm" for the first time and the subsequent reconstitution of a farm made thereafter will include all land operated by an individual entity or joint operation as a single farming unit except that it may not include:

(1) Land under separate ownership unless the owners agree in writing or have previously agreed in writing and the labor, equipment, accounting system, and management are operated in common by the operator, but separate from other tracts;

(2) Land under a lease agreement of less than 1 year duration;

(3) Federally owned land unless it is rangeland on which no crops are planted and on which there are no crop base acres established;

(4) State-owned wildlife lands unless the former owner has possession of the land under a leasing agreement;

(5) Land constituting a farm that is declared ineligible to be enrolled in a program under the regulations governing the program;

(6) For base acre crops, land located in counties that are not contiguous except where:

(i) Counties are divided by a river;

(ii) Counties do not share a common border because of a correction line adjustment; or

(iii) The land is within 20 miles, by road, of other land that will be a part of the farming unit;

(7) Land subject to either a default election or a valid election made under part 1412 of this title for each and all covered commodities constituted with land that has a different default election or valid election for each and all covered commodities, irrespective of whether or not any of the land has base acres; or

(8) Land subject to an election of individual coverage under the Agriculture Risk Coverage Program (ARC-IC) in any State constituted with any land in another State.

(b)(1) If all land on the farm is physically located in one county, the farm shall be administratively located in such county. If there is no FSA office in the county or the county offices have been consolidated, the farm shall be administratively located in the contiguous county most convenient for the farm operator.

(2) If the land on the farm is located in more than one county, the farm shall be administratively located in either of such counties as the county committees and the farm operator agree. If no agreement can be reached,

the farm shall be administratively located in the county where the principal dwelling is situated, or where the major portion of the farm is located if there is no dwelling.

(c) A reconstitution of a farm either by division or by combination is required whenever:

(1) A change has occurred in the operation of the land since the last constitution or reconstitution and as a result of such change the farm does not meet the conditions for constitution of a farm as specified in paragraph (a) of this section, except that no reconstitution will be made if the county committee determines that the primary purpose of the change in operation is to establish eligibility to transfer allotments subject to sale or lease, or increase the amount of program benefits received;

(2) The farm was not properly constituted the previous time;

(3) An owner requests in writing that the land no longer be included in a farm composed of tracts under separate ownership;

(4) The county committee determines that the farm was reconstituted on the basis of false information;

(5) The county committee determines that tracts included in a farm are not being operated as a single farming unit.

(d) An owner can file a written request to have FSA reconstitute from original tracts areas that are less than 10 DCP cropland acres and less than 5 percent of the original tract, if such request is accompanied by sufficient data from which FSA can determine the political county and State of land in both the original tract and the proposed tract. Any owner-initiated requests for tract divisions for physical location will be performed and effective prospectively from date of request and approval by FSA.

(e) Reconstitution shall not be approved if the county committee determines that the primary purpose of the reconstitution is to:

(1) Circumvent the provisions of part 12 of this title; or

(2) Circumvent any other chapter of this title.

[68 FR 16178, Apr. 3, 2003, as amended at 80 FR 41996, July 16, 2015; 84 FR 45887, Sept. 3, 2019]

§718.202 Determining the land constituting a farm.

(a) In determining the constitution of a farm, consideration shall be given to provisions such as ownership and operation. For purposes of this part, the following rules shall be applicable to determining what land is to be included in a farm.

(b) A minor shall be considered to be the same owner or operator as the parent, court-appointed guardian, or other person responsible for the minor child, unless the parent or guardian has no interest in the minor's farm or production from the farm, and the minor:

(1) Is a producer on a farm;

(2) Maintains a separate household from the parent or guardian;

(3) Personally carries out the farming activities; and

(4) Maintains a separate accounting for the farming operation.

(c) A minor shall not be considered to be the same owner or operator as the parent or court-appointed guardian if the minor's interest in the farming operation results from being the beneficiary of an irrevocable trust and ownership of the property is vested in the trust or the minor.

(d) A life estate tenant shall be considered to be the owner of the property for their life.

(e) A trust shall be considered to be an owner with the beneficiary of the trust; except a trust can be considered a separate owner or operator from the beneficiary, if the trust:

(1) Has a separate and distinct interest in the land or crop involved;

(2) Exercises separate responsibility for the separate and distinct interest; and

(3) Maintains funds and accounts separate from that of any other individual or entity for the interest.

(f) The county committee shall require specific proof of ownership.

(g) Land owned by different persons of an immediate family living in the same household and operated as a single farming unit shall be considered as being under the same ownership in determining a farm.

(h) All land operated as a single unit and owned and operated by a parent corporation and subsidiary corporations of which the parent corporation owns more than 50 percent of the value of the outstanding stock, or where the parent is owned and operated by subsidiary corporations, shall be constituted as one farm.

§718.203 County committee action to reconstitute a farm.

Action to reconstitute a farm may be initiated by the county committee, the farm owner, or the operator with the concurrence of the owner of the farm. Any request for a farm reconstitution shall be filed with the county committee.

§718.204 Reconstitution of base acres.

(a) Farms will be reconstituted in accordance with this subpart when it is determined that the land areas are not properly constituted and, to the extent practicable as determined by county committee, the reconstitution will be based on the facts and conditions existing at the time the change requiring the reconstitution occurred.

(b) Reconstitutions will be effective for the calendar year if initiated by August 1 of that year. Any reconstitution initiated after August 1 will not be effective for that year; it will be effective for the subsequent year.

(c) The Deputy Administrator may approve an exception to permit a reconstitution initiated after August 1 to be effective for the same year, if FSA determines that the failure is due to administrative problems as determined by FSA at the local or national level. Producers have no right to seek an exception under this paragraph. When such situations exist, FSA will establish procedures under which reconstitutions will be accepted and when those reconstitutions will become effective.

[79 FR 57714, Sept. 26, 2014, as amended at 84 FR 45887, Sept. 3, 2019]

§718.205 Substantive change in farming operation, and changes in related legal entities.

(a) Land that is properly constituted as a farm shall not be reconstituted if:

(1) The reconstitution request is based upon the formation of a newly established legal entity which owns or operates the farm or any part of the farm and the county committee determines there is not a substantive change in the farming operation;

(2) The county committee determines that the primary purpose of the request for reconstitution is to:

(i) Obtain additional benefits under one or more commodity programs;

(ii) Avoid damages or penalties under a contract or statute;

(iii) Correct an erroneous acreage report; or

(iv) Circumvent any other program provisions. In addition, no farm shall remain as constituted when the county committee determines that a substantive change in the farming operation has occurred which would require a reconstitution, except as otherwise approved by the State committee with the concurrence of the Deputy Administrator.

(b) In determining whether a substantive change has occurred with respect to a farming operation, the county committee shall consider factors such as the composition of the legal entities having an interest in the farming operation with respect to management, financing, and accounting. The county committee shall also consider the use of land, labor, and equipment available to the farming operations and any other relevant factors that bear on the determination.

(c) Unless otherwise approved by the State committee with the concurrence of the Deputy Administrator, when the county committee determines that a corporation, trust, or other legal entity is formed primarily for the purpose of obtaining additional benefits under the commodity programs of this title, the farm shall remain as constituted, or shall be reconstituted, as applicable, when the farm is owned or operated by:

(1) A corporation having more than 50 percent of the stock owned by members of the same family living in the same household;

(2) Corporations having more than 50 percent of the stock owned by stockholders common to more than one corporation; or

(3) Trusts in which the beneficiaries and trustees are family members living in the same household.

(d) Application of the provisions of paragraph (c) of this section shall not limit or affect the application of paragraphs (a) and (b) of this section.

§718.206 Determining farms, tracts, and base acres when reconstitution is made by division.

(a) The methods for dividing farms, tracts, and base acres are, in order of precedence: Estate, designation by landowner, cropland, and default. The proper method will be determined on a crop-by-crop basis.

(b) The estate method for reconstitution is the pro-rata distribution of base acres for a parent farm among the heirs in settling an estate. If the estate sells a tract of land before the farm is divided among the heirs, the base acres for that tract will be determined according to paragraphs (c) through (e) of this section.

(1) Base acres must be divided in accordance with a will, but only if the county committee determines that the terms of the will are such that a division can reasonably be made by the estate method.

(2) If there is no will or the county committee determines that the terms of a will are not clear as to the division of base acres, the base acres will be apportioned in the manner agreed to in writing by all interested heirs or devisees who acquire an interest in the property for which base acres have been established. An agreement by the administrator or executor will not be accepted in lieu of an agreement by the heirs or devisees.

(3) If base acres are not apportioned as specified in paragraph (b)(1) or (2) of this section, the base acres must be divided as specified in paragraph (d) or (e) of this section, as applicable.

(c) If the ownership of a tract of land is transferred from a parent farm, the transferring owner may request that the county committee divide the base acres, including historical acreage that has been double cropped, between the parent farm and the transferred tract, or between the various tracts if the entire farm is sold to two or more purchasers.

(1) If the county committee determines that base acres cannot be divided in the manner designated by the owner because the owner's designation does not meet the requirements of paragraph (c)(2) of this section, FSA will notify the owner and permit the owner to revise the designation to meet the requirements. If the owner does not furnish a revised designation of base acres within a reasonable time after such notification, or if the revised designation does not meet the requirements, the county committee will divide the base acres in a pro-rata manner in accordance with paragraph (d) or (e) of this section.

(2) The landowner may designate a manner in which base acres are divided by filing a signed written memorandum of understanding of the designation of base acres with the county committee before the transfer of ownership of the land. Both the transferring owner and transferee must sign the written designation of base acres.

(i) Within 30 days after a prescribed form, letter, or notice of base acres is issued by FSA following the reconstitution of a farm but before any subsequent transfer of ownership of the land, all owners in existence at time of the reconstitution request may seek a different manner of base acre designation by agreeing in writing by executing a form CCC–517 or other designated form.

(ii) The landowner must designate the base acres that will be permanently reduced when the sum of the base acres exceeds the effective cropland plus double-cropped acres for the farm.

(iii) When the part of the farm from which the ownership is being transferred was owned for less than 3 years, the designation by landowner method of designating base acres cannot be used unless the county committee determines that the primary purpose of the ownership transfer was other than to retain or to sell base acres. In the absence of such a determination, and if the farm contains land that has been owned for less than 3 years, the part of the farm that has been owned for less than 3 years will be considered as a separate farm and the base acres must be assigned to that farm in accordance with paragraph (d) or (e) of this sec-

tion. Such apportionment will be made prior to any designation of base acres with respect to the part that has been owned for 3 years or more.

(3) The designation by landowner method may be applied, at the owner's request, to land owned by an Indian Tribal Council that is leased to two or more producers for the production of any crop of a commodity for which base acres have been established. If the land is leased to two or more producers, an Indian Tribal Council may request that the county committee divide the base acres between the applicable tracts in the manner designated by the Council. The use of this method is not subject to the requirements specified in paragraph (c)(2) of this section.

(d) The cropland method for reconstitution is the pro-rata distribution of base acres to the resulting tracts in the same proportion that each resulting tract bears to the cropland for the parent tract. This method of division will be used if paragraphs (b) and (c) of this section do not apply.

(e) The default method for reconstitution is the separation of tracts from a farm with each tract maintaining the base acres attributed to the tract when the reconstitution is initiated.

(f) Farm program payment yields calculated for the resulting farms of a division may be increased or decreased if the county committee determines the method used did not provide an equitable distribution considering available land, cultural operations, and changes in the type of farming conducted on the farm. Any increase in the farm program payment yield on a resulting farm will be offset by a corresponding decrease on another resulting farm of the division.

[80 FR 41997, July 16, 2015]

§ 718.207 Determining base acres when reconstitution is made by combination.

(a) When two or more farms or tracts are combined for a year, that year's base acres, with respect to the combined farm or tract, as required by applicable program regulations, will not be greater than the sum of the base acres for each of the farms or tracts comprising the combination, subject to the provisions of § 718.204.

(b) [Reserved]

[80 FR 41998, July 16, 2015]

Subpart D—Equitable Relief From Ineligibility

SOURCE: 67 FR 66307, Oct. 31, 2002, unless otherwise noted.

§ 718.301 Applicability.

(a) This subpart is applicable to programs administered by the Farm Service Agency under chapters VII and XIV of this title, except for an agricultural credit program carried out under the Consolidated Farm and Rural Development Act (7 U.S.C. 1921 et seq.), as amended. Administration of this subpart shall be under the supervision of the Deputy Administrator, except that such authority shall not limit the exercise of authority allowed State Executive Directors of the Farm Service agency as provided for in § 718.307.

(b) Section 718.306 does not apply to a function performed under either section 376 of the Consolidated Farm and Rural Development Act (7 U.S.C. 1921 et seq.), or a conservation program administered by the Natural Resources Conservation Service of the United States Department of Agriculture.

(c) The relief provisions of this part cannot be used to extend a benefit or assistance not otherwise available under law or not otherwise available to others who have satisfied or complied with every eligibility or compliance requirement of the provisions of law or regulations governing the program benefit or assistance.

[67 FR 66307, Oct. 31, 2002, as amended at 80 FR 41998, July 16, 2015]

§ 718.302 Definitions and abbreviations.

In addition to the definitions provided in § 718.2 of this part, the following terms apply to this subpart:

Covered program means a program specified in § 718.301 of this subpart.

FSA means the Farm Service Agency of the United States Department of Agriculture.

OGC means the Office of the General Counsel of the United States Department of Agriculture.

SED means, for activities within a particular state, the State Executive Director of the United States Department of Agriculture, FSA, for that state.

[67 FR 66307, Oct. 31, 2002, as amended at 80 FR 41998, July 16, 2015]

§ 718.303 Reliance on incorrect actions or information.

(a) Notwithstanding any other law, if an action or inaction by a participant is based upon good faith reliance on the action or advice of an authorized representative of an FSA county or State committee, and that action or inaction results in the participant's noncompliance with the requirements of a covered program that is to the detriment of the participant, then that action or inaction still may be approved by the Deputy Administrator as meeting the requirements of the covered program, and benefits may be extended or payments made in as specified in § 718.305.

(b) This section applies only to a participant who:

(1) Relied in good faith upon the action of, or information provided by, an FSA county or State committee or an authorized representative of such committee regarding a covered program;

(2) Acted, or failed to act, as a result of the FSA action or information; and

(3) Was determined to be not in compliance with the requirements of that covered program.

(c) This section does not apply to cases where the participant had sufficient reason to know that the action or information upon which they relied was improper or erroneous or where the participant acted in reliance on their own misunderstanding or misinterpretation of program provisions, notices or information.

[80 FR 41998, July 16, 2015]

§ 718.304 Failure to fully comply.

(a) When the failure of a participant to fully comply with the terms and conditions of a covered program precludes the providing of payments or benefits, relief may be authorized as specified in § 718.305 if the participant made a good faith effort to comply fully with the requirements of the covered program.

(b) This section only applies to participants who are determined by FSA to have made a good faith effort to comply fully with the terms and conditions of the covered program and have performed substantial actions required for program eligibility.

[80 FR 41998, July 16, 2015]

§718.305 Forms of relief.

(a) The Administrator of FSA, Executive Vice President of CCC, or their designee, may authorize a participant in a covered program to:

(1) Retain loans, payments, or other benefits received under the covered program;

(2) Continue to receive loans, payments, and other benefits under the covered program;

(3) Continue to participate, in whole or in part, under any contract executed under the covered program;

(4) In the case of a conservation program, re-enroll all or part of the land covered by the program; and

(5) Receive such other equitable relief as determined to be appropriate.

(b) As a condition of receiving relief under this subpart, the participant may be required to remedy their failure to meet the program requirement, or mitigate its affects.

§718.306 Finality.

(a) A determination by an FSA State or county committee (or employee of such committee) becomes final on an application for benefits and binding 90 days from the date the application for benefits has been filed, and supporting documentation required to be supplied by the producer as a condition for eligibility for the particular program has been filed, unless any of the following exceptions exist:

(1) The participant has requested an administrative review of the determination in accordance with part 780 of this chapter;

(2) The determination was in any way based on erroneous, innocent, or purposeful misrepresentation; false statement; fraud; or willful misconduct by or on behalf of the participant;

(3) The determination was modified by the Administrator, FSA, or in the case of CCC programs conducted under Chapter XIV of this title, the Executive Vice President, CCC; or

(4) The participant knew or had reason to know that the determination was erroneous.

(b) Should an erroneous determination become final under the provisions of this section, the erroneous decision will be corrected according to paragraph (c) of this section.

(1) If, as a result of the erroneous decision, payment was issued, no action will be taken by FSA, CCC, or a State or county committee to recover unearned payment amounts unless one or more of the exceptions in paragraph (a) of this section applies;

(2) If payment was not issued before the error was discovered, the payment will not be issued. FSA and CCC are under no obligation to issue payments or render decisions that are contrary to law or regulation.

(c) FSA and CCC will modify and correct determinations when errors are discovered. As specified in paragraph (b) of this section, FSA or CCC may be precluded from recovering unearned payments that issued as a result of the erroneous decision. FSA or CCC's inability to recover or demand refunds of unearned amounts as specified in paragraph (b) will only be effective through the year in which the error was found and communicated to the participant.

[67 FR 66307, Oct. 31, 2002, as amended at 80 FR 41998, July 16, 2015]

§718.307 Special relief approval authority for State Executive Directors.

(a) *General nature of the special authority.* Notwithstanding provisions in this subpart providing supervision and relief authority to other officials, an SED, after consultation with and approval from OGC but without further review by other officials (other than the Secretary) may grant relief to a participant under the provisions of §§718.303 through 718.305 as if the SED were the final arbiter within the agency of such matters so long as:

(1) The program matter with respect to which the relief is sought is a program matter in a covered program

which is operated within the State under the control of the SED;

(2) The total amount of relief which will be provided to the participant (that is, to the individual or entity that applies for the relief) by that SED under this special authority for errors during that year is less than $20,000 (including in that calculation, any loan amount or other benefit of any kind payable for that year and any other year);

(3) The total amount of such relief which has been previously provided to the participant using this special authority for errors, as calculated above, is not more than $5,000;

(4) The total amount of loans, payments, and benefits of any kind for which relief is provided to similarly situated participants by an SED for errors for any year under the authority provided in this section, as calculated above, is not more than $1,000,000.

(b) *Report of the exercise of the power.* A grant of relief shall be considered to be under this section and subject to the special finality provided in this section only if the SED grants the relief in writing when granting the relief to the party who will receive the benefit of such relief and only if, in that document, the SED declares that they are exercising that power. The SED must report the exercise of that power to the Deputy Administrator so that a full accounting may be made in keeping with the limitations of this section. Absent such a report, relief will not be considered to have been made under this section.

(c) *Additional limits on the authority.* The authority provided under this section does not extend to:

(1) The administration of payment limitations under part 1400 of this chapter (§§ 1001 to 1001F of 7 U.S.C. 1308 *et seq.*);

(2) The administration of payment limitations under a conservation program administered by the Secretary; or

(3) Highly erodible land and wetland conservation requirements under subtitles B or C of Title XII of the Food Security Act of 1985 (16 U.S.C. 3811 *et seq.*) as administered under 7 CFR part 12.

(d) Relief may not be provided by the SED under this section until a written opinion or written acknowledgment is obtained from OGC that grounds exist for determination that requirements for granting relief under § 718.303 or § 718.304 have been met, that the form of relief is authorized under § 718.305, and that the granting of the relief is within the lawful authority of the SED.

(e) *Relation to other authorities.* The authority provided under this section is in addition to any other applicable authority that may allow relief.

[67 FR 66307, Oct. 31, 2002, as amended at 80 FR 41998, July 16, 2015]

SUBCHAPTER C [RESERVED]

SUBCHAPTER D—SPECIAL PROGRAMS

PART 750—SOIL BANK

EDITORIAL NOTE: Part 750 (formerly part 485 of title 6), published at 21 FR 6289, Aug. 22, 1956, and redesignated at 26 FR 5788, June 29, 1961, is no longer carried in the Code of Federal Regulations. This deletion does not relieve any person of any obligation or liability incurred under these regulations, nor deprive any person of any rights received or accrued under the provisions of this part. For FEDERAL REGISTER citations affecting this part, see the "List of CFR Sections Affected, 1949–1963, 1964–1972, and 1973–1985," published in seven separate volumes.

PART 755—REIMBURSEMENT TRANSPORTATION COST PAYMENT PROGRAM FOR GEOGRAPHICALLY DISADVANTAGED FARMERS AND RANCHERS

AUTHORITY: 7 U.S.C. 8792.

SOURCE: 75 FR 34340, June 17, 2010, unless otherwise noted.

§ 755.1 Administration.

(a) This part establishes the terms and conditions under which the Reimbursement Transportation Cost Payment (RTCP) Program for geographically disadvantaged farmers and ranchers will be administered.

(b) The RTCP Program will be administered under the general supervision of the FSA Administrator, or a designee, and will be carried out in the field by FSA State and county committees and FSA employees.

(c) FSA State and county committees, and representatives and employees thereof, do not have the authority to modify or waive any of the provisions of the regulations of this part, except as provided in paragraph (e) of this section.

(d) The FSA State committee will take any action required by the provisions of this part that has not been taken by the FSA county committee. The FSA State committee will also:

(1) Correct or require an FSA county committee to correct any action taken by the county committee that is not in compliance with the provisions of this part.

(2) Require an FSA county committee to not take an action or implement a decision that is not in compliance with the provisions of this part.

(e) No provision or delegation of this part to an FSA State committee or a county committee will preclude the FSA Administrator, or a designee, from determining any question arising under the program or from reversing or modifying any determination made by a State committee or a county committee.

(f) The Deputy Administrator for Farm Programs, FSA, may waive or modify program requirements of this part in cases where failure to meet requirements does not adversely affect the operation of the program and where the requirement is not statutorily mandated.

§ 755.2 Definitions.

The following definitions apply to this part. The definitions in parts 718 and 1400 of this title also apply, except where they may conflict with the definitions in this section.

Actual transportation rate means the transportation rate that reflects the actual transportation costs incurred and can be determined by supporting documentation.

Agricultural commodity means any agricultural commodity (including horticulture, aquaculture, and floriculture), food, feed, fiber, livestock (including elk, reindeer, bison, horses, or deer), or insects, and any product thereof.

Agricultural operation means a parcel or parcels of land; or body of water applicable to aquaculture, whether contiguous or noncontiguous, constituting a cohesive management unit for agricultural purposes. An agricultural operation will be regarded as located in the county in which the principal dwelling is situated, or if there is no dwelling thereon, it will be regarded to be in the county in which the major portion of the land or applicable body of water is located.

Application period means the period established by the Deputy Administrator for geographically disadvantaged farmers and ranchers to apply for program benefits.

County office or FSA county office means the FSA offices responsible for administering FSA programs in a specific area, sometimes encompassing more than one county, in a State.

Department or USDA means the U.S. Department of Agriculture.

Eligible reimbursement amount means the reported costs incurred to transport an agricultural commodity or input used to produce an agricultural commodity in an insular area, Alaska, or Hawaii, over a distance of more than 30 miles. The amount is calculated by multiplying the number of units of the reported transportation amount times the applicable transportation fixed, set, or actual rate times the applicable FY allowance (COLA).

Farm Service Agency or FSA means the Farm Service Agency of the USDA.

Fiscal year or FY means the year beginning October 1 and ending the following September 30. The fiscal year will be designated for this part by year reference to the calendar year in which it ends. For example, FY 2010 is from October 1, 2009, through September 30, 2010 (inclusive).

Fixed transportation rate means the per unit transportation rate determined by FSA to reflect the transportation cost applicable to an agricultural commodity or input used to produce an agricultural commodity in a particular region.

FY allowance (COLA) means the nonforeign area cost of living allowance or post differential, as applicable, for that FY set by Office of Personnel Management for Federal employees stationed in Alaska, Hawaii, and other insular areas, as authorized by 5 U.S.C. 5941 and E.O. 10000 and specified in 5 CFR part 591, subpart B, appendices A and B.

Geographically disadvantaged farmer or rancher means a farmer or rancher in an insular area, Alaska, or Hawaii.

Input transportation costs means those transportation costs of inputs used to produce an agricultural commodity including, but not limited to, air freight, ocean freight, and land freight of chemicals, feed, fertilizer, fuel, seeds, plants, supplies, equipment parts, and other inputs as determined by FSA.

Insular area means the Commonwealth of Puerto Rico; Guam; American Samoa; the Commonwealth of the Northern Mariana Islands; the Federated States of Micronesia; the Republic of the Marshall Islands; the Republic of Palau; and the Virgin Islands of the United States.

Payment amount means the amount due a producer that is the sum of all eligible reimbursement amounts, as calculated by FSA subject to the availability of funds, and subject to an $8,000 cap per producer per FY.

Producer means any geographically disadvantaged farmer or rancher who is an individual, group of individuals, partnership, corporation, estate, trust, association, cooperative, or other business enterprise or other legal entity, as defined in § 1400.3 of this title, who is, or whose members are, a citizen of or legal resident alien in the United States, and who, as determined by the Secretary, shares in the risk of producing an agricultural commodity in substantial commercial quantities, and who is entitled to a share of the agricultural commodity from the agricultural operation.

Reported transportation amount means the reported number of units (such as pounds, bushels, pieces, or parts) applicable to an agricultural commodity or input used to produce an agricultural commodity, which is used in calculating the eligible reimbursement amount.

Set transportation rate means the transportation rate established by FSA for a commodity or input for which there is not a fixed transportation rate or supporting documentation of the actual transportation rate.

United States means the 50 States of the United States of America, the District of Columbia, the Commonwealths of Puerto Rico and the Northern Mariana Islands, and any other territory or possession of the United States.

Verifiable records means evidence that is used to substantiate the amount of eligible reimbursements by geographically disadvantaged farmers and ranchers in an agricultural operation that can be verified by FSA through an independent source.

§755.3 Time and method of application.

(a) To be eligible for payment, producers must obtain and submit a completed application for payment and meet other eligibility requirements specified in this part. Producers may obtain an application in person, by mail, or by facsimile from any county FSA office. In addition, producers may download a copy of the application at *http://www.sc.egov.usda.gov.*

(b) An application for payment must be submitted on a completed application form. Applications and any other supporting documentation must be submitted to the FSA county office serving the county where the agricultural operation is located, but, in any case, must be received by the FSA county office by the close of business on the last day of the application period established by the Deputy Administrator.

(c) All producers who incurred transportation costs for eligible reimbursements and who share in the risk of an agricultural operation must certify to the information on the application before the application will be considered complete. FSA may require the producer to provide documentation to support all verifiable records.

(d) Each producer requesting payment under this part must certify to the accuracy and truthfulness of the information provided in their application and any supporting documentation. All information provided is subject to verification by FSA. Refusal to allow FSA or any other agency of the Department of Agriculture to verify any information provided will result in a denial of eligibility. Furnishing the information is voluntary; however, without it program benefits will not be approved. Providing a false certification to the Federal Government may be punishable by imprisonment, fines and other penalties or sanctions.

(e) To ensure all producers are provided an opportunity to submit actual costs for reimbursement at the actual cost rate, applicants will have 30 days after the end of the FY to provide supporting documentation of actual transportation costs to the FSA County Office. The actual costs documented in supporting documentation will override previously reported costs of eligible reimbursable costs at the fixed or set rate made during the application period.

(f) If verifiable records are not provided to FSA, the producer will be ineligible for payment.

(g) If supporting documentation is provided within 30 days after the end of the FY, but an application was not submitted to the applicable FSA County Office before the end of the application period, the producer is not eligible for payment.

(h) Producers who submit applications after the application period are not entitled to any payment consideration or determination of eligibility. Regardless of the reason why an application is not submitted to or received by FSA, any application received after the close of business on such date will not be eligible for benefits under this program.

§755.4 Eligibility.

(a) To be eligible to receive payments under this part, a geographically disadvantaged farmer or rancher must:

(1) Be a producer of an eligible agricultural commodity in substantial commercial quantities;

(2) Incur transportation costs for the transportation of the agricultural commodity or input used to produce the agricultural commodity;

(3) Submit an accurate and complete application for payment as specified in §755.3; and

(4) Be in compliance with the wetland and highly erodible conservation requirements in part 12 of this title and meet the adjusted gross income and

pay limit eligibility requirements in part 1400 of this title, as applicable, except that the $8,000 cap provided for in this rule is a per producer cap, not a per person cap. For example, a partnership of four individuals would be considered one producer, not four persons, for the purposes of this cap and thus the partnership could only generate a single $8,000 payment under this program if the cap holds because of full subscription of the program.

(b) Individual producers in an agricultural operation that is an entity are only eligible for a payment based on their share of the operation. A producer is not eligible for payment based on the share of production of any other producer.

(c) Multiple producers, such as the buyer and seller of a commodity (for example, a producer of hay and a livestock operation that buys the hay), are not eligible for payments for the same eligible transportation cost. Unless the multiple producers agree otherwise, only the last buyer will be eligible for the payment.

(d) A person or entity determined to be a "foreign person" under part 1400 of this title is not eligible to receive benefits under this part, unless that person provides land, capital, and a substantial amount of active personal labor in the production of crops on such farm.

(e) State and local governments and their political subdivisions and related agencies are not eligible for RTCP payments.

§ 755.5 Proof of eligible reimbursement costs incurred.

(a) To be eligible for reimbursement based on FSA fixed or set rates as specified in § 755.7, the requirements specified in paragraphs (b) and (c) of this section must be met at the time of the application. To be eligible for reimbursement of actual costs, the requirements of paragraph (d) must also be met, within 30 days after the end of the applicable fiscal year.

(b) Eligible verifiable records to support eligible reimbursement costs include, but are not limited to:

(1) Invoices;

(2) Account statements;

(3) Contractual Agreements; or

(4) Bill of Lading.

(c) Verifiable records must show:

(1) Name of producer(s);

(2) Commodity and unit of measure;

(3) Type of input(s) associated with transportation costs;

(4) Date(s) of service;

(5) Name of person or entity providing the service, as applicable, and;

(6) Retail sales receipts with verifiable records handwritten as applicable.

(d) To be eligible for reimbursement based on actual costs, the producer must provide supporting documentation that documents the specific costs incurred for transportation of each commodity or input. Such documentation must:

(1) Show transportation costs for each specific commodity or input, and

(2) Show the units of measure for each commodity or input, such that FSA can determine the transportation cost per unit.

§ 755.6 Availability of funds.

(a) Payments under this part are subject to the availability of funds.

(b) A reserve will be created to handle appeals and errors.

§ 755.7 Transportation rates.

(a) Payments may be based on fixed, set, or actual transportation rates. Fixed and set transportation rates will be established by FSA, based on available data for transportation costs for that commodity or input in the applicable State or insular region.

(b) Fixed transportation rates will establish per unit transportation costs for each eligible commodity or input used to produce the eligible commodity.

(c) Set transportation rates will be established for those transportation costs that are not on the FSA list of fixed rates and for which an actual rate cannot be documented. The set transportation rate will be set by FSA, based on available data of transportation costs for similar commodities and inputs.

(d) Actual transportation rates will be determined based on supporting documentation.

§755.8 Calculation of individual payments.

(a) Transportation cost for each commodity or input will be calculated by multiplying the number of reported eligible units (the reported transportation amount) times the fixed, set, or actual transportation rate, as applicable.

(b) Eligible reimbursement amounts will be calculated by multiplying the result of paragraph (a) of this section times the appropriate FY COLA percentage, as provided in this part.

(c) If transported inputs are used for both eligible and ineligible commodities, the eligible reimbursable costs will be determined on a revenue share of eligible commodities times input cost, as determined by FSA, and transportation may be allowed only for those commodities which were produced for the commercial market.

(d) The total payment amount for a producer is the sum of all eligible reimbursable amounts determined in paragraph (b) of this section for all commodities and inputs used to produce the eligible commodities listed on the application.

(e) Payment amounts are subject to $8,000 cap per FY per producer as defined in this part, not per "person" or "legal entity" as those terms might be defined in part 1400 of this title.

(f) In the event that approval of all calculated payment amounts would result in expenditures in excess of the amount available, FSA will recalculate the payment amounts in a manner that FSA determines to be fair and reasonable.

§755.9 Misrepresentation and scheme or device.

(a) In addition to other penalties, sanctions or remedies as may apply, a producer will be ineligible to receive payments under this part if the producer is determined by FSA to have:

(1) Adopted any scheme or device that tends to defeat the purpose of this part;

(2) Made any fraudulent representation; or

(3) Misrepresented any fact affecting a program determination.

(b) Any payment to any producer engaged in a misrepresentation, scheme, or device, must be refunded with interest together with such other sums as may become due. Any producer engaged in acts prohibited by this section and receiving payment under this part will be jointly and severally liable with other producers involved in such claim for benefits for any refund due under this section and for related charges. The remedies provided in this part will be in addition to other civil, criminal, or administrative remedies that may apply.

§755.10 Death, incompetence, or disappearance.

(a) In the case of the death, incompetency, or disappearance of a person or the dissolution of an entity that is eligible to receive a payment in accordance with this part, such alternate person or persons specified in part 707 of this chapter may receive such payment, as determined appropriate by FSA.

(b) Payments may be made to an otherwise eligible producer who is now deceased or to a dissolved entity if a representative who currently has authority to enter into an application for the producer or the producer's estate signs the application for payment. Proof of authority over the deceased producer's estate or a dissolved entity must be provided.

(c) If a producer is now a dissolved general partnership or joint venture, all members of the general partnership or joint venture at the time of dissolution or their duly authorized representatives must be identified in the application for payment.

§755.11 Maintaining records.

Persons applying for payment under this part must maintain records and accounts to document all eligibility requirements specified in this part. Such records and accounts must be retained for 3 years after the date of payment to the producer under this part.

§755.12 Refunds; joint and several liability.

(a) Any producer that receives excess payment, payment as the result of erroneous information provided by any person, or payment resulting from a failure to comply with any requirement or condition for payment under this part, must refund the amount of that payment to FSA.

(b) Any refund required will be due from the date of the disbursement by the agency with interest determined in accordance with paragraph (d) of this section and late payment charges as provided in part 1403 of this title.

(c) Each producer that has an interest in the agricultural operation will be jointly and severally liable for any refund and related charges found to be due to FSA.

(d) Interest will be applicable to any refunds to FSA required in accordance with parts 792 and 1403 of this title except as otherwise specified in this part. Such interest will be charged at the rate that the U.S. Department of the Treasury charges FSA for funds, and will accrue from the date FSA made the payment to the date the refund is repaid.

(e) FSA may waive the accrual of interest if it determines that the cause of the erroneous payment was not due to any action of the person or entity, or was beyond the control of the person or entity committing the violation. Any waiver is at the discretion of FSA alone.

§ 755.13 Miscellaneous provisions and appeals.

(a) *Offset.* FSA may offset or withhold any amount due to FSA from any benefit provided under this part in accordance with the provisions of part 1403 of this title.

(b) *Claims.* Claims or debts will be settled in accordance with the provisions of part 1403 of this title.

(c) *Other interests.* Payments or any portion thereof due under this part will be made without regard to questions of title under State law and without regard to any claim or lien against the eligible reimbursable costs thereof, in favor of the owner or any other creditor except agencies and instrumentalities of the U.S. Government.

(d) *Assignments.* Any producer entitled to any payment under this part may assign any payments in accordance with the provisions of part 1404 of this title.

(e) *Violations regarding controlled substances.* The provisions of § 718.6 of this chapter, which generally limit program payment eligibility for persons who have engaged in certain offenses with respect to controlled substances, will apply to this part.

(f) *Appeals.* The appeal regulations specified in parts 11 and 780 of this chapter apply to determinations made under this part.

PART 756—ORIENTAL FRUIT FLY PROGRAM

AUTHORITY: Sec. 778, Pub. L. 116–6, 133 Stat. 91.

SOURCE: 86 FR 70699, Dec. 13, 2021, unless otherwise noted.

§ 756.1 Applicability.

(a) The Oriental Fruit Fly (OFF) Program will provide payments to eligible producers who suffered losses due to the Oriental fruit fly quarantine in Miami-Dade County, Florida, in accordance with Public Law 116–6 (the Consolidated Appropriations Act, 2019).

(b) The regulations in this part are applicable to crops affected by the Oriental fruit fly quarantine.

(c) In any case in which money must be refunded to the Farm Service Agency (FSA) in connection with this part, interest will be due to run from the date of disbursement of the sum to be refunded. This paragraph (c) will apply, unless waived by the Deputy Administrator for Farm Programs, FSA, irrespective of any other regulation in this part.

§ 756.2 Administration.

(a) The OFF Program will be administered under the general supervision of the Administrator, FSA, and the Deputy Administrator for Farm Programs, FSA. The OFF Program is carried out by FSA State committees and FSA county committees with instructions issued by the Deputy Administrator.

(b) FSA State committees and FSA county committees, and representatives and their employees, do not have authority to modify or waive any of the provisions of the regulations in this part, except as provided in paragraph (e) of this section.

(c) The FSA State committee will take any required action not taken by the FSA county committee. The FSA State committee will also:

(1) Correct or require correction of an action taken by an FSA county committee that is not in compliance with this part; or

(2) Require an FSA county committee to not take an action or implement a decision that is not under the regulations of this part.

(d) The Deputy Administrator for Farm Programs, FSA, or a designee, may determine any question arising under these programs, or reverse or modify a determination made by an FSA State committee or FSA county committee.

(e) The Deputy Administrator for Farm Programs, FSA, may authorize FSA State committees and FSA county committees to waive or modify nonstatutory deadlines and other program requirements in cases where lateness or failure to meet such other requirements does not adversely affect the operation of the OFF Program.

(f) A representative of FSA may execute applications and related documents only under the terms and conditions determined and announced by FSA. Any document not executed under such terms and conditions, including any purported execution before the date authorized by FSA, will be null and void.

(g) Items of general applicability to program participants, including, but not limited to, application periods, application deadlines, internal operating guidelines issued to State and county offices, prices, and payment factors established by the OFF Program, are not subject to appeal.

§ 756.3 Definitions.

The definitions in this section apply for all purposes of OFF Program administration.

Administrative county office is the FSA county office where a producer's FSA records are maintained.

APHIS means Animal Plant Health and Inspection Service, U.S. Department of Agriculture.

Application period means the dates established by the Deputy Administrator for producers to apply for OFF Program benefits.

Calendar year means January 1st through December 31st.

Deputy Administrator means the Deputy Administrator for Farm Programs, FSA.

FSA means the Farm Service Agency, U.S. Department of Agriculture.

NAP means Non-insured Crop Disaster Assistance Program.

OFF Program means the Oriental Fruit Fly Program.

OFF quarantine period means August 28, 2015, through February 13, 2016.

Oriental fruit fly quarantine means the quarantine put in place during the OFF quarantine period in the quarantine area to protect against the entry and spread of the Oriental fruit fly by requiring strict adherence to treatment or destruction of the host crop.

Prevented planting means when producers chose not to plant an annual crop during the 2015 through 2016 season due to the Oriental fruit fly quarantine.

Producer means a person, partnership, association, corporation, estate, trust, or other legal entity that produces an eligible crop as a landowner, landlord, tenant, or sharecropper.

Program year means the relevant application year. The program year for OFF will be 2015 and include total revenue losses for calendar year 2015 and calendar year 2016.

Quarantine area means the area mapped by The Florida Department of Agriculture and Consumer Services Division, Division of Plant Industry (FDACS–DPI). The map identifies areas where the Oriental Fruit Fly was detected and the associated boundaries of the area quarantined by APHIS. The map is available by contacting FDACS–DPI, The Doyle Conner Building, 1911 SW 34th St., Gainesville, FL 32608–7100 or *https://www.fdacs.gov/Divisions-Offices/Plant-Industry*.

Reliable documentation means evidence provided by the participant that is used to substantiate the amount of revenue reported when verifiable documentation is not available, including copies of receipts, ledgers of income, income statements of deposit slips, register tapes, invoices for custom harvesting, and records to verify production costs, contemporaneous measurements truck scale tickets, and contemporaneous diaries that are determined acceptable by the FSA county committee. To determine whether the records are acceptable, the FSA county committee will consider whether they are consistent with the records of other producers of the crop in that area.

Revenue means the gross income from crop sales received during the applicable calendar years for the crops that suffered a loss due to the Oriental fruit fly quarantine. Revenue does not mean revenue received for crops grown under contract for crop owners unless the grower had an ownership share of the crop.

RMA means Risk Management Agency.

Secretary means the Secretary of the United States Department of Agriculture, or the Secretary's delegate.

Verifiable documentation means evidence that can be verified by FSA through an independent source.

§ 756.4   Qualifying disaster event.

The OFF Program will provide assistance to eligible producers who suffered revenue losses due to the State of Florida and APHIS implemented quarantine that took place from August 28, 2015, through February 13, 2016, in Miami-Dade County, Florida.

§ 756.5   Eligible producers.

(a) To be an eligible producer, the producer must:

(1) Be an individual person that is a U.S. Citizen or Resident Alien, or a partnership, association, corporation, estate, trust, or other legal entity consisting solely of U.S. Citizens or Resident Aliens that produces an eligible crop as a landowner, landlord, tenant, or sharecropper; and

(2) Comply with all provisions of this part and, as applicable;

(i) 7 CFR part 3—Debt Management;

(ii) 7 CFR part 12—Highly Erodible Land and Wetland Conservation;

(iii) 7 CFR 400.680, Controlled substance;

(iv) 7 CFR part 1400, adjusted gross income (AGI) provisions:

(A) Program year 2015 will be used to determine AGI for the OFF Program, therefore the AGI will be the average of tax years 2013, 2012, and 2011; and

(B) The OFF Program allows an exception to the $900,000 average AGI limitation if at least 75 percent of the average AGI was derived from farming, ranching, or forestry operations. CCC-942 is used to collect the producer and certified public accountant (CPA) or attorney certification statements;

(v) 7 CFR part 707—Payments Due Persons Who Have Died, Disappeared, or Have Been Declared Incompetent;

(vi) 7 CFR part 718—Provisions Applicable to Multiple Programs; and

(vii) 7 CFR part 1400—Payment Limitation and Payment Eligibility.

(b) A receiver or trustee of an insolvent or bankrupt debtor's estate, an executor or an administrator of a deceased person's estate, a guardian of an estate of a ward or an incompetent person, and trustees of a trust is considered to represent the insolvent or bankrupt debtor, the deceased person, the ward or incompetent, and the beneficiaries of a trust, respectively. The production of the receiver, executor, administrator, guardian, or trustee is the production of the person or estate represented by the receiver, executor, administrator, guardian, or trustee. OFF Program documents executed by any such person will be accepted by FSA only if they are legally valid and such person has the authority to sign the applicable documents.

(c) A minor who is otherwise an eligible producer is eligible to receive an OFF Program payment only if the minor meets one of the following requirements:

(1) The right of majority has been conferred on the minor by court proceedings or by statute.

(2) A guardian has been appointed to manage the minor's property and the applicable OFF Program documents are signed by the guardian.

(3) Any OFF Program application signed by the minor is cosigned by a person determined by the FSA county committee to be financially responsible.

(d) Foreign person rules in 7 CFR part 1400, subpart E, are not applicable to the OFF Program.

(e) Producers will not be required to be in the business of producing and marketing agricultural products at the time of OFF Program application.

(f) The producer must have been actively producing and marketing agricultural products during the OFF quarantine period.

§756.6 Eligible and ineligible causes of revenue loss.

(a) To be eligible for payments under this part the producer must have suffered a loss of revenue due to the Oriental fruit fly quarantine of one or more of the following types:

(1) Revenue loss on crop(s) planted or prevented from being planted within the Oriental Fruit Fly quarantine area during the OFF quarantine period. Crops that suffered a revenue loss due to prevented planting must have a prior history of being planted or be able to provide verifiable or reliable documentation demonstrating legitimate intent to plant the crop during the OFF quarantine period;

(2) Pre or post-harvest treatment costs;

(3) Transportation costs to a post-harvest treatment facility;

(4) Crop quality loss;

(5) Crop spoilage;

(6) Crop drop; or

(7) Reduced post-harvest shelf life.

(b) An ineligible cause of revenue loss under this part will apply to the following:

(1) Losses determined by FSA to be the result of poor management decisions or poor farming practices, such as using non-optimal chemical application, over-tilling, monoculture (growing of same crop year after year), allowing soil erosion, nonoptimal planting time, or poor quality seed selection.

(2) Losses due to conditions or events occurring outside of the applicable growing season for the crop.

(3) Losses due to failure of a power supply or lack of irrigation.

(4) Losses to crops not intended for harvest.

(5) Losses to home gardens for personal use and not intended to market.

(6) Losses to non-fruit bearing ornamental nursery.

(7) Losses caused by theft.

(8) Losses caused by disease or pest infestation other than the Oriental fruit fly.

(9) Losses to purchased crops.

§756.7 Time and method of application.

(a) An application for OFF Program payment under this part must be submitted in person, by mail, email, or facsimile to the FSA county office serving as the farm's administrative county office by the close of business 60 calendar days after the signup start date announced by FSA. A National Special Program (SP) Notice will be issued providing OFF program details including signup start date and program requirements.

(b) An application will include only the producer's share of revenue for the crops negatively affected by the Oriental fruit fly quarantine for the applicable calendar years.

(c) Once signed by a producer, the application for payment is considered to contain information and certifications of and pertaining to the producer regardless of who entered the information on the application.

(d) The producer applying for the OFF Program under this part certifies the accuracy and truthfulness of the information provided in the application as well as any documentation filed with or in support of the application.

(1) All information is subject to verification or spot check by FSA at

any time, either before or after payment is issued. Refusal to allow FSA or any agency of the Department of Agriculture to verify any information provided will result in the participant's forfeiting eligibility for the OFF Program. FSA may at any time, including before, during, or after processing and paying an application, require the producer to submit any additional information necessary to implement or determine any eligibility provision of this part. Furnishing required information is voluntary; however, without it, FSA is under no obligation to act on the application or approve payment.

(2) Providing a false certification will result in ineligibility and can also be punishable by imprisonment, fines, and other penalties.

(e) The application submitted· in accordance with paragraph (a) of this section is not considered valid and complete for issuance of payment under this part unless FSA determines all the applicable eligibility provisions have been satisfied and the participant has submitted all required documentation by the application deadline date announced by FSA.

(f) Applicants must submit all eligibility forms as listed on the FSA-438 Oriental Fruit Fly Program (OFF) Application within 60 calendar days from the date of submitting the application if not already on file with FSA.

§ 756.8 Calculating OFF Program payments.

(a) A revenue loss calculation and factor will determine the OFF Program payment.

(1) A factor will be applied to reduce the participant's payment to ensure that total OFF Program payments are no more than 70 percent of the total revenue losses by all eligible OFF Program participants.

(2) If necessary, at the close of the OFF Program sign-up period, a national payment factor may be determined by the Secretary and announced if full payment of all approved OFF Program applications would result in payments in excess of available OFF Program funds, less a reserve amount of 3 percent. A Price Support Division SP Notice will be issued to announce

the issuance of OFF and, if applicable, the factored rate.

(b)(1) The OFF Program payment calculation is:

(Calendar year 2014 producer certified gross revenue
− Calendar year 2015 producer certified gross . revenue)
+ (Calendar year 2014 producer certified gross revenue
− Calendar year 2016 producer certified gross revenue)
= Total revenue loss for calendar year 2015 and calendar year 2016
× 70%
= OFF Program payment (subject to proration after sign-up, see paragraph (a)(2) of this section)

(2) If the producer did not have 2014 revenue, then 2019 revenue will be used, and the calculation will be:

(Calendar year 2019 producer certified gross revenue
− Calendar year 2015 producer certified gross revenue)
+ (Calendar year 2019 producer certified gross revenue
− Calendar year 2016 producer certified gross revenue)
= Total revenue loss for calendar year 2015 and calendar year 2016
× 70%
= OFF Program Payment (subject to proration after sign-up, see paragraph (a)(2) of this section)

(c) If there is no gross revenue loss determined for calendar year 2015 or calendar year 2016, the payment will be zero.

§ 756.9 Availability of funds and timing of payments.

The total available program funds are $9 million as provided by Public Law 116-6 (the Consolidated Appropriations Act, 2019). OFF Program payments will be issued after all applications are received and FSA has approved the application.

§ 756.10 Miscellaneous provisions.

(a) Producers who are approved for OFF Program payment will not be required to purchase future NAP or crop insurance for those crops affected by the quarantine as is often required by other disaster programs, because the Oriental fruit fly quarantine was not an eligible covered loss by NAP, and RMA does not offer quarantine as an endorsement in Florida.

(b) All persons with a financial interest in a legal entity receiving payments under this part are jointly and severally liable for any refund, including related charges, that is determined to be due to FSA for any reason.

(c) In the event that any application under this part resulted from erroneous information or a miscalculation, the payment will be recalculated and any excess refunded to FSA with interest to be calculated from the date of disbursement.

(d) Any payment to any participant under this part will be made without regard to questions of title under State law, and without regard to any claim or lien against the commodity, or proceeds in favor of the owner or any other creditor except agencies of the U.S. Government. The regulations governing offsets and withholding in part 3 of this title apply to payments under this part.

(e) Any participant entitled to any payment may assign any payment(s) in accordance with regulations governing the assignment of payment in part 3 of this title.

(f) The regulations in part 11 of this title and part 780 of this chapter apply to determinations under this part.

§756.12 Payment limitation.

(a) For the program year 2015, direct or indirect payments made to an eligible person or legal entity, other than a joint venture or general partnership, will not exceed $125,000.

(b) The attribution of payment provisions in 7 CFR 1400.105 will be used to attribute payments to persons and legal entities for payment limitation determinations.

§756.13 Estates and trusts; minors.

(a) A receiver of an insolvent debtor's estate and the trustee of a trust estate will, for the purpose of this part, be considered to represent the insolvent affected producer or manufacturer and the beneficiaries of the trust, respectively.

(1) The production of the receiver or trustee will be considered to be the production of the represented person.

(2) Program documents executed by any such person will be accepted only if they are legally valid and such person has the authority to sign the applicable documents.

(b) [Reserved]

§756.14 Misrepresentation, scheme, or device.

(a) A producer will be ineligible to receive assistance under the OFF Program if the producer is determined by the FSA State committee or FSA county committee to have knowingly:

(1) Adopted any scheme or device that tends to defeat the purpose of the OFF Program;

(2) Made any fraudulent representation; or

(3) Misrepresented any fact affecting a determination under the OFF Program, then FSA will notify the appropriate investigating agencies of the United States and take steps deemed necessary to protect the interests of the Government.

(b) Any funds disbursed pursuant to this part to any person or operation engaged in a misrepresentation, scheme, or device, will be refunded to FSA. The remedies provided in this part are in addition to other civil, criminal, or administrative remedies that may apply.

§756.15 Death, incompetency, or disappearance.

In the case of the death, incompetency, or disappearance of any affected producer who would otherwise receive an OFF Program payment, such payment may be made to the person or persons specified in the regulations in part 707 of this chapter. The person requesting such payment must file Form FSA–325, "Application for Payment of Amounts Due Persons Who Have Died, Disappeared, or Have Been Declared Incompetent," as provided in part 707.

§756.16 Maintenance and inspection of records.

(a) Producers randomly selected for compliance spot checks by FSA must, in accordance with program notice instructions issued by the Deputy Administrator, provide adequate reports of revenue as applicable. The producer must report documentary evidence of crop revenue to FSA together with any supporting documentation to verify information entered on the application.

Verifiable documentation is preferred. If verifiable documentation is not available, FSA will accept reliable documentation, if determined to be acceptable by the FSA county committee.

(b) If supporting documentation is not presented to the county FSA office requesting the information within 30 calendar days of the request, producers will be determined ineligible for OFF Program benefits.

(c) The producer must maintain any existing books, records, and accounts supporting any information furnished in an approved OFF Program application for 3 years following the end of the year during which the application for payment was filed.

(d) The producer must permit authorized representatives of the Department of Agriculture and the General Accounting Office, during regular business hours, to inspect, examine, and make copies of such books, records, and accounts.

§ 756.17 Appeals.

Any producer who is dissatisfied with a determination made pursuant to this part may make a request for reconsideration or appeal of such determination in accordance with the appeal regulations in 7 CFR parts 11 and 780.

PART 759—DISASTER DESIGNATIONS AND NOTIFICATIONS

Sec.
759.1 Administration.
759.2 Purpose.
759.3 Abbreviations and definitions.
759.5 Secretarial disaster area determination and notification process.
759.6 EM to be made available.

AUTHORITY: 5 U.S.C. 301, 7 U.S.C. 1961 and 1989.

SOURCE: 77 FR 41254, July 13, 2012, unless otherwise noted.

§ 759.1 Administration.

(a) This part will be administered under the general supervision and direction of the Administrator, Farm Service Agency (FSA).

(b) FSA representatives do not have authority to modify or waive any of the provisions of the regulations of this part as amended or supplemented.

(c) The Administrator will take any action required by the regulations of this part that the Administrator determines has not already been taken. The Administrator will also:

(1) Correct or require correction of any action taken that is not in accordance with the regulations of this part; or

(2) Require withholding taking any action that is not in accordance with this part.

(d) No provision or delegation in these regulations will preclude the Administrator or a designee or other such person, from determining any question arising under this part, or from reversing or modifying any determination made under this part.

(e) Absent a delegation to the contrary, this part will be administered by the Deputy Administrator for Farm Programs of FSA on behalf of the Administrator of FSA or the Secretary, but nothing in this part will inhibit the ability of the Administrator of FSA or the person holding the equivalent position in the event of a reorganization to delegate the functions of DAFP under these regulations to another person. Likewise, nothing shall inhibit the ability of the Secretary to reassign any duties with respect to the designations of disasters under this part.

§ 759.2 Purpose.

(a) This part specifies the types of incidents that can result in an area being determined a disaster area, which under other regulations makes qualified farmers in such areas eligible for Emergency loans (EM) or eligible for such other assistance that may be available, based on Secretarial disaster designations. Nothing in this part overrides provision of those regulations that govern the actual administration and availability of the disaster assistance regulations.

(b) This part specifies the responsibility of the County Emergency Board (CEB), State Emergency Board (SEB), and the State Executive Director (SED) in regard to Secretarial Designations with regards to disasters. It also addresses matters relating to the handling of a Presidential declaration of disaster or the imposition of a USDA quarantine by the Secretary with respect to triggering the availability of EM loans.

§759.3 **Abbreviations and definitions.**

(a) *Abbreviations.* The following abbreviations apply to this part.

CEB means the County Emergency Board.

CED means the County Executive Director.

DAFP means the Deputy Administrator for Farm Programs of the Farm Service Agency.

EM means Emergency loan administered under 7 CFR part 764.

FSA means the Farm Service Agency.

LAR means the Loss Assessment Report.

SEB means the State Emergency Board.

SED means the State Executive Director.

USDA means the United States Department of Agriculture.

(b) *Definitions.* The following definitions apply to this part.

Administrator means the Administrator of FSA.

Contiguous county is used in reference to a primary county as defined in this section. A contiguous county is any county whose boundary touches at any point with that of the primary county. For programs other than the EM Program, disaster assistance regulations will specify whether benefits will be available only in the primary counties or also in the contiguous counties. For the EM Program that issue is addressed in §759.6, unless specified otherwise in the disaster assistance regulations for other programs or in §759.6 for the EM Program, only the "primary" county will be considered the qualifying "disaster county." Therefore, if the disaster assistance regulations specify that they cover the disaster area and contiguous counties, then the only eligible counties would be the primary county and those contiguous to that county. Coverage would not include coverage of those counties that are in turn contiguous to those counties that are contiguous to the primary county.

County is used when referring to a geographical area, a local administrative subdivision of a State or a similar political subdivision of the United States generally considered to be in county usage, for example, it includes an area referred to as a "county" or "parish." Except where otherwise specified, the use of the term county or similar political subdivision is for administrative purposes only.

CEB is comprised of the representatives of several USDA agencies that have responsibilities for reporting the occurrence of, and assessing the damage caused by, a natural disaster, and for requesting approval in declaring a county a disaster area.

CED is the person in charge of administering the local FSA county office for a particular county.

Disaster area is the county or counties declared or designated as a disaster area as a result of natural disaster related losses. The disaster area only includes the primary counties, but benefits may be available in the counties contiguous to the primary county if so provided by the disaster assistance regulations or, in the case of the EM Program, in §759.6.

LAR is a loss assessment report prepared by the CEB relating to the State and county where the potential disaster occurred and for which county or counties the CEB is responsible. The LAR includes as applicable, but is not limited to, starting and ending dates of the disaster, crop year affected, type of disaster incident, area of county affected by disaster; total number of farms affected, crop loss or pasture loss data associated with the applicable disaster (or both types of losses), livestock destroyed, and other property losses.

Natural disaster is a disaster in which unusual and adverse weather conditions or other natural phenomena have substantially affected farmers by causing severe physical losses, severe production losses, or both.

Primary county is a county determined to be a disaster area.

Presidential declaration is a declaration of a disaster by the President under the Robert T. Stafford Disaster Relief and Emergency Assistance Act (42 U.S.C. 5121-2) requiring Federal emergency assistance to supplement State and local efforts to save lives and protect property, public health and safety, or to avert or lessen the threat of a disaster.

Production losses (severe) within a county are those in which there has been a reduction county-wide of at least a 30 percent or more loss of production of at least one crop in the county.

SEB means the State Emergency Board which is comprised of the representatives of several USDA agencies having emergency program responsibilities at the State level. The board is required to respond to emergencies and carry out the Secretary's emergency preparedness responsibilities.

SED is the person who serves as the Chairperson of the USDA SEB in each State, is responsible for providing the leadership and coordination for all USDA emergency programs at the State level, and is subject to the supervision of DAFP.

Severe physical losses means, for the purpose of determining an Administrator's declaration of physical loss, losses that consist of severe damage to, or destruction of: Physical farm property including farmland (except sheet erosion); structures on the land including, but not limited to, building, fences, dams; machinery, equipment, supplies, and tools; livestock, livestock products, poultry and poultry products; harvested crops and stored crops.

Substantially affected when used to refer to producers and to the relationship of a particular producer to a particular disaster means a producer who has sustained qualifying physical or production losses, as defined in this section, as a result of the natural disaster.

U.S. Drought Monitor is a system for classifying drought severity according to a range of abnormally dry to exceptional drought. It is a collaborative effort between Federal and academic partners that is produced on a weekly basis to synthesize multiple indices, outlooks, and drought impacts on a map and in narrative form. This synthesis of indices is reported by the National Drought Mitigation Center.

United States means each of the several States, the Commonwealth of Puerto Rico, the Virgin Islands of the United States, Guam, American Samoa, and the Commonwealth of the Northern Mariana Islands. Extension of disaster assistance, following a disaster designation, to insular areas of the United States not covered by this definition of "United States" will be only as authorized by law, and as determined by the Administrator on behalf of the Secretary to be appropriate.

§ 759.5 Secretarial disaster area determination and notification process.

(a) *U.S. Drought Monitor.* With respect to drought and without requiring an LAR:

(1) If any portion of a county is physically located in an area with a Drought Monitor Intensity Classification value of D3 (drought-extreme) or higher during any part of the growing season of the crops affected by the disaster in the county, then the county will be designated a disaster area by the Secretary.

(2) If any portion of a county meets the threshold Drought Monitor Intensity Classification value of D2 (drought-severe) for at least 8 consecutive weeks during the growing season of affected crops, then the county will be designated a disaster area by the Secretary.

(b) *CEB and SEB recommendations.* In instances where counties have been impacted by a disaster but the county has not been designated a disaster area under the provisions of paragraph (a) of this section, CEB will make a disaster designation recommendation request to SEB when a disaster has resulted in severe production losses. The determination of the sufficiency of the production losses will be governed by the provisions in paragraph (c) of this section. The CEB may make such efforts as are needed to identify counties that have been impacted and had such production losses. A farmer, Indian Tribal Council, or local governing body may initiate the process by reporting production losses or drought conditions to CEB and suggesting that there be a recommendation in favor of designating a county as a disaster area. Recommendations by a CEB in favor of a disaster designation by a CEB under this paragraph are subject to the following:

(1) A LAR is required as part of a CEB disaster designation request. CEB

will submit a disaster designation request with a LAR to SEB for review and recommendation for approval by the Secretary. CEB's written request and SEB recommendation must be submitted within three months of the last day of the occurrence of a natural disaster.

(2) If SEB determines a qualifying natural disaster and loss have occurred, SEB will forward the recommendation to the Administrator. The natural disaster may include drought conditions that were not sufficiently severe to meet the criteria in paragraph (a) of this section. Since the U.S. Drought Monitor tracks only drought conditions, not specifically agricultural losses resulting from those conditions, it is possible for a drought that does not meet the criteria in paragraph (a) of this section to result in production losses that constitute a natural disaster.

(3) The Secretary or the Secretary's designee will make disaster area determinations. The Secretary may delegate the authority to the SED. In such case, the SED will act on behalf of the Secretary, subject to review by DAFP as may be appropriate and consistent with the delegation. The delegation of authority to the SED may be revoked by the authority making that delegation or by other authorized person. In all cases, DAFP may reverse any SED determination made in accordance with this section unless the delegation to the SED specifies that such review is not allowed.

(c) *Eligible production losses.* For purposes of making determinations under paragraph (b) of this section, in order for an area to be declared a disaster area under paragraph (b) of this section based on production losses, the county must have had production losses of 30 percent of at least one crop in the county as the result of a natural disaster.

(d) *Discretionary exception to production losses for designating a county as a disaster county.* For purposes of the EM program only, unless otherwise specified in the designation, a county may be designated by DAFP as a designated disaster county even though the conditions specified in paragraphs (a) through (c) of this section are not present so long as the disaster has otherwise produced such significant production losses, or other such extenuating circumstances so as to justify, in the opinion of the Secretary, the designation of a county as a disaster area. In making this determination, the Secretary may consider all relevant factors including such factors as the nature and extent of production losses; the number of farmers who have sustained qualifying production losses; the number of farmers that other lenders in the county indicate they will not be in position to provide emergency financing; whether the losses will cause undue hardship to a certain segment of farmers in the county; whether damage to particular crops has resulted in undue hardship; whether other Federal or State benefit programs, which are being made available due to the same disaster, will consequently lessen undue hardship and the demand for EM; and any other factors considered relevant.

§759.6 EM to be made available.

(a) For purposes of the EM Program under part 764, subpart I, of this chapter, a county will be considered an eligible disaster area as designated by FSA for coverage of the EM Program as follows:

(1) *Secretarial designations.* When production losses meet the requirements in §759.5 and the county has been designated as a disaster area for that reason, or when the discretionary exception to production losses for EM under §759.5(d) has been exercised, the primary and contiguous counties will be areas in which otherwise eligible producers can receive EM loans.

(2) *Physical loss notification.* When only qualifying physical losses occur, the SED will submit a request to the FSA Administrator to make a determination that a natural disaster has occurred in a county, resulting in severe physical losses. If the FSA Administrator determines that such a natural disaster has occurred, then EM can be made available to eligible farmers for physical losses only in the primary county (the county that was the subject of that determination) and the counties contiguous to that county.

(3) *USDA quarantine.* Any quarantine imposed by the Secretary of Agriculture under the Plant Protection Act or the animal quarantine laws, as defined in section 2509 of the Food, Agriculture, Conservation, and Trade Act of 1990, automatically authorizes EM for production and physical losses resulting from the quarantine in a primary county (the county in which the quarantine was in force) and (where the quarantine effects extend beyond that county) the counties contiguous to that primary county.

(4) *Presidential declaration.* Whenever the President declares a Major Disaster Declaration or an Emergency Declaration, FSA will make EM available to eligible applicants in declared and contiguous counties, provided:

(i) The Presidential declaration is not solely for Category A or Category B Public Assistance or Hazard Mitigation Grant Assistance, and

(ii) The Presidential Major Disaster declaration is for losses due to severe, general disaster conditions including but not limited to conditions such as flood, hurricane, or earthquake.

(b) [Reserved]

PART 760—INDEMNITY PAYMENT PROGRAMS

Subpart A—Dairy Indemnity Payment Program

Subpart B—General Provisions for Supplemental Agricultural Disaster Assistance Programs

Subpart C—Emergency Assistance for Livestock, Honeybees, and Farm-Raised Fish Program

Subpart R—Quality Loss Adjustment Program

AUTHORITY: 7 U.S.C. 4501 and 1531; 16 U.S.C. 3801, note; 19 U.S.C. 2497; Title III, Pub. L. 109–234, 120 Stat. 474; Title IX, Pub. L. 110–28, 121 Stat. 211; Sec. 748, Pub. L. 111–80, 123 Stat. 2131; Title I, Pub. L. 115–123, 132 Stat. 65; Title I, Pub. L. 116–20, 133 Stat. 871; and Division B, Title VII, Pub. L. 116–94, 133 Stat. 2658.

Subpart A—Dairy Indemnity Payment Program

AUTHORITY: 7 U.S.C. 450j-1.

SOURCE: 43 FR 10535, Mar. 14, 1978, unless otherwise noted.

PROGRAM OPERATIONS

§ 760.1 Administration.

This indemnity payment program will be carried out by FSA under the direction and supervision of the Deputy Administrator. In the field, the program will be administered by the State and county committees.

§ 760.2 Definitions.

For purposes of this subject, the following terms shall have the meanings specified:

Affected farmer means a person who produces whole milk which is removed from the commercial market any time from:

(1) Pursuant to the direction of a public agency because of the detection of pesticide residues in such whole milk by tests made by a public agency or under a testing program deemed adequate for the purpose by a public agency, or

(2) Pursuant to the direction of a public agency because of the detection of other residues of chemicals or toxic substances residues, or contamination from nuclear radiation or fallout in such whole milk by tests made by a public agency or under a testing program deemed adequate for the purpose by a public agency.

Affected manufacturer means a person who manufactures dairy products which are removed from the commercial market pursuant to the direction of a public agency because of the detection of pesticide residue in such dairy products by tests made by a public agency or under a testing program deemed adequate for the purpose by a public agency.

Application period means any period during which an affected farmer's whole milk is removed from the commercial market pursuant to direction of a public agency for a reason specified in paragraph (k) of this section and for which application for payment is made.

Base period means the calendar month or 4-week period immediately preceding removal of milk from the market.

Chemicals or Toxic Substances means any chemical substance or mixture as defined in the Toxic Substances Control Act (15 U.S.C. 2602).

Commercial market means (1) the market to which the affected farmer normally delivers his whole milk and from which it was removed because of detection therein of a residue of a violating substance(s) or (2) the market to which the affected manufacturer normally delivers his dairy products and from which they were removed because of detection therein of pesticide residue.

Contaminated milk means milk containing elevated levels of any violating substance that may affect public health based on tests made by the applicable public agency and resulting in the removal of the milk from the commercial market.

County committee means the FSA county committee.

Depopulation means, consistent with the American Veterinary Medical Association (AVMA)[1] definition, the rapid destruction of a population of cows with as much consideration given to the welfare of the animals as practicable.

Deputy Administrator means the Deputy Administrator for Farm Programs, FSA.

FSA means the Farm Service Agency, U.S. Department of Agriculture.

Milk handler means the marketing agency to or through which the affected dairy farmer marketed his whole milk at the time he was directed by the public agency to remove his whole milk from the commercial market.

Not marketable means no commercial market is available for affected cows to be slaughtered, processed, and marketed through the food chain system as determined by the Deputy Administrator.

Nuclear Radiation or Fallout means contamination from nuclear radiation or fallout from any source.

Pay period means (1) in the case of an affected farmer who markets his whole milk through a milk handler, the period used by the milk handler in settling with the affected farmer for his whole milk, usually biweekly or monthly, or (2) in the case of an affected farmer whose commercial market consists of direct retail sales to consumers, a calendar month.

Payment subject to refund means a payment which is made by a milk handler to an affected farmer, and which such farmer is obligated to refund to the milk handler.

Person means an individual, partnership, association, corporation, trust, estate, or other legal entity.

Pesticide means an economic poison which was registered pursuant to the provisions of the Federal Insecticide, Fungicide, and Rodenticide Act, as amended (7 U.S.C. 135 through 135k), and approved for use by the Federal Government.

----

[1] The AVMA Guidelines for the Depopulation of Animals is available at: *https://www.avma.org/sites/default/files/resources/AVMA-Guidelines-for-the-Depopulation-of-Animals.pdf.*

Public agency means any Federal, State or local public regulatory agency.

Removed from the commercial market means (1) produced and destroyed or fed to livestock, (2) produced and delivered to a handler who destroyed it or disposed of it as salvage (such as separating whole milk, destroying the fat, and drying the skim milk), or (3) produced and otherwise diverted to other than the commercial market.

Same loss means the event or trigger that caused the milk to be removed from the commercial market. For example, if milk is contaminated, the original cause of the contamination was the trigger and any loss related to that contamination would be considered the same loss.

Secretary means the Secretary of Agriculture of the United States or any officer or employee of the U.S. Department of Agriculture to whom he has delegated, or to whom he may hereafter delegate, authority to act in his stead.

State committee means the FSA State committee.

Violating substance means one or more of the following, as defined in this section: Pesticide, chemicals or toxic substances, or nuclear radiation or fallout.

Whole milk means milk as it is produced by cows.

[43 FR 10535, Mar. 14, 1978, as amended by Amdt. 1, 44 FR 36360, July 22, 1979; 52 FR 17935, May 13, 1987; 53 FR 44001, Nov. 1, 1988; 56 FR 1358, Jan. 14, 1991; 61 FR 18485, Apr. 26, 1996; 71 FR 27190, May 10, 2006; 84 FR 28176, June 18, 2019; 86 FR 70702, Dec. 13, 2021]

PAYMENTS TO DAIRY FARMERS FOR MILK

§ 760.3 Indemnity payments on milk.

(a) The amount of an indemnity payment for milk, including, but not limited to organic milk, made to an affected farmer who is determined by the county committee to be in compliance with all the terms and conditions of this subpart will be in the amount of the fair market value of the farmer's normal marketings for the application period, as determined in accordance with §§ 760.4 and 760.5, less:

(1) Any amount the affected farmer received for whole milk marketed during the application period; and

(2) Any payment not subject to refund that the affected farmer received from a milk handler with respect to milk removed from the commercial market during the application period.

(b) The eligible period for Dairy Indemnity Payment Program (DIPP) benefits for milk for the same loss is limited to 3 calendar months from when the first claim for milk benefits is approved. Upon written request from an affected farmer on the milk indemnity form authorized by the Deputy Administrator, the Deputy Administrator may authorize, at the Deputy Administrator's discretion, additional months of benefits for the affected farmer for milk due to extenuating circumstances, which may include allowing additional time for public agency approval of a removal plan for cow indemnification and confirmation of site disposal for affected cows. Additionally, the Deputy Administrator has discretion to approve additional months based on issues that are beyond the control of the affected farmer who is seeking cow indemnification, as well as when the affected farmer is following a plan to reduce chemical residues in milk, cows, and heifers to marketable levels.

[86 FR 70703, Dec. 13, 2021]

§760.4 Normal marketings of milk.

(a) The county committee shall determine the affected farmer's normal marketings which, for the purposes of this subpart, shall be the sum of the quantities of whole milk which such farmer would have sold in the commercial market in each of the pay periods in the application period but for the removal of his whole milk from the commercial market because of the detection of a residue of a violating substance.

(b) Normal marketings for each pay period are based on the average daily production during the base period.

(c) Normal marketings determined in paragraph (b) of this section are adjusted for any change in the daily average number of cows milked during each pay period the milk is off the market compared with the average number of cows milked daily during the base period.

(d) If only a portion of a pay period falls within the application period, normal marketings for such pay period shall be reduced so that they represent only that part of such pay period which is within the application period.

[43 FR 10535, Mar. 14, 1978, as amended by Amdt. 1, 44 FR 36360, July 22, 1979]

§760.5 Fair market value of milk.

(a) The county committee shall determine the fair market value of the affected farmer's normal marketings, which, for the purposes of this subpart, shall be the sum of the net proceeds such farmer would have received for his normal marketings in each of the pay periods in the application period.

(b) The county committee shall determine the net proceeds the affected farmer would have received in each of the pay periods in the application period (1) in the case of an affected farmer who markets his whole milk through a milk handler, by multiplying the affected farmer's normal marketings for each such pay period by the average net price per hundred-weight of whole milk paid during the pay period by such farmer's milk handler in the same area for whole milk similar in quality and butterfat test to that marketed by the affected farmer in the base period used to determine his normal marketings, or (2) in the case of an affected farmer whose commercial market consists of direct retail sales to consumers, by multiplying the affected farmer's normal marketings for each such pay period by the average net price per hundredweight of whole milk, as determined by the county committee, which other producers in the same area who marketed their whole milk through milk handlers received for whole milk similar in quality and butterfat test to that marketed by the affected farmer during the base period used to determine his normal marketings.

(c) In determining the net price for whole milk, the county committee shall deduct from the gross price therefor any transportation, administrative, and other costs of marketing which it determines are normally incurred by the affected farmer but which were not incurred because of the removal of his whole milk from the commercial market.

§ 760.6 Information to be furnished.

The affected farmer shall furnish to the county committee complete and accurate information sufficient to enable the county committee or the Deputy Administrator to make the determinations required in this subpart. Such information shall include, but is not limited to:

(a) A copy of the notice from, or other evidence of action by, the public agency which resulted in the removal of the affected farmer's whole milk from the commercial market.

(b) The specific name of the violating substance causing the removal of his whole milk from the commercial market, if not included in the notice or other evidence of action furnished under paragraph (a) of this section.

(c) The quantity and butterfat test of whole milk produced and marketed during the base period. This information must be a certified statement from the affected farmer's milk handler or any other evidence the county committee accepts as an accurate record of milk production and butterfat tests during the base period.

(d) The average number of cows milked during the base period and during each pay period in the application.

(e) If the affected farmer markets his whole milk through a milk handler, a statement from the milk handler showing, for each pay period in the application period, the average price per hundred-weight of whole milk similar in quality to that marketed by the affected farmer during the base period used to determine his normal marketings. If the milk handler has information as to the transportation, administrative, and other costs of marketing which are normally incurred by producers who market through the milk handler but which the affected farmer did not incur because of removal of his whole milk from the market, the average price stated by the milk handler shall be the average gross price paid producers less any such costs. If the milk handler does not have such information, the affected farmer shall furnish a statement setting forth such costs, if any.

(f) The amount of proceeds, if any, received by the affected farmer from the marketing of whole milk produced during the application period.

(g) The amount of any payments not subject to refund made to the affected farmer by the milk handler with respect to the whole milk produced during the application period and remove from the commercial market.

(h) To the extent that such information is available to the affected farmer, the name of any pesticide, chemical, or toxic substance used on the farm within 24 months prior to the application period, the use made of the pesticide, chemical, or toxic substance, the approximate date of such use, and the name of the manufacturer and the registration number, if any, on the label on the container of the pesticide, chemical, or toxic substance.

(i) To the extent possible, the source of the pesticide, chemical, or toxic substance that caused the contamination of the whole milk, the results of any laboratory tests on the feed supply, and the monthly milk testing results that detail the chemical residue levels.

(j) Such other information as the county committee may request to enable the county committee or the Deputy Administrator to make the determinations required in this subpart.

[43 FR 10535, Mar. 14, 1978, as amended by Amdt. 1, 44 FR 36360, June 22, 1979; 86 FR 70703, Dec. 13, 2021]

§ 760.7 Conditions required for milk or cow indemnity.

(a) An indemnity payment for milk or cows (dairy cows including, but not limited to, bred and open heifers) may be made under this subpart to an affected farmer under the conditions in this section.

(b) If the pesticide, chemical, or toxic substance, in the contaminated milk was used by the affected farmer, the affected farmer must establish that each of the conditions in this section are met:

(1) That the pesticide, chemical, or toxic substance, when used, was registered (if applicable) and approved for use as provided in § 760.2(f);

(2) That the contaminated milk was not the result of the affected farmer's failure to use the pesticide, chemical, or toxic substance, according to the directions and limitations stated on the label; and

(3) That the contaminated milk was not otherwise the affected farmer's fault.

(c) If the violating substance in the contaminated milk was not used by the affected farmer, the affected farmer must establish that each of the conditions in this section are met:

(1) The affected farmer did not know or have reason to believe that any purchased feed contained a violating substance;

(2) None of the milk was produced by dairy cattle that the affected farmer knew, or had reason to know at the time they were acquired, had elevated levels of a violating substance; and

(3) The contaminated milk was not otherwise the affected farmer's fault.

(d) The affected farmer has adopted recommended practices and taken action to eliminate or reduce chemical residues of violating substances from the milk as soon as practicable following the initial discovery of the contaminated milk.

[86 FR 70703, Dec. 13, 2021]

§760.8 Application for payments for milk.

The affected farmer or his legal representative, as provided in §§760.25 and 760.29, must sign and file an application for payment on a form which is approved for that purpose by the Deputy Administrator. The form must be filed with the county FSA office for the county where the farm headquarters are located no later than December 31 following the end of the fiscal year in which the loss occurred, or such later date as the Deputy Administrator may specify. The application for payment shall cover application periods of at least 28 days, except that, if the entire application period, or the last application period, is shorter than 28 days, applications for payment may be filed for such shorter period. The application for payment shall be accompanied by the information required by §760.6 as well as any other information which will enable the county committee to determine whether the making of an indemnity payment is precluded for any of the reasons set forth in §760.7. Such information shall be submitted on forms approved for the purpose by the Deputy Administrator.

[43 FR 10535, Mar. 14, 1978, as amended at 51 FR 12986, Apr. 17, 1986; 52 FR 17935, May 13, 1987]

§760.9 Payments for the same loss.

(a) No indemnity payment shall be made for contaminated milk resulting from residues of chemicals or toxic substances if, within 30 days after receiving a complete application, the Deputy Administrator determines that other legal recouse is available to the farmer. An application shall not be deemed complete unless it contains all information necessary to make a determination as to whether other legal recourse is available to the farmer. However, notwithstanding such a determination, the Deputy Administrator may reopen the case at a later date and make a new determination on the merits of the case as may be just and equitable.

(b) In the event that a farmer receives an indemnity payment under this subpart, and such farmer is later compensated for the same loss by the person (or the representative or successor in interest of such person) responsible for such loss, the indemnity payment shall be refunded by the farmer to the Department of Agriculture: *Provided,* That the amount of such refund shall not exceed the amount of other compensation received by the farmer.

(c) For any affected farmer that exceeded 3 months of milk indemnity payments before December 13, 2021 no further payments for milk indemnity will be made for the same loss except as provided in §760.3(b) and the affected farmer may apply for cow indemnity as specified in this subpart.

(d) An affected farmer that has an approved application for cow indemnity is no longer eligible for milk indemnity payments for the same loss.

(e) Cows purchased or bred after the initial discovery of the milk contamination are not eligible for DIPP benefits due to the same loss.

[Amdt. 1, 44 FR 36361, June 22, 1979, as amended at 84 FR 28176, June 18, 2019; 86 FR 70703, Dec. 13, 2021]

§ 760.10 Indemnity payments for cows.

(a) The Deputy Administrator for Farm Programs (DAFP) will determine eligibility for DIPP indemnification based on if the cows of the affected farmer are likely to be not marketable for 3 months or longer [from the date the affected farmer submits an application for cow indemnification per § 760.13]. The Deputy Administrator will review the following factors in making that determination:

(1) Milk testing results;

(2) Non marketability of affected cows through commercial marketing facilities;

(3) Type and source of chemical residues impacting the milk and animal tissues; and

(4) Projected duration for chemical residue reduction including the actions taken by the affected farmer to reduce the chemical residues to marketable levels since the affected cows were discovered.

(b) See § 760.11 for indemnity payment eligibility for bred and open heifers.

(c) Affected farmers applying for indemnification of cows, including heifers, must develop a removal plan both to permanently remove the affected cows by depopulating the cows.

(1) The removal plan for affected cows for which an affected farmer applies for indemnification under DIPP must be approved by the applicable public agency where the cows are located and must be in accordance with any applicable Environmental Protection Agency (EPA) and public agency depopulation and animal disposal requirements and guidelines, including contaminant disposal requirements, in the State where the affected cows are located.

(2) The approved removal plan must be submitted with the application for indemnification.

(d) The amount of an indemnity payment for cows to an affected farmer who is determined by the Deputy Administrator to be eligible for indemnification and by the county committee to be in compliance with all the terms and conditions of this subpart will be based on the national average fair market value of the cows. DIPP cow indemnification will be based on the 100 percent value of the Livestock Indemnity Program (LIP) rates as applicable for the calendar year for milk indemnification established for dairy cows, per head. For example, for a 100-cow farm: 100 cows multiplied by $1,300 (2021 LIP rate based on 100 percent value of average cow) = $130,000 payment.

(e) For any cow indemnification payment under this section or § 760.11, the affected farmer has the option to receive 50 percent of calculated payment in advance after application approval with the remaining fifty percent paid after the affected cows have been depopulated and removed. Otherwise, the affected farmer may choose to receive 100 percent of payment after cows have been depopulated and removed. Documented records of depopulation and removal of affected cows must be provided to FSA to the satisfaction of the county committee, before the final payment will be made.

(f) Upon written request from an affected farmer on a form authorized by the Deputy Administrator, the Deputy Administrator may approve, at the Deputy Administrator's discretion, indemnification of additional affected cows as specified in paragraphs (f)(1) through (3) of this section.

(1) The affected cows were depopulated or died above normal mortality rates for cows between approval of the affected farmer's application for the first month of milk indemnity and public agency approval of the affected farmer's removal plan for cow indemnification. Normal mortality rates established annually by the FSA State committee for their state for the following cow and heifer weight groups will be used:

(i) Dairy, nonadult less than 400 pounds;

(ii) Dairy, nonadult 400 pounds or more; and

(iii) Dairy, adult cow.

(2) This request may include both cows that were included in applications

for milk indemnity and heifers that were affected from the same loss.

(3) An affected farmer making such a request must submit the information specified in §760.12(c).

(g) Affected cows that are marketed as cull or for breeding are not eligible for indemnification.

[86 FR 70703, Dec. 13, 2021]

§760.11 **Indemnity payments for bred and open heifers.**

(a) Bred (young dairy female in gestation) and open (young dairy female not in gestation) heifers that contain elevated levels of chemical residues as the result of the same loss may be eligible for indemnification through DIPP. For affected bred and open heifers participating affected farmers may receive indemnification if the farmer's dairy cows were determined to be likely not marketable for three months or longer according to §760.10(a) and the Deputy Administrator determines the bred and open heifers to be eligible under paragraph (b) of this section. Except as provided in this section or otherwise stated in this subpart, the provisions in this subpart for cow indemnity apply equally to bred and open heifers, for example the removal requirements in §760.10(b).

(b) The county committee will make the recommendation to the Deputy Administrator to determine if eligible bred and open heifers that have been affected by the same loss will likely be not marketable for 3 months or longer from the date the affected farmer submits an application for cow indemnification per §760.13 because of elevated levels of chemical residues that will pass through milk once lactating. Affected farmers must provide the information specified in §760.12(a) and (b) for the county committee to make a recommendation of eligibility to the Deputy Administrator. The Deputy Administrator will take into consideration the recommendation of the county committee in making its eligibility determination.

(c) The amount of the cow indemnity for bred and open heifers will be based on the national average fair market value of the non-adult heifers. DIPP bred and open heifer indemnification will be based on the 100 percent value of the Livestock Indemnity Program (LIP) rates as applicable for the calendar year of milk indemnification established for non-adult dairy, by weight range, per head. For example, for an affected farmer with 40 bred or open heifers at different weight ranges: 10 bred heifers at 800 pounds or more multiplied by $986.13 ($9861.30), 10 bred or open heifers at 400 to 799 pounds multiplied by $650.00 ($6500.00), 10 open heifers at 250 to 399 pounds multiplied by $325.00 ($3250.00), and 10 open heifers 250 pounds or less multiplied by $57.65 ($576.50) = $20,187.80 payment.

[86 FR 70704, Dec. 13, 2021]

§760.12 **Information to be furnished for payment on dairy cows, and bred and open heifers.**

(a) To apply for DIPP for affected cows, the affected farmer must provide the county committee complete and accurate information to enable the Deputy Administrator to make the determinations required in this subpart in addition to providing the information requested in §760.6(a), (b), (h), and (i), if not previously provided to FSA in a milk indemnity application. The information specified in this section must be submitted as part of the cow indemnity application and includes, but is not limited to, the following items:

(1) An inventory of all dairy cows as of the date of application including lactating cows, bred heifers, and open heifers on the farm;

(2) A detailed description and timeline of how, where, and when cows will be depopulated and permanently removed from the farm (the removal plan);

(3) Documentation of public agency approval of the removal plan for cow depopulation and cow and contaminate disposal in accordance with any applicable EPA and public agency disposal requirements and guidelines;

(4) Documentation from 2 separate commercial markets stating that such market declined to accept the affected cows through a cull cow market, slaughter facility, or processing facility due to elevated levels of chemical residues;

(5) Documentation of any projected timelines for reducing chemical residues, any actions the affected farmer has taken to reduce chemical residues to marketable levels including any documents verifying steps undertaken, and any professional assistance obtained, including, discussion of strategy with the public agencies; and

(6) Any other documentation that may support the determination that the affected cows or milk from such cows is likely to be not marketable for longer than 3 months; and other documentation as requested or determined to be necessary by the county committee or the Deputy Administrator.

(b) To apply for DIPP for bred and open heifers the affected farmer must provide the information specified in paragraph (a) of this section and: veterinarian records, blood test results, and other testing information requested by the county committee for the recommendation specified in § 760.11(b) and eligibility for indemnification.

(c) To request consideration for indemnification of affected cows and heifers under § 760.10(e), the affected farmer must submit the information specified in paragraphs (c)(1) and (2) of this section to provide an accounting of affected cows and heifers that were depopulated or died above normal mortality rates for cows between approval of the affected farmer's application for the first month of milk indemnity and the public agency approval of the affected farmer's removal plan for cow indemnification.

(1) Herd health record documenting cow and heifer deaths; and

(2) Farm inventory or other record identifying the loss of dairy cows and heifers.

(d) The affected farmer certifies at application that once the cow indemnity application is approved, the affected farmer will dry off all lactating cows in a reasonable timeframe and discontinue milking.

[86 FR 70704, Dec. 13, 2021]

§ 760.13 Application for payment of cows.

(a) Any affected farmer may apply for cow indemnity under §§ 760.10 and 760.11. To apply for DIPP for affected cows, the affected farmer must sign and file an application for payment on a form that is approved for that purpose by the Deputy Administrator and provide the information described in § 760.12.

(b) The form must be filed with the FSA county office for the county where the farm headquarters is located by December 31 following the fiscal year end in which the affected farmer's milk was removed from the commercial market, except that affected farmers that have received 3 months of milk indemnity payments prior to December 13, 2021, must file the form within 120 days after December 13, 2021. Upon written request from an affected farmer and at Deputy Administrator's discretion, the deadline for that affected farmer may be extended.

[86 FR 70705, Dec. 13, 2021]

PAYMENTS TO MANUFACTURERS
AFFECTED BY PESTICIDES

§ 760.20 Payments to manufacturers of dairy products.

An indemnity payment may be made to the affected manufacturer who is determined by the Deputy Administrator to be in compliance with all the terms and conditions of this subpart in the amount of the fair market value of the product removed from the commercial market because of pesticide residues, less any amount the manufacturer receives for the product in the form of salvage.

NOTE: Manufacturers are not eligible for payment when dairy products are contaminated by chemicals, toxic substances (other than pesticides) or nuclear radiation or fallout.

[43 FR 10535, Mar. 14, 1978, as amended at 47 FR 24689, June 8, 1982]

§ 760.21 Application for payments by manufacturers.

The affected manufacturer, or his legal representatives, shall file an application for payment with the Deputy Administrator, FSA, Washington, D.C., through the county office serving the county where the contaminated product is located. The application for payment may be in the form of a letter or memorandum. Such letter or memorandum, however, must be accompanied by acceptable documentation to support such application for payment.

§ 760.22 Information to be furnished by manufacturer.

The affected manufacturer shall furnish the Deputy Administrator, through the county committee, complete and accurate information sufficient to enable him to make the determination as to the manufacturer's eligibility to receive an indemnity payment. Such information shall include, but is not limited to:

(a) A copy of the notice or other evidence of action by the public agency which resulted in the product being removed from the commercial market.

(b) The name of the pesticide causing the removal of the product from the commercial market and, to the extent possible, the source of the pesticide.

(c) A record of the quantity of milk or butterfat used to produce the product for which an indemnity payment is requested.

(d) The identity of any pesticide used by the affected manufacturer.

(e) Such other information as the Deputy Administrator may request to enable him to make the determinations required in this subpart.

§ 760.23 Other requirements for manufacturers.

An indemnity payment may be made under this subpart to an affected manufacturer only under the following conditions:

(a) If the pesticide contaminating the product was used by the affected manufacturer, he establishes each of the following: (1) That the pesticide, when used, was registered and recommended for such use as provided in § 760.2(f); (2) that the contamination of his product was not the result of his failure to use the pesticide in accordance with the directions and limitations stated on the label of the pesticide; and (3) that the contamination of his product was not otherwise his fault.

(b) If the pesticide contaminating the product was not used by the affected manufacturer: (1) He did not know or have reason to believe that the milk from which the product was processed contained a harmful level of pesticide residue, and (2) the contamination of his product was not otherwise his fault.

(c) In the event that a manufacturer receives an indemnity payment under this subpart, and such manufacturer is later compensated for the same loss by the person (or the representative or successor in interest of such person) responsible for such loss, the indemnity payment shall be refunded by the manufacturer to the Department of Agriculture: *Provided*, That the amount of such refund shall not exceed the amount of other compensation received by the manufacturer.

[43 FR 10535, Mar. 14, 1978, as amended at 47 FR 24689, June 8, 1982; 51 FR 12987, Apr. 17, 1986; 52 FR 17935, May 13, 1987]

GENERAL PROVISIONS

§ 760.24 Limitation of authority.

(a) County executive directors and State and county committees do not have authority to modify or waive any of the provisions of the regulations in this subpart.

(b) The State committee may take any action authorized or required by the regulations in this subpart to be taken by the county committee when such action has not been taken by the county committee. The State committee may also:

(1) Correct, or require a county committee to correct, any action taken by such county committee which is not in accordance with the regulations in this subpart, or (2) require a county committee to withhold taking any action which is not in accordance with the regulations in this subpart.

(c) No delegation herein to a State or county committee shall preclude the Deputy Administrator or his designee from determining any question arising under the regulations in this subpart or from reversing or modifying any determination made by a State or county committee.

§ 760.25 Estates and trusts; minors.

(a) A receiver of an insolvent debtor's estate and the trustee of a trust estate shall, for the purpose of this subpart, be considered to represent an insolvent affected farmer or manufacturer and

77

the beneficiaries of a trust, respectively, and the production of the receiver or trustee shall be considered to be the production of the person or manufacturer he represents. Program documents executed by any such person will be accepted only if they are legally valid and such person has the authority to sign the applicable documents.

(b) An affected dairy farmer or manufacturer who is a minor shall be eligible for indemnity payments only if he meets one of the following requirements:

(1) The right of majority has been conferred on him by court proceedings or by statute;

(2) A guardian has been appointed to manage his property and the applicable program documents are signed by the guardian; or

(3) A bond is furnished under which the surety guarantees any loss incurred for which the minor would be liable had he been an adult.

§ 760.26 Appeals.

The appeal regulations issued by the Administrator, FSA, part 780 of this chapter, shall be applicable to appeals by dairy farmers or manufacturers from determinations made pursuant to the regulations in this subpart.

§ 760.27 Setoffs.

(a) If the affected farmer or manufacturer is indebted to any agency of the United States and such indebtedness is listed on the county debt record, indemnity payments due the affected farmer or manufacturer under the regulations in this part shall be applied, as provided in the Secretary's setoff regulations, part 13 of this title, to such indebtedness.

(b) Compliance with the provisions of this section shall not deprive the affected farmer or manufacturer of any right he would otherwise have to contest the justness of the indebtedness involved in the setoff action, either by administrative appeal or by legal action.

§ 760.28 Overdisbursement.

If the indemnity payment disbursed to an affected farmer or to a manufacturer exceeds the amount authorized under the regulations in this subpart,

the affected farmer or manufacturer shall be personally liable for repayment of the amount of such excess.

§ 760.29 Death, incompetency, or disappearance.

In the case of the death, incompetency, or disappearance of any affected farmer or manufacturer who would otherwise receive an indemnity payment, such payment may be made to the person or persons specified in the regulations contained in part 707 of this chapter. The person requesting such payment shall file Form FSA-325, "Application for Payment of Amounts Due Persons Who Have Died, Disappeared, or Have Been Declared Incompetent," as provided in that part.

[43 FR 10535, Mar. 14, 1978, as amended at 47 FR 24689, June 8, 1982]

§ 760.30 Records and inspection thereof.

(a) The affected farmer, as well as his milk handler and any other person who furnished information to such farmer or to the county committee for the purpose of enabling such farmer to receive a milk indemnity payment under this subpart, shall maintain any existing books, records, and accounts supporting any information so furnished for 3 years following the end of the year during which the application for payment was filed. The affected farmer, his milk handler, and any other person who furnishes such information to the affected farmer or to the county committee shall permit authorized representatives of the Department of Agriculture and the General Accounting Office, during regular business hours, to inspect, examine, and make copies of such books, records, and accounts.

(b) The affected manufacturer or any other person who furnishes information to the Deputy Administrator for the purposes of enabling such manufacturer to receive an indemnity payment under this subpart shall maintain any books, records, and accounts supporting any information so furnished for 3 years following the end of the year during which the application for payment was filed. The affected manufacturer or any other person who furnishes such information to the Deputy Administrator shall permit authorized representatives of the Department of Agriculture and the General Accounting Office, during regular business hours, to inspect, examine, and make copies of such books, records, and accounts.

§760.31 Assignment.

No assignment shall be made of any indemnity payment due or to come due under the regulations in this subpart. Any assignment or attempted assignment of any indemnity payment due or to come due under this subpart shall be null and void.

§760.32 Instructions and forms.

The Deputy Administrator shall cause to be prepared such forms and instructions as are necessary for carrying out the regulations in this subpart. Affected farmers and manufacturers may obtain information necessary to make application for a dairy indemnity payment from the county FSA office. Form FSA–373—Application for Indemnity Payment, is available at the county ASC office.

[43 FR 10535, Mar. 14, 1978, as amended at 47 FR 24689, June 8, 1982]

§760.33 Availability of funds.

(a) Payment of indemnity claims will be contingent upon the availability of FSA funds to pay such claims. Claims will be, to the extent practicable within funding limits, paid from available funds, on a first-come, first-paid basis, based on the date FSA approves the application, until funds available in that fiscal year have been expended.

(b) DIPP claims received in a fiscal year after all available funds have been expended will not receive payment for such claims.

[75 FR 41367, July 16, 2010]

Subpart B—General Provisions for Supplemental Agricultural Disaster Assistance Programs

SOURCE: 74 FR 31571, July 2, 2009, unless otherwise noted.

§760.101 Applicability.

(a) This subpart establishes general conditions for this subpart and subparts C through H of this part and applies only to those subparts. Subparts C through H cover the following programs provided for in the "2008 Farm Bill" (Pub. L. 110–246):

(1) Emergency Assistance for Livestock, Honey Bees, and Farm-Raised Fish Program (ELAP);

(2) Livestock Forage Disaster Program (LFP);

(3) Livestock Indemnity Payments Program (LIP);

(4) Supplemental Revenue Assistance Payments Program (SURE); and

(5) Tree Assistance Program (TAP).

(b) To be eligible for payments under these programs, participants must comply with all provisions under this subpart and the relevant particular subpart for that program. All other provisions of law also apply.

§760.102 Administration of ELAP, LFP, LIP, SURE, and TAP.

(a) The programs in subparts C through H of this part will be administered under the general supervision and direction of the Administrator, Farm Service Agency (FSA), and the Deputy Administrator for Farm Programs, FSA (who is referred to as the "Deputy Administrator" in this part).

(b) FSA representatives do not have authority to modify or waive any of the provisions of the regulations of this part as amended or supplemented, except as specified in paragraph (e) of this section.

(c) The State FSA committee will take any action required by the regulations of this part that the county FSA committee has not taken. The State FSA committee will also:

(1) Correct, or require a county FSA committee to correct, any action taken by such county FSA committee that is not in accordance with the regulations of this part or

(2) Require a county FSA committee to withhold taking any action that is not in accordance with this part.

(d) No provision or delegation to a State or county FSA committee will preclude the Administrator, the Deputy Administrator for Farm Programs, or a designee or other such person, from determining any question arising under the programs of this part, or from reversing or modifying any determination made by a State or county FSA committee.

(e) The Deputy Administrator for Farm Programs may authorize State and county FSA committees to waive or modify non-statutory deadlines, or other program requirements of this part in cases where lateness or failure to meet such requirements does not adversely affect operation of the programs in this part. Participants have no right to seek an exception under this provision. The Deputy Administrator's refusal to consider cases or circumstances or decision not to exercise this discretionary authority under this provision will not be considered an adverse decision and is not appealable.

§ 760.103 Eligible producer.

(a) In general, the term "eligible producer" means, in addition to other requirements as may apply, an individual or entity described in paragraph (b) of this section that, as determined by the Secretary, assumes the production and market risks associated with the agricultural production of crops or livestock on a farm either as the owner of the farm, when there is no contract grower, or a contract grower of the livestock when there is a contract grower.

(b) To be eligible for benefits, an individual or entity must be a:

(1) Citizen of the United States;

(2) Resident alien; for purposes of this part, resident alien means "lawful alien" as defined in 7 CFR part 1400;

(3) Partnership of citizens of the United States; or

(4) Corporation, limited liability corporation, or other farm organizational structure organized under State law.

§ 760.104 Risk management purchase requirements.

(a) To be eligible for program payments under:

(1) ELAP, SURE, and TAP, eligible producers for any commodity at any location for which the producer seeks benefits must have for every commodity on every farm in which the producer has an interest for the relevant program year:

(i) In the case of an "insurable commodity," (which for this part means a commodity for which the Deputy Administrator determines catastrophic coverage is available from the USDA Risk Management Agency (RMA)) obtained catastrophic coverage or better under a policy or plan of insurance administered by RMA under the Federal Crop Insurance Act (FCIA) (7 U.S.C. 1501-1524), except that this obligation will not include crop insurance pilot programs so designated by RMA or to forage crops intended for grazing, and

(ii) In the case of a "noninsurable commodity," (which is any commodity for which, as to the particular production in question, is not an "insurable commodity," but for which coverage is available under the Noninsured Crop Disaster Assistance Program (NAP) operated under 7 CFR part 1437), have obtained NAP coverage by filing the proper paperwork and fee within the relevant deadlines, except that this requirement will not include forage on grazing land.

(2) LFP, with respect to those grazing lands incurring losses for which assistance is being requested, eligible livestock producers must have:

(i) Obtained a policy or plan of insurance for the forage crop under FCIA, or

(ii) Filed the required paperwork and paid the administrative fee by the applicable State filing deadline for NAP coverage for that grazing land.

(b) Producers who did not purchase a policy or plan of insurance administered by RMA in accordance with FCIA (7 U.S.C. 1501-1524), or NAP coverage for their applicable crops, will not be eligible for assistance under ELAP, LFP, SURE, and TAP, as provided in paragraph (a) of this section unless the producer is one of the classes of farmers for which an exemption under § 760.107 apply, is exempt under the "buy-in" provisions of this subpart, or is granted relief from that requirement by the Deputy Administrator under some other provision of this part.

(c) Producers who have obtained insurance by a written agreement as specified in § 400.652(d) of this title even though that production would not normally be considered an "insurable commodity" under the rules of this subpart, will be considered to have met the risk management purchase requirement of this subpart with respect to such production. The commodity to which the agreement applies will be considered for purposes of this subpart to be an "insurable commodity."

(d) Producers by an administrative process who were granted NAP coverage for the relevant period as a form of relief in an administrative proceeding, or who were awarded NAP coverage for the relevant period through an appeal through the National Appeals Division (NAD), will be considered as having met the NAP eligibility criteria of this section for that crop as long as the applicable NAP service fee has been paid.

(e) The risk management purchase requirement for programs specified under this part will be determined based on the initial intended use of a crop at the time a policy or plan of insurance or NAP coverage was purchased and as reported on the acreage report.

[74 FR 31571, July 2, 2009, as amended at 74 FR 46673, Sept. 11, 2009]

§760.105 Waiver for certain crop years; buy-in.

(a) For the 2008 crop year, the insurance or NAP purchase requirements of §760.104 (this is referred to as the "purchase" requirement) will be waived for eligible producers for losses during the 2008 crop year if the eligible producer paid a fee (buy-in fee) equal to the applicable NAP service fee or catastrophic risk protection plan fee to the Secretary by September 16, 2008. Payment of a buy-in fee under this section is for the sole purpose of becoming eligible for participation in ELAP, LFP, SURE, and TAP. Payment of a buy-in fee does not provide any actual insurance or NAP coverage or assistance.

(b) For the 2009 crop year, the purchase requirement will be waived for purchases where the closing date for coverage occurred prior to August 14, 2008, so long as the buy-in fee set by the Secretary of Agriculture was paid by January 12, 2009.

(c) Any producer of 2008 commodities who is otherwise ineligible because of the purchase requirement and who did not meet the conditions of paragraph (a) of this section may still be covered for ELAP, SURE, or TAP assistance if the producer paid the applicable fee de-scribed in paragraph (d) of this section no later than May 18, 2009, provided that in the case of each:

(1) Insurable commodity, excluding grazing land, the eligible producers on the farm agree to obtain a policy or plan of insurance under FCIA (7 U.S.C. 1501–1524), excluding a crop insurance pilot program under that subtitle, for the next insurance year for which crop insurance is available to the eligible producers on the farm at a level of coverage equal to 70 percent or more of the recorded or appraised average yield indemnified at 100 percent of the expected market price, or an equivalent coverage, and

(2) Noninsurable commodity, the eligible producers on the farm must agree to file the required paperwork, and pay the administrative fee by the applicable State filing deadline, for NAP for the next year for which a policy is available.

(d) For producers seeking eligibility under paragraph (c) of this section, the applicable buy-in fee for the 2008 crop year was the catastrophic risk protection plan fee or the applicable NAP service fee in effect prior to NAP service fee adjustments specified in the 2008 Farm Bill.

§760.106 Equitable relief.

(a) The Secretary may provide equitable relief on a case-by-case basis for the purchase requirement to eligible participants that:

(1) Are otherwise ineligible or provide evidence, satisfactory to FSA, that the failure to meet the requirements of §760.104 for one or more eligible crops on the farm was unintentional and not because of any fault of the participant, as determined by the Secretary, or

(2) Failed to meet the requirements of §760.104 due to the enactment of the 2008 Farm Bill after the:

(i) Applicable sales closing date for a policy or plan of insurance in accordance with the FCIA (7 U.S.C. 1501–1524) or

(ii) Application closing date for NAP.

(b) Equitable relief will not be granted to participants in instances of:

(1) A scheme or device that had the effect or intent of defeating the purposes of a program of insurance, NAP,

or any other program administered under this part or elsewhere in this title,

(2) An intentional decision to not meet the purchase or buy-in requirements,

(3) Producers against whom sanctions have been imposed by RMA or FSA prohibiting the purchase of coverage or prohibiting the receipt of payments otherwise payable under this part,

(4) Violations of highly erodible land and wetland conservation provisions of 7 CFR part 12,

(5) Producers who are ineligible under any provisions of law, including regulations, relating to controlled substances (see for example 7 CFR 718.6), or

(6) A producer's debarment by a federal agency from receiving any federal government payment if such debarment included payments of the type involved in this matter.

(c) In general, no relief that is discretionary will be allowed except upon a finding by the Deputy Administrator or the Deputy Administrator's designee that the person seeking the relief acted in good faith as determined in accordance with such rules and procedures as may be set by the Deputy Administrator.

[74 FR 31571, July 2, 2009, as amended at 76 FR 54075, Aug. 31, 2011]

§ 760.107 Socially disadvantaged, limited resource, or beginning farmer or rancher.

(a) Risk management purchase requirements, as provided in § 760.104, will be waived for a participant who, as specified in paragraphs (b)(1) through (3) of this section, is eligible to be considered a "socially disadvantaged farmer or rancher," a "limited resource farmer or rancher," or a "beginning farmer or rancher."

(b) To qualify for this section as a "socially disadvantaged farmer or rancher," "limited resource farmer or rancher," or "beginning farmer or rancher," participants must meet eligibility criteria as follows:

(1) A "socially disadvantaged farmer or rancher" is, for this section, a farmer or rancher who is a member of a socially disadvantaged group whose members have been subjected to racial

or ethnic prejudice because of their identity as members of a group without regard to their individual qualities. Gender is not included as a covered group. Socially disadvantaged groups include the following and no others unless approved in writing by the Deputy Administrator:

(i) American Indians or Alaskan Natives,

(ii) Asians or Asian-Americans,

(iii) Blacks or African Americans,

(iv) Native Hawaiians or other Pacific Islanders, and

(v) Hispanics.

(2) A "limited resource farmer or rancher" means for this section a producer who is both:

(i) A producer whose direct or indirect gross farm sales do not exceed $100,000 in both of the two calendar years that precede the calendar year that corresponds to the relevant program year, adjusted upwards for any general inflation since fiscal year 2004, inflation as measured using the Prices Paid by Farmer Index compiled by the National Agricultural Statistics Service (NASS), and

(ii) A producer whose total household income is at or below the national poverty level for a family of four, or less than 50 percent of the county median household income for the same two calendar years referenced in paragraph (b)(2)(i) of this section, as determined annually using Commerce Department data. (Limited resource farmer or rancher status can be determined using a Web site available through the Limited Resource Farmer and Rancher Online Self Determination Tool through the National Resource and Conservation Service at *http:// www.lrftool.sc.egov.usda.gov/tool.asp.*)

(3) A "beginning farmer or rancher" means for this section a person or legal entity who for a program year both:

(i) Has never previously operated a farm or ranch, or who has not operated a farm or ranch in the previous 10 years, applicable to all members (shareholders, partners, beneficiaries, etc., as fits the circumstances) of an entity, and

(ii) Will have or has had for the relevant period materially and substantially participated in the operation of a farm or ranch.

(c) If a legal entity requests to be considered a "socially disadvantaged," "limited resource," or "beginning" farmer or rancher, at least 50 percent of the persons in the entity must in their individual capacities meet the definition as provided in paragraphs (b)(1) through (3) of this section and it must be clearly demonstrated that the entity was not formed for the purposes of avoiding the purchase requirements or formed after the deadline for the purchase requirement.

[74 FR 31571, July 2, 2009, as amended at 76 FR 54075, Aug. 31, 2011]

§760.108   Payment limitation.

(a) For 2008, no person, as defined and determined under the provisions in part 1400 of this title in effect for 2008 may receive more than:

(1) $100,000 total for the 2008 program year under ELAP, LFP, LIP, and SURE combined or

(2) $100,000 for the 2008 program year under TAP.

(b) For 2009 and subsequent program years, no person or legal entity, excluding a joint venture or general partnership, as determined by the rules in part 1400 of this title may receive, directly or indirectly, more than:

(1) $100,000 per program year total under ELAP, LFP, LIP, and SURE combined; or

(2) $100,000 per program year under TAP.

(c) The Deputy Administrator may take such actions as needed, whether or not specifically provided for, to avoid a duplication of benefits under the multiple programs provided for in this part, or duplication of benefits received in other programs, and may impose such cross-program payment limitations as may be consistent with the intent of this part.

(1) FSA will review ELAP payments after the funding factor as specified in §760.208 is determined to be 100 percent. FSA will ensure that total ELAP payments provided to a participant in a year, together with any amount provided to the same participant for the same loss as a result of any Federal crop insurance program, the Noninsured Crop Disaster Assistance Program, or any other Federal disaster program, plus the value of the com-

modity that was not lost, is not more than 95 percent of the value of the commodity in the absence of the loss, as estimated by FSA.

(2) [Reserved]

(d) In applying the limitation on average adjusted gross income (AGI) for 2008, an individual or entity is ineligible for payment under ELAP, LFP, LIP, SURE, and TAP if the individual's or entity's average adjusted gross income (AGI) exceeds $2.5 million for 2007, 2006, and 2005 under the provisions in part 1400 of this title in effect for 2008.

(e) For 2009 through 2011, the average AGI limitation provisions in part 1400 of this title relating to limits on payments for persons or legal entities, excluding joint ventures and general partnerships, with certain levels of average adjusted gross income (AGI) will apply under this subpart and will apply to each applicant for ELAP, LFP, LIP, SURE, and TAP. Specifically, for 2009 through 2011, a person or legal entity with an average adjusted gross nonfarm income, as defined in §1404.3 of this title, that exceeds $500,000 will not be eligible to receive benefits under this part.

(f) The direct attribution provisions in part 1400 of this title apply to ELAP, LFP, LIP, SURE, and TAP for 2009 and subsequent years. Under those rules, any payment to any legal entity will also be considered for payment limitation purposes to be a payment to persons or legal entities with an interest in the legal entity or in a sub-entity. If any such interested person or legal entity is over the payment limitation because of direct payment or their indirect interests or a combination thereof, then the payment to the actual payee will be reduced commensurate with the amount of the interest of the interested person in the payee. Likewise, by the same method, if anyone with a direct or indirect interest in a legal entity or sub-entity of a payee entity exceeds the AGI levels that would allow a participant to directly receive a payment under this part, then the payment to the actual payee will be reduced commensurately with that interest. For all purposes under this section, unless otherwise specified in part 1400 of this title, the AGI figure that will be relevant for a person or legal entity will be an average AGI for the three taxable years that precede the most immediately preceding complete taxable year, as determined by CCC.

[74 FR 31571, July 2, 2009, as amended at 74 FR 46673, Sept. 11, 2009]

§ 760.109 Misrepresentation and scheme or device.

(a) A participant who is determined to have deliberately misrepresented any fact affecting a program determination made in accordance with this part, or otherwise used a scheme or device with the intent to receive benefits for which the participant would not otherwise be entitled, will not be entitled to program payments and must refund all such payments received, plus interest as determined in accordance with part 792 of this chapter. The participant will also be denied program benefits for the immediately subsequent period of at least 2 crop years, and up to 5 crop years. Interest will run from the date of the original disbursement by FSA.

(b) A participant will refund to FSA all program payments, plus interest, as determined in accordance with part 792 of this chapter, provided however, that in any case it will run from the date of the original disbursement, received by such participant with respect to all contracts or applications, as may be applicable, if the participant is determined to have knowingly done any of the following:

(1) Adopted any scheme or device that tends to defeat the purpose of the program,

(2) Made any fraudulent representation, or

(3) Misrepresented any fact affecting a program determination.

§ 760.110 Appeals.

(a) *Appeals.* Appeal regulations set forth at parts 11 and 780 of this title apply to this part.

(b) *Determinations not eligible for administrative review or appeal.* FSA determinations that are not in response to a specific individual participant's application are not to be construed to be individual program eligibility determinations or adverse decisions and are, therefore, not subject to administrative review or appeal under parts 11 or 780 of this title. Such determinations include, but are not limited to, application periods, deadlines, coverage periods, crop years, fees, prices, general statutory or regulatory provisions that apply to similarly situated participants, national average payment prices, regions, crop definition, average yields, and payment factors established by FSA for any of the programs for which this subpart applies or similar matters requiring FSA determinations.

§ 760.111 Offsets, assignments, and debt settlement.

(a) Any payment to any participant under this part will be made without regard to questions of title under State law, and without regard to any claim or lien against the commodity, or proceeds, in favor of the owner or any other creditor except agencies of the U.S. Government. The regulations governing offsets and withholdings in part 792 of this title apply to payments made under this part.

(b) Any participant entitled to any payment may assign any payment(s) in accordance with regulations governing the assignment of payments in part 1404 of this title.

§ 760.112 Records and inspections.

(a) Any participant receiving payments under any program in ELAP, LFP, LIP, SURE, or TAP, or any other legal entity or person who provides information for the purposes of enabling a participant to receive a payment under ELAP, LFP, LIP, SURE, or TAP, must:

(1) Maintain any books, records, and accounts supporting the information for 3 years following the end of the year during which the request for payment was submitted, and

(2) Allow authorized representatives of USDA and the Government Accountability Office, during regular business hours, to inspect, examine, and make copies of such books or records, and to enter the farm and to inspect and verify all applicable livestock and acreage in which the participant has an interest for the purpose of confirming the accuracy of information provided by or for the participant.

(b) [Reserved]

§760.113 Refunds; joint and several liability.

(a) In the event that the participant fails to comply with any term, requirement, or condition for payment or assistance arising under ELAP, LFP, LIP, SURE, or TAP and if any refund of a payment to FSA will otherwise become due in connection with this part, the participant must refund to FSA all payments made in regard to such matter, together with interest and late-payment charges as provided for in part 792 of this chapter provided that interest will in all cases run from the date of the original disbursement.

(b) All persons with a financial interest in an operation or in an application for payment will be jointly and severally liable for any refund, including related charges, that is determined to be due FSA for any reason under this part.

§760.114 Minors.

A minor child is eligible to apply for program benefits under ELAP, LFP, LIP, SURE, or TAP if all the eligibility requirements are met and the provision for minor children in part 1400 of this title are met.

§760.115 Deceased individuals or dissolved entities.

(a) Payments may be made for eligible losses suffered by an eligible participant who is now a deceased individual or is a dissolved entity if a representative, who currently has authority to enter into a contract, on behalf of the participant, signs the application for payment.

(b) Legal documents showing proof of authority to sign for the deceased individual or dissolved entity must be provided.

(c) If a participant is now a dissolved general partnership or joint venture, all members of the general partnership or joint venture at the time of dissolution or their duly authorized representatives must sign the application for payment.

§760.116 Miscellaneous.

(a) As a condition to receive benefits under ELAP, LFP, LIP, SURE, or TAP, a participant must have been in compliance with the provisions of parts 12 and 718 of this title, and must not otherwise be precluded from receiving benefits under those provisions or under any law.

(b) Rules of the Commodity Credit Corporation that are cited in this part will be applied to this subpart in the same manner as if the programs covered in this subpart were programs funded by the Commodity Credit Corporation.

Subpart C—Emergency Assistance for Livestock, Honeybees, and Farm-Raised Fish Program

SOURCE: 74 FR 46673, Sept. 11, 2009, unless otherwise noted.

§760.201 Applicability.

(a) This subpart establishes the terms and conditions under which the Emergency Assistance for Livestock, Honeybees, and Farm-Raised Fish Program (ELAP) will be administered.

(b) Eligible producers of livestock, honeybees, and farm-raised fish will be compensated to reduce eligible losses that occurred in the calendar year for which the producer requests benefits. The eligible loss must have been a direct result of eligible adverse weather or eligible loss conditions as determined by the Deputy Administrator, including, but not limited to, blizzards, wildfires, disease, and insect infestation. ELAP does not cover losses that are covered under LFP, LIP, or SURE.

§760.202 Definitions.

The following definitions apply to this subpart and to the administration of ELAP. The definitions in parts 718 and 1400 of this title also apply, except where they conflict with the definitions in this section.

Adult beef bull means a male beef breed bovine animal that was used for breeding purposes that was at least 2 years old before the beginning date of the eligible adverse weather or eligible loss condition.

Adult beef cow means a female beef breed bovine animal that had delivered one or more offspring before the beginning date of the eligible adverse weather or eligible loss condition. A first-time bred beef heifer is also considered an adult beef cow if it was pregnant on or by the beginning date of the eligible adverse weather or eligible loss condition.

Adult buffalo and beefalo bull means a male animal of those breeds that was used for breeding purposes and was at least 2 years old before the beginning date of the eligible adverse weather or eligible loss condition.

Adult buffalo and beefalo cow means a female animal of those breeds that had delivered one or more offspring before the beginning date of the eligible adverse weather or eligible loss condition. A first-time bred buffalo or beefalo heifer is also considered an adult buffalo or beefalo cow if it was pregnant by the beginning date of the eligible adverse weather or eligible loss condition.

Adult dairy bull means a male dairy breed bovine animal that was used primarily for breeding dairy cows and was at least 2 years old by the beginning date of the eligible adverse weather or eligible loss condition.

Adult dairy cow means a female bovine dairy breed animal used for the purpose of providing milk for human consumption that had delivered one or more offspring by the beginning date of the eligible adverse weather or eligible loss condition. A first-time bred dairy heifer is also considered an adult dairy cow if it was pregnant by the beginning date of the eligible adverse weather or eligible loss condition.

Agricultural operation means a farming operation.

Application means FSA form used to apply for either the emergency loss assistance for livestock or emergency loss assistance for farm-raised fish or honeybees.

Aquatic species means any species of aquatic organism grown as food for human consumption, fish raised as feed for fish that are consumed by humans, or ornamental fish propagated and reared in an aquatic medium by a commercial operator on private property in water in a controlled environment. Catfish and crawfish are both defined as aquatic species for ELAP. However, aquatic species do not include reptiles or amphibians.

Bait fish means small fish caught for use as bait to attract large predatory fish. For ELAP, it also must meet the definition of aquatic species and not be raised as food for fish; provided, however, that only bait fish produced in a controlled environment can generate claims under ELAP.

Buck means a male goat.

Commercial use means used in the operation of a business activity engaged in as a means of livelihood for profit by the eligible producer.

Contract means, with respect to contracts for the handling of livestock, a written agreement between a livestock owner and another individual or entity setting the specific terms, conditions, and obligations of the parties involved regarding the production of livestock or livestock products.

Controlled environment means an environment in which everything that can practicably be controlled by the participant with structures, facilities, and growing media (including, but not limited to, water and nutrients) was in fact controlled by the participant at the time of the eligible adverse weather or eligible loss condition.

County committee or county office means the respective FSA committee or office.

Deputy Administrator or DAFP means the Deputy Administrator for Farm Programs, Farm Service Agency, U.S. Department of Agriculture or the designee.

Eligible adverse weather or eligible loss condition means any disease, adverse weather, or other loss condition as determined by the Deputy Administrator. The eligible adverse weather or eligible loss condition would have resulted in agricultural losses not covered by other programs in this part for which the Deputy Administrator determines financial assistance needs to be provided to producers. The disease, adverse weather, or other conditions may include, but are not limited to, blizzards, wildfires, water shortages, and other factors. Specific eligible adverse weather and eligible loss conditions may vary based on the type of loss. Identification of eligible adverse weather and eligible loss conditions will include locations (National, State, or county-level) and start and end dates.

Equine animal means a domesticated horse, mule, or donkey.

Ewe means a female sheep.

Farming operation means a business enterprise engaged in producing agricultural products.

Farm-raised fish means any aquatic species that is propagated and reared in a controlled environment.

FSA means the Farm Service Agency.

Game or sport fish means fish pursued for sport by recreational anglers; provided, however, that only game or sport fish produced in a controlled environment can generate claims under ELAP.

Goat means a domesticated, ruminant mammal of the genus Capra, including Angora goats. Goats are further delineated into categories by sex (bucks and nannies) and age (kids).

Kid means a goat less than 1 year old.

Lamb means a sheep less than 1 year old.

Livestock owner, for death loss purposes, means one having legal ownership of the livestock for which benefits are being requested on the day such livestock died due to an eligible adverse weather or eligible loss condition. For all other purposes of loss under ELAP, "livestock owner" means one having legal ownership of the livestock for which benefits are being requested during the 60 days prior to the beginning date of the eligible adverse weather or eligible loss condition.

Nanny means a female goat.

Non-adult beef cattle means a beef breed bovine animal that does not meet the definition of adult beef cow or bull. Non-adult beef cattle are further delineated by weight categories of either less than 400 pounds or 400 pounds or more at the time they died. For a loss other than death, means a bovine animal less than 2 years old that that weighed 500 pounds or more on or before the beginning date of the eligible adverse weather or eligible loss condition.

· *Non-adult buffalo or beefalo* means an animal of those breeds that does not meet the definition of adult buffalo or beefalo cow or bull. Non-adult buffalo or beefalo are further delineated by weight categories of either less than

400 pounds or 400 pounds or more at the time of death. For a loss other than death, means an animal of those breeds that is less than 2 years old that weighed 500 pounds or more on or before the beginning date of the eligible adverse weather or eligible loss condition.

Non-adult dairy cattle means a bovine dairy breed animal used for the purpose of providing milk for human consumption that does not meet the definition of adult dairy cow or bull. Non-adult dairy cattle are further delineated by weight categories of either less than 400 pounds or 400 pounds or more at the time they died. For a loss other than death, means a bovine dairy breed animal used for the purpose of providing milk for human consumption that is less than 2 years old that weighed 500 pounds or more on or before the beginning date of the eligible adverse weather or eligible loss condition.

Normal grazing period, with respect to a county, means the normal grazing period during the calendar year with respect to each specific type of grazing land or pastureland in the county.

Normal mortality means the numerical amount, computed by a percentage, as established for the area by the FSA State Committee, of expected livestock deaths, by category, that normally occur during a calendar year for a producer.

Poultry means domesticated chickens, turkeys, ducks, and geese. Poultry are further delineated into categories by sex, age, and purpose of production as determined by FSA.

Ram means a male sheep.

Secretary means the Secretary of Agriculture or a designee of the Secretary.

Sheep means a domesticated, ruminant mammal of the genus Ovis. Sheep are further defined by sex (rams and ewes) and age (lambs) for purposes of dividing into categories for loss calculations.

State committee, State office, county committee, or county office means the respective FSA committee or office.

Swine means a domesticated omnivorous pig, hog, or boar. Swine for purposes of dividing into categories for loss calculations are further delineated into categories by sex and weight as determined by FSA.

United States means all 50 States of the United States, the Commonwealth of Puerto Rico, the Virgin Islands of the United States, Guam, and the District of Columbia.

§ 760.203 Eligible losses, adverse weather, and other loss conditions.

(a) An eligible loss covered under this subpart is a loss that an eligible producer or contract grower of livestock, honeybees, or farm-raised fish incurs due to an eligible adverse weather or eligible loss condition, as determined by the Deputy Administrator, (including, but not limited to, blizzards and wildfires).

(b) A loss covered under LFP, LIP, or SURE is not eligible for ELAP.

(c) To be eligible, the loss must have occurred:

(1) During the calendar year for which payment is being requested and

(2) Due to an eligible adverse weather event or loss condition that occurred on or after January 1, 2008, and before October 1, 2011.

(d) For a livestock feed loss to be considered an eligible loss, the livestock feed loss must be one of the following:

(1) Loss of purchased forage or feedstuffs that was intended for use as feed for the participant's eligible livestock that was physically located in the county where the eligible adverse weather or eligible loss condition occurred on the beginning date of the eligible adverse weather or eligible loss condition. The loss must be due to an eligible adverse weather or eligible loss condition, as determined by the Deputy Administrator, including, but not limited to, blizzard, flood, hurricane, tidal surge, tornado, volcanic eruption, wildfire on non-Federal land, or lightning;

(2) Loss of mechanically harvested forage or feedstuffs intended for use as feed for the participant's eligible livestock that was physically located in the county where the eligible adverse weather or eligible loss condition occurred on the beginning date of the eligible adverse weather or eligible loss condition. The loss must have occurred after harvest due to an eligible adverse weather or eligible loss condition, as

determined by the Deputy Administrator, including, but not limited to, blizzard, flood, hurricane, tidal surge, tornado, volcanic eruption, wildfire on non-Federal land, or lightning;

(3) A loss resulting from the additional cost incurred for providing or transporting livestock feed to eligible livestock due to an eligible adverse weather or eligible loss condition as determined by the Deputy Administrator, including, but not limited to, costs associated with equipment rental fees for hay lifts and snow removal. The additional costs incurred must have been incurred for losses suffered in the county where the eligible adverse weather or eligible loss condition occurred;

(4) A loss resulting from the additional cost of purchasing additional livestock feed, above normal quantities, required to maintain the eligible livestock during an eligible adverse weather or eligible loss condition, until additional livestock feed becomes available, as determined by the Deputy Administrator. To be eligible, the additional feed purchased above normal quantities must be feed that is fed to maintain livestock in the county where the eligible adverse weather or eligible loss condition occurred.

(e) For a grazing loss to be considered eligible, the grazing loss must have been incurred on eligible grazing lands physically located in the county where the eligible adverse weather or eligible loss condition occurred. The grazing loss must be due to an eligible adverse weather or eligible loss condition, as determined by the Deputy Administrator, including, but not limited to, flood, freeze, hurricane, hail, tidal surge, volcanic eruption, and wildfire on non-Federal land. The grazing loss will not be eligible if it is due to an adverse weather condition covered by LFP as specified in subpart D, such as drought or wildfire on federally managed land where the producer is prohibited by the Federal agency from grazing the normally permitted livestock on the managed rangeland due to a fire.

(f) For a loss due to livestock death to be considered eligible, the livestock death must have occurred in the county where the eligible loss condition occurred. The livestock death must be due to an eligible loss condition determined as eligible by the Deputy Administrator and not related to an eligible adverse weather event as specified in Subpart E for LIP.

(g) For honeybee or farm-raised fish feed losses to be considered eligible, the honeybee or farm-raised fish feed producer must have incurred the loss in the county where the eligible adverse weather or eligible loss condition occurred. The honeybee or farm-raised fish feed losses must be for feed that was intended as feed for the honeybees or farm-raised fish that was damaged or destroyed due to an eligible adverse weather or eligible loss condition, as determined by the Deputy Administrator, including, but not limited to, earthquake, excessive wind, flood, hurricane, tidal surge, tornado, volcanic eruption, and wildfire.

(h) For honeybee colony or honeybee hive losses to be considered eligible, the honeybee colony or honeybee hive producer must have incurred the loss in the county where the eligible adverse weather or eligible loss condition occurred. The honeybee colony or honeybee hive losses must be due to an eligible adverse weather or eligible loss condition, as determined by the Deputy Administrator, including, but not limited to, earthquake, excessive wind, flood, hurricane, tornado, volcanic eruption, and wildfire. To be eligible for a loss of honeybees due to colony collapse disorder, the eligible honeybee producer must provide acceptable documentation to support that the loss was due to colony collapse disorder. Except for 2008 and 2009 honeybee losses, acceptable documentation must include an acceptable colony collapse disorder certification by an independent third party as determined by the Deputy Administrator, plus any other documentation requested by FSA. For 2008 and 2009 honeybee losses such an independent certification is not required in all cases, but rather a self-certification by the honeybee producer as determined acceptable by the Deputy Administrator may be allowed in addition to whatever other documentation might be requested.

(i) For a death loss for bait fish or game fish to be considered eligible, the producer must have incurred the loss in the county where the eligible adverse weather or eligible loss condition occurred. The bait fish or game fish death must be due to an eligible adverse weather or eligible loss condition as determined by the Deputy Administrator including, but not limited to, an earthquake, flood, hurricane, tidal surge, tornado, and volcanic eruption.

[74 FR 46673, Sept. 11, 2009, as amended at 75 FR 19188, Apr. 14, 2010; 76 FR 54075, Aug. 31, 2010]

§ 760.204 Eligible livestock, honeybees, and farm-raised fish.

(a) To be considered eligible livestock for livestock feed losses and grazing losses, livestock must meet all the following conditions:

(1) Be alpacas, adult or non-adult dairy cattle, adult or non-adult beef cattle, adult or non-adult buffalo, adult or non-adult beefalo, deer, elk, emus, equine, goats, llamas, poultry, reindeer, sheep, or swine;

(2) Be livestock that would normally have been grazing the eligible grazing land or pastureland during the normal grazing period for the specific type of grazing land or pastureland for the county;

(3) Be livestock that is owned, cash-leased, purchased, under contract for purchase, or been raised by a contract grower or an eligible livestock producer, during the 60 days prior to the beginning date of the eligible adverse weather or eligible loss condition;

(4) Be livestock that has been maintained for commercial use as part of the producer's farming operation on the beginning date of the eligible adverse weather or eligible loss condition;

(5) Be livestock that has not been produced and maintained for reasons other than commercial use as part of a farming operation; and

(6) Be livestock that was not in a feedlot, on the beginning date of the eligible adverse weather or eligible loss condition, as a part of the normal business operation of the producer, as determined by the Deputy Administrator.

(b) The eligible livestock types for feed losses and grazing losses are:

(1) Adult beef cows or bulls,

(2) Adult buffalo or beefalo cows or bulls,

(3) Adult dairy cows or bulls,

(4) Alpacas,

(5) Deer,

(6) Elk,

(7) Emus,

(8) Equine,

(9) Goats,

(10) Llamas,

(11) Non-adult beef cattle,

(12) Non-adult buffalo or beefalo,

(13) Non-adult dairy cattle,

(14) Poultry,

(15) Reindeer,

(16) Sheep, and

(17) Swine;

(c) Ineligible livestock for feed losses and grazing losses include, but are not limited to:

(1) Livestock that were or would have been in a feedlot, on the beginning date of the eligible adverse weather or eligible loss condition, as a part of the normal business operation of the producer, as determined by FSA;

(2) Yaks;

(3) Ostriches;

(4) All beef and dairy cattle, and buffalo and beefalo that weighed less than 500 pounds on the beginning date of the eligible adverse weather or eligible loss condition;

(5) Any wild free roaming livestock, including horses and deer;

(6) Livestock produced or maintained for reasons other than commercial use as part of a farming operation, including, but not limited to, livestock produced or maintained exclusively for recreational purposes, such as:

(i) Roping,

(ii) Hunting,

(iii) Show,

(iv) Pleasure,

(v) Use as pets, or

(vi) Consumption by owner.

(d) For death losses for livestock owners to be eligible, the livestock must meet all of the following conditions:

(1) Be alpacas, adult or non-adult dairy cattle, beef cattle, beefalo, buffalo, deer, elk, emus, equine, goats, llamas, poultry, reindeer, sheep, or swine, and meet all the conditions in paragraph (f) of this section.

(2) Be one of the following categories of animals for which calculations of

eligibility for payments will be calculated separately for each producer with respect to each category:

(i) Adult beef bulls;

(ii) Adult beef cows;

(iii) Adult buffalo or beefalo bulls;

(iv) Adult buffalo or beefalo cows;

(v) Adult dairy bulls;

(vi) Adult dairy cows;

(vii) Alpacas;

(viii) Chickens, broilers, pullets;

(ix) Chickens, chicks;

(x) Chickens, layers, roasters;

(xi) Deer;

(xii) Ducks;

(xiii) Ducks, ducklings;

(xiv) Elk;

(xv) Emus;

(xvi) Equine;

(xvii) Geese, goose;

(xviii) Geese, gosling;

(xix) Goats, bucks;

(xx) Goats, nannies;

(xxi) Goats, kids;

(xxii) Llamas;

(xxiii) Non-adult beef cattle;

(xxiv) Non-adult buffalo or beefalo;

(xxv) Non-adult dairy cattle;

(xxvi) Reindeer;

(xxvii) Sheep, ewes;

(xxviii) Sheep, lambs;

(xxix) Sheep, rams;

(xxx) Swine, feeder pigs under 50 pounds;

(xxxi) Swine, sows, boars, barrows, gilts 50 to 150 pounds;

(xxxii) Swine, sows, boars, barrows, gilts over 150 pounds;

(xxxiii) Turkeys, poults; and

(xxxiv) Turkeys, toms, fryers, and roasters.

(e) Under ELAP, "contract growers" will only be deemed to include producers of livestock, other than feedlots, whose income is dependent on the actual weight gain and survival of the livestock. For death losses for contract growers to be eligible, the livestock must meet all of the following conditions:

(1) Be poultry or swine, as defined in § 760.202, and meet all the conditions in paragraph (f) of this section.

(2) Be one of the following categories of animals for which calculations of eligibility for payments will be calculated separately for each contract grower with respect to each category:

(i) Chickens, broilers, pullets;

(ii) Chickens, layers, roasters;

(iii) Geese, goose;

(iv) Swine, boars, sows;

(v) Swine, feeder pigs;

(vi) Swine, lightweight barrows, gilts;

(vii) Swine, sows, boars, barrows, gilts; and

(viii) Turkeys, toms, fryers, and roasters.

(f) For livestock death losses to be considered eligible livestock for the purpose of generating payments under this subpart, livestock must meet all of the following conditions:

(1) They must have died:

(i) On or after the beginning date of the eligible loss condition; and

(ii) On or after January 1, 2008, and no later than 60 calendar days from the ending date of the eligible loss condition, but before November 30, 2011; and

(iii) As a direct result of an eligible loss condition that occurs on or after January 1, 2008, and before October 1, 2011; and

(iv) In the calendar year for which payment is being requested; and

(2) Been maintained for commercial use as part of a farming operation on the day the livestock died; and

(3) Before dying, not have been produced or maintained for reasons other than commercial use as part of a farming operation, such non-eligible uses being understood to include, but not be limited to, any uses of wild free roaming animals or use of the animals for recreational purposes, such as pleasure, hunting, roping, pets, or for show.

(g) For honeybee losses to be eligible, the honeybee colony must meet the following conditions:

(1) Been maintained for the purpose of producing honey or pollination for commercial use in a farming operation on the beginning date of the eligible adverse weather or eligible loss condition;

(2) Been physically located in the county where the eligible adverse weather or eligible loss condition occurred on the beginning date of the eligible adverse weather or eligible loss condition;

(3) Been a honeybee colony in which the participant has a risk in the honey production or pollination farming operation on the beginning date of the eligible adverse weather or eligible loss condition;

(4) Been a honeybee colony for which the producer had an eligible loss of a honeybee colony, honeybee hive, or honeybee feed; the feed must have been intended as feed for honeybees.

(h) For fish to be eligible to generate payments under ELAP, the fish must be produced in a controlled environment so to be considered "farm raised fish" as defined in this subpart, and the farm-raised fish must:

(1) For feed losses:

(i) Be an aquatic species that is propagated and reared in a controlled environment;

(ii) Be maintained and harvested for commercial use as part of a farming operation; and

(iii) Be physically located in the county where the eligible adverse weather or eligible loss condition occurred on the beginning date of the eligible adverse weather or eligible loss condition.

(2) For death losses:

(i) Be bait fish or game fish that are propagated and reared in a controlled environment;

(ii) Been maintained for commercial use as part of a farming operation; and

(iii) Been physically located in the county where the eligible loss adverse weather or eligible loss condition occurred on the beginning date of the eligible adverse weather or eligible loss condition.

[74 FR 46673, Sept. 11, 2009, as amended at 76 FR 54075, Aug. 31, 2011]

§760.205 Eligible producers, owners, and contract growers.

(a) To be considered an eligible livestock producer for livestock feed losses and to receive payments, the participant must have owned, cash-leased, purchased, entered into a contract to purchase, or been a contract grower of eligible livestock during the 60 days prior to the beginning date of the eligible adverse weather or eligible loss condition and must have had a loss that is determined to be eligible as specified in §760.203(d), and the producer's eligible livestock must have been livestock that would normally have been grazing the eligible grazing land or pastureland during the normal grazing period for the specific type of grazing land or pastureland for the county as specified in paragraph (b)(1)(i) or (ii) of this section.

(b) To be considered an eligible livestock producer for grazing losses and to receive payments, the participant must have:

(1) Owned, cash-leased, purchased, entered into a contract to purchase, or been a contract grower of eligible livestock during the 60 days prior to the beginning date of the eligible adverse weather or eligible loss condition, must have had a loss that is determined to be eligible as specified in § 760.203(e), and the loss must have occurred on land that is:

(i) Native or improved pastureland with permanent vegetative cover or

(ii) Planted to a crop planted specifically for the purpose of providing grazing for covered livestock;

(2) Have had eligible livestock that would normally have been grazing the eligible grazing land or pastureland during the normal grazing period for the specific type of grazing land or pastureland for the county as specified in paragraph (b)(1)(i) or (ii) of this section;

(3) Provided for the eligible livestock pastureland or grazing land, including cash leased pastureland or grazing land for covered livestock that is physically located in the county where the eligible adverse weather or loss condition occurred during the normal grazing period for the county.

(c) For livestock death losses to be eligible the producer must have had a loss that is determined to be eligible as specified in § 760.203(f) and in addition to other eligibility rules that may apply to be eligible as a:

(1) Livestock owner for the payment with respect to the death of an animal under this subpart, the applicant must have had legal ownership of the livestock on the day the livestock died and under conditions in which no contract grower could have been eligible for ELAP payment with respect to the animal. Eligible types of animal categories for which losses can be calculated for an owner are specified in § 760.204(d).

(2) Contract grower for ELAP payment with respect to the death of an animal, the animal must be in one of the categories specified in § 760.204(e), and the contract grower must have had:

(i) A written agreement with the owner of eligible livestock setting the specific terms, conditions, and obligations of the parties involved regarding the production of livestock;

(ii) Control of the eligible livestock on the day the livestock died; and

(iii) A risk of loss in the animal.

(d) To be considered an eligible honeybee producer, a participant must have an interest and risk in an eligible honeybee colony, as specified in § 760.204(g), for the purpose of producing honey or pollination for commercial use as part of a farming operation and must have had a loss that is determined to be eligible as specified in § 760.203(g) or (h).

(e) To be considered an eligible farm-raised fish producer for feed loss purposes, the participant must have produced eligible farm-raised fish, as specified in § 760.204(h)(1), with the intent to harvest for commercial use as part of a farming operation and must have had a loss that is determined to be eligible as specified in § 760.203(g);

(f) A producer seeking payments must not be ineligible under the restrictions applicable to foreign persons contained in § 760.103(b) and must meet all other requirements of subpart B and other applicable USDA regulations.

§ 760.206 Notice of loss and application process.

(a) To apply for ELAP, the participant that suffered eligible livestock, honeybee, or farm-raised fish losses must submit, to the FSA administrative county office that maintains the participant's farm records for the agricultural operation, the following:

(1) A notice of loss to FSA as specified in § 760.207(a),

(2) A completed application as specified in § 760.207(b) for one or both of the following:

(i) For livestock feed, grazing and death losses, the participant must submit a completed Emergency Loss Assistance for Livestock Application;

(ii) For honeybee feed, honeybee colony, honeybee hive, or farm-raised fish feed or death losses, the participant must submit a completed Emergency Loss Assistance for Farm-Raised Fish or Honeybees Application;

(3) A report of acreage;

(4) A copy of the participant's grower contract, if the participant is a contract grower; and

(5) Other supporting documents required for FSA to determine eligibility of the participant, livestock, and loss.

(b) For livestock, honeybee, or farm-raised fish feed losses, participant must provide verifiable documentation of:

(1) Purchased feed intended as feed for livestock, honeybees, or farm-raised fish that was lost, or additional feed purchased above normal quantities to sustain livestock, honeybees, and farm-raised fish for a short period of time until additional feed becomes available, due to an eligible adverse weather or eligible loss condition. To be considered acceptable documentation, the participant must provide original feed receipts and each feed receipt must include the date of feed purchase, name, address, and telephone number of feed vendor, type and quantity of feed purchased, cost of feed purchased, and signature of feed vendor if the vendor does not have a license to conduct this type of transaction.

(2) Harvested feed intended as feed for livestock, honeybees, or farm-raised fish that was lost due to an eligible adverse weather or eligible loss condition. Documentation may include, but is not limited to, weight tickets, truck scale tickets, contemporaneous diaries used to verify that the crop was stored with the intent to feed the crop to livestock, honeybees, or farm-raised fish, and custom harvest documents that clearly identify the amount of feed produced from the applicable acreage. Documentation must clearly identify the acreage from which the feed was produced.

(c) For eligible honeybee colony and honeybee hive losses and eligible farm-raised fish losses, the participant must also provide documentation of inventory on the beginning date of the eligible adverse weather or loss condition and the ending inventory. Documentation may include, but is not limited to, any combination of the following:

(1) A report of acreage,

(2) Loan records,

(3) Private insurance documents,

(4) Property tax records,

(5) Sales and purchase receipts,

(6) State colony registration documentation, and

(7) Chattel inspections.

(d) For the loss of honeybee colonies due to colony collapse disorder, the participant must also provide acceptable documentation or certification that the loss of the honeybee colony was due to colony collapse disorder. Except for 2008 and 2009 honeybee colony losses, acceptable documentation must include an independent third party certification determined acceptable by the Deputy Administrator, plus such additional information and documentation as may be requested. For 2008 and 2009 honeybee colony losses a self-certification may be accepted by FSA together with any additional information demanded by FSA as determined appropriate by the Deputy Administrator.

(e) For livestock death losses, the participant must provide evidence of loss, current physical location of livestock in inventory, and physical location of claimed livestock at the time of death. The participant must provide:

(1) Documentation listing the quantity and kind of livestock that died as a direct result of the eligible loss condition during the calendar year for which payment is being requested, which must include: Purchase records, veterinarian records, bank or other loan papers, rendering truck receipts, Federal Emergency Management Agency records, National Guard records, written contracts, production records, Internal Revenue Service records, property tax records, private insurance documents, or other similar verifiable documents as determined by FSA.

(2) Adequate proof that the death of the eligible livestock occurred as a direct result of an eligible loss condition in the calendar year for which payment is requested.

(3) If adequate verifiable proof of death documentation is not available, the participant must provide reliable records, in conjunction with verifiable beginning and ending inventory records, as proof of death. Reliable records may include: Contemporaneous

93

producer records, dairy herd improvement records, brand inspection records, vaccination records, pictures, and other similar reliable documents, as determined by FSA.

(4) Certification of livestock deaths by third parties will be acceptable for eligibility determination only if verifiable proof of death records or reliable proof of death records in conjunction with verifiable beginning and ending inventory records are not available and both of the following conditions are met:

(i) The livestock owner or livestock contract grower, as applicable, certifies in writing:

(A) That there is no other verifiable or reliable documentation of death available;

(B) The number of livestock, by category as determined by FSA, was in inventory at the time the applicable loss condition occurred;

(C) The physical location of the livestock, by category, in inventory when the deaths occurred; and

(D) Any other details required for FSA to determine the certification acceptable; and

(ii) The third party is an independent source who is not affiliated with the farming operation such as a hired hand and is not a "family member," defined as a person to whom a member in the farming operation or their spouse is related as a lineal ancestor, lineal descendant, sibling, spouse, or otherwise by marriage, and provides their telephone number, address, and a written statement containing specific details about:

(A) Their knowledge of the livestock deaths;

(B) Their affiliation with the livestock owner;

(C) The accuracy of the deaths claimed by the livestock owner or contract grower including, but not limited to, the number and kind or type of the participant's livestock that died because of the eligible loss condition; and

(D) Any other information required for FSA to determine the certification acceptable.

(f) FSA will use the data furnished by the participant and the third party to determine eligibility for program payment. Furnishing the data is voluntary; however, without all required data program, payment will not be approved or provided.

[74 FR 46673, Sept. 11, 2009, as amended at 75 FR 19188, Apr. 14, 2010]

§ 760.207 Notice of loss and application period.

(a) In addition to submitting an application for payment at the appropriate time, the participant that suffered eligible livestock, honeybee, or farm-raised fish losses that create or could create a claim for benefits must:

(1) For losses during calendar year 2008 and in calendar year 2009 prior to September 11, 2009, provide a notice of loss to FSA no later than December 10, 2009;

(2) For losses on or after September 11, 2009, the participant must provide a notice of loss to FSA within the earlier of:

(i) 30 calendar days of when the loss is apparent to the participant or

(ii) 30 calendar days after the end of the calendar year in which the loss occurred.

(3) The participant must submit the notice of loss required in paragraphs (a)(1) and (a)(2) of this section to the administrative FSA county office

(b) In addition to the notices of loss required in paragraph (a) of this section, a participant must also submit a completed application for payment no later than:

(1) 30 calendar days after the end of the calendar year in which the loss occurred or

(2) December 10, 2009 for losses that occurred during 2008.

§ 760.208 Availability of funds.

By law, "up to" $50 million per year for the years in question may be approved for use by the Secretary and accordingly, within that cap, the only funds that will be considered available to pay claims will be that amount approved by the Secretary. Nothing in these regulations will limit the ability of the Secretary to restrict the availability of funds for the program as permitted by the relevant legislation. Payments will not be made for claims arising out of a particular year until, for all claims for that year, the time for applying for a payment has passed.

In the event that, within the limits of the funding made available by the Secretary within the statutory cap, approval of eligible applications would result in expenditures in excess of the amount available, FSA will prorate the available funds by a national factor to reduce the total expected payments to the amount made available by the Secretary. FSA will make payments based on the factor for the national rate determined by FSA. FSA will prorate the payments in such manner as it determines appropriate and reasonable. Claims that are unpaid or prorated for a calendar year for any reason will not be carried forward for payment under other funds for later years or otherwise, but will be considered, as to any unpaid amount, void and nonpayable.

§ 760.209 Livestock payment calculations.

(a) Payments for an eligible livestock producer will be calculated based on losses for no more than 90 days during the calendar year. Payment calculations for feed losses will be based on 60 percent of the producer's actual cost for:

(1) Livestock feed that was purchased forage or feedstuffs intended for use as feed for the participant's eligible livestock that was physically damaged or destroyed due to the direct result of an eligible adverse weather or eligible loss condition, as provided in § 760.203(d)(1);

(2) Livestock feed that was mechanically harvested forage or feedstuffs intended for use as feed for the participant's eligible livestock that was physically damaged or destroyed after harvest due to the direct result of an eligible adverse weather or eligible loss condition, as provided in § 760.203(d)(2);

(3) The additional cost incurred for providing or transporting livestock feed to eligible livestock due to an eligible adverse weather or eligible loss condition, as provided in § 760.203(d)(3); or

(4) The additional cost of purchasing additional livestock feed above normal, to maintain the eligible livestock during an eligible adverse weather or eligible loss condition until additional livestock feed becomes available, as provided in § 760.203(d)(4).

(b) Payments for an eligible livestock producer for grazing losses, except for losses due to wildfires on non-Federal land, will be calculated based on 60 percent of the lesser of:

(1) The total value of the feed cost for all covered livestock owned by the eligible livestock producer based on the number of days grazing was lost, not to exceed 90 days of daily feed cost for all covered livestock, or

(2) The total value of grazing lost for all eligible livestock based on the normal carrying capacity, as determined by the Secretary, of the eligible grazing land of the eligible livestock producer for the number of grazing days lost, not to exceed 90 days of lost grazing.

(c) The total value of feed cost to be used in the calculation for paragraph (b)(1) of this section is based on the number of days grazing was lost and equals the product obtained by multiplying:

(1) A payment quantity equal to the feed grain equivalent, as determined in paragraph (d) of this section;

(2) A payment rate equal to the corn price per pound, as determined in paragraph (e) of this section;

(3) The number of all covered livestock owned by the eligible producer converted to an animal unit basis;

(4) The number of days grazing was lost, not to exceed 90 calendar days during the normal grazing period for the specific type of grazing land; and

(5) The producer's ownership share in the livestock.

(d) The feed grain equivalent to be used in the calculation for paragraph (c)(1) of this section equals, in the case of:

(1) An adult beef cow, 15.7 pounds of corn per day or

(2) Any other type or weight of livestock, an amount determined by the Secretary that represents the average number of pounds of corn per day necessary to feed that specific type of livestock.

(e) The corn price per pound to be used in the calculation for paragraph (c)(2) of this section equals the quotient obtained by dividing:

(1) The higher of:

(i) The national average corn price per bushel of corn for the 12-month period immediately preceding March 1 of the calendar year for which payments are calculated; or

(ii) The national average corn price per bushel of corn for the 24-month period immediately preceding March 1 of the calendar year for which payments are calculated; by

(2) 56.

(f) The total value of grazing lost to be used in the calculation for paragraph (b)(2) of this section equals the product obtained by multiplying:

(1) A payment quantity equal to the feed grain equivalent of 15.7 pounds of corn per day;

(2) A payment rate equal to the corn price per pound, as determined in paragraph (e) of this section;

(3) The number of animal units the eligible livestock producer's grazing land or pastureland can sustain during the normal grazing period in the county for the specific type of grazing land or pastureland, in the absence of an eligible adverse weather or eligible loss condition, determined by dividing the:

(i) Number of eligible grazing land or pastureland acres of the specific type of grazing land or pastureland by

(ii) The normal carrying capacity of the specific type of eligible grazing land or pastureland; and

(4) The number of days grazing was lost, not to exceed 90 calendar days during the normal grazing period for the specific type of grazing land.

(g) Payments for an eligible livestock producer for grazing losses due to a wildfire on non-Federal land will be calculated by multiplying:

(1) The result of dividing:

(i) The number of acres of grazing land or pastureland acres affected by the fire by

(ii) The normal carrying capacity of the specific type of eligible grazing land or pastureland; times

(2) The daily value of grazing as calculated by FSA under this section; times

(3) The number of days grazing was lost due to fire, not to exceed 180 calendar days; times

(4) 50 percent.

(h) Payments for an eligible livestock producer for eligible livestock death losses due to an eligible loss condition will be based on the following:

(1) Payments will be calculated by multiplying:

(i) The national payment rate for each livestock category times

(ii) The number of eligible livestock that died in each category as a result of an eligible loss condition in excess of normal mortality, as determined in paragraph (d)(2) of this section;

(2) Normal mortality for each livestock category as determined by FSA on a statewide basis using local data sources including, but not limited to, State livestock organizations and the Cooperative Extension Service for the State.

(3) National payment rates to be used in the calculation for paragraph (b)(1) of this section for eligible livestock owners and eligible livestock contract growers are:

(i) A national payment rate for eligible livestock owners that is based on 75 percent of the average fair market value of the applicable livestock as computed using nationwide prices for the previous calendar year unless some other price is approved by the Deputy Administrator.

(ii) A national payment rate for eligible livestock contract growers that is based on 75 percent of the relevant average income loss sustained by the contract grower, with respect to the dead livestock.

(i) Payments calculated in this section are subject to the adjustments and limits provided for in this part.

§ 760.210 Honeybee payment calculations.

(a) An eligible honeybee producer may receive payments for honeybee feed losses due to an eligible adverse weather or loss condition, as provided in § 760.203(g), based on 60 percent of the producer's actual cost for honeybee feed that was:

(1) Damaged or destroyed due to an eligible adverse weather or eligible loss condition and

(2) Intended as feed for an eligible honeybee colony, as provided in § 760.204(g);

(b) An eligible honeybee producer may receive payments for honeybee colony losses due to an eligible adverse weather or eligible loss condition, as provided in §760.203(h), based on 60 percent of the average fair market value for the number of honeybee colonies that were damaged or destroyed due to an eligible adverse weather or eligible loss condition, as computed using nationwide prices unless some other price data is approved for use by the Deputy Administrator, for losses in excess of normal honeybee mortality, as determined by the Deputy Administrator.

(c) An eligible honeybee producer may receive payments for honeybee hive losses due to an eligible adverse weather or eligible loss condition, as provided in §760.203(h), based on 60 percent of the average fair market value for the number of honeybee hives that were damaged or destroyed due to an eligible adverse weather or eligible loss condition, as computed using nationwide prices unless some other price data is approved for use by the Deputy Administrator.

(d) Payments calculated in this section are subject to the adjustments and limits provided for in this part.

[74 FR 46673, Sept. 11, 2009, as amended at 75 FR 19188, Apr. 14, 2010]

§760.211 Farm-raised fish payment calculations.

(a) An eligible farm-raised fish producer may receive payments for fish feed losses due to an eligible adverse weather or eligible loss condition, as provided in §760.203(g), based on 60 percent of the producer's actual replacement cost for the fish feed that was:

(1) Damaged or destroyed due to an eligible adverse weather or eligible loss condition and

(2) Intended as feed for the eligible farm-raised fish, as provided in §760.204(h)(1).

(b) An eligible producer of farm-raised game or sport fish may receive payments for death losses of farm-raised fish due to an eligible adverse weather or eligible loss condition, as provided in §760.203(i), based on 60 percent of the average fair market value of the game fish or sport fish that died as a direct result of an eligible adverse weather or eligible loss condition, as computed using nationwide prices unless some other price data is approved for use by the Deputy Administrator.

(c) Payments calculated in this section or elsewhere with respect to ELAP are subject to the adjustments and limits provided for in this part and are also subject to the payment limitations and average adjusted gross income limitations that are contained in subpart B.

[74 FR 46673, Sept. 11, 2009, as amended at 75 FR 19189, Apr. 14, 2010]

Subpart D—Livestock Forage Disaster Program

Source: 74 FR 46680, Sept. 11, 2009, unless otherwise noted.

§760.301 Applicability.

(a) This subpart establishes the terms and conditions under which the Livestock Forage Disaster Program (LFP) will be administered.

(b) Eligible livestock producers will be compensated for eligible grazing losses for covered livestock that occur due to a qualifying drought or fire that occurs:

(1) On or after January 1, 2008, and before October 1, 2011, and

(2) In the calendar year for which benefits are being requested.

§760.302 Definitions.

The following definitions apply to this subpart and to the administration of LFP. The definitions in parts 718 and 1400 of this title also apply, except where they conflict with the definitions in this section.

Adult beef bull means a male beef breed bovine animal that was at least 2 years old and used for breeding purposes on or before the beginning date of a qualifying drought or fire.

Adult beef cow means a female beef breed bovine animal that had delivered one or more offspring. A first-time bred beef heifer is also considered an adult beef cow if it was pregnant on or before the beginning date of a qualifying drought or fire.

Adult buffalo and beefalo bull means a male animal of those breeds that was at least 2 years old and used for breeding purposes on or before the beginning date of a qualifying drought or fire.

Adult buffalo and beefalo cow means a female animal of those breeds that had delivered one or more offspring. A first-time bred buffalo or beefalo heifer is also considered an adult buffalo or beefalo cow if it was pregnant on or before the beginning date of a qualifying drought or fire.

Adult dairy bull means a male dairy breed bovine animal at least 2 years old used primarily for breeding dairy cows on or before the beginning date of a qualifying drought or fire.

Adult dairy cow means a female dairy breed bovine animal used for the purpose of providing milk for human consumption that had delivered one or more offspring. A first-time bred dairy heifer is also considered an adult dairy cow if it was pregnant on or before the beginning date of a qualifying drought or fire.

Agricultural operation means a farming operation.

Application means the "Livestock Forage Disaster Program" form.

Commercial use means used in the operation of a business activity engaged in as a means of livelihood for profit by the eligible livestock producer.

Contract means, with respect to contracts for the handling of livestock, a written agreement between a livestock owner and another individual or entity setting the specific terms, conditions, and obligations of the parties involved regarding the production of livestock or livestock products.

Covered livestock means livestock of an eligible livestock producer that, during the 60 days prior to the beginning date of a qualifying drought or fire, the eligible livestock producer owned, leased, purchased, entered into a contract to purchase, was a contract grower of, or sold or otherwise disposed of due to a qualifying drought during the current production year. It includes livestock that the producer otherwise disposed of due to drought in one or both of the two production years immediately preceding the current production year as determined by the Secretary. Notwithstanding the foregoing portions of this definition, covered livestock for "contract growers" will not include livestock in feedlots. "Contract growers" under LFP will only include producers of livestock not in feedlots whose income is dependent

on the actual weight gain and survival of the livestock.

Equine animal means a domesticated horse, mule, or donkey.

Farming operation means a business enterprise engaged in producing agricultural products.

Federal Agency means, with respect to the control of grazing land, an agency of the Federal government that manages rangeland on which livestock is generally permitted to graze. For the purposes of this section, it includes, but is not limited to, the U.S. Department of the Interior (DOI) Bureau of Indian Affairs (BIA), DOI Bureau of Land Management (BLM), and USDA Forest Service (FS).

Goat means a domesticated, ruminant mammal of the genus *Capra*, including Angora goats.

Non-adult beef cattle means a beef breed bovine animal that weighed 500 pounds or more on or before the beginning date of a qualifying drought or fire but that does not meet the definition of adult beef cow or bull.

Non-adult buffalo or beefalo means an animal of those breeds that weighed 500 pounds or more on or before the beginning date of a qualifying drought or fire, but does not meet the definition of adult buffalo or beefalo cow or bull.

Non-adult dairy cattle means a bovine animal, of a breed used for the purpose of providing milk for human consumption, that weighed 500 pounds or more on or before the beginning date of a qualifying drought or fire, but that does not meet the definition of adult dairy cow or bull.

Normal carrying capacity means, with respect to each type of grazing land or pastureland in a county, the normal carrying capacity that would be expected from the grazing land or pastureland for livestock during the normal grazing period in the county, in the absence of a drought or fire that diminishes the production of the grazing land or pastureland.

Normal grazing period means, with respect to a county, the normal grazing period during the calendar year with respect to each specific type of grazing land or pastureland in the county served by the applicable county committee.

Owner means one who had legal ownership of the livestock for which benefits are being requested during the 60 days prior to the beginning of a qualifying drought or fire.

Poultry means a domesticated chicken, turkey, duck, or goose. Poultry are further delineated by sex, age, and purpose of production, as determined by FSA.

Sheep means a domesticated, ruminant mammal of the genus *Ovis.*

Swine means a domesticated omnivorous pig, hog, or boar. Swine are further delineated by sex and weight, as determined by FSA.

U.S. Drought Monitor is a system for classifying drought severity according to a range of abnormally dry to exceptional drought. It is a collaborative effort between Federal and academic partners, produced on a weekly basis, to synthesize multiple indices, outlooks, and drought impacts on a map and in narrative form. This synthesis of indices is reported by the National Drought Mitigation Center at *http://www.drought.unl.edu/dm/monitor.html.*

§ 760.303 **Eligible livestock producer.**

(a) To be considered an eligible livestock producer, the eligible producer on a farm must:

(1) During the 60 days prior to the beginning date of a qualifying drought or fire, own, cash or share lease, or be a contract grower of covered livestock or

(2) Provide pastureland or grazing land for covered livestock, including cash-leased pastureland or grazing land, that is:

(i) Physically located in a county affected by a qualifying drought during the normal grazing period for the county or

(ii) Rangeland managed by a Federal agency for which the otherwise eligible livestock producer is prohibited by the Federal agency from grazing the normal permitted livestock due to a qualifying fire.

(b) The eligible livestock producer must have certified that the livestock producer has suffered a grazing loss due to a qualifying drought or fire to be eligible for LFP payments.

(c) An eligible livestock producer does not include any owner, cash or share lessee, or contract grower of livestock that rents or leases pastureland or grazing land owned by another person on a rate-of-gain basis. (That is, where the lease or rental agreement calls for payment based in whole or in part on the amount of weight gained by the animals that use the pastureland or grazing land.)

(d) A producer seeking payment must not be ineligible for payments under the restrictions applicable to foreign persons contained in § 760.103(b) and must meet all other requirements of subpart B and other applicable USDA regulations.

(e) If a contract grower is an eligible livestock producer for covered livestock, the owner of that livestock is not eligible for payment.

§ 760.304 **Covered livestock.**

(a) To be considered covered livestock for LFP payments, livestock must meet all the following conditions:

(1) Be adult or non-adult beef cattle, adult or non-adult beefalo, adult or non-adult buffalo, adult or non-adult dairy cattle, alpacas, deer, elk, emus, equine, goats, llamas, poultry, reindeer, sheep, or swine;

(2) Be livestock that would normally have been grazing the eligible grazing land or pastureland in the county:

(i) During the normal grazing period for the specific type of grazing land or pastureland for the county or

(ii) When the Federal agency prohibited the eligible livestock producer from using the managed rangeland for grazing due to a fire;

(3) Be livestock that the eligible livestock producer:

(i) During the 60 days prior to the beginning date of a qualifying drought or fire:

(A) Owned,

(B) Leased,

(C) Purchased,

(D) Entered into a contract to purchase, or

(E) Was a contract grower of; or

(ii) Sold or otherwise disposed of due to qualifying drought during:

(A) The current production year or

(B) 1 or both of the 2 production years immediately preceding the current production year;

(4) Been maintained for commercial use as part of the producer's farming

operation on the beginning date of the qualifying drought or fire;

(5) Not have been produced and maintained for reasons other than commercial use as part of a farming operation. Such excluded uses include, but are not limited to, any uses of wild free roaming animals or use of the animals for recreational purposes, such as pleasure, roping, hunting, pets, or for show; and

(6) Not have been livestock that were or would have been in a feedlot, on the beginning date of the qualifying drought or fire, as a part of the normal business operation of the eligible livestock producer, as determined by the Secretary.

(b) The covered livestock categories are:

(1) Adult beef cows or bulls,

(2) Adult buffalo or beefalo cows or bulls,

(3) Adult dairy cows or bulls,

(4) Alpacas,

(5) Deer,

(6) Elk,

(7) Emu,

(8) Equine,

(9) Goats,

(10) Llamas,

(11) Non-adult beef cattle,

(12) Non-adult buffalo or beefalo,

(13) Non-adult dairy cattle,

(14) Poultry,

(15) Reindeer,

(16) Sheep, and

(17) Swine.

(c) Livestock that are not covered include, but are not limited to:

(1) Livestock that were or would have been in a feedlot, on the beginning date of the qualifying drought or fire, as a part of the normal business operation of the eligible livestock producer, as determined by the Secretary;

(2) Yaks;

(3) Ostriches;

(4) All beef and dairy cattle, and buffalo and beefalo that weighed less than 500 pounds on the beginning date of the qualifying drought or fire;

(5) Any wild free roaming livestock, including horses and deer; and

(6) Livestock produced or maintained for reasons other than commercial use as part of a farming operation, including, but not limited to, livestock produced or maintained for recreational purposes, such as:

(i) Roping,

(ii) Hunting,

(iii) Show,

(iv) Pleasure,

(v) Use as pets, or

(vi) Consumption by owner.

[74 FR 46680, Sept. 11, 2009, as amended at 75 FR 19189, Apr. 14, 2010]

§ 760.305   Eligible grazing losses.

(a) A grazing loss due to drought is eligible for LFP only if the grazing loss for the covered livestock occurs on land that:

(1) Is native or improved pastureland with permanent vegetative cover or

(2) Is planted to a crop planted specifically for the purpose of providing grazing for covered livestock; and

(3) Is grazing land or pastureland that is owned or leased by the eligible livestock producer that is physically located in a county that is, during the normal grazing period for the specific type of grazing land or pastureland for the county, rated by the U.S. Drought Monitor as having a:

(i) D2 (severe drought) intensity in any area of the county for at least 8 consecutive weeks during the normal grazing period for the specific type of grazing land or pastureland for the county, as determined by the Secretary, or

(ii) D3 (extreme drought) or D4 (exceptional drought) intensity in any area of the county at any time during the normal grazing period for the specific type of grazing land or pastureland for the county, as determined by the Secretary. (As specified elsewhere in this subpart, the amount of potential payment eligibility will be higher than under (a)(3)(i) of this section where the D4 trigger applies or where the D3 condition as determined by the Secretary lasts at least 4 weeks during the normal grazing period for the specific type of grazing land or pastureland for the county.)

(b) A grazing loss is not eligible for LFP if the grazing loss due to drought on land used for haying or grazing under the Conservation Reserve Program established under subchapter B of chapter 1 of subtitle D of title XII of the Food Security Act of 1985 (16 U.S.C. 3831–3835a).

(c) A fire qualifies for LFP only if:

(1) The grazing loss occurs on range-land that is managed by a Federal agency and

(2) The eligible livestock producer is prohibited by the Federal agency from grazing the normal permitted livestock on the managed rangeland due to a fire.

(d) An eligible livestock producer may be eligible for LFP payments only on those grazing lands incurring losses for which the livestock producer:

(1) Meets the risk management purchase requirements specified in §760.104; or

(2) Does not meet the risk management purchase requirements specified in §760.104 because the risk management purchase requirement is waived according to §§760.105, 760.106, or 760.107.

§760.306 Application for payment.

(a) To apply for LFP, the participant that suffered eligible grazing losses:

(1) During 2008, must submit a completed application for payment and required supporting documentation to the administrative FSA county office no later than December 10, 2009 or

(2) During 2009 and later years, must submit a completed application for payment and required supporting documentation to the administrative FSA county office no later than 30 calendar days after the end of the calendar year in which the grazing loss occurred.

(b) A participant must also provide a copy of the grower contract, if a contract grower, and other supporting documents required for determining eligibility as an applicant at the time the participant submits the completed application for payment. Supporting documents must include:

(1) Evidence of loss,

(2) Current physical location of livestock in inventory,

(3) Evidence of meeting risk management purchase requirements as specified in subpart B,

(4) Evidence that grazing land or pastureland is owned or leased,

(5) A report of acreage according to part 718 of this chapter for the grazing lands incurring losses for which assistance is being requested under this subpart;

(6) Adequate proof, as determined by FSA that the grazing loss:

(i) Was for the covered livestock;

(ii) If the loss of grazing occurred as the result of a fire that the:

(A) Loss was due to a fire and

(B) Participant was prohibited by the Federal agency from grazing the normal permitted livestock on the managed rangeland due to a fire;

(iii) Occurred on or after January 1, 2008, and before October 1, 2011; and

(iv) Occurred in the calendar year for which payments are being requested;

(7) Adequate proof, absent an appropriate waiver (if there is a waiver, it itself must be documented by the producer), as determined by FSA, that the participant had obtained, for the grazing land incurring the losses for which assistance is being requested, one or both of the following:

(i) A policy or plan of insurance under the Federal Crop Insurance Act (7 U.S.C. 1501–1524); or

(ii) Filed the required paperwork, and paid the administrative fee by the applicable State filing deadline, for the noninsured crop disaster assistance program;

(8) Any other supporting documentation as determined by FSA to be necessary to make a determination of eligibility of the participant. Supporting documents include, but are not limited to: Verifiable purchase and sales records; grower contracts; veterinarian records; bank or other loan papers; rendering truck receipts; Federal Emergency Management Records; National Guard records; written contracts; production records; private insurance documents; sales records; and similar documents determined acceptable to FSA.

(c) Data furnished by the participant will be used to determine eligibility for program benefits. Furnishing the data is voluntary; however, without all required data, program benefits will not be approved or provided.

§760.307 Payment calculation.

(a) An eligible livestock producer will be eligible to receive payments for grazing losses for qualifying drought as specified in §760.305(a) equal to one, two, or three times the monthly payment rate specified in paragraphs (e) or (f) of this section. Total LFP payments

101

to an eligible livestock producer in a calendar year for grazing losses due to qualifying drought will not exceed three monthly payments for the same livestock. Payments calculated in this section or elsewhere with respect to LFP are subject to the adjustments and limits provided for in this part and are also subject to the payment limitations and average adjusted gross income provisions that are contained in subpart B. Payment may only be made to the extent that eligibility is specifically provided for in this subpart. Hence, with respect to drought, payments will be made only as a "one month" payment, a "two month" payment, or a "three month" payment based on the provisions of paragraphs (b), (c), and (d) of this section.

(b) To be eligible to receive a one month payment, that is a payment equal to the monthly feed cost as determined under paragraph (g) of this section, the eligible livestock producer must own or lease grazing land or pastureland that is physically located in a county that is rated by the U.S. Drought Monitor as having at least a D2 severe drought (intensity) in any area of the county for at least 8 consecutive weeks during the normal grazing period for the specific type of grazing land or pastureland in the county.

(c) To be eligible to receive a two month payment, that is a payment equal to twice the monthly feed cost as determined under paragraph (g) of this section, the eligible livestock producer must own or lease grazing land or pastureland that is physically located in a county that is rated by the U.S. Drought Monitor as having at least a D3 (extreme drought) intensity in any area of the county at any time during the normal grazing period for the specific type of grazing land or pastureland for the county.

(d) To be eligible to receive a three month payment, that is a payment equal to three times the monthly feed cost as determined under paragraph (g) of this section, the eligible livestock producer must own or lease grazing land or pastureland that is physically located in a county that is rated by the U.S. Drought Monitor as having at least a D3 (extreme drought) intensity in any area of the county for at least 4

weeks during the normal grazing period for the specific type of grazing land or pastureland for the county, or is rated as having a D4 (exceptional drought) intensity in any area of the county at any time during the normal grazing period for the specific type of grazing land or pastureland for the county.

(e) The monthly payment rate for LFP for grazing losses due to a qualifying drought, except as provided in paragraph (f) of this section, will be equal to 60 percent of the lesser of:

(1) The monthly feed cost for all covered livestock owned or leased by the eligible livestock producer, as determined in paragraph (g) of this section or

(2) The monthly feed cost calculated by using the normal carrying capacity of the eligible grazing land of the eligible livestock producer, as determined in paragraph (j) of this section.

(f) In the case of an eligible livestock producer that sold or otherwise disposed of covered livestock due to a qualifying drought in 1 or both of the 2 production years immediately preceding the current production year, the payment rate is 80 percent of the monthly payment rate calculated in paragraph (e) of this section.

(g) The monthly feed cost for covered livestock equals the product obtained by multiplying:

(1) 30 days;

(2) A payment quantity equal to the amount referred to in paragraph (h) of this section as the "feed grain equivalent", as determined under paragraph (h) of this section; and

(3) A payment rate equal to the corn price per pound, as determined in paragraph (i) of this section.

(h) The feed grain equivalent equals, in the case of:

(1) An adult beef cow, 15.7 pounds of corn per day or

(2) In the case of any other type or weight of covered livestock, an amount determined by the Secretary that represents the average number of pounds of corn per day necessary to feed that specific type of livestock.

(i) The corn price per pound equals the quotient obtained by dividing:

(1) The higher of:

(i) The national average corn price per bushel for the 12-month period immediately preceding March 1 of the calendar year for which LFP payment is calculated or

(ii) The national average corn price per bushel for the 24-month period immediately preceding March 1 of the calendar year for which LFP payment is calculated

(2) By 56.

(j) The monthly feed cost using the normal carrying capacity of the eligible grazing land equals the product obtained by multiplying:

(1) 30 days;

(2) A payment quantity equal to the feed grain equivalent of 15.7 pounds of corn per day;

(3) A payment rate equal to the corn price per pound, as determined in paragraph (i) of this section; and

(4) The number of animal units the eligible livestock producer's grazing land or pastureland can sustain during the normal grazing period in the county for the specific type of grazing land or pastureland, in the absence of a drought or fire, determined by dividing the:

(i) Number of eligible grazing land or pastureland acres of the specific type of grazing land or pastureland by

(ii) The normal carrying capacity of the specific type of eligible grazing land or pastureland as determined under this subpart.

(k) An eligible livestock producer will be eligible to receive payments for grazing losses due to a fire as specified in §760.305(c):

(1) For the period, subject to paragraph (l)(2) of this section:

(i) Beginning on the date on which the Federal Agency prohibits the eligible livestock producer from using the managed rangeland for grazing and

(ii) Ending on the earlier of the last day of the Federal lease of the eligible livestock producer or the day that would make the period a 180 day period and

(2) For grazing losses that occur on not more than 180 days per calendar year.

(3) For 50 percent of the monthly feed cost, as determined under §760.308(g), pro-rated to a daily rate, for the total number of livestock covered by the Federal lease of the eligible livestock producer.

Subpart E—Livestock Indemnity Program

SOURCE: 74 FR 31575, July 2, 2009, unless otherwise noted.

§760.401 Applicability.

(a) This subpart establishes the terms and conditions under which the Livestock Indemnity Program (LIP) will be administered under Titles XII and XV of the 2008 Farm Bill (Pub. L. 110–246).

(b) Eligible livestock owners and contract growers will be compensated in accordance with §760.406 for eligible livestock deaths in excess of normal mortality that occurred in the calendar year for which benefits are being requested as a direct result of an eligible adverse weather event. An "eligible adverse weather event" is one, as determined by the Secretary, occurring in the program year that could and did, even when normal preventative or corrective measures were taken and good farming practices were followed, directly result in the death of livestock. Because feed can be purchased or otherwise obtained in the event of a drought, drought is not an eligible adverse weather event except when anthrax, resulting from drought, causes the death of eligible livestock.

§760.402 Definitions.

The following definitions apply to this subpart. The definitions in parts 718 and 1400 of this title also apply, except where they conflict with the definitions in this section.

Adult beef bull means a male beef breed bovine animal that was at least 2 years old and used for breeding purposes before it died.

Adult beef cow means a female beef breed bovine animal that had delivered one or more offspring before dying. A first-time bred beef heifer is also considered an adult beef cow if it was pregnant at the time it died.

Adult buffalo and beefalo bull means a male animal of those breeds that was at least 2 years old and used for breeding purposes before it died.

103

Adult buffalo and beefalo cow means a female animal of those breeds that had delivered one or more offspring before dying. A first-time bred buffalo or beefalo heifer is also considered an adult buffalo or beefalo cow if it was pregnant at the time it died.

Adult dairy bull means a male dairy breed bovine animal at least 2 years old used primarily for breeding dairy cows before it died.

Adult dairy cow means a female bovine dairy breed animal used for the purpose of providing milk for human consumption that had delivered one or more offspring before dying. A first-time bred dairy heifer is also considered an adult dairy cow if it was pregnant at the time it died.

Adverse weather means damaging weather events, including, but not limited to, hurricanes, floods, blizzards, disease, wildfires, extreme heat, and extreme cold.

Agricultural operation means a farming operation.

Application means the "Livestock Indemnity Program" form.

Buck means a male goat.

Commercial use means used in the operation of a business activity engaged in as a means of livelihood for profit by the eligible producer.

Contract means, with respect to contracts for the handling of livestock, a written agreement between a livestock owner and another individual or entity setting the specific terms, conditions, and obligations of the parties involved regarding the production of livestock or livestock products.

Deputy Administrator or DAFP means the Deputy Administrator for Farm Programs, Farm Service Agency, U.S. Department of Agriculture or the designee.

Equine animal means a domesticated horse, mule, or donkey.

Ewe means a female sheep.

Farming operation means a business enterprise engaged in producing agricultural products.

FSA means the Farm Service Agency.

Goat means a domesticated, ruminant mammal of the genus Capra, including Angora goats. Goats are further defined by sex (bucks and nannies) and age (kids).

Kid means a goat less than 1 year old.

Lamb means a sheep less than 1 year old.

Livestock owner means one having legal ownership of the livestock for which benefits are being requested on the day such livestock died.

Nanny means a female goat.

Non-adult beef cattle means a beef breed bovine animal that does not meet the definition of adult beef cow or bull. Non-adult beef cattle are further delineated by weight categories of either less than 400 pounds or 400 pounds or more at the time they died.

Non-adult buffalo or beefalo means an animal of those breeds that does not meet the definition of adult buffalo or beefalo cow or bull. Non-adult buffalo or beefalo are further delineated by weight categories of either less than 400 pounds or 400 pounds or more at the time of death.

Non-adult dairy cattle means a dairy breed bovine animal, of a breed used for the purpose of providing milk for human consumption, that does not meet the definition of adult dairy cow or bull. Non-adult dairy cattle are further delineated by weight categories of either less than 400 pounds or 400 pounds or more at the time they died.

Normal mortality means the numerical amount, computed by a percentage, as established for the area by the FSA State Committee, of expected livestock deaths, by category, that normally occur during a calendar year for a producer.

Poultry means domesticated chickens, turkeys, ducks, and geese. Poultry are further delineated by sex, age, and purpose of production as determined by FSA.

Ram means a male sheep.

Secretary means the Secretary of Agriculture or a designee of the Secretary.

Sheep means a domesticated, ruminant mammal of the genus Ovis. Sheep are further defined by sex (rams and ewes) and age (lambs) for purposes of dividing into categories for loss calculations.

State committee, State office, county committee, or county office means the respective FSA committee or office.

Swine means a domesticated omnivorous pig, hog, or boar. Swine for purposes of dividing into categories for loss calculations are further delineated by sex and weight as determined by FSA.

United States means all fifty States of the United States, the Commonwealth of Puerto Rico, the Virgin Islands, Guam, and the District of Columbia.

§760.403 Eligible owners and contract growers.

(a) In addition to other eligibility rules that may apply, to be eligible as a:

(1) Livestock owner for benefits with respect to the death of an animal under this subpart, the applicant must have had legal ownership of the eligible livestock on the day the livestock died and under conditions in which no contract grower could have been eligible for benefits with respect to the animal. Eligible types of animal categories for which losses can be calculated for an owner are specified in §760.404(a).

(2) Contract grower for benefits with respect to the death of an animal, the animal must be in one of the categories specified on §760.404(b), and the contract grower must have had

(i) A written agreement with the owner of eligible livestock setting the specific terms, conditions, and obligations of the parties involved regarding the production of livestock;

(ii) Control of the eligible livestock on the day the livestock died; and

(iii) A risk of loss in the animal.

(b) A producer seeking payment must not be ineligible under the restrictions applicable to foreign persons contained in §760.103(b) and must meet all other requirements of subpart B and other applicable USDA regulations.

§760.404 Eligible livestock.

(a) To be considered eligible livestock for livestock owners, the kind of livestock must be alpacas, adult or non-adult dairy cattle, beef cattle, buffalo, beefalo, elk, emus, equine, llamas, sheep, goats, swine, poultry, deer, or reindeer and meet all the conditions in paragraph (c) of this section.

(b) To be considered eligible livestock for contract growers, the kind of livestock must be poultry or swine as defined in §760.402 and meet all the conditions in paragraph (c) of this section.

(c) To be considered eligible livestock for the purpose of generating payments under this subpart, livestock must meet all of the following conditions:

(1) Died as a direct result of an eligible adverse weather event that occurred on or after January 1, 2008, and before October 1, 2011;

(2) Died no later than 60 calendar days from the ending date of the applicable adverse weather event, but before November 30, 2011;

(3) Died in the calendar year for which benefits are being requested;

(4) Been maintained for commercial use as part of a farming operation on the day they died; and

(5) Before dying, not have been produced or maintained for reasons other than commercial use as part of a farming operation, such non-eligible uses being understood to include, but not be limited to, any uses of wild, free roaming animals or use of the animals for recreational purposes, such as pleasure, hunting, roping, pets, or for show.

(d) The following categories of animals owned by a livestock owner are eligible livestock and calculations of eligibility for payments will be calculated separately for each producer with respect to each category:

(1) Adult beef bulls;

(2) Adult beef cows;

(3) Adult buffalo or beefalo bulls;

(4) Adult buffalo or beefalo cows;

(5) Adult dairy bulls;

(6) Adult dairy cows;

(7) Alpacas;

(8) Chickens, broilers, pullets;

(9) Chickens, chicks;

(10) Chickens, layers, roasters;

(11) Deer;

(12) Ducks;

(13) Ducks, ducklings;

(14) Elk;

(15) Emus;

(16) Equine;

(17) Geese, goose;

(18) Geese, gosling;

(19) Goats, bucks;

(20) Goats, nannies;

(21) Goats, kids;

(22) Llamas;

(23) Non-adult beef cattle;

(24) Non-adult buffalo or beefalo;

(25) Non-adult dairy cattle;

(26) Reindeer;

(27) Sheep, ewes;

(28) Sheep, lambs;

(29) Sheep, rams;

(30) Swine, feeder pigs under 50 pounds;

(31) Swine, sows, boars, barrows, gilts 50 to 150 pounds;

(32) Swine, sows, boars, barrows, gilts over 150 pounds;

(33) Turkeys, poults; and

(34) Turkeys, toms, fryers, and roasters.

(e) The following categories of animals are eligible livestock for contract growers and calculations of eligibility for payments will be calculated separately for each producer with respect to each category:

(1) Chickens, broilers, pullets;

(2) Chickens, layers, roasters;

(3) Geese, goose;

(4) Swine, boars, sows;

(5) Swine, feeder pigs;

(6) Swine, lightweight barrows, gilts;

(7) Swine, sows, boars, barrows, gilts; and

(8) Turkeys, toms, fryers, and roasters.

[74 FR 31575, July 2, 2009, as amended at 76 FR 54075, Aug. 31, 2011]

§ 760.405   Application process.

(a) In addition to submitting an application for payment at the appropriate time, a producer or contract grower that suffered livestock losses that create or could create a claim for benefits must:

(1) For losses during 2008 and losses in 2009, prior to July 13, 2009, provide a notice of loss to FSA no later than September 13, 2009.

(2) For losses on or after July 13, 2009, provide a notice of loss to FSA within the earlier of:

(i) 30 calendar days of when the loss of livestock is apparent to the participant or

(ii) 30 calendar days after the end of the calendar year in which the loss of livestock occurred.

(3) The participant must submit the notice of loss required in paragraphs (a)(1) and (a)(2) to the FSA administrative county office that maintains the participant's farm records for the agricultural operation.

(b) In addition to the notices of loss required in paragraph (a) of this section, a participant must also submit a completed application for payment no later than

(1) 30 calendar days after the end of the calendar year in which the loss of livestock occurred or

(2) September 13, 2009 for losses during 2008.

(c) Applicants must submit supporting documentation with their application. For contract growers, the information must include a copy of the grower contract and other documents establishing their status. In addition, for all applicants, including contract growers, supporting documents must show:

(1) Evidence of loss,

(2) Current physical location of livestock in inventory,

(3) Physical location of claimed livestock at the time of death, and

(4) Inventory numbers and other inventory information necessary to establish actual mortality as required by FSA.

(d) The participant must provide adequate proof that the death of the eligible livestock occurred as a direct result of an eligible adverse weather event in the calendar year for which benefits are requested. The quantity and kind of livestock that died as a direct result of the eligible adverse weather event during the calendar year for which benefits are being requested may be documented by: purchase records; veterinarian records; bank or other loan papers; rendering-plant truck receipts; Federal Emergency Management Agency records; National Guard records; written contracts; production records; Internal Revenue Service records; property tax records; private insurance documents; and other similar verifiable documents as determined by FSA.

(e) If adequate verifiable proof of death documentation is not available, the participant may provide reliable records, in conjunction with verifiable beginning and ending inventory records, as proof of death. Reliable records may include contemporaneous producer records, dairy herd improvement records, brand inspection records, vaccination records, pictures, and other similar reliable documents as determined by FSA.

(f) Certification of livestock deaths by third parties may be accepted only if verifiable proof of death records or reliable proof of death records in conjunction with verifiable beginning and ending inventory records are not available and both of the following conditions are met:

(1) The livestock owner or livestock contract grower, as applicable, certifies in writing:

(i) That there is no other verifiable or reliable documentation of death available;

(ii) The number of livestock, by category identified in this subpart and by FSA were in inventory at the time the applicable adverse weather event occurred;

(iii) The physical location of the livestock, by category, in inventory when the deaths occurred; and

(iv) Other details required for FSA to determine the certification acceptable; and

(2) The third party is an independent source who is not affiliated with the farming operation such as a hired hand and is not a "family member," defined as a person whom a member in the farming operation or their spouse is related as lineal ancestor, lineal descendant, sibling, spouse, and provides their telephone number, address, and a written statement containing specific details about:

(i) Their knowledge of the livestock deaths;

(ii) Their affiliation with the livestock owner;

(iii) The accuracy of the deaths claimed by the livestock owner or contract grower including, but not limited to, the number and kind or type of the participant's livestock that died because of the eligible adverse weather event; and

(iv) Other information required by FSA to determine the certification acceptable.

(g) Data furnished by the participant and the third party will be used to determine eligibility for program benefits. Furnishing the data is voluntary; however, without all required data program benefits will not be approved or provided.

§760.406 Payment calculation.

(a) Under this subpart, separate payment rates for eligible livestock owners and eligible livestock contract growers are specified in paragraphs (b) and (c) of this section, respectively. Payments for LIP are calculated by multiplying the national payment rate for each livestock category by the number of eligible livestock in excess of normal mortality in each category that died as a result of an eligible adverse weather event. Normal mortality for each livestock category will be determined by FSA on a State-by-State basis using local data sources including, but not limited to, State livestock organizations and the Cooperative Extension Service for the State. Adjustments will be applied as specified in paragraph (d) of this section.

(b) The LIP national payment rate for eligible livestock owners is based on 75 percent of the average fair market value of the applicable livestock as computed using nationwide prices for the previous calendar year unless some other price is approved by the Deputy Administrator.

(c) The LIP national payment rate for eligible livestock contract growers is based on 75 percent of the average income loss sustained by the contract grower with respect to the dead livestock.

(d) The LIP payment calculated for eligible livestock contract growers will be reduced by the amount the participant received from the party who contracted with the producer to raise the livestock for the loss of income from the dead livestock.

Subpart F—Tree Assistance Program

SOURCE: 75 FR 25108, May 7, 2010, unless otherwise noted.

§760.500 Applicability.

(a) This subpart establishes the terms and conditions under which the Tree Assistance Program (TAP) will be administered under Titles XII and XV of the Food, Conservation, and Energy Act of 2008 (Pub. L. 110–246, the 2008 Farm Bill).

(b) Eligible orchardists and nursery tree growers will be compensated as specified in § 760.506 for eligible tree, bush, and vine losses in excess of 15 percent mortality, or, where applicable, 15 percent damage, adjusted for normal mortality and normal damage, that occurred in the calendar year for which benefits are being requested and as a direct result of a natural disaster.

§ 760.501 Administration.

The program will be administered as specified in § 760.102 and in this subpart.

§ 760.502 Definitions.

The following definitions apply to this subpart. The definitions in parts 718 and 1400 of this title also apply, except where they conflict with the definitions in this section.

Bush means, a low, branching, woody plant, from which at maturity of the bush, an annual fruit or vegetable crop is produced for commercial purposes, such as a blueberry bush. The definition does not cover plants that produce a bush after the normal crop is harvested such as asparagus.

Commercial use means used in the operation of a business activity engaged in as a means of livelihood for profit by the eligible producer.

County committee means the respective FSA committee.

County office means the FSA or U.S. Department of Agriculture (USDA) Service Center that is responsible for servicing the farm on which the trees, bushes, or vines are located.

Cutting means a piece of a vine which was planted in the ground to propagate a new vine for the commercial production of fruit, such as grapes, kiwi fruit, passion fruit, or similar fruit.

Deputy Administrator or DAFP means the Deputy Administrator for Farm Programs, FSA, USDA, or the designee.

Eligible nursery tree grower means a person or legal entity that produces nursery, ornamental, fruit, nut, or Christmas trees for commercial sale.

Eligible orchardist means a person or legal entity that produces annual crops from trees, bushes, or vines for commercial purposes.

FSA means the Farm Service Agency.

Lost means, with respect to the extent of damage to a tree or other plant, that the plant is destroyed or the damage is such that it would, as determined by FSA, be more cost effective to replace the tree or other plant than to leave it in its deteriorated, low-producing state.

Natural disaster means plant disease, insect infestation, drought, fire, freeze, flood, earthquake, lightning, or other natural occurrence of such magnitude or severity so as to be considered disastrous, as determined by the Deputy Administrator.

Normal damage means the percentage, as established for the area by the FSA State Committee, of trees, bushes, or vines in the individual stand that would normally be damaged during a calendar year for a producer.

Normal mortality means percentage, as established for the area by the FSA State Committee, of expected lost trees, bushes, or vines in the individual stand that normally occurs during a calendar year for a producer. This term refers to the number of whole trees, bushes, or vines that are destroyed or damaged beyond rehabilitation. Mortality does not include partial damage such as lost tree limbs.

Seedling means an immature tree, bush, or vine that was planted in the ground or other growing medium to grow a new tree, bush, or vine for commercial purposes.

Stand means a contiguous acreage of the same type of trees (including Christmas trees, ornamental trees, nursery trees, and potted trees), bushes (including shrubs), or vines.

State committee means the respective FSA committee.

Tree means a tall, woody plant having comparatively great height, and a single trunk from which an annual crop is produced for commercial purposes, such as a maple tree for syrup, papaya tree, or orchard tree. Trees used for pulp or timber are not considered eligible trees under this subpart.

Vine means a perennial plant grown under normal conditions from which an annual fruit crop is produced for commercial market for human consumption, such as grape, kiwi, or passion fruit, and that has a flexible stem supported by climbing, twining, or creeping along a surface. Perennials that are normally propagated as annuals such as tomato plants, biennials such as the plants that produce strawberries, and annuals such as pumpkins, squash, cucumbers, watermelon, and other melons, are excluded from the term vine in this subpart.

§ 760.503 Eligible losses.

(a) To be considered an eligible loss under this subpart:

(1) Eligible trees, bushes, or vines must have been lost or damaged as a result of natural disaster as determined by the Deputy Administrator;

(2) The individual stand must have sustained a mortality loss or damage, as the case may be, loss in excess of 15 percent after adjustment for normal mortality or damage;

(3) The loss could not have been prevented through reasonable and available measures; and

(4) The trees, bushes, or vines, in the absence of a natural disaster, would not normally have required rehabilitation or replanting within the 12-month period following the loss.

(b) The damage or loss must be visible and obvious to the county committee representative. If the damage is no longer visible, the county committee may accept other evidence of the loss as it determines is reasonable.

(c) The county committee may require information from a qualified expert, as determined by the county committee, to determine extent of loss in the case of plant disease or insect infestation.

(d) The Deputy Administrator will determine the types of trees, bushes, and vines that are eligible.

(e) An individual stand that did not sustain a sufficient loss as specified in paragraph (a)(2) of this section is not eligible for payment, regardless of the amount of loss sustained.

§ 760.504 Eligible orchardists and nursery tree growers.

(a) To be eligible for TAP payments, the eligible orchardist or nursery tree grower must:

(1) Have planted, or be considered to have planted (by purchase prior to the loss of existing stock planted for commercial purposes) trees, bushes, or vines for commercial purposes, or have a production history, for commercial purposes, of planted or existing trees, bushes, or vines;

(2) Have suffered eligible losses of eligible trees, bushes, or vines occurring between January 1, 2008, and September 30, 2011, as a result of a natural disaster or related condition;

(3) Meet the risk management purchase requirement as specified in § 760.104 or the waiver requirements in § 760.105 or § 760.107; and

(4) Have continuously owned the stand from the time of the disaster until the time that the TAP application is submitted.

(b) A new owner of an orchard or nursery who does not meet the requirements of paragraph (a) of this section may receive TAP payments approved for the previous owner of the orchard or nursery and not paid to the previous owner, if the previous owner of the orchard or nursery agrees to the succession in writing and if the new owner:

(1) Acquires ownership of trees, bushes, or vines for which benefits have been approved;

(2) Agrees to complete all approved practices that the original owner has not completed; and

(3) Otherwise meets and assumes full responsibility for all provisions of this part, including refund of payments made to the previous owner, if applicable.

(c) A producer seeking payment must not be ineligible under the restrictions applicable to citizenship and foreign corporations contained in § 760.103(b) and must meet all other requirements of subpart B of this part.

(d) Federal, State, and local governments and agencies and political subdivisions thereof are not eligible for payment under this subpart.

§ 760.505 Application.

(a) To apply for TAP, a producer that suffered eligible tree, bush, or vine losses that occurred:

(1) During calendar years 2008, 2009, or 2010, prior to May 7, 2010, must provide an application for payment and supporting documentation to FSA no later than July 6, 2010.

(2) On or after May 7, 2010, must provide an application for payment and supporting documentation to FSA within 90 calendar days of the disaster event or date when the loss of trees, bushes, or vines is apparent to the producer.

(b) The producer must submit the application for payment within the time specified in paragraph (a) of this section to the FSA administrative county office that maintains the producer's farm records for the agricultural operation.

(c) A complete application includes all of the following:

(1) A completed application form provided by FSA;

(2) An acreage report for the farming operation as specified in part 718, subpart B, of this chapter;

(3) Subject to verification and a loss amount determined appropriate by the county committee, a written estimate of the number of trees, bushes, or vines lost or damaged that is certified by the producer or a qualified expert, including the number of acres on which the loss occurred; and

(4) Sufficient evidence of the loss to allow the county committee to calculate whether an eligible loss occurred.

(d) Before requests for payment will be approved, the county committee:

(1) Must make an eligibility determination based on a complete application for assistance;

(2) Must verify actual qualifying losses and the number of acres involved by on-site visual inspection of the land and the trees, bushes, or vines;

(3) May request additional information and may consider all relevant information in making its determination; and

(4) Must verify actual costs to complete the practices, as documented by the producer.

§ 760.506 Payment calculations.

(a) Payment to an eligible orchardist or nursery tree grower for the cost of replanting or rehabilitating trees, bushes, or vines damaged or lost due to a natural disaster, in excess of 15 percent damage or mortality (adjusted for normal damage or mortality), will be calculated as follows:

(1) For the cost of planting seedlings or cuttings, to replace lost trees, bushes, or vines, the lesser of:

(i) 70 percent of the actual cost of the practice, or

(ii) The amount calculated using rates established by the Deputy Administrator for the practice.

(2) For the cost of pruning, removal, and other costs incurred for salvaging damaged trees, bushes, or vines, or in the case of mortality, to prepare the land to replant trees, bushes, or vines, the lesser of:

(i) 50 percent of the actual cost of the practice, or

(ii) The amount calculated using rates established by the Deputy Administrator for the practice.

(b) An orchardist or nursery tree grower that did not plant the trees, bushes, or vines, but has a production history for commercial purposes on planted or existing trees and lost the trees, bushes, or vines as a result of a natural disaster, in excess of 15 percent damage or mortality (adjusted for normal damage or mortality), will be eligible for the salvage, pruning, and land preparation payment calculation as specified in paragraph (a)(2) of this section. To be eligible for the replanting payment calculation as specified in paragraph (a)(1) of this section, the orchardist or nursery grower who did not plant the stock must be a new owner who meets all of the requirements of § 760.504(b) or be considered the owner of the trees under provisions appearing elsewhere in this subpart.

(c) Eligible costs for payment calculation include costs for:

(1) Seedlings or cuttings, for tree, bush, or vine replanting;

(2) Site preparation and debris handling within normal horticultural practices for the type of stand being re-established, and necessary to ensure successful plant survival;

(3) Pruning, removal, and other costs incurred to salvage damaged trees, bushes, or vines, or, in the case of tree mortality, to prepare the land to replant trees, bushes, or vines;

(4) Chemicals and nutrients necessary for successful establishment;

(5) Labor to plant seedlings or cuttings as determined reasonable by the county committee; and

(6) Labor used to transplant existing seedlings established through natural regeneration into a productive tree stand.

(d) The following costs are not eligible:

(1) Costs for fencing, irrigation, irrigation equipment, protection of seedlings from wildlife, general improvements, re-establishing structures, and windscreens.

(2) Any other costs not listed in paragraphs (c)(1) through (c)(6) of this section, unless specifically determined eligible by the Deputy Administrator.

(e) Producers must provide the county committee documentation of actual costs to complete the practices, such as receipts for labor costs, equipment rental, and purchases of seedlings or cuttings.

(f) When lost stands are replanted, the types planted may be different from those originally planted. The alternative types will be eligible for payment if the new types have the same general end use, as determined and approved by the county committee. Payments for alternative types will be based on the lesser of rates established to plant the types actually lost or the cost to establish the alternative used. If the type of plantings, seedlings, or cuttings differs significantly from the types lost, the costs may not be approved for payment.

(g) When lost stands are replanted, the types planted may be planted on the same farm in a different location than the lost stand. To be eligible for payment, site preparation costs for the new location must not exceed the cost to re-establish the original stand in the original location.

(h) Eligible orchardists or nursery tree growers may elect not to replant the entire eligible stand. If so, the county committee will calculate payment based on the number of qualifying trees, bushes, or vines actually replanted.

(i) If a practice, such as site preparation, is needed to both replant and rehabilitate trees, bushes, or vines, the producer must document the expenses attributable to replanting versus rehabilitation. The county committee will determine whether the documentation of expenses detailing the amounts attributable to replanting versus rehabilitation is acceptable. In the event that the county committee determines the documentation does not include acceptable detail of cost allocation, the county committee will pro-rate payment based on physical inspection of the loss, damage, replanting, and rehabilitation.

(j) The cumulative total quantity of acres planted to trees, bushes, or vines for which a producer may receive payment under this part for losses that occurred between January 1, 2008, and September 30, 2011, will not exceed 500 acres.

§ 760.507 Obligations of a participant.

(a) Eligible orchardists and nursery tree growers must execute all required documents and complete the TAP-funded practice within 12 months of application approval.

(b) Eligible orchardist or nursery tree growers must allow representatives of FSA to visit the site for the purposes of certifying compliance with TAP requirements.

(c) Producers who do not meet all applicable requirements and obligations will not be eligible for payment.

Subpart G—Supplemental Revenue Assistance Payments Program

SOURCE: 74 FR 68490, Dec. 28, 2009, unless otherwise noted.

§ 760.601 Applicability.

(a) This subpart specifies the terms and conditions of the Supplemental Revenue Assistance Payments Program (SURE).

(b) Assistance in the form of SURE payments is available for crop losses occurring in the crop year 2008 through September 30, 2011, caused by disaster as determined by the Secretary. Crop losses must have occurred in crop year 2008 or subsequent crop years due to an eligible disaster event that occurs on or before September 30, 2011.

(c) SURE provides disaster assistance to eligible participants on farms in:

(1) Disaster counties designated by the Secretary, which also includes counties contiguous to such declared

disaster counties, if the participant incurred actual production losses of at least 10 percent to at least one crop of economic significance on the farm; and

(2) Any county, if the participant incurred eligible total crop losses of greater than or equal to 50 percent of the normal production on the farm, as measured by revenue, including a loss of at least 10 percent to at least one crop of economic significance on the farm.

(d) Subject to the provisions in subpart B of this part, SURE payments will be issued on 60 percent of the difference between the SURE guarantee and total farm revenue, calculated using the National Average Market Price as specified in this subpart.

[74 FR 68490, Dec. 28, 2009, as amended at 76 FR 54075, Aug. 31, 2011]

§ 760.602 Definitions.

(a) The following definitions apply to all determinations made under this subpart.

(b) The terms defined in parts 718, 1400, and 1437 of this title and subpart B of this part will be applicable, except where those definitions conflict with the definitions set forth in this section In the event that a definition in any of those parts conflicts with the definitions set forth in this subpart, the definitions in this subpart apply. Any additional conflicts will be resolved by the Deputy Administrator.

Actual crop acreage means all acreage for each crop planted or intended to be planted on the farm.

Actual production history yield means the average of the actual production history yields for each insurable or noninsurable crop as calculated under the Federal Crop Insurance Act (FCIA) (7 U.S.C. 1501–1524) or Noninsured Crop Disaster Assistance Program (NAP) as set forth in part 1437 of this title, respectively. FSA will use the actual production history yield data provided for crop insurance or NAP, if available, in the SURE payment calculation.

Actual production on the farm means, unless the Deputy Administrator determines that the context requires otherwise, the sum obtained by adding:

(1) For each insurable crop on the farm, excluding value loss crops, the product obtained by multiplying:

(i) 100 percent of the per unit price for the crop used to calculate a crop insurance indemnity for the applicable crop insurance if a crop insurance indemnity is triggered. If a price is not available, then the price is 100 percent of the NAP established price for the crop, times

(ii) The relevant per unit quantity of the crop produced on the farm, adjusted for quality losses, plus

(2) For each noninsurable crop on the farm, excluding value loss crops, the product obtained by multiplying:

(i) 100 percent of the per unit NAP established price for the crop, times

(ii) The relevant per unit quantity of the crop produced on the farm, adjusted for quality losses, plus

(3) For value loss crops, the value of inventory immediately after the disaster.

Adjusted actual production history yield means a yield that will not be less than the participant's actual production history yield for a year and:

(1) In the case of an eligible participant on a farm that has at least 4 years of actual production history for an insurable crop that are established other than pursuant to section 508(g)(4)(B) of FCIA, the average of the production history for the eligible participant without regard to any yields established under that section;

(2) In the case of an eligible participant on a farm that has less than 4 years of actual production history for an insurable crop, of which one or more were established pursuant to section 508(g)(4)(B) of FCIA, the average of the production history for the eligible participant as calculated without including the lowest of the yields established pursuant to section 508(g)(4)(B) of FCIA; or

(3) In all other cases, the actual production history yield of the eligible participant on a farm.

Adjusted NAP yield means a yield that will not be less than the participant's actual production history yield for NAP for a year and:

(1) In the case of an eligible participant on a farm that has at least 4 years of actual production history under NAP that are not replacement yields, the average of the production history

without regard to any replacement yields;

(2) In the case of an eligible participant on a farm that has less than 4 years of actual production history under NAP that are not replacement yields, the average of the production history without including the lowest of replacement yields; or

(3) In all other cases, the actual production history yield of the eligible participant on the farm under NAP.

Administrative fee means a fixed fee payable by a participant for NAP or crop insurance coverage, including buy-in fees, based on the number of covered crops under NAP or insurance under FCIA.

Appraised production means production determined by FSA, or an insurance provider approved by FCIC, that was unharvested, but which was determined to reflect the crop's yield potential at the time of appraisal. An appraisal may be provided in terms of a potential value of the crop.

Aquaculture means the reproduction and rearing of aquatic species as specified in part 1437 of this title in controlled or selected environments.

Brownout means a disruption of electrical or other similar power source for any reason. A brownout, although it may indirectly have an adverse effect on crops, is not a disaster for the purposes of this subpart and losses caused by a brownout will not be considered a qualifying loss.

Catastrophic risk protection (CAT) means the minimum level of coverage offered by the Risk Management Agency (RMA) for crop insurance. CAT is further specified in parts 402 and 1437 of this title.

Counter-cyclical program payment yield means the weighted average payment yield established under part 1412, subpart C of this title.

County expected yield means an estimated yield, expressed in a specific unit of measure equal to the average of the most recent five years of official county yields established by FSA, excluding the years with the highest and lowest yields, respectively.

Crop insurance indemnity means, for the purpose of this subpart, the net payment to a participant excluding the value of the premium for crop losses covered under crop insurance administered in accordance with FCIA by RMA.

Crop of economic significance means any crop, as defined in this subpart that contributed, or, if the crop is not successfully produced, would have contributed or is expected to contribute, 5 percent or more of the total expected revenue from all of a participant's crops on a farm.

Crop year means as determined by the Deputy Administrator for a commodity on a nationwide basis the calendar year in which the crop is normally harvested or, where more than one calendar year is involved, the calendar year in which the majority of the crop would have been harvested. For crops on which catastrophic risk protection, as defined in this section, is available, the crop year will be as defined as in such coverage. Crop year determinations by the Deputy Administrator will be final in all cases and, because these are matters of general applicability, will not be considered by the Farm Service Agency to be subject to administrative appeal.

Determined acreage or determined production means the amount of acres or production for a farm established by a representative of FSA by use of appropriate means such as official acreage, digitizing and planimetering areas on the photograph or other photographic image, or computations from scaled dimensions or ground measurements. In the case of production, any production established by a representative of FSA through audit, review, measurement, appraisal, or other acceptable means of determining production, as determined by FSA.

Disaster means damaging weather, including drought, excessive moisture, hail, freeze, tornado, hurricane, typhoon, excessive wind, excessive heat, weather-related saltwater intrusion, weather-related irrigation water rationing, or any combination thereof and adverse natural occurrences such as earthquakes or volcanic eruptions. Disaster includes a related condition that occurs as a result of the damaging weather or adverse natural occurrence and exacerbates the condition of the crop, such as disease and insect infestation. It does not include brownouts or power failures.

Disaster county means a county included in the geographic area covered by a qualifying natural disaster designation under section 321(a) of the Consolidated Farm and Rural Development Act (7 U.S.C. 1961(a)) and for SURE, the term "disaster county" also includes a county contiguous to a county declared a disaster by the Secretary; however, farms not in a disaster county may qualify under SURE where for the relevant period, as determined under this subpart, the actual production on a farm is less than 50 percent of the normal production on the farm.

Double-cropping means, as determined by the Deputy Administrator on a regional basis, planting for harvest a crop of a different commodity on the same acres in cycle with another crop in a 12-month period in an area where such double-cropping is considered normal, or could be considered to be normal, for all growers and under normal growing conditions and normal agricultural practices for the region and being able to repeat the same cycle in the following 12-month period.

Farm means, for the purposes of determining SURE eligibility, the entirety of all crop acreage in all counties that a producer planted or intended to be planted for harvest for normal commercial sale or on-farm livestock feeding, including native and improved grassland intended for haying. In the case of aquaculture, except for species for which an Aquaculture Grant Program payment was received, the term "farm" includes all acreage used for all aquatic species being produced in all counties that the producer intended to harvest for normal commercial sale. In the case of honey, the term "farm" means all bees and beehives in all counties that the participant intended to be harvested for a honey crop for normal commercial sale.

FCIC means the Federal Crop Insurance Corporation, a wholly owned Government Corporation operated and managed by USDA RMA.

FSA means the Farm Service Agency.

Harvested means:

(1) For insurable crops, harvested is as defined according to the applicable crop insurance policy administered in accordance with FCIA by RMA;

(2) For NAP-covered single harvest crops, a mature crop that has been removed from the field, either by hand or mechanically;

(3) For noninsurable crops with potential multiple harvests in one year or one crop harvested over multiple years, that the participant has, by hand or mechanically, removed at least one mature crop from the field during the crop year; or

(4) For mechanically harvested noninsurable crops, that the mature crop has been removed from the field and placed in or on a truck or other conveyance, except hay is considered harvested when in the bale, whether removed from the field or not. Grazing of land will not be considered harvested for the purpose of determining an unharvested or prevented planting payment factor.

Initial crop means a first crop planted for which assistance is provided under this subpart.

Insurable crop means an agricultural commodity (excluding livestock) for which the participant on a farm is eligible to obtain a policy or plan of crop insurance administered in accordance with FCIA by RMA. Such a crop for which the participant purchased insurance from RMA is referred to as an insured crop.

Insurance is available means when crop information is contained in RMA's county actuarial documents for a particular crop and a policy or plan of insurance administered in accordance with FCIA by RMA. If the Adjusted Gross Revenue Plan of crop insurance was the only plan of insurance available for the crop in the county in the applicable crop year, insurance is considered not available for that crop. If an AGR plan or a pilot plan was the only plan available, producers are not required to purchase it to meet the risk management purchase requirement, but it will satisfy the risk management purchase requirement. In that case, the other ways to meet the requirement would be, if all the requirements of this subpart are met, a buy-in or NAP.

114

Intended use means the original use for which a crop or a commodity is grown and produced.

Marketing year means the 12 months immediately following the established final harvest date of the crop of a commodity, as determined by the Deputy Administrator, and not an individual participant's final harvest date. FSA will use the marketing year determined by NASS, when available.

Maximum average loss level means the maximum level of crop loss that will be used in calculating SURE payments for a participant without reliable or verifiable production records as defined in this section. Loss levels are expressed in either a percent of loss or a yield per acre, and reflect the amount of production that a participant should have produced considering the eligible disaster conditions in the area or county, as determined by the FSA county committee in accordance with instructions issued by the Deputy Administrator.

Multi-use crop means a crop intended for more than one use during the calendar year such as grass harvested for seed, hay, or grazing.

Multiple planting means the planting for harvest of the same crop in more than one planting period in a crop year on the same or different acreage. This is also sometimes referred in this rule as multiple cropping.

NAMP means the national average market price determined in accordance with §§ 760.640 and 760.641.

NASS is the USDA National Agricultural Statistics Service.

Noninsurable crop means a commercially produced crop for which the eligible participants on a farm may obtain coverage under NAP.

Noninsured Crop Disaster Assistance Program or NAP means the FSA program carried out under 7 U.S.C. 7333, as specified in part 1437 of this title.

Normal production on the farm means, for purposes of the revenue calculations of this subpart, the sum of the expected revenue for all crops on the farm. It is stated in terms of revenue, because different crops may have different units of measure.

Planted acreage means land in which seed, plants, or trees have been placed, appropriate for the crop and planting method, at a correct depth, into a seed bed that has been properly prepared for the planting method and production practice normal to the area, as determined by the FSA county committee.

Prevented planting means the inability to plant an eligible crop with proper equipment during the planting period as a result of a disaster, as determined by FSA. All prevented planted cropland must meet conditions provided in § 718.103 of this chapter. Additionally, all insured crops must satisfy the provisions of prevented planting provided in § 457.8 of this title.

Price election means, for an insured crop, the crop insurance price elected by the participant multiplied by the percentage of price elected by the participant.

Production means quantity of a crop or commodity produced on the farm expressed in a specific unit of measure including, but not limited to, bushels or pounds and used to determine the normal production on a farm. Normal production for the whole farm is stated in terms of revenue, because different crops may have different units of measure.

Qualifying loss means a 10 percent loss of at least one crop of economic significance due to disaster and on a farm that is either:

(1) Located in a disaster county (a county for which a Secretarial disaster designation has been issued or in a county contiguous to a county that has received a Secretarial disaster designation), or

(2) If not located in any disaster county or county contiguous to such a county, but has an overall loss greater than or equal to 50 percent of normal production on the farm (expected revenue for all crops on the farm) due to disaster.

Qualifying natural disaster designation means a natural disaster designated by the Secretary for production losses under section 321(a) of the Consolidated Farm and Rural Development Act (7 U.S.C. 1961(a)).

Related condition means, with respect to a disaster, a condition that causes deterioration of a crop such as insect infestation, plant disease, or aflatoxin that is accelerated or exacerbated as a result of damaging weather, as determined by the Deputy Administrator.

Reliable production records means evidence provided by the participant to the FSA county office that FSA determines is adequate to substantiate the amount of production reported when verifiable records are not available, including copies of receipts, ledgers of income, income statements, deposit slips, register tapes, invoices for custom harvesting, records to verify production costs, contemporaneous measurements, truck scale tickets, and contemporaneous diaries. When the term "acceptable production records" is used in this rule, it may be either reliable or verifiable production records, as defined in this section.

Reported acreage or production means information obtained from the participant or the participant's agent, on a form prescribed by FSA or through insurance records.

RMA means the Risk Management Agency.

Salvage value means the dollar amount or equivalent for the quantity of the commodity that cannot be marketed or sold in any recognized market for the crop.

Secretary means the Secretary of Agriculture.

State means a State; the District of Columbia, the Commonwealth of Puerto Rico, and any other territory or possession of the United States.

Subsequent crop means any crop planted after an initial crop, on the same land, during the same crop year.

SURE means the Supplemental Revenue Assistance Payments Program.

Unit of measure means:

(1) For all insurable crops, the FCIC established unit of measure;

(2) For all noninsurable crops, if available, the established unit of measure used for the NAP price and yield;

(3) For aquatic species, a standard unit of measure such as gallons, pounds, inches or pieces, established by the FSA State committee for all aquatic species or varieties;

(4) For turfgrass sod, a square yard;

(5) For maple sap, a gallon; and

(6) For all other crops, the smallest unit of measure that lends itself to the greatest level of accuracy, as determined by the FSA State committee.

USDA means United States Department of Agriculture.

Value loss crop has the meaning specified in part 1437, subpart D of this title. Unless otherwise announced by FSA, value loss crops for SURE include aquaculture, floriculture, ornamental nursery, Christmas trees, mushrooms, ginseng, and turfgrass sod.

Verifiable production records mean evidence that is used to substantiate the amount of production reported and that can be verified by FSA through an independent source.

Volunteer stand means plants that grow from seed residue or are indigenous or are not planted. Volunteer plants may sprout from seeds left behind during a harvest of a previous crop; be unintentionally introduced to land by wind, birds, or fish; or be inadvertently mixed into a crop's growing medium.

§ 760.610 Participant eligibility.

(a) In addition to meeting the eligibility requirements of § 760.103, a participant must meet all of the following conditions:

(1) All insurable crops on the participant's farm must be covered by crop insurance administered by RMA in accordance with FCIA, and all noninsured crops must be covered under NAP, as specified in § 760.104, unless the participant meets the requirements in either § 760.105 or § 760.107. At the discretion of FSA, the equitable relief provisions in § 760.106 may apply.

(2) Crop losses must have occurred in crop year 2008 or subsequent crop years due to an eligible disaster event that occurred on or before September 30, 2011.

(i) For insured crops, the coverage period, as defined in the insurance policy, must have begun on or before September 30, 2011;

(ii) For NAP crops, the coverage period must have begun on or before September 30, 2011; and

(iii) The final planting date for that crop according to the Federal crop insurance or NAP policy must have been on or before September 30, 2011.

(3) A qualifying loss as defined in § 760.602 must have occurred.

(4) The participant must have been in compliance with the Highly Erodible

Land Conservation and Wetland Conservation provisions of part 12 of this title, for 2008 and subsequent crop years through September 30, 2011, as applicable, and must not otherwise be barred from receiving benefits or payments under part 12 of this title or any other law.

(5) The participant must not be ineligible or otherwise barred from the requisite risk management insurance programs or NAP because of past violations where those insurance programs or NAP would otherwise be available absent such violations.

(6) The participant must have an entitlement to an ownership share of the crop and also assume production and market risks associated with the production of the crop. In the event the crop was planted but not produced, participants must have an ownership share of the crop that would have been produced.

(i) Any verbal or written contract that precludes the grower from having an ownership share renders the grower ineligible for payments under this subpart.

(ii) Growers growing eligible crops under contract are not eligible participants under this subpart unless the grower has an ownership share of the crop.

(b) In the event that a producer is determined not to be an eligible producer of a crop in accordance with this section, such crop will be disregarded in determining the producer's production or eligibility for payments under this subpart. However, any insurance, farm program, or NAP payments received by the producer on such crop will count as farm revenue if that producer is an eligible participant as a producer of other crops.

(c) Participants may not receive payments with respect to volunteer stands of crops. Volunteer stands will not be considered in either the calculation of revenue or of the SURE guarantee.

(d) A deceased applicant or an applicant that is a dissolved entity that suffered losses prior to the death or the dissolution that met all eligibility criteria prior to death or dissolution may be eligible for payments for such losses if an authorized representative signs the application for payment. Proof of authority to sign for the deceased participant or dissolved entity must be provided. If a participant is now a dissolved general partnership or joint venture, all members of the general partnership or joint venture at the time of dissolution or their duly authorized representatives must sign the application for payment. Eligibility of such participant will be determined, as it is for other participants, based upon ownership share and risk in producing the crop.

(e) Participants receiving payments under the Emergency Assistance for Livestock, Honey Bees, and Farm-Raised Fish Program (ELAP) as specified in subpart C of this part are not eligible to receive payments under SURE for the same loss.

(f) Participants with a farming interest in multiple counties who apply for SURE payment based on a Secretarial disaster designation must have a 10 percent loss of a crop of economic significance located in at least one disaster county, as defined in this subpart, to be eligible for SURE.

[74 FR 68490, Dec. 28, 2009, as amended at 76 FR 54075, Aug. 31, 2011]

§760.611 Qualifying losses, eligible causes and types of loss.

(a) Eligible causes of loss are disasters which cause types of losses where the crop could not be planted or where crop production was adversely affected in quantity, quality, or both. A qualifying loss, as defined in this subpart, must be the result of a disaster.

(b) A loss will not be considered a qualifying loss if any of the following apply:

(1) The cause of the loss was not the result of disaster;

(2) The cause of loss was due to poor management decisions or poor farming practices, as determined by the FSA county committee on a case-by-case basis;

(3) The cause of loss was due to failure of the participant to re-seed or replant to the same crop in a county where it is customary to re-seed or replant after a loss before the final planting date;

(4) The cause of loss was due to water contained or released by any governmental, public, or private dam or reservoir project if an easement exists on the acreage affected by the containment or release of the water;

(5) The cause of loss was due to conditions or events occurring outside of the applicable crop year growing season; or

(6) The cause of loss was due to a brownout.

(c) The following types of loss, regardless of whether they were the result of a disaster, are not qualifying losses:

(1) Losses to crops not intended for harvest in the applicable crop year;

(2) Losses of by-products resulting from processing or harvesting a crop, such as, but not limited to, cotton seed, peanut shells, wheat or oat straw, or corn stalks or stovers;

(3) Losses to home gardens; or to a crop subject to a de minimis election according to § 760.613;

(4) Losses of crops that were grazed or, if prevented from being planted, had the intended use of grazing; or

(5) Losses of first year seeding for forage production, or immature fruit crops.

(d) The following losses of ornamental nursery stock are not a qualifying loss:

(1) Losses caused by a failure of power supply or brownout as defined in § 760.602;

(2) Losses caused by the inability to market nursery stock as a result of quarantine, boycott, or refusal of a buyer to accept production;

(3) Losses caused by fires that are not the result of disaster;

(4) Losses affecting crops where weeds and other forms of undergrowth in the vicinity of nursery stock have not been controlled; or

(5) Losses caused by the collapse or failure of buildings or structures.

(e) The following losses for honey, where the honey production by colonies or bees was diminished, are not a qualifying loss:

(1) Losses caused by the unavailability of equipment or the collapse or failure of equipment or apparatus used in the honey operation;

(2) Losses caused by improper storage of honey;

(3) Losses caused by bee feeding;

(4) Losses caused by the application of chemicals;

(5) Losses caused by theft or fire not caused by a natural condition including, but not limited to, arson or vandalism;

(6) Losses caused by the movement of bees by the participant or any other legal entity or person;

(7) Losses caused by disease or pest infestation of the colonies, unless approved by the Secretary;

(8) Losses of income from pollinators; or

(9) Losses of equipment or facilities.

§ 760.613 De minimis exception.

(a) Participants seeking the de minimis exception to the risk management purchase requirements of this subpart, must certify:

(1) That a specific crop on the farm is not a crop of economic significance on the farm; or

(2) That the administrative fee required for the purchase of NAP coverage for a crop exceeds 10 percent of the value of that coverage.

(b) To be eligible for a de minimis exception to the risk management purchase requirement in § 760.104, the participant must elect such exception at the same time the participant files the application for payment and the certification of interests, as specified in § 760.620, and specify the crop or crops for which the participant is requesting such exception.

(c) FSA will not consider the value of any crop elected under paragraph (b) of this section in calculating both the SURE guarantee and the total farm revenue.

(d) All provisions of this subpart apply in the event a participant does not obtain an exception according to this section.

§ 760.614 Lack of access.

In addition to other provisions for eligibility provided for in this part, the Deputy Administrator may provide assistance to participants who suffered 2008 production losses that meet the lack of access provisions in 19 U.S.C.

2497(g)(7)(F), where deemed appropriate, and consistent with the statutory provision. Such a determination to exercise that authority, and the terms on which to exercise that authority, will be considered to be a determination of general effect, not a "relief" determination, and will not be considered by the Farm Service Agency to be appealable administratively either within FSA or before the National Appeals Division.

§ 760.620 Time and method of application and certification of interests.

(a) Each producer interested in obtaining a SURE payment must file an application for payment and provide an accurate certification of interests. The application will be on a form prescribed by FSA and will require information or certifications from the producer regarding any other assistance, payment, or grant benefit the producer has received for any of the producer's crops or interests on a farm as defined in this subpart; regardless of whether the crop or interest is covered in the farm's SURE guarantee according to § 760.631. The producer's certification of interests will help FSA establish whether the producer is an eligible participant.

(b) Eligible participants with a qualifying loss as defined in this subpart must submit an application for payment and certification of interests by March 1 of the calendar year that is two years after the relevant corresponding calendar year for the crop year which benefits are sought to be eligible for payment (for example, the final date to submit an application for a SURE payment for the 2009 crop year will be March 1, 2011). Producers who do not submit the application by that date will not be eligible for payment.

(c) To the extent available and practicable, FSA will assist participants with information regarding their interests in a farm, as of the date of certification, based on information already available to FSA from various sources. However, the participant is solely responsible for providing an accurate certification from which FSA can determine the participant's farm interests for the purposes of this program. As determined appropriate by FSA, failure of a participant to provide an accurate certification of interests as part of the application may render the participant ineligible for any assistance under SURE.

(d) To elect a de minimis exception to the risk management purchase requirement for a crop or crops, the participant must meet the requirements specified in § 760.613. When electing a de minimis exception, the participant must specify the crops for which the exception is requested and provide the certification and supporting documentation for that exception at the time the application and certification of interests is filed with FSA.

§ 760.621 Requirement to report acreage and production.

(a) As a condition of eligibility for payment under this subpart, participants must submit an accurate and timely report of all cropland, non-cropland, prevented planting, and subsequent crop acreage and production for the farm in all counties.

(b) Acreage and production reports that have been submitted to FSA for NAP or to RMA for crop insurance purposes may satisfy the requirement of paragraph (a) of this section provided that the participant's certification of interests submitted as required by § 760.620 corresponds to the report requirements in paragraph (a) of this section, as determined by the FSA county committee.

(c) Reports of production submitted for NAP or FCIA purposes must satisfy the requirements of NAP or FCIA, as applicable. In all other cases, in order for production reports or appraisals to be considered acceptable for SURE, production reports and appraisals must meet the requirements set forth in part 1437 of this title.

(d) In any case where production reports or an appraisal is not acceptable, maximum loss provisions apply as specified in § 760.637.

§ 760.622 Incorrect or false producer production evidence.

(a) If production evidence, including but not limited to acreage and production reports, provided by a participant is false or incorrect, as determined by the FSA county committee at any time after an application for payment is made, the FSA county committee will determine whether:

(1) The participant submitting the production evidence acted in good faith or took action to defeat the purposes of the program, such that the information provided was intentionally false or incorrect.

(2) The same false, incorrect, or unacceptable production evidence was submitted for payment(s) under crop insurance or NAP, and if so, for NAP covered crops, make any NAP program adjustments according to § 1437.15 of this title.

(b) If the FSA county committee determines that the production evidence submitted is false, incorrect, or unacceptable, and the participant who submitted the evidence did not act in good faith or took action to defeat the purposes of the program, the provisions of § 760.109, including a denial of future program benefits, will apply. The Deputy Administrator may take further action, including, but not limited to, making further payment reductions or requiring refunds or taking other legal action.

(c) If the FSA county committee determines that the production evidence is false, incorrect, or unacceptable, but the participant who submitted the evidence acted in good faith, payment may be adjusted and a refund may be required.

§ 760.631 SURE guarantee calculation.

(a) Except as otherwise provided in this part, the SURE guarantee for a farm is the sum obtained by adding the dollar amounts calculated in paragraphs (a)(1) through (a)(3) of this section.

(1) For each insurable crop on the farm except for value loss crops, 115 percent of the product obtained by multiplying together:

(i) The price election. If a price election was not made or a participant is eligible as specified in §§ 760.105, 760.106, or 760.107, then the percentage of price will be 55 percent of the NAP established price;

(ii) The payment acres determined according to § 760.632;

(iii) The SURE yield as calculated according to § 760.638; and

(iv) The coverage level elected by the participant. If a coverage level was not elected or a participant is eligible as specified in § 760.105, § 760.106, or § 760.107, a coverage level of 50 percent will be used in the calculation.

(2) For each noninsurable crop on a farm except for value loss crops, 120 percent of the product obtained by multiplying:

(i) 100 percent of the NAP established price for the crop;

(ii) The payment acres determined according to § 760.632;

(iii) The SURE yield calculated according to § 760.638; and

(iv) 50 percent.

(3) The guarantee for value loss crops as calculated according to § 760.634.

(4) In the case of an insurable crop for which crop insurance provides for an adjustment in the guarantee liability, or indemnity, such as in the case of prevented planting, that adjustment will be used in determining the guarantee for the insurable crop.

(5) In the case of a noninsurable crop for which NAP provides for an adjustment in the level of assistance, such as in the case of unharvested crops, that adjustment will be used for determining the guarantee for the noninsurable crop.

(b) Those participants who are eligible according to § 760.105, § 760.106, or § 760.107 who do not have crop insurance or NAP coverage will have their SURE guarantee calculated based on catastrophic risk protection or NAP coverage available for those crops.

(c) FSA will not include in the SURE guarantee the value of any crop that has a de minimis exception, according to § 760.613.

(d) For crops where coverage may exist under both crop insurance and NAP, such as for pasture, rangeland, and forage, adjustments to the guarantee will be the product obtained by multiplying the county expected yield for that crop times:

(1) 115 percent;

(2) 100 percent of the NAP established price;

(3) The payment acres determined according to § 760.632;

(4) The SURE yield calculated according to § 760.638; and

(5) The coverage level elected by the participant.

(e) Participants who do not have a SURE yield as specified in §760.638 will have a yield determined for them by the Deputy Administrator.

(f) The SURE guarantee may not be greater than 90 percent of the sum of the expected revenue for each of the crops on a farm, as determined by the Deputy Administrator.

§ 760.632 Payment acres.

(a) Payment acres as calculated in this section are used in determining both total farm revenue and the SURE guarantee for a farm. Payment acreage will be calculated using the lesser of the reported or determined acres shown to have been planted or prevented from being planted to a crop.

(b) Initial crop acreage will be the payment acreage for SURE, unless the provisions for subsequent crops in this section are met. Subsequently planted or prevented planted acre acreage is considered acreage for SURE only if the provisions of this section are met. All plantings of an annual or biennial crop are considered the same as a planting of an initial crop in tropical regions as defined in part 1437, subpart F, of this title.

(c) In cases where there is double cropped acreage, each crop may be included in the acreage for SURE only if the specific crops are either insured crops eligible for double cropping according to RMA or approved by the FSA State committee as eligible double cropping practices in accordance with procedures approved by the Deputy Administrator.

(d) Except for insured crops, participants with double cropped acreage not meeting the criteria in paragraph (c) of this section may have such acreage included in the acreage for SURE on more than one crop only if the participant submits verifiable records establishing a history of carrying out a successful double cropping practice on the specific crops for which payment is requested.

(e) Participants having multiple plantings may have each planting included in the SURE guarantee only if the planting meets the requirements of part 1437 of this title and all other provisions of this subpart are satisfied.

(f) Provisions of part 718 of this title specifying what is considered prevented planting and how it must be documented and reported will apply to this payment acreage for SURE.

(g) Subject to the provisions of this subpart, the FSA county committee will:

(1) Use the most accurate data available when determining planted and prevented planted acres; and

(2) Disregard acreage of a crop produced on land that is not eligible for crop insurance or NAP.

(h) For any crop acreage for which crop insurance or NAP coverage is canceled, those acres will no longer be considered the initial crop and will, therefore, no longer be eligible for SURE.

(i) Notwithstanding any other provisions of these or other applicable regulations that relate to tolerance in part 718 of this title, if a farm has a crop that has both FSA and RMA acreage for insured crops, payment acres for the SURE guarantee calculation will be based on acres for which an indemnity was received if RMA acres do not differ from FSA acres by more than the larger of 5 percent or 10 acres not to exceed 50 acres. If the difference between FSA and RMA acres is more than the larger of 5 percent or 10 acres not to exceed 50 acres, then the payment acres for the SURE guarantee will be calculated using RMA acres. In that case, the participant will be notified of the discrepancy and that refunds of unearned payments may be required after FSA and RMA reconcile acreage data.

§ 760.633 2008 SURE guarantee calculation.

(a) For a participant who is eligible due to the 2008 buy-in waiver for risk management purchase under the provisions of §760.105(c), the SURE guarantee for their farm for the 2008 crop will be calculated according to §760.631, or according to §760.634 for value loss crops, with the exception that the:

(1) Price election in §760.631(a)(1)(i) is 100 percent of the NAP established price for the crop;

(2) Coverage level in §760.631(a)(1)(iv) is 70 percent; and

(3) The percent specified in § 760.631(a)(2)(iv) is 70 percent instead of 50 percent; and

(4) Coverage level used in § 760.634(a)(1)(ii) is 70 percent; and

(5) The percent specified in § 760.634(a)(2)(ii) is 70 percent instead of 50 percent.

(b) For those 2008 crops that meet the requirements of §§ 760.104, 760.105(a), 760.106, or 760.107, the SURE guarantee will be the higher of:

(1) The guarantee calculated according to § 760.631, or according to § 760.634 for value loss crops, with the exception that the percent specified in §§ 760.631(a)(1) and 760.634(a)(1) will be 120 percent instead of 115 percent;

(2) The guarantee calculated according to § 760.631, or according to § 760.634 for value loss crops, will be used with the exception that the:

(i) Price election in § 760.631(a)(1)(i) is 100 percent of the NAP established price for the crop; and

(ii) Coverage level in §§ 760.631(a)(1)(iv) and 760.634(a)(1)(ii) will be 70 percent; and

(iii) The percent specified in §§ 760.631(a)(2)(iv) and 760.634(a)(2)(ii) will be 70 percent instead of 50 percent.

§ 760.634 SURE guarantee for value loss crops.

(a) The SURE guarantee for value loss crops will be the sum of the amounts calculated in paragraphs (a)(1) and (a)(2) of this section, except as otherwise specified.

(1) For each insurable crop on the farm, 115 percent of the product obtained by multiplying:

(i) The value of inventory immediately prior to disaster, and

(ii) The coverage level elected by the participant. If a coverage level was not elected or a participant is eligible as specified in §§ 760.106 or 760.107, a coverage level of 27.5 percent will be used in the calculation.

(2) For each noninsurable crop on the farm, 120 percent of the product obtained by multiplying:

(i) The value of inventory immediately prior to a disaster, and

(ii) 50 percent.

(b) Aquaculture participants who received assistance under the Aquaculture Grant Program (Pub. L. 111-5)

will not be eligible for SURE assistance on those species for which a grant benefit was received under the Aquaculture Grant Program for feed losses associated with that species.

(c) In the case of an insurable value loss crop for which crop insurance provides for an adjustment in the guarantee, liability, or indemnity, such as in the case of inventory exceeding peak inventory value, the adjustment will be used in determining the SURE guarantee for the insurable crop.

(d) In the case of a noninsurable value loss crop for which NAP provides for an adjustment in the level of assistance, such as in the case of unharvested field grown inventory, the adjustment will be used in determining the SURE guarantee for the noninsurable crop.

§ 760.635 Total farm revenue.

(a) For the purpose of SURE payment calculation, total farm revenue will equal the sum obtained by adding the amounts calculated in paragraphs (a)(1) through (a)(12) of this section.

(1) The estimated actual value for each crop produced on a farm, except for value loss crops, which equals the product obtained by multiplying:

(i) The actual production of the payment acres for each crop on a farm for purposes of determining losses under FCIA or NAP; and

(ii) NAMP, as calculated for the marketing year as specified in § 760.640 and as adjusted if required as specified in § 760.641.

(2) The estimated actual value for each value loss crop produced on a farm that equals the value of inventory immediately after disaster.

(3) 15 percent of the amount of any direct payments made to the participant under part 1412 of this title.

(4) The total amount of any counter-cyclical and average crop revenue election payments made to the participant under part 1412 of this title.

(5) The total amount of any loan deficiency payments, marketing loan gains, and marketing certificate gains made to the participant under parts 1421 and 1434 of this title.

(6) The amount of payments for prevented planting.

(7) The amount of crop insurance indemnities.

(8) The amount of NAP payments received.

(9) The value of any guaranteed payments made to a participant in lieu of production pursuant to an agreement or contract, if the crop is included in the SURE guarantee.

(10) Salvage value for any crops salvaged.

(11) The value of any other disaster assistance payments provided by the Federal Government for the same loss for which the eligible participant applied for SURE.

(12) For crops for which the eligible participant received a waiver under the provisions of § 760.105(c) or obtained relief according to § 760.106, the value determined by FSA based on what the participant would have received, irrespective of any other provision, if NAP or crop insurance coverage had been obtained.

(b) Sale of plant parts or by-products, such as straw, will not be counted as farm revenue.

(c) For value loss crops:

(1) Other inventory on hand or marketed at some time other than immediately prior to and immediately after the disaster event are irrelevant for revenue purposes and will not be counted as revenue for SURE.

(2) Revenue will not be adjusted for market loss.

(3) Quality losses will not be considered in determining revenue.

(4) In no case will market price declines in value loss crops, due to any cause, be considered in the calculation of payments for those crops.

§ 760.636 Expected revenue.

The expected revenue for each crop on a farm is:

(a) For each insurable crop, except value loss crops, the product obtained by multiplying:

(1) The SURE yield as specified in § 760.638;

(2) The payment acres as specified in § 760.632; and

(3) 100 percent of the price for the crop used to calculate a crop insurance indemnity for an applicable policy of insurance if a crop insurance indemnity is triggered. If a price is not avail-

able, then the price is 100 percent of the NAP established price for the crop, and

(b) For each noninsurable crop, except value loss crops, the product obtained by multiplying

(1) The SURE yield as specified in § 760.638;

(2) The payment acres as specified in § 760.632; and

(3) 100 percent of the NAP price.

(c) For each value loss crop, the value of inventory immediately prior to the disaster.

§ 760.637 Determination of production.

(a) Except for value loss crops, production for the purposes of this part includes all harvested, appraised, and assigned production for the payment acres determined according to § 760.632.

(b) The FSA county committee will use the best available data to determine production, including RMA and NAP loss records and yields for insured and noninsured crops.

(c) The production of any eligible crop harvested more than once in a crop year will include the total harvested production from all harvests.

(d) Crop production losses occurring in tropical regions, as defined in part 1437, subpart F of this chapter, will be based on a crop year beginning on January 1 and ending on December 31 of the same calendar year. All crop harvests in tropical regions that take place between those dates will be considered a single crop.

(e) Any record of an appraisal of crop production conducted by RMA or FSA through a certified loss adjustor will be used if available. Unharvested appraised production will be included in the calculation of revenue under SURE. If the unharvested appraised crop is subsequently harvested for the original intended use, the larger of the actual or appraised production will be used to determine payment.

(1) If no appraisal is available, the participant is required to submit verifiable or reliable production evidence.

(2) If the participant does not have verifiable or reliable production evidence, the FSA county committee will use the higher of the participant's crop certification or the maximum average

loss level to determine the participant's crop production losses.

(f) Production will be adjusted based on a whole grain equivalent, as established by FSA, for all crops with an intended use of grain, but harvested as silage, cobbage, or hay, cracked, rolled, or crimped.

(g) For crops sold in a market that is not a recognized market for that crop and has no established county expected yield and NAMP, the quantity of such crops will not be considered production; rather, 100 percent of the salvage value will be included in the revenue calculation.

(h) Production from different counties that is commingled on the farm before it was a matter of record and cannot be separated by using records or other means acceptable to FSA will have the NAMP prorated to each respective county by FSA. Commingled production may be attributed to the applicable county, if the participant made the location of production of a crop a matter of record before commingling, if the participant does either of the following:

(1) Provides copies of verifiable documents showing that production of the crop was purchased, acquired, or otherwise obtained from the farm in that county; or

(2) Had the farm's production in that county measured in a manner acceptable to the FSA county committee.

(i) The FSA county committee will assign production for the purpose of NAMP for the farm if the FSA county committee determines that the participant failed to provide verifiable or reliable production records.

(j) If RMA loss records are not available, or if the FSA county committee determines that the RMA loss records as reported by the insured participant appear to be questionable or incomplete, or if the FSA county committee makes inquiry, then participants are responsible for:

(1) Retaining and providing, when required, the best available verifiable and reliable production records available for the crops;

(2) Summarizing all the production evidence;

(3) Accounting for the total amount of production for the crop on a farm, whether or not records reflect this production;

(4) Providing the information in a manner that can be easily understood by the FSA county committee; and

(5) Providing supporting documentation if the FSA county committee has reason to question the disaster event or that all production has been taken into account.

(k) The participant must supply verifiable or reliable production records to substantiate production to the FSA county committee. If the eligible crop was sold or otherwise disposed of through commercial channels, acceptable production records include: Commercial receipts; settlement sheets; warehouse ledger sheets or load summaries; or appraisal information from a loss adjuster acceptable to FSA. If the eligible crop was farm-stored, sold, fed to livestock, or disposed of by means other than commercial channels, acceptable production records for these purposes include: Truck scale tickets; appraisal information from a loss adjuster acceptable to FSA; contemporaneous reliable diaries; or other documentary evidence, such as contemporaneous reliable measurements. Determinations of reliability with respect to this paragraph will take into account, as appropriate, the ability of the agency to verify the evidence as well as the similarity of the evidence to reports or data received by FSA for the crop or similar crops. Other factors deemed relevant may also be taken into account.

(l) If no verifiable or reliable production records are available, the FSA county committee will use the higher of the participant's certification or the maximum average loss level to determine production.

(m) Participants must provide all records for any production of a crop that is grown with an arrangement, agreement, or contract for guaranteed payment.

(n) FSA may verify the production evidence submitted with records on file at the warehouse, gin, or other entity that received or may have received the reported production.

§760.638 Determination of SURE yield.

(a) Except for value loss crops as specified in §760.634, a SURE yield will be determined for each crop, type, and intended use on a farm, using the higher of the participant's weighted:

(1) Adjusted actual production history yield as determined in paragraph (b) of this section; or

(2) Counter-cyclical yield as determined in paragraph (c) of this section.

(b) The adjusted actual production history yield, as defined in §760.602, will be weighted by the applicable crop year total planted and prevented planted acres, by crop, type, and intended use for each county. RMA data will be used for calculating the SURE yield for insured crops.

(c) The counter-cyclical yield for a crop on a SURE farm will be weighted in such manner as FSA deems fit taking into account a desire for a consistent system and FSA's ability to make timely yield determinations.

(d) Participants who do not purchase crop insurance or NAP coverage, but who are otherwise eligible for payment, will have a SURE yield determined by the FSA county committee as follows:

(1) A weighted yield, based on planted and prevented planted acres, the location county, crop type, and intended use, will be determined at 65 percent of the county expected yield for each crop.

(2) The SURE yield will be the higher of the yield calculated using the method in paragraph (d)(1) of this section or 65 percent of the weighted counter-cyclical yield as determined in paragraph (c) of this section.

(e) For those participants with crop insurance but without an adjusted actual production history yield, a SURE yield will be determined by the applicable FSA county committee. This paragraph will apply in the case where the insurance policy does not require an actual production history yield, or where a participant has no production history.

[74 FR 68490, Dec. 28, 2009, as amended at 75 FR 19189, Apr. 14, 2010]

§760.640 National average market price.

(a) The Deputy Administrator will establish the National Average Market Price (NAMP) using the best sources available, as determined by the Deputy Administrator, which may include, but are not limited to, data from NASS, Cooperative Extension Service, Agricultural Marketing Service, crop insurance, and NAP.

(b) NAMP may be adjusted by the FSA State committee, in accordance with instructions issued by the Deputy Administrator and as specified in §760.641, to recognize average quality loss factors that are reflected in the market by county or part of a county.

(c) With respect to a crop for which an eligible participant on a farm receives assistance under NAP, the NAMP will not exceed the price of the crop established under NAP.

(d) To the extent practicable, the NAMP will be established on a harvested basis without the inclusion of transportation, storage, processing, marketing, or other post-harvest expenses, as determined by FSA.

(e) NAMP may be adjusted by the FSA State committee, as authorized by The Deputy Administrator, to reflect regional variations in price consistent with those prices established under the FCIA or NAP.

§760.641 Adjustments made to NAMP to reflect loss of quality.

(a) The Deputy Administrator will authorize FSA county committees, with FSA State committee concurrence, to adjust NAMP for a county or part of a county:

(1) To reflect the average quality discounts applied to the local or regional market price of a crop due to a reduction in the intrinsic characteristics of the production resulting from adverse weather, as determined annually by the State office of the FSA; or

(2) To account for a crop for which the value is reduced due to excess moisture resulting from a disaster related condition.

(3) For adjustments specified in paragraphs (a)(1) and (a)(2) of this section, an adjustment factor that represents the regional or local price received for

the crop in the county will be calculated by the FSA State committee. The adjustment factor will be based on the average actual market price compared to NAMP.

(b) For adjustments made under paragraph (a) of this section, participants must provide verifiable evidence of actual or appraised production, clearly indicating an average loss of value caused by poor quality or excessive moisture that meets or exceeds the quality adjustment for the county or part of a county established in paragraph (a)(3) of this section to be eligible to receive the quality-adjusted NAMP as part of their SURE payment calculation. In order to be considered at all for the purpose of quality adjustments, the verifiable evidence of production must itself detail the extent of the quality loss for a specific quantity. With regard to test evidence, in addition to meeting all the requirements of this section, tests must have been completed by January 1 of the year following harvest.

§ 760.650 Calculating SURE.

(a) Subject to the provision of this subpart, SURE payments for crop losses in crop year 2008 and subsequent crop years will be calculated as the amount equal to 60 percent of the difference between:

(1) The SURE guarantee, as specified in § 760.631, § 760.633 or § 760.634 of this subpart, and

(2) The total farm revenue, as specified in § 760.635.

(b) In addition to the other provisions of this subpart and subpart B of this part, SURE payments may be adjusted downward as necessary to insure compliance with the payment limitations in subpart B and to insure that payments do not exceed the maximum amount specified in § 760.108(a)(1) or (b)(1) or otherwise exceed the perceived intent of 19 U.S.C. 2497(j). Such adjustments can include, but are not limited to, adjustments to insure that there is no duplication of benefits as specified in § 760.108(c).

Subpart H—Crop Assistance Program

AUTHORITY: 7 U.S.C. 612c.

SOURCE: 75 FR 65428, Oct. 25, 2010, unless otherwise noted.

§ 760.701 Applicability.

(a) This subpart specifies the eligibility requirements and payment calculations for the Crop Assistance Program (CAP), which will be administered using funds authorized by Section 32 of the Agricultural Adjustment Act of 1935 (7 U.S.C. 612c, as amended).

(b) CAP, within the limits of the funds made available by the Secretary for this program, is intended to help reestablish purchasing power to producers of long grain rice, medium or short grain rice, upland cotton, soybeans, and sweet potatoes who suffered a five percent or greater loss in the 2009 crop year due to disaster.

(c) Only producers who have a share in a farm located in a disaster county (a county that is the primary county that is the subject of a Secretarial disaster designation for 2009 crop year due to excessive moisture and related conditions, as determined by FSA) are eligible for CAP benefits.

§ 760.702 Definitions.

The following definitions apply to CAP. The definitions in parts 718, 760, and 1400 of this title also apply, except where they conflict with the definitions in this section.

Acceptable production records means verifiable or reliable production records deemed acceptable by FSA.

Application means the CAP application form.

Application period means the 45-day period established by the Deputy Administrator for producers on farms in disaster counties to apply for CAP that ends December 9, 2010.

Approved yield means the amount of production per acre, computed in accordance with FCIC's Actual Production History (APH) Program at part 400, subpart G of this title or, for crops not included under part 400, subpart G of this title, the yield used to determine the guarantee. For crops covered under NAP, the approved yield is established according to part 1437 of this title.

Considered planted means acreage approved as prevented planted or failed in accordance with §718.103 of this chapter.

Crop means the reported or determined 2009 crop year planted and considered planted acres of long grain rice, medium or short grain rice, upland cotton, soybean, or sweet potatoes as reflected on 2009 crop year form FSA–578, Report of Acreage, for a producer in a disaster county as of October 22, 2010. Subsequent crops, replacement crops, reseeded crops, and replanted crops are not eligible crops under this part and no revision of the Report of Acreage that would increase an eligibility for payment will be permitted to produce that effect.

Crop year means for 2009:

(1) For insurable crops, the crop year as defined according to the applicable crop insurance policy;

(2) For NAP covered crops, the crop year as provided in part 1437 of this title.

Disaster means excessive moisture or related condition, resulting from any of the following: flood, flash flooding, excessive rain, moisture, humidity, severe storms, thunderstorms, ground saturation or standing water, hail, winter storms, ice storms, snow, blizzard, hurricane, typhoons, tropical storms, and cold wet weather. A disaster does not include brownouts or power failures.

Disaster county means a county included in the geographic area covered by a qualifying natural disaster designation under section 321(a) of the Consolidated Farm and Rural Development Act (7 U.S.C. 1961(a)). For CAP, the term "disaster county" is limited to those primary counties declared a disaster by the Secretary for excessive moisture or a related condition, which are limited to designations based on any of the following: flood, flash flooding, excessive rain, moisture, humidity, severe storms, thunderstorms, ground saturation or standing water, hail, winter storms, ice storms, snow, blizzard, hurricane, typhoons, tropical storms, and cold wet weather.

Expected production means, for a producer on a farm who attempts to determine what the producer might produce for an eligible crop on a farm, the historic yield multiplied by the producer's share of planted and considered planted acres of the crop for the farm. Expected production may be used to assist producers in determining whether the producer has a crop or crops that suffered a qualifying loss of five percent and to determine whether that crop is eligible for CAP benefits.

Historic yield means, for a producer on a farm, the higher of the county average yield or the producer's approved yields for eligible crops on the farm.

(1) An insured producer's yield will be the higher of the county average yield listed or the approved federal crop insurance APH, for the disaster year.

(2) A NAP producer's yield will be the higher of the county average yield or NAP approved yield for the disaster year.

Replacement crop means the planting or approved prevented planting of any crop for harvest following the failed planting or prevented planting of a crop of long grain rice, medium or short grain rice, upland cotton, soybeans, or sweet potatoes not in a recognized double-cropping sequence. Replacement crops are not eligible for CAP.

Reseeded or replanted crop means the second planting of a crop of long grain rice, medium or short grain rice, upland cotton, soybeans, or sweet potatoes on the same acreage after the first planting of that same crop that failed.

§ 760.703 **Producer eligibility requirements.**

(a) A producer must meet all of the requirements in this subpart to be eligible for a CAP payment.

(b) To be eligible, a producer must be an individual or entity who is entitled to an ownership share of an eligible crop and who has the production and market risks associated with the agricultural production of the crop on a farm. An eligible producer must be a:

(1) Citizen of the United States;

(2) Resident alien, which for purposes of this subpart means "lawful alien" as defined in 7 CFR part 1400;

(3) Partnership of citizens of the United States; or

(4) Corporation, limited liability corporation, or other farm organizational structure organized under State law.

(c) To be eligible, a producer must have:

(1) Produced a 2009 crop year planted or considered planted long grain rice, medium or short grain rice, upland cotton, soybean, or sweet potato crop in a 2009 eligible disaster county, and

(2) Suffered a five percent or greater loss in an eligible disaster county in 2009. A list of the disaster counties for CAP is available on the FSA Web site and at FSA county offices.

§ 760.704 Time and method of application.

(a) To request a CAP payment, the producer must submit a CAP application on the form designated by FSA to the FSA county office responsible for administration of the farm.

(b) Producers submitting an application for a crop must certify that they suffered a five percent or greater loss of the crop on the farm in a disaster county and that they have documentation to support that certification as required in § 760.713.

(c) Once submitted by a producer, the application is considered to contain information and certifications of and pertaining to the producer's crop and farm regardless of who entered the information on the application.

(d) Producers requesting benefits under CAP must certify the accuracy and truthfulness of the information provided in the application as well as with any documentation that may be provided with the application or documentation that will be provided to FSA in substantiation of the application. All certifications and information are subject to verification by FSA.

(e) Producers applying for CAP must certify that they have an eligible ownership share interest in the 2009 crop acreage that sustained a five percent or greater loss. The determination and certification by a producer that a crop suffered the requisite five percent or greater farm crop loss is the expected quantity of production of the crop less the actual production of the crop.

(f) In the event that the producer does not submit documentation in response to any request of FSA to support the producer's application or documentation furnished does not show a crop loss of at least five percent as claimed, the application for that crop will be disapproved in its entirety. For quantity losses, producers need to apply a standard similar to the historic yield provisions used under previous ad hoc disaster programs. Those provisions provided that a historic yield was the higher of a county average yield or a producer's approved yield. Thus, if an applicant is determining whether a farm has a crop that suffered a loss of five percent or greater on the farm's planted and considered planted acreage, the applicant could compare the amount successfully produced in 2009 from those planted and considered planted acres to what the participant expected to produce from that acreage using either the county average yield (which may be obtained from FSA by request) or based on analysis of approved actual production history yields that may exist for producers of the crop on the farm.

(g) Unless otherwise determined necessary by FSA, producers will not be required to submit documentation of farm crop production or loss at time of application. FSA's decision not to require proof, documentation, or evidence in support of any application at time of application is not to be construed as a determination of a producer's eligibility.

(h) Producers who apply are required to retain documentation in support of their application for three years after the date of application in accordance with § 760.713.

(i) The application submitted in accordance with this section is not considered valid and complete for issuance of payment under this part unless FSA determines all the applicable eligibility provisions have been satisfied and the producer has submitted all the required forms. In addition to the completed, certified application form, if the information for the following forms or certifications is not on file in the FSA county office or is not current for 2009, the producer must also submit:

(1) Farm operating plan for individual or legal entity;

(2) Average adjusted gross income statement for 2009; and

(3) Highly erodible land conservation (HELC) and wetland conservation certification.

(j) Application approval and payment by FSA does not relieve a producer from having to submit any form, records, or documentation required, but not filed at the time of application or payment, according to paragraph (h) of this section.

§ 760.705 **Payment rates and calculation of payments.**

(a) CAP payments will be calculated by multiplying the total number of reported or determined acres of an eligible crop by the per acre payment rate for that crop. Payment rates are as follows:

(1) Long grain rice, $31.93 per acre;
(2) Medium or short grain rice, $52.46 per acre;
(3) Upland cotton, $17.70 per acre;
(4) Soybeans, $15.62 per acre; and
(5) Sweet potatoes, $155.41 per acre.

(b) Payments will be calculated based on the 2009 crop year reported or determined planted or considered planted acres of an eligible crop on a farm in a disaster county as reflected on a form FSA–578, Report of Acreage, on file in FSA as of October 22, 2010.

§ 760.706 **Availability of funds.**

(a) Payments specified in this subpart are subject to the availability of funds. The total available program funds are $550 million. In order to keep payments within available funds, the Deputy Administrator may pro-rate payments, to the extent the Deputy Administrator determines that necessary.

(b) Funds for CAP are being made available only for the 2009 crop year reported and determined eligible crop acreage in disaster counties as reflected on a form FSA–578, Report of Acreage, as of October 22, 2010.

§ 760.707 **Proof of loss.**

(a) All certifications, applications, and documentation are subject to spot check and verification by FSA. Producers must submit documentation to FSA if and when FSA requests documentation to substantiate any certified application.

(b) Producers are responsible for retaining or providing, when required, verifiable or reliable production or loss records available for the crop. Producers are also responsible for summarizing all the production or loss evidence and providing the information in a manner that can be understood by the county committee.

(c) Any producer receiving payment under this subpart agrees to maintain any books, records, and accounts supporting any information or certification made according to this part for 3 years after the end of the year following application.

(d) Producers receiving payments or any other person who furnishes such information to FSA must permit FSA or authorized representatives of USDA and the General Accounting Office during regular business hours to inspect, examine, and to allow such persons to make copies of such books, records or other items for the purpose of confirming the accuracy of the information provided by the producer.

§ 760.708 **Miscellaneous provisions and limitations.**

(a) A person ineligible under § 1437.15(c) of this title concerning violations of the Noninsured Crop Disaster Assistance Program for the 2009 crop year is ineligible for benefits under this subpart.

(b) A person ineligible under § 400.458 of this title for the 2009 crop year concerning violations of crop insurance regulations is ineligible for CAP.

(c) In the event that any request for CAP payment resulted from erroneous information or a miscalculation, the payment will be recalculated and the producer must refund any excess to FSA with interest to be calculated from the date of the disbursement to the producer. If for whatever reason the producer signing a CAP application overstates the loss level of the crop when the actual loss level determined by FSA for the crop is less than the level claimed, or where the CAP payment would exceed the producer's actual loss, the application will be disapproved for the crop and the full CAP payment for that crop will be required to be refunded with interest from date of disbursement. The CAP payment cannot exceed the producer's actual loss.

(d) The liability of anyone for any penalty or sanction under or in connection with this subpart, or for any refund to FSA or related charge is in addition to any other liability of such person under any civil or criminal fraud statute or any other provision of law including, but not limited to: 18 U.S.C. 286, 287, 371, 641, 651, 1001, and 1014; 15 U.S.C. 714; and 31 U.S.C. 3729.

(e) The regulations in parts 11 and 780 of this title apply to determinations under this subpart.

(f) Any payment to any person under this subpart will be made without regard to questions of title under State law and without regard to any claim or lien against the crop, or its proceeds.

(g) Any payment made under this subpart will be considered farm revenue for 2009 for the Supplemental Revenue Assistance Payments Program.

(h) The average AGI limitation provisions in part 1400 of this title relating to limits on payments for persons or legal entities, excluding joint ventures and general partnerships, with certain levels of average adjusted gross income (AGI) apply to each applicant for CAP. Specifically, a person or legal entity with an average adjusted gross nonfarm income, as defined in § 1404.3 of this title, that exceeds $500,000 is not eligible to receive CAP payments.

(i) No person or legal entity, excluding a joint venture or general partnership, as determined by the rules in part 1400 of this title may receive, directly or indirectly, more than $100,000 in payments under this subpart.

(j) The direct attribution provisions in part 1400 of this title apply to CAP. Under those rules, any payment to any legal entity will also be considered for payment limitation purposes to be a payment to persons or legal entities with an interest in the legal entity or in a sub-entity. If any such interested person or legal entity is over the payment limitation because of direct payment or their indirect interests or a combination thereof, then the payment to the actual payee will be reduced commensurate with the amount of the interest of the interested person in the payee. Likewise, by the same method, if anyone with a direct or indirect interest in a legal entity or sub-entity of a payee entity exceeds the AGI levels that would allow a producer to directly receive a CAP payment, then the payment to the actual payee will be reduced commensurately with that interest. For CAP, unless otherwise specified in part 1400 of this title, the AGI amount will be that person's or legal entity's average AGI for the three taxable years that precede the 2008 taxable year (that is 2005, 2006, and 2007).

(k) For the purposes of the effect of lien on eligibility for Federal programs (28 U.S.C. 3201(e)), FSA waives the restriction on receipt of funds under CAP but only as to beneficiaries who, as a condition of such waiver, agree to apply the CAP payments to reduce the amount of the judgment lien.

(l) For CAP, producers are either eligible or ineligible. Therefore, the provisions of § 718.304 of this chapter, "Failure to Fully Comply," do not apply to this subpart.

(m) The regulations in subpart B apply to CAP. In addition to those regulations that specifically include subpart H or apply to this part, the following sections specifically apply to this subpart: §§ 760.113(a), 760.114, and 760.116(a).

Subpart I—2005–2007 Crop Disaster Program

Source: 72 FR 72867, Dec. 21, 2007, unless otherwise noted.

§ 760.800 Applicability.

This part sets forth the terms and conditions for the 2005–2007 Crop Disaster Program (2005–2007 CDP). CDP makes emergency financial assistance available to producers who have incurred crop losses in quantity or quality for eligible 2005, 2006, or 2007 crop years due to disasters as determined by the Secretary under provisions of Title IX of the U.S. Troop Readiness, Veterans' Care, Katrina Recovery, and Iraq Accountability Appropriations Act, 2007 (Pub. L. 110–28). However, to be eligible for assistance, the crop subject to the loss must have been planted or existed before February 28, 2007, or, in the case of prevented planting, would have been planted before February 28, 2007.

§760.801 Administration.

(a) The program will be administered under the general supervision of the Deputy Administrator for Farm Programs and will be carried out in the field by FSA State and county committees.

(b) State and county committees and representatives do not have the authority to modify or waive any of the provisions of this part.

(c) The State committee will take any action required by this part that has not been taken by a county committee. The State committee will also:

(1) Correct, or require a county committee to correct, any action taken by that FSA county committee that is not in accordance with this part; and

(2) Require a county committee to withhold taking or reverse any action that is not in accordance with this part.

(d) No provision or delegation to a State or county committee will prevent the Deputy Administrator for Farm Programs from determining any question arising under the program or from reversing or modifying any determination made by a State or county committee.

(e) The Deputy Administrator for Farm Programs may authorize State and county committees to waive or modify non-statutory deadlines or other program requirements in cases where lateness or failure to meet such does not adversely affect the operation of the program.

§760.802 Definitions.

The following definitions apply to this part. The definitions in parts 718 and 1400 of this title also apply, except where they conflict with the definitions in this section.

Actual production means the total quantity of the crop appraised, harvested, or assigned, as determined by the FSA State or county committee in accordance with instructions issued by the Deputy Administrator for Farm Programs.

Administrative fee means an amount the producer must pay for Noninsured Crop Disaster Assistance Program (NAP) enrollment for non-insurable crops.

Affected production means, with respect to quality losses, the harvested production of an eligible crop that has a documented quality reduction of 25 percent or more on the verifiable production record.

Appraised production means production determined by FSA, or a company reinsured by the Federal Crop Insurance Corporation (FCIC), that was unharvested but was determined to reflect the crop's yield potential at the time of appraisal.

Approved yield means the amount of production per acre, computed in accordance with FCIC's Actual Production History (APH) Program at part 400, subpart G of this title or, for crops not included under part 400, subpart G of this title, the yield used to determine the guarantee. For crops covered under NAP, the approved yield is established according to part 1437 of this title. Only the approved yields based on production evidence submitted to FSA prior to May 25, 2007 will be used for purposes of the 2005–2007 CDP.

Aquaculture means a value loss crop for the reproduction and rearing of aquatic species in controlled or selected environments including, but not limited to, ocean ranching, except private ocean ranching of Pacific salmon for profit in those States where such ranching is prohibited by law.

Aquaculture facility means any land or structure including, but not limited to, a laboratory, concrete pond, hatchery, rearing pond, raceway, pen, incubator, or other equipment used in aquaculture.

Aquaculture species means any aquaculture species as defined in part 1437 of this title.

Average market price means the price or dollar equivalent on an appropriate basis for an eligible crop established by FSA, or CCC, or RMA, as applicable, for determining payment amounts. Such price will be based on historical data of the harvest basis excluding transportation, storage, processing, packing, marketing, or other post-harvesting expenses. Average market prices are generally applicable to all similarly situated participants and are not established in response to individual participants. Accordingly, the established average market prices are

131

not appealable under parts 11 or 780 of this title.

Catastrophic risk protection means the minimum level of coverage offered by FCIC.

CCC means the Commodity Credit Corporation.

Controlled environment means, with respect to those crops for which a controlled environment is expected to be provided, including but not limited to ornamental nursery, aquaculture (including ornamental fish), and floriculture, an environment in which everything that can practically be controlled with structures, facilities, growing media (including, but not limited to, water, soil, or nutrients) by the producer, is in fact controlled by the producer.

Crop insurance means an insurance policy reinsured by FCIC under the provisions of the Federal Crop Insurance Act, as amended.

Crop year means:

(1) For insured crops, the crop year as defined according to the applicable crop insurance policy;

(2) For NAP covered crops, as provided in part 1437 of this title.

Damaging weather means drought, excessive moisture, hail, freeze, tornado, hurricane, typhoon, excessive wind, excessive heat, weather-related saltwater intrusion, weather-related irrigation water rationing, and earthquake and volcanic eruptions, or any combination. It also includes a related condition that occurs as a result of the damaging weather and exacerbates the condition of the crop, such as crop disease, and insect infestation.

Deputy Administrator means the Deputy Administrator for Farm Programs, Farm Service Agency, U.S. Department of Agriculture or designee.

Eligible crop means a crop insured by FCIC as defined in part 400 of this title, or included under NAP as defined under part 1437 of this title for which insurance or NAP coverage was obtained timely for the year which CDP benefits are sought.

End use means the purpose for which the harvested crop is used, such as grain, hay, or seed.

Expected production means, for an agricultural unit, the historic yield multiplied by the number of planted or prevented acres of the crop for the unit.

FCIC means the Federal Crop Insurance Corporation, a wholly owned Government Corporation within USDA.

Final planting date means the latest date, established by the Risk Management Agency (RMA) for insured crops, by which the crop must initially be planted in order to be insured for the full production guarantee or amount of insurance per acre. For NAP covered crops, the final planting date is as provided in part 1437 of this title.

Flood prevention means:

(1) For aquaculture species, placing the aquaculture facility in an area not prone to flood;

(2) In the case of raceways, devices or structures designed for the control of water level; and

(3) With respect to nursery crops, placing containerized stock in a raised area above expected flood level and providing draining facilities, such as drainage ditches or tile, gravel, cinder, or sand base.

Good nursery growing practices means utilizing flood prevention, growing media, fertilization to obtain expected production results, irrigation, insect and disease control, weed, rodent and wildlife control, and over winterization storage facilities.

Ground water means aqueous supply existing in an aquifer subsurface that is brought to the surface and made available for irrigation by mechanical means such as by pumps and irrigation wells.

Growing media means:

(1) For aquaculture species, media that provides nutrients necessary for the production of the aquaculture species and protects the aquaculture species from harmful species or chemicals or

(2) For nursery crops, a well-drained media with a minimum 20 percent air pore space and pH adjustment for the type of plant produced designed to prevent "root rot."

Harvested means:

(1) For insured crops, harvested as defined according to the applicable crop insurance policy;

(2) For NAP covered single harvest crops, that a crop has been removed

132

from the field, either by hand or mechanically, or by grazing of livestock;

(3) For NAP covered crops with potential multiple harvests in 1 year or harvested over multiple years, that the producer has, by hand or mechanically, removed at least one mature crop from the field during the crop year;

(4) For mechanically-harvested NAP covered crops, that the crop has been removed from the field and placed in a truck or other conveyance, except hay is considered harvested when in the bale, whether removed from the field or not. Grazed land will not be considered harvested for the purpose of determining an unharvested or prevented planting payment factor. A crop that is intended for mechanical harvest, but subsequently grazed and not mechanically harvested, will have an unharvested factor applied.

Historic yield means, for a unit, the higher of the county average yield or the participant's approved yield.

(1) An insured participant's yield will be the higher of the county average yield listed or the approved federal crop insurance APH, for the disaster year.

(2) NAP participant's yield will be the higher of the county average or approved NAP APH for the disaster year.

Insurable crop means an agricultural crop (excluding livestock) for which the producer on a farm is eligible to obtain a policy or plan of insurance under the Federal Crop Insurance Act (7 U.S.C. 1501–1524).

Marketing contract means a legally binding written contract between a purchaser and grower for the purpose of marketing a crop.

Market value means:

(1) The price(s) designated in the marketing contract; or

(2) If not designated in a marketing contract, the rate established for quantity payments under § 760.811.

Maximum average loss level means the maximum average level of crop loss to be attributed to a participant without acceptable production records (verifiable or reliable). Loss levels are expressed in either a percent of loss or yield per acre, and are intended to reflect the amount of production that a participant would have been expected to make if not for the eligible disaster

conditions in the area or county, as determined by the county committee in accordance with instructions issued by the Deputy Administrator.

Multi-use crop means a crop intended for more than one end use during the calendar year such as grass harvested for seed, hay, and grazing.

Multiple cropping means the planting of two or more different crops on the same acreage for harvest within the same crop year.

Multiple planting means the planting for harvest of the same crop in more than one planting period in a crop year on different acreage.

NASS means the National Agricultural Statistics Service.

Net crop insurance indemnity means the indemnity minus the producer paid premium.

NAP covered means a crop for which the participants obtained assistance under section 196 of the Federal Agriculture Improvement and Reform Act of 1996 (7 U.S.C. 7333).

Normal mortality means the percentage of dead aquaculture species that would normally occur during the crop year.

Person means person as defined in part 1400 of this title, and all rules with respect to the determination of a person found in that part are applicable to this part. However, the determinations made in this part in accordance with part 1400, subpart B, Person Determinations, of this title will also take into account any affiliation with any entity in which an individual or entity has an interest, regardless of whether or not such entities are considered to be actively engaged in farming.

Planted acreage means land in which seed, plants, or trees have been placed, appropriate for the crop and planting method, at a correct depth, into a seedbed that has been properly prepared for the planting method and production practice normal to the USDA plant hardiness zone as determined by the county committee.

Prevented planting means the inability to plant an eligible crop with proper equipment during the planting period as a result of an eligible cause of loss, as determined by FSA.

Production means quantity of the crop or commodity produced expressed in a specific unit of measure including, but not limited to, bushels or pounds.

Rate means price per unit of the crop or commodity.

Recording county means, for a producer with farming interests in only one county, the FSA county office in which the producer's farm is administratively located or, for a producer with farming interests that are administratively located in more than one county, the FSA county office designated by FSA to control the payments received by the producer.

Related condition means, with respect to a disaster, a condition that causes deterioration of a crop, such as insect infestation, plant disease, or aflatoxin, that is accelerated or exacerbated as a result of damaging weather, as determined in accordance with instructions issued by the Deputy Administrator.

Reliable production records means evidence provided by the participant that is used to substantiate the amount of production reported when verifiable records are not available, including copies of receipts, ledgers of income, income statements of deposit slips, register tapes, invoices for custom harvesting, and records to verify production costs, contemporaneous measurements, truck scale tickets, and contemporaneous diaries that are determined acceptable by the county committee.

Repeat crop means, with respect to production, a commodity that is planted or prevented from being planted in more than one planting period on the same acreage in the same crop year.

RMA means the Risk Management Agency.

Salvage value means the dollar amount or equivalent for the quantity of the commodity that cannot be marketed or sold in any recognized market for the crop.

Secondary use means the harvesting of a crop for a use other than the intended use.

Secondary use value means the value determined by multiplying the quantity of secondary use times the FSA or CCC-established price for that use.

State committee means the FSA State committee.

Surface irrigation water means aqueous supply anticipated for irrigation of agricultural crops absent an eligible disaster condition impacting either the aquifer or watershed. Surface irrigation water may result from feral sources or from irrigation districts.

Tropical crops has the meaning assigned in part 1437 of this title.

Tropical region has the meaning assigned in part 1437 of this title.

Unharvested factor means a percentage established for a crop and applied in a payment formula to reduce the payment for reduced expenses incurred because commercial harvest was not performed. Unharvested factors are generally applicable to all similarly situated participants and are not established in response to individual participants. Accordingly established unharvested factors are not appealable under parts 11 and 780 of this title.

Unit means, unless otherwise determined by the Deputy Administrator, basic unit as defined in part 457 of this title that, for ornamental nursery production, includes all eligible plant species and sizes.

Unit of measure means:

(1) For all insured crops, the FCIC-established unit of measure;

(2) For all NAP covered crops, the established unit of measure, if available, used for the 2005, 2006, or 2007 NAP price and yield;

(3) For aquaculture species, a standard unit of measure such as gallons, pounds, inches, or pieces, established by the State committee for all aquaculture species or varieties;

(4) For turfgrass sod, a square yard;

(5) For maple sap, a gallon;

(6) For honey, pounds; and

(7) For all other crops, the smallest unit of measure that lends itself to the greatest level of accuracy with minimal use of fractions, as determined by the State committee.

United States means all 50 States of the United States, the Commonwealth of Puerto Rico, the Virgin Islands of the United States, and to the extent the Deputy Administrator determines it to be feasible and appropriate, Guam, American Samoa, the Commonwealth of the Northern Mariana Islands, and the former Trust Territory of the Pacific Islands, which include Palau, Federated States of Micronesia, and the Marshall Islands.

USDA means the United States Department of Agriculture.

USDA Plant Hardiness Zone means 11 regions or planting zones as defined by a 10 degree Fahrenheit difference in the average annual minimum temperature.

Value loss crop has the meaning assigned in part 1437 of this title.

Verifiable production record means:

(1) For quantity losses, evidence that is used to substantiate the amount of production reported and that can be verified by FSA through an independent source; or

(2) For quality losses, evidence that is used to substantiate the amount of production reported and that can be verified by FSA through an independent source including determined quality factors and the specific quantity covered by those factors.

Yield means unit of production, measured in bushels, pounds, or other unit of measure, per area of consideration, usually measured in acres.

§760.803 Eligibility.

(a) Participants will be eligible to receive disaster benefits under this part only if they incurred qualifying quantity or quality losses for the 2005, 2006, or 2007 crops, as further specified in this part, as a result of damaging weather or any related condition. Participants may not receive benefits with respect to volunteer stands of crops.

(b) Payments may be made for losses suffered by an eligible participant who, at the time of application, is a deceased individual or is a dissolved entity if a representative, who currently has authority to enter into a contract for the participant, signs the 2005, 2006, or 2007 Crop Disaster Program application. Participants must provide proof of the authority to sign legal documents for the deceased individual or dissolved entity. If a participant is now a dissolved general partnership or joint venture, all members of the general partnership or joint venture at the time of dissolution or their duly authorized representatives must sign the application for payment.

(c) As a condition to receive benefits under this part, the Participant must have been in compliance with the Highly Erodible Land Conservation and Wetland Conservation provisions of part 12 of this title for the 2005, 2006, or 2007 crop year, as applicable, and must not otherwise be precluded from receiving benefits under parts 12 or 1400 of this title or any law.

§760.804 Time and method of application.

(a) The 2005, 2006, 2007 Crop Disaster Program application must be submitted on a completed FSA–840, or such other form designated for such application purpose by FSA, in the FSA county office in the participant's control county office before the close of business on a date that will be announced by the Deputy Administrator.

(b) Once signed by a participant, the application for benefits is considered to contain information and certifications of and pertaining to the participant regardless of who entered the information on the application.

(c) The participant requesting benefits under this program certifies the accuracy and truthfulness of the information provided in the application as well as any documentation filed with or in support of the application. All information is subject to verification by FSA. For example, as specified in §760.818(f), the participant may be required to provide documentation to substantiate and validate quality standards and marketing contract prices. Refusal to allow FSA or any agency of the Department of Agriculture to verify any information provided will result in the participant's forfeiting eligibility under this program. Furnishing required information is voluntary; however without it, FSA is under no obligation to act on the application or approve benefits. Providing a false certification to the government is punishable by imprisonment, fines, and other penalties.

(d) FSA may require the participant to submit any additional information it deems necessary to implement or determine any eligibility provision of this part. For example, as specified in §760.818(f), the participant may be required to provide documentation to substantiate and validate quality standards and marketing contract prices.

(e) The application submitted in accordance with paragraph (a) of this section is not considered valid and complete for issuance of payment under this part unless FSA determines all the applicable eligibility provisions have been satisfied and the participant has submitted all of following completed forms:

(1) If Item 16 on FSA-840 is answered "YES," FSA-840M, Crop Disaster Program for Multiple Crop—Same Acreage Certification;

(2) CCC-502, Farm Operating Plan for Payment Eligibility;

(3) CCC-526, Payment Eligibility Average Adjusted Gross Income Certification;

(4) AD-1026, Highly Erodible Land Conservation (HELC) and Wetland Conservation Certification; and

(5) FSA-578, Report of Acreage.

(f) Application approval and payment by FSA does not relieve a participant from having to submit any form required, but not filed, according to paragraph (e) of this section.

§ 760.805 Limitations on payments and other benefits.

(a) A participant may receive benefits for crop losses for only one of the 2005, 2006, or 2007 crop years as specified under this part.

(b) Payments will not be made under this part for grazing losses.

(c) Payments determined to be issued are considered due and payable not later than 60 days after a participant's application is completed with all information necessary for FSA to determine producer eligibility for benefits.

(d) FSA may divide and classify crops based on loss susceptibility, yield, and other factors.

(e) No person, as defined by part 1400 subpart B of this title, may receive more than a total of $80,000 in disaster benefits under this part. In applying the $80,000 per person payment limitation, regardless of whether 2005, 2006, or 2007 crop year benefits are at issue or sought, the most restrictive "person" determination for the participant in the years 2005, 2006, and 2007, will be used to limit benefits.

(f) No participant may receive disaster benefits under this part in an amount that exceeds 95 percent of the value of the expected production for the relevant period as determined by FSA. Accordingly, the sum of the value of the crop not lost, if any; the disaster payment received under this part; and any crop insurance payment or payments received under the NAP for losses to the same crop, cannot exceed 95 percent of what the crop's value would have been if there had been no loss.

(g) An individual or entity whose adjusted gross income is in excess of $2.5 million, as defined by and determined under part 1400 subpart G of this title, is not eligible to receive disaster benefits under this part.

(h) Any participant in a county eligible for either of the following programs must complete a duplicate benefits certification. If the participant received a payment authorized by either of the following, the amount of that payment will be reduced from the calculated 2005-2007 CDP payment:

(1) The Hurricane Indemnity Program (subpart B of this part);

(2) The Hurricane Disaster Programs (subparts D, E, F, and G of part 1416 of this title);

(3) The 2005 Louisiana Sugarcane Hurricane Disaster Assistance Program; or

(4) The 2005 Crop Florida Sugarcane Disaster Program.

§ 760.806 Crop eligibility requirements.

(a) A participant on a farm is eligible for assistance under this section with respect to losses to an insurable commodity or NAP if the participant:

(1) In the case of an insurable commodity, obtained a policy or plan of insurance under the Federal Crop Insurance Act for the crop incurring the losses; or

(2) In the case of a NAP covered crop, filed the required paperwork and paid the administrative fee by the applicable filing deadline, for the noninsurable commodity under section 196 of the Federal Agriculture Improvement and Reform Act of 1996 for the crop incurring the losses.

(b) The reasons a participant either elected not to have coverage or did not have coverage mentioned in paragraphs (a)(1) or (2) of this section are not relevant to the determination of the participant's ineligibility under this section. In addition, such reasons for not having crop insurance coverage have no bearing for consideration under part 718, subpart D of this chapter.

§ 760.807 Miscellaneous provisions.

(a) A person is not eligible to receive disaster assistance under this part if it is determined by FSA that the person has:

(1) Adopted any scheme or other device that tends to defeat the purpose of this part;

(2) Made any fraudulent representation;

(3) Misrepresented any fact affecting a program determination;

(4) Is ineligible under §1400.5 of this title; or

(5) Does not have entitlement to an ownership share of the crop.

(i) Growers growing eligible crops under contract for crop owners are not eligible unless the grower can be determined to have a share of the crop.

(ii) Any verbal or written contract that precludes the grower from having an ownership share renders the grower ineligible for benefits under this part.

(b) A person ineligible under §1437.15(c) of this title for any year is likewise ineligible for benefits under this part for that year or years.

(c) A person ineligible under §400.458 of this title for any year is likewise ineligible for benefits under this part for that year or years.

(d) All persons with a financial interest in the operation receiving benefits under this part are jointly and severally liable for any refund, including related charges, which is determined to be due FSA for any reason.

(e) In the event that any request for assistance or payment under this part resulted from erroneous information or a miscalculation, the assistance or payment will be recalculated and any excess refunded to FSA with interest to be calculated from the date of the disbursement to the producer.

(f) The liability of anyone for any penalty or sanction under or in connection with this part, or for any refund to FSA or related charge is in addition to any other liability of such person under any civil or criminal fraud statute or any other provision of law including, but not limited to: 18 U.S.C. 286, 287, 371, 641, 651, 1001, and 1014; 15 U.S.C. 714; and 31 U.S.C. 3729.

(g) The regulations in parts 11 and 780 of this title apply to determinations under this part.

(h) Any payment to any person will be made without regard to questions of title under State law and without regard to any claim or lien against the crop, or its proceeds.

(i) For the purposes of the effect of lien on eligibility for Federal programs (28 U.S.C. 3201(e)), FSA waives the restriction on receipt of funds or benefits under this program but only as to beneficiaries who, as a condition of such waiver, agree to apply the benefits received under this part to reduce the amount of the judgment lien.

(j) Under this program, participants are either eligible or ineligible. Participants in general, do not render performance or need to comply. They either suffered eligible losses or they did not. Accordingly, the provisions of §718.304 of this chapter do not apply to this part.

§ 760.808 General provisions.

(a) For calculations of loss, the participant's existing unit structure will be used as the basis for the calculation established in accordance with:

(1) For insured crops, part 457 of this title; or

(2) For NAP covered crops, part 1437 of this title.

(b) County average yield for loss calculations will be the average of the 2001 through 2005 official county yields established by FSA, excluding the years with the highest and lowest yields, respectively.

(c) County committees will assign production or reduce the historic yield when the county committee determines:

(1) An acceptable appraisal or record of harvested production does not exist;

(2) The loss is due to an ineligible cause of loss or practices, soil type, climate, or other environmental factors that cause lower yields than those upon which the historic yield is based;

(3) The participant has a contract providing a guaranteed payment for all or a portion of the crop; or

(4) The crop was planted beyond the normal planting period for the crop.

(d) The county committee will establish a maximum average loss level that reflects the amount of production producers would have produced if not for the eligible damaging weather or related conditions in the area or county for the same crop. The maximum average loss level for the county will be expressed as either a percent of loss or yield per acre. The maximum average loss level will apply when:

(1) Unharvested acreage has not been appraised by FSA, or a company reinsured by FCIC; or

(2) Acceptable production records for harvested acres are not available from any source.

(e) Assignment of production or reduction in yield will apply for practices that result in lower yields than those for which the historic yield is based.

§ 760.809   Eligible damaging conditions.

(a) Except as provided in paragraphs (b) and (c) of this section, to be eligible for benefits under this part the loss of the crop, or reduction in quality, or prevented planting must be due to damaging weather or related conditions as defined in § 760.802.

(b) Benefits are not available under this part for any losses in quantity or quality, or prevented planting due to:

(1) Poor farming practices;

(2) Poor management decisions; or

(3) Drifting herbicides.

(c) With the exception of paragraph (d) of this section, in all cases, the eligible damaging condition must have directly impacted the specific crop or crop acreage during its planting or growing period.

(d) If FSA has determined that there has been an eligible loss of surface irrigation water due to drought and such loss of surface irrigation water impacts eligible crop acreage, FSA may approve assistance to the extent permitted by section 760.814.

§ 760.810   Qualifying 2005, 2006, or 2007 quantity crop losses.

(a) To receive benefits under this part, the county committee must determine that because of eligible damaging weather or related condition specifically impacting the crop or crop acreage, the participant with respect to the 2005, 2006, or 2007 crop:

(1) Was prevented from planting a crop;

(2) Sustained a loss in excess of 35 percent of the expected production of a crop; or

(3) Sustained a loss in excess of 35 percent of the value for value loss crops.

(b) Qualifying losses under this part do not include losses:

(1) For the 2007 crop, those acres planted, or in the case of prevented planting, would have been planted, on or after February 28, 2007;

(2) That are determined by FSA to be the result of poor management decisions, poor farming practices, or drifting herbicides;

(3) That are the result of the failure of the participant to re-seed or replant the same crop in the county where it is customary to re-seed or replant after a loss;

(4) That are not as a result of a damaging weather or a weather related condition specifically impacting the crop or crop acreage;

(5) To crops not intended for harvest in crop year 2005, 2006, or 2007;

(6) Of by-products resulting from processing or harvesting a crop, such as cottonseed, peanut shells, wheat, or oat straw;

(7) To home gardens;

(8) That are a result of water contained or released by any governmental, public, or private dam or reservoir project if an easement exists on the acreage affected for the containment or release of the water; or

(9) If losses could be attributed to conditions occurring outside of the applicable crop year growing season.

(c) Qualifying losses under this part for nursery stock will not include losses:

(1) For the 2007 crop, that nursery inventory acquired on or after February 28, 2007;

(2) Caused by a failure of power supply or brownouts;

(3) Caused by the inability to market nursery stock as a result of lack of

compliance with State and local commercial ordinances and laws, quarantine, boycott, or refusal of a buyer to accept production;

(4) Caused by fire unless directly related to an eligible natural disaster;

(5) Affecting crops where weeds and other forms of undergrowth in the vicinity of the nursery stock have not been controlled; or

(6) Caused by the collapse or failure of buildings or structures.

(d) Qualifying losses under this part for honey, where the honey production by colonies or bees was diminished, will not include losses:

(1) For the 2007 crop, for production from those bees acquired on or after February 28, 2007;

(2) Where the inability to extract was due to the unavailability of equipment, the collapse or failure of equipment, or apparatus used in the honey operation;

(3) Resulting from storage of honey after harvest;

(4) To honey production because of bee feeding;

(5) Caused by the application of chemicals;

(6) Caused by theft, fire, or vandalism;

(7) Caused by the movement of bees by the producer or any other person; or

(8) Due to disease or pest infestation of the colonies.

(e) Qualifying losses for other value loss crops, except nursery, will not include losses for the 2007 crop that were acquired on or after February 28, 2007.

(f) Loss calculations will take into account other conditions and adjustments provided for in this part.

§ 760.811 Rates and yields; calculating payments.

(a)(1) Payments made under this part to a participant for a loss of quantity on a unit with respect to yield-based crops are determined by multiplying the average market price times 42 percent, times the loss of production which exceeds 35 percent of the expected production, as determined by FSA, of the unit.

(2) Payments made under this part to a participant for a quantity loss on a unit with respect to value-based crops are determined by multiplying the payment rate established for the crop by

FSA times the loss of value that exceeds 35 percent of the expected production value, as determined by FSA, of the unit.

(3) As determined by FSA, additional quality loss payments may be made using a 25 percent quality loss threshold. The quality loss threshold is determined according to § 760.817.

(b) Payment rates for the 2005, 2006, or 2007 year crop losses will be 42 percent of the average market price.

(c) Separate payment rates and yields for the same crop may be established by the State committee as authorized by the Deputy Administrator, when there is supporting data from NASS or other sources approved by FSA that show there is a significant difference in yield or value based on a distinct and separate end use of the crop. Despite potential differences in yield or values, separate rates or yields will not be established for crops with different cultural practices, such as those grown organically or hydroponically.

(d) Production from all end uses of a multi-use crop or all secondary uses for multiple market crops will be calculated separately and summarized together.

(e) Each eligible participant's share of a disaster payment will be based on the participant's ownership entitlement share of the crop or crop proceeds, or, if no crop was produced, the share of the crop the participant would have received if the crop had been produced. If the participant has no ownership share of the crop, the participant is ineligible for assistance under this part.

(f) When calculating a payment for a unit loss:

(1) An unharvested payment factor will be applied to crop acreage planted but not harvested;

(2) A prevented planting factor will be applied to any prevented planted acreage eligible for payment; and

(3) Unharvested payment factors may be adjusted if costs normally associated with growing the crop are not incurred.

§ 760.812 Production losses; participant responsibility.

(a) Where available and determined accurate by FSA, RMA loss records will be used for insured crops.

(b) If RMA loss records are not available, or if the FSA county committee determines the RMA loss records are inaccurate or incomplete, or if the FSA county committee makes inquiry, participants are responsible for:

(1) Retaining or providing, when required, the best verifiable or reliable production records available for the crop;

(2) Summarizing all the production evidence;

(3) Accounting for the total amount of unit production for the crop, whether or not records reflect this production;

(4) Providing the information in a manner that can be easily understood by the county committee; and

(5) Providing supporting documentation if the county committee has reason to question the damaging weather event or question whether all production has been accounted for.

(c) In determining production under this section, the participant must supply verifiable or reliable production records to substantiate production to the county committee. If the eligible crop was sold or otherwise disposed of through commercial channels, production records include: commercial receipts; settlement sheets; warehouse ledger sheets; load summaries; or appraisal information from a loss adjuster acceptable to FSA. If the eligible crop was farm-stored, sold, fed to livestock, or disposed of in means other than commercial channels, production records for these purposes include: truck scale tickets; appraisal information from a loss adjuster acceptable to FSA; contemporaneous diaries; or other documentary evidence, such as contemporaneous measurements.

(d) Participants must provide all records for any production of a crop that is grown with an arrangement, agreement, or contract for guaranteed payment.

§ 760.813 Determination of production.

(a) Production under this part includes all harvested production, unharvested appraised production, and assigned production for the total planted acreage of the crop on the unit.

(b) The harvested production of eligible crop acreage harvested more than once in a crop year includes the total harvested production from all these harvests.

(c) If a crop is appraised and subsequently harvested as the intended use, the actual harvested production must be taken into account to determine benefits. FSA will analyze and determine whether a participant's evidence of actual production represents all that could or would have been harvested.

(d) For all crops eligible for loan deficiency payments or marketing assistance loans with an intended use of grain but harvested as silage, ensilage, cobbage, hay, cracked, rolled, or crimped, production will be adjusted based on a whole grain equivalent as established by FSA.

(e) For crops with an established yield and market price for multiple intended uses, a value will be calculated by FSA with respect to the intended use or uses for disaster purposes based on historical production and acreage evidence provided by the participant and FSA will determine the eligible acres for each use.

(f) For crops sold in a market that is not a recognized market for the crop with no established county average yield and average market price, 42 percent of the salvage value received will be deducted from the disaster payment.

(g) If a participant does not receive compensation based upon the quantity of the commodity delivered to a purchaser, but has an agreement or contract for guaranteed payment for production, the determination of the production will be the greater of the actual production or the guaranteed payment converted to production as determined by FSA.

(h) Production that is commingled between units before it was a matter of record or combination of record and cannot be separated by using records or other means acceptable to FSA will be prorated to each respective unit by FSA. Commingled production may be attributed to the applicable unit, if the participant made the unit production of a commodity a matter of record before commingling and does any of the following, as applicable:

(1) Provides copies of verifiable documents showing that production of the commodity was purchased, acquired, or otherwise obtained from beyond the unit;

(2) Had the production measured in a manner acceptable to the county committee; or

(3) Had the current year's production appraised in a manner acceptable to the county committee.

(i) The county committee will assign production for the unit when the county committee determines that:

(1) The participant has failed to provide adequate and acceptable production records;

(2) The loss to the crop is because of a disaster condition not covered by this part, or circumstances other than natural disaster, and there has not otherwise been an accounting of this ineligible cause of loss;

(3) The participant carries out a practice, such as multiple cropping, that generally results in lower yields than the established historic yields;

(4) The participant has a contract to receive a guaranteed payment for all or a portion of the crop;

(5) A crop was late-planted;

(6) Unharvested acreage was not timely appraised; or

(7) Other appropriate causes exist for such assignment as determined by the Deputy Administrator.

(j) For peanuts, the actual production is all peanuts harvested for nuts, regardless of their disposition or use, as adjusted for low quality.

(k) For tobacco, the actual production is the sum of the tobacco: marketed or available to be marketed; destroyed after harvest; and produced but unharvested, as determined by an appraisal.

§ 760.814 Calculation of acreage for crop losses other than prevented planted.

(a) Payment acreage of a crop is limited to the lesser of insured acreage or NAP covered acreage of the crop, as applicable, or actual acreage of the crop planted for harvest.

(b) In cases where there is a repeat crop or a multiple planted crop in more than one planting period, or if there is multiple cropped acreage meeting criteria established in paragraph (c) or (d) of this section, each of these crops may be considered separate crops if the county committee determines that all of the following conditions are met:

(1) Were planted with the intent to harvest;

(2) Were planted within the normal planting period for that crop;

(3) Meet all other eligibility provisions of this part including good farming practices; and

(4) Could reach maturity if each planting was harvested or would have been harvested.

(c) In cases where there is multiple-cropped acreage, each crop may be eligible for disaster assistance separately if both of the following conditions are met:

(1) The specific crops are approved by the State committee as eligible multiple-cropping practices in accordance with procedures approved by the Deputy Administrator and separately meet all requirements, including insurance or NAP requirements ; and

(2) The farm containing the multiple-cropped acreage has a history of successful multiple cropping more than one crop on the same acreage in the same crop year, in the year previous to the disaster, or at least 2 of the 4 crop years immediately preceding the disaster crop year based on timely filed crop acreage reports.

(d) A participant with multiple-cropped acreage not meeting the criteria in paragraph (c) of this section may be eligible for disaster assistance on more than one crop if the participant has verifiable records establishing a history of carrying out a successful multiple-cropping practice on the specific crops for which assistance is requested. All required records acceptable to FSA as determined by the Deputy Administrator must be provided before payments are issued.

(e) A participant with multiple-cropped acreage not meeting the criteria in paragraphs (c) or (d) of this section must select the crop for which assistance will be requested. If more than one participant has an interest in the multiple cropped acreage, all participants must agree to the crop designated for payment by the end of the application period or no payment will be approved for any crop on the multiple-cropped acreage.

(f) Benefits under this part apply to irrigated crops where, in cases determined by the Deputy Administrator, acreage was affected by a lack of surface irrigation water due to drought or contamination of ground water or surface irrigation water due to saltwater intrusion. In no case is a loss of ground water, for any reason, an eligible cause of loss.

§760.815 Calculation of prevented planted acreage.

(a) When determining losses under this part, prevented planted acreage will be considered separately from planted acreage of the same crop.

(b) For insured crops, or NAP covered crops, as applicable, disaster payments under this part for prevented planted acreage will not be made unless RMA or FSA, as applicable, documentation indicates that the eligible participant received a prevented planting payment under either NAP or the RMA-administered program.

(c) The participant must prove, to the satisfaction of the county committee, an intent to plant the crop and that such crop could not be planted because of an eligible disaster. The county committee must be able to determine the participant was prevented from planting the crop by an eligible disaster that:

(1) Prevented other producers from planting on acreage with similar characteristics in the surrounding area;

(2) Occurred after the previous planting period for the crop; and

(3) Unless otherwise approved by the Deputy Administrator, began no earlier than the planting season for that crop.

(d) Prevented planted disaster benefits under this part do not apply to:

(1) Acreage not insured or NAP covered;

(2) Any acreage on which a crop other than a cover crop was harvested, hayed, or grazed during the crop year;

(3) Any acreage for which a cash lease payment is received for the use of

the acreage the same crop year, unless the county committee determines the lease was for haying and grazing rights only and was not a lease for use of the land;

(4) Acreage for which the participant or any other person received a prevented planted payment for any crop for the same acreage, excluding share arrangements;

(5) Acreage for which the participant cannot provide verifiable proof to the county committee that inputs such as seed, chemicals, and fertilizer were available to plant and produce a crop with the expectation of producing at least a normal yield; and

(6) Any other acreage for which, for whatever reason, there is cause to question whether the crop could have been planted for a successful and timely harvest, or for which prevented planting credit is not allowed under the provisions of this part.

(e) Prevented planting payments are not provided on acreage that had either a previous or subsequent crop planted in the same crop year on the acreage, unless the county committee determines that all of the following conditions are met:

(1) There is an established practice of planting two or more crops for harvest on the same acreage in the same crop year;

(2) Both crops could have reached maturity if each planting was harvested or would have been harvested;

(3) Both the initial and subsequent planted crops were planted or prevented planting within the normal planting period for that crop;

(4) Both the initial and subsequent planted crops meet all other eligibility provisions of this part including good farming practices; and

(5) The specific crops meet the eligibility criteria for a separate crop designation as a repeat or approved multiple cropping practice set out in §760.814.

(f)(1) Disaster benefits under this part do not apply to crops where the prevented planted acreage was affected by a disaster that was caused by drought unless on the final planting date or the late planting period for non-irrigated acreage, the area that was prevented from being planted had insufficient soil moisture for germination of seed and progress toward crop maturity because of a prolonged period of dry weather;

(2) Verifiable information collected by sources whose business or purpose is to record weather conditions, including, but not limited to, local weather reporting stations of the U.S. National Weather Service.

(g) Prevented planting benefits under this part apply to irrigated crops where adequate irrigation facilities were in place before the eligible disaster and the acreage was prevented from being planted due to a lack of water resulting from drought conditions or contamination by saltwater intrusion of an irrigation supply resulting from drought conditions.

(h) For NAP covered crops, prevented planting provisions apply according to part 718 of this chapter.

(i) Late-filed crop acreage reports for prevented planted acreage in previous years are not acceptable for CDP purposes.

§ 760.816 Value loss crops.

(a) Notwithstanding any other provisions of this part, this section applies to value loss crops and tropical crops. Unless otherwise specified, all the eligibility provisions of part 1437 of this title apply to value loss crops and tropical crops under this part.

(b) For value loss crops, benefits under this part are calculated based on the loss of value at the time of the damaging weather or related condition, as determined by FSA.

(c) For tropical crops:

(1) CDP benefits for 2005 are calculated according to general provisions of part 1437, but not subpart F, of this title.

(2) CDP benefits for 2006 and 2007 are calculated according to part 1437, subpart F of this title.

§ 760.817 Quality losses for 2005, 2006, and 2007 crops.

(a) Subject to other provisions of this part, assistance will be made available to participants determined eligible under this section for crop quality losses of 25 percent or greater of the value that all affected production of the crop would have had if the crop had not suffered a quality loss.

(b) The amount of payment for a quality loss will be equal to 65 percent of the quantity of the crop affected by the quality loss, not to exceed expected production based on harvested acres, multiplied by 42 percent of the per unit average market value based on percentage of quality loss for the crop as determined by the Deputy Administrator.

(c) This section applies to all crops eligible for 2005, 2006, and 2007 crop disaster assistance under this part, with the exceptions of value loss crops, honey, and maple sap, and applies to crop production that has a reduced economic value due to the reduction in quality.

(d) Participants may not be compensated under this section to the extent that such participants have received assistance under other provisions of this part, attributable in whole or in part to diminished quality.

§ 760.818 Marketing contracts.

(a) A marketing contract must meet all of the conditions outlined in paragraphs (b), (c), and (d) of this section.

(b) A marketing contract, at a minimum, must meet all of the following conditions:

(1) Be a legal contract in the State where executed;

(2) Specify the commodity under contract;

(3) Specify crop year;

(4) Be signed by both the participant, or legal representative, and the purchaser of the specified commodity;

(5) Include a commitment to deliver the contracted quantity;

(6) Include a commitment to purchase the contracted quantity that meets specified minimum quality standards and other criteria as specified;

(7) Define a determinable quantity by containing either a:

(i) Specified production quantity or

(ii) A specified acreage for which production quantity can be calculated;

(8) Define a determinable price by containing either a:

(i) Specified price or

(ii) Method to determine such a price;

(9) Contain a relationship between the price and the quality using either:

(i) Specified quality standards or

(ii) A method to determine such quality standards from published third party data; and

(10) Have been executed within 10 days after:

(i) End of insurance period for insured crops or

(ii) Normal harvest date for NAP covered crops as determined by FSA.

(c) The purchaser of the commodity specified in the marketing contract must meet at least one of the following:

(1) Be a licensed commodity warehouseman;

(2) Be a business enterprise regularly engaged in the processing of a commodity, that possesses all licenses and permits for marketing the commodity required by the State in which it operates, and that possesses or has contracted for facilities with enough equipment to accept and process the commodity within a reasonable amount of time after harvest; or

(3) Is able to physically receive the harvested production.

(d) In order for the commodity specified in the marketing contract to be considered sold pursuant to the marketing contract, the commodity must have been produced by the participant in the crop year specified in the contract, and at least one of the following conditions must be met:

(1) Commodity was sold under the terms of the contract or

(2) Participant attempted to deliver the commodity to the purchaser, but the commodity was rejected due to quality factors as specified in the contract.

(e) The amount of payment for affected production, as determined in § 760.817(b), sold pursuant to one or more marketing contracts will take into consideration the marketing contract price as determined by FSA.

(f) County committees have the authority to require a participant to provide necessary documentation, which may include, but is not limited to, previous marketing contracts fulfilled, to substantiate and validate quality standards in paragraph (b)(9) of this section and marketing contract price

received for the commodity for which crop quality loss assistance is requested. In cases where the county committee has reason to believe the participant lacks the capacity or history to fulfill the quality provisions of the marketing contract the county committee will require such documentation.

§ 760.819 Misrepresentation, scheme, or device.

(a) A person is ineligible to receive assistance under this part if it is determined that such person has:

(1) Adopted any scheme or device that tends to defeat the purpose of this program;

(2) Made any fraudulent representation under this program;

(3) Misrepresented any fact affecting a program or person determination; or

(4) Has violated or been determined ineligible under § 1400.5 of this title.

§ 760.820 Offsets, assignments, and debt settlement.

(a) Except as provided in paragraph (b) of this section, any payment to any person will be made without regard to questions of title under State law and without regard to any claim or lien against the crop, or proceeds, in favor of the owner or any other creditor except agencies of the U.S. Government. The regulations governing offsets and withholdings found at part 1403 of this title apply to any payments made under this part.

(b) Any participant entitled to any payment may assign any payments in accordance with regulations governing the assignment of payments found at part 1404 of this title.

(c) A debt or claim may be settled according to part 792 of this chapter.

§ 760.821 Compliance with highly erodible land and wetland conservation.

(a) The highly erodible land and wetland conservation provisions of part 12 of this title apply to the receipt of disaster assistance for 2005, 2006, and 2007 crop losses made available under this authority.

(b) Eligible participants must be in compliance with the highly erodible land and wetland conservation compliance provisions for the year for which financial assistance is requested.

Subpart J—2005–2007 Livestock Indemnity Program

SOURCE: 72 FR 72867, Dec. 21, 2007, unless otherwise noted.

§760.900 Administration.

(a) The regulations in this subpart specify the terms and conditions applicable to the 2005–2007 Livestock Indemnity Program (2005–2007 LIP), which will be administered under the general supervision and direction of the Administrator, FSA.

(b) FSA representatives do not have authority to modify or waive any of the provisions of the regulations of this subpart.

(c) The State FSA committee will take any action required by the regulations of this subpart that the county FSA committee has not taken. The State FSA committee will also:

(1) Correct, or require a county committee to correct, any action taken by such county committee that is not in accordance with the regulations of this subpart; or

(2) Require a county committee to withhold taking any action that is not in accordance with this subpart.

(d) No delegation to a State or county FSA committee will preclude the Deputy Administrator for Farm Programs from determining any question arising under the program or from reversing or modifying any determination made by a State or county FSA committee.

§760.901 Applicability.

(a) This subpart establishes the terms and conditions under which the 2005–2007 LIP will be administered under Title IX of the U.S. Troop Readiness, Veterans' Care, Katrina Recovery, and Iraq Accountability Appropriations Act, 2007 (Pub. L. 110–28) for eligible counties as specified in §760.902(a).

(b) Eligible livestock owners and contract growers will be compensated in accordance with §760.909 for eligible livestock deaths that occurred in eligible counties as a direct result of an eligible disaster event. Drought is not an eligible disaster event except when anthrax, as a related condition that occurs as a result of drought, results in the death of eligible livestock.

§760.902 Eligible counties and disaster periods.

Counties are eligible for agricultural assistance under the 2005–2007 LIP if they received a timely Presidential designation, a timely Secretarial declaration, or a qualifying Administrator's Physical Loss Notice (APLN) determination in a county otherwise the subject of a timely Presidential declaration, or are counties contiguous to such counties. Presidential designations and Secretarial declarations will be considered timely only if made after January 1, 2005, and before February 28, 2007. Eligible counties, disaster events, and disaster periods are listed at *http:// disaster.fsa.usda.gov*.

§760.903 Definitions.

The following definitions apply to this subpart. The definitions in parts 718 and 1400 of this title also apply, except where they conflict with the definitions in this section.

Adult beef bull means a male beef bovine animal that was at least 2 years old and used for breeding purposes before it died.

Adult beef cow means a female beef bovine animal that had delivered one or more offspring before dying. A first-time bred beef heifer is also considered an adult beef cow if it was pregnant at the time it died.

Adult buffalo and beefalo bull means a male animal of those breeds that was at least 2 years old and used for breeding purposes before it died.

Adult buffalo and beefalo cow means a female animal of those breeds that had delivered one or more offspring before dying. A first-time bred buffalo or beefalo heifer is also considered an adult buffalo or beefalo cow if it was pregnant at the time it died.

Adult dairy bull means a male dairy breed bovine animal at least 2 years old used primarily for breeding dairy cows before it died.

Adult dairy cow means a female bovine animal used for the purpose of providing milk for human consumption that had delivered one or more off-spring before dying. A first-time bred dairy heifer is also considered an adult dairy cow if it was pregnant at the time it died.

Agricultural operation means a farming operation.

Application means the "2005–2007 Livestock Indemnity Program" form.

Application period means the date established by the Deputy Administrator for Farm Programs for participants to apply for program benefits.

Buck means a male goat.

Catfish means catfish grown as food for human consumption by a commercial operator on private property in water in a controlled environment.

Commercial use means used in the operation of a business activity engaged in as a means of livelihood for profit by the eligible producer to apply for program benefits.

Contract means, with respect to contracts for the handling of livestock, a written agreement between a livestock owner and another individual or entity setting the specific terms, conditions, and obligations of the parties involved regarding the production of livestock or livestock products.

Controlled environment means an environment in which everything that can practicably be controlled by the participant with structures, facilities, and growing media (including, but not limited to, water and nutrients) and was in fact controlled by the participant at the time of the disaster.

Crawfish means crawfish grown as food for human consumption by a commercial operator on private property in water in a controlled environment.

. *Deputy Administrator* means the Deputy Administrator for Farm Programs, Farm Service Agency, U.S. Department of Agriculture or the designee.

Doe means a female goat.

Equine animal means a domesticated horse, mule, or donkey.

Ewe means a female sheep.

Farming operation means a business enterprise engaged in producing agricultural products.

Goat means a domesticated, ruminant mammal of the genus *Capra*, including Angora goats. Goats are further defined by sex (bucks and does) and age (kids).

Kid means a goat less than 1 year old.

Lamb means a sheep less than 1 year old.

Livestock owner means one having legal ownership of the livestock for which benefits are being requested on the day such livestock died due to an eligible disaster.

Non-adult beef cattle means a bovine that does not meet the definition of adult beef cow or bull. Non-adult beef cattle are further delineated by weight categories of less than 400 pounds, and 400 pounds or more at the time they died.

Non-adult buffalo or beefalo means an animal of those breeds that does not meet the definition of adult buffalo/beefalo cow or bull. Non-adult buffalo or beefalo are further delineated by weight categories of less than 400 pounds, and 400 pounds or more at the time of death.

Non-adult dairy cattle means a bovine livestock, of a breed used for the purpose of providing milk for human consumption, that do not meet the definition of adult dairy cow or bull. Non-adult dairy cattle are further delineated by weight categories of less than 400 pounds, and 400 pounds or more at the time they died.

Poultry means domesticated chickens, turkeys, ducks, and geese. Poultry are further delineated by sex, age, and purpose of production as determined by FSA.

Ram means a male sheep.

Sheep means a domesticated, ruminant mammal of the genus *Ovis*. Sheep are further defined by sex (rams and ewes) and age (lambs).

Swine means a domesticated omnivorous pig, hog, and boar. Swine are further delineated by sex and weight as determined by FSA.

§ 760.904   Limitations on payments and other benefits.

(a) A participant may receive benefits for livestock losses for only one of the 2005, 2006, or 2007 calendar years as specified under this part.

(b) A "person" as determined under part 1400 of this title may receive no more than $80,000 under this subpart. In applying the $80,000 per person payment limitation, regardless of whether 2005, 2006, or 2007 calendar year benefits are at issue or sought, the most restrictive "person" determination for the participant in the years 2005, 2006, and 2007, will be used to limit benefits.

(c) The provisions of part 1400, subpart G, of this title relating to limits to payments for individuals or entities with certain levels of adjusted gross income apply to this program.

(d) As a condition to receive benefits under this subpart, a participant must have been in compliance with the provisions of parts 12 and 718 of this title and must not otherwise be precluded from receiving benefits under any law.

(e) An individual or entity determined to be a foreign person under part 1400 of this title is not eligible to receive benefits under this subpart.

§760.905 Eligible owners and contract growers.

(a) To be considered eligible, a livestock owner must have had legal ownership of the eligible livestock, as provided in §760.906(a), on the day the livestock died.

(b) To be considered eligible, a contract grower on the day the livestock died must have had:

(1) A written agreement with the owner of eligible livestock setting the specific terms, conditions, and obligations of the parties involved regarding the production of livestock; and

(2) Control of the eligible livestock, as provided in §760.906(b), on the day the livestock died.

§760.906 Eligible livestock.

(a) To be considered eligible livestock for livestock owners, livestock must be adult or non-adult dairy cattle, beef cattle, buffalo, beefalo, catfish, crawfish, equine, sheep, goats, swine, poultry, deer, or reindeer and meet all the conditions in paragraph (c) of this section.

(b) To be considered eligible livestock for contract growers, livestock must be poultry or swine as defined in §760.903 and meet all the conditions in paragraph (c) of this section.

(c) To be considered eligible, livestock must meet all of the following conditions:

(1) Died in an eligible county as a direct result of an eligible disaster event;

(i) After January 1, 2005, but before February 28, 2007;

(ii) No later than 60 calendar days from the ending date of the applicable disaster period, but before February 28, 2007; and

(iii) In the calendar year for which benefits are being requested.

(2) The disaster event that caused the loss must be the same event for which a natural disaster was declared or designated.

(3) Been maintained for commercial use as part of a farming operation on the day they died; and

(4) Before dying, not have been produced or maintained for reasons other than commercial use as part of a farming operation, including, but not limited to, wild free roaming animals or animals used for recreational purposes, such as pleasure, hunting, roping, pets, or for show.

(d) In those counties in §760.902, the following types of animals owned by a livestock owner are eligible livestock:

(1) Adult beef bulls;
(2) Adult beef cows;
(3) Adult buffalo or beefalo bulls;
(4) Adult buffalo or beefalo cows;
(5) Adult dairy bulls;
(6) Adult dairy cows;
(7) Catfish;
(8) Chickens, broilers, pullets;
(9) Chickens, chicks;
(10) Chickens, layers, roasters;
(11) Crawfish;
(12) Deer;
(13) Ducks;
(14) Ducks, ducklings;
(15) Equine;
(16) Geese, goose;
(17) Geese, gosling;
(18) Goats, bucks;
(19) Goats, does;
(20) Goats, kids;
(21) Non-adult beef cattle;
(22) Non-adult buffalo/beefalo;
(23) Non-adult dairy cattle;
(24) Reindeer
(25) Sheep, ewes;
(26) Sheep, lambs;
(27) Sheep, rams;
(28) Swine, feeder pigs under 50 pounds;
(29) Swine, sows, boars, barrows, gilts 50 to 150 pounds;
(30) Swine, sows, boars, barrows, gilts over 150 pounds;

(31) Turkeys, poults; and

(32) Turkeys, toms, fryers, and roasters.

(e) In those counties in § 760.902, the following types of animals are eligible livestock for contract growers:

(1) Chickens, broilers, pullets;

(2) Chickens, layers, roasters;

(3) Geese, goose;

(4) Swine, boars, sows;

(5) Swine, feeder pigs;

(6) Swine, lightweight barrows, gilts;

(7) Swine, sows, boars, barrows, gilts; and

(8) Turkeys, toms, fryers, and roasters.

§ 760.907   Application process.

(a) To apply for 2005–2007 LIP, submit a completed application to the administrative county FSA office that maintains the farm records for your agricultural operation, a copy of your grower contract, if you are a contract grower, and other supporting documents required for determining your eligibility as an applicant. Supporting documents must show:

(1) Evidence of loss,

(2) Current physical location of livestock in inventory, and

(3) Physical location of claimed livestock at the time of death.

(b) The application must be filed during the application period announced by the Deputy Administrator.

(c) A minor child is eligible to apply for program benefits if all eligibility requirements are met and one of the following conditions exists:

(1) The right of majority has been conferred upon the minor by court proceedings or statute;

(2) A guardian has been appointed to manage the minor's property, and the applicable program documents are executed by the guardian; or

(3) A bond is furnished under which a surety guarantees any loss incurred for which the minor would be liable had the minor been an adult.

(d) The participant must provide adequate proof that the death of the eligible livestock occurred in an eligible county as a direct result of an eligible disaster event during the applicable disaster period. The quantity and kind of livestock that died as a direct result of the eligible disaster event may be documented by: purchase records; veterinarian records; bank or other loan papers; rendering truck receipts; Federal Emergency Management Agency records; National Guard records; written contracts; production records; Internal Revenue Service records; property tax records; private insurance documents; and other similar verifiable documents as determined by FSA.

(e) Certification of livestock deaths by third parties may be accepted only if both the following conditions are met:

(1) The livestock owner or livestock contract grower, as applicable, certifies in writing:

(i) That there is no other documentation of death available;

(ii) The number of livestock, by category determined by FSA, were in inventory at the time the applicable disaster event occurred; and

(iii) Other details required for FSA to determine the certification acceptable; and

(2) The third party provides their telephone number, address, and a written statement containing:

(i) Specific details about their knowledge of the livestock deaths;

(ii) Their affiliation with the livestock owner;

(iii) The accuracy of the deaths claimed by the livestock owner; and

(iv) Other details required by FSA to determine the certification acceptable.

(f) Data furnished by the participant will be used to determine eligibility for program benefits. Furnishing the data is voluntary; however, without all required data program benefits will not be approved or provided.

§ 760.908   Deceased individuals or dissolved entities.

(a) Payments may be made for eligible losses suffered by an eligible participant who is now a deceased individual or is a dissolved entity if a representative, who currently has authority to enter into a contract, on behalf of the participant, signs the application for payment.

(b) Legal documents showing proof of authority to sign for the deceased individual or dissolved entity must be provided.

(c) If a participant is now a dissolved general partnership or joint venture, all members of the general partnership or joint venture at the time of dissolution or their duly authorized representatives must sign the application for payment.

§760.909 Payment calculation.

(a) Under this subpart separate payment rates are established for eligible livestock owners and eligible livestock contract growers in accordance with paragraphs (b) and (c) of this section. Payments for the 2005–2007 LIP are calculated by multiplying the national payment rate for each livestock category, as determined in paragraphs (b) and (c) of this section, by the number of eligible livestock in each category, as provided in §760.906. Adjustments will be applied in accordance with paragraphs (d) and (e) of this section.

(b) The 2005–2007 LIP national payment rate for eligible livestock owners is based on 26 percent of the average fair market value of the livestock.

(c) The 2005–2007 LIP national payment rate for eligible livestock contract growers is based on 26 percent of the average income loss sustained by the contract grower with respect to the dead livestock.

(d) The 2005 payment calculated under 2005–2007 LIP for eligible livestock owners will be reduced by the amount the participant received under:

(1) The Livestock Indemnity Program (subpart E of this part);

(2) The Aquaculture Grant Program (subpart G of this part); and

(3) The Livestock Indemnity Program II (part 1416, subpart C of this title).

(e) The 2005 payment calculated under 2005–2007 LIP for eligible livestock contract growers will be reduced by the amount the participant received:

(1) Under the Livestock Indemnity Program (subpart E of this part);

(2) For the loss of income from the dead livestock from the party who contracted with the producer to grow the livestock; and

(3) Under the Livestock Indemnity Program II (part 1416, subpart C of this title).

§760.910 Appeals.

The appeal regulations set forth at parts 11 and 780 of this title apply to determinations made pursuant to this subpart.

§760.911 Offsets, assignments, and debt settlement.

(a) Any payment to any participant will be made without regard to questions of title under State law and without regard to any claim or lien against the commodity, or proceeds, in favor of the owner or any other creditor except agencies of the U.S. Government. The regulations governing offsets and withholdings found at part 792 of this chapter apply to payments made under this subpart.

(b) Any participant entitled to any payment may assign any payment in accordance with regulations governing the assignment of payments found at part 1404 of this title.

§760.912 Records and inspections.

Participants receiving payments under this subpart or any other person who furnishes information for the purposes of enabling such participant to receive a payment under this subpart must maintain any books, records, and accounts supporting any information so furnished for 3 years following the end of the year during which the application for payment was filed. Participants receiving payments or any other person who furnishes such information to FSA must allow authorized representatives of USDA and the General Accountability Office, during regular business hours, to inspect, examine, and make copies of such books or records, and to enter upon, inspect and verify all applicable livestock and acreage in which the participant has an interest for the purpose of confirming the accuracy of information provided by or for the participant.

§760.913 Refunds; joint and several liability.

In the event there is a failure to comply with any term, requirement, or condition for payment or assistance arising under this subpart, and if any refund of a payment to FSA will otherwise become due in connection with

149

this subpart, all payments made in regard to such matter must be refunded to FSA together with interest and late-payment charges as provided for in part 792 of this chapter.

Subpart K—General Provisions for 2005–2007 Livestock Compensation and Catfish Grant Programs

SOURCE: 72 FR 72881, Dec. 21, 2007, unless otherwise noted.

§ 760.1000 Applicability.

(a) This subpart establishes the terms and conditions under which the following programs will be administered under Title IX of the U.S. Troop Readiness, Veterans' Care, Katrina Recovery, and Iraq Accountability Appropriations Act, 2007 for participants affected by eligible disaster events and located in counties that are eligible as specified in § 760.1001:

(1) The 2005–2007 Livestock Compensation Program (2005–2007 LCP); and

(2) The 2005–2007 Catfish Grant Program (2005–2007 CGP).

(b) Farm Service Agency (FSA) funds as are necessary for the programs in subparts L and M of this part are available under Title IX of the U.S. Troop Readiness, Veterans' Care, Katrina Recovery, and Iraq Accountability Appropriations Act, 2007.

§ 760.1001 Eligible counties, disaster events, and disaster periods.

(a) Except as provided in this subpart, FSA will provide assistance under the programs listed in § 760.1000 to eligible participants who have suffered certain losses due to eligible disaster events in eligible disaster counties provided in paragraph (c) of this section.

(b) The "Disaster Period" is the time period in which losses occurred for the particular disaster that may be considered eligible for the programs under subparts L and M of this part. The start and end dates for each eligible disaster period are specified at *http://disaster.fsa.usda.gov.*

(c) Eligible counties are those primary counties declared by the Secretary or designated for the applicable loss by the President, including counties contiguous to those counties, between January 1, 2005, and February 28, 2007 (that is after January 1, 2005 and before February 28, 2007). The listing is provided at *http://disaster.fsa.usda.gov.* For counties where there was an otherwise timely Presidential declaration, but the declarations do not cover agricultural physical loss, the subject counties may still be eligible if the counties were the subject of an approved Administrator's Physical Loss Notice (APLN) when the APLN applies to a natural disaster timely designated by the President.

§ 760.1002 Definitions.

The following definitions apply to the programs in subpart L and M of this part. The definitions in parts 718 and 1400 of this title also apply, except where they conflict with the definitions in this section.

Commercial use means a use performed as part of the operation of a business activity engaged in as a means of livelihood for profit by the eligible producer.

Farming operation means a business enterprise engaged in producing agricultural products.

§ 760.1003 Limitations on payments and other benefits.

(a) A participant may receive benefits for eligible livestock feed losses, including additional feed costs, for only one of the 2005, 2006, or 2007 calendar years under 2005–2007 LCP, subpart L of this part, or under the CGP of subpart M of this part.

(b) As specified in § 760.1106(c), the payment under the 2005–2007 LCP may not exceed the smaller of the calculated payment in § 760.1106(a) or the value of the producer's eligible feed loss, increased feed costs, or forage or grazing loss.

(c) A person may receive no more than $80,000 under 2005–2007 LCP, subpart L of this part. In applying the $80,000 per person payment limitation, regardless of whether the 2005, 2006, or 2007 calendar year benefits are at issue or sought, the most restrictive "person" determination for the participant in the years 2005, 2006, and 2007, will be used to limit benefits. The rules and definitions of part 1400 of this title apply in construing who is a qualified separate "person" for purposes of this limit. All payment eligibility requirements of part 1400 as they apply to any other payments, also apply to payments under subpart L of this part.

(d) For payments under 2005–2007 CGP, a farming operation may receive no more than $80,000, except for general partnerships and joint ventures, in which case assistance will not exceed $80,000 times the number of eligible members of the general partnership or joint venture. This limit must be enforced by the state government administering the grant program.

(e) The provisions of part 1400, subpart G, of this title apply to these programs. That is the rules that limit the eligibility for benefits of those individuals or entities with an adjusted gross income greater than a certain limit will be applied in the same manner to payments under subparts L and M of this part.

(f) As a condition to receive benefits under subparts L and M of this part, a participant must have been in compliance with the provisions of parts 12 and 718 of this title for the calendar year for which benefits are being requested and must not otherwise be precluded from receiving benefits under any law.

(g) An individual or entity determined to be a foreign person under part 1400 of this title is not eligible to receive benefits under subparts L and M of this part.

(h) In addition to limitations provided in subparts L and M of this part, participants cannot receive duplicate benefits under subparts L and M of this part for the same loss or any similar loss under:

(1) An agricultural disaster assistance provision contained in the announcement of the Secretary on January 26, 2006, or August 29, 2006;

(2) The Emergency Supplemental Appropriations Act for Defense, the Global War on Terror, and Hurricane Recovery, 2006 (Pub. L. 109–234; 120 Stat. 418); or

(3) Any other disaster assistance program.

Subpart L—2005–2007 Livestock Compensation Program

SOURCE: 72 FR 72881, Dec. 21, 2007, unless otherwise noted.

§ 760.1100 Applicability.

This subpart sets forth the terms and conditions applicable to the 2005–2007 Livestock Compensation Program (LCP).

§ 760.1101 Administration.

(a) This program is administered under the general supervision of the Administrator, Farm Service Agency (FSA).

(b) FSA representatives do not have authority to modify or waive any of the provisions of the regulations of this subpart.

(c) The State FSA committee must take any action required by the regulations of this subpart that the county FSA committee has not taken. The State committee must also:

(1) Correct, or require a county committee to correct, any action taken by such county committee that is not in accordance with the regulations of this subpart; or

(2) Require a county committee to withhold taking any action that is not in accordance with this subpart.

(d) No provision or delegation to a State or county FSA committee will preclude the FSA Deputy Administrator for Farm Programs (Deputy Administrator), or a designee of such, from determining any question arising under the program or from reversing or modifying any determination made by a State or county FSA committee.

(e) The Deputy Administrator for Farm Programs may authorize state and county committees to waive or modify nonstatutory deadlines or other program requirements in cases where lateness or failure to meet such does not adversely affect the operation of the program.

§ 760.1102 Definitions.

The following definitions apply to this subpart.

Adult beef bull means a male beef bovine animal that was at least 2 years old and used for breeding purposes on the beginning date of the disaster period.

Adult beef cow means a female beef bovine animal that had delivered one or more offspring before the disaster period. A first-time bred beef heifer is also considered an adult beef cow if it was pregnant on the beginning date of the disaster period.

Adult buffalo and beefalo bull means a male animal of those breeds that was at least 2 years old and used for breeding purposes on the beginning date of the disaster period.

Adult buffalo and beefalo cow means a female animal of those breeds that had delivered one or more offspring before the beginning date of the applicable disaster period. A first-time bred buffalo or beefalo heifer is also considered to be an adult buffalo or beefalo cow if it was pregnant on the beginning date of the disaster period.

Adult dairy bull means a male dairy bovine breed animal at least 2 years old used primarily for breeding dairy cows on the beginning date of the disaster period.

Adult dairy cow means a female bovine animal used for the purpose of providing milk for human consumption that had delivered one or more offspring before the beginning date of the applicable disaster period. A first-time bred dairy heifer is also considered an adult dairy cow if it was pregnant on the beginning date of the disaster period.

Agricultural operation means a farming operation.

Application means the "2005/2006/2007 Livestock Compensation Program" form.

Application period means the date established by the Deputy Administrator for Farm Programs for participants to apply for program benefits.

Disaster period means the applicable disaster period specified in § 760.1001.

Equine animal means a domesticated horse, mule, or donkey.

Goat means a domesticated, ruminant mammal of the genus *Capra*, including Angora goats.

Non-adult beef cattle means a bovine animal that weighed 500 pounds or more on the beginning date of the disaster period, but does not meet the definition of an adult beef cow or bull.

Non-adult buffalo/beefalo means an animal of those breeds that weighed 500 pounds or more on the beginning date of the disaster period, but does not meet the definition of an adult buffalo or beefalo cow or bull.

Non-adult dairy cattle means a bovine livestock, of a breed used for the purpose of providing milk for human consumption, that weighed 500 pounds or more on the beginning date of the disaster period, but does not meet the definition of an adult dairy cow or bull.

Owner means one who had legal ownership of the livestock for which benefits are being requested under this subpart on the beginning date of the applicable disaster period as set forth in § 760.1001.

Poultry means a domesticated chicken, turkey, duck, or goose. Poultry are further delineated by sex, age and purpose of production, as determined by FSA.

Sheep means a domesticated, ruminant mammal of the genus *Ovis*.

Swine means a domesticated omnivorous pig, hog, and boar. Swine are further delineated by sex and weight as determined by FSA.

§ 760.1103 Eligible livestock and producers.

(a) To be considered eligible livestock to generate benefits under this subpart, livestock must meet all the following conditions:

(1) Be adult or non-adult dairy cattle, beef cattle, buffalo, beefalo, equine, poultry, elk, reindeer, sheep, goats, swine, or deer;

(2) Been physically located in the eligible disaster county on the beginning date of the disaster period;

(3) Been maintained for commercial use as part of the producer's farming operation on the beginning date of the disaster period; and

(4) Not have been produced and maintained for reasons other than commercial use as part of a farming operation. Such excluded uses include, but are not limited to, wild free roaming animals or animals used for recreational purposes, such as pleasure, roping, hunting, pets, or for show.

(b) To be considered an eligible livestock producer, the participant's eligible livestock must have been located in the eligible disaster county on the beginning date of the disaster period. To be eligible, also, the livestock producer must have:

(1) Owned or cash-leased eligible livestock on the beginning date of the disaster period (provided that if there is a cash lease, only the cash lessee and not the owner will be eligible); and

(2) Suffered any of the following:

(i) A grazing loss on eligible grazing lands physically located in the eligible disaster county, where the forage was damaged or destroyed by an eligible disaster event, and intended for use as feed for the participant's eligible livestock;

(ii) A loss of feed from forage or feedstuffs physically located in the eligible disaster county, that was mechanically harvested and intended for use as feed for the participant's eligible livestock, that was damaged or destroyed after harvest as the result of an eligible disaster event;

(iii) A loss of feed from purchased forage or feedstuffs physically located in the eligible disaster county, intended for use as feed for the participant's eligible livestock, that was damaged or destroyed by an eligible disaster event; or

(iv) Increased feed costs incurred in the eligible disaster county, due to an eligible disaster event, to feed the participant's eligible livestock.

(c) The eligible livestock categories are:

(1) Adult beef cows or bulls;

(2) Non-adult beef cattle;

(3) Adult buffalo or beefalo cows or bulls;

(4) Non-adult buffalo or beefalo;

(5) Adult dairy cows or bulls;

(6) Non-adult dairy cattle;

(7) Goats;

(8) Sheep;

(9) Equine;

(10) Reindeer;

(11) Elk;

(12) Poultry; and

(13) Deer.

(d) Ineligible livestock include, but are not limited to, livestock:

(1) Livestock that were or would have been in a feedlot regardless of whether there was a disaster or where such livestock were in a feedlot as part of a participant's normal business operation, as determined by FSA;

(2) Emus;

(3) Yaks;

(4) Ostriches;

(5) Llamas;

(6) All beef and dairy cattle, and buffalo and beefalo that weighed less than 500 pounds on the beginning date of the disaster period;

(7) Any wild free roaming livestock, including horses and deer;

(8) Livestock produced or maintained for reasons other than commercial use as part of a farming operation, including, but not limited to, livestock produced or maintained for recreational purposes, such as:

(i) Roping,

(ii) Hunting,

(iii) Show,

(iv) Pleasure,

(v) Use as pets, or

(vi) Consumption by owner.

§760.1104   Application for payment.

(a) To apply for 2005–2007 LCP, an application and required supporting documentation must be submitted to the administrative county FSA office.

(b) The application must be filed during the application period announced by the Deputy Administrator for Farm Programs.

(c) Payments may be made for eligible losses suffered by an eligible livestock producer who is now a deceased individual or is a dissolved entity if a representative who currently has authority to enter into a contract, on behalf of the livestock producer, signs the application for payment. Legal documents showing proof of authority to sign for the deceased individual or dissolved entity must be provided. If a participant is now a dissolved general partnership or joint venture, all members of the general partnership or joint venture at the time of dissolution or their duly authorized representatives must sign the application for payment.

(d) Data furnished by the participant will be used to determine eligibility for program benefits. Furnishing the data is voluntary; however, without all required data program benefits will not be approved or provided.

(e) A minor child is eligible to apply for program benefits if all eligibility requirements are met and one of the following conditions exists:

(1) The right of majority has been conferred upon the minor by court proceedings or statute;

(2) A guardian has been appointed to manage the minor's property, and the applicable program documents are executed by the guardian; or

(3) A bond is furnished under which a surety guarantees any loss incurred for which the minor would be liable had the minor been an adult.

§ 760.1105 Application process.

(a) Participants must submit to FSA:

(1) A completed application in accordance with § 760.1104;

(2) Adequate proof, as determined by FSA, that the feed lost:

(i) Was for the claimed eligible livestock;

(ii) Was lost as a direct result of an eligible disaster event during an eligible disaster period specified in § 760.1001;

(iii) Was lost after January 1, 2005, but before February 28, 2007; and

(iv) Occurred in the calendar year for which benefits are being requested; and

(3) Any other supporting documentation as determined by FSA to be necessary to make a determination of eligibility of the participant. Supporting documents include, but are not limited to: verifiable purchase records; veterinarian records; bank or other loan papers; rendering truck receipts; Federal Emergency Management Agency records; National Guard records; written contracts; production records; Internal Revenue Service records; property tax records; private insurance documents; sales records, and similar documents determined acceptable by FSA.

(b) [Reserved]

§ 760.1106 Payment calculation.

(a) Preliminary, unadjusted LCP payments are calculated for a producer by multiplying the national payment rate for each livestock category, as provided in paragraph (c) of this section, by the number of eligible livestock for the producer in each category. The national payment rate represents the cost of the amount of corn needed to maintain the specific livestock for 30 days, as determined by FSA. As pro-

vided in subpart K of this part, a producer may receive benefits for only one of the three program years, 2005, 2006, or 2007. The producer must indicate which year has been chosen. Payments are available only with respect to disaster-related fees losses in the period from January 2, 2005 through February 27, 2007, in eligible counties for losses during the times specified for the disaster periods as specified in § 760.1001(b).

(b) The preliminary LCP payment calculated in accordance with paragraph (a) of this section:

(1) For 2005 LCP provided for under this subpart will be reduced by the amount the participant received for the specific livestock under the Feed Indemnity Program in accordance with subpart D of this part and LCP for the 2005 hurricanes under subpart B of part 1416 of this title; and

(2) For 2006 LCP under this subpart will be reduced by the amount the participant received for the same or similar loss under the Livestock Assistance Grant Program in accordance with subpart H of this part.

(c) Subject to such other limitations as may apply, including those in paragraph (b) of this section, the payment under the 2005–2007 LCP may not exceed for the relevant year chosen by the producer the smaller of either the:

(1) Payment calculated in paragraph (a) of this section for that year; or

(2) Value of the producer's eligible feed loss, increased feed costs, or forage or grazing loss as determined by FSA for that year.

(d) The actual payment to the producer will be the amount provided for in paragraph (c) of this section subject to the adjustments and limits provided for in this section or in this part.

§ 760.1107 Appeals.

The appeal regulations in parts 11 and 780 of this title apply to determinations made under this subpart.

§ 760.1108 Offsets, assignments, and debt settlement.

(a) Any payment to any participant will be made without regard to any claim or lien against the commodity, or proceeds, in favor of the owner or any other creditor except agencies of the U.S. Government. The regulations governing offsets and withholdings in parts 792 and 1403 of this title apply to payments made under this subpart.

(b) Any participant entitled to any payment may assign any payments in accordance with regulations governing the assignment of payments in part 1404 of this chapter.

§760.1109 Recordkeeping and inspections.

Participants receiving payments under this subpart or any other person who furnishes information for the purposes of enabling the participant to receive a payment under this subpart must maintain any books, records, and accounts supporting that information for a minimum of 3 years following the end of the year during which the application for payment was filed. Participants receiving payments or any other person who furnishes the information to FSA must allow authorized representatives of USDA and the General Accounting Office, during regular business hours, and to enter upon, inspect, examine, and make copies of the books or records, and to inspect and verify all applicable livestock and acreage in which the participant has an interest for the purpose of confirming the accuracy of the information provided by or for the participant.

§760.1110 Refunds; joint and several liability.

In the event there is a failure to comply with any term, requirement, or condition for payment or assistance arising under this subpart, and if any refund of a payment to FSA will otherwise become due in connection with this subpart, all payments made in regard to such matter must be refunded to FSA together with interest and late-payment charges as provided for in part 792 of this title, provided that interest will run from the date of the disbursement of the refund to the producer.

Subpart M—2005–2007 Catfish Grant Program

SOURCE: 72 FR 72881, Dec. 21, 2007, unless otherwise noted.

§760.1200 Administration.

FSA will administer a limited 2005–2007 CGP to provide assistance to catfish producers in eligible counties that suffered catfish feed and related losses between January 1, 2005, and February 28, 2007, that is after January 1, 2005, and before February 28, 2007. Under the 2005–2007 CGP, FSA will provide grants to State governments in those States that have catfish producers that are located in eligible counties and that have agreed to participate in the 2005–2007 CGP. The amount of each grant will be based on the total value of catfish feed and related losses suffered in eligible counties in the subject state. Each State must submit a work plan providing a summary of how the State will implement the 2005–2007 CGP.

§760.1201 Application for payment.

Application procedures for 2005–2007 CGP will be as determined by the State governments.

§760.1202 Eligible producers.

(a) To be considered an eligible catfish producer, an participant must:

(1) Raise catfish in a controlled environment and be physically located in an eligible county on the beginning date of the disaster period;

(2) Maintain the catfish for commercial use as part of a farming operation;

(3) Have a risk in production of such catfish; and

(4) Have suffered one of the following types of losses relating to catfish feed as a direct result of the county's disaster event that occurred in that year:

(i) Physical loss of feed that was damaged or destroyed,

(ii) Cost to the extent allowed by FSA, associated with lost feeding days, or

(iii) Cost associated with increased feed prices.

(b) [Reserved]

§760.1203 Payment calculation.

(a) Producers must be paid for feed losses of higher costs only for one of the three years, 2005, 2006, or 2007, and the loss must be for eligible catfish feed losses in an eligible county, as determined pursuant to subpart K of this part. Further, the feed loss or higher costs must be caused by the disaster that caused the county to qualify as an eligible county. The loss, moreover, to qualify for payment, must have occurred during the allowable time period provided in this part, namely the period beginning on January 2, 2005 and ending February 27, 2007. The producer must pick the year of the benefits sought.

(b) Subject to all adjustments and limits provided for in this part the amount of assistance provided to each participant from the State will be equal to the smaller of:

(1) Depending on the year chosen by the producer, the value of the participant's 2005, 2006, or 2007 catfish feed and related losses as a direct result of an eligible disaster event, as determined by the State or

(2) Result of multiplying:

(i) Total tons of catfish feed purchased by the participant in depending on the year chosen by the producer 2005 (entire year), 2006 (entire year), or 2007 (through February 27, 2007, only), times,

(ii) Catfish feed payment rate for 2005, 2006, or 2007, as applicable, as set by FSA.

(c) The catfish feed rate represents 61 percent of the normal cost of a ton of feed for a year divided by six to reflect the normal feeding price for catfish.

Subpart N—Dairy Economic Loss Assistance Payment Program

SOURCE: 74 FR 67808, Dec. 21, 2009, unless otherwise noted.

§ 760.1301 Administration.

(a) This subpart establishes, subject to the availability of funds, the terms and conditions under which the Dairy Economic Loss Assistance Payments (DELAP) program as authorized by section 10104 of the Farm Security and Rural Investment Act of 2002 (Pub. L. 107–171) will be administered with respect to funds appropriated under Section 748 of the Agriculture, Rural Development, Food and Drug Administration, and Related Agencies Appropriations Act, 2010 (2010 Agriculture Appropriations Bill, Pub. L. 111–80).

(b) The DELAP program will be administered under the general supervision of the Administrator, FSA, and

the Deputy Administrator for Farm Programs, FSA (who is referred to as the "Deputy Administrator" in this part), and will be carried out by FSA's Price Support Division (PSD) and Kansas City Management Office (KCMO).

(c) FSA representatives do not have authority to modify or waive any of the provisions of the regulations of this subpart, except as provided in paragraph (d) of this section.

(d) The State committee will take any action required by the provisions of this subpart that has not been taken by the county committee. The State committee will also:

(1) Correct or require the county committee to correct any action taken by the county committee that is not in compliance with the provisions of this subpart.

(2) Require a county committee to not take an action or implement a decision that is not in compliance with the provisions of this subpart.

(e) No provision or delegation of this subpart to PSD, KCMO, a State committee, or a county committee will preclude the Administrator, FSA, or a designee, from determining any question arising under the program or from reversing or modifying any determination made by PSD, KCMO, a State committee, or a county committee.

(f) The Deputy Administrator may waive or modify non-statutory deadlines and other program requirements of this part in cases where lateness or failure to meet other requirements does not adversely affect the operation of the program. Participants have no right to seek an exception under this provision. The Deputy Administrator's refusal to consider cases or circumstances or decision not to exercise the discretionary authority of this provision will not be considered an adverse decision and is not appealable.

§ 760.1302 Definitions and acronyms.

The following definitions apply to this subpart. The definitions in parts 718 and 1400 of this title also apply, except where they may conflict with the definitions in this section.

County office or FSA county office means the FSA offices responsible for administering FSA programs in a specific areas, sometimes encompassing more than one county, in a State.

Dairy operation means any person or group of persons who, as a single unit, as determined by FSA, produce and market milk commercially produced from cows, and whose production facilities are located in the United States. In any case, however, dairy operation may be given by the agency the same meaning as the definition of dairy operation as found in part 1430 of this title for other dairy assistance programs.

Department or USDA means the U. S. Department of Agriculture.

Deputy Administrator means the Deputy Administrator for Farm programs (DAFP), FSA, or a designee.

Eligible production means milk from cows that was produced during February through July 2009, by a dairy producer in the United States and marketed commercially by a producer in a participating State.

Farm Service Agency or FSA means the Farm Service Agency of the USDA.

Fiscal year or FY means the year beginning October 1 and ending the following September 30. The fiscal year will be designated for this subpart by year reference to the calendar year in which it ends. For example, FY 2009 is from October 1, 2008, through September 30, 2009 (inclusive).

Marketed commercially means sold to the market to which the dairy operation normally delivers whole milk and receives a monetary amount and in any case this term will be construed to allow the use of MILC records in making DELAP payments.

Milk handler means the marketing agency to or through which the dairy operation commercially markets whole milk.

Milk marketing means a marketing of milk for which there is a verifiable sales or delivery record of milk marketed for commercial use.

Participating State means each of the 50 States in the United States of America, the District of Columbia, and the Commonwealth of Puerto Rico, or any other territory or possession of the United States.

Payment quantity means the pounds of milk production for which an operation is eligible to be paid under this subpart.

Producer means any individual, group of individuals, partnership, corporation, estate, trust association, cooperative, or other business enterprise or other legal entity, as defined in 7 CFR 1400.3, who is, or whose members are, a citizen of or legal resident alien in the United States, and who directly or indirectly, as determined by the Secretary, shares in the risk of producing milk, and who is entitled to a share of the commercial production available for marketing from the dairy operation. This term, and other terms in this subpart, will in any case be applied in a way that allows MILC records to be used to make DELAP payments.

United States means the 50 States of the United States of America, the District of Columbia, the Commonwealth of Puerto Rico, and any other territory or possession of the United States.

Verifiable production records means evidence that is used to substantiate the amount of production marketed commercially by a dairy operation and its producers and that can be verified by FSA through an independent source.

§760.1303 Requesting benefits.

(a) If as a dairy operation or producer, your records are currently available in the FSA county office from previous participation in a fiscal year 2009 dairy program administered by FSA, you do not need to request benefits under this subpart to receive payments. FSA will make payments as specified in this subpart to eligible dairy producers based on production data maintained by the FSA county office for the months of February through July 2009.

(b) If records are not available in the FSA county office, dairy producers may request benefits. The request for benefits may be a letter or email; no specific form is required.

(1) Submit your request for DELAP to: Deputy Administrator for Farm Programs, FSA, USDA, STOP 0512, 1400 Independence Avenue, SW., Washington, DC 20250–0512; Attention: DELAP Program. Or you may send your request for DELAP via fax to (202) 690–1536 or e-mail to *Danielle.Cooke@wdc.usda.gov.*

(2) The complete request as described in this subpart must be received by FSA by the close of business on January 19, 2010.

(3) The complete request for benefits must include all of the following:

(i) The name and location of the dairy operation;

(ii) Contact information for the dairy operation, including telephone number;

(iii) Name, percentage share, and tax identification number for the entity or individual producer's receiving a share of the payment; and

(iv) Proof of production (acceptable documentation as specified in § 760.1305).

(4) Requests for benefits and related documents not provided to FSA as required by this subpart, will not be approved.

(5) If not already provided and available to FSA, the dairy producer or dairy operation must provide documentation to support:

(i) The amount (quantity in pounds) of milk produced by the dairy operation during the months of February 2009 through July 2009;

(ii) Percentage share of milk production during February through July 2009 attributed to each producer in the dairy operation; and

(iii) Average adjusted gross income for each individual or entity with a share in the operation and any additional entities or individuals as needed to apply the adjusted gross income rules of these regulations.

(6) Each dairy producer requesting benefits under this subpart is responsible for providing accurate and truthful information and any supporting documentation. If the dairy operation provides the required information, each dairy producer who shares in the risk of a dairy operation's total production is responsible for the accuracy and truthfulness of the information submitted for the request for benefits before the request will be considered complete. Providing a false statement, request, or certification to the Government may be punishable by imprisonment, fines, other penalties, or sanctions.

(c) All information provided by the dairy producer or dairy operation is subject to verification, spot check, and audit by FSA. Further verification information may be obtained from the dairy operation's milk handler or marketing cooperative if necessary for FSA to verify provided information. Refusal to allow FSA or any other USDA agency to verify any information provided or the inability of FSA to verify such information will result in a determination of ineligibility for benefits under this subpart.

(d) Data furnished by dairy producers and dairy operations, subject to verification, will be used to determine eligibility for program benefits. Although participation in the DELAP program is voluntary, program benefits will not be provided unless a producer or operation furnishes all requested data or such data is already recorded at the FSA county office.

§ 760.1304 Eligibility.

(a) Payment under DELAP will only be made to producers, but the dairy "operation" must first qualify its production within limits provided for in this subpart in order to have the individuals or entities that qualify as "producers" receive payment subject to whatever additional limits (such as the adjusted gross income provisions of these regulations) apply. As needed the agency may construe the terms of this regulation in any manner needed to facilitate and expedite payments using existing data and records from other assistance programs. Further, those parties (State and local governments and their political subdivisions and related agencies) excluded from the MILC program will not be eligible for DELAP payments notwithstanding any other provision of these regulation. That said, to be eligible to receive payments under this subpart, a dairy producer in the United States must:

(1) Have produced milk in the United States and commercially marketed the milk produced any time during February 2009 through July 2009;

(2) Be a producer, as defined in § 760.1302;

(3) Provide FSA with proof of milk production commercially marketed by all dairy producers in the dairy operation during February 2009 through July 2009; and

(4) Submit an accurate and complete request for benefits as specified in §760.1303, if production data is not available in the FSA county office.

(b) To be eligible to receive a payment, each producer in an eligible dairy operation must meet the average adjusted gross income eligibility requirements of 7 CFR part 1400. No person or entity will be eligible to receive any payment or direct or indirect benefit under this subpart if their annual average adjusted nonfarm income is over $500,000 as determined under 7 CFR part 1400. In the case of indirect benefits, direct benefits to other parties will be reduced accordingly. This will mean that all of the attribution rules of part 1400 will apply. For example if Individual A is over the limit and owns 100 percent of Corporation C which had a 20 percent interest in Corporation B which had a 50 percent interest in milk producer Corporation A, the AGI of Individual A would result in a 10 percent (100 percent times 20 percent times 50 percent) loss in benefits to Corporation A. For DELAP, the relevant period for the annual average adjusted nonfarm income is 2005 through 2007.

(1) Individual dairy producers in a dairy operation that is an entity are only eligible for a payment based on their share of the dairy operation.

(2) No payment will be made to any other producer based on the share of any dairy producer who exceeds the income limit or who, because of the attribution rules, has their payment reduced.

§760.1305 Proof of production.

(a) Dairy producers requesting benefits must, as required by this subpart, provide adequate proof of the dairy operation's eligible production during the months of February through July 2009, if those records are not already available at the FSA county office. The dairy operation must also provide proof that the eligible production was also commercially marketed during the same period.

(b) To be eligible for payment, dairy producers marketing milk during February through July 2009 must provide any required supporting documents to assist FSA in verifying production.

Supporting documentation may be provided by either the dairy producer or by the dairy operation for each of its producers. Examples of supporting documentation may include, but are not limited to: Milk marketing payment stubs, tank records, milk handler records, daily milk marketings, copies of any payments received as compensation from other sources, or any other documents available to confirm the production and production history of the dairy operation. Dairy operations and producers may also be required to allow FSA to examine the herd of cattle as production evidence. If supporting documentation requested is not presented to FSA, the request for benefits will be denied.

§760.1306 Availability of funds.

(a) Payments under this subpart are subject to the availability of funds. The total available program funds are $290,000,000.

(b) FSA will prorate the available funds by a national factor to ensure payments do not exceed $290,000,000. The payment will be made based on the national payment rate as determined by FSA. FSA will prorate the payments based on the amount of milk production eligible for payments in a fair and reasonable manner.

(c) A reserve will be created to handle new applications, appeals, and errors.

§760.1307 Dairy operation payment quantity.

(a) A dairy operation's payment quantity (the quantity of milk on which the "operation" can generate payments for "producers" involved in the operation) will be determined by FSA, based on the pounds of production of commercially marketed milk during the months of February 2009 through July 2009, multiplied by two.

(b) The maximum payment quantity for which a dairy operation can generate payments for its dairy producers under this subpart will be 6,000,000 pounds.

(c) The dairy operation's payment quantity will be used to determine the amount of DELAP payments made to dairy producers.

§760.1308 Payment rate.

(a) A national per-hundredweight payment rate will be calculated by dividing the available funding, less a reserve established by FSA, by the total pounds of eligible production approved for payment.

(b) Each eligible dairy producer's payment with respect to an operation will be calculated by multiplying the payment rate determined in paragraph (a) of this section by the dairy producer's share in the dairy operation's eligible production payment quantity as determined in accordance with section §760.1307.

(c) In the event that approval of all eligible requests for benefits would result in expenditures in excess of the amount available, FSA will reduce the payment rate in a manner that FSA determines to be fair and reasonable.

§760.1309 Appeals.

The appeal regulations set forth at 7 CFR parts 11 and 780 apply to determinations made under this subpart.

§760.1310 Misrepresentation and scheme or device.

(a) In addition to other penalties, sanctions or remedies as may apply, a dairy producer or operation will be ineligible to receive benefits under this subpart if the producer or operation is determined by FSA to have:

(1) Adopted any scheme or device that tends to defeat the purpose of this subpart;

(2) Made any fraudulent representation; or

(3) Misrepresented any fact affecting a program determination.

(b) Any payment to any person or operation engaged in a misrepresentation, scheme, or device, must be refunded with interest together with such other sums as may become due. Any dairy operation or person engaged in acts prohibited by this section and receiving payment under this subpart will be jointly and severally liable with other producers or operations involved in such claim for benefits for any refund due under this section and for related charges. The remedies provided in this subpart will be in addition to other civil, criminal, or administrative remedies that may apply.

§760.1311 Death, incompetence, or disappearance.

(a) In the case of the death, incompetency, or disappearance of a person or the dissolution of an entity that is eligible to receive benefits in accordance with this subpart, such alternate person or persons specified in 7 CFR part 707 may receive such benefits, as determined appropriate by FSA.

(b) Payments may be made to an otherwise eligible dairy producer who is now deceased or to a dissolved entity if a representative who currently has authority to enter into an application for the producer or the producer's estate makes the request for benefits as specified in §760.1303. Proof of authority over the deceased producer's estate or a dissolved entity must be provided.

(c) If a dairy producer is now a dissolved general partnership or joint venture, all members of the general partnership or joint venture at the time of dissolution or their duly authorized representatives must be identified in the request for benefits.

§760.1312 Maintaining records.

(a) Persons requesting benefits under this subpart must maintain records and accounts to document all eligibility requirements specified in this subpart. Such records and accounts must be retained for 3 years after the date of payment to the dairy producer under this subpart.

(b) Destruction of the records after 3 years from the date of payment will be at the decision and risk of the party undertaking the destruction.

§760.1313 Refunds; joint and several liability.

(a) Any dairy producer that receives excess payment, payment as the result of erroneous information provided by any person, or payment resulting from a failure to comply with any requirement or condition for payment under this subpart, must refund the amount of that payment to FSA.

(b) Any refund required will be due from the date of the disbursement by the agency with interest determined in accordance with paragraph (d) of this section and late payment charges as provided in 7 CFR part 1403.

(c) Each dairy producer that has an interest in the dairy operation will be jointly and severally liable for any refund and related charges found to be due to FSA.

(d) Interest will be applicable to any refunds to FSA required in accordance with 7 CFR parts 792 and 1403. Such interest will be charged at the rate that the U.S. Department of the Treasury charges FSA for funds, and will accrue from the date FSA made the payment to the date the refund is repaid.

(e) FSA may waive the accrual of interest if it determines that the cause of the erroneous payment was not due to any action of the person or entity, or was beyond the control of the person or entity committing the violation. Any waiver is at the discretion of FSA alone.

§ 760.1314 Miscellaneous provisions.

(a) *Offset.* FSA may offset or withhold any amount due to FSA from any benefit provided under this subpart in accordance with the provisions of 7 CFR part 1403.

(b) *Claims.* Claims or debts will be settled in accordance with the provisions of 7 CFR part 1403.

(c) *Other interests.* Payments or any portion thereof due under this subpart will be made without regard to questions of title under State law and without regard to any claim or lien against the milk production, or proceeds thereof, in favor of the owner or any other creditor except agencies and instrumentalities of the U.S. Government.

(d) *Assignments.* Any dairy producer entitled to any payment under this part may assign any payments in accordance with the provisions of 7 CFR part 1404.

(e) *Violations of highly erodible land and wetland conservation provisions.* The provisions of part 12 of this title apply to this subpart. That part sets out certain conservation requirements as a general condition for farm benefits.

(f) *Violations regarding controlled substances.* The provisions of §718.6 of this title, which generally limit program payment eligibility for persons who have engaged in certain offenses with respect to controlled substances, will apply to this subpart.

Subpart O—Agricultural Disaster Indemnity Programs

SOURCE: 83 FR 33801, July 18, 2018, unless otherwise noted.

§ 760.1500 Applicability.

(a) This subpart specifies the terms and conditions for the 2017 Wildfires and Hurricanes Indemnity Program (2017 WHIP) and the Wildfires and Hurricanes Indemnity Program Plus (WHIP+).

(b) The 2017 WHIP provides disaster assistance for necessary expenses related to crop, tree, bush, and vine losses related to the consequences of wildfires, hurricanes, and Tropical Storm Cindy that occurred in calendar year 2017, and for losses of peach and blueberry crops in calendar year 2017 due to extreme cold, and blueberry productivity losses in calendar year 2018 due to extreme cold and hurricane damage in calendar year 2017.

(c) WHIP+ provides disaster assistance for necessary expenses related to losses of crops, trees, bushes, and vines, as a consequence of Hurricanes Michael and Florence, other hurricanes, floods, tornadoes, typhoons, volcanic activity, snowstorms, wildfires, excessive moisture, and qualifying drought occurring in calendar years 2018 and 2019.

[84 FR 48528, Sept. 13, 2019, as amended at 86 FR 445, Jan. 6, 2021]

§ 760.1501 Administration.

(a) Programs under this subpart are administered under the general supervision of the Administrator, Farm Service Agency (FSA), and the Deputy Administrator for Farm Programs, FSA. Programs under this subpart are carried out by FSA State and county committees with instructions issued by the Deputy Administrator.

(b) FSA State and county committees, and representatives and their employees, do not have authority to modify or waive any of the provisions of the regulations in this subpart or instructions issued by the Deputy Administrator.

(c) The FSA State committee will take any action required by the regulations in this subpart that the FSA county committee has not taken. The FSA State committee will also:

(1) Correct, or require an FSA county committee to correct, any action taken by the FSA county committee that is not in accordance with the regulations in this subpart; or

(2) Require an FSA county committee to withhold taking any action that is not in accordance with this subpart.

(d) No delegation to an FSA State or county committee precludes the FSA Administrator, the Deputy Administrator, or a designee, from determining any question arising under this subpart or from reversing or modifying any determination made by an FSA State or county committee.

(e) The Deputy Administrator has the authority to permit State and county committees to waive or modify a non-statutory deadline specified in this part.

(f) Items of general applicability to program participants, including, but not limited to, application periods, application deadlines, internal operating guidelines issued to FSA State and county offices, prices, yields, and payment factors established under this subpart, are not subject to appeal in accordance with part 780 of this chapter.

[83 FR 33801, July 18, 2018, as amended 84 FR 48528, Sept. 13, 2019]

§ 760.1502 Definitions.

The following definitions apply to this subpart. The definitions in §§ 718.2 and 1400.3 of this title also apply, except where they conflict with the definitions in this section. In the event of conflict, the definitions in this section apply.

2017 WHIP factor means the factor in § 760.1511, determined by the Deputy Administrator, that is based on the crop insurance or NAP coverage level elected by the *2017* WHIP participant for a crop for which a payment is being requested; or, as applicable, the factor that applies for a crop of a crop year where the participant had no insurance or NAP coverage.

2017 WHIP yield means, for a unit:

(1) For an insured crop, excluding crops located in Puerto Rico, the ap-

proved federal crop insurance APH, for the disaster year;

(2) For a NAP covered crop, excluding crops located in Puerto Rico, the approved yield for the disaster year;

(3) For a crop located in Puerto Rico or an uninsured crop, excluding citrus crops located in Florida, the county expected yield for the disaster year; and

(4) For citrus crops located in Florida, the yield based on documentation submitted according to § 760.1511(c)(3), or if documentation is not submitted, the county expected yield.

Actual production means the total quantity of the crop appraised, harvested, or assigned, as determined by the FSA State or county committee in accordance with instructions issued by the Deputy Administrator.

Administrative county office means the FSA county office designated to make determinations, handle official records, and issue payments for the farm as specified in accordance part 718 of this title.

Appraised production means the amount of production determined by FSA, or a company reinsured by the Federal Crop Insurance Corporation (FCIC), that was unharvested but was determined to reflect the crop's yield potential at the time of appraisal.

Approved yield means the amount of production per acre, computed as specified in FCIC's Actual Production History (APH) Program in part 400, subpart G of this title or, for crops not included in part 400, subpart G of this title, the yield used to determine the guarantee. For crops covered under NAP, the approved yield is established according to part 1437 of this title.

Average adjusted gross farm income means the average of the portion of adjusted gross income of the person or legal entity that is attributable to activities related to farming, ranching, or forestry. The relevant tax years are:

(1) For 2017 WHIP, 2013, 2014, and 2015; and

(2) For WHIP+, 2015, 2016, and 2017.

Average adjusted gross income means the average of the adjusted gross income as defined under 26 U.S.C. 62 or comparable measure of the person or legal entity. The relevant tax years are:

(1) For 2017 WHIP, 2013, 2014, and 2015; and

(2) For WHIP+, 2015, 2016, and 2017.

Bush means, a low, branching, woody plant, from which at maturity of the bush, an annual fruit or vegetable crop is produced for commercial market for human consumption, such as a blueberry bush. The definition does not cover nursery stock or plants that produce a bush after the normal crop is harvested.

Buy-up NAP coverage means NAP coverage at a payment amount that is equal to an indemnity amount calculated for buy-up coverage computed under section 508(c) or (h) of the Federal Crop Insurance Act and equal to the amount that the buy-up coverage yield for the crop exceeds the actual yield for the crop.

Catastrophic coverage has the meaning as defined in §1437.3 of this title.

Citrus crops and citrus trees include grapefruit, lemon, lime, Mandarin, Murcott, orange (all types), pummelo (pomelo), tangelo, tangerine, tangor.

County disaster yield means the average yield per acre calculated for a county or part of a county for the applicable crop year based on disaster events, and is intended to reflect the amount of production that a participant would have been expected to make based on the eligible disaster conditions in the county or area, as determined by the FSA county committee in accordance with instructions issued by the Deputy Administrator.

County expected yield has the meaning assigned in §1437.102(b) of this title.

Coverage level means the percentage determined by multiplying the elected yield percentage under a crop insurance policy or NAP coverage by the elected price percentage.

Crop insurance means an insurance policy reinsured by FCIC under the provisions of the Federal Crop Insurance Act, as amended. It does not include private plans of insurance.

Crop insurance indemnity means, for the purpose of this subpart, the payment to a participant for crop losses covered under crop insurance administered by RMA in accordance with the Federal Crop Insurance Act (7 U.S.C. 1501–1524).

Crop year means:

(1) For insurable crops, trees, bushes, and vines, the crop year as defined according to the applicable crop insurance policy;

(2) For NAP eligible crops, the crop year as defined in §1437.3 of this title;

(3) For uninsurable trees, bushes, and vines, the calendar year in which the qualifying disaster event occurred.

Damage factor means a percentage of the value lost when a tree, bush, or vine is damaged and requires rehabilitation but is not completely destroyed, as determined by the Deputy Administrator.

Eligible crop means a crop for which coverage was available either from FCIC under part 400 of this title, or through NAP under §1437.4 of this title, that was affected by a qualifying disaster event.

Eligible disaster event means a disaster event that was:

(1) For insured crops, an eligible cause of loss under the applicable crop insurance policy for the crop year;

(2) For NAP covered crops and uninsured crops, an eligible cause of loss as specified in §1437.10 of this title.

End use means the purpose for which the harvested crop is used, such as grain, hay, or seed.

Expected production means, for an agricultural unit, the historic yield multiplied by the number of planted or prevented planted acres of the crop for the unit.

FCIC means the Federal Crop Insurance Corporation, a wholly owned Government Corporation of USDA, administered by RMA.

Final planting date means the latest date, established by RMA for insurable crops, by which the crop must initially be planted in order to be insured for the full production guarantee or amount of insurance per acre. For NAP eligible crops, the final planting date is as defined in §1437.3 of this title.

Growth stage means a classification system for trees, bushes, and vines based on a combination of age and production capability, determined by:

(1) The applicable insurance policy for insurable trees, bushes, and vines; or

(2) The Deputy Administrator for trees, bushes, and vines for which RMA does not offer an insurance policy.

Harvested means:

(1) For insurable crops, harvested as defined according to the applicable crop insurance policy;

(2) For NAP eligible single harvest crops, that a crop has been removed from the field, either by hand or mechanically;

(3) For NAP eligible crops with potential multiple harvests in 1 year or harvested over multiple years, that the producer has, by hand or mechanically, removed at least one mature crop from the field during the crop year;

(4) For mechanically-harvested NAP eligible crops, that the crop has been removed from the field and placed in a truck or other conveyance, except that hay is considered harvested when in the bale, whether removed from the field or not. Grazed land will not be considered harvested for the purpose of determining an unharvested or prevented planting payment factor.

Insurable crop means an agricultural crop (excluding livestock) for which the producer on a farm is eligible to obtain a policy or plan of insurance under the Federal Crop Insurance Act (7 U.S.C. 1501–1524).

Multi-use crop means a crop intended for more than one end use during the crop year such as grass harvested for seed, hay, and grazing.

Multiple cropping means the planting of two or more different crops on the same acreage for harvest within the same crop year.

Multiple planting means the planting for harvest of the same crop in more than one planting period in a crop year on different acreage.

NASS means the National Agricultural Statistics Service.

NAP means the Noninsured Crop Disaster Assistance Program under section 196 of the Federal Agriculture Improvement and Reform Act of 1996 (7 U.S.C. 7333) and part 1437 of this title.

NAP covered crop means a crop for which the producer on a farm obtained NAP coverage.

NAP eligible crop means an agricultural crop for which the producer on a farm is eligible to obtain NAP coverage.

NAP service fee means the amount the producer must pay to obtain NAP coverage.

Planted acreage means land in which seed, plants, or trees have been placed, appropriate for the crop and planting method, at a correct depth, into a seedbed that has been properly prepared for the planting method and production practice normal to the USDA plant hardiness zone as determined by the county committee.

Prevented planting means the inability to plant an eligible crop with proper equipment during the planting period as a result of an eligible cause of loss, as determined by FSA.

Price means price per unit of the crop or commodity and will be:

(1) For an insured crop under a crop insurance policy that establishes a price, and under WHIP+, the price for a crop for which the producer obtained a revenue plan of insurance is the greater of the projected price or the harvest price to determine liability, that established price;

(2) For an insured crop under a crop insurance policy that does not establish a price to determine crop insurance liability, the county average price, as determined by FSA;

(3) For a NAP covered crop or uninsured crop, the average market price determined in § 1437.12 of this title; or

(4) For a tree, bush, or vine, the price determined by the Deputy Administrator based on the species of tree, bush, or vine and its growth stage.

Production means quantity of the crop or commodity produced expressed in a specific unit of measure including, but not limited to, bushels or pounds. Production under this subpart includes all harvested production, unharvested appraised production, and assigned production for the total planted acreage of the crop on the unit.

Qualifying disaster event means:

(1) For 2017 WHIP, a hurricane, wildfire, or Tropical Storm Cindy or related condition that occurred in the 2017 calendar year; extreme cold in calendar year 2017 for losses of peach and blueberry crops in calendar year 2017; and extreme cold and hurricane damage in calendar year 2017 for blueberry productivity losses in calendar year 2018; and

(2) For WHIP+, a hurricane, flood, tornado, typhoon, volcanic activity,

snowstorm, wildfire, excessive moisture, qualifying drought, or related condition that occurred in the 2018 or 2019 calendar year.

Qualifying drought means an area within the county was rated by the U.S. Drought Monitor as having a D3 (extreme drought) or higher level of drought intensity during the applicable calendar year.

Related condition means damaging weather or an adverse natural occurrence that occurred as a direct result of a specified qualifying disaster event, as determined by FSA, such as excessive rain, high winds, flooding, mudslides, and heavy smoke, as determined by the Deputy Administrator.

Repeat crop means, with respect to production, a commodity that is planted or prevented from being planted in more than one planting period on the same acreage in the same crop year.

RMA means the Risk Management Agency.

Salvage value means the dollar amount or equivalent for the quantity of the commodity that cannot be marketed or sold in any recognized market for the crop.

Secondary use means the harvesting of a crop for a use other than the intended use.

Secondary use value means the value determined by multiplying the quantity of secondary use times the FSA-established price for that use.

Tree means a tall, woody plant having comparatively great height, and a single trunk from which an annual crop is produced for commercial market for human consumption, such as a maple tree for syrup, or papaya or orchard tree for fruit. It includes immature trees that are intended for commercial purposes. Nursery stock, banana and plantain plants, and trees used for pulp or timber are not considered eligible trees under this subpart.

Tropical crops is defined in §1437.501 of this title.

Tropical region is defined in §1437.502 of this title.

Unharvested payment factor means a percentage established by FSA for a crop and applied in a payment formula to reduce the payment for reduced expenses incurred because commercial harvest was not performed.

Uninsured means a crop that was not covered by crop insurance or NAP for the crop year for which a payment is being requested under this subpart.

Unit means, unless otherwise determined by the Deputy Administrator, basic unit as defined in part 457 or §1437.9 of this title, for ornamental nursery production, includes all eligible plant species and sizes.

Unit of measure means:

(1) For insurable crops, the FCIC-established unit of measure; and

(2) For NAP eligible crops, the established unit of measure used for the NAP price and yield.

USDA means the U.S. Department of Agriculture.

USDA Plant Hardiness Zone means the 11 regions or planting zones as defined by a 10 degree Fahrenheit difference in the average annual minimum temperature.

U.S. drought monitor is a system for classifying drought severity according to a range of abnormally dry to exceptional drought. It is a collaborative effort between Federal and academic partners, produced on a weekly basis, to synthesize multiple indices, outlooks, and drought impacts on a map and in narrative form. This synthesis of indices is reported by the National Drought Mitigation Center at *http://droughtmonitor.unl.edu.*

Value loss crop has the meaning specified in subpart D, of part 1437 of this title.

Vine means a perennial plant grown under normal conditions from which an annual fruit crop is produced for commercial market for human consumption, such as grape, kiwi, or passion fruit, and that has a flexible stem supported by climbing, twining, or creeping along a surface. Nursery stock, perennials that are normally propagated as annuals such as tomato plants, biennials such as strawberry plants, and annuals such as pumpkin, squash, cucumber, watermelon, and other melon plants, are excluded from the term vine in this subpart.

WHIP+ factor means the factor in §760.1511, determined by the Deputy Administrator, that is based on the crop insurance or NAP coverage level elected by the WHIP+ participant for a crop for which a payment is being requested; or, as applicable, the factor that applies for a crop during a crop year in which the participant had no insurance or NAP coverage.

WHIP+ yield means, for a unit:

(1) For an insured crop, excluding crops located in Puerto Rico, the approved federal crop insurance APH, for the crop year;

(2) For a NAP covered crop, excluding crops located in Puerto Rico, the approved yield for the crop year;

(3) For a crop located in Puerto Rico or an uninsured crop, excluding select crops, the county expected yield for the crop year; and

(4) For select crops, the yield based on documentation submitted according to § 760.1511(c)(3), or if documentation is not submitted, the county expected yield.

Yield means unit of production, measured in bushels, pounds, or other unit of measure, per area of consideration, usually measured in acres.

[83 FR 33801, July 18, 2018, as amended 84 FR 48529, Sept. 13, 2019; 86 FR 445, Jan. 6, 2021]

§ 760.1503 Eligibility.

(a) Participants will be eligible to receive a payment under this subpart only if they incurred a loss to an eligible crop, tree, bush, or vine due to a qualifying disaster event, as further specified in this subpart.

(b) To be an eligible participant under this subpart a producer who is a person or legal entity must be a:

(1) Citizen of the United States;

(2) Resident alien; for purposes of this subpart, resident alien means "lawful alien;"

(3) Partnership consisting of solely of citizens of the United States or resident aliens; or

(4) Corporation, limited liability company, or other organizational structure organized under State law consisting solely of citizens of the United States or resident aliens.

(c) If any person who would otherwise be eligible to receive a payment dies before the payment is received, payment may be released as specified in § 707.3 of this title. Similarly, if any person or legal entity who would otherwise been eligible to apply for a payment dies or is dissolved, respectively, before the payment is applied for, payment may be released in accordance with this subpart if a timely application is filed by an authorized representative. Proof of authority to sign for the deceased producer or dissolved entity must be provided. If a participant is now a dissolved general partnership or joint venture, all members of the general partnership or joint venture at the time of dissolution or their duly authorized representatives must sign the application for payment. Eligibility of such participant will be determined, as it is for other participants, based upon ownership share and risk in producing the crop.

(d) Growers growing eligible crops under contract for crop owners are not eligible unless the grower is also determined to have an ownership share of the crop. Any verbal or written contract that precludes the grower from having an ownership share renders the grower ineligible for payments under this subpart.

(e) A person or legal entity is not eligible to receive disaster assistance under this subpart if it is determined by FSA that the person or legal entity:

(1) Adopted any scheme or other device that tends to defeat the purpose of this subpart or any of the regulations applicable to this subpart;

(2) Made any fraudulent representation; or

(3) Misrepresented any fact affecting a program determination under any or all of the following: This subpart and parts 12, 400, 1400, and 1437 of this title.

(g) A person ineligible for crop insurance or NAP under §§ 400.458 or 1437.16 of this title, respectively, for any year is ineligible for payments under this subpart for the same year.

(h) The provisions of § 718.11 of this title, providing for ineligibility for payments for offenses involving controlled substances, apply.

(i) As a condition of eligibility to receive payments under this subpart, the participant must have been in compliance with the Highly Erodible Land Conservation and Wetland Conservation provisions of part 12 of this title for the applicable crop year for which the producer is applying for benefits under this subpart, and must not otherwise be precluded from receiving payments under parts 12, 400, 1400, or 1437 of this title or any law.

(j) Members of cooperative processors are not eligible for WHIP+ assistance for sugar beet losses.

[83 FR 33801, July 18, 2018, as amended 84 FR 48529, Sept. 13, 2019; 86 FR 445, Jan. 6, 2021]

§ 760.1504 Miscellaneous provisions.

(a) All persons with a financial interest in the legal entity receiving payments under this subpart are jointly and severally liable for any refund, including related charges, which is determined to be due to FSA for any reason.

(b) In the event that any application for payment under this subpart resulted from erroneous information or a miscalculation, the payment will be recalculated and any excess refunded to FSA with interest to be calculated from the date of the disbursement.

(c) Any payment to any participant under this subpart will be made without regard to questions of title under State law, and without regard to any claim or lien against the commodity, or proceeds, in favor of the owner or any other creditor except agencies of the U.S. Government. The regulations governing offsets and withholdings in part 792 of this chapter apply to payments made under this subpart.

(d) Any participant entitled to any payment may assign any payment(s) in accordance with regulations governing the assignment of payments in part 792 of this chapter.

(e) The regulations in parts 11 and 780 of this title apply to determinations under this subpart.

§ 760.1505 General provisions.

(a) For loss calculations, the participant's unit structure will be:

(1) For an insured crop, the participant's existing unit structure established in accordance with part 457 of this title;

(2) For a crop with NAP coverage, the participant's existing unit structure established in accordance with part 1437 of this title;

(3) For an uninsured crop, the participant's unit structure established in accordance with part 1437 of this title.

(b) FSA county committees will make the necessary adjustments to assign production or reduce the 2017 WHIP yield or WHIP+ yield when the county committee determines:

(1) An acceptable appraisal or record of harvested production does not exist;

(2) The loss is due to an ineligible cause of loss;

(3) The loss is due to practices, soil type, climate, or other environmental factors that cause lower yields than those upon which the historic yield is based;

(4) The participant has a contract providing a guaranteed payment for all or a portion of the crop; or

(5) The crop was planted beyond the normal planting period for the crop.

(c) Assignment of production or reduction in yield will apply for practices that result in lower yields than those for which the historic yield is based.

(d) Eligibility and payments under this subpart will be determined based on a unit's:

(1) Physical location county for insured crops; and

(2) Administrative county for NAP covered crops and uninsured crops.

(e) FSA may separate or combine types and varieties as a crop for eligibility and payment purposes under this subpart when specific credible information as determined by FSA shows the crop of a specific type or variety has a significantly different or similar value, respectively, when compared to other types or varieties, as determined by the Deputy Administrator.

(f) Unless otherwise specified, all the eligibility provisions of part 1437 of this title apply to value loss crops and tropical crops under this subpart.

(g) The quantity or value of a crop will not be reduced for any quality consideration unless a zero value is established based on a total loss of quality, except as specified in § 760.1513(i).

(h) FSA will use the most reliable data available at the time payments under this subpart are calculated. If additional data or information is provided or becomes available after a payment is issued, FSA will recalculate the payment amount and the producer must return any overpayment amount to FSA. In all cases, payments can only issue based on the payment formula for losses that affirmatively occurred.

(i) A participant who received a payment for a loss under 2017 WHIP cannot:

(1) Be paid for the same loss under WHIP+; or

(2) Refund the 2017 WHIP payment to be eligible for payment for that loss under WHIP+.

[83 FR 33801, July 18, 2018, as amended 84 FR 48529, Sept. 13, 2019]

§ 760.1506 Availability of funds and timing of payments.

(a) For 2017 WHIP:

(1) An initial payment will be issued for 50 percent of each 2017 WHIP payment calculated according to this subpart, as determined by the Secretary. The remainder of the calculated 2017 WHIP payment will be paid to a participant only after the application period has ended and any crop insurance indemnity or NAP payment the participant is entitled to receive for the crop has been calculated and reported to FSA, and then only if there are funds available for such payment as discussed in this subpart.

(2) In the event that, within the limits of the funding made available by the Secretary, approval of eligible applications would result in payments in excess of the amount available, FSA will prorate payments by a national factor to reduce the payments to an amount that is less than available funds as determined by the Secretary. FSA will prorate the payments in such manner as it determines equitable.

(3) Applications and claims that are unpaid or prorated for any reason will not be carried forward for payment under other funds for later years or otherwise, but will be considered, as to any unpaid amount, void and nonpayable.

(b) For WHIP:

(1) For the 2018 crop year, the calculated WHIP+ payment will be paid at 100 percent.

(2) For the 2019 and 2020 crop years, an initial payment will be issued for 50 percent of each WHIP+ payment calculated according to this subpart, as determined by the Secretary. Up to the remaining 50 percent of the calculated

WHIP+ payment will be paid only to the extent that there are funds available for such payment as discussed in this subpart.

(3) In the event that, within the limits of the funding made available by the Secretary, approval of eligible applications would result in payments in excess of the amount available, FSA will prorate 2019 and 2020 payments by a national factor to reduce the payments to the remaining available funds, as determined by the Secretary. FSA will prorate the payments accordingly.

(4) Applications and claims that are unpaid or prorated for aforementioned reasons of fund availability will not be carried forward for payment and will be considered, as to any unpaid amount, void and non-payable.

[83 FR 33801, July 18, 2018, as amended 84 FR 48529, Sept. 13, 2019]

§ 760.1507 Payment limitation.

(a) For any 2017 WHIP payments for the 2017 or 2018 crop year combined, a person or legal entity, other than a joint venture or general partnership, is eligible to receive, directly or indirectly, 2017 WHIP payments of not more than:

(1) $125,000, if less than 75 percent of the person or legal entity's average adjusted gross income is average adjusted gross farm income; or

(2) $900,000, if not less than 75 percent of the average adjusted gross income of the person or legal entity is average adjusted gross farm income.

(b) For any WHIP+ payments, a person or legal entity, other than a joint venture or general partnership, is eligible to receive, directly or indirectly, WHIP+ payments of not more than:

(1) $125,000 combined for the 2018, 2019, and 2020 crop years, if less than 75 percent of the person or legal entity's average adjusted gross income is average adjusted gross farm income; or

(2) $250,000 for each of the 2018, 2019, and 2020 crop years, if 75 percent or more of the average adjusted gross income of the person or legal entity is average adjusted gross farm income, and such payments cannot exceed a total of $500,000 combined for all of the 2018, 2019, and 2020 crop years.

(c) A person or legal entity's average adjusted gross income and average adjusted gross farm income are determined based on the:

(1) 2013, 2014, and 2015 tax years for 2017 WHIP;

(2) 2015, 2016, and 2017 tax years for WHIP+.

(d) To be eligible for more than $125,000 in payments for the applicable period specified in this section, a person or legal entity must submit FSA–892 and provide a certification in the manner prescribed by FSA from a certified public accountant or attorney that at least 75 percent of the person or legal entity's average adjusted gross income was average adjusted gross farm income. Persons or legal entities who fail to provide FSA–892 and the required certification may not receive a 2017 WHIP payment, directly or indirectly, of more than $125,000.

(e) The direct attribution provisions in part 1400 of this chapter apply to payments under this subpart for both payment limitation as well as in determining average AGI as defined and used in this rule.

[83 FR 33801, July 18, 2018, as amended 84 FR 48529, Sept. 13, 2019]

§ 760.1508 Qualifying disaster events.

(a) A producer will be eligible for payments under this subpart for a crop, tree, bush, or vine loss only if the producer suffered a loss to the crop, tree, bush, or vine on the unit due to a qualifying disaster event.

(b) For a loss due to hurricane and conditions related to hurricanes, the crop, tree, bush, or vine loss must have occurred on acreage that was physically located in a county that received a:

(1) Presidential Emergency Disaster Declaration authorizing public assistance for categories C through G or individual assistance due to a hurricane occurring in the 2017 calendar year; or

(2) Secretarial Disaster Designation for a hurricane occurring in the 2017 calendar year.

(c) A producer with crop, tree, bush, or vine losses on acreage not located in a physical location county that was eligible under paragraph (b) of this section will be eligible for 2017 WHIP for losses due to hurricane and related conditions only if the producer provides supporting documentation that is acceptable to FSA from which the FSA county committee determines that the loss of the crop, tree, bush, or vine on the unit was reasonably related to a qualifying disaster event as specified in this subpart. Supporting documentation may include furnishing climatological data from a reputable source or other information substantiating the claim of loss due to a qualifying disaster event.

(d) For a loss due to wildfires and conditions related to wildfire in the 2017 calendar year, all counties where wildfires occurred, as determined by FSA county committees, are eligible for 2017 WHIP; a Presidential Emergency Disaster Declaration or Secretarial Disaster Designation for wildfire is not required. The loss of the crop, tree, bush, or vine must be reasonably related to wildfire and conditions related to wildfire, as specified in this subpart's definition of qualifying disaster event.

(e) For WHIP+, for a loss due to a qualifying disaster event, the crop, tree, bush, or vine loss must have occurred on acreage that was physically located in a county that received a:

(1) Presidential Emergency Disaster Declaration authorizing public assistance for categories C through G or individual assistance due to a qualifying disaster event occurring in the 2018 or 2019 calendar years; or

(2) Secretarial Disaster Designation for a qualifying disaster event occurring in the 2018 or 2019 calendar years.

(f) A producer with crop, tree, bush, or vine losses on acreage not located in a physical location county that was eligible under paragraph (e) of this section will be eligible for WHIP+ for losses due to qualifying disaster events only if the producer provides supporting documentation that is acceptable to FSA from which the FSA county committee determines that the loss of the crop, tree, bush, or vine on the unit was reasonably related to a qualifying disaster event as specified in this subpart. Supporting documentation may include furnishing climatological data from a reputable source or other information substantiating the claim of loss due to a qualifying disaster event.

[83 FR 33801, July 18, 2018, as amended 84 FR 48530, Sept. 13, 2019; 86 FR 445, Jan. 6, 2021]

§ 760.1509   Eligible and ineligible losses.

(a) Except as provided in paragraphs (b) through (e) of this section, to be eligible for payments under this subpart the unit must have suffered a loss of the crop, tree, bush, or vine, or prevented planting of a crop, due to a qualifying disaster event.

(b) A loss will not be eligible under this subpart if any of the following apply:

(1) The cause of loss is determined by FSA to be the result of poor management decisions, poor farming practices, or drifting herbicides;

(2) The cause of loss was due to failure of the participant to re-seed or re-plant to the same crop in a county where it is customary to re-seed or re-plant after a loss before the final planting date;

(3) The cause of loss was due to water contained or released by any governmental, public, or private dam or reservoir project if an easement exists on the acreage affected by the containment or release of the water;

(4) The cause of loss was due to conditions or events occurring outside of the applicable growing season for the crop, tree, bush, or vine;

(5) The cause of loss was due to failure of a power supply or brownout; or

(6) FSA or RMA have previously disapproved a notice of loss for the crop and disaster event unless that notice of loss was disapproved solely because it was filed after the applicable deadline.

(c) The following types of loss, regardless of whether they were the result of an eligible disaster event, are not eligible losses:

(1) Losses to crops intended for grazing;

(2) Losses to crops for which FCIC coverage or NAP coverage is unavailable;

(3) Losses to volunteer crops;

(4) Losses to crops not intended for harvest;

(5) Losses of by-products resulting from processing or harvesting a crop, such as, but not limited to, cotton seed, peanut shells, wheat or oat straw, or corn stalks or stovers;

(6) Losses to home gardens;

(7) Losses of first year seeding for forage production, or immature fruit crops; or

(8) Losses to crops that occur after harvest.

(d) The following losses of ornamental nursery stock are not eligible losses:

(1) Losses caused by the inability to market nursery stock as a result of lack of compliance with State and local commercial ordinances and laws, quarantine, boycott, or refusal of a buyer to accept production;

(2) Losses affecting crops where weeds and other forms of undergrowth in the vicinity of nursery stock have not been controlled; or

(3) Losses caused by the collapse or failure of buildings or structures.

(e) The following losses for honey, as a crop, where the honey production by colonies or bees was diminished, are not eligible losses:

(1) Losses caused by the unavailability of equipment or the collapse or failure of equipment or apparatus used in the honey operation;

(2) Losses caused by improper storage of honey;

(3) Losses caused by bee feeding;

(4) Losses caused by the application of chemicals;

(5) Losses caused by theft;

(6) Losses caused by the movement of bees by or for the participant;

(7) Losses caused by disease or pest infestation of the colonies, unless approved by the Deputy Administrator;

(8) Losses of income from pollinators; or

(9) Losses of equipment or facilities.

(f) Qualifying losses for trees, bushes, and vines will not include losses:

(1) That could have been prevented through reasonable and available measures; and

(2) To trees, bushes, or vines that were abandoned or were not in use or intended for commercial operation at the time of the loss.

[83 FR 33801, July 18, 2018, as amended 84 FR 48530, Sept. 13, 2019]

§760.1510 Application for payment.

(a) An application for payment under this subpart must be submitted to the FSA county office serving as the farm's administrative county office by the close of business on October 30, 2020. Producers must submit:

(1) For 2017 WHIP, a completed form FSA–890, Wildfires and Hurricanes Indemnity Program Application; or

(2) For WHIP+, a completed form FSA–894, Wildfires and Hurricanes Indemnity Program + Application.

(b) Once signed by a producer, the application for payment is considered to contain information and certifications of and pertaining to the producer regardless of who entered the information on the application.

(c) The producer applying for payment under this subpart certifies the accuracy and truthfulness of the information provided in the application as well as any documentation filed with or in support of the application. All information is subject to verification or spot check by FSA at any time, either before or after payment is issued. Refusal to allow FSA or any agency of the Department of Agriculture to verify any information provided will result in the participant's forfeiting eligibility for payment under this subpart. FSA may at any time, including before, during, or after processing and paying an application, require the producer to submit any additional information necessary to implement or determine any eligibility provision of this subpart. Furnishing required information is voluntary; however, without it FSA is under no obligation to act on the application or approve payment. Providing a false certification will result in ineligibility and can also be punishable by imprisonment, fines, and other penalties.

(d) The application submitted in accordance with paragraph (a) of this section is not considered valid and complete for issuance of payment under this subpart unless FSA determines all the applicable eligibility provisions have been satisfied and the participant has submitted all of following completed forms and information:

(1) Report of all acreage for the crop for the unit for which payments under this subpart are requested, on FSA–578, Report of Acreage, or in another format acceptable to FSA;

(2) AD–1026, Highly Erodible Land Conservation (HELC) and Wetland Conservation Certification; and

(3) For 2017 WHIP:

(i) FSA–891, Crop Insurance and/or NAP Coverage Agreement;

(ii) FSA–892, Request for an Exception to the WHIP Payment Limitation of $125,000, if the applicant is requesting 2017 WHIP payments in excess of the $125,000 payment limitation; and

(iii) FSA–893, 2018 Citrus Actual Production History and Approved Yield Record, Florida Only, for participants applying for payment for a citrus crop located in Florida;

(4) For WHIP+:

(i) FSA–895, Crop Insurance and/or NAP Coverage Agreement;

(ii) FSA–896, Request for an Exception to the WHIP Payment Limitation of $125,000, if 75 percent or more of an applicant's average AGI is attributable to activities related to farming, ranching, or forestry and the applicant wants to be eligible to receive WHIP+ payments of more than $125,000, up to the $250,000 payment limitation per crop year, with an overall WHIP+ limit of $500,000; and

(iii) FSA–897, Actual Production History and Approved Yield Record (WHIP+ Select Crops Only), for applicants requesting payments for select crops.

(e) Application approval and payment by FSA does not relieve a participant from having to submit any form required, but not filed, according to paragraph (d) of this section.

[83 FR 33801, July 18, 2018, as amended 84 FR 48530, Sept. 13, 2019; 86 FR 446, Jan. 6, 2021]

§760.1511 Calculating payments for yield-based crop losses.

(a) Payments made under this subpart to a participant for a loss to yield-based crops, including losses due to prevented planting, subject to §760.1514(i) and (j), are determined for a unit by:

(1) Multiplying the eligible acres by the 2017 WHIP yield in paragraph (c) of this section or the WHIP+ yield in paragraph (d) of this section by the price;

(2) Multiplying the result from paragraph (a)(1) of this section by the applicable 2017 WHIP factor or WHIP+ factor in paragraph (b) of this section;

(3) Multiplying the applicable production in paragraph (d) of this section by the price;

(4) Subtracting the result from paragraph (a)(3) of this section from the result of paragraph (a)(2) of this section;

(5) Multiplying the result from paragraph (a)(4) of this section by the participant's share in paragraph (e) of this section;

(6) Multiplying the result from paragraph (a)(5) of this section by the applicable payment factor in paragraph (g) of this section;

(7) Subtracting the amount of the gross insurance indemnity or NAP payment from the result from paragraph (a)(6) of this section;

(8) Subtracting the secondary use or salvage value of the crop from the result from paragraph (a)(7) of this section; and

(b) If the NAP or crop insurance coverage is at the coverage level listed in the first column, then the 2017 WHIP factor is listed in the second column, and the WHIP+ factor is listed in the third column:

TABLE 1 TO § 760.1511(b)

Coverage level	2017 WHIP factor (percent)	WHIP+ factor (percent)
(1) No crop insurance or No NAP coverage	65	70
(2) Catastrophic coverage	70	75
(3) More than catastrophic coverage but less than 55 percent	72.5	77.5
(4) At least 55 percent but less than 60 percent	75	80
(5) At least 60 percent but less than 65 percent	77.5	82.5
(6) At least 65 percent but less than 70 percent	80	85
(7) At least 70 percent but less than 75 percent	85	87.5
(8) At least 75 percent but less than 80 percent	90	92.5
(9) At least 80 percent	95	95

(c) The 2017 WHIP yield is:

(1) The producer's APH for insured crops under a crop insurance policy that has an associated yield and for NAP covered crops, excluding all crops located in Puerto Rico;

(2) The county expected yield for crops located in Puerto Rico and uninsured crops, excluding citrus crops located in Florida; or

(3) For uninsured citrus crops located in Florida:

(i) Determined based on information provided on FSA-893 and supported by evidence that meets the requirements of § 760.1513(c), or

(ii) If FSA-893 and supporting documentation are not submitted, the county expected yield.

(d) The WHIP+ yield is:

(1) The producer's APH for insured crops under a crop insurance policy that has an associated yield and for NAP covered crops, excluding all crops located in Puerto Rico;

(2) The county expected yield for crops located in Puerto Rico and uninsured crops, excluding select crops; or

(3) For select crops:

(i) Determined based on information provided on FSA-897 and supported by evidence that meets the requirements of § 760.1513(c), or

(ii) If FSA-897 and supporting documentation are not submitted, the county expected yield.

(e) The production used to calculate a payment under this subpart will be determined as specified in § 760.1513.

(f) The eligible participant's share of a payment under this subpart is based on the participant's ownership entitlement share of the crop or crop proceeds, or, if no crop was produced, the share of the crop the participant would have received if the crop had been produced. If the participant has no ownership share of the crop, the participant is ineligible for payment.

(g) Payment factors will be used to calculate payments for crops produced with significant and variable production and harvesting expenses that are not incurred because the crop acreage was prevented planted, or planted but not harvested, as determined by FSA.

The use of payment factors is based on whether the crop acreage was unharvested or prevented planted, not whether a participant actually incurs or does not incur expenses. Payment factors are generally applicable to all similarly situated participants and are not established in response to individual participants. Accordingly established payment factors are not appealable under parts 11 and 780 of this title. A crop that is intended for mechanical harvest, but subsequently grazed and not mechanically harvested, will have an unharvested payment factor applied.

(h) Production from all end uses of a multi-use crop will be calculated separately and summarized together.

[83 FR 33801, July 18, 2018, as amended 84 FR 48530, Sept. 13, 2019; 86 FR 446, Jan. 6, 2021]

§760.1512 Production losses; participant responsibility.

(a) For any record submitted along with the certification of production, the record must be either a verifiable or reliable record that substantiates the certification to the satisfaction of the FSA county committee. If the eligible crop was sold or otherwise disposed of through commercial channels, a record of that disposition must be provided to FSA with the certification.

(1) Acceptable production records include:

(i) RMA or NAP records, if accurate and complete;

(ii) Commercial receipts;

(iii) Settlement sheets;

(iv) Warehouse ledger sheets or load summaries; or

(v) Appraisal information from a loss adjuster acceptable to FSA.

(2) If the eligible crop was farm-stored, sold, fed to livestock, or disposed of by means other than verifiable commercial channels, acceptable records for these purposes include:

(i) Truck scale tickets;

(ii) Appraisal information from a loss adjuster acceptable to FSA;

(iii) Contemporaneous reliable diaries; or

(iv) Other documentary evidence, such as contemporaneous reliable measurements.

(3) Determinations of reliability with respect to this paragraph will take into account, as appropriate, the ability for FSA to review and verify or compare the evidence against the similarity of the evidence or reports or data received by FSA for the crop or similar crops. Other factors deemed relevant may also be taken into account.

(b) If RMA or NAP records are not available, or if the FSA county committee determines the RMA or NAP records as reported by the insured or covered participant appear to be questionable or incomplete, or if the FSA county committee makes inquiry, the participant is responsible for:

(1) Retaining and providing, at time of application and whenever required by FSA, the best available verifiable or reliable or other production records for the crop;

(2) Summarizing all the production evidence;

(3) Accounting for the total amount of unit production for the crop, whether or not records reflect this production;

(4) Providing the information in a manner that can be easily understood by the FSA county committee; and

(5) Providing supporting documentation if the FSA county committee has reason to question the disaster event or that all production has been taken into account.

(c) FSA may verify the production evidence submitted with records on file at the warehouse, gin, or other entity that received or may have received the reported production.

(d) Participants must provide all records for any production of a crop that is grown with an arrangement, agreement, or contract for guaranteed payment.

(e) Under WHIP+, participants requesting payments for losses to adulterated wine grapes must submit verifiable sales tickets that document that the reduced price received was due to adulteration due to a qualifying disaster event. For adulterated wine grapes that have not been sold, participants must submit verifiable records obtained by testing or analysis to establish that the wine grapes were adulterated due to a qualifying disaster event and the price they would receive due to adulteration.

[83 FR 33801, July 18, 2018, as amended 84 FR 48531, Sept. 13, 2019]

§ 760.1513 Determination of production.

(a) The harvested production of eligible crop acreage harvested more than once in a crop year includes the total harvested production from all the harvests in the crop year.

(b) If a crop is appraised and subsequently harvested as the intended use, the actual harvested production must be taken into account to determine payments. FSA will analyze and determine whether a participant's evidence of actual production represents all that could or would have been harvested.

(c) For all crops eligible for loan deficiency payments or marketing assistance loans (see parts 1421 and 1434 of this title) with an intended use of grain but harvested as silage, ensilage, cabbage, hay, cracked, rolled, or crimped, production will be converted to a whole grain equivalent based on conversion factors as previously established by FSA.

(d) If a participant does not receive compensation based upon the quantity of the commodity delivered to a purchaser, but has an agreement or contract for guaranteed payment for production, the determination of the production will be the greater of the actual production or the guaranteed payment converted to production as determined by FSA.

(e) Production that is commingled between crop years, units, ineligible and eligible acres, or different practices before it was a matter of record or combination of record and cannot be separated by using records or other means acceptable to FSA will be prorated to each respective year, unit, type of acreage, or practice, respectively. Commingled production may be attributed to the applicable unit, if the participant made the unit production of a commodity a matter of record before commingling and does any of the following, as applicable:

(1) Provides copies of verifiable documents showing that production of the commodity was purchased, acquired, or otherwise obtained from beyond the unit;

(2) Had the production measured in a manner acceptable to the FSA county committee; or

(3) Had the current year's production appraised in a manner acceptable to the FSA county committee.

(f) The FSA county committee will assign production for the unit when the FSA county committee determines that:

(1) The participant has failed to provide adequate and acceptable production records;

(2) The loss to the crop is because of a disaster condition not covered by this subpart, or circumstances other than natural disaster, and there has not otherwise been an accounting of this ineligible cause of loss;

(3) The participant carries out a practice, such as multiple cropping, that generally results in lower yields than the established historic yields;

(5) A crop was late-planted;

(6) Unharvested acreage was not timely appraised; or

(7) Other appropriate causes exist for such assignment as determined by the Deputy Administrator.

(g) The FSA county committee will establish a county disaster yield that reflects the amount of production producers would have produced considering the eligible disaster events in the county or area for the same crop. The county disaster yield for the county or area will be expressed as either a percent of loss or yield per acre. The county disaster yield will apply when:

(1) Unharvested acreage has not been appraised by FSA or a company reinsured by FCIC; or

(2) Acceptable production records for harvested acres are not available from any source.

(h) In no case will the production amount of any applicant be less than the producer's certified loss.

(i) Under WHIP+, production for eligible adulterated wine grapes will be adjusted for quality deficiencies due to a qualifying disaster event. Wine grapes are eligible for production adjustment only if adulteration occurred prior to harvest and as a result of a qualifying disaster event or as a result of a related condition (such as application of fire retardant). Losses due to all other causes of adulteration (such as addition of artificial flavoring or chemicals for economic purposes) are not eligible for WHIP+. Production will be eligible for quality adjustment if, due to a qualifying disaster event, it has a value of less than 75 percent of the average market price of undamaged grapes of the same or similar variety. The value per ton of the qualifying damaged production and the average market price of undamaged grapes will be determined on the earlier of the date the damaged production is sold or the date of final inspection for the unit. Grape production that is eligible for quality adjustment will be reduced by:

(1) Dividing the value per ton of the damaged grapes by the value per ton for undamaged grapes; and

(2) Multiplying this result (not to exceed 1.000) by the number of tons of the eligible damaged grapes.

[83 FR 33801, July 18, 2018, as amended 84 FR 48531, Sept. 13, 2019]

§ 760.1514 Eligible acres.

(a) Eligible acreage will be calculated using the lesser of the reported or determined acres shown to have been planted or prevented from being planted to a crop.

(b) Initial crop acreage will be the payment acreage for under this subpart, unless the provisions for subsequent crops in this section are met. Subsequently planted or prevented planted acre acreage is considered acreage for under this subpart only if the provisions of this section are met. All plantings of an annual or biennial crop are considered the same as a planting of an initial crop in tropical regions as defined in part 1437, subpart F, of this title.

(c) In cases where there is double cropped acreage, each crop may be included in the acreage only if the specific crops are approved by the FSA State committee as eligible double cropping practices in accordance with procedures approved by the Deputy Administrator.

(d) Except for insured crops, participants with double cropped acreage not meeting the criteria in paragraph (c) of this section may have such acreage included in the acreage on more than one crop only if the participant submits verifiable records establishing a history of carrying out a successful double cropping practice on the specific crops for which payment is requested.

(e) Participants having multiple plantings may receive payments for each planting included only if the planting meets the requirements of part 1437 of this title and all other provisions of this subpart are satisfied.

(f) Losses due to prevented planting are eligible under this subpart only if the loss was due to a qualifying disaster event. Provisions of parts 718 and 1437 of this title specifying what is considered prevented planting and how it must be documented and reported apply. Crops located in tropical regions are not eligible for prevented planting.

(g) Subject to the provisions of this subpart, the FSA county committee will:

(1) Use the most accurate data available when determining planted and prevented planted acres; and

(2) Disregard acreage of a crop produced on land that is not eligible for crop insurance or NAP.

(h) If a farm has a crop that has both FSA and RMA acreage for insured crops, eligible acres will be based on the lesser of RMA or FSA acres.

(i) For 2017 WHIP, prevented planting acres will be considered eligible acres if they meet all requirements of this subpart.

(j) For WHIP+:

(1) 2018 and 2020 crop year prevented planting acres and 2019 crop year uninsured and NAP-covered prevented planting acres will be eligible acres if they meet all requirements of this subpart; and

(2) 2019 crop year insured prevented planting acres will not be eligible acres.

[83 FR 33801, July 18, 2018, as amended 84 FR 48531, Sept. 13, 2019]

§ 760.1515 Calculating payments for value loss crops.

(a) Payments made under this subpart to a participant for a loss on a unit with respect to value loss crops are determined by:

(1) Multiplying the field market value of the crop immediately before the qualifying disaster event by the *2017* WHIP factor or WHIP+ factor specified in § 760.1511(b);

(2) Subtracting the sum of the field market value of the crop immediately after the qualifying disaster event and the value of the crop lost due to ineligible causes of loss from the result from paragraph (a)(1) of this section;

(3) Multiplying the result from paragraph (a)(2) of this section by the participant's share;

(4) Multiplying the result from paragraph (a)(3) of this section by the applicable payment factor;

(5) Subtracting the gross insurance indemnity or NAP payment from the result from paragraph (a)(4) of this section;

(6) Subtracting the secondary use or salvage value of the crop from the result from paragraph (a)(5) of this section; and

(7) Subtracting the amount of any payment for future economic losses received under the Florida Citrus Recovery Block Grant Program.

(b) In the case of an insurable value loss crop for which crop insurance provides for an adjustment in the guarantee, liability, or indemnity, such as in the case of inventory exceeding peak inventory value, the adjustment will be used in determining the payment under this subpart for the crop.

(c) In the case of a NAP eligible value loss crop for which NAP provides for an adjustment in the level of assistance, such as in the case of unharvested field grown inventory, the adjustment will be used in determining the payment for the crop.

[83 FR 33801, July 18, 2018, as amended 84 FR 48531, Sept. 13, 2019]

§ 760.1516 Calculating payments for tree, bush, and vine losses.

(a) Payments will be calculated separately based on the growth stage of the trees, bushes, or vines, as determined by the Deputy Administrator.

(b) Payments made under this subpart to a participant for a loss on a unit with respect to tree, bush, and vine losses are determined by:

(1) Multiplying the expected value (see paragraph (c) of this section) of the trees, bushes, or vines immediately before the qualifying disaster event by the *2017* WHIP factor or WHIP+ factor specified in § 760.1511(b);

(2) Subtracting the actual value (see paragraph (d) of this section) of the trees, bushes, or vines immediately after the qualifying disaster event from the result of paragraph (b)(1) of this section;

(3) Multiplying the result of paragraph (b)(2) of this section by the participant's share;

(4) Subtracting the amount of any insurance indemnity received from the result of paragraph (b)(3) of this section; and

(5) Subtracting the value of any secondary use or salvage value from the result of paragraph (b)(4) of this section.

(c) Expected value is determined by multiplying the total number of trees, bushes, or vines that were damaged or destroyed by a qualifying disaster event by the price.

(d) Actual value is determined by:

(1) Multiplying the number of trees, bushes, or vines damaged by a qualifying disaster event by the damage factor;

(2) Adding the result of paragraph (d)(1) of this section and the number of trees, bushes, or vines destroyed by a qualifying disaster event;

(3) Multiplying the result of paragraph (d)(2) of this section by the price; and

(4) Subtracting the result of paragraph (d)(3) of this section from the expected value from paragraph (c) of this section.

(e) The FSA county committee will adjust the number of damaged and destroyed trees, bushes, and vines, if it determines that the number of damaged or destroyed trees, bushes, or vines certified by the participant is inaccurate.

(f) Citrus trees located in Florida are ineligible for payment under 2017 WHIP.

[83 FR 33801, July 18, 2018, as amended 84 FR 48532, Sept. 13, 2019]

§ 760.1517 Requirement to purchase crop insurance or NAP coverage.

(a) For the first 2 consecutive crop years for which crop insurance or NAP coverage is available after the enrollment period for 2017 WHIP or WHIP+ ends, subject to paragraph (c) of this section, a participant who receives payment under this subpart for a crop loss in a county must obtain:

(1) For an insurable crop, crop insurance with at least a 60 percent coverage level for that crop in that county; or

(2) For a NAP eligible crop:

(i) NAP coverage with a coverage level of 60 percent, if available for the applicable crop year, or NAP catastrophic coverage if NAP coverage is not offered at a 60 percent coverage level for that crop year.

(ii) Participants who exceed the average adjusted gross income limitation for NAP payment eligibility[1] for the applicable crop year may meet the purchase requirement specified in paragraph (a)(2)(i) of this section by purchasing Whole-Farm Revenue Protection crop insurance coverage, if eligible, or paying the NAP service fee and premium even though the participant will not be eligible to receive a NAP payment due to the average adjusted gross income limit but will be eligible for the WHIP payment.

(b) For the first 2 consecutive insurance years for which crop insurance is available after the enrollment period for 2017 WHIP ends, subject to paragraph (c) of this section, any participant who receives 2017 WHIP payments for a tree, bush, or vine loss must purchase a plan of insurance for the tree, bush, or vine with at least a 60 percent coverage level.

(c) The final crop year to purchase crop insurance or NAP coverage to meet the requirements of paragraphs (a) and (b) of this section is the:

(1) 2021 crop year for 2017 WHIP payment eligibility, except as provided in paragraph (c)(2) of this section;

(2) 2023 crop year for:

(i) WHIP+ payment eligibility; and

(ii) 2017 WHIP payment eligibility for losses due to Tropical Storm Cindy, losses of peach and blueberry crops in calendar year 2017 due to extreme cold, and blueberry productivity losses in calendar year 2018 due to extreme cold

and hurricane damage in calendar year 2017.

(d) If a producer fails to obtain crop insurance or NAP coverage as required in paragraphs (a) and (b) of this section, the producer must reimburse FSA for the full amount of 2017 WHIP payment or WHIP+ payment plus interest that the producer received for that crop, tree, bush, or vine loss. A producer will only be considered to have obtained NAP coverage for the purposes of this section if the participant applied and payed the requisite NAP service fee and paid any applicable premium by the applicable deadline and completed all program requirements, including filing an acreage report as may be required under such coverage agreement.

[83 FR 33801, July 18, 2018, as amended 84 FR 48532, Sept. 13, 2019]

Subpart P—On-Farm Storage Loss Program

SOURCE: 84 FR 48532, Sept. 13, 2019, unless otherwise noted.

§ 760.1600 Applicability.

(a) This subpart specifies the terms and conditions for the On-Farm Storage Loss Program. The On-Farm Storage Loss Program will provide payments to eligible producers who suffered uncompensated losses of harvested commodities stored in on farm structures as a result from hurricanes, floods, tornadoes, typhoons, volcanic activity, snowstorms, and wildfires that occurred in the 2018 and 2019 calendar years.

(b) The regulations in this subpart are applicable to crops of barley, small and large chickpeas, corn, grain sorghum, lentils, oats, dry peas, peanuts, rice, wheat, soybeans, oilseeds, hay and other crops designated by Commodity Credit Corporation (CCC) stored in on-farm structures. These regulations specify the general provisions under which the On-Farm Storage Loss Program will be administered by CCC. In any case in which money must be refunded to CCC in connection with this part, interest will be due to run from the date of disbursement of the sum to be refunded. This provision will apply,

---

[1] See §§ 1400.500(a) and 1400.1(a)(4) of this title.

unless waived by the Deputy Administrator, irrespective of any other rule.

(c) Eligible on-farm structures include all on-farm structures deemed acceptable by the Deputy Administrator for Farm Programs.

(d) Adjusted Gross Income (AGI) and payment limitation provisions specified in part 760.1607 of this chapter apply to this subpart.

§ 760.1601 Administration.

(a) The On-Farm Storage Loss Program will be administered under the general supervision of the Executive Vice President, CCC and will be carried out in the field by FSA State and county committees, respectively.

(b) State and county committees, and representatives and their employees, do not have authority to modify or waive any of the provisions of the regulations, except as provided in paragraph (e) of this section.

(c) The FSA State committee will take any required action not taken by the FSA county committee. The FSA State committee will also:

(1) Correct or require correction of an action taken by a county committee that is not in compliance with this part; or

(2) Require a county committee to not take an action or implement a decision that is not under the regulations of this part.

(d) The Executive Vice President, CCC, or a designee, may determine any question arising under these programs, or reverse or modify a determination made by a State or county committee.

(e) The Deputy Administrator for Farm Programs, FSA, may authorize State and county committees to waive or modify non-statutory deadlines and other program requirements in cases where lateness or failure to meet such other requirements does not adversely affect the operation of the On-Farm Storage Loss Program.

(f) A representative of CCC may execute applications and related documents only under the terms and conditions determined and announced by CCC. Any document not executed under such terms and conditions, including any purported execution before the date authorized by CCC, will be null and void.

(g) Items of general applicability to program participants, including, but not limited to, application periods, application deadlines, internal operating guidelines issued to State and county offices, prices, and payment factors established by the On-Farm Storage Loss Program, are not subject to appeal.

§ 760.1602 Definitions.

The definitions in this section apply for all purposes of program administration. Terms defined in §§ 760.1502 and 760.1421 of this chapter also apply, except where they conflict with the definitions in this section.

Administrative County Office is the FSA County Office where a producer's FSA records are maintained.

CCC means the Commodity Credit Corporation.

COC means the FSA county committee.

Covered commodity means wheat, oats, and barley (including wheat, oats, and barley used for haying), corn, grain sorghum, long grain rice, medium grain rice, seed cotton, pulse crops, soybeans, other oilseeds, and peanuts as specified in 7 CFR 1412 and produced and mechanically harvested in the United States.

Crop means with respect to a year, commodities harvested in that year. Therefore, the referenced crop year of a commodity means commodities that when planted were intended for harvest in that calendar year.

Crop year means the relevant contract or application year. For example, the 2014 crop year is the year that runs from October 1, 2013, through September 30, 2014, and references to payments for that year refer to payments made under contracts or applications with the compliance year that runs during those dates.

FSA means the Farm Service Agency of the United States Department of Agriculture.

Oilseeds means any crop of sunflower seed, canola, rapeseed, safflower, flaxseed, mustard seed, crambe, sesame seed, and other oilseeds as designated by CCC or the Secretary.

Qualifying disaster event means a hurricane, flood, tornado, typhoon, volcanic activity, snowstorm, or wildfire or related condition that occurred in the 2018 or 2019 calendar year.

Recording FSA County Office is the FSA County Office that records eligibility data for producers designated as multi-county producers.

Related condition means damaging weather or an adverse natural occurrence that occurred as a direct result of a hurricane or wildfire qualifying disaster event, such as excessive rain, high winds, flooding, mudslides, and heavy smoke.

Secretary means the Secretary of the United States Department of Agriculture, or the Secretary's delegate.

STC means the FSA State committee.

§760.1603 Eligible producers.

(a) To be an eligible producer, the producer must:

(1) Be a person, partnership, association, corporation, estate, trust, or other legal entity that produces an eligible commodity as a landowner, landlord, tenant, or sharecropper, or in the case of rice, furnishes land, labor, water, or equipment for a share of the rice crop.

(2) Comply with all provisions of this part and, as applicable:

(i) 7 CFR part 12—Highly Erodible Land and Wetland Conservation;

(ii) 7 CFR part 707—Payments Due Persons Who Have Died, Disappeared, or Have Been Declared Incompetent;

(iii) 7 CFR part 718—Provisions Applicable to Multiple Programs;

(v) 7 CFR part 1400—Payment Limitation & Payment Eligibility; and

(vii) 7 CFR part 1403—Debt Settlement Policies and Procedures.

(b) A receiver or trustee of an insolvent or bankrupt debtor's estate, an executor or an administrator of a deceased person's estate, a guardian of an estate of a ward or an incompetent person, and trustees of a trust is considered to represent the insolvent or bankrupt debtor, the deceased person, the ward or incompetent, and the beneficiaries of a trust, respectively. The production of the receiver, executor, administrator, guardian, or trustee is considered to be the production of the person or estate represented by the receiver, executor, administrator, guardian, or trustee. On-Farm Storage Loss

Program documents executed by any such person will be accepted by CCC only if they are legally valid and such person has the authority to sign the applicable documents.

(c) A minor who is otherwise an eligible producer is eligible to receive a program payment only if the minor meets one of the following requirements:

(1) The right of majority has been conferred on the minor by court proceedings or by statute;

(2) A guardian has been appointed to manage the minor's property and the applicable program documents are signed by the guardian;

(3) Any program application signed by the minor is cosigned by a person determined by the FSA county committee to be financially responsible; or

(e) A producer must meet the requirements of actively engaged in farming, cash rent tenant, and member contribution as specified in 7 CFR part 1400 to be eligible for program payments.

§760.1604 Eligible commodities.

(a) Commodities eligible to be compensated for loss made under this part are:

(1) Covered Commodities;

(2) Hay; and

(3) Stored in an on-farm structure that under normal circumstances, would have maintained the quality of the commodity throughout harvest until marketing or feed if not for the qualifying weather event.

(b) A commodity produced on land owned or otherwise in the possession of the United States that is occupied without the consent of the United States is not an eligible commodity.

§760.1605 Miscellaneous provisions.

(a) All persons with a financial interest in the legal entity receiving payments under this subpart are jointly and severally liable for any refund, including related charges, which is determined to be due to FSA for any reason.

(b) In the event that any application for payment under this subpart resulted from erroneous information or a miscalculation, the payment will be recalculated and any excess refunded to FSA with interest to be calculated from the date of the disbursement.

(c) Any payment to any participant under this subpart will be made without regard to questions of title under State law, and without regard to any claim or lien against the commodity, or proceeds, in favor of the owner or any other creditor except agencies of the U.S. Government. The regulations governing offsets and withholdings in part 792 of this chapter apply to payments made under this subpart.

(d) Any participant entitled to any payment may assign any payment(s) in accordance with regulations governing the assignment of payments in part 792 of this chapter.

(e) The regulations in 7 CFR parts 11 and 780 apply to determinations under this subpart.

§ 760.1606 General provisions.

Losses will be determined total production in storage at time of loss. Eligibility and payments will be based on physical location of storage. Payments will be made on commodities that were completely lost or destroyed while in storage due to the qualifying weather related event.

§ 760.1607 Availability of funds and timing of payments.

For the On-Farm Storage Loss Program, payments will be issued as applications are approved.

§ 760.1608 Payment limitation and AGI.

(a) Per loss year, a person or legal entity, other than a joint venture or general partnership, is eligible to receive, directly or indirectly payments of not more than $125,000.

(b) The direct attribution provisions in § 760.1507 of this part apply for payment limitation as defined and used in this rule.

§ 760.1609 Qualifying disaster events.

(a) The On-Farm Storage Loss Program will provide a payment to eligible producers who suffered losses of harvested commodities while such commodities were stored in on farm structures as a result from hurricanes, floods, tornadoes, typhoons, volcanic activity, snowstorms, and wildfires that occurred in the 2018 and 2019 calendar years.

(b) For a loss due to or related to an event specified in paragraph (a) of this section, the loss must have occurred on acreage that was physically located in a county that received a:

(1) Presidential Emergency Disaster Declaration authorizing public assistance for categories C through G or individual assistance due to a hurricane occurring in the 2018 or 2019 calendar year; or

(2) Secretarial Disaster Designation for a hurricane occurring in the 2018 or 2019 calendar year.

(c) A producer with a loss not located in a physical location county that was eligible under paragraph (b)(1) of this section will be eligible for a program payment for losses due to hurricane and related conditions only if the producer provides supporting documentation that is acceptable to FSA from which the FSA county committee determines that the loss of the commodity was reasonably related to a qualifying disaster event as specified in this subpart and meets all other eligibility conditions. Supporting documentation may include furnishing climatological data from a reputable source or other information substantiating the claim of loss due to a qualifying disaster event.

(d) For a loss due to wildfires and conditions related to wildfire in the 2018 or 2019 calendar year, all counties where wildfires occurred, as determined by FSA county committees, are eligible program payments; a Presidential Emergency Disaster Declaration or Secretarial Disaster Designation for wildfire is not required. The loss must be reasonably related to wildfire and conditions related to wildfire, as specified in this subpart's definition of qualifying disaster event.

(e) For a loss due to floods, tornadoes, typhoons, volcanic activity, snowstorms or any other directly related weather disaster event, the loss must be reasonably related to the disaster event as specified in this subpart's definition of qualifying disaster event.

§ 760.1610 Eligible and ineligible losses.

(a) Except as provided in paragraphs (b) of this section, to be eligible for payments under this subpart the commodity stored in an eligible structure must have suffered a loss due to a qualifying disaster event.

(b) A loss will not be eligible for the On-Farm Storage Loss Program this subpart if any of the following apply:

(1) The cause of loss is determined by FSA to be the result of poor management decisions, poor farming practices, or previously damaged structures;

(2) The cause of loss was due to failure of the participant to store the commodity in an eligible structure before the qualifying disaster event; or

(3) The cause of loss was due to water contained or released by any governmental, public, or private dam or reservoir project if an easement exists on the acreage affected by the containment or release of the water.

(c) The following types of loss, regardless of whether they were the result of an eligible disaster event, are not eligible losses:

(1) Losses to crops that have not been harvested.

(2) Losses to crops not intended for harvest;

(4) Losses caused by improper storage;

(5) Losses caused by the application of chemicals; and

(6) Losses caused by theft.

§760.1611 Application for payment.

(a) An application for payment under this subpart must be submitted to the FSA county office serving as the farm's administrative county office by the close of business on a date that will be announced by the Deputy Administrator.

(b) Once signed by a producer, the application for payment is considered to contain information and certifications of and pertaining to the producer regardless of who entered the information on the application.

(c) The producer applying for the On-Farm Storage Loss Program under this subpart certifies the accuracy and truthfulness of the information provided in the application as well as any documentation filed with or in support of the application. All information is subject to verification or spot check by FSA at any time, either before or after payment is issued. Refusal to allow

FSA or any agency of the Department of Agriculture to verify any information provided will result in the participant's forfeiting eligibility for this program. FSA may at any time, including before, during, or after processing and paying an application, require the producer to submit any additional information necessary to implement or determine any eligibility provision of this subpart. Furnishing required information is voluntary; however, without it FSA is under no obligation to act on the application or approve payment. Providing a false certification will result in ineligibility and can also be punishable by imprisonment, fines, and other penalties.

(d) The application submitted in accordance with paragraph (a) of this section is not considered valid and complete for issuance of payment under this subpart unless FSA determines all the applicable eligibility provisions have been satisfied and the participant has submitted all required documentation.

(e) Application approval and payment by FSA does not relieve a participant from having to submit any form required, but not filed.

§760.1612 Calculating payments on-farm storage losses.

(a) Payments made under this subpart to a participant for loss of stored commodities are calculated, except hay or silage, by:

(1) Multiplying the eligible quantity of the eligible commodity by the RMA determined price;

(2) Multiplying the result from paragraph (a)(1) of this section by a 75 percent factor.

(b) Payments made under this subpart to a participant for loss of stored hay or silage, by:

(1) Multiplying the eligible quantity of the eligible commodity by a price as determined by the Secretary;

(2) Multiplying the result from paragraph (b)(1) of this section by a 75 percent factor.

Subpart Q—Milk Loss Program

SOURCE: 84 FR 48534, Sept. 13, 2019, unless otherwise noted.

§ 760.1700 Applicability

This subpart specified the terms and conditions for the Milk Loss Program. The Milk Loss Program will provide payments to dairy operations for milk that was dumped or removed without compensation from the commercial milk market due to the results from hurricanes, floods, tornadoes, typhoons, volcanic activity, snowstorms, and wildfires that occurred in the 2018 and 2019 calendar year.

§ 760.1701 Administration.

This milk loss payment program will be carried out by FSA under the direction and supervision of the Deputy Administrator. In the field, the program will be administered by the State and county committees.

§ 760.1702 Definitions.

The following definitions apply to the Milk Loss Program.

Affected farmer means a person who produces whole milk which is removed from the commercial market any time or who produces but was unable to deliver milk to a commercial market as a result of a qualifying event limited to:

(1) Weather-related event prevented transportation of the milk,

(2) Weather-related event caused a power outage or structural damage causing milk to be unmerchantable.

Application period means any period during calendar year 2018 and 2019 which an affected farmer's whole milk is dumped or removed without compensation from the commercial market due to a qualified disaster event for which application for payment is made.

Base period means the calendar month or 4-week period immediately preceding when the producer was unable to deliver milk to a commercial market as a result of a qualifying disaster event.

Claim period means the calendar month, or months, in which milk was dumped or removed and usually is the calendar month immediately following the base period.

Commercial market means:

(1) The market to which the affected farmer normally delivers his whole milk and from which it was removed; or

(2) The market to which the affected manufacturer normally delivers his dairy products and from which they were removed.

County committee means the FSA county committee.

Deputy Administrator means the Deputy Administrator for Farm Programs, FSA.

FSA means the Farm Service Agency, U.S. Department of Agriculture.

Milk handler means the marketing agency to or through which the affected dairy farmer marketed his whole milk at the time he dumped milk or was unable to deliver milk to the commercial market due to a qualifying weather related event.

Pay period means:

(1) In the case of an affected farmer who markets his whole milk through a milk handler, the period used by the milk handler in settling with the affected farmer for his whole milk, usually biweekly or monthly; or

(2) In the case of an affected farmer whose commercial market consists of direct retail sales to consumers, a calendar month.

Payment subject to refund means a payment which is made by a milk handler to an affected farmer, and which such farmer is obligated to refund to the milk handler.

Person means an individual, partnership, association, corporation, trust, estate, or other legal entity.

Qualifying disaster event means a hurricane, flood, tornado, typhoon, volcanic activity, snowstorm, or wildfire or related condition that occurred in the 2018 or 2019 calendar year.

Removed from the commercial market means:

(1) Produced and destroyed or fed to livestock;

(2) Produced and delivered to a handler who destroyed it or disposed of it as salvage (such as separating whole milk, destroying the fat, and drying the skim milk); or

(3) Produced and otherwise diverted to other than the commercial market.

Same loss means the event or trigger that caused the milk to be removed from the commercial market.

Secretary means the Secretary of Agriculture of the United States or any officer or employee of the U.S. Department of Agriculture to whom the Secretary delegates authority to act as the Secretary.

State committee means the FSA State committee.

Whole milk means milk as it is produced by cows.

§ 760.1703 Payments to dairy farmers for milk.

A milk loss payment may be made to an affected farmer who is determined by the FSA county committee to be in compliance with all the terms and conditions of this subpart in the amount equal to 75 percent of the fair market value of the farmer's normal marketings for the application period, less:

(a) Any amount he received for whole milk marketed during the applications period; and

(b) Any payment not subject to refund which he received from a milk handler with respect to whole milk removed from the commercial market during the application period.

§ 760.1704 Normal marketings of milk.

(a) The FSA county committee will determine the affected farmer's dumped milk normal marketings which, for the purposes of this subpart, will be the sum of the quantities of whole milk for which the farmer would have sold in the commercial market in each of the pay periods in the application period be it not for the removal of his whole milk from the commercial market as a result of a qualifying disaster event.

(b) Normal marketings for each pay period are based on the average daily production during the base period.

(c) Normal marketings determined in paragraph (b) of this section are adjusted for any change in the daily average number of cows milked during each pay period the milk is off the market compared with the average number of cows milked daily during the base period.

(d) If only a portion of a pay period falls within the application period, normal marketings for such pay period will be reduced so that they represent only that part of such pay period which is within the application period.

§ 760.1705 Fair market value of milk.

(a) The FSA county committee will determine the fair market value of the affected farmer's dumped milk normal marketings, which, for the purposes of this subpart, will be the sum of the net proceeds such farmer would have received for his normal marketings in each of the pay periods in the application period but for the qualifying disaster event.

(b) The FSA county committee will determine the net proceeds the affected farmer would have received in each of the pay periods in the application period:

(1) In the case of an affected farmer who markets his whole milk through a milk handler, by multiplying the affected farmer's normal marketings for each such pay period by the average net price per hundred-weight of whole milk paid during the pay period by such farmer's milk handler in the same area for whole milk similar in quality and butterfat test to that marketed by the affected farmer in the base period used to determine his normal marketings; or

(2) In the case of an affected farmer whose commercial market consists of direct retail sales to consumers, by multiplying the affected farmer's normal marketings for each such pay period by the average net price per hundredweight of whole milk, as determined by the FSA county committee, which other producers in the same area who marketed their whole milk through milk handlers received for whole milk similar in quality and butterfat test to that marketed by the affected farmer during the base period used to determine his normal marketings.

(c) In determining the net price for whole milk, the FSA county committee will deduct from the gross price any transportation, administrative, and other costs of marketing which it determines are normally incurred by the affected farmer but which were not incurred because of the removal of his whole milk from the commercial market.

§ 760.1706 Information to be furnished.

The affected farmer must furnish to the FSA county committee complete and accurate information sufficient to enable the FSA county committee or the Deputy Administrator to make the determinations required in this subpart. Such information must include, but is not limited to:

(a) A copy of the notice from, or other evidence of action by, the public agency which resulted in the dumping or removal of the affected farmer's whole milk from the commercial market.

(b) The specific weather or disaster event and its results on milk marketing for the loss period.

(c) The quantity and butterfat test of whole milk produced and marketed during the base period. This information must be a certified statement from the affected farmer's milk handler or any other evidence the FSA county committee accepts as an accurate record of milk production and butterfat tests during the base period.

(d) The average number of dry cows, bred heifers, and cows milked during the base period and during each pay period in the application.

(e) If the affected farmer markets his whole milk through a milk handler, a statement from the milk handler showing, for each pay period in the application period, the average price per hundred-weight of whole milk similar in quality to that marketed by the affected farmer during the base period used to determine his normal marketings. If the milk handler has information as to the transportation, administrative, and other costs of marketing which are normally incurred by producers who market through the milk handler but which the affected farmer did not incur because of the dumping or removal of his whole milk from the market, the average price stated by the milk handler will be the average gross price paid producers less any such costs. If the milk handler does not have such information, the affected farmer will furnish a statement setting forth such costs, if any.

(f) The amount of proceeds, if any, received by the affected farmer from the marketing of whole milk produced during the application period.

(g) The amount of any payments not subject to refund made to the affected farmer by the milk handler with respect to the whole milk produced during the application period and remove from the commercial market.

(h) Such other information as the FSA county committee may request to enable the FSA county committee or the Deputy Administrator to make the determinations required in this subpart.

§ 760.1707 Application for payments for milk loss.

(a) The affected farmer or his legal representative must sign and file an application for payment on a form which is approved for that purpose by the Deputy Administrator. The form must be filed with the county FSA office for the county where the farm headquarters are located no later than 60 days after the designated deadline announced by the Secretary for 2018 and 2019 losses.

(b) The application for payment will cover application periods of at least 30 days, except that, if the entire application period, or the last application period, is shorter than 30 days, applications for payment may be filed for such shorter period. The application for payment must be accompanied by the information required for the Milk Loss Program as any other information which will enable the FSA county committee to determine whether the making of this payment is precluded for any of the reasons as determined ineligible by the Deputy Administrator.

§ 760.1708 Payment limitation and AGI.

(a) Per loss year, a person or legal entity, other than a joint venture or general partnership, is eligible to receive, directly or indirectly payments of not more than $125,000.

(b) The direct attribution provisions in § 760.1507 apply for payment limitation as defined and used in this subpart.

§ 760.1709 Limitation of authority.

(a) FSA county executive directors and State and county committees do not have authority to modify or waive any of the provisions of the regulations in this subpart.

(b) The FSA State committee may take any action authorized or required by the regulations in this subpart to be taken by the FSA county committee when such action has not been taken by the FSA county committee. The FSA State committee may also:

(1) Correct, or require a county committee to correct, any action taken by such county committee which is not in accordance with the regulations in this subpart; or

(2) Require a county committee to withhold taking any action which is not in accordance with the regulations in this subpart.

(c) No delegation herein to a State or county committee will preclude the Deputy Administrator or his designee from determining any question arising under the regulations in this subpart or from reversing or modifying any determination made by a State or county committee.

§ 760.1710 Estates and trusts; minors.

(a) A receiver of an insolvent debtor's estate and the trustee of a trust estate will, for the purpose of this subpart, be considered to represent an insolvent affected farmer or manufacturer and the beneficiaries of a trust, respectively, and the production of the receiver or trustee will be considered to be the production of the person or manufacturer he represents. Program documents executed by any such person will be accepted only if they are legally valid and such person has the authority to sign the applicable documents.

(b) An affected dairy farmer or manufacturer who is a minor will be eligible for milk loss payments only if he meets one of the following requirements:

(1) The right of majority has been conferred on him by court proceedings or by statute;

(2) A guardian has been appointed to manage his property and the applicable program documents are signed by the guardian; or

(3) A bond is furnished under which the surety guarantees any loss incurred for which the minor would be liable had he been an adult.

§ 760.1711 Setoffs.

(a) If the affected farmer or manufacturer is indebted to any agency of the United States and such indebtedness is listed on the county debt record, milk loss payments due the affected farmer the regulations in this part will be applied, as provided in the Secretary's setoff regulations, 7 CFR part 13, to such indebtedness.

(b) Compliance with the provisions of this section will not deprive the affected farmer of any right he would otherwise have to contest the justness of the indebtedness involved in the setoff action, either by administrative appeal or by legal action.

§ 760.1712 Overdisbursement.

If the milk loss payment disbursed to an affected farmer exceeds the amount authorized under the regulations in this subpart, the affected farmer or manufacturer will be personally liable for repayment of the amount of such excess.

§ 760.1713 Death, incompetency, or disappearance.

In the case of the death, incompetency, or disappearance of any affected farmer who would otherwise receive a milk loss payment, such payment may be made to the person or persons specified in the regulations contained in part 707 of this chapter. The person requesting such payment must file Form FSA–325, "Application for Payment of Amounts Due Persons Who Have Died, Disappeared, or Have Been Declared Incompetent," as provided in that part.

§ 760.1714 Records and inspection of records.

(a) The affected farmer, as well as his milk handler and any other person who furnished information to such farmer or to the FSA county committee for the purpose of enabling such farmer to receive a milk loss payment under this subpart, must maintain any existing books, records, and accounts supporting any information so furnished for 3 years following the end of the year during which the application for payment was filed.

(b) The affected farmer, his milk handler, and any other person who furnishes such information to the affected

185

farmer or to the FSA county committee must permit authorized representatives of the Department of Agriculture and the General Accounting Office, during regular business hours, to inspect, examine, and make copies of such books, records, and accounts.

§ 760.1715 Assignment.

No assignment will be made of any milk loss payment due or to come due under the regulations in this subpart. Any assignment or attempted assignment of any indemnity payment due or to come due under this subpart will be null and void.

§ 760.1716 Instructions and forms.

Affected farmers may obtain information necessary to make application for a milk loss payment from the county FSA office.

§ 760.1717 Availability of funds.

Milk loss program payments will be made on a first-come, first-served basis. Applications received after all funds are used will not be paid.

§ 760.1718 Calculating payments for milk losses.

(a) Payments made under this subpart to a participant for loss of milk as a result of a qualifying disaster event are calculated as follows:

(1) Amount of the fair market value of the farmer's normal marketings for the application period; less

(2) Any amount the farmer received for whole milk marketed during the applications period; and

(3) Any payment not subject to refund which the farmer received from a milk handler with respect to whole milk removed from the commercial market during the application period;

(4) Multiplied by a program factor of 75 percent.

(b) [Reserved]

Subpart R—Quality Loss Adjustment Program

SOURCE: 86 FR 446, Jan. 6, 2021, unless otherwise noted.

§ 760.1800 Applicability.

This subpart specifies the terms and conditions for the Quality Loss Adjustment (QLA) Program. The QLA Program provides disaster assistance for crop quality losses that were a consequence of hurricanes, excessive moisture, floods, qualifying drought, tornadoes, typhoons, volcanic activity, snowstorms, and wildfires occurring in calendar years 2018 and 2019.

§ 760.1801 Administration.

(a) The QLA Program is administered under the general supervision of the Administrator, Farm Service Agency (FSA), and the Deputy Administrator for Farm Programs, FSA. The QLA Program is carried out by FSA State and county committees with instructions issued by the Deputy Administrator.

(b) FSA State and county committees, and representatives and their employees, do not have authority to modify or waive any of the provisions of the regulations in this subpart or instructions issued by the Deputy Administrator.

(c) The FSA State committee will take any action required by the regulations in this subpart that the FSA county committee has not taken. The FSA State committee will also:

(1) Correct, or require an FSA county committee to correct, any action taken by the FSA county committee that is not in accordance with the regulations in this subpart; or

(2) Require an FSA county committee to withhold taking any action that is not in accordance with this subpart.

(d) No delegation to an FSA State or county committee precludes the FSA Administrator or the Deputy Administrator from determining any question arising under this subpart or from reversing or modifying any determination made by an FSA State or county committee.

(e) The Deputy Administrator has the authority to:

(1) Permit State and county committees to waive or modify a non-statutory deadline specified in this subpart; and

(2) Delegate authority to FSA State or county committees to make determinations under § 760.1812(f) and (g).

(f) Items of general applicability to program participants, including, but not limited to, application periods, application deadlines, internal operating guidelines issued to FSA State and county offices, prices, and payment factors established under this subpart, are not subject to appeal in accordance with part 780 of this chapter.

§ 760.1802 Definitions.

The following definitions apply to this subpart. The definitions in §§ 718.2 and 1400.3 of this title also apply, except where they conflict with the definitions in this section. In the event of conflict, the definitions in this section apply.

Affected production means the producer's ownership share of harvested production of an eligible crop, adjusted to standard moisture as established by the U.S. Grains Standards Act, a State regulatory agency, or industry standard, that had both:

(1) A quality loss due to a qualifying disaster event; and

(2) At least a 5 percent quality loss due to all eligible disaster events.

Average market price means the average market price determined according to § 1437.12 of this title.

Coverage level means the percentage determined by multiplying the elected yield percentage under a crop insurance policy or NAP coverage by the elected price percentage.

Crop insurance means an insurance policy reinsured by FCIC under the provisions of the Federal Crop Insurance Act, as amended. It does not include private plans of insurance.

Crop insurance indemnity means, for the purpose of this subpart, the payment to a participant for crop losses covered under crop insurance administered by RMA in accordance with the Federal Crop Insurance Act (7 U.S.C. 1501–1524).

Crop year means:

(1) For insurable crops, the crop year as defined according to the applicable crop insurance policy; and

(2) For NAP-eligible crops, the crop year as defined in § 1437.3 of this title.

Eligible crop means a crop for which coverage was available either from FCIC under part 400 of this title, or through NAP under § 1437.4 of this title.

Eligible disaster event means a disaster event that is an eligible cause of loss specified in § 1437.10 of this title, excluding insect infestation.

FCIC means the Federal Crop Insurance Corporation, a wholly owned Government Corporation of USDA, administered by RMA.

FSA means the Farm Service Agency, an agency of USDA.

Grading factor means a factor that describes the physical condition or a feature that is evaluated to determine the quality of the production, such as broken kernels and low-test weight.

Good farming practices means the cultural practices generally recognized as compatible with agronomic and weather conditions and used for the crop to make normal progress toward maturity, as determined by FSA. These practices are:

(1) For conventional farming practices, those generally recognized by agricultural experts for the area, which could include one or more counties; or

(2) For organic farming practices, those generally recognized by the organic agricultural experts for the area or contained in the organic system plan that is in accordance with the National Organic Program specified in part 205 of this title.

Harvested means:

(1) For insurable crops, harvested as defined according to the applicable crop insurance policy;

(2) For NAP-eligible single harvest crops, that a crop has been removed from the field, either by hand or mechanically;

(3) For NAP-eligible crops with potential multiple harvests in 1 year or harvested over multiple years, that the producer has, by hand or mechanically, removed at least 1 mature crop from the field during the crop year; and

(4) For mechanically harvested NAP-eligible crops, that the crop has been removed from the field and placed in a truck or other conveyance, except hay is considered harvested when in the bale, whether removed from the field or not.

Insurable crop means an agricultural crop (excluding livestock) for which the producer on a farm is eligible to obtain a policy or plan of insurance under the Federal Crop Insurance Act (7 U.S.C. 1501–1524).

Multiple market crop means a crop that is delivered to a single market but can have fresh and processed prices based on grading. For example, a producer may intend to sell all production of an apple crop as fresh production; however, based on grading of the crop at the market, the producer is compensated for some production at the fresh price and for some production at the processing price.

Multiple planting means the planting for harvest of the same crop in more than one planting period in a crop year on different acreage.

NAP means the Noninsured Crop Disaster Assistance Program under section 196 of the Federal Agriculture Improvement and Reform Act of 1996 (7 U.S.C. 7333) and part 1437 of this title.

NAP-eligible crop means an agricultural crop for which the producer on a farm is eligible to obtain NAP coverage.

NAP service fee means the amount specified in § 1437.7 of this title that the producer must pay to obtain NAP coverage.

Nutrient factor means a factor determined by a test that measures the nutrient value of a crop to be fed to livestock. Examples include, but are not limited to, relative feed value and total digestible nutrients.

Production means quantity of the crop produced, which is expressed in a specific unit of measure including, but not limited to, bushels or pounds.

QLA Program means the Quality Loss Adjustment Program.

Qualifying disaster event means a hurricane, flood, tornado, typhoon, volcanic activity, snowstorm, wildfire, excessive moisture, qualifying drought, or a related condition that occurred in the 2018 or 2019 calendar year.

Qualifying drought means an area within the county was rated by the U.S. Drought Monitor as having a D3 (extreme drought) or higher level of drought intensity during the applicable calendar year.

Quality loss means:

(1) For forage crops, a reduction in an applicable nutrient factor for the crop; and

(2) For crops other than forage, a reduction in the total dollar value of the crop due to reduction in the physical condition of the crop indicated by an applicable grading factor.

Related condition means damaging weather or an adverse natural occurrence that occurred as a direct result of a specified qualifying disaster event, such as excessive rain, high winds, flooding, mudslides, and heavy smoke, as determined by the Deputy Administrator. The term does not include insect infestation.

Reliable production record means evidence provided by the participant that is used to substantiate the amount of production reported when verifiable records are not available, including copies of receipts, ledgers of income, income statements of deposit slips, register tapes, invoices for custom harvesting, and records to verify production costs, contemporaneous measurements, truck scale tickets, and contemporaneous diaries that are determined acceptable by the FSA county committee. To determine whether the records are acceptable, the FSA county committee will consider whether they are consistent with the records of other producers of the crop in that area.

RMA means the Risk Management Agency, an agency of USDA.

Salvage value means the dollar amount or equivalent for the quantity of the commodity that cannot be marketed or sold in any recognized market for the crop.

Secondary use means the harvesting of a crop for a use other than the intended use.

Unit of measure means:

(1) For insurable crops, the FCIC-established unit of measure; and

(2) For NAP-eligible crops, the established unit of measure used for the NAP price and yield.

USDA means the U.S. Department of Agriculture.

U.S. Drought Monitor is a system for classifying drought severity according to a range of abnormally dry to exceptional drought. It is a collaborative effort between Federal and academic partners, produced on a weekly basis, to synthesize multiple indices, outlooks, and drought impacts on a map and in narrative form. This synthesis of indices is reported by the National Drought Mitigation Center at *http://droughtmonitor.unl.edu.*

Value loss crop has the meaning specified in subpart D of part 1437 of this title.

Verifiable documentation means evidence that can be verified by FSA through an independent source.

Verifiable percentage of loss is the percentage of loss determined by comparing the applicable nutrient factors for a producer's affected production of a forage crop with the average of such nutrient factors from the 3 preceding crop years, as documented on FSA–899, Historical Nutritional Value Weighted Average Worksheet.

WHIP+ means the Wildfires and Hurricanes Indemnity Program Plus under subpart O of this part.

§760.1803 Participant eligibility.

(a) Participants will be eligible to receive a payment under this subpart only if they incurred a loss to an eligible crop due to a qualifying disaster event, as further specified in this subpart.

(b) To be an eligible participant under this subpart, a person or legal entity must be a:

(1) Citizen of the United States;

(2) Resident alien; for purposes of this subpart, resident alien means "lawful alien" (see §1400.3 of this title);

(3) Partnership consisting solely of citizens of the United States or resident aliens; or

(4) Corporation, limited liability company, or other similar organizational structure organized under State law consisting solely of citizens or resident aliens of the United States.

(c) If any person who would otherwise be eligible to receive a payment dies before the payment is received, payment may be released as specified in §707.3 of this chapter. Similarly, if any person or legal entity who would otherwise have been eligible to apply for a payment dies or is dissolved, respectively, before the payment is applied for, payment may be released in accordance with this subpart if a timely application is filed by an authorized representative. Proof of authority to sign for the deceased producer or dissolved entity must be provided. If a participant is now a dissolved general partnership or joint venture, all members of the general partnership or joint venture at the time of dissolution or their duly authorized representatives must sign the application. Eligibility of such participant will be determined, as it is for other participants, based upon ownership share and risk in producing the crop.

(d) An ownership share is required to be eligible for a payment under this subpart. Growers growing eligible crops under contract for crop owners are not eligible for a payment under this subpart unless the grower is also determined to have an ownership share of the crop. Any verbal or written contract that precludes the grower from having an ownership share renders the grower ineligible for payments under this subpart.

(e) A person or legal entity is not eligible to receive assistance under this subpart if FSA determines that the person or legal entity:

(1) Adopted any scheme or other device that tends to defeat the purpose of this subpart or any of the regulations applicable to this subpart;

(2) Made any fraudulent representation; or

(3) Misrepresented any fact affecting a program determination under any or all of the following: This subpart and parts 12, 400, 1400, and 1437 of this title.

(f) A person who is ineligible for crop insurance or NAP under §400.458 or §1437.16 of this title, respectively, for any year is ineligible for payments under this subpart for the same year.

(g) The provisions of §718.11 of this chapter, providing for ineligibility for payments for offenses involving controlled substances, apply.

(h) As a condition of eligibility to receive payments under this subpart, the participant must have been in compliance with the Highly Erodible Land Conservation and Wetland Conservation provisions of part 12 of this title for the applicable crop year for which the producer is applying for benefits under this subpart, and must not otherwise be precluded from receiving payments under part 12, 400, 1400, or 1437 of this title or any law.

§ 760.1804 Eligibility of affected production.

(a) To be eligible for the QLA Program, an eligible crop's affected production must have suffered a quality loss due to a qualifying disaster event and had at least a 5 percent quality loss due to all eligible disaster events. Whether affected production of a crop had a 5 percent loss will be determined separately for crops with different crop types, intended uses, certified organic or conventional status, county, and crop year.

(b) Affected production of the following is not eligible for the QLA Program:

(1) Crops that were not grown commercially;

(2) Crops that were intended for grazing or were grazed;

(3) Crops not intended for harvest;

(4) Volunteer crops;

(5) Value loss crops;

(6) Maple sap;

(7) Honey;

(8) By-products resulting from processing or harvesting a crop, such as, but not limited to, cotton seed, peanut shells, wheat or oat straw, or corn stalks or stovers;

(9) First-year seeding for forage production;

(10) Immature fruit crops;

(11) Crops for which FCIC coverage or NAP coverage is unavailable;

(12) Multiple market crops for which the producer previously received a crop insurance indemnity or WHIP+ payment for a quality loss;

(13) Crops for which production used to calculate a crop insurance indemnity or WHIP+ payment was adjusted based on a comparison of the producer's sale price to FCIC established price;

(14) Crops that received a crop insurance indemnity, NAP payment, or WHIP+ payment based on the quantity of production that was considered unmarketable;

(15) Crops for which the producer previously received a crop insurance indemnity, NAP payment, or WHIP+ payment for which production was reported as salvage value or secondary use;

(16) Sugar beets for which a member of a cooperative processor received a payment through a cooperative agreement; and

(17) Crops that were destroyed.

(c) Only affected production from initial crop acreage will be eligible for a QLA Program payment, unless the provisions for subsequent crops in this section are met. All plantings of an annual or biennial crop are considered the same as a planting of an initial crop in tropical regions as defined in part 1437, subpart F, of this title.

(d) In cases where there is double cropped acreage, affected production of each crop may be eligible only if the specific crops are approved by the FSA State committee as eligible double cropping practices in accordance with procedures approved by the Deputy Administrator.

(e) Participants having affected production from multiple plantings may receive payments for each planting only if the planting meets the requirements of part 1437 of this title and all other provisions of this subpart are satisfied.

§ 760.1805 Qualifying disaster events.

(a) A producer is eligible for payments under this subpart only if the producer's affected production of an eligible crop suffered a crop quality loss due to a qualifying disaster event.

(b) A crop quality loss due to a qualifying disaster event must have occurred on acreage that was physically located in a county that received a:

(1) Presidential Emergency Disaster Declaration authorizing public assistance for categories C through G or individual assistance due to a qualifying disaster event occurring in the 2018 or 2019 calendar years; or

(2) Secretarial Disaster Designation for a qualifying disaster event occurring in the 2018 or 2019 calendar years.

(c) A producer with a crop quality loss on acreage not physically located in a county that was eligible under paragraph (b) of this section will be eligible for the QLA Program for losses due to qualifying disaster events only if the producer provides supporting documentation from which the FSA county committee determines that the crop quality loss on the unit was reasonably related to a qualifying disaster event as specified in this subpart. Supporting documentation may include furnishing climatological data from a reputable source or other information substantiating the claim of loss due to a qualifying disaster event.

§ 760.1806 Ineligible losses.

(a) A loss is not eligible under this subpart if any of the following apply:

(1) The cause of loss is determined by FSA to be the result of poor management decisions, poor farming practices, or drifting herbicides;

(2) The loss could have been mitigated using good farming practices, including losses due to high moisture content that could be mitigated by following best practices for drying and storing the crop;

(3) The qualifying disaster event occurred after the crop was harvested;

(4) FSA or RMA have previously disapproved a notice of loss for the crop and disaster event, unless that notice of loss was disapproved solely because it was filed after the applicable deadline; or

(5) The cause of loss was due to:

(i) Conditions or events occurring outside of the applicable growing season for the crop;

(ii) Insect infestation;

(iii) Water contained or released by any governmental, public, or private dam or reservoir project if an easement exists on the acreage affected by the containment or release of the water;

(iv) Failure of a power supply or brownout; or

(v) Failure to harvest or market the crop due to lack of a sufficient plan or resources.

(b) [Reserved]

§ 760.1807 Miscellaneous provisions.

(a) All persons with a financial interest in a legal entity receiving payments under this subpart are jointly and severally liable for any refund, including related charges, that is determined to be due to FSA for any reason.

(b) In the event that any application under this subpart resulted from erroneous information or a miscalculation, the payment will be recalculated and any excess refunded to FSA with interest to be calculated from the date of the disbursement.

(c) Any payment to any participant under this subpart will be made without regard to questions of title under State law, and without regard to any claim or lien against the commodity, or proceeds, in favor of the owner or any other creditor except agencies of the U.S. Government. The regulations governing offsets and withholdings in part 3 of this chapter apply to payments made under this subpart.

(d) Any participant entitled to any payment may assign any payment(s) in accordance with regulations governing the assignment of payments in part 3 of this chapter.

(e) The regulations in parts 11 and 780 of this title apply to determinations under this subpart.

§ 760.1808 General provisions.

(a) Eligibility and payments under this subpart will be determined based on the county where the affected production was harvested.

(b) FSA county committees will make any necessary adjustments to the applicant's affected production and other information on the application form used to calculate a payment when the county committee determines:

(1) Additional documentation has been requested by FSA but has not been provided by the participant;

(2) The loss is due to an ineligible cause; or

(3) The participant has a contract providing a guaranteed payment for all or a portion of the crop.

(c) Unless otherwise specified, all eligibility provisions of part 1437 of this title also apply to tropical crops for eligibility under this subpart.

(d) FSA will use the most reliable data available at the time payments under this subpart are calculated. If additional data or information is provided or becomes available after a payment is issued, FSA will recalculate the payment amount and the producer must return any overpayment amount to FSA. In all cases, payments can only issue based on the payment formula for losses that affirmatively occurred.

(e) Production that is commingled between counties, crop years, or ineligible and eligible acres before it was a matter of record or combination of record and cannot be separated by using records or other means acceptable to FSA will be prorated to each respective year, county, or type of acreage, respectively.

§ 760.1809 Payment and adjusted gross income limitation.

(a) A person or legal entity, other than a joint venture or general partnership, is eligible to receive, directly or indirectly, payments of not more than $125,000 for each of the 2018, 2019, and 2020 crop years under this subpart.

(b) Payments made to a joint operation, including a joint venture or general partnership, cannot exceed the amount determined by multiplying the maximum payment amount specified in paragraph (a) of this section by the number of persons and legal entities, other than joint operations, that comprise the ownership of the joint operation.

(c) The direct attribution provisions in § 1400.105 of this title apply to payments under this subpart.

(d) The notification of interest provisions in § 1400.107 of this title apply to payments under this subpart.

(e) The provisions for recognizing persons added to a farming operation for payment limitation purposes as described in § 1400.104 of this title apply to payments under this subpart.

(f) The $900,000 average AGI limitation provisions in part 1400 of this title relating to limits on payments for persons or legal entities, excluding joint ventures and general partnerships, apply to each applicant for the QLA Program unless at least 75 percent of the person or legal entity's average AGI is derived from farming, ranching, or forestry-related activities. A person's or legal entity's average AGI for each of the program years 2018, 2019 or 2020, is determined by using the average of the adjusted gross incomes for the 3 taxable years preceding the most immediately preceding taxable year. If the person's or legal entity's average AGI is below $900,000 or at least 75 percent of the person or legal entity's average AGI is derived from farming,

ranching, or forestry-related activities, the person or legal entity, is eligible to receive payments under this subpart.

§ 760.1810 Time and method of application.

(a) A completed FSA–898, Quality Loss Adjustment (QLA) Program Application, must be submitted in person, by mail, email, or facsimile to any FSA county office by the close of business on March 5, 2020.

(b) An application submitted in accordance with paragraph (a) of this section is not considered valid and complete for issuance of payment under this subpart unless FSA determines all the applicable eligibility provisions have been satisfied and the producer has submitted all of following by March 19, 2020:

(1) Documentation required by § 760.1811;

(2) FSA–578, Report of Acreage, for all acreage for any crop for which payments under this subpart are requested;

(3) FSA–895, Crop Insurance and/or NAP Coverage Agreement; and

(4) For forage crops, FSA–899, Historical Nutritional Value Weighted Average Worksheet, if verifiable documentation of historical nutrient factors is available.

(c) In addition to the forms listed in paragraph (b) of this section, applicants must also submit all the following eligibility forms within 60 days from the date of signing the QLA Program application if not already on file with FSA:

(1) AD–1026, Highly Erodible Land Conservation (HELC) and Wetland Conservation Certification;

(2) CCC–902 Automated, Farm Operating Plan for Payment Eligibility 2009 and Subsequent Program Years;

(3) CCC–941 Average Adjusted Gross Income (AGI) Certification and Consent to Disclosure of Tax Information; and

(4) CCC–942 Certification of Income from Farming, Ranching and Forestry Operations, if applicable.

(d) Failure to submit all required forms by the applicable deadlines in paragraphs (b) and (c) of this section may result in no payment or a reduced payment.

(e) Application approval and payment by FSA does not relieve a participant from having to submit any form required, but not filed, according to this section.

(f) Once signed by a producer, the application is considered to contain information and certifications of and pertaining to the producer regardless of who entered the information on the application.

(g) The producer applying for payment under this subpart certifies the accuracy and truthfulness of the information provided in the application as well as any documentation filed with or in support of the application. All information is subject to verification or spot check by FSA at any time, either before or after payment is issued. Refusal to allow FSA or any agency of the USDA to verify any information provided will result in the participant's forfeiting eligibility for payment under this subpart. FSA may at any time, including before, during, or after processing and paying an application, require the producer to submit any additional information necessary to implement or determine any eligibility provision of this subpart. Furnishing information specified in this subpart is voluntary; however, FSA may choose not to act on the application or approve payment if the required information is not provided. Providing a false certification will result in ineligibility and can also be punishable by imprisonment, fines, and other penalties.

§760.1811 Required documentation and verification.

(a) If requested by FSA, an applicant must provide documentation that establishes the applicant's ownership share and value at risk in the crop.

(b) The applicant must provide acceptable documentation that is dated and contains all information required to substantiate the applicant's certification to the satisfaction of the FSA county committee. Verifiable documentation is required to substantiate the total dollar value loss and associated affected production, grading factors, and nutritional factors. FSA may verify the records with records on file at the warehouse, gin, or other entity that received or may have received the reported production. Reliable production records are required to substantiate the reported amount of affected production for applications not based on the total dollar value loss.

(c) To be considered acceptable, verifiable documentation for grain crops that were sold may come from any time between harvest and sale of the affected production, unless the FSA county committee determines the record is not representative of the condition within 30 days of harvest. For all other crops, the verifiable documentation must come from tests or analysis completed within 30 days of harvest, unless the FSA county committee determines that the record is representative of the condition of the affected production at time of harvest. Examples of acceptable records for purposes of this paragraph (c) include:

(1) Warehouse grading sheets;

(2) Settlement sheets;

(3) Sales receipts showing grade and price or disposition to secondary market due to quality; and

(4) Laboratory test results.

(d) For forage crops, producers must submit verifiable documentation showing the nutrient factors for the affected production. Producers must also submit verifiable documentation of the historical nutrient factors for the 3 preceding crop years if available. The nutrient factors that must be documented for a crop will determined by the FSA county committee based on the standard practice for the crop in that county.

(e) For all crops other than forage crops, producers must submit verifiable documentation of the total dollar value loss due to quality, if available, and verifiable documentation of grading factors due to quality.

(f) The participant is responsible for:

(1) Retaining, providing, and summarizing, at time of application and whenever required by FSA, the best available verifiable production records for the crop;

(2) Providing the information in a manner that can be easily understood by the FSA county committee; and

(3) Providing supporting documentation about the disaster event if the FSA county committee has reason to question the disaster event.

(e) Participants must provide all records for any production of a crop that is grown with an arrangement, agreement, or contract for guaranteed payment.

(f) Participants are required to retain documentation in support of their application for 3 years after the date of approval.

(g) Participants receiving QLA Program payments or any other person who furnishes such information to USDA must permit authorized representatives of USDA or the Government Accountability Office, during regular business hours, to enter the agricultural operation and to inspect, examine, and make copies of books, records, or other items for the purpose of confirming the accuracy of the information provided by the participant.

§ 760.1812   Payment calculation.

(a) Payments will be calculated separately for crops based on the crop type, intended use, certified organic or conventional status, county, and crop year.

(b) For forage crops with verifiable documentation of nutrient factors for the affected production and for the 3 preceding crop years, the payment will be equal to the producer's total affected production multiplied by the producer's verifiable percentage of loss, multiplied by the average market price, multiplied by 70 percent.

(c) For forage crops with verifiable documentation of nutrient factors for the affected production but not for the 3 preceding crop years, the payment will be equal to the producer's total affected production multiplied by the county average percentage of loss in paragraph (f) of this section, multiplied by the average market price, multiplied by 70 percent, multiplied by 50 percent.

(d) For crops other than forage with verifiable documentation of the total dollar value loss due to quality, the payment will be equal to the producer's total dollar value loss on the affected production, multiplied by 70 percent.

(e) For crops other than forage without verifiable documentation of the total dollar value loss but with verifiable documentation of grading factors, the payment will be equal to the producer's affected production multiplied by the county average loss per unit of measure in paragraph (g) of this section, multiplied by 70 percent, multiplied by 50 percent.

(f) The county average percentage of loss is the average percentage of loss from producers eligible for payment under paragraph (b) of this section if at least 5 producers in a county are eligible for payment for a crop under paragraph (b) of this section. If less than 5 producers in a county are eligible for payment for a crop under paragraph (b) of this section, the Deputy Administrator will:

(1) Determine a county average percentage of loss based on the best available data, including, but not limited to, evidence of losses in contiguous counties; or

(2) If a county average percentage of loss cannot be determined due to insufficient data, not issue payments to applicants under paragraph (c) of this section.

(g) The county average loss per unit of measure is based on the weighted average sales price from producers eligible for payment under paragraph (d) of this section if at least 5 producers in a county are eligible for payment for a crop under paragraph (d) of this section. If less than 5 producers are eligible for payment in a county under paragraph (d) of this section, the Deputy Administrator will:

(1) Determine a county average loss per unit of measure based on the best available data, including, but not limited to, evidence of losses in contiguous counties; or

(2) If a county average loss per unit of measure cannot be determined due to insufficient data, not issue payments to applicants under paragraph (e) of this section.

§ 760.1813   Availability of funds and timing of payments.

(a) Payments will be issued after the application period has ended and all applications have been reviewed by FSA.

(b) In the event that, within the limits of the funding made available by the Secretary, approval of eligible applications would result in payments in excess of the amount available, FSA will prorate payments by a national factor to reduce the payments to an amount that is less than available funds as determined by the Secretary.

(c) Applications and claims that are unpaid or prorated for any reason will not be carried forward for payment under other funds for later years or otherwise, but will be considered, as to any unpaid amount, void and nonpayable.

§ 760.1814 Requirement to purchase crop insurance or NAP coverage.

(a) For the first 2 consecutive crop years for which crop insurance or NAP coverage is available after the enrollment period for the QLA Program ends, a participant who receives payment under this subpart for a crop loss in a county must obtain:

(1) For an insurable crop, crop insurance with at least a 60 percent coverage level for that crop in that county; or

(2) For a NAP-eligible crop, NAP coverage with a coverage level of 60 percent.

(b) Participants who exceed the average adjusted gross income limitation for NAP payment eligibility[1] for the applicable crop year may meet the purchase requirement specified in paragraph (a)(2) of this section by purchasing Whole-Farm Revenue Protection crop insurance coverage, if eligible, or paying the NAP service fee and premium even though the participant will not be eligible to receive a NAP payment due to the average adjusted gross income limit.

(c) The final crop year to purchase crop insurance or NAP coverage to meet the requirements of paragraph (a) of this section is the 2023 crop year.

(d) A participant who obtained crop insurance or NAP coverage for the crop in accordance with the requirements for WHIP+ in § 760.1517 is considered to have met the requirement to purchase crop insurance or NAP coverage for the QLA Program.

(e) If a producer fails to obtain crop insurance or NAP coverage as required in this section, the producer must reimburse FSA for the full amount of

---

[1] See §§ 1400.500(a) and 1400.1(a)(4) of this title.

QLA Program payment plus interest. A producer will only be considered · to have obtained NAP coverage for the purposes of this section if the participant submitted a NAP application for coverage and paid the requisite NAP service fee and any applicable premium by the applicable deadline and completed all program requirements required under the coverage agreement, including filing an acreage report.

PART 761—FARM LOAN PROGRAMS; GENERAL PROGRAM ADMINISTRATION

Subpart A—General Provisions

Subpart B—Supervised Bank Accounts

Subpart C—Progression Lending

Subpart D—Allocation of Farm Loan Programs Funds to State Offices

AUTHORITY: 5 U.S.C. 301 and 7 U.S.C. 1989.

SOURCE: 72 FR 63285, Nov. 8, 2007, unless otherwise noted.

Subpart A—General Provisions

§ 761.1 Introduction.

(a) The Administrator delegates the responsibility to administer Farm Loan Programs of the Consolidated Farm and Rural Development Act (7 U.S.C. 1921 *et seq.*) to the Deputy Administrator for Farm Loan Programs subject to any limitations established in 7 CFR 2.16(a)(2) and 7 CFR 2.42.

(b) The Deputy Administrator:

(1) Delegates to each State Executive Director within the State Executive Director's jurisdiction the authority, and in the absence of the State Executive Director, the person acting in that position, to act for, on behalf of, and in the name of the United States of America or the Farm Service Agency to do and perform acts necessary in connection with making and guaranteeing loans, such as, but not limited to, making advances, servicing loans and other indebtedness, and obtaining, servicing, and enforcing or releasing security and other instruments related to the loan. For actions that do not result in a loss to the Farm Service Agency, a State Executive Director may redelegate authorities received under this paragraph to a Farm Loan Chief, Farm Loan Specialist, District Director, Farm Loan Manager, or Senior Farm Loan Officer, Farm Loan Officer, Loan Analyst, Loan Resolution Specialist, or Program Technician.

(2) May establish procedures for further redelegation or limitation of authority.

(c) Parts 761 through 767 describe the Agency's policies for its Farm Loan Programs. The objective of these programs is to provide supervised credit and management assistance to eligible farmers to become owners or operators, or both, of family farms, to continue such operations when credit is not available elsewhere, or to return to normal farming operations after sustaining substantial losses as a result of a designated or declared disaster. The programs are designed to allow those who participate to transition to private commercial credit or other sources of credit in the shortest period of time practicable through the use of supervised credit, including farm assessments, borrower training, market placement, and borrower graduation requirements. These regulations apply to loan applicants, borrowers, lenders, holders, Agency personnel, and other parties involved in making, guaranteeing, holding, servicing, or liquidating such loans.

(d) This part describes the Agency's general and administrative policies for its guaranteed and direct Farm Loan Programs. In general, this part addresses issues that affect both guaranteed and direct loan programs.

(e) Part 3 of this title and 31 CFR part 285 describe the policies and procedures the Agency will follow for noncentralized offset (including administrative offset) and referral to Treasury for centralized offset (TOP), Federal salary offset, Administrative Wage Garnishment, and collection through Treasury's private collection agencies (cross-servicing). Supplemental provisions for FLP purposes are described in part 761, subpart F of this title.

(f) Part 3 of this title and 31 CFR parts 900–904 describe the policies and procedures the Agency will follow for debt settlement authorities pursuant to the Federal Claims Collection Standards. Supplemental provisions for FLP purposes are described in part 761, subpart F of this title.

(g) Part 761, subpart F of this title describes the debt settlement policies and procedures for FLP debt pursuant to the Act.

[72 FR 63285, Nov. 8, 2007, as amended at 76 FR 5057, Jan. 28, 2011; 83 FR 11869, Mar. 19, 2018; 85 FR 36691, June 17, 2020]

§761.2 Abbreviations and definitions.

The following abbreviations and definitions are applicable to the Farm Loan Programs addressed in parts 761 through 767 and 769 unless otherwise noted.

(a) *Abbreviations.*

ARA Alternative Repayment Agreement.

CL Conservation Loan.

CLP Certified Lender Program.

DSA Disaster Set-Aside.

EE Economic Emergency loan.

EM Emergency loan.

FCCS Federal Claims Collection Standards.

FLP Farm Loan Programs.

FO Farm Ownership loan.

FSA Farm Service Agency, an Agency of the USDA, including its personnel and any successor Agency.

HPRP The Heirs' Property Relending Program.

LIBOR London Interbank Offered Rate.

ML Microloan.

MLP Micro Lender Program.

NRCS National Resources and Conservation Service, USDA.

OIG Office of the Inspector General, USDA.

OGC Office of the General Counsel of the USDA.

OL Operating loan.

PLP Preferred Lender Program.

RHF Rural Housing loan for farm service buildings.

RL Recreation loan.

SAA Shared Appreciation Agreement.

SA Shared Appreciation loan.

SEL Standard Eligible Lender.

ST Softwood Timber loan.

SW Soil and Water loan.

USDA United States Department of Agriculture.

USPAP Uniform Standards of Professional Appraisal Practice.

(b) *Definitions.*

Abandoned security property is security property that a borrower is not occupying, is not in possession of, or has relinquished control of and has not made arrangements for its care or sale.

Accrued deferred interest is unpaid interest from past due installments posted to a borrower's loan account.

Act is the Consolidated Farm and Rural Development Act (7 U.S.C. 1921 *et seq.*).

Additional security is property which provides security in excess of the amount of security value equal to the loan amount.

Adequate security is property which is required to provide security value at least equal to the direct loan amount.

Adjustment means the settlement of an FLP debt for less than the total amount owed. The adjusted amount is collected through a series of payments that are scheduled over time. An adjustment is not a final settlement until all scheduled payments have been made. After applying all payments pursuant to the adjustment agreement, any remaining balance is canceled. The amount canceled is reported to the IRS pursuant to §3.90 of this title and applicable IRS requirements.

Administrative appraisal review is a review of an appraisal to determine if the appraisal:

(1) Meets applicable Agency requirements; and

(2) Is accurate outside the requirements of standard 3 of USPAP.

Agency is the FSA.

Agreement for the use of proceeds is an agreement between the borrower and the Agency for each production cycle that reflects the proceeds from the sale of normal income security that will be used to pay scheduled FLP loan installments, including any past due installments, during the production cycle covered by the agreement.

Agricultural commodity means livestock, grains, cotton, oilseeds, dry beans, tobacco, peanuts, sugar beets, sugar cane, fruit, vegetable, forage, nursery crops, nuts, aquacultural species, and the products resulting from: livestock, tree farming, and other plant or animal production as determined by the Agency.

Allonge is an attachment or an addendum to a promissory note.

Allowable costs are those costs for replacement or repair that are supported by acceptable documentation, including, but not limited to, written estimates, invoices, and bills.

Alternative repayment agreement is a written repayment agreement accepted by both the borrower and the Agency as specified in §§ 3.42(b) and 3.80 of this title. The agreement may allow for payments to be made from the borrower to the Agency as an alternative to collecting the payment amounts through administrative offset, or Federal salary offset.

Applicant is the individual or entity applying for a loan or loan servicing under either the direct or guaranteed loan program.

Apprentice means an individual who receives applied guidance and input from an individual with the skills and knowledge pertinent to the successful operation of the farm enterprise being financed.

Aquaculture is the husbandry of any aquatic organisms (including fish, mollusks, crustaceans or other invertebrates, amphibians, reptiles, or aquatic plants) raised in a controlled or selected environment of which the applicant has exclusive rights to use.

Assignment of guaranteed portion is a process by which the lender transfers the right to receive payments or income on a guaranteed loan to another party, usually in return for payment in the amount of the loan's guaranteed principal. The lender retains the unguaranteed portion in its portfolio and receives a fee from the purchaser or assignee to service the loan and receive and remit payments according to a written assignment agreement. This assignment can be reassigned or sold multiple times.

Assignment of indemnity is the transfer of rights to compensation under an insurance contract.

Assistance is financial assistance in the form of a direct or guaranteed loan or interest subsidy or servicing action.

Assumption is the act of agreeing to be legally responsible for another party's indebtedness.

Assumption agreement is a written agreement on the appropriate Agency form to pay the FLP debt incurred by another.

Basic part of an applicant's total farming operation is any single agricultural commodity or livestock production enterprise of an applicant's farming operation which normally generates sufficient income to be considered essential to the success of such farming operation.

Basic security is all farm machinery, equipment, vehicles, foundation and breeding livestock herds and flocks, including replacements, and real estate that serves as security for a loan made or guaranteed by the Agency.

Beginning farmer is an individual or entity who:

(1) Meets the loan eligibility requirements for a direct or guaranteed CL, FO, or OL, as applicable;

(2) Has not operated a farm for more than 10 years. This requirement applies to all members of an entity;

(3) Will materially and substantially participate in the operation of the farm:

(i) In the case of a loan made to an individual, individually or with the family members, material and substantial participation requires that the individual provide substantial day-to-day labor and management of the farm, consistent with the practices in the county or State where the farm is located.

(ii) In the case of a loan made to an entity, all members must materially and substantially participate in the operation of the farm. Material and substantial participation requires that the member provide some amount of the management, or labor and management necessary for day-to-day activities, such that if the individual did not provide these inputs, operation of the farm would be seriously impaired;

(4) Agrees to participate in any loan assessment and borrower training required by Agency regulations;

(5) Except for an OL applicant, does not own real farm property or who, directly or through interests in family farm entities owns real farm property, the aggregate acreage of which does not exceed 30 percent of the average farm acreage of the farms in the county where the property is located. If the farm is located in more than one county, the average farm acreage of the county where the applicant's residence is located will be used in the calculation. If the applicant's residence is not located on the farm or if the applicant is an entity, the average farm acreage of the county where the major portion

of the farm is located will be used. The average county farm acreage will be determined from the most recent Census of Agriculture;

(6) Demonstrates that the available resources of the applicant and spouse (if any) are not sufficient to enable the applicant to enter or continue farming on a viable scale; and

(7) In the case of an entity:

(i) All the members are related by blood or marriage; and

(ii) All the members are beginning farmers.

Borrower (or debtor) is an individual or entity that has an outstanding obligation to the Agency or to a lender under any direct or guaranteed FLP loan, without regard to whether the loan has been accelerated. The term "borrower" includes all parties liable for such obligation, including collection-only borrowers, except for debtors whose total loans and accounts have been voluntarily or involuntarily foreclosed, sold, or conveyed, or who have been discharged of all such obligations owed to the Agency or guaranteed lender.

Cancellation means the final resolution of an FLP debt without receiving payment in full. Any amounts still owed, after applying payments in accordance with approved adjustment and compromise agreements, is canceled. The amount canceled is reported to the IRS pursuant to §3.90 of this title and applicable IRS requirements.

Cash flow budget is a projection listing all anticipated cash inflows (including all farm income, nonfarm income and all loan advances) and all cash outflows (including all farm and nonfarm debt service and other expenses) to be incurred during the period of the budget. Advances and principal repayments of lines of credit may be excluded from a cash flow budget. Cash flow budgets for guaranteed loans under $125,000 do not require income and expenses itemized by categories. A cash flow budget may be completed either for a 12-month period, a typical production cycle, or the life of the loan, as appropriate. It may also be prepared with a breakdown of cash inflows and outflows for each month of the review period and include the expected outstanding operating credit balance for the end of each month. The latter type is referred to as a "monthly cash flow budget."

Chattel or real estate essential to the operation is chattel or real estate that would be necessary for the applicant to continue operating the farm after the disaster in a manner similar to the manner in which the farm was operated immediately prior to the disaster, as determined by the Agency.

Chattel security is property that may consist of, but is not limited to: Crops; livestock; aquaculture species; farm equipment; inventory; accounts; contract rights; general intangibles; and supplies that are covered by financing statements and security agreements, chattel mortgages, and other security instruments.

Civil action is a court proceeding to protect the Agency's financial interests. A civil action does not include bankruptcy and similar proceedings to impound and distribute the bankrupt's assets to creditors, or probate or similar proceedings to settle and distribute estates of incompetents or decedents, and pay claims of creditors.

Closing agent is the attorney or title insurance company selected by the applicant and approved by the Agency to provide closing services for the proposed loan or servicing action. Unless a title insurance company provides loan closing services, the term "title company" does not include "title insurance company."

Coastal barrier is an area of land identified as part of the national Coastal Barrier Resources System under the Coastal Barrier Resources Act of 1980.

Compromise is the settlement of an FLP debt or claim by a lump-sum payment of less than the total amount owed in satisfaction of the debt or claim.

Conditional commitment is the Agency's commitment to a lender that the material the lender has submitted is approved subject to the completion of all listed conditions and requirements.

Conservation Contract is a contract under which a borrower agrees to set aside land for conservation, recreation or wildlife purposes in exchange for reduction of a portion of an outstanding FLP debt.

Conservation Contract review team is comprised by the appropriate offices of FSA, the Natural Resources Conservation Service, U.S. Fish and Wildlife Service, State Fish and Wildlife Agencies, Conservation Districts, National Park Service, Forest Service, State Historic Preservation Officer, State Conservation Agencies, State Environmental Protection Agency, State Natural Resource Agencies, adjacent public landowner, and any other entity that may have an interest and qualifies to be a management authority for a proposed conservation contract.

Conservation loan means a loan made to eligible applicants to cover the costs to the applicant of carrying out a qualified conservation project.

Conservation plan means an NRCS-approved written record of the land user's decisions and supporting information, for treatment of a land unit or water as a result of the planning process, that meets NRCS Field Office Technical Guide (FOTG) quality criteria for each natural resource (soil, water, air, plants, and animals) and takes into account economic and social considerations. The conservation plan describes the schedule of operations and activities needed to solve identified natural resource problems and takes advantage of opportunities at a conservation management system level. This definition only applies to the direct loans and guaranteed loans for the Conservation Loan Program.

Conservation practice means a specific treatment, such as a structural or vegetative measure, or management technique, commonly used to meet specific needs in planning and implementing conservation, for which standards and specifications have been developed. Conservation practices are contained in the appropriate NRCS Field Office Technical Guide (FOTG), which is based on the National Handbook of Conservation Practices (NHCP).

Conservation project means conservation measures that address provisions of a conservation plan or Forest Stewardship Management Plan.

Consolidation is the process of combining the outstanding principal and interest balance of two or more loans of the same type made for operating purposes.

Construction is work such as erecting, repairing, remodeling, relocating, adding to, or salvaging any building or structure, and the installing, repairing, or adding to heating and electrical systems, water systems, sewage disposal systems, walks, steps, and driveways.

Controlled is when a director or an employee has more than a 50 percent ownership in an entity or, the director or employee, together with relatives of the director or employee, have more than a 50 percent ownership.

Controlled substance is the term as defined in 21 U.S.C. 812.

Cooperative is an entity that has farming as its purpose, whose members have agreed to share the profits of the farming enterprise, and is recognized as a farm cooperative by the laws of the state in which the entity will operate a farm.

Corporation is a private domestic corporation created and organized under the laws of the state in which it will operate a farm.

Cosigner is a party, other than the applicant, who joins in the execution of a promissory note to assure its repayment. The cosigner becomes jointly and severally liable to comply with the repayment terms of the note, but is not authorized to severally receive loan servicing available under 7 CFR parts 765 and 766. In the case of an entity applicant, the cosigner cannot be a member of the entity.

County is a local administrative subdivision of a State or similar political subdivision of the United States.

County average yield is the historical average yield for an agricultural commodity in a particular political subdivision, as determined or published by a government entity or other recognized source.

Criminal action is the prosecution by the United States to exact punishment in the form of fines or imprisonment for alleged violation of criminal statutes.

Crop allotment or quota is a farm's share of an approved national tobacco or peanut allotment or quota.

Current market value buyout is the termination of a borrower's loan obligations to the Agency in exchange for payment of the current appraised value of the borrower's security property and non-essential assets, less any prior liens.

Debt forgiveness means the reduction or termination of a debt under the Act in a manner that results in a loss to the Agency, through:

(i)(A) Writing down or writing off a debt pursuant to 7 U.S.C. 2001;

(B) Cancellation of remaining amounts owed after compromising, adjusting, reducing, or charging off a debt or claim pursuant to 7 U.S.C. 1981;

(C) Paying a loss pursuant to 7 U.S.C. 2005 on a FLP loan guaranteed by the Agency;

(D) Discharging a debt as a result of bankruptcy; or

(E) Releases of liability which result in a loss to the Agency.

(ii) Debt forgiveness does not include:

(A) Debt reduction through a conservation contract;

(B) Any writedown provided as part of the resolution of a discrimination complaint against the Agency;

(C) Prior debt forgiveness that has been repaid in its entirety;

(D) Consolidation, rescheduling, reamortization, or deferral of a loan; and

(E) Forgiveness of a YL debt due to circumstances beyond the borrower's control.

Debt settlement is a compromise, adjustment, or cancellation of an FLP debt.

Debt service margin is the difference between all of the borrower's expected expenditures in a planning period (including farm operating expenses, capital expenses, essential family living expenses, and debt payments) and the borrower's projected funds available to pay all expenses and payments.

Debt writedown is the reduction of the borrower's debt to that amount the Agency determines to be collectible based on an analysis of the security value and the borrower's ability to pay.

Default is the failure of a borrower to observe any agreement with the Agency, or the lender in the case of a guaranteed loan, as contained in promissory notes, security instruments, and similar or related instruments.

Deferral is a postponement of the payment of interest or principal, or both.

Delinquent borrower, for loan servicing purposes, is a borrower who has failed to make all scheduled payments by the due date.

Direct loan is a loan funded and serviced by the Agency as the lender.

Disaster is an event of unusual and adverse weather conditions or other natural phenomena, or quarantine, that has substantially affected the production of agricultural commodities by causing physical property or production losses in a county, or similar political subdivision, that triggered the inclusion of such county or political subdivision in the disaster area as designated by the Agency.

Disaster area is the county or counties declared or designated as a disaster area for EM loan assistance as a result of disaster related losses. This area includes counties contiguous to those counties declared or designated as disaster areas.

Disaster set-aside is the deferral of payment of an annual loan installment to the Agency to the end of the loan term in accordance with part 766, subpart B of this chapter.

Disaster yield is the per-acre yield of an agricultural commodity for the operation during the production cycle when the disaster occurred.

Down payment loan is a type of FO loan made to beginning farmers and socially disadvantaged farmers to finance a portion of a real estate purchase under part 764, subpart E of this chapter.

Economic Emergency loan is a loan that was made or guaranteed to an eligible applicant to allow for continuation of the operation during an economic emergency which was caused by a lack of agricultural credit or an unfavorable relationship between production costs and prices received for agricultural commodities. EE loans are not currently funded; however, such outstanding loans are serviced by the Agency or the lender in the case of a guaranteed EE loan.

Embedded entity means an entity that has a direct or indirect interest, as a stockholder, member, beneficiary, or otherwise, in another entity.

Emergency loan is a loan made to eligible applicants who have incurred substantial financial losses from a disaster.

Entity means a corporation, partnership, joint operation, cooperative, limited liability company, trust, or other legal business organization, as determined by the Agency, that is authorized to conduct business in the state in which the organization operates. Organizations operating as non-profit entities under Internal Revenue Code 501 (26 U.S.C. 501) and estates are not considered eligible entities for Farm Loan Programs purposes.

Entity member means all individuals and all embedded entities, as well as the individual members of the embedded entities, having an ownership interest in the assets of the entity.

Essential family living and farm operating expenses:

(1) Are those that are basic, crucial or indispensable.

(2) Are determined by the Agency based on the following considerations:

(i) The specific borrower's operation;

(ii) What is typical for that type of operation in the area; and

(iii) What is an efficient method of production considering the borrower's resources.

(3) Include, but are not limited to, essential: Household operating expenses; food, including lunches; clothing and personal care; health and medical expenses, including medical insurance; house repair and sanitation; school and religious expenses; transportation; hired labor; machinery repair; farm building and fence repair; interest on loans and credit or purchase agreement; rent on equipment, land, and buildings; feed for animals; seed, fertilizer, pesticides, herbicides, spray materials and other necessary farm supplies; livestock expenses, including medical supplies, artificial insemination, and veterinarian bills; machinery hire; fuel and oil; taxes; water charges; personal, property and crop insurance; auto and truck expenses; and utility payments.

Established farmer means a farmer who operates the farm (in the case of an entity, its members as a group) who meets all of the following conditions:

(1) Actively participated in the operation and the management, including, but not limited to, exercising control over, making decisions regarding, and establishing the direction of, the farm-

ing operation at the time of the disaster;

(2) Spends a substantial portion of time in carrying out the farming operation;

(3) Planted the crop, or purchased or produced the livestock on the farming operation;

(4) In the case of an entity, is primarily engaged in farming and has over 50 percent of its gross income from all sources from its farming operation based on the operation's projected cash flow for the next crop year or the next 12-month period, as mutually determined;

(5) Is not an integrated livestock, poultry, or fish processor who operates primarily and directly as a commercial business through contracts or business arrangements with farmers, except a grower under contract with an integrator or processor may be considered an established farmer, provided the farming operation is not managed by an outside full-time manager or management service and Agency loans will be based on the applicant's share of the agricultural production as specified in the contract; and

(6) Does not employ a full time farm manager.

EZ Guarantee means a type of OL or FO of $100,000 or less made using a simplified loan application. As part of the simplified application process, EZ Guarantees are processed using a streamlined underwriting method to determine financial feasibility.

False information is information provided by an applicant, borrower or other source to the Agency that the applicant or borrower knows to be incorrect.

Family farm is a business operation that:

(1) Produces agricultural commodities for sale in sufficient quantities so that it is recognized as a farm rather than a rural residence;

(2) Has both physical labor and management provided as follows:

(i) The majority of day-to-day, operational decisions, and all strategic management decisions are made by:

(A) The borrower, with input and assistance allowed from persons who are either related to the borrower by blood or marriage, or are a relative, for an individual borrower; or

(B) The members responsible for operating the farm, in the case of an entity.

(ii) A substantial amount of labor to operate the farm is provided by:

(A) The borrower, with input and assistance allowed from persons who are either related to the borrower by blood or marriage, or are a relative, for an individual borrower; or

(B) The members responsible for operating the farm, in the case of an entity.

(3) May use full-time hired labor in amounts only to supplement family labor.

(4) May use reasonable amounts of temporary labor for seasonal peak workload periods or intermittently for labor intensive activities.

Family living expenses are the costs of providing for the needs of family members and those for whom the borrower has a financial obligation, such as alimony, child support, and care expenses of an elderly parent.

Family members are the immediate members of the family residing in the same household with the borrower.

Farm is a tract or tracts of land, improvements, and other appurtenances that are used or will be used in the production of crops, livestock, or aquaculture products for sale in sufficient quantities so that the property is recognized as a farm rather than a rural residence. The term "farm" also includes the term "ranch." It may also include land and improvements and facilities used in a non-eligible enterprise or the residence which, although physically separate from the farm acreage, is ordinarily treated as part of the farm in the local community.

Farmer is an individual, corporation, partnership, joint operation, cooperative, trust, or limited liability company that is the operator of a farm.

Farm income is the proceeds from the sale of agricultural commodities that are normally sold annually during the regular course of business, such as crops, feeder livestock, and other farm products.

Farm Loan Programs are Agency programs to make, guarantee, and service loans to family farmers authorized under the Act or Agency regulations.

Farm Ownership loan is a loan made to eligible applicants to purchase, enlarge, or make capital improvements to family farms, or to promote soil and water conservation and protection. It also includes the Downpayment loan.

Farm Program payments are benefits received from FSA for any commodity, disaster, or cost share program.

Feasible plan is when an applicant or borrower's cash flow budget or farm operating plan indicates that there is sufficient cash inflow to pay all cash outflow. If a loan approval or servicing action exceeds one production cycle and the planned cash flow budget or farm operating plan is atypical due to cash or inventory on hand, new enterprises, carryover debt, atypical planned purchases, important operating changes, or other reasons, a cash flow budget or farm operating plan must be prepared that reflects a typical cycle. If the request is for only one cycle, a feasible plan for only one production cycle is required for approval.

Financially distressed borrower is a borrower unable to develop a feasible plan for the current or next production cycle.

Financially viable operation, for the purposes of considering a waiver of OL term limits under §764.252 of this chapter, is a farming operation that, with Agency assistance, is projected to improve its financial condition over a period of time to the point that the operator can obtain commercial credit without further Agency assistance. Such an operation must generate sufficient income to:

(1) Meet annual operating expenses and debt payments as they become due;

(2) Meet essential family living expenses to the extent they are not met by dependable non-farm income;

(3) Provide for replacement of capital items; and

(4) Provide for long-term financial growth.

Fixture is an item of personal property attached to real estate in such a way that it cannot be removed without defacing or dismantling the structure, or damaging the item itself.

Floodplains are lowland and relatively flat areas adjoining inland and coastal waters, including flood-prone areas of offshore islands, including at a minimum, that area subject to a one percent or greater chance of flooding in any given year. The base floodplain is used to designate the 100-year floodplain (one percent chance floodplain). The critical floodplain is defined as the 500-year floodplain (0.2 percent chance floodplain).

Foreclosed is the completed act of selling security either under the power of sale in the security instrument or through judicial proceedings.

Foreclosure sale is the act of selling security either under the power of sale in the security instrument or through judicial proceedings.

Forest Stewardship Management Plan means a property-specific, long-term, multi-resource plan that addresses private landowner objectives while recommending a set and schedule of management practices designed to achieve a desired future forest condition developed and approved through the USDA Forest Service or its agent.

Good faith is when an applicant or borrower provides current, complete, and truthful information when applying for assistance and in all past dealings with the Agency, and adheres to all written agreements with the Agency including, but not limited to, loan agreement, security instruments, farm operating plans, and agreements for use of proceeds. The Agency considers a borrower to act in good faith, however, if the borrower's inability to adhere to all agreements is due to circumstances beyond the borrower's control. In addition, the Agency will consider fraud, waste, or conversion actions, when substantiated by a legal opinion from OGC, when determining if an applicant or borrower has acted in good faith.

Graduation means the payment in full of all direct FLP loans, except for CLs, made for operating, real estate, or both purposes by refinancing with other credit sources either with or without an Agency guarantee.

Guaranteed loan is a loan made and serviced by a lender for which the Agency has entered into a Lender's Agreement and for which the Agency has issued a Loan Guarantee. This term also includes guaranteed lines of credit except where otherwise indicated.

Guarantor is a party not included in the farming operation who assumes responsibility for repayment in the event of default.

Hazard insurance is insurance covering fire, windstorm, lightning, hail, explosion, riot, civil commotion, aircraft, vehicles, smoke, builder's risk, public liability, property damage, flood or mudslide, workers compensation, or any similar insurance that is available and needed to protect the Agency security or that is required by law.

Hearing official. For the purposes of salary offset, the hearing official is an Administrative Law Judge of the USDA or another individual not under the supervision or control of the USDA. For the purposes of administrative wage garnishment, the hearing official is selected pursuant to part 3, subpart E of this title.

Heirs' property means a farm that is jointly held by multiple heirs as tenants in common as a result of inheriting title from a relative.

Highly erodible land is land as determined by Natural Resources Conservation Service to meet the requirements provided in section 1201 of the Food Security Act of 1985.

Holder is a person or organization other than the lender that holds all or a part of the guaranteed portion of an Agency guaranteed loan but has no servicing responsibilities. When the lender assigns a part of the guaranteed loan by executing an Agency assignment form, the assignee becomes a holder.

Homestead protection is the previous owner's right to lease with an option to purchase the principal residence and up to 10 acres of adjoining land which secured an FLP direct loan.

Homestead protection property is the principal residence that secured an FLP direct loan and is subject to homestead protection.

Household contents are essential household items necessary to maintain viable living quarters. Household contents exclude all luxury items such as jewelry, furs, antiques, paintings, etc.

HPRP loan agreement means the signed agreement between FSA and the intermediary that specifies the terms and conditions of the HPRP loan.

HPRP loan funds means cash proceeds of a loan obtained through HPRP, including the portion of an HPRP revolving loan fund directly provided by the HPRP loan as well as the proceeds advanced to an ultimate recipient. HPRP loan funds are Federal funds.

HPRP revolving loan fund means a group of assets, obtained through or related to an HPRP loan and recorded by the intermediary in a bookkeeping account or set of accounts and accounted for, along with related liabilities, revenues, and expenses, as an entity or enterprise separate from the intermediary's other assets and financial activities.

Inaccurate information is incorrect information provided by an applicant, borrower, lender, or other source without the intent of fraudulently obtaining benefits.

Indian reservation is all land located within the limits of any Indian reservation under the jurisdiction of the United States, notwithstanding the issuance of any patent, and including rights-of-way running through the reservation; trust or restricted land located within the boundaries of a former reservation of a Federally recognized Indian Tribe in the State of Oklahoma; or all Indian allotments the Indian titles to which have not been extinguished if such allotments are subject to the jurisdiction of a Federally recognized Indian Tribe.

In-house expenses are expenses associated with credit management and loan servicing by the lender and the lender's contractor. In-house expenses include, but are not limited to, employee salaries, staff lawyers, travel, supplies, and overhead.

Interest Assistance Agreement is the appropriate Agency form executed by the Agency and the lender containing the terms and conditions under which the Agency will make interest assistance payments to the lender on behalf of the guaranteed loan borrower.

Intermediary means the entity requesting or receiving HPRP loan funds for establishing a revolving loan fund and relending to ultimate recipients.

Inventory property is real estate or chattel property and related rights that formerly secured an FLP loan and to which the Federal Government has acquired title.

Joint financing arrangement is an arrangement in which two or more lenders make separate loans simultaneously to supply the funds required by one applicant.

Joint operation is an operation run by individuals who have agreed to operate a farm or farms together as an entity, sharing equally or unequally land, labor, equipment, expenses, or income, or some combination of these items. The real and personal property is owned separately or jointly by the individuals.

Land contract is an installment contract executed between a buyer and a seller for the sale of real property, in which complete fee title ownership of the property is not transferred until all payments under the contract have been made.

Leasehold is a right to use farm property for a specific period of time under conditions provided for in a lease agreement.

Lender is the organization making and servicing a loan, or advancing and servicing a line of credit that is guaranteed by the Agency. The lender is also the party requesting a guarantee.

Lender's Agreement is the appropriate Agency form executed by the Agency and the lender setting forth their loan responsibilities when the Loan Guarantee is issued.

Lien is a legally enforceable claim against real or chattel property of another obtained as security for the repayment of indebtedness or an encumbrance on property to enforce payment of an obligation.

Limited resource interest rate is an interest rate normally below the Agency's regular interest rate, which is available to applicants unable to develop a feasible plan at regular rates and are requesting:

(1) FO or OL loan assistance under part 764 of this title; or

(2) Primary loan servicing on an FO, OL, or SW loan under part 766 of this title.

Line of Credit Agreement is a contract between the borrower and the lender that contains certain lender and borrower conditions, limitations, and responsibilities for credit extension and acceptance where loan principal balance may fluctuate throughout the term of the contract.

Liquidation is the act of selling security for recovery of amounts owed to the Agency or lender.

Liquidation expenses are the costs of an appraisal, due diligence evaluation, environmental assessment, outside attorney fees, and other costs incurred as a direct result of liquidating the security for a direct or guaranteed loan. Liquidation expenses do not include internal Agency expenses for a direct loan or in-house expenses for a guaranteed loan.

Livestock is a member of the animal kingdom, or product thereof, as determined by the Agency.

Loan Agreement is a contract between the borrower and the lender that contains certain lender and borrower agreements, conditions, limitations, and responsibilities for credit extension and acceptance.

Loan servicing programs include any primary loan servicing program, conservation contract, current market value buyout, and homestead protection.

Loan transaction is any loan approval or servicing action.

Loss claim is a request made to the Agency by a lender to receive a reimbursement based on a percentage of the lender's loss on a loan covered by an Agency guarantee.

Loss rate is the net amount of loan loss claims paid on FSA guaranteed loans made in the previous 7 years divided by the total loan amount of all such loans guaranteed during the same period.

Major deficiency is a deficiency that directly affects the soundness of the loan.

Majority interest is more than a 50 percent interest in an entity held by an individual or group of individuals.

Market value is the amount that an informed and willing buyer would pay an informed and willing, but not forced, seller in a completely voluntary sale.

Microloan means a type of OL or FO of $50,000 or less made using a reduced loan application. Direct MLs are made under modified eligibility and security requirements.

Mineral right is an ownership interest in minerals in land, with or without ownership of the surface of the land.

Minor deficiency is a deficiency that violates Agency regulations, but does not affect the soundness of the loan.

Mortgage is a legal instrument giving the lender a security interest or lien on real or personal property of any kind. The term "mortgage" also includes the terms "deed of trust" and "security agreement."

Natural disaster is unusual and adverse weather conditions or natural phenomena that have substantially affected farmers by causing severe physical or production, or both, losses.

Negligent servicing is servicing that fails to include those actions that are considered normal industry standards of loan management or comply with the lender's agreement or the guarantee. Negligent servicing includes failure to act or failure to act in a timely manner consistent with actions of a reasonable lender in loan making, servicing, and collection.

Negotiated sale is a sale in which there is a bargaining of price or terms, or both.

Net recovery value of security is the market value of the security property, assuming that the lender in the case of a guaranteed loan, or the Agency in the case of a direct loan, will acquire the property and sell it for its highest and best use, less the lender's or the Agency's costs of property acquisition, retention, maintenance, and liquidation.

Net recovery value of non-essential assets is the appraised market value of the non-essential assets less any prior liens and any selling costs that may include such items as taxes due, commissions, and advertising costs. However, no deduction is made for maintenance of the property while in inventory.

Non-capitalized interest is accrued interest on a loan that was not reclassified as principal at the time of restructuring. Between October 10, 1988, and November 27, 1990, the Agency did not capitalize interest that was less than 90 days past due when restructuring a direct loan.

Non-eligible enterprise is a business that meets the criteria in any one of the following categories:

(1) Produces exotic animals, birds, or aquatic organisms or their products which may be agricultural in nature, but are not normally associated with agricultural production, e.g., there is no established or stable market for them or production is speculative in nature.

(2) Produces non-farm animals, birds, or aquatic organisms ordinarily used for pets, companionship, or pleasure and not typically associated with human consumption, fiber, or draft use.

(3) Markets non-farm goods or provides services which might be agriculturally related, but are not produced by the farming operation.

(4) Processes or markets farm products when the majority of the commodities processed or marketed are not produced by the farming operation.

Non-essential assets are assets in which the borrower has an ownership interest, that:

(1) Do not contribute to:

(i) Income to pay essential family living expenses, or

(ii) The farming operation; and

(2) Are not exempt from judgment creditors or in a bankruptcy action.

Non-monetary default means a situation where a borrower is not in compliance with the covenants or requirements of the loan documents, program requirements, or loan.

Non-program loan is a loan on terms more stringent than terms for a program loan that is an extension of credit for the convenience of the Agency, because the applicant does not qualify for program assistance or the property to be financed is not suited for program purposes. Such loans are made or continued only when it is in the best interest of the Agency.

Normal income security is all security not considered basic security, including crops, livestock, poultry products, other property covered by Agency liens that is sold in conjunction with the operation of a farm or other business, and FSA Farm Program payments.

Normal production yield as used in 7 CFR part 764 for EM loans, is:

(1) The per acre actual production history of the crops produced by the farming operation used to determine Federal crop insurance payments or payment under the Noninsured Crop Disaster Assistance Program for the production year during which the disaster occurred;

(2) The applicant's own production records, or the records of production on which FSA Farm Program payments are made contained in the applicant's Farm Program file, if available, for the previous 3 years, when the actual production history in paragraph (1) of this definition is not available;

(3) The county average production yield, when the production records outlined in paragraphs (1) and (2) of this definition are not available.

Operating loan is a loan made to an eligible applicant to assist with the financial costs of operating a farm. The term also includes a Youth loan.

Operator is the individual or entity that provides the labor, management, and capital to operate the farm. The operator can be either an owner-operator or tenant-operator. Under applicable State law, an entity may have to receive authorization from the State in which the farm is located to be the owner and/or operator of the farm. Operating-only entities may be considered owner-operators when the individuals who own the farm real estate own at least 50 percent of the family farm operation.

Participated in the business operations of a farm requires that an applicant has:

(1) Been the owner, manager or operator of a farming operation for the year's complete production cycle as evidenced by tax returns, FSA farm records or similar documentation;

(2) Been employed as a farm manager or farm management consultant for the year's complete production cycle; or

(3) Participated in the operation of a farm by virtue of being raised on a farm or having worked on a farm (which can include a farm-related apprenticeship, internship, or similar educational program with applied work experience) with significant responsibility for the day-to-day decisions for the year's complete production cycle,

which may include selection of seed varieties, weed control programs, input suppliers, or livestock feeding programs or decisions to replace or repair equipment.

Partnership is any entity consisting of two or more individuals who have agreed to operate a farm as one business unit. The entity must be recognized as a partnership by the laws of the State in which the partnership will operate a farm. It also must be authorized to own both real and personal property and to incur debt in its own name.

Past due is when a payment is not made by the due date.

Physical loss is verifiable damage or destruction with respect to real estate or chattel, excluding annual growing crops.

Potential liquidation value is the amount of a lender's protective bid at a foreclosure sale. Potential liquidation value is determined by an independent appraiser using comparables from other forced liquidation sales.

Present value is the present worth of a future stream of payments discounted to the current date.

Presidentially-designated emergency is a major disaster or emergency designated by the President under the Robert T. Stafford Disaster Relief and Emergency Assistance Act (42 U.S.C. 5121 *et seq.*).

Primary loan servicing programs include:

(1) Loan consolidation and rescheduling, or reamortization;

(2) Interest rate reduction, including use of the limited resource rate program;

(3) Deferral;

(4) Write-down of the principal or accumulated interest; or

(5) Any combination of paragraphs (1) through (4) of this definition.

Production cycle is the time it takes to produce an agricultural commodity from the beginning of the production process until it is normally disposed of or sold.

Production loss is verifiable damage or destruction with respect to annual growing crops.

Program loans include CL, FO, OL, and EM. In addition, for loan servicing purposes the term includes existing loans for the following programs no longer funded: SW, RL, EE, ST, and RHF.

Promissory note is a written agreement to pay a specified sum on demand or at a specified time to the party designated. The terms "promissory note" and "note" are interchangeable.

Prospectus consists of a transmittal letter, a current balance sheet and projected year's budget which is sent to commercial lenders to determine their interest in financing or refinancing specific Agency direct loan applicants and borrowers.

Protective advance is an advance made by the Agency or a lender to protect or preserve the collateral from loss or deterioration.

Quarantine is a quarantine imposed by the Secretary under the Plant Protection Act or animal quarantine laws (as defined in section 2509 of the Food, Agriculture, Conservation and Trade Act of 1990).

Reamortization is the rewriting of rates or terms, or both, of a loan made for real estate purposes.

Reasonable rates and terms are those commercial rates and terms that other farmers are expected to meet when borrowing from a commercial lender or private source for a similar purpose and similar period of time. The "similar period of time" of available commercial loans will be measured against, but need not be the same as, the remaining or original term of the loan.

Recoverable cost is a loan cost expense chargeable to either a borrower or property account.

Recreation loan is a loan that was made to eligible applicants to assist in the conversion of all or a portion of the farm they owned or operated to outdoor income producing recreation enterprises to supplement or supplant farm income. RL's are no longer funded, however, such outstanding loans are serviced by the Agency.

Redemption right is a Federal or state right to reclaim property for a period of time established by law, by paying the amount paid at the involuntary sale plus accrued interest and costs.

Related by blood or marriage is being connected to one another as husband, wife, parent, child, brother, sister, uncle, aunt, or grandparent.

Relative is the spouse and anyone having one of the following relationships to an applicant or borrower: parent, son, daughter, sibling, stepparent, stepson, stepdaughter, stepbrother, stepsister, half brother, half sister, uncle, aunt, nephew, niece, cousin, grandparent, grandson, granddaughter, or the spouses of the foregoing.

Repossessed property is security property in the Agency's custody.

Rescheduling is the rewriting of the rates or terms, or both, of a loan made for operating purposes.

Restructuring is changing the terms of a debt through rescheduling, reamortization, deferral, writedown, or a combination thereof.

Revolved funds means the cash portion of an HPRP revolving loan fund that is not composed of HPRP loan funds, including funds that are repayments of HPRP loans and including fees and interest collected on such loans.

Rural youth is a person who has reached the age of 10 but has not reached the age of 21 and resides in a rural area or any city or town with a population of 50,000 or fewer people.

Security is property or right of any kind that is subject to a real or personal property lien. Any reference to "collateral" or "security property" will be considered a reference to the term "security."

Security instrument includes any document giving the Agency a security interest on real or personal property.

Security value is the market value of real estate or chattel property (less the value of any prior liens) used as security for an Agency loan.

Shared Appreciation Agreement is an agreement between the Agency, or a lender in the case of a guaranteed loan, and a borrower on the appropriate Agency form that requires the borrower who has received a writedown on a direct or guaranteed loan to repay the Agency or the lender some or all of the writedown received, based on a percentage of any increase in the value of the real estate securing an SAA at a future date.

Socially disadvantaged applicant or farmer is an individual or entity who is a member of a socially disadvantaged group. For an entity, the majority interest must be held by socially disadvantaged individuals. For married couples, the socially disadvantaged individual must have at least 50 percent ownership in the farm business and make most of the management decisions, contribute a significant amount of labor, and generally be recognized as the operator of the farm.

Socially disadvantaged group is a group whose members have been subject to racial, ethnic, or gender prejudice because of their identity as members of a group without regard to their individual qualities. These groups consist of: American Indians or Alaskan Natives, Asians, Blacks or African Americans, Native Hawaiians or other Pacific Islanders, Hispanics, and women.

Softwood Timber Program loan was available to eligible financially distressed borrowers who would take marginal land, including highly erodible land, out of production of agricultural commodities other than the production of softwood timber. ST loans are no longer available, however, such outstanding loans are serviced by the Agency.

Soil and Water loan is a loan that was made to an eligible applicant to encourage and facilitate the improvement, protection, and proper use of farmland by providing financing for soil conservation, water development, conservation, and use; forestation; drainage of farmland; the establishment and improvement of permanent pasture; pollution abatement and control; and other related measures consistent with all Federal, State and local environmental standards. SW loans are no longer funded, however, such outstanding loans are serviced by the Agency.

Streamlined Conservation Loan means a direct or guaranteed CL made to eligible applicants based on reduced documentation.

Subordination is a creditor's temporary relinquishment of all or a portion of its lien priority to another party providing the other party with a priority lien on the collateral.

Subsequent loan is any FLP loan processed by the Agency after an initial loan of the same type has been made to the same borrower.

Succession plan means a general plan to address the continuation of the farm, which may include specific intrafamily succession agreements or strategies to address business asset transfer planning to create opportunities for farmers and ranchers.

Supervised bank account is an account with a financial institution established through a deposit agreement entered into between the borrower, the Agency, and the financial institution.

Technical appraisal review is a review of an appraisal to determine if such appraisal meets the requirements of USPAP pursuant to standard 3 of USPAP.

Transfer and assumption is the conveyance by a debtor to an assuming party of the assets, collateral, and liabilities of a loan in return for the assuming party's binding promise to pay the debt outstanding or the market value of the collateral.

Trust is an entity that under applicable state law meets the criteria of being a trust of any kind but does not meet the criteria of being a farm cooperative, private domestic corporation, partnership, or joint operation.

Ultimate recipient means an entity or individual that receives a loan from an intermediaries' HPRP revolving loan fund.

Unaccounted for security is security for a direct or guaranteed loan that was misplaced, stolen, sold, or otherwise missing, where replacement security was not obtained or the proceeds from its sale have not been applied to the loan.

Unauthorized assistance is any loan, loan servicing action, lower interest rate, loan guarantee, or subsidy received by a borrower, or lender, for which the borrower or lender was not eligible, which was not made in accordance with all Agency procedures and requirements, or which the Agency obligated from the wrong appropriation or fund. Unauthorized assistance may result from borrower, lender, or Agency error.

Undivided ownership interest means a common interest in the whole parcel of land that is owned by two or more people. Undivided ownership interest does not include those who own a specific piece of a parcel of land; rather they own a percentage interest in a parcel of land as a whole.

Uniform Standards of Professional Appraisal Practice are standards governing the preparation, reporting, and reviewing of appraisals established by the Appraisal Foundation pursuant to the Financial Institutions Reform, Recovery, and Enforcement Act of 1989.

United States is any of the 50 States, the Commonwealth of Puerto Rico, the Virgin Islands of the United States, Guam, American Samoa, the Commonwealth of the Northern Mariana Islands, Republic of Palau, Federated States of Micronesia, and the Republic of the Marshall Islands.

U. S. Attorney is an attorney for the United States Department of Justice.

Veteran is any person who served in the military, naval, or air service during any war as defined in section 101(12) of title 38, United States Code.

Veteran farmer is a farmer who has served in the Armed Forces (as defined in 38 U.S.C. 101(10)) and who—

(1) has not operated a farm; or

(2) has operated a farm for not more than 10 years.

Wetlands are those lands or areas of land as determined by the Natural Resources Conservation Service to meet the requirements provided in section 1201 of the Food Security Act of 1985.

Working capital is cash available to conduct normal daily operations including, but not limited to, paying for feed, seed, fertilizer, pesticides, farm supplies, cooperative stock, and cash rent.

Youth loan is an operating type loan made to an eligible rural youth applicant to finance a modest income-producing agricultural project.

[72 FR 63285, Nov. 8, 2007; 72 FR 74153, Dec. 31, 2007, as amended at 73 FR 74344, Dec. 8, 2008; 75 FR 54012, Sept. 3, 2010; 76 FR 75430, Dec. 2, 2011; 77 FR 15938, May 18, 2012; 78 FR 3834, Jan. 16, 2013; 78 FR 14005, Mar. 4, 2013; 78 FR 65529, Nov. 1, 2013; 79 FR 60743, Oct. 8, 2014; 79 FR 78693, Dec. 31, 2014; 80 FR 74970, Dec. 1, 2015; 81 FR 3292, Jan. 21, 2016; 81 FR 72690, Oct. 21, 2016; 85 FR 36691, June 17, 2020; 86 FR 43390, Aug. 9, 2021]

§ 761.3　Civil rights.

Part 15d of this title contains applicable regulations pertaining to civil rights and filing of discrimination complaints by program participants.

§761.4 Conflict of interest.

The Agency enforces conflict of interest policies to maintain high standards of honesty, integrity, and impartiality in the making and servicing of direct and guaranteed loans. These requirements are established in 5 CFR parts 2635 and 8301.

§761.5 Restrictions on lobbying.

A person who applies for or receives a loan made or guaranteed by the Agency must comply with the restrictions on lobbying in 2 CFR part 418.

[72 FR 63285, Nov. 8, 2007, as amended at 79 FR 75996, Dec. 19, 2014]

§761.6 Appeals.

Except as provided in 7 CFR part 762, appeal of an adverse decision made by the Agency will be handled in accordance with 7 CFR parts 11 and 780.

§761.7 Appraisals.

(a) *General.* This section describes Agency requirements for:

(1) Real estate and chattel appraisals made in connection with the making and servicing of direct FLP and Non-program loans; and

(2) Appraisal reviews conducted on appraisals made in connection with the making and servicing of direct and guaranteed FLP and Non-program loans.

(b) *Appraisal standards.* (1) Real estate appraisals, technical appraisal reviews and their respective forms must comply with the standards contained in USPAP, as well as applicable Agency regulations and procedures for the specific FLP activity involved. Applicable appraisal procedures and regulations are available for review in each Agency State Office.

(2) When a chattel appraisal is required, it must be completed on an applicable Agency form (available in each Agency State Office) or other format containing the same information.

(c) *Use of an existing real estate appraisal.* Except where specified elsewhere, when a real estate appraisal is required, the Agency will use the existing real estate appraisal to reach loan making or servicing decisions under either of the following conditions:

(1) The appraisal was completed within the previous 18 months and the Agency determines that:

(i) The appraisal meets the provisions of this section and the applicable Agency loan making or servicing requirements; and

(ii) Market values have remained stable since the appraisal was completed; or

(2) The appraisal was not completed in the previous 18 months, but has been updated by the appraiser or appraisal firm that completed the appraisal, and both the update and the original appraisal were completed in accordance with USPAP.

(d) *Appraisal reviews.* (1) With respect to a real estate appraisal, the Agency may conduct a technical appraisal review or an administrative appraisal review, or both.

(2) With respect to a chattel appraisal, the Agency may conduct an administrative appraisal review.

(e) *Appraisal appeals.* Challenges to an appraisal used by the Agency are limited as follows:

(1) When an applicant or borrower challenges a real estate appraisal used by the Agency for any loan making or loan servicing decision, except primary loan servicing decisions as specified in §766.115 of this chapter, the issue for review is limited to whether the appraisal used by the Agency complies with USPAP. The applicant or borrower must submit a technical appraisal review prepared by a State Certified General Appraiser that will be used to determine whether the Agency's appraisal complies with USPAP. The applicant or borrower is responsible for obtaining and paying for the technical appraisal review.

(2) When an applicant or borrower challenges a chattel appraisal used by the Agency for any loan making or loan servicing decision, except for primary loan servicing decisions as specified in §766.115 of this chapter, the issue for review is limited to whether the appraisal used by the Agency is consistent with present market values of similar items in the area. The applicant or borrower must submit an independent appraisal review that will be

used to determine whether the appraisal is consistent with present market values of similar items in the area. The applicant or borrower is responsible for obtaining and paying for the independent appraisal.

[72 FR 63285, Nov. 8, 2007, as amended at 78 FR 65529, Nov. 1, 2013; 79 FR 78693, Dec. 31, 2014; 81 FR 72690, Oct. 21, 2016; 86 FR 43390, Aug. 9, 2021]

§ 761.8 Loan Limitations.

(a) *Dollar limits.* The outstanding principal balances for an applicant or anyone who will sign the promissory note cannot exceed any of the following at the time of loan closing or assumption of indebtedness. If the outstanding principal balance exceeds any of the limits at the time of approval, the farm operating plan must reflect that funds will be available to reduce the indebtedness prior to loan closing or assumption of indebtedness.

(1) Farm Ownership, Down payment loans, Conservation loans, and Soil and Water loans:

(i) Direct—$300,000;

(ii) Guaranteed—$700,000 (for fiscal year 2000 and increased at the beginning of each fiscal year in accordance with paragraph (b) of this section);

(iii) Any combination of a direct Farm Ownership loan, direct Conservation loan, direct Soil and Water loan, guaranteed Farm Ownership loan, guaranteed Conservation loan, and guaranteed Soil and Water loan-$700,000 (for fiscal year 2000 and increased each fiscal year in accordance with paragraph (b) of this section);

(2) Operating loans:

(i) Direct—$300,000;

(ii) Guaranteed—$700,000 (for fiscal year 2000 and increased each fiscal year in accordance with paragraph (b) of this section);

(iii) Any combination of a direct Operating loan and guaranteed Operating loan—$700,000 (for fiscal year 2000 and increased each fiscal year in accordance with paragraph (b) of this section);

(3) Any combination of guaranteed Farm Ownership loan, guaranteed Conservation loan, guaranteed Soil and Water loan, and guaranteed Operating loan-$700,000 (for fiscal year 2000 and increased each fiscal year in accordance with paragraph (b) of this section);

(4) Any combination of direct Farm Ownership loan, direct Conservation loan, direct Soil and Water loan, direct Operating loan, guaranteed Farm Ownership loan, guaranteed Conservation loan, guaranteed Soil and Water loan, and guaranteed Operating loan-the amount in paragraph (a)(1)(ii) of this section plus $300,000;

(5) Emergency loans—$500,000;

(6) Any combination of direct Farm Ownership loan, direct Conservation loan, direct Soil and Water loan, direct Operating loan, guaranteed Farm Ownership, guaranteed Conservation loan, guaranteed Soil and Water loan, guaranteed Operating loan, and Emergency loan-the amount in paragraph (a)(1)(ii) of this section plus $800,000.

(b) *Guaranteed loan limit.* The dollar limits of guaranteed loans will be increased each fiscal year based on the percentage change in the Prices Paid by Farmers Index as compiled by the National Agricultural Statistics Service, USDA. The maximum loan limits for the current fiscal year are available in any FSA office and on the FSA website at *http://www.fsa.usda.gov.*

(c) *Line of credit advances.* The total dollar amount of guaranteed line of credit advances and income releases cannot exceed the total estimated expenses, less interest expense, as indicated on the borrower's cash flow budget, unless the cash flow budget is revised and continues to reflect a feasible plan.

[72 FR 63285, Nov. 8, 2007, as amended at 73 FR 74345, Dec. 8, 2008; 75 FR 54012, Sept. 3, 2010; 86 FR 43390, Aug. 9, 2021]

§ 761.9 Interest rates for direct loans.

Interest rates for all direct loans are set in accordance with the Act. A copy of the current interest rates may be obtained in any Agency office.

§ 761.10 Planning and performing construction and other development.

(a) *Purpose.* This section describes Agency policies regarding the planning and performing of construction and other development work performed with:

(1) Direct FLP loan funds; or

(2) Insurance or other proceeds resulting from damage or loss to direct loan security.

(b) *Funds for development work.* The applicant or borrower:

(1) Must provide the Agency with an estimate of the total cash cost of all planned development prior to loan approval;

(2) Must show proof of sufficient funds to pay for the total cash cost of all planned development at or before loan closing;

(3) Must not incur any debts for materials or labor or make any expenditures for development purposes prior to loan closing with the expectation of being reimbursed from Agency loan funds.

(c) *Scheduling, planning, and completing development work.* The applicant or borrower:

(1) Is responsible for scheduling and planning development work in a manner acceptable to the Agency and must furnish the Agency information fully describing the planned development, the proposed schedule, and the manner in which it will be accomplished;

(2) Is responsible for obtaining all necessary State and local construction approvals and permits prior to loan closing;

(3) Must ensure that all development work meets the environmental requirements established in part 799 of this chapter;

(4) Must schedule development work to start as soon as feasible after the loan is closed and complete work as quickly as practicable;

(5) Is responsible for obtaining any required technical services from qualified technicians, tradespeople, and contractors.

(d) *Construction and repair standards.* (1) The construction of a new building and the alteration or repair of an existing building must conform with industry-acceptable construction practices and standards.

(2) All improvements to a property must conform to applicable laws, ordinances, codes, and regulations.

(3) The applicant or borrower is responsible for selecting a design standard that meets all applicable local and state laws, ordinances, codes, and regulations, including building, plumbing, mechanical, electrical, water, and waste management.

(4) The Agency will require drawings, specifications, and estimates to fully describe the work as necessary to protect the Agency's financial interests. The drawings and specifications must identify any specific development standards being used. Such information must be sufficiently complete to avoid any misunderstanding as to the extent, kind, and quality of work to be performed.

(5) The Agency will require technical data, tests, or engineering evaluations to support the design of the development as necessary to protect its financial interests.

(6) The Agency will require the applicant or borrower to provide written certification that final drawings and specifications conform with the applicable development standard as necessary to protect its financial interests. Certification must be obtained from individuals or organizations trained and experienced in the compliance, interpretation, or enforcement of the applicable development standards, such as licensed architects, professional engineers, persons certified by a relevant national model code organization, authorized local building officials, or national code organizations.

(e) *Inspection.* (1) The applicant or borrower is responsible for inspecting development work as necessary to protect their interest.

(2) The applicant or borrower must provide the Agency written certification that the development conforms to the plans and good construction practices, and complies with applicable laws, ordinances, codes, and regulations.

(3) The Agency will require the applicant or borrower to obtain professional inspection services during construction as necessary to protect its financial interests.

(4) Agency inspections do not create or imply any duty or obligation of the Agency to the applicant or borrower.

(f) *Warranty and lien waivers.* The applicant or borrower must obtain and submit all lien waivers on any construction before the Agency will issue final payment.

(g) *Surety.* The Agency will require surety to guarantee both payment and performance for construction contracts as necessary to protect its financial interests.

(h) *Changing the planned development.* An applicant or borrower must request, in writing, Agency approval for any change to a planned development. The Agency will approve a change if all of the following are met:

(1) It will not reduce the value of the Agency's security;

(2) It will not adversely affect the soundness of the farming operation;

(3) It complies with all applicable laws and regulations;

(4) It is for an authorized loan purpose;

(5) It is within the scope of the original loan proposal;

(6) If required, documentation that sufficient funding for the full amount of the planned development is approved and available;

(7) If required, surety to cover the full revised development amount has been provided; and

(8) The modification is certified in accordance with paragraph (d) (6) of this section.

[72 FR 63285, Nov. 8, 2007, as amended at 81 FR 51284, Aug. 3, 2016]

§§ 761.11–761.50 [Reserved]

Subpart B—Supervised Bank Accounts

§ 761.51 Establishing a supervised bank account.

(a) Supervised bank accounts will be used to:

(1) Assure correct use of funds are planned and released for capital purchases, construction projects, site development work, debt refinancing, or proceeds from the sale of basic security, and perfection of the Agency's security interest in assets purchased or refinanced when electronic funds transfer or treasury check processes are not practicable;

(2) Protect the Agency's security interest in insurance indemnities or other loss compensation resulting from loss or damage to loan security; or

(3) Assist borrowers with limited financial skills with cash management, subject to the following conditions:

(i) Use of a supervised bank for this purpose will be temporary and infrequent;

(ii) The need for a supervised bank account in this situation will be determined on a case-by-case basis; and

(iii) The borrower agrees to the use of a supervised bank account for this purpose by executing the deposit agreement.

(b) The borrower may select the financial institution in which the account will be established, provided the institution is Federally insured. If the borrower does not select an institution, the Agency will choose one.

(c) Only one supervised bank account will be established for any borrower.

(d) If both spouses sign an FLP note and security agreement, the supervised bank account will be established as a joint tenancy account with right of survivorship from which either borrower can withdraw funds.

(e) If the funds to be deposited into the account cause the balance to exceed the maximum amount insurable by the Federal Government, the financial institution must agree to pledge acceptable collateral with the Federal Reserve Bank for the excess over the insured amount, before the deposit is made.

(1) If the financial institution is not a member of the Federal Reserve System, the institution must pledge acceptable collateral with a correspondent bank that is a member of the Federal Reserve System. The correspondent bank must inform the Federal Reserve Bank that it is holding securities pledged for the supervised bank account in accordance with 31 CFR part 202 (Treasury Circular 176).

(2) When the balance in the account has been reduced, the financial institution may request a release of part or all of the collateral, as applicable, from the Agency.

[72 FR 63285, Nov. 8, 2007, as amended at 76 FR 5057, Jan. 28, 2011; 86 FR 43390, Aug. 9, 2021]

§761.52 Deposits into a supervised bank account.

(a) Checks or money orders may be deposited into a supervised bank account provided they are not payable:

(1) Solely to the Federal Government or any agency thereof; or

(2) To the Treasury of the United States as a joint payee.

(b) Loan proceeds may be deposited electronically.

§761.53 Interest bearing accounts.

(a) A supervised bank account, if possible, will be established as an interest bearing deposit account provided that the funds will not be immediately disbursed, and the account is held jointly by the borrower and the Agency if this arrangement will benefit the borrower.

(b) Interest earned on a supervised bank account will be treated as normal income security.

§761.54 Withdrawals from a supervised bank account.

(a) The Agency will authorize a withdrawal from the supervised bank account for an approved purpose after ensuring that:

(1) Sufficient funds in the supervised bank account are available;

(2) No loan proceeds are disbursed prior to confirmation of proper lien position, except to pay for lien search if needed;

(3) No checks are issued to "cash;" and

(4) The use of funds is consistent with the current farm operating plan or other agreement with the Agency.

(b) A check must be signed by the borrower with countersignature of the Agency, except as provided in paragraph (c) of this section. All checks must bear the legend "countersigned, not as co-maker or endorser."

(c) The Agency will withdraw funds from a supervised bank account without borrower counter-signature only for the following purposes:

(1) For application on Agency indebtedness;

(2) To refund Agency loan funds;

(3) To protect the Agency's lien or security;

(4) To accomplish a purpose for which such advance was made; or

(5) In the case of a deceased borrower, to continue to pay necessary farm expenses to protect Agency security in conjunction with the borrower's estate.

§761.55 Closing a supervised bank account.

(a) If the supervised bank account is no longer needed and the loan account is not paid in full, the Agency will determine the source of the remaining funds in the supervised bank account. If the funds are determined to be:

(1) Loan funds:

(i) From any loan type, except Youth loan, and the balance is less than $1,000, the Agency will provide the balance to the borrower to use for authorized loan purposes;

(ii) From a Youth loan, and the balance is less than $100, the Agency will provide the balance to the borrower to use for authorized loan purposes;

(2) Loan funds:

(i) From any loan type, except Youth loan, and the balance is $1,000 or greater, the Agency will apply the balance to the FLP loan;

(ii) From a Youth loan, and the balance is $100 or greater, the Agency will apply the balance to the FLP loan;

(3) Normal income funds, the Agency will apply the balance to the remaining current year's scheduled payments and pay any remaining balance to the borrower; and

(4) Basic security funds, the Agency will apply the balance to the FLP loan as an extra payment or the borrower may apply the balance toward the purchase of basic security, provided the Agency obtains a lien on such security and its security position is not diminished.

(b) If the borrower is uncooperative in closing a supervised bank account, the Agency will make written demand to the financial institution for the balance and apply it in accordance with paragraph (a) of this section.

(c) In the event of a borrower's death, the Agency may:

(1) Apply the balance to the borrower's FLP loan;

(2) Continue with a remaining borrower, provided the supervised bank account was established as a joint tenancy with right of survivorship account;

(3) Refund unobligated balances from other creditors in the supervised bank account for specific operating purposes in accordance with any prior written agreement between the Agency and the deceased borrower; or

(4) Continue to pay expenses from the supervised bank account in conjunction with the borrower's estate.

§§ 761.56–761.100 [Reserved]

Subpart C—Progression Lending

§ 761.101 Applicability.

This subpart applies to all direct applicants and borrowers, except borrowers with only Non-program loans.

§ 761.102 Borrower recordkeeping, reporting, and supervision.

(a) A borrower must maintain accurate records sufficient to make informed management decisions and to allow the Agency to render loan making and servicing decisions in accordance with Agency regulations. These records must include the following:

(1) Production (e.g., total and per unit for livestock and crops);

(2) Revenues, by source;

(3) Other sources of funds, including borrowed funds;

(4) Operating expenses;

(5) Interest;

(6) Family living expenses;

(7) Profit and loss;

(8) Tax-related information;

(9) Capital expenses;

(10) Outstanding debt; and

(11) Debt repayment.

(b) A borrower also must agree in writing to:

(1) Cooperate with the Agency and comply with all supervisory agreements, farm assessments, farm operating plans, year-end analyses, and all other loan-related requirements and documents;

(2) Submit financial information and an updated farm operating plan when requested by the Agency;

(3) Immediately notify the Agency of any proposed or actual significant change in the farming operation, any significant changes in family income, expenses, or the development of problem situations, or any losses or proposed significant changes in security.

(c) If the borrower fails to comply with these requirements, unless due to reasons outside the borrower's control, the non-compliance may adversely impact future requests for assistance.

§ 761.103 Farm assessment.

(a) The Agency, in collaboration with the applicant, will assess the farming operation to:

(1) Determine the applicant's financial condition, organizational structure, and management strengths and weaknesses;

(2) Identify and prioritize training and supervisory needs; and

(3) Develop a progressive lending plan to assist the borrower in achieving financial viability and transitioning to private commercial credit or other sources of credit in the shortest time practicable, except for CL.

(b) Except for ML, the initial assessment must evaluate, at a minimum, the:

(1) Farm organization and key personnel qualifications;

(2) Type of farming operation;

(3) Goals for the operation;

(4) Adequacy of real estate, including facilities, to conduct the farming operation;

(5) Adequacy of chattel property used to conduct the farming operation;

(6) Historical performance, except for streamlined CL;

(7) Farm operating plan;

(8) Progression lending plan, except for streamlined CL;

(9) Training plan; and

(10) Graduation plan, except for CL.

(c) For ML, the Agency will complete a narrative that will evaluate, at a minimum, the:

(1) Type of farming operation and adequacy of resources;

(2) Amount of assistance necessary to cover expenses to carry out the proposed farm operating plan, including building an adequate equity base;

(3) The goals of the operation;

(4) The financial viability of the entire operation, including a marketing plan, and available production history, as applicable;

(5) Progression lending plan; and

(6) Training plan.

(d) An assessment update must be prepared for each subsequent loan. The update must include a farm operating plan and any other items discussed in paragraph (b) of this section that have significantly changed since the initial assessment.

(e) The Agency reviews the assessment to determine a borrower's progress at least annually, combining any required classification and graduation reviews as part of the review. For streamlined CLs, the borrower must provide a current balance sheet and income tax records. Any negative trends noted between the previous years' and the current years' information must be evaluated and addressed in the assessment of the streamlined CL borrower.

(f) If a CL borrower becomes financially distressed, delinquent, or receives any servicing options available under part 766 of this chapter, all elements of the assessment in paragraph (b) of this section must be addressed.

[72 FR 63285, Nov. 8, 2007, as amended at 75 FR 54012, Sept. 3, 2010; 76 FR 5057, Jan. 28, 2011; 78 FR 3835, Jan. 17, 2013; 78 FR 65529, Nov. 1, 2013; 86 FR 43391, Aug. 9, 2021]

§761.104 Developing the farm operating plan.

(a) An applicant or borrower must submit a farm operating plan to the Agency, upon request, for loan making or servicing purposes.

(b) An applicant or borrower may request Agency assistance in developing the farm operating plan.

(c) The farm operating plan will be based on accurate and verifiable information.

(1) Historical information will be used as a guide.

(2) Positive and negative trends, mutually agreed upon changes and improvements, and current input prices will be taken into consideration when arriving at reasonable projections.

(3) Projected yields will be calculated according to the following priorities:

(i) The applicant or borrower's own production records for the previous 3 years;

(ii) The per-acre actual production history of the crops produced by the farming operation used to determine Federal crop insurance payments, if available;

(iii) FSA Farm Program actual yield records;

(iv) County averages;

(v) State averages.

(4) If the applicant or borrower's production history has been substantially affected by a disaster declared by the President or designated by the Secretary of Agriculture, or the applicant or borrower has had a qualifying loss from such disaster but the farming operation was not located in a declared or designated disaster area, the applicant or borrower may:

(i) Use county average yields, or state average yields if county average yields are not available, in place of the disaster year yields when the county or state average yields are realistic and reasonable compared to the applicant's actual non-disaster year yields, as determined by the agency approval official; or

(ii) Exclude the production year with the lowest actual or county average yield if their yields were affected by disasters during at least 2 of the 3 years.

(d) Unit prices for agricultural commodities established by the Agency will generally be used. Applicants and borrowers that provide evidence that they will receive a premium price for a commodity may use a price above the price established by the Agency.

(e) For MLs, when projected yields and unit prices cannot be determined as specified in paragraphs (c) and (d) of this section because the data is not available or practicable, other documentation from other reliable sources may be used to assist in developing the applicant's farm operating plan.

(f) Except as provided in paragraph (g) of this section, the applicant or borrower must sign the final farm operating plan prior to approval of any loan or servicing action.

(g) If the Agency believes the applicant or borrower's farm operating plan is inaccurate, or the information upon which it is based cannot be verified, the Agency will discuss and try to resolve the concerns with the applicant or borrower. If an agreement cannot be reached, the Agency will make loan approval and servicing determinations based on the Agency's revised farm operating plan.

[72 FR 63285, Nov. 8, 2007, as amended at 78 FR 3835, Jan. 17, 2013; 86 FR 43391, Aug. 9, 2021]

§ 761.105 Year-end analysis.

(a) The Agency conducts a year-end analysis at its discretion or if the borrower:

(1) Is being considered for a new direct loan or subordination;

(2) Is financially distressed or delinquent;

(3) Has a loan deferred, excluding deferral of an installment under subpart B of part 766; or

(4) Is receiving a limited resource interest rate on any loan, in which case the review will be completed at least every 2 years.

(b) To the extent practicable, the year-end analysis will be completed within 60 days after the end of the business year or farm budget planning period and must include:

(1) An analysis comparing actual income, expenses, and production to projected income, expenses, and production for the preceding production cycle; and

(2) An updated farm operating plan.

[72 FR 63285, Nov. 8, 2007, as amended at 75 FR 54013, Sept. 3, 2010; 86 FR 43391, Aug. 9, 2021]

§§ 761.106–761.200 [Reserved]

Subpart D—Allocation of Farm Loan Programs Funds to State Offices

§ 761.201 Purpose.

(a) This subpart addresses:

(1) The allocation of funds for direct and guaranteed FO, CL, and OL loans;

(2) The establishment of socially disadvantaged target participation rates; and

(3) The reservation of loan funds for beginning farmers.

(b) The Agency does not allocate EM loan funds to State Offices but makes funds available following a designated or declared disaster. EM loan funds are available on a first-come first-served basis.

(c) State funding information is available for review in any State Office.

[72 FR 63285, Nov. 8, 2007, as amended at 75 FR 54013, Sept. 3, 2010]

§ 761.202 Timing of allocations.

The Agency's National Office allocates funds for FO, CL, and OL loans to the State Offices on a fiscal year basis, as made available by the Office of Management and Budget. However, the National Office will retain control over the funds when funding or administrative constraints make allocation to State Offices impractical.

[72 FR 63285, Nov. 8, 2007, as amended at 75 FR 54013, Sept. 3, 2010]

§ 761.203 National reserves for Farm Ownership and Operating loans.

(a) *Reservation of funds.* At the start of each fiscal year, the National Office reserves a portion of the funds available for each direct and guaranteed loan program. These reserves enable the Agency to meet unexpected or justifiable program needs during the fiscal year.

(b) *Allocation of reserved funds.* The National Office distributes funds from the reserve to one or more State Offices to meet a program need or Agency objective.

§ 761.204 Methods of allocating funds to State Offices.

FO, CL, and OL loan funds are allocated to State Offices using one or more of the following allocation methods:

(a) Formula allocation, if data, as specified in § 761.205, is available to use the formula for the State.

(b) Administrative allocation, if the Agency cannot adequately meet program objectives with a formula allocation. The National Office determines the amount of an administrative allocation on a case-by-case basis.

(c) Base allocation, to ensure funding for at least one loan in each State, District, or County Office. In making a base allocation, the National Office may use criteria other than those used in the formula allocation, such as historical Agency funding information.

[72 FR 63285, Nov. 8, 2007, as amended at 75 FR 54013, Sept. 3, 2010]

§ 761.205 Computing the formula allocation.

(a) The formula allocation for FO, CL, or OL loan funds is equal to:

(1) The amount available for allocation by the Agency minus the amounts held in the National Office reserve and distributed by base and administrative allocation, multiplied by

(2) The State Factor, which represents the percentage of the total amount of the funds for a loan program that the National Office allocates to a State Office.

formula allocation = (amount available for allocation – national reserve – base allocation – administrative allocation) × State Factor

(b) To calculate the State Factor, the Agency:

(1) Uses the following criteria, data sources, and weights:

Criteria	Loan type criterion is used for	Data source	Weight for FO loans (percent)	Weight for OL loans (percent)
Farm operators with sales of $2,500–$39,999 and less than 200 days work off the farm.	FO, CL, and OL loans	U.S. Census of Agriculture.	15	15
Farm operators with sales of $40,000 or more and less than 200 days work off farm.	FO, CL, and OL loans	U.S. Census of Agriculture.	35	35
Tenant farm operators	FO, CL, and OL loans	U.S. Census of Agriculture.	25	20
3-year average net farm income	FO, CL, and OL loans	USDA Economic Research Service.	15	15
Value of farm real estate assets	FOs and CLs	USDA Economic Research Service.	10	N/A
Value of farm non-real estate assets	OL loans	USDA Economic Research Service.	N/A	15

(2) Determines each State's percentage of the national total for each criterion;

(3) Multiplies the percentage for each State determined in paragraph (b)(2) of this section by the applicable weight for that criterion;

(4) Sums the weighted criteria for each State to obtain the State factor.

[72 FR 63285, Nov. 8, 2007, as amended at 75 FR 54013, Sept. 3, 2010]

§ 761.206 Pooling of unobligated funds allocated to State Offices.

The Agency periodically pools unobligated FO, CL, and OL loan funds that have been allocated to State Offices. When pooling these funds, the Agency places all unobligated funds in the appropriate National Office reserve. The pooled funds may be retained in the national reserve or reallocated to the States.

[72 FR 63285, Nov. 8, 2007, as amended at 75 FR 54013, Sept. 3, 2010]

§ 761.207 Distribution of loan funds by State Offices.

A State Office may distribute its allocation of loan funds to District or County level using the same allocation methods that are available to the National Office. State Offices may reserve a portion of the funds to meet unexpected or justifiable program needs during the fiscal year.

§ 761.208 Target participation rates for socially disadvantaged groups.

(a) *General.* (1) The Agency establishes target participation rates for providing FO, CL, and OL loans to members of socially disadvantaged groups.

(2) The Agency sets the target participation rates for State and County levels annually.

(3) When distributing loan funds in counties within Indian reservations, the Agency will allocate the funds on a reservation-wide basis.

(4) The Agency reserves and allocates sufficient loan funds to achieve these target participation rates. The Agency

may also use funds that are not reserved and allocated for socially disadvantaged groups to make or guarantee loans to members of socially disadvantaged groups.

(b) *FO and CL, loans based on ethnicity or race.* The FO and CL, loan target participation rate based on ethnicity or race in each:

(1) State is equal to the percent of the total rural population in the State who are members of such socially disadvantaged groups.

(2) County is equal to the percent of rural population in the county who are members of such socially disadvantaged groups.

(c) *OL loans based on ethnicity or race.* The OL loan target participation rate based on ethnicity or race in each:

(1) State is equal to the percent of the total number of farmers in the State who are members of such socially disadvantaged groups.

(2) County is equal to the percent of the total number of farmers in the county who are members of socially disadvantaged ethnic groups.

(d) *Women farmers.* (1) The target participation rate for women farmers in each:

(i) State is equal to the percent of farmers in the State who are women.

(ii) County is equal to the percent of farmers in the county who are women.

(2) In developing target participation rates for women, the Agency will consider the number of women who are current farmers and potential farmers.

[72 FR 63285, Nov. 8, 2007, as amended at 75 FR 54013, Sept. 3, 2010]

§ 761.209 Loan funds for beginning farmers.

Each fiscal year, the Agency reserves a portion of direct and guaranteed FO and OL loan funds for beginning farmers in accordance with section 346(b)(2) of the Act.

§ 761.210 CL funds.

(a) The following applicants and conservation projects will receive priority for CL funding:

(1) Beginning farmer or socially disadvantaged farmer,

(2) An applicant who will use the loan funds to convert to a sustainable or organic agriculture production system as evidenced by one of the following:

(i) A conservation plan that states the applicant is moving toward a sustainable or organic production system, or

(ii) An organic plan, approved by a certified agent and the State organic certification program, or

(iii) A grant awarded by the Sustainable Agriculture Research and Education (SARE) program of the National Institute of Food and Agriculture, USDA.

(3) An applicant who will use the loan funds to build conservation structures or establish conservation practices to comply with 16 U.S.C. 3812 (section 1212 of the Food Security Act of 1985) for highly erodible land.

(b) [Reserved]

[75 FR 54013, Sept. 3, 2010]

§ 761.211 Transfer of funds.

If sufficient unsubsidized guaranteed OL funds are available, then beginning on:

(a) August 1 of each fiscal year, the Agency will use available unsubsidized guaranteed OL loan funds to make approved direct FO loans to beginning farmers and socially disadvantaged farmers under the Down payment loan program; and

(b) September 1 of each fiscal year the Agency will use available unsubsidized guaranteed OL loan funds to make approved direct FO loans to beginning farmers.

[72 FR 63285, Nov. 8, 2007, as amended at 73 FR 74345, Dec. 8, 2008. Redesignated at 75 FR 54013, Sept. 3, 2010, as amended at 86 FR 43391, Aug. 9, 2021]

Subpart F—Farm Loan Programs Debt Settlement

SOURCE: 85 FR 36691, June 17, 2020, unless otherwise noted.

§ 761.401 Purpose.

(a) This subpart describes the Agency's policies for debt settlement as authorized by the Consolidated Farm and Rural Development Act (CONACT) (7 U.S.C. 1921, 7 U.S.C. 1981, 1981a, 1981d, and 2008h).

(b) FLP debts that cannot be debt settled using CONACT debt settlement authority such as when a borrower has received previous debt forgiveness on another direct loan made under the CONACT, will be processed as specified in 31 U.S.C. chapter 37 and 31 CFR parts 900 through 904.

§761.402 Abbreviations and definitions.

(a) Abbreviations and definitions for terms used in this subpart are provided in 7 CFR part 3 and §761.2.

(b) Definitions used only in this subpart include:

(1) *Third party converter* means an individual or entity who:

(i) Is in possession of agency security property, or money from the sale of security, in relation to a loan or other debt that the individual or entity was not liable for; or

(ii) Assists, or participates knowingly or unknowingly, in the transportation or sale of agency security, in relation to a loan or other debt that the individual or entity was not liable for; or

(iii) Assists, or participates knowingly or unknowingly, in temporarily or permanently relocating or concealing the location of agency security property, or money from the sale of agency security, in relation to a loan or other debt that the individual or entity was not liable for.

(2) [Reserved]

§761.403 General.

(a) The Agency will settle debts that result from, except as otherwise specified in this section:

(1)(i) Farm Ownership loans (part 764, subpart D of this chapter), including down payment loans (764, subpart E of this chapter);

(ii) Operating loans (part 764, subpart G of this chapter), including microloans part 764 of this chapter), and youth loans (part 764, subpart H of this chapter);

(iii) Emergency loans (part 764, subpart I of this chapter);

(iv) Conservation loans (part 764, subpart F of this chapter);

(v) Economic Emergency loans (serviced under parts 761 through 767 of this chapter); softwood timber loans; Soil

and Water loans; Individual Recreation Loans; Irrigation and Drainage loans; and Shift-in-land-use (Grazing Association) loans;

(2) Costs associated with servicing a borrower's account including, but not limited to, Uniform Commercial Code filing fees, surveys, appraisals, protective advances, and liquidation expenses;

(3) Debts reduced to judgment;

(4) Non-Program Loans;

(5) Amounts the Agency is authorized to recapture through agreements such as the Shared Appreciation Agreement (part 766, subpart E of this chapter);

(6) Loss claims paid on guaranteed loans (part 762 of this chapter);

(7) Unauthorized assistance;

(8) Amounts the Agency may collect from third party converters, or other individuals or entities having possession of security for FLP loans or monies obtained through the sale of FLP loan security; and

(9) Debt returned to the Agency from the Treasury cross-servicing program.

(b) The debtor's signature is not required to process some debt settlement actions. These cases include, but are not limited to, debts discharged in bankruptcy and debts returned from Treasury's cross-servicing program with amounts still owing when no further collection can be taken.

(c) FSA will not engage in settlement of a debt if:

(1) Foreclosure of security has been initiated and is pending with Justice, unless Justice has advised FSA that it does not object to the settlement; or

(2) Debts that have been referred to Justice for a judgment, or a judgement has been obtained by the United States Attorney or Justice, unless Justice closes its file and releases the judgement back to FSA for continued servicing; or

(3) The debtor's account is involved in a fiscal irregularity investigation in which final action has not been taken or the account shows evidence that a shortage may exist and an investigation will be requested.

(d) The Agency will consider settlement of a debt only when:

(1) All security has been liquidated and the proceeds, less any prior lien amounts, have been applied to the

debt; or the Agency received a lump sum payment equal to the security's current market value, less any prior lien amounts, and

(2) Payment is received based on the Agency's determination of the amount the borrower can pay to resolve the remaining balance owed on the unsecured debt.

(3) The lump sum payment made under paragraph (d)(1) of this section for the security's market value may be submitted by the borrower, an individual authorized to act for the borrower pursuant to a power of attorney document or court order, or an individual who is not an obligor on the debt but who has an ownership interest in the security.

(e) If an FLP loan has been accelerated and all security has been liquidated, and the agency has approved an adjustment debt settlement offer in accordance with this subpart, voluntary payments and involuntary payments (such as offsets) will be applied in the following order, as applicable:

(1) Recoverable costs and protective advances plus interest;

(2) Loan principal;

(3) Deferred non-capitalized interest;

(4) Accrued deferred interest; and

(5) Interest accrual to date of payment.

(f) Settlement of FLP debt referred to Treasury's cross-servicing program and returned to the Agency as uncollectible will not be processed for the borrower until all FLP debts referred to the cross-servicing program for that borrower have been returned, with or without payment agreements.

[85 FR 36691, June 17, 2020, as amended at 86 FR 10441, Feb. 22, 2021]

§ 761.404 Eligibility.

(a) A borrower is eligible for debt settlement if the borrower:

(1) Meets the requirements for the particular type of debt settlement under this part; and

(2) Submits a complete application for debt settlement as specified in § 761.405.

(b) All parties liable for the debt must submit a complete application with the following exceptions:

(1) The applicable information required in § 761.405 can be provided by the administrator or executor of the Estate, heir, or other authorized person who can sign the debt settlement application; or compiled by FSA staff when a signature cannot be obtained.

(2) The debt may be settled when the borrower has no known assets or income from which collection can be made, has disappeared and cannot be located without undue expense, and there is no security remaining for the debt.

(3) In cases where the full amount of the unsecured debt cannot be collected in a reasonable time by legal action or through enforced collection proceedings, the Agency may consider a debt settlement offer submitted by a borrower without requiring a complete application. When evaluating these offers, the Agency will consider the likelihood of the debtor obtaining a larger income or additional assets, including inheritance prospects within 5 years, from which legal or enforced collection could be made.

(c) A borrower is not eligible for debt settlement if:

(1) The borrower is indebted on another active FLP loan that the borrower cannot or will not debt settle; or

(2) The debt has been referred to the OIG, OGC, or Justice because of suspected civil or criminal violation, unless investigation was declined or advice was provided that the debt can be canceled, compromised, or adjusted.

§ 761.405 Application.

(a) A borrower requesting debt settlement must submit complete and accurate information from which the Agency can make a full determination of the borrower's financial circumstances and repayment ability. Except for the situations listed in § 761.404(b), each liable party, must submit the following:

(1) One completed original debt settlement application on the applicable Agency form signed by all parties liable for the debt;

(2) A current financial statement;

(3) A cash flow projection for the next production or earnings period;

(4) Verification of employment or other earned income, including verification of a nondebtor spouse's income which will be included as available to pay family living expenses;

(5) Verification of assets including, but not limited to, cash, checking accounts, savings accounts, certificates of deposit, individual retirement accounts, retirement and pension funds, mutual funds, stocks, bonds, and accounts receivable;

(6) Verification of debts exceeding an amount determined by the Agency;

(7) Copies of complete Federal income tax returns for the previous 3 years; and

(8) Any other items requested by the Agency to evaluate the debtor's financial condition.

(b) [Reserved]

[85 FR 36691, June 17, 2020, as amended at 86 FR 43391, Aug. 9, 2021]

§ 761.406 Types of debt settlement.

(a) *Compromise.* The Agency may compromise a debt owed to the Agency if the requirements of this subpart are met and:

(1) The borrower pays a lump sum as a compromise for the remaining unsecured debt; and

(2) The amount is reasonable based on the Agency's determination of what the borrower can pay to settle the debt.

(b) *Adjustment.* The Agency may settle a debt owed to the Agency through an adjustment agreement if the requirements of this subpart are met and:

(1) The borrower agrees to pay the adjustment amount for a period of time not to exceed 5 years; and

(2) The amount is reasonable based on the Agency's determination of what the borrower can pay to settle the. debt; and

(3) The borrower provides documentation that funds are, or will be, available to pay the adjustment offer through its term.

(c) *Cancellation.* The Agency may cancel a debt owed to the Agency if the requirements of this subpart are met and the application and supporting documents indicate that the borrower is unable to pay a compromise or adjustment offer.

§ 761.407 Failure to pay.

(a) Failure to pay any compromise amount approved by FSA by the date

agreed will result in cancellation of the compromise agreement.

(b) Failure to pay debt adjustment amounts approved by FSA by the dates agreed will result in cancellation of the adjustment agreement.

(c) A debtor who has entered into an agreement under this subpart may request that FSA extend a repayment date for 90 days. The debtor must provide information that supports the basis for the request at the time the request is made.

(d) If a debtor is delinquent under the terms of an adjustment agreement and FSA determines the debtor is likely to be financially unable to meet the terms of the agreement, the existing agreement may be cancelled and the debtor may be allowed to apply for a different type of settlement more consistent with the debtor's repayment ability.

(e) If an agreement is cancelled, any payments received will be retained as payments on the debt owed.

§ 761.408 Administrator authority.

On an individual case basis, the Agency may consider granting an exception to any requirement of this part if:

(a) The exception is not inconsistent with the authorizing statute or other applicable law; and

(b) The Agency's financial interest would be adversely affected by acting in accordance with this part and granting an exception would resolve or eliminate the adverse effect upon its financial interest.

PART 762—GUARANTEED FARM LOANS

AUTHORITY: 5 U.S.C. 301 and 7 U.S.C. 1989.

SOURCE: 64 FR 7378, Feb. 12, 1999, unless otherwise noted.

EDITORIAL NOTE: Nomenclature changes to part 762 appear at 72 FR 63297, Nov. 8, 2007.

§§ 762.1–762.100 [Reserved]

§ 762.101 Introduction.

(a) *Scope.* This subpart contains regulations governing Operating loans, Farm Ownership loans, and Conservation loans guaranteed by the Agency. This subpart applies to lenders, holders, borrowers, Agency personnel, and other parties involved in making, guaranteeing, holding, servicing, or liquidating such loans.

(b) *Lender list.* The Agency maintains a current list of lenders who express a desire to participate in the guaranteed loan program. This list is made available to farmers upon request.

(c) *Lender classification.* Lenders who participate in the Agency guaranteed loan program will be classified into one of the following categories:

(1) Standard Eligible Lender under § 762.105;

(2) Certified Lender;

(3) Preferred Lender under § 762.106; or

(4) Micro Lender under § 762.107.

(d) *Type of guarantee.* Guarantees are available for both a loan note or a line of credit. A loan note is used for a loan of fixed amount and term. A line of credit has a fixed term, but no fixed amount. The principal amount outstanding at any time, however, may not exceed the line of credit ceiling contained in the contract. Both guarantees are evidenced by the same loan guarantee form.

(e) *Termination of loan guarantee.* The loan guarantee will automatically terminate as follows:

(1) Upon full payment of the guaranteed loan. A zero balance within the period authorized for advances on a line of credit will not terminate the guarantee;

(2) Upon payment of a final loss claim; or

(3) Upon written notice from the lender to the Agency that a guarantee is no longer desired provided the lender holds all of the guaranteed portion of the loan. The loan guarantee will be returned to the Agency office for cancellation within 30 days of the date of the notice by the lender.

[64 FR 7378, Feb. 12, 1999, as amended at 72 FR 63297, Nov. 8, 2007; 75 FR 54013, Sept. 3, 2010; 81 FR 72690, Oct. 21, 2016]

§ 762.102 Abbreviations and definitions.

Abbreviations and definitions for terms used in this part are provided in § 761.2 of this chapter.

[72 FR 63297, Nov. 8, 2007]

§ 762.103 Full faith and credit.

(a) *Fraud and misrepresentation.* The loan guarantee constitutes an obligation supported by the full faith and credit of the United States. The Agency may contest the guarantee only in cases of fraud or misrepresentation by a lender or holder, in which:

(1) The lender or holder had actual knowledge of the fraud or misrepresentation at the time it became the lender or holder, or

(2) The lender or holder participated in or condoned the fraud or misrepresentation.

(b) *Lender violations.* The loan guarantee cannot be enforced by the lender, regardless of when the Agency discovers the violation, to the extent that the loss is a result of:

(1) Violation of usury laws;

(2) Negligent servicing;

(3) Failure to obtain the required security; or,

(4) Failure to use loan funds for purposes specifically approved by the Agency.

(c) *Enforcement by holder.* The guarantee and right to require purchase will be directly enforceable by the holder even if:

(1) The loan guarantee is contestable based on the lender's fraud or misrepresentation; or

(2) The loan note guarantee is unenforceable by the lender based on a lender violation.

§762.104 Appeals.

(a) A decision made by the lender adverse to the borrower is not a decision by the Agency, whether or not concurred in by the Agency, and may not be appealed.

(b) The lender or Agency may request updated information from the borrower to implement an appeal decision.

(c) Appeals will be handled in accordance with parts 11 and 780 of this title.

[64 FR 7378, Feb. 12, 1999, as amended at 72 FR 63297, Nov. 8, 2007]

§762.105 Eligibility and substitution of lenders.

(a) *General.* To participate in FSA guaranteed farm loan programs, a lender must meet the eligibility criteria in this part. The standard eligible lender must demonstrate eligibility and provide such evidence as the Agency may request.

(b) *Standard eligible lender eligibility criteria.* (1) A lender must have experience in making and servicing agricultural loans and have the capability to make and service the loan for which a guarantee is requested;

(2) The lenders must not have losses or deficiencies in processing and servicing guaranteed loans above a level which would indicate an inability to properly process and service a guaranteed agricultural loan.

(3) A lender must be subject to credit examination and supervision by an acceptable State or Federal regulatory agency;

(4) The lender must maintain an office near enough to the collateral's location so it can properly and efficiently discharge its loan making and loan servicing responsibilities or use Agency approved agents, correspondents, branches, or other institutions or persons to provide expertise to assist in carrying out its responsibilities. The lender must be a local lender unless it:

(i) Normally makes loans in the region or geographic location in which the applicant's operation being financed is located, or

(ii) Demonstrates specific expertise in making and servicing loans for the proposed operation.

(5) The lender, its officers, or agents must not be debarred or suspended from participation in Government contracts or programs or be delinquent on a Government debt.

(c) *Substitution of lenders.* A new eligible lender may be substituted for the original lender, upon the original lender's concurrence, under the following conditions:

(1) The Agency approves of the substitution in writing by executing a modification of the guarantee to identify the new lender, the amount of debt at the time of the substitution and any new loan terms if applicable.

(2) The new lender agrees in writing to:

(i) Assume all servicing and other responsibilities of the original lender and to acquire the unguaranteed portion of the loan;

(ii) Execute a lender's agreement if one is not in effect;

(iii) [Reserved]

(iv) Give any holder written notice of the substitution. If the rate and terms are changed, written concurrence from the holder is required.

(3) The original lender will:

(i) Assign their promissory note, lien instruments, loan agreements, and other documents to the new lender.

(ii) If the loan is subject to an existing interest assistance agreement, submit a request for subsidy for the partial year that it has owned the loan.

(d) *Lender name or ownership changes.* (1) When a lender begins doing business under a new name or undergoes an ownership change the lender will notify the Agency.

(2) The lender's CLP, PLP, or MLP status is subject to reconsideration when ownership changes.

(3) The lender will execute a new lender's agreement when ownership changes.

[64 FR 7378, Feb. 12, 1999, as amended at 66 FR 7567, Jan. 24, 2001; 81 FR 72690, Oct. 21, 2016]

§762.106 Preferred and certified lender programs.

(a) *General.* (1) Lenders who desire PLP or CLP status must prepare a written request addressing:

(i) The States in which they desire to receive PLP or CLP status and their branch offices which they desire to be considered by the Agency for approval; and

(ii) Each item of the eligibility criteria for PLP or CLP approval in this section, as appropriate.

(2) The lender may include any additional supporting evidence or other information the lender believes would be helpful to the Agency in making its determination.

(3) The lender must send its request to the Agency State office for the State in which the lender's headquarters is located.

(4) The lender must provide any additional information requested by the Agency to process a PLP or CLP request if the lender continues with the approval process.

(b) *CLP criteria.* The lender must meet the following requirements to obtain CLP status:

(1) Qualify as a standard eligible lender under §762.105;

(2) Have a lender loss rate not in excess of the maximum CLP loss rate established by the Agency and published periodically in a FEDERAL REGISTER Notice. The Agency may waive the loss rate criteria for those lenders whose loss rate was substantially affected by a disaster as defined in §761.2(b) and part 759 of this chapter.

(3) Have proven an ability to process and service Agency guaranteed loans by showing that the lender:

(i) Submitted substantially complete and correct guaranteed loan applications; and

(ii) Serviced all guaranteed loans according to Agency regulations;

(4) Have made the minimum number of guaranteed OL, FO, CL, or SW loans established by the Agency and published periodically in a FEDERAL REGISTER Notice.

(5) Not be under any regulatory enforcement action such as a cease and desist order, written agreement, or an appointment of conservator or receiver, based upon financial condition;

(6) Designate a qualified person or persons to process and service Agency guaranteed loans for each of the lender offices which will process CLP loans. To be qualified, the person must meet the following conditions:

(i) Have attended Agency sponsored training in the past 12 months or will attend training in the next 12 months; and

(ii) Agree to attend Agency sponsored training each year;

(7) Use forms acceptable to the Agency for processing, analyzing, securing, and servicing Agency guaranteed loans and lines of credit;

(c) *PLP criteria.* The lender must meet the following requirements to obtain PLP status:

(1) Meet the CLP eligibility criteria under this section.

(2) Have a credit management system, satisfactory to the Agency, based on the following:

(i) The lender's written credit policies and underwriting standards;

(ii) Loan documentation requirements;

(iii) Exceptions to policies;

(iv) Analysis of new loan requests;

(v) Credit file management;

(vi) Loan funds and collateral management system;

(vii) Portfolio management;

(viii) Loan reviews;

(ix) Internal credit review process;

(x) Loan monitoring system; and

(xi) The board of director's responsibilities.

(3) Have made the minimum number of guaranteed OL, FO, CL, or SW loans established by the Agency and published periodically in a FEDERAL REGISTER Notice.

(4) Have a lender loss rate not in excess of the rate of the maximum PLP loss rate established by the Agency and

published periodically in a FEDERAL REGISTER Notice. The Agency may waive the loss rate criteria for those lenders whose loss rate was substantially affected by a disaster as defined in §761.2(b) and part 759 of this chapter.

(5) Show a consistent practice of submitting applications for guaranteed loans containing accurate information supporting a sound loan proposal.

(6) Show a consistent practice of processing Agency guaranteed loans without recurring major or minor deficiencies.

(7) Demonstrate a consistent, above average ability to service guaranteed loans based on the following:

(i) Borrower supervision and assistance;

(ii) Timely and effective servicing; and

(iii) Communication with the Agency.

(8) Designate a person or persons, either by name, title, or position within the organization, to process and service PLP loans for the Agency.

(d) *CLP and PLP approval.* (1) If a lender applying for CLP or PLP status is or has recently been involved in a merger or acquisition, all loans and losses attributed to both lenders will be considered in the eligibility calculations.

(2) The Agency will determine which branches of the lender have the necessary experience and ability to participate in the CLP or PLP program based on the information submitted in the lender application and on Agency experience.

(3) Lenders who meet the criteria will be granted CLP or PLP status for a period not to exceed 5 years.

(4) PLP status will be conditioned on the lender carrying out its credit management system as proposed in its request for PLP status and any additional loan making or servicing requirements agreed to and documented the PLP lender's agreement. If the PLP lender's agreement does not specify any agreed upon process for a particular action, the PLP lender will act according to regulations governing CLP lenders.

(e) *Monitoring CLP and PLP lenders.* CLP and PLP lenders will provide information and access to records upon Agency request to permit the Agency to audit the lender for compliance with these regulations.

(f) *Renewal of CLP or PLP status.* (1) PLP or CLP status will expire within a period not to exceed 5 years from the date the lender's agreement is executed, unless a new lender's Agreement is executed.

(2) Renewal of PLP or CLP status is not automatic. A lender must submit a written request for renewal of a lender's agreement with PLP or CLP status which includes information:

(i) Updating the material submitted in the initial application; and,

(ii) Addressing any new criteria established by the Agency since the initial application.

(3) PLP or CLP status will be renewed if the applicable eligibility criteria under this section are met, and no cause exists for denying renewal under paragraph (g) of this section.

(g) *Revocation of PLP or CLP status.* (1) The Agency may revoke the lender's PLP or CLP status at any time during the 5 year term for cause.

(2) Any of the following instances constitute cause for revoking or not renewing PLP or CLP status:

(i) Violation of the terms of the lender's agreement;

(ii) Failure to maintain PLP or CLP eligibility criteria. The Agency may allow a PLP lender with a loss rate which exceeds the maximum PLP loss rate, to retain its PLP status for a two-year period, if:

(A) The lender documents in writing why the excessive loss rate is beyond their control;

(B) The lender provides a written plan that will reduce the loss rate to the PLP maximum rate within two years from the date of the plan, and

(C) The Agency determines that exceeding the maximum PLP loss rate standard was beyond the control of the lender. Examples include, but are not limited to, a freeze with only local impact, economic downturn in a local area, drop in local land values, industries moving into or out of an area, loss of access to a market, and biological or chemical damage.

(D) The Agency will revoke PLP status if the maximum PLP loss rate is not met at the end of the two-year period, unless a second two year extension is granted under this subsection.

(iii) Knowingly submitting false or misleading information to the Agency;

(iv) Basing a request on information known to be false;

(v) Deficiencies that indicate an inability to process or service Agency guaranteed farm loan programs loans in accordance with this subpart;

(vi) Failure to correct cited deficiencies in loan documents upon notification by the Agency;

(vii) Failure to submit status reports in a timely manner;

(viii) Failure to use forms, or follow credit management systems (for PLP lenders) accepted by the Agency; or

(ix) Failure to comply with the reimbursement requirements of § 762.144(c)(7) and (c)(8).

(3) A lender which has lost PLP or CLP status must be reconsidered for eligibility to continue as a Standard Eligible Lender (for former PLP and CLP lenders), or as a CLP lender (for former PLP lenders) in submitting loan guarantee requests. They may reapply for CLP or PLP status when the problem causing them to lose their status has been resolved.

[64 FR 7378, Feb. 12, 1999; 64 FR 38298, July 16, 1999, as amended at 70 FR 56107, Sept. 26, 2005; 71 FR 43957, Aug. 3, 2006; 75 FR 54013, Sept. 3, 2010; 77 FR 41256, July 13, 2012]

§ 762.107 Micro Lender Program.

(a) *General.* The lenders must submit the following items:

(1) To request MLP Status, a lender must submit an application form to any local FSA office.

(2) The lender must provide any additional information requested by the Agency to process an MLP request, if the lender continues with the approval process.

(3) MLP lender authorities are limited to originating and servicing EZ Guarantee loans.

(b) *MLP criteria.* An MLP lender must satisfy the following requirements to obtain MLP Status:

(1) Have experience in making and servicing business loans.

(2) Have the staff and resources to properly and efficiently discharge its loan making and loan servicing responsibilities that may include use of Agency approved agents.

(3) Be subject to oversight as established and announced by the Agency on the FSA Web site (*www.fsa.usda.gov*).

(4) Have a loss rate not in excess of the maximum MLP loss rate established and announced by the Agency on the FSA Web site (*www.fsa.usda.gov*).

(5) Have made the minimum number of loans as established and announced by the Agency on the FSA Web site (*www.fsa.usda.gov*).

(6) Not be debarred or suspended from participation in Government contracts or programs or be delinquent on a Government debt. This includes the lender's officers and agents.

(c) *Renewal of MLP Status.* MLP Status will expire within a period not to exceed 5 years from the date the lender's agreement is executed, unless a new lender's agreement is executed.

(1) Renewal of MLP Status is not automatic. A lender must submit a new application for renewal.

(2) MLP Status will be renewed if the applicable eligibility criteria under this section are met, and no cause exists for denying renewal under paragraph (d)(1) of this section.

(d) *Revocation of MLP Status.* The Agency may revoke the lender's MLP Status at any time during the 5 year term for cause as specified in paragraph (d)(1) of this section.

(1) Any of the following instances constitutes cause for revoking or not renewing MLP Status:

(i) Violation of the terms of the lender's agreement;

(ii) Failure to maintain MLP eligibility criteria;

(iii) Knowingly submitting false or misleading information to the Agency;

(iv) Deficiencies that indicate an inability to process or service Agency guaranteed farm loan programs loans in accordance with this subpart;

(v) Failure to correct cited deficiencies in loan documents upon notification by the Agency;

(vi) Failure to submit status reports in a timely manner; or

(vii) Failure to comply with the reimbursement requirements of § 762.144(c)(7) and (c)(8).

(2) A lender that has lost MLP Status may reapply for MLP Status once the

problem that caused the MLP Status to be revoked has been resolved.

[81 FR 72690, Oct. 21, 2016]

§§ 762.108–762.109 [Reserved]

§ 762.110 Loan application.

(a) *General.* This paragraph (a) specifies the general requirements for guaranteed loan applications:

(1) Lenders must perform at least the same level of evaluation and documentation for a guaranteed loan that the lender typically performs for non-guaranteed loans of a similar type and amount.

(2) The application thresholds in this section apply to any single loan, or package of loans submitted for consideration at any one time. A lender must not split a loan into two or more parts in order to fall below the threshold in order to avoid additional documentation.

(3) The Agency may require lenders with a lender loss rate in excess of the rate for CLP lenders to assemble additional documentation specified in paragraph (d) of this section.

(b) *EZ Guarantee loans.* MLP lenders may submit an EZ Guarantee application for loans up to $50,000. All other lenders may submit EZ Guarantee applications for loans up to $100,000. Lenders must submit:

(1) An EZ Guarantee application form.

(2) If the loan fails to pass the underwriting criteria for EZ Guarantee approval in § 762.125(d), or the responses in the application are insufficient for the Agency to make a loan decision, the lender must provide additional information as requested by the Agency.

(c) *Loans up to $125,000.* Lenders must submit the following items for loans up to $125,000 (other than EZ Guarantees):

(1) The application form;

(2) Loan narrative, including a plan for servicing the loan;

(3) Balance sheet;

(4) Cash flow budget; and

(5) Credit report.

(d) *Loans over $125,000.* A complete application for loans over $125,000 will require items specified in paragraph (c) of this section, plus the following items:

(1) Verification of income;

(2) Verification of debts exceeding an amount determined by the Agency;

(3) Three years financial history;

(4) Three years of production history (for standard eligible lenders only);

(5) Proposed loan agreements; and,

(6) If construction or development is planned, a copy of the plans, specifications, and development schedule.

(e) *Applications from PLP lenders.* Notwithstanding paragraphs (c) and (d) of this section, a complete application for PLP lenders will consist of at least:

(1) An application form;

(2) A loan narrative;

(3) Any other items agreed to during the approval of the PLP lender's status and contained in the PLP lender agreement.

(f) *CL Guarantees.* In addition to the other requirements in this section, the following items apply when a lender is requesting a CL guarantee:

(1) Lenders must submit a copy of the conservation plan or Forest Stewardship Management Plan;

(2) Lenders must submit plans to transition to organic or sustainable agriculture when the funds requested will be used to facilitate the transition and the lender is requesting consideration for priority funding;

(3) When CL guarantee applicants meet all the following criteria, the cash flow budget requirement in this section will be waived:

(i) Be current on all payments to all creditors including the Agency (if currently an Agency borrower);

(ii) Debt to asset ratio is 40 percent or less;

(iii) Balance sheet indicates a net worth of 3 times the requested loan amount or greater; and

(iv) FICO credit score is at least 700; for entity applicants, the FICO credit score of the majority of the individual members of the entity must be at least 700.

(g) *Submitting applications.* (1) All lenders must compile and maintain in their files a complete application for each guaranteed loan.

(2) The Agency will notify CLP lenders which items to submit to the Agency.

(3) PLP lenders will submit applications in accordance with their agreement with the Agency for PLP status.

229

(4) All lenders must certify that the required items, not submitted, are in their files.

(5) The Agency may request additional information from any lender or review the lender's loan file as needed to make eligibility and approval decisions.

(h) *Incomplete applications.* If the lender does not provide the information needed to complete its application by the deadline established in an Agency request for the information, the application will be considered withdrawn by the lender.

(i) *Conflict of interest.* (1) When a lender submits the application for a guaranteed loan, the lender will inform the Agency in writing of any relationship which may cause an actual or potential conflict of interest.

(2) Relationships include:

(i) The lender or its officers, directors, principal stockholders (except stockholders in a Farm Credit System institution that have stock requirements to obtain a loan), or other principal owners having a financial interest (other than lending relationships in the normal course of business) in the applicant or borrower.

(ii) The applicant or borrower, a relative of the applicant or borrower, anyone residing in the household of the applicant or borrower, any officer, director, stockholder or other owner of the applicant or borrower holds any stock or other evidence of ownership in the lender.

(iii) The applicant or borrower, a relative of the applicant or borrower, or anyone residing in the household of the applicant or borrower is an Agency employee.

(iv) The officers, directors, principal stockholders (except stockholders in a Farm Credit System institution that have stock requirements to obtain a loan), or other principal owners of the lender have substantial business dealings (other than in the normal course of business) with the applicant or borrower.

(v) The lender or its officers, directors, principal stockholders, or other principal owners have substantial business dealings with an Agency employee.

(3) The lender must furnish additional information to the Agency upon request.

(4) The Agency will not approve the application until the lender develops acceptable safeguards to control any actual or potential conflicts of interest.

(j) *Market placement program.* Except for CL guarantees, when the Agency determines that a direct applicant or borrower may qualify for guaranteed credit, the Agency may submit the applicant or borrower's financial information to one or more guaranteed lenders. If a lender indicates interest in providing financing to the applicant or borrower through the guaranteed loan program, the Agency will assist in completing the application for a guarantee.

[64 FR 7378, Feb. 12, 1999, as amended at 68 FR 7695, Feb. 18, 2003; 72 FR 63297, Nov. 8, 2007; 75 FR 54013, Sept. 3, 2010; 77 FR 15938, Mar. 19, 2012; 81 FR 72691, Oct. 21, 2016; 86 FR 43391, Aug. 9, 2021]

§§ 762.111–762.119 [Reserved]

§ 762.120 Applicant eligibility.

Unless otherwise provided, applicants must meet all of the following requirements to be eligible for a guaranteed OL, FO, or CL.

(a) *Agency loss.* (1) Except as provided in paragraph (a)(2) of this section, the applicant, and anyone who will execute the promissory note, has not caused the Agency a loss by receiving debt forgiveness on all or a portion of any direct or guaranteed loan made under the authority of the Act by debt write-down or write-off; compromise, adjustment, reduction, or charge-off under the provisions of section 331 of the Act; discharge in bankruptcy; or through payment of a guaranteed loss claim on:

(i) More than three occasions on or prior to April 4, 1996; or

(ii) Any occasion after April 4, 1996.

(2) The applicant may receive a guaranteed OL to pay annual farm operating and family living expenses, provided the applicant meets all other requirements for the loan, if the applicant and anyone who will execute the promissory note:

(i) Received a write-down under section 353 of the Act;

(ii) Is current on payments under a confirmed reorganization plan under chapter 11, 12, or 13 of title 11 of the United States Code; or

(iii) Received debt forgiveness on not more than one occasion after April 4, 1996, resulting directly and primarily from a Presidentially-designated emergency for a county or contiguous county in which the applicant operates. Only applicants who were current on all existing direct and guaranteed FSA loans prior to the beginning date of the incidence period for a Presidentially-designated emergency and received debt forgiveness on that debt within three years after the designation of such emergency meet this exception.

(b) *Delinquent Federal debt.* The applicant, and anyone who will execute the promissory note, is not delinquent on any Federal debt, other than a debt under the Internal Revenue Code of 1986. (Any debt under the Internal Revenue Code of 1986 may be considered by the lender in determining cash flow and creditworthiness.)

(c) *Outstanding judgments.* The applicant, and anyone who will execute the promissory note, have no outstanding unpaid judgment obtained by the United States in any court. Such judgments do not include those filed as a result of action in the United States Tax Courts.

(d) *Citizenship.* (1) The applicant must be a citizen of the United States, a United States non-citizen national, or a qualified alien under applicable Federal immigration laws. For an entity applicant, the majority interest of the entity must be held by members who are United States citizens, United States non-citizen nationals, or qualified aliens under applicable Federal immigration laws.

(2) United States non-citizen nationals and qualified aliens must provide the appropriate documentation as to their immigration status as required by the United States Department of Homeland Security, Bureau of Citizenship and Immigration Services.

(e) *Legal capacity.* The applicant and all borrowers on the loan must possess the legal capacity to incur the obligations of the loan.

(f) *False or misleading information.* The applicant, in past dealings with the Agency, must not have provided the Agency with false or misleading documents or statements.

(g) *Credit history.* (1) The individual or entity applicant and all entity members must have acceptable credit history demonstrated by debt repayment.

(2) A history of failures to repay past debts as they came due when the ability to repay was within their control will demonstrate unacceptable credit history.

(3) Unacceptable credit history will not include:

(i) Isolated instances of late payments which do not represent a pattern and were clearly beyond their control; or,

(ii) Lack of credit history.

(h) *Test for credit.* Except for CL guarantees,

(1) The applicant is unable to obtain sufficient credit elsewhere without a guarantee to finance actual needs at reasonable rates and terms.

(2) The potential for sale of any significant nonessential assets will be considered when evaluating the availability of other credit.

(3) Ownership interests in property and income received by an individual or entity applicant, and any entity members as individuals will be considered when evaluating the availability of other credit to the applicant.

(i) *For OLs:*

(1) The individual or entity applicant must be an operator of not larger than a family farm after the loan is closed.

(2) In the case of an entity borrower:

(i) The entity must be authorized to operate, and own if the entity is also an owner, a farm in the State or States in which the farm is located; and

(ii) If the entity members holding a majority interest are related by marriage or blood, at least one member of the entity must operate the family farm; or,

(iii) If the entity members holding a majority interest are not related by marriage or blood, the entity members holding at least 50 percent interest must also operate the family farm.

(j) *For FOs:*

(1) The individual must be the operator of not larger than a family farm and the owner of a farm after the loan is closed. Ownership of the family farm

operation or the farm real estate may be held either directly in the individual's name or indirectly through interest in a legal entity.

(2) In the case of an entity borrower:

(i) An ownership entity must be authorized to own a farm in the state or states in which the farm is located. An operating entity must be authorized to operate a farm in the state or states in which the farm is located; and

(ii) If the entity members holding a majority interest are related by marriage or blood, at least one member of the entity must operate the family farm and at least one member of the entity or the entity must own the farm; or

(iii) If the entity members holding a majority interest are not related by marriage or blood, the entity members holding at least 50 percent interest must operate the family farm and the entity members holding at least 50 percent or the entity must own the farm.

(3) If the entity is an operator-only entity, the individuals that own the farm (real estate) must own at least 50 percent of the family farm (operating entity).

(4) All ownership may be held either directly in the individual's name or indirectly through interest in a legal entity.

(k) *For entity applicants.* Except for CL, entity applicants must meet the following additional eligibility criteria:

(1) Each entity member's ownership interest may not exceed the family farm definition limits;

(2) The collective ownership interest of all entity members may exceed the family farm definition limits only if the following conditions are met:

(i) All of the entity members are related by blood or marriage;

(ii) All of the members are or will be operators of the entity; and,

(iii) The majority interest holders of the entity must meet the requirements of paragraphs (d), (f), (g), and (i) through (j) of this section;

(3) The entity must be controlled by farmers engaged primarily and directly in farming in the United States after the loan is made; and

(4) If the applicant has one or more embedded entities, at least 75 percent of the individual ownership interests of each embedded entity must be owned by members actively involved in managing or operating the family farm.

(l) *For CL entity applicants.* Entity applicants for CL guarantees must meet the following eligibility criteria:

(1) The majority interest holders of the entity must meet the requirements of paragraph (d), (f), and (g) of this section;

(2) The entity must be controlled by farmers engaged primarily and directly in farming in the United States after the loan is made;

(3) If the applicant has one or more embedded entities, at least 75 percent of the individual ownership interests of each embedded entity must be owned by members actively involved in managing or operating the family farm; and

(4) The entity must be authorized to operate a farm in the State or States in which the farm is located.

(m) *For CL individual applicants.* Individual applicants for CL guarantees must be farmers in the United States.

(n) *Controlled substances.* The applicant, and anyone who will sign the promissory note, must not be ineligible as a result of a conviction for controlled substances according to 7 CFR part 718 of this chapter. If the lender uses the lender's Agency approved forms, the certification may be an attachment to the form.

[64 FR 7378, Feb. 12, 1999, as amended at 68 FR 62223, Nov. 3, 2003; 69 FR 5262, Feb. 4, 2004; 72 FR 63297, Nov. 8, 2007; 75 FR 54013, Sept. 3, 2010; 78 FR 65529, Nov. 1, 2013; 79 FR 60743, Oct. 8, 2014; 86 FR 43391, Aug. 9, 2021]

§ 762.121 Loan purposes.

(a) *Operating Loan purposes.* (1) Loan funds disbursed under an OL guarantee may only be used for the following purposes:

(i) Payment of costs associated with reorganizing a farm to improve its profitability;

(ii) Purchase of livestock, including poultry, and farm equipment or fixtures, quotas and bases, and cooperative stock for credit, production, processing or marketing purposes;

(iii) Payment of annual farm operating expenses, examples of which include feed, seed, fertilizer, pesticides, farm supplies, repairs and improvements which are to be expensed, cash rent and family subsistence;

(iv) Payment of scheduled principal and interest payments on term debt provided the debt is for authorized FO or OL purposes;

(v) Other farm needs;

(vi) Payment of costs associated with land and water development for conservation or use purposes;

(vii) Refinancing indebtedness incurred for any authorized OL purpose, when the lender and applicant can demonstrate the need to refinance;

(viii) Payment of loan closing costs;

(ix) Payment of costs associated with complying with Federal or State-approved standards under the Occupational Safety and Health Act of 1970 (29 U.S.C. 655, 667). This purpose is limited to applicants who demonstrate that compliance or non-compliance with the standards will cause them substantial economic injury; and

(x) Payment of training costs required or recommended by the Agency.

(2) Loan funds under a line of credit may be advanced only for the following purposes:

(i) Payment of annual operating expenses, family subsistence, and purchase of feeder animals;

(ii) Payment of current annual operating debts advanced for the current operating cycle; (Under no circumstances can carry-over operating debts from a previous operating cycle be refinanced);

(iii) Purchase of routine capital assets, such as replacement of livestock, that will be repaid within the operating cycle;

(iv) Payment of scheduled, non-delinquent, term debt payments provided the debt is for authorized FO or OL purposes.

(v) Purchase of cooperative stock for credit, production, processing or marketing purposes; and

(vi) Payment of loan closing costs.

(b) *Farm ownership loan purposes.* Guaranteed FO are authorized only to:

(1) Acquire or enlarge a farm; examples include, but are not limited to, providing down payments, purchasing easements for the applicant's portion of land being subdivided, and partici-

pating in the down payment FO program under part 764 of this chapter;

(2) Make capital improvements; examples include, but are not limited to, the construction, purchase, and improvement of a farm dwelling, service buildings and facilities that can be made fixtures to the real estate, (Capital improvements to leased land may be financed subject to the limitations in §762.122);

(3) Promote soil and water conservation and protection; examples include the correction of hazardous environmental conditions, and the construction or installation of tiles, terraces and waterways;

(4) Pay closing costs, including but not limited to, purchasing stock in a cooperative and appraisal and survey fees; and

(5) Refinancing indebtedness incurred for authorized FO and OL purposes, provided the lender and applicant demonstrate the need to refinance the debt.

(c) *CL purposes.* Loan funds disbursed under a CL guarantee may be used for any conservation activities included in a conservation plan or Forestry Stewardship Management Plan including, but not limited to:

(1) The installation of conservation structures to address soil, water, and related resources;

(2) The establishment of forest cover for sustained yield timber management, erosion control, or shelter belt purposes;

(3) The installation of water conservation measures;

(4) The installation of waste management systems;

(5) The establishment or improvement of permanent pasture;

(6) Other purposes including the adoption of any other emerging or existing conservation practices, techniques, or technologies; and

(7) Refinancing indebtedness incurred for any authorized CL purpose, when refinancing will result in additional conservation benefits.

(d) *Highly erodible land or wetlands conservation.* Loans may not be made for any purpose which contributes to excessive erosion of highly erodible land or to the conversion of wetlands to produce an agricultural commodity. A decision by the Agency to reject an application for this reason may be appealable. An appeal questioning whether the presence of a wetland, converted wetland, or highly erodible land on a particular property must be filed directly with the USDA agency making the determination in accordance with the agency's appeal procedures.

(e) *Judgment debts.* Loans may not be used to satisfy judgments obtained in the United States District courts. However, Internal Revenue Service judgment liens may be paid with loan funds.

[64 FR 7378, Feb. 12, 1999, as amended at 72 FR 63297, Nov. 8, 2007; 73 FR 74345, Dec. 8, 2008; 75 FR 54014, Sept. 3, 2010; 77 FR 15938, Mar. 19, 2012; 78 FR 65529, Nov. 1, 2013; 86 FR 43391, Aug. 9, 2021]

§ 762.122 Loan limitations.

(a) *Dollar limits.* The Agency will not guarantee any loan that would result in the applicant's total indebtedness exceeding the limits established in § 761.8 of this chapter.

(b) *Leased land.* When FO or CL funds are used for improvements to leased land the terms of the lease must provide reasonable assurance that the applicant will have use of the improvement over its useful life, or provide compensation for any unexhausted value of the improvement if the lease is terminated.

(c) *Tax-exempt transactions.* The Agency will not guarantee any loan made with the proceeds of any obligation the interest on which is excluded from income under section 103 of the Internal Revenue Code of 1986. Funds generated through the issuance of tax-exempt obligations may not be used to purchase the guaranteed portion of any Agency guaranteed loan. An Agency guaranteed loan may not serve as collateral for a tax-exempt bond issue.

(d) *Floodplain restrictions.* The Agency will not guarantee any loan to purchase, build, or expand buildings located in a special 100 year floodplain as defined by FEMA flood hazard area maps unless flood insurance is available and purchased.

[64 FR 7378, Feb. 12, 1999; 64 FR 38298, July 16, 1999, as amended at 66 FR 7567, Jan. 24, 2001; 72 FR 63297, Nov. 8, 2007; 73 FR 74345, Dec. 8, 2008; 75 FR 54014, Sept. 3, 2010; 79 FR 78693, Dec. 31, 2014]

§ 762.123 Insurance and farm inspection requirements.

(a) *Insurance.* (1) Lenders must require borrowers to maintain adequate property, public liability, and crop insurance to protect the lender and Government's interests.

(2) By loan closing, applicants must either:

(i) Obtain at least the catastrophic risk protection (CAT) level of crop insurance coverage, if available, for each crop of economic significance, as defined by § 400.651 of this title, or

(ii) Waive eligibility for emergency crop loss assistance in connection with the uninsured crop. EM loan assistance under part 764 of this chapter is not considered emergency crop loss assistance for purposes of this waiver and execution of the waiver does not render the borrower ineligible for EM loans.

(3) Applicants must purchase flood insurance if buildings are or will be located in a special flood hazard area as defined by FEMA flood hazard area maps and if flood insurance is available.

(4) Insurance, including crop insurance, must be obtained as required by the lender or the Agency based on the strengths and weaknesses of the loan.

(b) *Farm inspections.* Before submitting an application the lender must make an inspection of the farm to assess the suitability of the farm and to determine any development that is needed to make it a suitable farm.

[64 FR 7378, Feb. 12, 1999, as amended at 70 FR 56107, Sept. 26, 2005; 72 FR 63297, Nov. 8, 2007; 86 FR 43391, Aug. 9, 2021]

§ 762.124 Interest rates, terms, charges, and fees.

(a) *Interest rates.* (1) The interest rate on a guaranteed loan or line of credit may be fixed or variable as agreed upon between the borrower and the lender. The lender may charge different rates on the guaranteed and the non-guaranteed portions of the note. The guaranteed portion may be fixed while the unguaranteed portion may be variable, or vice versa. If both portions are variable, different bases may be used.

(2) If a variable rate is used, it must be tied to an index or rate specifically agreed to between the lender and borrower in the loan instruments and the

rate adjustments must be in accordance with normal practices of the lender for unguaranteed loans. Upon request, the lender must provide the Agency with copies of its written rate adjustment practices.

(3) At the time of loan closing or loan restructuring, the interest rate on both the guaranteed portion and the unguaranteed portion of a fixed or variable rate OL or FO loan may not exceed the following, as applicable:

(i) For lenders using risk-based pricing practices, the risk tier at least one tier lower (representing lower risk) than that borrower would receive without a guarantee. The lender must provide the Agency with copies of its written pricing practices, upon request.

(ii) For lenders not using risk-based pricing practices, for variable rate loans or fixed rate loans with rates fixed for less than five years, 650 basis points (6.5 percentage points) above the 3-month LIBOR.

(iii) For lenders not using risk-based pricing practices, for loans with rates fixed for five or more years, 550 basis points (5.5 percentage points) above the 5-year Treasury note rate.

(4) In the event the 3-month LIBOR is below 2 percent, the maximum rates specified in paragraph (a)(3) of this section do not apply. In that case, at the time of loan closing or loan restructuring, the interest rate on both the guaranteed portion and the unguaranteed portion of an OL or FO loan may not exceed 750 basis points above the 3-month LIBOR for variable rate loans and 650 basis points above the 5-year Treasury rate for fixed rate loans.

(5) Interest must be charged only on the actual amount of funds advanced and for the actual time the funds are outstanding. Interest on protective advances made by the lender to protect the security will be charged at the note rate but limited to paragraph (a)(3) of this section.

(6) The lender and borrower may collectively obtain a temporary reduction in the interest rate through the interest assistance program in accordance with §762.150.

(b) *OL terms.* (1) Loan funds or advances on a line of credit used to pay annual operating expenses will be repaid when the income from the year's operation is received, except when the borrower is establishing a new enterprise, developing a farm, purchasing feed while feed crops are being established, or recovering from disaster or economic reverses.

(2) The final maturity date for each loan cannot exceed 7 years from the date of the promissory note or line of credit agreement. Advances for purposes other than for annual operating expenses will be scheduled for repayment over the minimum period necessary considering the applicant's ability to repay and the useful life of the security, but not in excess of 7 years.

(3) All advances on a line of credit must be made within 5 years from the date of the Loan Guarantee.

(c) *FO terms.* Each loan must be scheduled for repayment over a period not to exceed 40 years from the date of the note or such shorter period as may be necessary to assure that the loan will be adequately secured, taking into account the probable depreciation of the security.

(d) *CL terms.* Each loan must be scheduled for repayment over a period not to exceed 30 years from the date of the note or such shorter period as may be necessary to assure that the loan will be adequately secured, taking into account the probable depreciation of the security.

(e) *Balloon installments under loan note guarantee.* Balloon payment terms are permitted on FO, OL, or CL subject to the following:

(1) Extended repayment schedules may include equal, unequal, or balloon installments if needed on any guaranteed loan to establish a new enterprise, develop a farm, or recover from a disaster or an economical reversal.

(2) Loans with balloon installments must have adequate collateral at the time the balloon installment comes due. Crops, livestock other than breeding livestock, or livestock products produced are not sufficient collateral for securing such a loan.

(3) The borrower must be projected to be able to refinance the remaining debt at the time the balloon payment comes due based on the expected financial

condition of the operation, the depreciated value of the collateral, and the principal balance on the loan.

(f) *Charges and fees.* (1) The lender may charge the applicant and borrower fees for the loan provided they are no greater than those charged to unguaranteed customers for similar transactions. Similar transactions are those involving the same type of loan requested (for example, operating loans or farm real estate loans).

(2) Late payment charges (including default interest charges) are not covered by the guarantee. These charges may not be added to the principal and interest due under any guaranteed note or line of credit. However, late payment charges may be made outside of the guarantee if they are routinely made by the lender in similar types of loan transactions.

(3) Lenders may not charge a loan origination and servicing fee greater than 1 percent of the loan amount for the life of the loan when a guaranteed loan is made in conjunction with a down payment FO under part 764 of this chapter.

[64 FR 7378, Feb. 12, 1999, as amended at 72 FR 17358, Apr. 9, 2007; 72 FR 63297, Nov. 8, 2007; 73 FR 74345, Dec. 8, 2008; 75 FR 54014, Sept. 3, 2010; 77 FR 15938, Mar. 19, 2012; 78 FR 14005, Mar. 4, 2013]

§ 762.125 Financial feasibility.

(a) *General.* Except for streamlined CL guarantees (see § 762.110(f)), the following requirements must be met:

(1) Notwithstanding any other provision of this section, PLP lenders will follow their internal procedures on financial feasibility as agreed to by the Agency during PLP certification.

(2) The applicant's proposed operation must project a feasible plan.

(3) For standard eligible lenders, the projected income and expenses of the borrower and operation used to determine a feasible plan must be based on the applicant's proven record of production and financial management.

(4) For CLP lenders, the projected income and expenses of the borrower and the operation must be based on the applicant's financial history and proven record of financial management.

(5) For those farmers without a proven history, a combination of any actual history and any other reliable source of information that are agreeable with the lender, the applicant, and the Agency will be used.

(6) The cash flow budget analyzed to determine a feasible plan must represent the predicted cash flow of the operating cycle.

(7) Lenders must use price forecasts that are reasonable and defensible. Sources must be documented by the lender and acceptable to the Agency.

(8) When a feasible plan depends on income from other sources in addition to income from owned land, the income must be dependable and likely to continue.

(9) The lender will analyze business ventures other than the farm operation to determine their soundness and contribution to the operation. Except for CL, guaranteed loan funds will not be used to finance a nonfarm enterprise. Nonfarm enterprises include, but are not limited to: raising earthworms, exotic birds, tropical fish, dogs, or horses for nonfarm purposes; welding shops; boarding horses; and riding stables.

(10) When the applicant has or will have a cash flow budget developed in conjunction with a proposed or existing Agency direct loan, the two cash flow budgets must be consistent.

(b) *Estimating production.* Except for streamlined CL guarantees (see § 762.110(f)), the following requirements must be met:

(1) Standard eligible lenders must use the best sources of information available for estimating production in accordance with this subsection when developing cash flow budgets.

(2) Deviations from historical performance may be acceptable, if specific to changes in operation and adequately justified and acceptable to the Agency.

(3) For existing farmers, actual production for the past 3 years will be utilized.

(4) For those farmers without a proven history, a combination of any actual history and any other reliable source of information that are agreeable with the lender, the applicant, and the Agency will be used.

(5) When the production of a growing commodity can be estimated, it must be considered when projecting yields.

Farm Service Agency, USDA

§ 762.126

(6) When the applicant's production history has been so severely affected by a declared disaster that an accurate projection cannot be made, the following applies:

(i) County average yields are used for the disaster year if the applicant's disaster year yields are less than the county average yields. If county average yields are not available, State average yields are used. Adjustments can be made, provided there is factual evidence to demonstrate that the yield used in the farm plan is the most probable to be realized.

(ii) To calculate a historical yield, the crop year with the lowest actual or county average yield may be excluded, provided the applicant's yields were affected by disasters at least 2 of the previous 5 consecutive years.

(c) *Refinancing.* Loan guarantee requests for refinancing must ensure that a reasonable chance for success still exists. The lender must demonstrate that problems with the applicant's operation that have been identified, can be corrected, and the operation returned to a sound financial basis.

(d) *EZ Guarantee feasibility.* Notwithstanding any other provision of this section:

(1) The Agency will evaluate EZ Guarantee application financial feasibility using criteria determined and announced by the Agency on the FSA Web site (*www.fsa.usda.gov*).

(2) *EZ Guarantee* applications that satisfy the criteria will be determined to meet the financial feasibility standards in this section.

(3) *EZ Guarantee* applications that do not satisfy the criteria will require further documentation as determined by the Agency and announced on the FSA Web site (*www.fsa.usda.gov*).

[64 FR 7378, Feb. 12, 1999, as amended at 66 FR 7567, Jan. 24, 2001; 75 FR 54014, Sept. 3, 2010; 81 FR 72691, Oct. 21, 2016]

§ 762.126 **Security requirements.**

(a) *General.* (1) The lender is responsible for ensuring that proper and adequate security is obtained and maintained to fully secure the loan, protect the interest of the lender and the Agency, and assure repayment of the loan or line of credit.

(2) The lender will obtain a lien on additional security when necessary to protect the Agency's interest.

(b) *Guaranteed and unguaranteed portions.* (1) All security must secure the entire loan or line of credit. The lender may not take separate security to secure only that portion of the loan or line of credit not covered by the guarantee.

(2) The lender may not require compensating balances or certificates of deposit as means of eliminating the lender's exposure on the unguaranteed portion of the loan or line of credit. However, compensating balances or certificates of deposit as otherwise used in the ordinary course of business are allowed for both the guaranteed and unguaranteed portions.

(c) *Identifiable security.* The guaranteed loan must be secured by identifiable collateral. To be identifiable, the lender must be able to distinguish the collateral item and adequately describe it in the security instrument.

(d) *Type of security.* (1) Guaranteed loans may be secured by any property if the term of the loan and expected life of the property will not cause the loan to be undersecured.

(2) For loans with terms greater than 7 years, a lien must be taken on real estate.

(3) Loans can be secured by a mortgage on leasehold properties if the lease has a negotiable value and is subject to being mortgaged.

(4) The lender or Agency may require additional personal and corporate guarantees to adequately secure the loan. These guarantees are separate from, and in addition to, the personal obligations arising from members of an entity signing the note as individuals.

(e) *Lien position.* All guaranteed loans will be secured by the best lien obtainable. Provided that:

(1) Any chattel-secured guaranteed loan must have a higher lien priority (including purchase money interest) than an unguaranteed loan secured by the same chattels and held by the same lender.

(2) Junior lien positions are acceptable only if the total amount of debt with liens on the security, including the debt in junior lien position, is less than or equal to 85 percent of the value

237

of the security. Junior liens on crops or livestock products will not be relied upon for security unless the lender is involved in multiple guaranteed loans to the same borrower and also has the first lien on the collateral.

(3) When taking a junior lien, prior lien instruments will not contain future advance clauses (except for taxes, insurance, or other reasonable costs to protect security), or cancellation, summary forfeiture, or other clauses that jeopardize the Government's or the lender's interest or the borrower's ability to pay the guaranteed loan, unless any such undesirable provisions are limited, modified, waived or subordinated by the lienholder for the benefit of the Agency and the lender.

(f) *Additional security*, or any loan of $10,000 or less may be secured by the best lien obtainable on real estate without title clearance or legal services normally required, provided the lender believes from a search of the county records that the applicant can give a mortgage on the farm and provided that the lender would, in the normal course of business, waive the title search. This exception to title clearance will not apply when land is to be purchased.

(g) *Multiple owners.* If security has multiple owners, all owners must execute the security documents for the loan.

(h) *Exceptions.* The Deputy Administrator for Farm Loan Programs has the authority to grant an exception to any of the requirements involving security, if the proposed change is in the best interest of the Government and the collection of the loan will not be impaired.

[64 FR 7378, Feb. 12, 1999, as amended at 70 FR 56107, Sept. 26, 2005]

§ 762.127 Appraisal requirements.

(a) *General.* The general requirements for an appraisal are:

(1) *Value of collateral.* The lender is responsible for ensuring that the value of chattel and real estate pledged as collateral is sufficient to fully secure the guaranteed loan.

(2) *Additional security.* The lender is not required to complete an appraisal or evaluation of collateral that will serve as additional security, but the lender must provide an estimated value.

(3) *Appraisal cost.* Except for authorized liquidation expenses, the lender is responsible for all appraisal costs, which may be passed on to the borrower or transferee in the case of a transfer and assumption.

(b) *Chattel security.* The requirements for chattel appraisals are:

(1) *Need for chattel appraisal.* A current appraisal (not more than 12 months old) of primary chattel security is required on all loans except loans or lines of credit for annual production purposes secured by crops, which require an appraisal only when the guarantee is requested late in the current production year and actual yields can be reasonably estimated. An appraisal is not required for loans of $50,000 or less if a strong equity position exists.

(2) *Basis of value.* The appraised value of chattel property will be based on public sales of the same or similar property in the market area. In the absence of such public sales, reputable publications reflecting market values may be used.

(3) *Appraisal form.* Appraisal reports may be on the Agency's appraisal of chattel property form or on any other appraisal form containing at least the same information.

(4) *Experience and training.* Chattel appraisals will be performed by appraisers who possess sufficient experience or training to establish market (not retail) values as determined by the Agency.

(c) *Real estate security.* The requirements for real estate appraisals are:

(1) *Loans of $250,000 or less.* The lender must document the value of the real estate by applying the same policies and procedures as their non-guaranteed loans.

(2) *Loans greater than $250,000.* The lender must document the value of real estate using a current appraisal (not more than 18 months old) completed by a State Certified General Appraiser. Real estate appraisals must be completed in accordance with USPAP. Restricted reports as defined in USPAP are not acceptable. The Agency may allow an appraisal more than 18

months old to be used only if documentation provided by the lender reflects each of the following:

(i) Market conditions have remained stable or improved based on sales of similar properties,

(ii) The property in question remains in the same or better condition, and

(iii) The value of the property has remained the same or increased.

(3) Agency determinations under paragraph (c)(2) of this section to permit appraisals more than 18 months old are not appealable.

[78 FR 65529, Nov. 1, 2013, as amended at 86 FR 43391, Aug. 9, 2021]

§762.128 Environmental and special laws.

(a) *Environmental requirements.* The requirements found in part 799 of this chapter must be met for guaranteed OL, FO, and CL. CLP, PLP, and MLP lenders may certify that they have documentation in their file to demonstrate compliance with paragraph (c) of this section. Standard eligible lenders must submit evidence supporting compliance with this section.

(b) *Determination.* The Agency determination of whether an environmental problem exists will be based on:

(1) The information supplied with the application;

(2) The Agency Official's personal knowledge of the operation;

(3) Environmental resources available to the Agency including, but not limited to, documents, third parties, and governmental agencies;

(4) A visit to the farm operation when the available information is insufficient to make a determination;

(5) Other information supplied by the lender or applicant upon Agency request. If necessary, information not supplied with the application will be requested by the Agency.

(c) *Special requirements.* Lenders will assist in the environmental review process by providing environmental information. In all cases, the lender must retain documentation of their investigation in the applicant's case file.

(1) A determination must be made as to whether there are any potential impacts to a 100 year floodplain as defined by Federal Emergency Management Agency floodplain maps, Natural Resources Conservation Service data, or other appropriate documentation.

(2) The lender will assist the borrower in securing any applicable permits or waste management plans. The lender may consult with the Agency for guidance on activities which require consultation with State regulatory agencies, special permitting or waste management plans.

(3) The lender will examine the security property to determine if there are any structures or archeological sites which are listed or may be eligible for listing in the National Register of Historic Places. The lender may consult with the Agency for guidance on which situations will need further review in accordance with the National Historical Preservation Act and part 799 of this chapter.

(4) The applicant must certify they will not violate the provisions of §363 of the Act, the Food Security Act of 1985, and Executive Order 11990 relating to Highly Erodible Land and Wetlands.

(5) All lenders are required to ensure that due diligence is performed in conjunction with a request for guarantee of a loan involving real estate. Due diligence is the process of evaluating real estate in the context of a real estate transaction to determine the presence of contamination from release of hazardous substances, petroleum products, or other environmental hazards and determining what effect, if any, the contamination has on the security value of the property. The Agency will accept as evidence of due diligence the most current version of the American Society of Testing Materials (ASTM) transaction screen questionnaire available from 100 Barr Harbor Drive, West Conshohocken, Pennsylvania 19428–2959, or similar documentation, approved for use by the Agency, supplemented as necessary by the ASTM phase I environmental site assessments form.

(d) *Equal opportunity and nondiscrimination.* (1) With respect to any aspect of a credit transaction, the lender will not discriminate against any applicant on the basis of race, color, religion, national origin, sex, marital status, or age, provided the applicant can execute a legal contract. Nor will the lender discriminate on the basis of whether all or a part of the applicant's income derives from any public assistance program, or whether the applicant in good faith, exercises any rights under the Consumer Protection Act.

(2) Where the guaranteed loan involves construction, the contractor or subcontractor must file all compliance reports, equal opportunity and non-discrimination forms, and otherwise comply with all regulations prescribed by the Secretary of Labor pursuant to Executive Orders 11246 and 11375.

(e) *Other Federal, State and local requirements.* Lenders are required to coordinate with all appropriate Federal, State, and local agencies and comply with special laws and regulations applicable to the loan proposal.

[64 FR 7378, Feb. 12, 1999, as amended at 72 FR 63297, Nov. 8, 2007; 75 FR 54014, Sept. 3, 2010; 81 FR 51284, Aug. 3, 2016; 81 FR 72691, Oct. 21, 2016]

§ 762.129 Percent of guarantee and maximum loss.

(a) *Percent of guarantee.* The percent of guarantee will not exceed 90 percent based on the credit risk to the lender and the Agency both before and after the transaction. The Agency will determine the percentage of guarantee. See paragraph (b) of this section for exceptions.

(b) *Exceptions.* The guarantee will be determined by the Agency except:

(1) For OLs and FOs, the guarantee will be issued at 95 percent if:

(i) The sole purpose of a guaranteed FO or OL is to refinance an Agency direct farm loan. When only a portion of the loan is used to refinance a direct Agency loan, a weighted percentage of a guarantee will be provided; or

(ii) When the purpose of a guaranteed FO is to participate in the down payment loan program; or

(iii) When a guaranteed OL is made to a farmer who is participating in the Agency's down payment loan program. The guaranteed OL must be made during the period that a borrower has the down payment loan outstanding; or

(iv) When a guaranteed OL is made to a farmer whose farm land is subject to the jurisdiction of an Indian tribe and whose loan is secured by one or more security instruments that are subject to the jurisdiction of an Indian tribe.

(2) For CLs, the guarantee will be issued at 80 percent; however, the guarantee will be issued at 90 percent if:

(i) The applicant is a qualified SDA farmer; or

(ii) The applicant is a qualified beginning farmer.

(c) *CLP and PLP guarantees.* All guarantees issued to CLP or PLP lenders will not be less than 80 percent.

(d) *Maximum loss.* The maximum amount the Agency will pay the lender under the loan guarantee will be any loss sustained by such lender on the guaranteed portion including:

(1) The pro rata share of principal and interest indebtedness as evidenced by the note or by assumption agreement;

(2) Any loan subsidy due and owing;

(3) The pro rata share of principal and interest indebtedness on secured protective and emergency advances made in accordance with this subpart; and

(4) Principal and interest indebtedness on recapture debt pursuant to a shared appreciation agreement. Provided that the lender has paid the Agency its pro rata share of the recapture amount due.

[64 FR 7378, Feb. 12, 1999, as amended at 68 FR 7695, Feb. 18, 2003; 72 FR 63297, Nov. 8, 2007; 75 FR 54014, Sept. 3, 2010; 79 FR 78693, Dec. 31, 2014; 86 FR 43391, Aug. 9, 2021]

§ 762.130 Loan approval and issuing the guarantee.

(a) *Processing timeframes.* (1) *Standard eligible lenders.* Complete applications from Standard Eligible Lenders will be approved or rejected, and the lender notified in writing, no later than 30 calendar days after receipt.

(2) CLP and PLP lenders.

(i) Complete applications from CLP or PLP lenders will be approved or rejected not later than 14 calendar days after receipt.

(ii) For PLP lenders, if the 14 day time frame is not met, the proposed guaranteed loan will automatically be approved, subject to funding, and receive an 80 or 95 percent guarantee for FO or OL loans, and 80 or 90 percent guarantee for CL, as appropriate.

(3) *Complete applications.* For purposes of determining the application processing timeframes, an application

will be not be considered complete until all information required to make an approval decision, including the information for an environmental review, is received by the Agency.

(4) The Agency will confirm the date an application is received with a written notification to the lender.

(b) *Funding preference.* Loans are approved subject to the availability of funding. When it appears that there are not adequate funds to meet the needs of all approved applicants, applications that have been approved will be placed on a preference list according to the date of receipt of a complete application. If approved applications have been received on the same day, the following will be given priority:

(1) An application from a veteran

(2) An application from an Agency direct loan borrower

(3) An application from a applicant who:

(i) Has a dependent family,

(ii) Is an owner of livestock and farm implements necessary to successfully carry out farming operations, or

(iii) Is able to make down payments.

(4) Any other approved application.

(c) *Conditional commitment.* (1) The lender must meet all of the conditions specified in the conditional commitment to secure final Agency approval of the guarantee.

(2) The lender, after reviewing the conditions listed on the conditional commitment, will complete, execute, and return the form to the Agency. If the conditions are not acceptable to the lender, the Agency may agree to alternatives or inform the lender and the applicant of their appeal rights.

(d) *Lender requirements prior to issuing the guarantee*—(1) *Lender certification.* The lender will certify as to the following on the appropriate Agency form:

(i) No major changes have been made in the lender's loan or line of credit conditions and requirements since submission of the application (except those approved in the interim by the Agency in writing);

(ii) Required hazard, flood, crop, worker's compensation, and personal life insurance (when required) are in effect;

(iii) Truth in lending requirements have been met;

(iv) All equal employment and equal credit opportunity and nondiscrimination requirements have been or will be met at the appropriate time;

(v) The loan or line of credit has been properly closed, and the required security instruments have been obtained, or will be obtained, on any acquired property that cannot be covered initially under State law;

(vi) The borrower has marketable title to the collateral owned by the borrower, subject to the instrument securing the loan or line of credit to be guaranteed and subject to any other exceptions approved in writing by the Agency. When required, an assignment on all USDA crop and livestock program payments has been obtained;

(vii) When required, personal, joint operation, partnership, or corporate guarantees have been obtained;

(viii) Liens have been perfected and priorities are consistent with requirements of the conditional commitment;

(ix) Loan proceeds have been, or will be disbursed for purposes and in amounts consistent with the conditional commitment and as specified on the loan application. In line of credit cases, if any advances have occurred, advances have been disbursed for purposes and in amounts consistent with the conditional commitment and line of credit agreements;

(x) There has been no material adverse change in the borrower's condition, financial or otherwise, since submission of the application; and

(xi) All other requirements specified in the conditional commitment have been met.

(2) *Inspections.* The lender must notify the Agency of any scheduled inspections during construction and after the guarantee has been issued. The Agency may attend these field inspections. Any inspections or review performed by the Agency, including those with the lender, are solely for the benefit of the Agency. Agency inspections do not relieve any other parties of their inspection responsibilities, nor can these parties rely on Agency inspections for any purpose.

(3) *Execution of lender's agreement.* The lender must execute the Agency's

241

lender's agreement and deliver it to the Agency.

(4) *Closing report and guarantee fees.* (i) The lender must complete an Agency closing report form and return it to the Agency along with any guarantee fees.

(ii) The guarantee fee is established by the Agency at the time the guarantee is obligated. The current fee schedule is available at *http:// www.fsa.usda.gov* and any FSA office. Guaranteed fees may be adjusted annually based on factors that affect program costs. The nonrefundable fee is paid to the Agency by the lender. The fee may be passed on to the borrower and included in loan funds. The guarantee fee for the loan type will be calculated as follows:

(A) FO guarantee fee = Loan Amount × % guaranteed × (FO percentage established by FSA).

(B) OL guarantee fee = Loan Amount × % guaranteed × (OL percentage established by FSA).

(C) CL guarantee fee = Loan Amount × % guaranteed × (CL percentage established by FSA).

(iii) The following guaranteed loan transactions are not charged a fee:

(A) Loans involving interest assistance;

(B) Loans where a majority of the funds are used to refinance an Agency direct loan; and

(C) Loans to beginning or socially disadvantaged farmers involved in the direct Down payment Loan Program or beginning farmers participating in a qualified State Beginning Farmer Program.

(e) *Promissory notes, line of credit agreements, mortgages, and security agreements.* The lender will use its own promissory notes, line of credit agreements, real estate mortgages (including deeds of trust and similar instruments), and security agreements (including chattel mortgages), provided:

(1) The forms meet Agency requirements;

(2) Documents comply with State law and regulation;

(3) The principal and interest repayment schedules are stated clearly in the notes and are consistent with the conditional commitment;

(4) The note is executed by the individual liable for the loan. For entity applicants, the promissory note will be executed to evidence the liability of the entity, any embedded entities, and the individual liability of all entity members. Individual liability can be waived by the Agency for members holding less than 10 percent ownership in the entity if the collectability of the loan will not be impaired; and

(5) When the loan purpose is to refinance or restructure the lender's own debt, the lender may continue to use the existing debt instrument and attach an allonge that modifies the terms of the original note.

(f) *Replacement of loan guarantee, or assignment guarantee agreement.* If the guarantee or assignment guarantee agreements are lost, stolen, destroyed, mutilated, or defaced, except where the evidence of debt was or is a bearer instrument, the Agency will issue a replacement to the lender or holder upon receipt of acceptable documentation including a certificate of loss and an indemnity bond.

[64 FR 7378, Feb. 12, 1999, as amended at 72 FR 63297, Nov. 8, 2007; 73 FR 74345, Dec. 8, 2008; 75 FR 54014, Sept. 3, 2010; 76 FR 58094, Sept. 20, 2011; 79 FR 60744, Oct. 8, 2014; 79 FR 78693, Dec. 31, 2014; 86 FR 43391, Aug. 9, 2021]

§§ 762.131–762.139   [Reserved]

§ 762.140 General servicing responsibilities.

(a) *General.* (1) Lenders are responsible for servicing the entire loan in a reasonable and prudent manner, protecting and accounting for the collateral, and remaining the mortgagee or secured party of record.

(2) The lender cannot enforce the guarantee to the extent that a loss results from a violation of usury laws or negligent servicing.

(b) *Borrower supervision.* The lender's responsibilities regarding borrower supervision include, but are not limited to the following:

(1) Ensuring loan funds are not used for unauthorized purposes.

(2) Ensuring borrower compliance with the covenants and provisions contained in the promissory note, loan agreement, mortgage, security instruments, any other agreements, and this

part. Any violations which indicate non-compliance on the part of the borrower must be reported, in writing, to both the Agency and the borrower.

(3) Ensuring the borrower is in compliance with all laws and regulations applicable to the loan, the collateral, and the operations of the farm.

(4) Receiving all payments of principal and interest on the loan as they fall due and promptly disbursing to any holder its pro-rata share according to the amount of interest the holder has in the loan, less only the lender's servicing fee.

(5) Performing an annual analysis of the borrower's financial condition to determine the borrower's progress for all term loans with aggregate balances greater than $100,000 and all line of credit loans. The annual analysis will include:

(i) For loans secured by real estate only, the analysis for standard eligible lenders must include an analysis of the borrower's balance sheet. CLP lenders will determine the need for the annual analysis based on the financial strength of the borrower and document the file accordingly. PLP lenders will perform an annual analysis in accordance with the requirements established in the lender's agreement.

(ii) For loans secured by chattels, all lenders will review the borrower's progress regarding business goals, trends and changes in financial performance, and compare actual to planned income and expenses for the past year.

(iii) An account of the whereabouts or disposition of all collateral.

(iv) A discussion of any observations about the farm business with the borrower.

(v) For borrowers with an outstanding loan balance for existing term loans of $100,000 or less, the need for an annual analysis will be determined by the Agency for SEL, CLP, and MLP lenders. The annual analysis for PLP lenders will be in accordance with requirements in lender's credit management system (CMS).

(c) *Monitoring of development.* The lender's responsibilities regarding the construction, repairs, or other development include, but are not limited to:

(1) Determining that all construction is completed as proposed in the loan application;

(2) Making periodic inspections during construction to ensure that any development is properly completed within a reasonable period of time; and

(3) Verification that the security is free of any mechanic's, materialmen's, or other liens which would affect the lender's lien or result in a different lien priority from that proposed in the request for guarantee.

(d) *Loan installments.* When a lender receives a payment from the sale of encumbered property, loan installments will be paid in the order of lien priority. When a payment is received from the sale of unencumbered property or other sources of income, loan installments will be paid in order of their due date. Agency approval is required for any other proposed payment plans.

[64 FR 7378, Feb. 12, 1999, as amended at 69 FR 44579, July 27, 2004; 81 FR 72692, Oct. 21, 2016]

§762.141 Reporting requirements.

Lenders are responsible for providing the local Agency credit officer with all of the following information on the loan and the borrower:

(a) When the guaranteed loan becomes 30 days past due, and following the lender's meeting or attempts to meet with the borrower, all lenders will submit the appropriate Agency form showing guaranteed loan borrower default status. The form will be resubmitted every 60 days until the default is cured either through restructuring or liquidation.

(b) All lenders will submit the appropriate guaranteed loan status reports as of March 31 and September 30 of each year;

(c) CLP lenders also must provide the following:

(1) A written summary of the lender's annual analysis of the borrower's operation. This summary should describe the borrower's progress and prospects for the upcoming operating cycle. This annual analysis may be waived or postponed if the borrower is financially strong. The summary will include a description of the reasons an analysis was not necessary.

(2) For lines of credit, an annual certification stating that a cash flow projecting at least a feasible plan has been developed, that the borrower is in compliance with the provisions of the line of credit agreement, and that the previous year income and loan funds and security proceeds have been accounted for.

(d) In addition to the requirements of paragraphs (a), (b), and (c) of this section, the standard eligible lender also will provide:

(1) Borrower's balance sheet, and income and expense statement for the previous year.

(2) For lines of credit, the cash flow for the borrower's operation that projects a feasible plan or better for the upcoming operating cycle. The standard eligible lender must receive approval from the Agency before advancing future years' funds.

(3) An annual farm visit report or collateral inspection.

(e) PLP lenders will submit additional reports as required in their lender's agreement.

(f) A lender receiving a final loss payment must complete and return an annual report on its collection activities for each unsatisfied account for 3 years following payment of the final loss claim.

§ 762.142 Servicing related to collateral.

(a) *General.* The lender's responsibilities regarding servicing collateral include, but are not limited to, the following:

(1) Obtain income and insurance assignments when required.

(2) Ensure the borrower has or obtains marketable title to the collateral.

(3) Inspect the collateral as often as deemed necessary to properly service the loan.

(4) Ensure the borrower does not convert loan security.

(5) Ensure the proceeds from the sale or other disposition of collateral are accounted for and applied in accordance with the lien priorities on which the guarantee is based or used for the purchase of replacement collateral.

(6) Ensure the loan and the collateral are protected in the event of foreclosure, bankruptcy, receivership, insolvency, condemnation, or other litigation.

(7) Ensure taxes, assessments, or ground rents against or affecting the collateral are paid.

(8) Ensure adequate insurance is maintained.

(9) Ensure that insurance loss payments, condemnation awards, or similar proceeds are applied on debts in accordance with lien priorities on which the guarantee was based, or used to rebuild or acquire needed replacement collateral.

(b) *Partial releases.* (1) A lender may release guaranteed loan security without FSA concurrence as follows:

(i) When the security item is being sold for market value and the proceeds will be applied to the loan in accordance with lien priorities. In the case of term loans, proceeds will be applied as extra payments and not as a regular installment on the loan.

(ii) The security item will be used as a trade-in or source of down payment funds for a like item that will be taken as security.

(iii) The security item has no present or prospective value.

(2) A partial release of security may be approved in writing by the Agency upon the lender's request when:

(i) Proceeds will be used to make improvements to real estate that increase the value of the security by an amount equal to or greater than the value of the security being released.

(ii) Security will be released outright with no consideration, but the total unpaid balance of the guaranteed loan is less than or equal to 75 percent of the value of the security for the loan after the release, excluding the value of growing crops or planned production, based on a current appraisal of the security.

(iii) Significant income generating property will not be released unless it is being replaced and business assets will not be released for use as a gift or any similar purpose.

(iv) Agency concurrence is provided in writing to the lender's written request. Standard eligible lenders and CLP lenders will submit the following to the Agency:

(A) A current balance sheet on the borrower; and

(B) A current appraisal of the security. Based on the level of risk and estimated equity involved, the Agency will determine what security needs to be appraised. Any required security appraisals must meet the requirements of § 762.127; and

(C) A description of the purpose of the release; and

(D) Any other information requested by the Agency to evaluate the proposed servicing action.

(3) The lender will provide the Agency copies of any agreements executed to carry out the servicing action.

(4) PLP lenders will request servicing approval in accordance with their agreement with the Agency at the time of PLP status certification.

(c) *Subordinations.* (1) The Agency may subordinate its security interest on a direct loan when a guaranteed loan is being made if the requirements of the regulations governing Agency direct loan subordinations are met and only in the following circumstances:

(i) To permit a guaranteed lender to advance funds and perfect a security interest in crops, feeder livestock, livestock offspring, or livestock products;

(ii) When the lender requesting the guarantee needs the subordination of the Agency's lien position to maintain its lien position when servicing or restructuring;

(iii) When the lender requesting the guarantee is refinancing the debt of another lender and the Agency's position on real estate security will not be adversely affected; or

(iv) To permit a line of credit to be advanced for annual operating expenses.

(2) The Agency may subordinate its basic security in a direct loan to permit guaranteed line of credit only when both of the following additional conditions are met:

(i) The total unpaid balance of the direct loans is less than or equal to 75 percent of the value of all of the security for the direct loans, excluding the value of growing crops or planned production, at the time of the subordination. The direct loan security value will be determined by an appraisal. The lender requesting the subordination

and guarantee is responsible for providing the appraisal and may charge the applicant a reasonable appraisal fee.

(ii) The applicant cannot obtain sufficient credit through a conventional guaranteed loan without a subordination.

(3) The lender may not subordinate its interest in property which secures a guaranteed loan except as follows:

(i) The lender may subordinate its security interest in crops, feeder livestock, livestock offspring, or livestock products when no funds have been advanced from the guaranteed loan for their production, so a lender can make a loan for annual production expenses; or

(ii) The lender may, with written Agency approval, subordinate its interest in basic security in cases where the subordination is required to allow another lender to refinance an existing prior lien, no additional debt is being incurred, and the lender's security position will not be adversely affected by the subordination.

(iii) The Agency's national office may provide an exception to the subordination prohibition if such action is in the Agency's best interest. However, in no case can the loan made under the subordination include tax exempt financing.

(d) *Transfer and assumption.* Transfers and assumptions are subject to the following conditions:

(1) For standard eligible and CLP lenders, the servicing action must be approved by the Agency in writing.

(2) For standard eligible and CLP lenders, the transferee must apply for a loan in accordance with § 762.110, including a current appraisal, unless the lien position of the guaranteed loan will not change, and any other information requested by the Agency to evaluate the transfer and assumption.

(3) PLP lenders may process transfers and assumptions in accordance with their agreement with the Agency.

(4) Any required security appraisals must meet the requirements of § 762.127.

(5) The Agency will review, approve or reject the request in accordance with the time frames in § 762.130.

(6) The transferee must meet the eligibility requirements and loan limitations for the loan being transferred, all requirements relating to loan rates and terms, loan security, feasibility, and environmental and other laws applicable to a applicant under this part.

(7) The lender will use its own assumption agreements or conveyance instruments, providing they are legally sufficient to obligate the transferee for the total outstanding debt. The lender will provide the Agency copies of any agreements executed to carry out the servicing action.

(8) The Agency approves the transfer and assumption by executing a modification of the guarantee to designate the party that assumed the guaranteed debt, the amount of debt at the time of the assumption, including interest that is being capitalized, and any new loan terms, if applicable.

(9) The lender must give any holder notice of the transfer. If the rate and terms are changed, written concurrence from the holder is required.

(10) The Agency will agree to releasing the transferor or any guarantor from liability only if the requirements of § 762.146(c) are met.

[64 FR 7378, Feb. 12, 1999, as amended at 66 FR 7567, Jan. 24, 2001; 69 FR 44579, July 27, 2004]

§ 762.143 Servicing distressed accounts.

(a) A borrower is in default when 30 days past due on a payment or in violation of provisions of the loan documents.

(b) In the event of a borrower default, SEL and CLP lenders will:

(1) Report to the Agency in accordance with § 762.141.

(2) Determine whether it will repurchase the guaranteed portion from the holder in accordance with § 762.144, if the guaranteed portion of the loan was sold on the secondary market.

(3) Arrange a meeting with the borrower within 15 days of default (45 days after payment due date for monetary defaults) to identify the nature of the delinquency and develop a course of action that will eliminate the delinquency and correct the underlying problems. Non-monetary defaults will be handled in accordance with the lender's note, loan agreements and any other applicable loan documents.

(i) The lender and borrower will prepare a current balance sheet and cash flow projection in preparation for the meeting. If the borrower refuses to cooperate, the lender will compile the best financial information available.

(ii) The lender or the borrower may request the attendance of an Agency official. If requested, the Agency official will assist in developing solutions to the borrower's financial problems.

(iii) The lender will summarize the meeting and proposed solutions on the Agency form for guaranteed loan borrower default status completed after the meeting. The lender will indicate the results on this form for the lender's consideration of the borrower for interest assistance in conjunction with rescheduling under § 762.145(b).

(iv) The lender must decide whether to restructure or liquidate the account within 90 days of default, unless the lender can document circumstances that justify an extension by the Agency.

(v) The lender may not initiate foreclosure action on the loan until 60 days after eligibility of the borrower to participate in the interest assistance programs has been determined by the Agency. If the lender or the borrower does not wish to consider servicing options under this section, this should be documented, and liquidation under § 762.149 should begin.

(vi) If a borrower is current on a loan, but will be unable to make a payment, a restructuring proposal may be submitted in accordance with § 762.145 prior to the payment coming due.

(c) PLP lenders will service defaulted loans according to their lender's agreement.

[64 FR 7378, Feb. 12, 1999, as amended at 72 FR 63297, Nov. 8, 2007]

§ 762.144 Repurchase of guaranteed portion from a secondary market holder.

(a) *Request for repurchase.* The holder may request the lender to repurchase the unpaid guaranteed portion of the loan when:

(1) The borrower has not made a payment of principal and interest due on the loan for at least 60 days; or

(2) The lender has failed to remit to the holder its pro-rata share of any payment made by the borrower within 30 days of receipt of a payment.

(b) *Repurchase by the lender.* (1) When a lender is requested to repurchase a loan from the holder, the lender must consider the request according to the servicing actions that are necessary on the loan. In order to facilitate servicing and simplified accounting of loan transactions, lenders are encouraged to repurchase the loan upon the holder's request.

(2) The repurchase by the lender will be for an amount equal to the portion of the loan held by the holder plus accrued interest.

(3) The guarantee will not cover separate servicing fees that the lender accrues after the repurchase.

(c) *Repurchase by the Agency.* (1) If the lender does not repurchase the loan, the holder must inform the Agency in writing that demand was made on the lender and the lender refused. Following the lender's refusal, the holder may continue as holder of the guaranteed portion of the loan or request that the Agency purchase the guaranteed portion. Within 30 days after written demand to the Agency from the holder with required attachments, the Agency will forward to the holder payment of the unpaid principal balance, with accrued interest to the date of repurchase. If the holder does not desire repurchase or purchase of a defaulted loan, the lender must forward the holder its pro-rata share of payments, liquidation proceeds and Agency loss payments.

(2) With its demand on the Agency, the holder must include:

(i) A copy of the written demand made upon the lender.

(ii) Originals of the guarantee and note properly endorsed to the Agency, or the original of the assignment of guarantee.

(iii) A copy of any written response to the demand of the holder by the lender.

(iv) An account to which the Agency can forward the purchase amount via electronic funds transfer.

(3) The amount due the holder from the Agency includes unpaid principal, unpaid interest to the date of demand, and interest which has accrued from the date of demand to the proposed payment date.

(i) Upon request by the Agency, the lender must furnish upon Agency request a current statement, certified by a bank officer, of the unpaid principal and interest owed by the borrower and the amount due the holder.

(ii) Any discrepancy between the amount claimed by the holder and the information submitted by the lender must be resolved by the lender and the holder before payment will be approved by the Agency. The Agency will not participate in resolution of any such discrepancy. When there is a discrepancy, the 30 day Agency payment requirement to the holder will be suspended until the discrepancy is resolved.

(iii) In the case of a request for Agency purchase, the Agency will only pay interest that accrues for up to 90 days from the date of the demand letter to the lender requesting the repurchase. However, if the holder requested repurchase from the Agency within 60 days of the request to the lender and for any reason not attributable to the holder and the lender, the Agency cannot make payment within 30 days of the holder's demand to the Agency, the holder will be entitled to interest to the date of payment.

(4) At the time of purchase by the Agency, the original assignment of guarantee will be assigned by the holder to the Agency without recourse, including all rights, title, and interest in the loan.

(5) Purchase by the Agency does not change, alter, or modify any of the lender's obligations to the Agency specified in the lender's agreement or guarantee; nor does the purchase waive any of the Agency's rights against the lender.

(6) The Agency succeeds to all rights of the holder under the Guarantee including the right of set-off against the lender.

(7) Within 180 days of the Agency's purchase, the lender will reimburse the Agency the amount of repurchase, with accrued interest, through one of the following ways:

(i) By liquidating the loan security and paying the Agency its pro-rata share of liquidation proceeds; or

(ii) Paying the Agency the full amount the Agency paid to the holder plus any accrued interest.

(8) The lender will be liable for the purchase amount and any expenses incurred by the Agency to maintain the loan in its portfolio or liquidate the security. While the Agency holds the guaranteed portion of the loan, the lender will transmit to the Agency any payment received from the borrower, including the pro-rata share of liquidation or other proceeds.

(9) If the borrower files for reorganization under the provisions of the bankruptcy code or pays the account current while the purchase by the Government is being processed, the Agency may hold the loan as long it determines this action to be in the Agency's interest. If the lender is not proceeding expeditiously to collect the loan or reimbursement is not waived under this paragraph, the Agency will demand payment by the lender and collect the purchase amount through administrative offset of any claims due the lender.

(10) The Agency may sell a purchased guaranteed loan on a non-recourse basis if it determines that selling the portion of the loan that it holds is in the Government's best interest. A non-recourse purchase from the Agency requires a written request to the Agency from the party that wishes to purchase it, and written concurrence from the lender;

(d) *Repurchase for servicing.* (1) If, due to loan default or imminent loan restructuring, the lender determines that repurchase is necessary to adequately service the loan, the lender may repurchase the guaranteed portion of the loan from the holder, with the written approval of the Agency.

(2) The lender will not repurchase from the holder for arbitrage purposes. With its request for Agency concurrence, the lender will notify the Agency of its plans to resell the guaranteed portion following servicing.

(3) The holder will sell the guaranteed portion of the loan to the lender

for an amount agreed to between the lender and holder.

[64 FR 7378, Feb. 12, 1999, as amended at 69 FR 44579, July 27, 2004]

§ 762.145 **Restructuring   guaranteed loans.**

(a) *General.* (1) To restructure guaranteed loans standard eligible lenders must:

(i) Obtain prior written approval of the Agency for all restructuring actions; and,

(ii) Provide the items in paragraph (b) and (e) of this section to the Agency for approval.

(2) If the standard eligible lender's proposal for servicing is not agreed to by the Agency, the Agency approval official will notify the lender in writing within 14 days of the lender's request.

(3) To restructure guaranteed loans CLP lenders must:

(i) Obtain prior written approval of the Agency only for debt write down under this section.

(ii) Submit all calculations required in paragraph (e) of this section for debt writedown.

(iii) For restructuring other than write down, provide FSA with a certification that each requirement of this section has been met, a narrative outlining the circumstances surrounding the need for restructuring, and copies of any applicable calculations.

(4) PLP lenders will restructure loans in accordance with their lender's agreement.

(5) All lenders will submit copies of any restructured notes or lines of credit to the Agency.

(b) *Requirements.* For any restructuring action, the following conditions apply:

(1) The borrower meets the eligibility criteria of § 762.120, except the provisions regarding prior debt forgiveness and delinquency on a federal debt do not apply.

(2) The borrower's ability to make the amended payment is documented by the following:

(i) A feasible plan.

(ii) Current financial statements from all liable parties.

(iii) Verification of nonfarm income.

(iv) Verification of all debts of exceeding an amount determined by the Agency.

(v) Applicable credit reports.

(vi) Financial history (and production history for standard eligible lenders) for the past 3 years to support the cash flow projections.

(3) A final loss claim may be reduced, adjusted, or rejected as a result of negligent servicing after the concurrence with a restructuring action under this section.

(4) Loans secured by real estate and/or equipment can be restructured using a balloon payment, equal installments, or unequal installments. Under no circumstances may livestock or crops alone be used as security for a loan to be rescheduled using a balloon payment. If a balloon payment is used, the projected value of the real estate and/or equipment security must indicate that the loan will be fully secured when the balloon payment becomes due. The projected value will be derived from a current appraisal adjusted for depreciation of depreciable property, such as buildings and other improvements, that occurs until the balloon payment is due. For equipment security, a current appraisal is required. The lender is required to project the security value of the equipment at the time the balloon payment is due based on the remaining life of the equipment, or the depreciation schedule on the borrower's Federal income tax return. Loans restructured with a balloon payment that are secured by real estate will have a minimum term of 5 years, and other loans will have a minimum term of 3 years before the scheduled balloon payment. If statutory limits on terms of loans prevent the minimum terms, balloon payments may not be used. If the loan is rescheduled with unequal installments, a feasible plan, as defined in §762.2(b), must be projected for when installments are scheduled to increase.

(5) If a borrower is current on a loan, but will be unable to make a payment, a restructuring proposal may be submitted prior to the payment coming due.

(6) The lender may capitalize the outstanding interest when restructuring the loan as follows:

(i) As a result of the capitalization of interest, a rescheduled promissory note may increase the amount of principal the borrower is required to pay. However, in no case will such principal amount exceed the statutory loan limits contained in §761.8 of this chapter.

(ii) When accrued interest causes the loan amount to exceed the statutory loan limits, rescheduling may be approved without capitalization of the amount that exceeds the limit. Noncapitalized interest may be scheduled for repayment over the term of the rescheduled note.

(iii) Only interest that has accrued at the rate indicated on the borrower's original promissory notes may be capitalized. Late payment fees or default interest penalties that have accrued due to the borrower's failure to make payments as agreed are not covered under the guarantee and may not be capitalized.

(iv) The Agency will execute a modification of guarantee form to identify the new loan principal and the guaranteed portion if greater than the original loan amounts, and to waive the restriction on capitalization of interest, if applicable, to the existing guarantee documents. The modification form will be attached to the original guarantee as an addendum.

(v) Approved capitalized interest will be treated as part of the principal and interest that accrues thereon, in the event that a loss should occur.

(7) The lender's security position will not be adversely affected because of the restructuring. New security instruments may be taken if needed, but a loan does not have to be fully secured in order to be restructured, unless it is restructured with a balloon payment. When a loan is restructured using a balloon payment the lender must take a lien on all assets and project the loan to be fully secured at the time the balloon payment becomes due, in accordance with paragraph (b)(4) of this section.

(8) Any holder agrees to any changes in the original loan terms. If the holder does not agree, the lender must repurchase the loan from the holder for any loan restructuring to occur.

(9) After a guaranteed loan is restructured, the lender must provide the

Agency with a copy of the restructured promissory note.

(10) For CL, the lender must ensure that the borrower is maintaining the practice for which the CL was made.

(c) *Rescheduling.* The following conditions apply when a guaranteed loan is rescheduled or reamortized:

(1) Payments will be rescheduled within the following terms:

(i) FO and existing SW may be amortized over the remaining term of the note or rescheduled with an uneven payment schedule. The maturity date cannot exceed 40 years from the date of the original note.

(ii) OL notes must be rescheduled over a period not to exceed 15 years from the date of the rescheduling. An OL line of credit may be rescheduled over a period not to exceed 7 years from the date of the rescheduling or 10 years from the date of the original note, whichever is less. Advances cannot be made against a line of credit loan that has had any portion of the loan rescheduled.

(iii) CL will be amortized over the remaining term or rescheduled with an uneven payment schedule. The maturity date cannot exceed 30 years from the date of the original note.

(2) The interest rate for a rescheduled loan is the negotiated rate agreed upon by the lender and the borrower at the time of the action, subject to the loan limitations for each type of loan.

(3) A new note is not necessary when rescheduling occurs. However, if a new note is not taken, the existing note or line of credit agreement must be modified by attaching an allonge or other legally effective amendment, evidencing the revised repayment schedule and any interest rate change. If a new note is taken, the new note must reference the old note and state that the indebtedness evidenced by the old note or line of credit agreement is not satisfied. The original note or line of credit agreement must be retained.

(d) *Deferrals.* The following conditions apply to deferrals:

(1) Payments may be deferred up to 5 years, but the loan may not be extended beyond the final due date of the note.

(2) The principal portion of the payment may be deferred either in whole or in part.

(3) Interest may be deferred only in part. Payment of a reasonable portion of accruing interest as indicated by the borrower's cash flow projections is required for multi-year deferrals.

(4) There must be a reasonable prospect that the borrower will be able to resume full payments at the end of the deferral period.

(e) *Debt writedown.* The following conditions apply to debt writedown:

(1) A lender may only write down a delinquent guaranteed loan or line of credit in an amount sufficient to permit the borrower to develop a feasible plan as defined in § 762.102(b).

(2) The lender will request other creditors to negotiate their debts before a writedown is considered.

(3) The borrower cannot develop a feasible plan after consideration is given to rescheduling and deferral under this section.

(4) The present value of the loan to be written down, based on the interest rate of the rescheduled loan, will be equal to or exceed the net recovery value of the loan collateral.

(5) The loan will be restructured with regular payments at terms no shorter than 5 years for a line of credit and OL term note; and no shorter than 20 years for FO and CL, unless required to be shorter by paragraphs (c)(1)(i) through (iii) of this section.

(6) No further advances may be made on a line of credit that is written down.

(7) Loans may not be written down with interest assistance. If a borrower's loan presently on interest assistance requires a writedown, the writedown will be considered without interest assistance.

(8) The writedown is based on writing down the shorter-term loans first.

(9) When a lender requests approval of a writedown for a borrower with multiple loans, the security for all of the loans will be cross-collateralized and continue to serve as security for the loan that is written down. If a borrower has multiple loans and one loan is written off entirely through debt writedown, the security for that loan will not be released and will remain as security for the other written down

debt. Additional security instruments will be taken if required to cross-collateralize security and maintain lien priority.

(10) The writedown will be evidenced by an allonge or amendment to the existing note or line of credit reflecting the writedown.

(11) The borrower executes an Agency shared appreciation agreement for loans which are written down and secured by real estate.

(i) The lender will attach the original agreement to the restructured loan document.

(ii) The lender will provide the Agency a copy of the executed agreement, and

(iii) Security instruments must ensure future collection of any appreciation under the agreement.

(12) The lender will prepare and submit the following to the Agency:

(i) A current appraisal of all security in accordance with §762.127.

(ii) A completed report of loss on the appropriate Agency form for the proposed writedown loss claim.

(iii) Detailed writedown calculations as follows:

(A) Calculate the present value.

(B) Determine the net recovery value.

(C) If the net recovery value exceeds the present value, writedown is unavailable; liquidation becomes the next servicing consideration. If the present value equals or exceeds the net recovery value, the debt may be written down to the present value.

(iv) The lender will make any adjustment in the calculations as requested by the Agency.

[64 FR 7378, Feb. 12, 1999; 64 FR 38298, July 16, 1999, as amended at 66 FR 7567, Jan. 24, 2001; 69 FR 44579, July 27, 2004; 70 FR 56107, Sept. 26, 2005; 72 FR 17358, Apr. 9, 2007; 75 FR 54014, Sept. 1, 2010; 77 FR 15938, Mar. 19, 2012; 78 FR 65530, Nov. 1, 2013; 86 FR 43391, Aug. 9, 2021]

§762.146 Other servicing procedures.

(a) *Additional loans and advances.* (1) Notwithstanding any provision of this section, the PLP lender may make additional loans or advances in accordance with the lender's agreement with the Agency.

(2) SEL and CLP lenders must not make additional loans or advances without prior written approval of the Agency, except as provided in the borrower's loan or line of credit agreement.

(3) In cases of a guaranteed line of credit, lenders may make an emergency advance when a line of credit has reached its ceiling. The emergency advance will be made as an advance under the line and not as a separate note. The lender's loan documents must contain sufficient language to provide that any emergency advance will constitute a debt of the borrower to the lender and be secured by the security instrument. The following conditions apply:

(i) The loan funds to be advanced are for authorized operating loan purposes;

(ii) The financial benefit to the lender and the Government from the advance will exceed the amount of the advance; and

(iii) The loss of crops or livestock is imminent unless the advance is made.

(4) Protective advance requirements are found in §762.149.

(b) *Release of liability upon withdrawal.* An individual who is obligated on a guaranteed loan may be released from liability by a lender, with the written consent of the Agency, provided the following conditions have been met:

(1) The individual to be released has withdrawn from the farming operation;

(2) A divorce decree or final property settlement does not hold the withdrawing party responsible for the loan payments;

(3) The withdrawing party's interest in the security is conveyed to the individual or entity with whom the loan will be continued;

(4) The ratio of the amount of debt to the value of the remaining security is less than or equal to .75, or the withdrawing party has no income or assets from which collection can be made; and

(5) Withdrawal of the individual does not result in legal dissolution of the entity to which the loans are made. Individually liable members of a general or limited partnership may not be released from liability.

(6) The remaining liable party projects a feasible plan (see §761.2(b) of this chapter).

(c) *Release of liability after liquidation.* After a final loss claim has been paid on the borrower's account, the lender may release the borrower or guarantor from liability if;

(1) The Agency agrees to the release in writing;

(2) The lender documents its consideration of the following factors concerning the borrower or guarantors:

(i) The likelihood that the borrower or guarantor will have a sufficient level of income in the reasonably near future to contribute to a meaningful reduction of the debt;

(ii) The prospect that the borrower or guarantor will inherit assets in the near term that may be attached by the Agency for payment of a significant portion of the debt;

(iii) Whether collateral has been properly accounted for, and whether liability should be retained in order to take action against the borrower or a third party for conversion of security;

(iv) The availability of other income or assets which are not security;

(v) The possibility that assets have been concealed or improperly transferred;

(vi) The effect of other guarantors on the loan; and

(vii) Cash consideration or other collateral in exchange for the release of liability.

(3) The lender will use its own release of liability documents.

(d) *Interest rate changes.* (1) The lender may change the interest rate on a performing (nondelinquent) loan only with the borrower's consent.

(2) If the loan has been sold on the secondary market, the lender must repurchase the loan or obtain the holder's written consent.

(3) To change a fixed rate of interest to a variable rate of interest or vice versa, the lender and the borrower must execute a legally effective allonge or amendment to the existing note.

(4) If a new note is taken, it will be attached to and refer to the original note.

(5) The lender will inform the Agency of the rate change.

(e) *Consolidation.* Two or more Agency guaranteed loans may be consolidated, subject to the following conditions:

(1) The borrower must project a feasible plan after the consolidation. See §761.2(b) of this chapter for definition of feasible plan.

(2) Only OL may be consolidated.

(3) Existing lines of credit may only be consolidated with a new line of credit if the final maturity date and conditions for advances of the new line of credit are made the same as the existing line of credit.

(4) Guaranteed OL may not be consolidated with a line of credit, even if the line of credit has been rescheduled.

(5) Guaranteed loans made prior to October 1, 1991, cannot be consolidated with those loans made on or after October 1, 1991.

(6) OL secured by real estate or with an outstanding interest assistance agreement or shared appreciation agreement cannot be consolidated.

(7) A new note or line of credit agreement will be taken. The new note or line of credit agreement must describe the note or line of credit agreement being consolidated and must state that the indebtedness evidenced by the note or line of credit agreement is not satisfied. The original note or line of credit agreement must be retained.

(8) The interest rate for a consolidated OL loan is the negotiated rate agreed upon by the lender and the borrower at the time of the action, subject to the loan limitations for each type of loan.

(9) The Agency approves the consolidation by executing a modification of guarantee. The modification will indicate the consolidated loan amount, new terms, and percentage of guarantee, and will be attached to the originals of the guarantees being consolidated. If loans with a different guarantee percentage are consolidated, the new guarantee will be at the lowest percentage of guarantee being consolidated

(10) Any holders must consent to the consolidation, or the guaranteed portion must be repurchased by the lender.

[64 FR 7378, Feb. 12, 1999, as amended at 66 FR 7567, Jan. 24, 2001; 78 FR 65530, Nov. 1, 2013]

§762.147 Servicing shared appreciation agreements.

(a) *Lender responsibilities.* The lender is responsible for:

(1) Monitoring the borrower's compliance with the shared appreciation agreement;

(2) Notifying the borrower of the amount of recapture due; and,

(3) Beginning October 1, 1999, a notice of the agreement's provisions not later than 12 months before the end of the agreement; and

(4) Reimbursing the Agency for its pro-rata share of recapture due.

(b) *Recapture.* (1) Recapture of any appreciation of real estate security will take place at the end of the term of the agreement, or sooner if the following occurs:

(i) On the conveyance of the real estate security (or a portion thereof) by the borrower.

(A) If only a portion of the real estate is conveyed, recapture will only be triggered against the portion conveyed. Partial releases will be handled in accordance with §762.142(b).

(B) Transfer of title to the spouse of the borrower on the death of such borrower will not be treated as a conveyance under the agreement.

(ii) On repayment of the loan; or

(iii) If the borrower ceases farming.

(2) Calculating recapture.

(i) The amount of recapture will be based on the difference between the value of the security at the time recapture is triggered and the value of the security at the time of writedown, as shown on the shared appreciation agreement.

(ii) Security values will be determined through appraisals obtained by the lender and meeting the requirements of §762.127.

(iii) All appraisal fees will be paid by the lender.

(iv) The amount of recapture will not exceed the amount of writedown shown on the shared appreciation agreement.

(v) If recapture is triggered within 4 years of the date of the shared appreciation agreement, the lender shall recapture 75 percent of any positive appreciation in the market value of the property securing the loan or line of credit agreement.

(vi) If recapture is triggered after 4 years from the date of the shared appreciation agreement, the lender shall recapture 50 percent of any positive appreciation in the market value of the property securing the loan or line of credit agreement.

(3) Servicing recapture debt.

(i) If recapture is triggered under the shared appreciation agreement and the borrower is unable to pay the recapture in a lump sum, the lender may:

(A) Reschedule the recapture debt with the consent of the Agency, provided the lender can document the borrower's ability to make amortized payments on the recapture debt, plus pay all other obligations. In such case, the recapture debt will not be covered by the guarantee;

(B) Pay the Agency its pro rata share of the recapture due. In such case, the recapture debt of the borrower will be covered by the guarantee; or

(C) Service the account in accordance with §762.149.

(ii) If recapture is triggered, and the borrower is able but unwilling to pay the recapture in a lump sum, the lender will service the account in accordance with §762.149.

(4) Paying the Agency. Any shared appreciation recaptured by the lender will be shared on a pro-rata basis between the lender and the Agency.

[64 FR 7378, Feb. 12, 1999, as amended at 75 FR 54014, Sept. 3, 2010]

§762.148 Bankruptcy.

(a) *Lender responsibilities.* The lender must protect the guaranteed loan debt and all collateral securing the loan in bankruptcy proceedings. The lender's responsibilities include, but are not limited to:

(1) Filing a proof of claim where required and all the necessary papers and pleadings;

(2) Attending, and where necessary, participating in meetings of the creditors and court proceedings;

(3) Protecting the collateral securing the guaranteed loan and resisting any adverse changes that may be made to the collateral;

(4) Seeking a dismissal of the bankruptcy proceeding when the operation as proposed by the borrower to the bankruptcy court is not feasible;

(5) When permitted by the bankruptcy code, requesting a modification of any plan of reorganization if it appears additional recoveries are likely.

(6) Monitor confirmed plans under chapters 11, 12 and 13 of the bankruptcy code to determine borrower compliance. If the borrower fails to comply, the lender will seek a dismissal of the reorganization plan; and

(7) Keeping the Agency regularly informed in writing on all aspects of the proceedings.

(i) The lender will submit a default status report when the borrower defaults and every 60 days until the default is resolved or a final loss claim is paid.

(ii) The default status report will be used to inform the Agency of the bankruptcy filing, the reorganization plan confirmation date and effective date, when the reorganization plan is complete, and when the borrower is not in compliance with the reorganization plan.

(b) *Bankruptcy expenses.* (1) Reorganization.

(i) Expenses, such as legal fees and the cost of appraisals incurred by the lender as a direct result of the borrower's chapter 11, 12, or 13 reorganization, are covered under the guarantee, provided they are reasonable, customary, and provide a demonstrated economic benefit to the lender and the Agency.

(ii) Lender's in-house expenses, which are those expenses which would normally be incurred for administration of the loan, including in-house lawyers, are not covered by the guarantee.

(2) Liquidation expenses in bankruptcy.

(i) Reasonable and customary liquidation expenses may be deducted from the proceeds of the collateral in liquidation bankruptcy cases.

(ii) In-house expenses are not considered customary liquidation expenses, may not be deducted from collateral proceeds, and are not covered by the guarantee.

(c) *Estimated loss claims in reorganization*—(1) *At confirmation.* The lender may submit an estimated loss claim upon confirmation of the reorganization plan in accordance with the following:

(i) The estimated loss payment will cover the guaranteed percentage of the principal and accrued interest written off, plus any allowable costs incurred as of the effective date of the plan.

(ii) The lender will submit supporting documentation for the loss claim, and any additional information requested by the Agency, including justification for the legal fees included on the claim.

(iii) The estimated loss payment may be revised as consistent with a court-approved reorganization plan.

(iv) Protective advances made and approved in accordance with § 762.149 may be included in an estimated loss claim associated with a reorganization, if:

(A) They were incurred in connection with the initiation of liquidation action prior to bankruptcy filing; or

(B) The advance is required to provide repairs, insurance, etc. to protect the collateral as a result of delays in the case, or failure of the borrower to maintain the security.

(2) Interest only losses. The lender may submit an estimated loss claim for interest only after confirmation of the reorganization plan in accordance with the following:

(i) The loss claims may cover interest losses sustained as a result of a court-ordered, permanent interest rate reduction.

(ii) The loss claims will be processed annually on the anniversary date of the effective date of the reorganization plan.

(iii) If the borrower performs under the terms of the reorganization plan, annual interest reduction loss claims will be submitted on or near the same date, beyond the period of the reorganization plan.

(3) Actual loss.

(i) Once the reorganization plan is complete, the lender will provide the Agency with documentation of the actual loss sustained.

(ii) If the actual loss sustained is greater than the prior estimated loss payment, the lender may submit a revised estimated loss claim to obtain payment of the additional amount owed by the Agency under the guarantee.

(iii) If the actual loss is less than the prior estimated loss, the lender will reimburse the Agency for the overpayment plus interest at the note rate from the date of the payment of the estimated loss.

(4) *Payment to holder.* In reorganization bankruptcy, if a holder makes demand upon the Agency, the Agency will pay the holder interest to the plan's effective date. Accruing interest thereafter will be based upon the provisions of the reorganization plan.

(d) *Liquidation under the bankruptcy code.* (1) Upon receipt of notification that a borrower has filed for protection under Chapter 7 of the bankruptcy code, or a liquidation plan under chapter 11, the lender must proceed according to the liquidation procedures of this part.

(2) If the property is abandoned by the trustee, the lender will conduct the liquidation according to §762.149.

(3) Proceeds received from partial sale of collateral during bankruptcy may be used by the lender to pay reasonable costs, such as freight, labor and sales commissions, associated with the partial sale. Reasonable use of proceeds for this purpose must be documented with the final loss claim in accordance with §762.149(i)(4).

[64 FR 7378, Feb. 12, 1999, as amended at 71 FR 43957, Aug. 3, 2006; 73 FR 32637, June 10, 2008; 75 FR 54014, Sept. 3, 2010]

§762.149 Liquidation.

(a) *Mediation.* When it has been determined that default cannot be cured through any of the servicing options available, or if the lender does not wish to utilize any of the authorities provided in this part, the lender must:

(1) Participate in mediation according to the rules and regulations of any State which has a mandatory farmer-creditor mediation program;

(2) Consider private mediation services in those States which do not have a mandatory farmer-creditor mediation program; and

(3) Not agree to any proposals to rewrite the terms of a guaranteed loan which do not comply with this part. Any agreements reached as a result of mediation involving defaults and or loan restructuring must have written concurrence from the Agency before they are implemented.

(b) *Liquidation plan.* If a default cannot be cured after considering servicing options and mediation, the lender will proceed with liquidation of the collateral in accordance with the following:

(1) Within 150 days after the payment due date, all lenders will prepare a liquidation plan. Standard eligible and CLP lenders will submit a written liquidation plan to the Agency which includes:

(i) Current balance sheets from all liable parties or, if the parties are not cooperative, the best information available, or in liquidation bankruptcies, a copy of the bankruptcy schedules or discharge notice;

(ii) A proposed method of maximizing the collection of debt which includes specific plans to collect any remaining loan balances on the guaranteed loan after loan collateral has been liquidated, including possibilities for judgment;

(A) If the borrower has converted loan security, the lender will determine whether litigation is cost effective. The lender must address, in the liquidation plan, whether civil or criminal action will be pursued. If the lender does not pursue the recovery, the reason must be documented when an estimated loss claim is submitted.

(B) Any proposal to release the borrower from liability will be addressed in the liquidation plan in accordance with §762.146(c)(2);

(iii) An independent appraisal report on all collateral securing the loan that meets the requirements of §762.127 and a calculation of the net recovery value of the security as defined in ;§761.2(b) of this chapter. The appraisal requirement may be waived by the Agency in the following cases:

(A) The bankruptcy trustee is handling the liquidation and the lender has submitted the trustee's determination of value;

(B) The lender's proposed method of liquidation rarely results in receipt of less than market value for livestock and used equipment; or

(C) A purchase offer has already been received for more than the debt;

(iv) An estimate of time necessary to complete the liquidation;

(v) An estimated loss claim must be filed no later than 150 days past the payment due date unless the account has been completely liquidated and then a final loss claim must be filed.

(vi) An estimate of reasonable liquidation expenses; and

(vii) An estimate of any protective advances.

(2) PLP lenders will submit a liquidation plan as required by their lender's agreement.

(c) *Agency approval of the liquidation plan.* (1) CLP lender's or standard eligible lender's liquidation plan, and any revisions of the plan, must be approved by the Agency.

(2) If, within 20 calendar days of the Agency's receipt of the liquidation plan, the Agency fails to approve it or fails to request that the lender make revisions, the lender may assume the plan is approved. The lender may then proceed to begin liquidation actions at its discretion as long as it has been at least 60 days since the borrower's eligibility for interest assistance was considered.

(3) At its option, the Agency may liquidate the guaranteed loan as follows:

(i) Upon Agency request, the lender will transfer to the Agency all rights and interests necessary to allow the Agency to liquidate the loan. The Agency will not pay the lender for any loss until after the collateral is liquidated and the final loss is determined; and

(ii) If the Agency conducts the liquidation, interest accrual will cease on the date the Agency notifies the lender in writing that it assumes responsibility for the liquidation.

(d) *Estimated loss claims.* An estimated loss claim must be submitted by all lenders no later than 150 days after the payment due date unless the account has been completely liquidated and then a final loss claim must be filed. The estimated loss will be based on the following:

(1) The Agency will pay the lender the guaranteed percentage of the total outstanding debt, less the net recovery value of the remaining security, less any unaccounted for security; and

(2) The lender will discontinue interest accrual on the defaulted loan at the time the estimated loss claim is paid

by the Agency. The Agency will not pay interest beyond 210 days from the payment due date. If the lender estimates that there will be no loss after considering the costs of liquidation, an estimated loss of zero will be submitted and interest accrual will cease upon the approval of the estimated loss and never later than 210 days from the payment due date. The following exceptions apply:

(i) In the case of a Chapter 7 bankruptcy, in cases where the lender filed an estimated loss claim, the Agency will pay the lender interest that accrues during and up to 45 days after the discharge on the portion of the chattel only secured debt that was estimated to be secured, but upon final liquidation was found to be unsecured, and up to 90 days after the date of discharge on the portion of real estate secured debt that was estimated to be secured, but was found to be unsecured upon final disposition.

(ii) The Agency will pay the lender interest that accrues during and up to 90 days after the time period the lender is unable to dispose of acquired property due to state imposed redemption rights on any unsecured portion of the loan during the redemption period, if an estimated loss claim was paid by the Agency during the liquidation action.

(3) Packager fees and outside consultant fees for servicing of guaranteed loans are not covered by the guarantee, and will not be paid in an estimated loss claim.

(e) *Protective advances.* (1) Prior written authorization from the Agency is required for all protective advances in excess of $5,000 for CLP lenders and $3,000 for standard eligible lenders. The dollar amount of protective advances allowed for PLP lenders will be specified when PLP status is awarded by the Agency or as contained in the lender's agreement.

(2) The lender may claim recovery for the guaranteed portion of any loss of monies advanced as protective advances as allowed in this part, plus interest that accrues on the protective advances.

(3) Payment for protective advances is made by the Agency when the final

loss claim is approved, except in bankruptcy actions.

(4) Protective advances are used only when the borrower is in liquidation, liquidation is imminent, or when the lender has taken title to real property in a liquidation action.

(5) Legal fees are not a protective advance.

(6) Protective advances may only be made when the lender can demonstrate the advance is in the best interest of the lender and the Agency.

(7) Protective advances must constitute a debt of the borrower to the lender and be secured by the security instrument.

(8) Protective advances must not be made in lieu of additional loans.

(f) *Unapproved loans or advances.* The amount of any payments made by the borrower on unapproved loans or advances outside of the guarantee will be deducted from any loss claim submitted by the lender on the guaranteed loan, if that loan or advance was paid prior to, and to the detriment of, the guaranteed loan.

(g) *Acceleration.* (1) If the borrower is not in bankruptcy, the lender shall send the borrower notice that the loan is in default and the entire debt has been determined due and payable immediately after other servicing options have been exhausted.

(2) The loan cannot be accelerated until after the borrower has been considered for interest assistance and the conclusion of mandatory mediation in accordance with §762.149.

(3) The lender will submit a copy of the acceleration notice or other document to the Agency.

(h) *Foreclosure.* (1) The lender is responsible for determining the necessary parties to any foreclosure action, or who should be named on a deed of conveyance taken in lieu of foreclosure.

(2) When the property is liquidated, the lender will apply the net proceeds to the guaranteed loan debt.

(3) When it is necessary to enter a bid at a foreclosure sale, the lender may bid the amount that it determines is reasonable to protect its and the Agency's interest. At a minimum, the lender will bid the lesser of the net recovery value or the unpaid guaranteed loan balance.

(i) *Final loss claims.* (1) Lenders must submit a final loss claim when the security has been liquidated and all proceeds have been received and applied to the account. All proceeds must be applied to principal first and then toward accrued interest if the interest is still accruing. The application of the loss claim payment to the account does not automatically release the borrower of liability for any portion of the borrower's debt to the lender. The lender will continue to be responsible for collecting the full amount of the debt and sharing these future recoveries with the Agency in accordance with paragraph (j) of this section.

(2) If a lender acquires title to property either through voluntary conveyance or foreclosure proceeding, the lender will submit a final loss claim after disposing of the property. The lender may pay reasonable maintenance expenses to protect the value of the property while it is owned by the lender. These may be paid as protective advances or deducted as liquidation expenses from the sales proceeds when the lender disposes of the property. The lender must obtain Agency written concurrence before incurring maintenance expenses which exceed the amounts allowed in §762.149(e)(1). Packager fees and outside consultant fees for servicing of guaranteed loans are not covered by the guarantee, and will not be paid in a final loss claim.

(3) The lender will make its records available to the Agency for the Agency's audit of the propriety of any loss payment.

(4) All lenders will submit the following documents with a final loss claim:

(i) An accounting of the use of loan funds;

(ii) An accounting of the disposition of loan security and its proceeds;

(iii) A copy of the loan ledger indicating loan advances, interest rate changes, protective advances, and application of payments, rental proceeds, and security proceeds, including a running outstanding balance total; and

(iv) Documentation, as requested by the Agency, concerning the lender's compliance with the requirements of this part.

(5) The Agency will notify the lender of any discrepancies in the final loss claim or, approve or reject the claim within 40 days. Failure to do so will result in additional interest being paid to the lender for the number of days over 40 taken to process the claim.

(6) The Agency will reduce a final loss claim based on its calculation of the dollar amount of loss caused by the lender's negligent servicing of the account. Loss claims may be reduced or rejected as a result of the following:

(i) A loss claim may be reduced by the amount caused by the lender's failure to secure property after a default, and will be reduced by the amount of interest that accrues when the lender fails to contact the borrower or takes no action to cure the default, once it occurs. Losses incurred as a result of interest accrual during excessive delays in collection, as determined by the Agency, will not be paid.

(ii) Unauthorized release of security proceeds, failure to verify ownership or possession of security to be purchased, or failure to inspect collateral as often required so as to ensure its maintenance.

(7) Losses will not be reduced for the following:

(i) Servicing deficiencies that did not contribute materially to the dollar amount of the loss.

(ii) Unaccounted security, as long as the lender's efforts to locate and recover the missing collateral was equal to that which would have been expended in the case of an unguaranteed loan in the lender's portfolio.

(8) Default interest, late charges, and loan servicing fees are not payable under the loss claim.

(9) The final loss will be the remaining outstanding balance after application of the estimated loss payment and the application of proceeds from the liquidation of the security.

(10) If the final loss is less than the estimated loss, the lender will reimburse the Agency for the overpayment, plus interest at the note rate from the date of the estimated loss payment.

(11) The lender will return the original guarantee marked paid after receipt of a final loss claim.

(j) *Future Recovery.* The lender will remit any recoveries made on the account after the Agency's payment of a final loss claim to the Agency in proportion to the percentage of guarantee, in accordance with the lender's agreement, until the account is paid in full or otherwise satisfied.

(k) *Overpayments.* The lender will repay any final loss overpayment determined by the Agency upon request.

(l) *Electronic funds transfer.* The lender will designate one or more financial institutions to which any Agency payments will be made via electronic funds transfer.

(m) *Establishment of Federal debt.* Any amounts paid by the Agency on account of liabilities of the guaranteed loan borrower will constitute a Federal debt owing to the Agency by the guaranteed loan borrower. In such case, the Agency may use all remedies available to it, including offset under the Debt Collection Improvement Act of 1996, to collect the debt from the borrower. Interest charges will be established at the note rate of the guaranteed loan on the date the final loss claim is paid.

[64 FR 7378, Feb. 12, 1999, as amended at 67 FR 44016, July 1, 2002; 69 FR 44580, July 27, 2004; 71 FR 43957, Aug. 3, 2006; 73 FR 32637, June 10, 2008; 78 FR 65530, Nov. 1, 2013]

§ 762.150 **Interest assistance program.**

(a) *Requests for interest assistance.* In addition to the loan application items required by § 762.110, to apply for interest assistance the lender's cash flow budget for the guaranteed applicant must reflect the need for interest assistance and the ability to cash flow with the subsidy. Interest assistance is available only on new guaranteed Operating Loans (OL).

(b) *Eligibility requirements.* The lender must document that the following conditions have been met for the applicant to be eligible for interest assistance:

(1) A feasible plan cannot be achieved without interest assistance, but can be achieved with interest assistance.

(2) If significant changes in the borrower's cash flow budget are anticipated after the initial 12 months, then the typical cash flow budget must demonstrate that the borrower will still

have a feasible plan following the anticipated changes, with or without interest assistance.

(3) The typical cash flow budget must demonstrate that the borrower will have a feasible plan throughout the term of the loan.

(4) The borrower, including members of an entity borrower, does not own any significant assets that do not contribute directly to essential family living or farm operations. The lender must determine the market value of any such non-essential assets and prepare a cash flow budget and interest assistance calculations based on the assumption that these assets will be sold and the market value proceeds used for debt reduction. If a feasible plan can then be achieved, the borrower is not eligible for interest assistance.

(5) A borrower may only receive interest assistance if their total debts (including personal debts) prior to the new loan exceed 50 percent of their total assets (including personal assets). An entity's debt to asset ratio will be based upon a financial statement that consolidates business and personal debts and assets of the entity and its members. Beginning farmers §761.2(b) of this chapter, as defined in §762.102, are excluded from this requirement.

(c) *Maximum assistance.* The maximum total guaranteed OL debt on which a borrower can receive interest assistance is $400,000, regardless of the number of guaranteed loans outstanding. This is a lifetime limit.

(d) *Maximum time for which interest assistance is available.* (1) A borrower may only receive interest assistance for one 5-year period. The term of the interest assistance agreement executed under this section shall not exceed 5 consecutive years from the date of the initial agreement signed by the applicant, including any entity members, or the outstanding term of the loan, whichever is less. This is a lifetime limit.

(2) Beginning farmers §761.2(b) of this chapter, as defined in §762.102, however, may be considered for two 5-year periods. The applicant must meet the definition of a beginning farmer and meet the other eligibility requirements outlined in paragraph (b) of this section at the onset of each 5-year period. A needs test will be completed in the fifth year

of IA eligibility for beginning farmers, to determine continued eligibility for a second 5-year period.

(3) Notwithstanding the limitation of paragraph (d)(1) of this section, a new interest assistance agreement may be approved for eligible borrowers to provide interest assistance through June 8, 2009, provided the total period does not exceed 10 years from the effective date of the original interest assistance agreement.

(e) *Multiple loans.* In the case of a borrower with multiple guaranteed loans with one lender, interest assistance can be applied to each loan, only to one loan or any distribution the lender selects, as necessary to achieve a feasible plan, subject to paragraph (c) of this section.

(f) *Terms.* The typical term of scheduled loan repayment will not be reduced solely for the purpose of maximizing eligibility for interest assistance. A loan must be scheduled over the maximum term typically used by lenders for similar type loans within the limits in §762.124. An OL for the purpose of providing annual operating and family living expenses will be scheduled for repayment when the income is scheduled to be received from the sale of the crops, livestock, and/or livestock products which will serve as security for the loan. An OL for purposes other than annual operating and family living expenses (i.e. purchase of equipment or livestock, or refinancing existing debt) will be scheduled over 7 years from the effective date of the proposed interest assistance agreement, or the life of the security, whichever is less.

(g) *Rate of interest.* The lender interest rate will be set according to §762.124(a).

(h) *Agreement.* The lender and borrower must execute an interest assistance agreement as prescribed by the Agency.

(i) *Interest assistance claims and payments.* To receive an interest assistance payment, the lender must prepare and submit a claim on the appropriate Agency form. The following conditions apply:

(1) Interest assistance payments will be four (4) percent of the average daily principal loan balance prorated over

the number of days the loan has been outstanding during the payment period. For loans with a note rate less than four (4) percent, interest assistance payments will be the weighted average interest rate multiplied by the average daily principal balance.

(2) The lender may select at the time of loan closing the date that they wish to receive an interest assistance payment. That date will be included in the interest assistance agreement.

(i) The initial and final claims submitted under an agreement may be for a period less than 12 months. All other claims will be submitted for a 12-month period, unless there is a lender substitution during the 12-month period in accordance with this section.

(ii) In the event of liquidation, the final interest assistance claim will be submitted with the estimated loss claim or the final loss claim if an estimated loss claim was not submitted. Interest will not be paid beyond the interest accrual cutoff dates established in the loss claims according to § 762.149(d)(2).

(3) A claim should be filed within 60 days of its due date. Claims not filed within 1 year from the due date will not be paid, and the amount due the lender will be permanently forfeited.

(4) All claims will be supported by detailed calculations of average daily principal balance during the claim period.

(5) Requests for continuation of interest assistance for agreements dated prior to June 8, 2007 will be supported by the lender's analysis of the applicant's farming operation and need for continued interest assistance as set out in their Interest Assistance Agreements. The following information will be submitted to the Agency:

(i) A summary of the operation's actual financial performance in the previous year, including a detailed income and expense statement.

(ii) A narrative description of the causes of any major differences between the previous year's projections and actual performance, including a detailed income and expense statement.

(iii) A current balance sheet.

(iv) A cash-flow budget for the period being planned. A monthly cash-flow budget is required for all lines of credit and operating loans made for annual operating purposes. All other loans may include either an annual or monthly cash-flow budget.

(v) A copy of the interest assistance needs analysis portion of the application form which has been completed based on the planned period's cash-flow budget.

(6) Interest Assistance Agreements dated June 8, 2007 or later do not require a request for continuation of interest assistance. The lender will only be required to submit an Agency IA payment form and the average daily principal balance for the claim period, with supporting documentation.

(7) Lenders may not charge or cause a borrower with an interest assistance agreement to be charged a fee for preparation and submission of the items required for an annual interest assistance claim.

(j) *Transfer, consolidation, and writedown.* Loans covered by interest assistance agreements cannot be consolidated. Such loans can be transferred only when the transferee was liable for the debt on the effective date of the interest assistance agreement. Loans covered by interest assistance can be transferred to an entity if the entity is eligible in accordance with §§ 762.120 and 762.150(b) and at least one entity member was liable for the debt on the effective date of the interest assistance agreement. Interest assistance will be discontinued as of the date of any writedown on a loan covered by an interest assistance agreement.

(k) *Rescheduling and deferral.* When a borrower defaults on a loan with interest assistance or the loan otherwise requires rescheduling or deferral, the interest assistance agreement will remain in effect for that loan at its existing terms. The lender may reschedule the loan in accordance with § 762.145. For Interest Assistance Agreements dated June 8, 2007 or later increases in the restructured loan amount above the amount originally obligated do not require additional funding; however, interest assistance is not available on that portion of the loan as interest assistance is limited to the original loan amount.

(1) *Bankruptcy.* In cases where the interest on a loan covered by an interest assistance agreement is reduced by court order in a reorganization plan under the bankruptcy code, interest assistance will be terminated effective on the date of the court order. Guaranteed loans which have had their interest reduced by bankruptcy court order are not eligible for interest assistance.

(m) *Termination of interest assistance payments.* Interest assistance payments will cease upon termination of the loan guarantee, upon reaching the expiration date contained in the agreement, or upon cancellation by the Agency under the terms of the interest assistance agreement. In addition, for loan guarantees sold into the secondary market, Agency purchase of the guaranteed portion of a loan will terminate the interest assistance.

(n) *Excessive interest assistance.* Upon written notice to the lender, borrower, and any holder, the Agency may amend or cancel the interest assistance agreement and collect from the lender any amount of interest assistance granted which resulted from incomplete or inaccurate information, an error in computation, or any other reason which resulted in payment that the lender was not entitled to receive.

(o) *Condition for cancellation.* The Interest Assistance Agreement is incontestable except for fraud or misrepresentation, of which the lender or borrower have actual knowledge at the time the interest assistance agreement is executed, or which the lender or borrower participates in or condones.

(p) *Substitution.* If there is a substitution of lender, the original lender will prepare and submit to the Agency a claim for its final interest assistance payment calculated through the effective date of the substitution. This final claim will be submitted for processing at the time of the substitution.

(1) Interest assistance will continue automatically with the new lender.

(2) The new lender must follow paragraph (i) of this section to receive their initial and subsequent interest assistance payments.

(q) *Exception Authority.* The Deputy Administrator for Farm Loan Programs has the authority to grant an exception to any requirement involving interest assistance if it is in the best interest of the Government and is not inconsistent with other applicable law.

[72 FR 17358, Apr. 9, 2007, as amended at 78 FR 14005, Mar. 4, 2013; 78 FR 65530, Nov. 1, 2013]

§§762.151–762.158 [Reserved]

§762.159 Pledging of guarantee.

A lender may pledge all or part of the guaranteed or unguaranteed portion of the loan as security to a Federal Home Loan Bank, a Federal Reserve Bank, a Farm Credit System Bank, or any other funding source determined acceptable by the Agency.

[70 FR 56107, Sept. 26, 2005]

§762.160 Assignment of guarantee.

(a) The following general requirements apply to assigning guaranteed loans:

(1) Subject to Agency concurrence, the lender may assign all or part of the guaranteed portion of the loan to one or more holders at or after loan closing, if the loan is not in default. However, a line of credit cannot be assigned. The lender must always retain the unguaranteed portion in their portfolio, regardless of how the loan is funded.

(2) The Agency may refuse to execute the Assignment of Guarantee and prohibit the assignment in case of the following:

(i) The Agency purchased and is holder of a loan that was assigned by the lender that is requesting the assignment.

(ii) The lender has not complied with the reimbursement requirements of §762.144(c)(7), except when the 180 day reimbursement or liquidation requirement has been waived by the Agency.

(3) The lender will provide the Agency with copies of all appropriate forms used in the assignment.

(4) The guaranteed portion of the loan may not be assigned by the lender until the loan has been fully disbursed to the borrower.

(5) The lender is not permitted to assign any amount of the guaranteed or unguaranteed portion of the loan to the applicant or borrower, or members of their immediate families, their officers, directors, stockholders, other

owners, or any parent, subsidiary, or affiliate.

(6) Upon the lender's assignment of the guaranteed portion of the loan, the lender will remain bound to all obligations indicated in the Guarantee, Lender's Agreement, the Agency program regulations, and to future program regulations not inconsistent with the provisions of the Lenders Agreement. The lender retains all rights under the security instruments for the protection of the lender and the United States.

(b) The following will occur upon the lender's assignment of the guaranteed portion of the loan:

(1) The holder will succeed to all rights of the Guarantee pertaining to the portion of the loan assigned.

(2) The lender will send the holder the borrower's executed note attached to the Guarantee.

(3) The holder, upon written notice to the lender and the Agency, may assign the unpaid guaranteed portion of the loan. The holder must assign the guaranteed portion back to the original lender if requested for servicing or liquidation of the account.

(4) The Guarantee or Assignment of Guarantee in the holder's possession does not cover:

(i) Interest accruing 90 days after the holder has demanded repurchase by the lender, except as provided in the Assignment of Guarantee and §762.144(c)(3)(iii).

(ii) Interest accruing 90 days after the lender or the Agency has requested the holder to surrender evidence of debt repurchase, if the holder has not previously demanded repurchase.

(c) Negotiations concerning premiums, fees, and additional payments for loans are to take place between the holder and the lender. The Agency will participate in such negotiations only as a provider of information.

[70 FR 56107, Sept. 26, 2005]

PART 763—LAND CONTRACT GUARANTEE PROGRAM

AUTHORITY: 5 U.S.C. 501 and 7 U.S.C. 1989.

SOURCE: 76 FR 75430, Dec. 2, 2011, unless otherwise noted.

§ 763.1 Introduction.

(a) *Purpose.* The Land Contract Guaranteed Program provides certain financial guarantees to the seller of a farm through a land contract sale to a beginning farmer or a socially disadvantaged farmer.

(b) *Types of guarantee.* The seller may request either of the following:

(1) *The prompt payment guarantee plan.* The Agency will guarantee an amount not to exceed three amortized annual installments plus an amount equal to the total cost of any related real estate taxes and insurance incurred during the period covered by the annual installment; or

(2) *The standard guarantee plan.* The Agency will guarantee an amount equal to 90 percent of the outstanding principal under the land contract.

(c) *Guarantee period.* The guarantee period is 10 years for either plan regardless of the term of the land contract.

§ 763.2 Abbreviations and definitions.

Abbreviations and definitions for terms used in this part are in § 761.2 of this chapter.

§763.3 Full faith and credit.

(a) The land contract guarantee constitutes an obligation supported by the full faith and credit of the United States. The Agency may contest the guarantee only in cases of fraud or misrepresentation by the seller, in which:

(1) The seller had actual knowledge of the fraud or misrepresentation at the time it because the seller, or

(2) The seller participated in or condoned the fraud or misrepresentation.

(b) Loss claims also may be reduced or denied to the extent that any negligence contributed to the loss under §763.22.

§763.4 Authorized land contract purpose.

The Agency will only guarantee the Contract installments, real estate taxes and insurance; or outstanding principal balance for an eligible seller of a family farm, through a land contract sale to an eligible beginning or socially disadvantaged farmer.

§763.5 Eligibility.

(a) *Seller eligibility requirements.* The private seller, and each entity member in the case of an entity seller, must:

(1) Possess the legal capacity to enter into a legally binding agreement;

(2) Not have provided false or misleading documents or statements during past or present dealings with the Agency;

(3) Not be ineligible due to disqualification resulting from Federal Crop Insurance violation, according to 7 CFR part 718; and

(4) Not be suspended or debarred under 2 CFR parts 180 and 417.

(b) *Buyer eligibility requirements.* The buyer must meet the following requirements to be eligible for the Land Contract Guarantee Program:

(1) Is a beginning farmer or socially disadvantaged farmer engaged primarily in farming in the United States after the guarantee is issued.

(2) Is the owner and operator of a family farm after the Contract is completed. Ownership of the family farm operation or the farm real estate may be held either directly in the individual's name or indirectly through interest in a legal entity. In the case of an entity buyer:

(i) Each entity member's ownership interest may not exceed the amount specified in the family farm definition in §761.2 of this chapter.

(ii) If the applicant has one or more embedded entities, at least 75 percent of the individual ownership interests of each embedded entity must be owned by members actively involved in managing or operating the family farm.

(iii) An ownership entity must be authorized to own a farm in the state or states in which the farm is located. An operating entity must be authorized to operate a farm in the state or states in which the farm is located.

(iv) If the entity members holding a majority interest are related by blood or marriage, at least one member of the entity must:

(A) Operate the family farm and

(B) Own the farm after the contract is completed;

(v) If the entity members holding a majority interest are not related by blood or marriage, the entity members holding a majority interest must:

(A) Operate the family farm; and

(B) Own the farm, or the entity itself must own the farm after the contract is completed;

(vi) If the entity is an operator-only entity, the individuals that own the farm (real estate) must own at least 50 percent of the family farm (operating entity).

(vii) All ownership may be held either directly in the individual's name or indirectly through interest in a legal entity.

(3) Must have participated in the business operations of a farm or ranch for at least 3 years out of the last 10 years prior to the date the application is submitted. Of those 3 years, 1 year can be substituted with the following experience:

(i) Postsecondary education in agriculture business, horticulture, animal science, agronomy, or other agricultural related fields,

(ii) Significant business management experience, or

(iii) Leadership or management experience while serving in any branch of the military.

(4) The buyer, and all entity members in the case of an entity, must not have caused the Agency a loss by receiving

debt forgiveness on all or a portion of any direct or guaranteed loan made under the authority of the Act by debt write-down or write-off; compromise, adjustment, reduction, or charge off under the provisions of section 331 of the Act; discharge in bankruptcy; or through payment of a guaranteed loss claim on more than three occasions on or prior to April 4, 1996 or any occasion after April 4, 1996. If the debt forgiveness is resolved by repayment of the Agency's loss, the Agency may still consider the debt forgiveness in determining the applicant's creditworthiness.

(5) The buyer, and all entity members in the case of an entity, must not be delinquent on any Federal debt, other than a debt under the Internal Revenue Code of 1986, when the guarantee is issued.

(6) The buyer, and all entity members in the case of an entity, may have no outstanding unpaid judgment awarded to the United States in any court. Such judgments do not include those filed as a result of action in the United States Tax Courts.

(7) The buyer, and all entity members in the case of an entity, must be a citizen of the United States, United States non-citizen national, or a qualified alien under applicable Federal immigration laws. United States non-citizen nationals and qualified aliens must provide the appropriate documentation as to their immigration status as required by the United States Department of Homeland Security, Bureau of Citizenship and Immigration Services.

(8) The buyer, and all entity members in the case of an entity, must possess the legal capacity to enter into a legally binding agreement.

(9) The buyer, and all entity members in the case of an entity, must not have provided false or misleading documents or statements during past or present dealings with the Agency.

(10) The buyer, and all entity members in the case of an entity, must not be ineligible as a result of a conviction for controlled substances according to 7 CFR part 718.

(11) The buyer, and all entity members in the case of an entity, must have an acceptable credit history demonstrated by satisfactory debt repayment.

(i) A history of failures to repay past debts as they came due when the ability to repay was within their control will demonstrate unacceptable credit history.

(ii) Unacceptable credit history will not include:

(A) Isolated instances of late payments which do not represent a pattern and were clearly beyond their control; or

(B) Lack of credit history.

(12) The buyer is unable to enter into a contract unless the seller obtains an Agency guarantee to finance the purchase of the farm at reasonable rates and terms.

(13) The buyer, and all entity members in the case of an entity, must not be ineligible due to disqualification resulting from Federal Crop Insurance violation, according to 7 CFR part 718.

(14) The buyer, and all entity members in the case of an entity, must not be suspended or debarred under 2 CFR parts 180 and 417.

[76 FR 75430, Dec. 2, 2011, as amended at 79 FR 60744, Oct. 8, 2014]

§ 763.6 Limitations.

(a) To qualify for a guarantee, the purchase price of the farm to be acquired through the land contract sale cannot exceed the lesser of:

(1) $500,000 or

(2) The current market value of the property.

(b) A guarantee will not be issued if the appraised value of the farm is greater than $500,000.

(c) Existing land contracts are not eligible for the Land Contract Guarantee Program.

(d) Guarantees may not be used to establish or support a non-eligible enterprise.

§ 763.7 Application requirements.

(a) *Seller application requirements.* A seller who contacts the Agency with interest in a guarantee under the Land Contract Guarantee Program will be sent the land contract letter of interest outlining specific program details. To formally request a guarantee on the proposed land contract, the seller, and each entity member in the case of an entity, must:

(1) Complete, sign, date, and return the land contract letter of interest to the Agency, and

(2) Provide the name, address, and telephone number of the chosen servicing or escrow agent.

(b) *Buyer application requirements.* A complete application from the buyer will include:

(1) The completed Agency application form;

(2) A current financial statement (not older than 90 days);

(3) If the buyer is an entity:

(i) A complete list of entity members showing the address, citizenship, principal occupation, and the number of shares and percentage of ownership or stock held in the entity by each member, or the percentage of interest in the entity held by each member;

(ii) A current financial statement for each member of the entity;

(iii) A current financial statement for the entity itself;

(iv) A copy of the entity's charter or any entity agreement, any articles of incorporation and bylaws, any certificate or evidence of current registration (in good standing), and a resolution adopted by the Board of Directors or entity members authorizing specified officers of the entity to apply for and obtain the land contract guarantee and execute required debt, security, and other instruments and agreements; and

(v) In the form of a married couple applying as a joint operation, items in paragraphs (b)(3)(i) and (b)(3)(iv) of this section will not be required. The Agency may request copies of the marriage license, prenuptial agreement, or similar documents as needed to verify loan eligibility and security. The information specified in paragraphs (b)(3)(ii) and (iii) of this section are only required to the extent needed to show the individual and joint finances of the husband and wife without duplication;

(4) A brief written description of the buyer's proposed operation;

(5) A farm operating plan;

(6) A brief written description of the buyer's farm training and experience;

(7) Three years of income tax and other financial records acceptable to the Agency, unless the buyer has been farming less than 3 years;

(8) Three years of farm production records, unless the buyer has been farming less than 3 years;

(9) Verification of income and off-farm employment if relied upon for debt repayment;

(10) Verification of all debts;

(11) Payment of the credit report fee;

(12) Documentation of compliance with the environmental regulations in part 799 of this chapter;

(13) A copy of the proposed land contract; and

(14) Any additional information deemed necessary by the Agency to effectively evaluate the applicant's eligibility and farm operating plan.

[76 FR 75430, Dec. 2, 2011, as amended at 79 FR 60744, Oct. 8, 2014; 81 FR 51284, Aug. 3, 2016]

§763.8 Incomplete applications.

(a) Within 10 days of receipt of an incomplete application, the Agency will provide the seller and buyer written notice of any additional information that must be provided. The seller or buyer, as applicable, must provide the additional information within 20 calendar days of the date of the notice.

(b) If the additional information is not received, the Agency will provide written notice that the application will be withdrawn if the information is not received within 10 calendar days of the date of the second notice.

§763.9 Processing complete applications.

Applications will be approved or rejected and all parties notified in writing no later than 30 calendar days after application is considered complete.

§763.10 Feasibility.

(a) The buyer's proposed operation as described in a form acceptable to the Agency must represent the operating cycle for the farm operation and must project a feasible plan as defined in §761.2(b) of this chapter.

(b) The projected income, expenses, and production estimates:

(1) Must be based on the buyer's last 3 years actual records of production and financial management unless the

buyer has been farming less than 3 years;

(2) For those farming less than 3 years, a combination of any actual history and other reliable sources of information may be used. Sources must be documented and acceptable to the Agency; and

(3) May deviate from historical performance if deviations are the direct result of specific changes in the operation, reasonable, justified, documented, and acceptable to the Agency.

(c) Price forecasts used in the plan must be reasonable, documented, and acceptable to the Agency.

(d) The Agency will analyze the buyer's business ventures other than the farm operation to determine their soundness and contribution to the operation.

(e) When a feasible plan depends on income from sources other than from owned land, the income must be dependable and likely to continue.

(f) When the buyer's farm operating plan is developed in conjunction with a proposed or existing Agency direct loan, the two farm operating plans must be consistent.

§ 763.11 Maximum loss amount, guarantee period, and conditions.

(a) *Maximum loss amount.* The maximum loss amount due to nonpayment by the buyer covered by the guarantee is based on the type of guarantee initially selected by the seller as follows:

(1) The prompt payment guarantee will cover:

(i) Three amortized annual installments; or

(ii) An amount equal to three annual installments (including an amount equal to the total cost of any tax and insurance incurred during the period covered by the annual installments).

(2) The standard guarantee will cover an amount equal to 90 percent of the outstanding principal balance.

(b) *Guarantee period.* The period of the guarantee will be 10 years from the effective date of the guarantee unless terminated earlier under § 763.23.

(c) *Conditions.* The seller will select an escrow agent to service a Land Contract Agreement if selecting the prompt payment guarantee plan, and a servicing agent to service a Land Contract Agreement if selecting the standard guarantee plan.

(1) An escrow agent must provide the Agency evidence of being a bonded title insurance company, attorney, financial institution or fiscally responsible institution.

(2) A servicing agent must provide the Agency evidence of being a bonded commercial lending institution or similar entity, registered and authorized to provide escrow and collection services in the State in which the real estate is located.

§ 763.12 Down payment, rates, terms, and installments.

(a) *Down payment.* The buyer must provide a minimum down payment of five percent of the purchase price of the farm.

(b) *Interest rate.* The interest rate charged by the seller must be fixed at a rate not to exceed the Agency's direct FO loan interest rate in effect at the time the guarantee is issued, plus three percentage points. The seller and buyer may renegotiate the interest rate for the remaining term of the contract following expiration of the guarantee.

(c) *Land contract terms.* The contract payments must be amortized for a minimum of 20 years and payments on the contract must be of equal amounts during the term of the guarantee.

(d) *Balloon installments.* Balloon payments are prohibited during the 10-year term of the guarantee.

§ 763.13 Fees.

(a) *Payment of fees.* The seller and buyer will be responsible for payment of any expenses or fees necessary to process the Land Contract Agreement required by the State or County to ensure that proper title is vested in the seller including, but not limited to, attorney fees, recording costs, and notary fees.

(b) [Reserved]

§ 763.14 Appraisals.

(a) *Standard guarantee plan.* For the standard guarantee plan, the value of real estate to be purchased will be established by an appraisal obtained at

Agency expense and completed as specified in §761.7 of this chapter. An appraisal is required prior to, or as a condition of, approval of the guarantee.

(b) *Prompt payment guarantee plan.* The Agency may, at its option and expense, obtain an appraisal to determine value of real estate to be purchased under the Prompt Payment Guarantee plan.

§763.15 Taxes and insurance.

(a) The seller will ensure that taxes and insurance on the real estate are paid timely and will provide the evidence of payment to the escrow or servicing agent.

(b) The seller will maintain flood insurance, if available, if buildings are located in a special 100-year floodplain as defined by FEMA flood hazard area maps.

(c) The seller will report any insurance claim and use of proceeds to the escrow or servicing agent.

§763.16 Environmental regulation compliance.

(a) *Environmental compliance requirements.* The environmental requirements contained in part 799 of this chapter must be met prior to approval of guarantee request.

(b) *Determination.* The Agency determination of whether an environmental problem exists will be based on:

(1) The information supplied with the application;

(2) Environmental resources available to the Agency including, but not limited to, documents, third parties, and government agencies;

(3) Other information supplied by the buyer or seller upon Agency request; and

(4) A visit to the farm.

[76 FR 75430, Dec. 2, 2011, as amended at 81 FR 51284, Aug. 3, 2016]

§763.17 Approving application and executing guarantee.

(a) Approval is subject to the availability of funds, meeting the requirements in this part, and the participation of an approved escrow or servicing agent, as applicable.

(b) Upon approval of the guarantee, all parties (buyer, seller, escrow or servicing agent, and Agency official) will execute the Agency's guarantee agreement.

(c) The "Land Contract Agreement for Prompt Payment Guarantee" or the "Land Contract Agreement for Standard Guarantee" will describe the conditions of the guarantee, outline the covenants and any agreements of the buyer, seller, escrow or servicing agent, and the Agency, and outline the process for payment of loss claims.

§763.18 General servicing responsibilities.

(a) For the prompt payment guarantee plan, the seller must use a third party escrow agent approved by the Agency. The escrow agent will:

(1) Provide the Agency a copy of the recorded Land Contract;

(2) Handle transactions relating to the Land Contract between the buyer and seller;

(3) Receive Land Contract installment payments from the buyer and send them to the seller;

(4) Provide evidence to the Agency that property taxes are paid and insurance is kept current on the security property;

(5) Send a notice of payment due to the buyer at least 30 days prior to the installment due date;

(6) Notify the Agency and the seller if the buyer defaults;

(7) Service delinquent accounts as specified in §763.20(a);

(8) Make demand on the Agency to pay missed payments;

(9) Send the seller any missed payment amount paid by the Agency under the guarantee;

(10) Notify the Agency on March 31 and September 30 of each year of the outstanding balance on the Land Contract and the status of payment; and

(11) Perform other duties as required by State law and as agreed to by the buyer and the seller;

(b) For the standard guarantee plan, the seller must use a third party servicing agent approved by the Agency. The servicing agent will:

(1) Provide the Agency a copy of the recorded Land Contract;

(2) Handle transactions relating to the Land Contract between the buyer and seller;

(3) Receive Land Contract installment payments from the buyer and send them to the seller;

(4) Provide evidence to the Agency that property taxes are paid and insurance is kept current on the security property;

(5) Perform a physical inspection of the farm each year during the term of the guarantee, and provide an annual inspection report to the Agency;

(6) Obtain from the buyer a current balance sheet, income statement, cash flow budget, and any additional information needed, perform, and provide the Agency an analysis of the buyer's financial condition on an annual basis;

(7) Notify the Agency on March 31 and September 30 of each year of the outstanding balance on the Land Contract and the status of payment;

(8) Send a notice of payment due to the buyer at least 30 days prior to the installment due date;

(9) Notify the Agency and the seller if the buyer defaults;

(10) Service delinquent accounts as specified in § 763.20(b); and

(11) Perform other duties as required by State law and as agreed to by the buyer and the seller.

§ 763.19 Contract modification.

(a) The seller and buyer may modify the land contract to lower the interest rate and corresponding amortized payment amount without Agency approval.

(b) With prior written approval from the Agency, the seller and buyer may modify the land contract provided that, in addition to a feasible plan for the upcoming operating cycle, a feasible plan can be reasonably projected throughout the remaining term of the guarantee. Such modifications may include but are not limited to:

(1) Deferral of installments,

(2) Leasing or subleasing, and

(3) Partial releases. All proceeds from a partial release or royalties from mineral extraction must be applied to a prior lien, if one exists, and in addition, the same amount must be credited to the principal balance of the land contract.

(4) Transfer and assumption. If the guarantee is to remain in effect, any transfer of the property and assump-

tion of the guaranteed debt must be made to an eligible buyer for the Land Contract Guarantee Program as specified in § 763.5(b), and must be approved by the Agency in writing. If an eligible buyer for transfer and assumption cannot be found, the Deputy Administrator for Farm Loan Programs may make an exception to this requirement when in the Government's best financial interests.

(5) Assignment. The seller may not assign the contract to another party without written consent of the Agency.

(c) Any contract modifications other than those listed above must be approved by the Deputy Administrator for Farm Loan Programs, and will only be approved if such action is determined permissible by law and in the Government's best financial interests.

§ 763.20 Delinquent servicing and collecting on guarantee.

(a) *Prompt payment guarantee plan.* If the buyer fails to pay an annual amortized installment or a portion of an installment on the contract or taxes or insurance when due, the escrow agent:

(1) Must make a written demand on the buyer for payment of the defaulted amount within 30 days of the missed payment, taxes, or insurance and send a copy of the demand letter to the Agency and to the seller; and

(2) Must make demand on the Agency within 90 days from the original payment, taxes, or insurance due date, for the missed payment in the event the buyer has not made the payment.

(b) *Standard guarantee plan.* If the buyer fails to pay an annual amortized installment or a portion of an installment on the contract, then the seller has the option of either liquidating the real estate, or having the amount of the loss established by the Agency by an appraisal of the real estate. For either option, the servicing agent:

(1) Must make a written demand on the buyer for payment of the defaulted amount within 30 days of the missed payment, and send a copy of the demand letter to the Agency and to the seller; and

(2) Must immediately inform the Agency which option the seller has chosen for establishing the amount of the loss, in the event the buyer does

not make the payment within 60 days of the demand letter.

(i) *Liquidation method.* If the seller chooses the liquidation method, the servicing agent will:

(A) Submit a liquidation plan to the Agency within 120 days from the missed payment for approval prior to any liquidation action. The Agency may require and pay for an appraisal prior to approval of the liquidation plan.

(B) Complete liquidation within 12 months of the missed installment unless prevented by bankruptcy, redemption rights, or other legal action.

(C) Credit an amount equal to the sale price received in a liquidation of the security property, with no deduction for expenses, to the principal balance of the land contract.

(D) File a loss claim immediately after liquidation, which must include a complete loan ledger.

(E) Base the loss claim amount on the appraisal method if the property is reacquired by the seller, through liquidation.

(ii) *Appraisal method.* If the seller chooses to have the loss amount established by appraisal rather than liquidation, the Agency will complete an appraisal on the real estate, and the loss claim amount will be based on the difference between the appraised value at the time the loss is calculated and the unpaid principal balance of the land contract at that time.

(A) The only administrative appeal allowed under §761.6 of this chapter related to the resulting appraisal amount will be a determination of whether the appraisal is Uniform Standards of Professional Appraisal Practice (USPAP) compliant.

(B) The seller will give the Agency a lien on the security property in the amount of the loss claim payment. If the property sells within 5 years from the date of the loss payment for an amount greater than the appraised value used to establish the loss claim amount, the seller must repay the difference, up to the amount of the loss claim. For purposes of determining the amount to be repaid (recapture), the market value of the property may be reduced by the value of certain capital improvements, as specified in

§766.202(a)(1)–(3) of this chapter, made by the seller to the property in the time period from the loss claim to final disposition. If the property is not sold within 5 years from the date of the loss payment, the Agency will release the lien and the seller will have no further obligation to the Agency.

§763.21 **Establishment of Federal debt and Agency recovery of loss claim payments.**

(a) Any amount paid by FSA as a result of an approved loss claim is immediately due and payable by the buyer after FSA notifies the buyer that a loss claim has been paid to the seller. If the debt is not restructured into a repayment plan or the obligation otherwise cured, FSA may use all remedies available, including offset as authorized by the Debt Collection Improvement Act of 1996, to collect the debt.

(1) Interest on the debt will be at the FLP non-program real property loan rate in effect at the time of the first Agency payment of a loss claim.

(2) The debt may be scheduled for repayment consistent with the buyer's repayment ability, not to exceed 7 years. Before any payment plan can be approved, the buyer must provide the Agency with the best lien obtainable on all of the buyer's assets. This includes the buyer's ownership interest in the real estate under contract for guarantees using the prompt payment guarantee plan. When the buyer is an entity, the best lien obtainable will be taken on all of the entity's assets, and all assets owned by individual members of the entity, including their ownership interest in the real estate under contract.

(b) Annually, buyers with an Agency approved repayment plan under this section will supply the Agency a current balance sheet, income statement, cash flow budget, complete copy of Federal income tax returns, and any additional information needed to analyze the buyer's financial condition.

(c) If a buyer fails to make required payments to the Agency as specified in the approved repayment plan, the debt will be treated as a non-program loan debt, and servicing will proceed as specified in §766.351(c) of this chapter.

§ 763.22 Negligence and negligent servicing.

(a) The Agency may deny a loss claim in whole or in part due to negligence that contributed to the loss claim. This could include, but is not limited to:

(1) The escrow and servicing agent failing to seek payment of a missed installment from the buyer within the prescribed timeframe or otherwise does not enforce the terms of the land contract;

(2) Losing the collateral to a third party, such as a taxing authority, prior lien holder, *etc;*

(3) Not performing the duties and responsibilities required of the escrow or servicing agent;

(4) The seller's failure to disclose environmental issues; or

(5) Any other action in violation of the land contract or guarantee agreement that does not terminate the guarantee.

(b) [Reserved]

§ 763.23 Terminating the guarantee.

(a) The guarantee and the Agency's obligations will terminate at the earliest of the following circumstances:

(1) Full payment of the land contract;

(2) Agency payment to the seller of 3 annual installments plus property taxes and insurance, if applicable, under the prompt payment guarantee plan, if not repaid in full by the buyer. An Agency approved repayment plan will not constitute payment in full until such time as the entire amount due for the Agency approved repayment plan is paid in full;

(3) Payment of a loss claim through the standard guarantee plan;

(4) Sale of real estate without guarantee being properly assigned;

(5) The seller terminates the land contract for reasons other than monetary default; or

(6) If for any reason the land contract becomes null and void.

(b) If none of the events in paragraph (a) of this section occur, the guarantee will automatically expire, without notice, 10 years from the effective date of the guarantee.

PART 764—DIRECT LOAN MAKING

Subpart A—Overview

Subpart B—Loan Application Process

Subpart C—Requirements for All Direct Program Loans

Subpart D—Farm Ownership Loan Program

Subpart E—Downpayment Loan Program

Subpart F—Conservation Loan Program

Subpart G—Operating Loan Program

AUTHORITY: 5 U.S.C. 301 and 7 U.S.C. 1989.

SOURCE: 72 FR 63298, Nov. 8, 2007, unless otherwise noted.

Subpart A—Overview

§764.1 Introduction.

(a) *Purpose.* This part describes the Agency's policies for making direct FLP loans.

(b) *Types of loans.* The Agency makes the following types of loans:

(1) FO, including ML and Downpayment loans;

(2) OL, including ML and Youth loans;

(3) EM; and

(4) CL.

[72 FR 63298, Nov. 8, 2007, as amended at 75 FR 54015, Sept. 3, 2010; 78 FR 3835, Jan. 17, 2013; 81 FR 3292, Jan. 21, 2016]

§764.2 Abbreviations and definitions.

Abbreviations and definitions for terms used in this part are provided in §761.2 of this chapter.

§§764.3–764.50 [Reserved]

Subpart B—Loan Application Process

§764.51 Loan application.

(a) A loan application must be submitted in the name of the actual operator of the farm. Two or more applicants applying jointly will be considered an entity applicant. The Agency will consider tax filing status and other business dealings as indicators of the operator of the farm.

(b) A complete loan application, except as provided in paragraphs (c) through (f) of this section, will include:

(1) The completed Agency application form;

(2) If the applicant is an entity:

(i) A complete list of entity members showing the address, citizenship, principal occupation, and the number of shares and percentage of ownership or stock held in the entity by each member, or the percentage of interest in the entity held by each member;

(ii) A current financial statement from each member of the entity;

(iii) A current financial statement from the entity itself;

(iv) A copy of the entity's charter or any entity agreement, any articles of incorporation and bylaws, any certificate or evidence of current registration (good standing), and a resolution adopted by the Board of Directors or entity members authorizing specified officers of the entity to apply for and obtain the desired loan and execute required debt, security and other loan instruments and agreements;

(v) In the form of married couples applying as a joint operation, items (i) and (iv) will not be required. The Agency may request copies of the marriage license, prenuptial agreement or similar documents as needed to verify loan eligibility and security. Items (ii) and (iii) are only required to the extent needed to show the individual and joint finances of the husband and wife without duplication.

(3) A written description of the applicant's farm training and experience, including each entity member who will be involved in managing or operating the farm. Farm experience of the applicant, without regard to any lapse of time between the farm experience and the new application, may be included in the applicant's written description. If farm experience occurred more than 5 years prior to the date of the new application, the applicant must demonstrate sufficient on-the-job training or education within the last 5 years to demonstrate managerial ability;

(4) The last 3 years of farm financial records, including tax returns, unless the applicant has been farming less than three years;

(5) The last 3 years of farm production records, unless the applicant has been farming less than 3 years;

(6) Except for CL, documentation that the applicant and each member of an entity applicant cannot obtain sufficient credit elsewhere on reasonable rates and terms, including a loan guaranteed by the Agency;

(7) Documentation of compliance with the Agency's environmental regulations contained in part 799 of this chapter;

(8) Verification of all non-farm income;

(9) A current financial statement and the operation's farm operating plan, including the projected cash flow budget reflecting production, income, expenses, and loan repayment plan;

(10) A legal description of the farm property owned or to be acquired and, if applicable, any leases, contracts, options, and other agreements with regard to the property;

(11) Payment to the Agency for ordering a credit report on the applicant;

(12) Verification of all debts;

(13) Any additional information deemed necessary by the Agency to effectively evaluate the applicant's eligibility and farm operating plan;

(14) For EM loans, a statement of loss or damage on the appropriate Agency form;

(15) For CL only, a conservation plan or Forest Stewardship Management Plan as defined in §761.2 of this chapter; and

(16) For CL only, and if the applicant wishes to request consideration for priority funding, plans to transition to organic or sustainable agriculture when the funds requested will be used to facilitate the transition.

(c) For an ML for OL purposes request, all of the following criteria must be met:

(1) The loan requested for OL purposes is:

(i) To pay annual or term operating expenses, and

(ii) $50,000 or less and the applicant's total outstanding Agency OL debt at the time of loan closing will be $50,000 or less,

(2) The applicant must submit the following:

(i) Items (1), (2), (3), (6), (7), (9), and (11) of paragraph (b) of this section;

(ii) Financial and production records for the most recent production cycle, if available, and practicable to project the cash flow of the operating cycle, and

(iv) Verification of all non-farm income relied upon for repayment; and

(3) The Agency may require an ML applicant to submit any other information listed in paragraph (b) of this section upon request when specifically needed to make a determination on the loan application.

(d) For an ML request for FO purposes, all of the following criteria must be met:

(1) The loan requested is:

(i) To pay for any authorized purpose under the FO Program, which are specified in §764.151; and

(ii) $50,000 or less and the applicant's total outstanding Agency FO debt at the time of loan closing will be $50,000 or less,

(2) The applicant must submit the following:

(i) Items specified in paragraphs (b)(1), (2), (3), (6), (7), (9), (10), and (11) of this section;

(ii) Financial and production records for the most recent production cycle, if available and practicable to project the cash flow of the operating cycle; and

(iv) Verification of all non-farm income relied upon for repayment; and

(v) Verification of applicant's farm experience;

(3) The Agency may require an ML applicant to submit any other information listed in paragraph (b) of this section upon request when necessary to make a determination on the loan application.

(e) For a CL Program streamlined application, the applicant must meet all of the following:

(1) Be current on all payments to all creditors including the Agency (if currently an Agency borrower).

(2) Have not received primary loan servicing on any FLP debt within the past 5 years.

(3) Have a debt to asset ratio that is 40 percent or less.

(4) Have a balance sheet that indicates a net worth of 3 times the requested loan amount or greater.

(5) Have a FICO credit score from the Agency obtained credit report of at least 700. For entity applicants, the FICO credit score of the majority of the individual members of the entity must be at least 700.

(6) Submit the following items:

(i) Items specified in paragraphs (b)(1), (b)(2), (b)(3), (b)(7), (b)(11), (b)(15), and (b)(16) of this section,

(ii) A current financial statement less than 90 days old, and

(iii) Upon Agency request, other information specified in paragraph (b) of this section necessary to make a determination on the loan application.

(f) For a youth loan request:

(1) The applicant must submit items (1), (7), and (9) of paragraph (b) of this section.

(2) Applicants 18 years or older, must also provide items (11) and (12) of paragraph (b) of this section.

(3) The Agency may require a youth loan applicant to submit any other information listed in paragraph (b) of this section as needed to make a determination on the loan application.

(g) The applicant need not submit any information under this section that already exists in the applicant's Agency file and is still current.

[72 FR 63298, Nov. 8, 2007, as amended at 75 FR 54015, Sept. 3, 2010; 76 FR 75434, Dec. 2, 2011; 77 FR 15938, Mar. 19, 2012; 78 FR 3835, Jan. 17, 2013; 79 FR 60744, Oct. 8, 2014; 81 FR 3292, Jan. 21, 2016; 81 FR 51284, Aug. 3, 2016]

§764.52 Processing an incomplete application.

(a) Within 7 calendar days of receipt of an incomplete application, the Agency will provide the applicant written notice of any additional information which must be provided. The applicant must provide the additional information within 15 calendar days of the date of this notice.

(b) If the additional information is not received, the Agency will provide written notice that the application will be withdrawn if the information is not received within 15 calendar days of the date of this second notice.

[72 FR 63298, Nov. 8, 2007, as amended at 86 FR 43391, Aug. 9, 2021]

§764.53 Processing the complete application.

Upon receiving a complete loan application, the Agency will:

(a) Consider the loan application in the order received, based on the date the application was determined to be complete.

(b) Provide written notice to the applicant that the application is complete.

(c) Within 60 calendar days after receiving a complete loan application, the Agency will complete the processing of the loan request and notify the applicant of the decision reached, and the reason for any disapproval.

(d) Except for CL requests, if based on the Agency's review of the application, it appears the applicant's credit needs could be met through the guaranteed loan program, the Agency will assist the applicant in securing guaranteed loan assistance under the market placement program as specified in §762.110(h) of this chapter.

(e) In the absence of funds for a direct loan, the Agency will keep an approved loan application on file until funding is available. At least annually, the Agency will contact the applicant to determine if the Agency should retain the application or if the applicant wants the application withdrawn.

(f) If funding becomes available, the Agency will resume processing of approved loans in accordance with this part.

[72 FR 63298, Nov. 8, 2007, as amended at 75 FR 54015, Sept. 3, 2010]

§764.54 Preferences when there is limited funding.

(a) *First priority*. When there is a shortage of loan funds, approved applications will be funded in the order of the date the application was received, whether or not complete.

(b) *Secondary priorities*. If two or more applications were received on the same date, the Agency will give preference to:

(1) First, an applicant who is a veteran of any war;

(2) Second, an applicant who is not a veteran, but:

(i) Has a dependent family;

(ii) Is able to make a down payment; or

(iii) Owns livestock and farm implements necessary to farm successfully.

(3) Third, to other eligible applicants.

[72 FR 63298, Nov. 8, 2007, as amended at 86 FR 43391, Aug. 9, 2021]

§§764.55–764.100 [Reserved]

Subpart C—Requirements for All Direct Program Loans

§764.101 General eligibility requirements.

The following requirements must be met unless otherwise provided in the eligibility requirements for the particular type of loan.

(a) *Controlled substances*. The applicant, and anyone who will sign the promissory note, must not be ineligible for loans as a result of a conviction for controlled substances according to 7 CFR part 718 of this chapter.

(b) *Legal capacity*. The applicant, and anyone who will sign the promissory note, must possess the legal capacity to incur the obligation of the loan. A Youth loan applicant will incur full personal liability upon execution of the promissory note without regard to the applicant's minority status.

(c) *Citizenship*. The applicant, and anyone who will sign the promissory note, must be a citizen of the United States, United States non-citizen national, or a qualified alien under applicable Federal immigration laws.

(d) *Credit history*. The applicant must have acceptable credit history demonstrated by debt repayment.

(1) As part of the credit history, the Agency will determine whether the applicant will carry out the terms and conditions of the loan and deal with the Agency in good faith. In making this determination, the Agency may examine whether the applicant has properly fulfilled its obligations to other parties, including other agencies of the Federal Government.

(2) When the applicant caused the Agency a loss by receiving debt forgiveness, the applicant may be ineligible for assistance in accordance with eligibility requirements for the specific loan type. If the debt forgiveness is cured by repayment of the Agency's loss, the Agency may still consider the debt forgiveness in determining the applicant's credit worthiness.

(3) A history of failures to repay past debts as they came due when the ability to repay was within the applicant's control will demonstrate unacceptable credit history. The following circumstances, for example, do not automatically indicate an unacceptable credit history:

(i) Foreclosures, judgments, delinquent payments of the applicant which occurred more than 36 months before the application, if no recent similar situations have occurred, or Agency delinquencies that have been resolved through loan servicing programs available under 7 CFR part 766.

(ii) Isolated incidents of delinquent payments which do not represent a general pattern of unsatisfactory or slow payment.

(iii) "No history" of credit transactions by the applicant.

(iv) Recent foreclosure, judgment, bankruptcy, or delinquent payment when the applicant can satisfactorily demonstrate that the adverse action or delinquency was caused by circumstances that were of a temporary nature and beyond the applicant's control; or the result of a refusal to make full payment because of defective goods or services or other justifiable dispute relating to the purchase or contract for goods or services.

(e) *Availability of credit elsewhere.* Except for CL, the applicant, and all entity members in the case of an entity, must be unable to obtain sufficient credit elsewhere to finance actual needs at reasonable rates and terms. The Agency will evaluate the ability to obtain credit based on factors including, but not limited to:

(1) Loan amounts, rates, and terms available in the marketplace; and

(2) Property interests, income, and significant non-essential assets.

(f) *Not in delinquent status on Federal debt.* As provided in 31 CFR part 285, except for EM loan applicants, the applicant, and anyone who will sign the promissory note, must not be in delinquent status on any Federal debt, other than a debt under the Internal Revenue Code of 1986 at the time of loan closing. All delinquent debts, however, will be considered in determining credit history and ability to repay under this part.

(g) *Outstanding judgments.* The applicant, and anyone who will sign the promissory note, must have no outstanding unpaid judgments obtained by the United States in any court. Such judgments do not include those filed as a result of action in the United States Tax Courts.

(h) *Federal crop insurance violation.* The applicant, and all entity members in the case of an entity, must not be ineligible due to disqualification resulting from Federal crop insurance violation according to 7 CFR part 718.

(i) *Managerial ability.* The applicant must have sufficient managerial ability to assure reasonable prospects of loan repayment, as determined by the Agency. The applicant must demonstrate this managerial ability by:

(1) *Education.* For example, the applicant obtained a 4-year college degree in agricultural business, horticulture, animal science, agronomy, or other agricultural-related field.

(2) *On-the-job training.* For example, the applicant is currently working on a farm as part of an apprenticeship program.

(3) *Farming experience.* For example, the applicant has been an owner, manager, or operator of a farm business for at least one entire production cycle or for MLs, made for OL purposes, the applicant may have obtained and successfully repaid one FSA Youth-OL. Farm experience of the applicant, without regard to any lapse of time between the farm experience and the new application, will be taken into consideration in determining loan eligibility. If farm experience occurred more than 5 years prior to the date of the new application, the applicant must demonstrate sufficient on-the-job training or education within the last 5 years to demonstrate managerial ability.

(4) *Alternatives for MLs made for OL purposes.* Applicants for MLs made for OL purposes, also may demonstrate managerial ability by one of the following:

(i) Certification of a past participation with an agriculture-related organization, such as, but not limited to, 4-H Club, FFA, beginning farmer and rancher development programs, or Community Based Organizations, that demonstrates experience in a related agricultural enterprise; or

(ii) A written description of a self-directed apprenticeship combined with either prior sufficient experience working on a farm or significant small business management experience. As a condition of receiving the loan, the self-directed apprenticeship requires that the applicant seek, receive, and apply guidance from a qualified person during the first cycle of production and marketing typical for the applicant's specific operation. The individual providing the guidance must be knowledgeable in production, management, and marketing practices that are pertinent to the applicant's operation, and agree to form a developmental partnership with the applicant to share knowledge, skills, information, and perspective of agriculture to foster the applicant's development of technical skills and management ability.

(j) *Borrower training.* The applicant must agree to meet the training requirements in subpart K of this part.

(k) *Operator of a family farm.* Except for CL:

(1) The applicant must be the operator of a family farm after the loan is closed.

(2) For an entity applicant, if the entity members holding a majority interest are:

(i) Related by blood or marriage, at least one member must be the operator of a family farm;

(ii) Not related by blood or marriage, the entity members holding at least 50 percent interest must be operators of a family farm.

(3) Except for EM loans, the collective interests of the members may be larger than a family farm only if:

(i) Each member's ownership interest is not larger than a family farm;

(ii) All of the members of the entity are related by blood or marriage; and

(iii) All of the members are or will become operators of the family farm; and .

(4) If the entity applicant has an operator and ownership interest for farm ownership loans and emergency loans for farm ownership loan purposes, in any other farming operation, that farming operation must not exceed the requirements of a family farm.

(1) *Entity composition.* If the applicant has one or more embedded entities, at least 75 percent of the individual ownnership interests of each embedded entity must be owned by members actively involved in managing or operating the family farm.

[72 FR 63298, Nov. 8, 2007, as amended at 75 FR 54015, Sept. 3, 2010; 76 FR 75434, Dec. 2, 2011; 78 FR 3835, Jan. 17, 2013; 79 FR 60744, Oct. 8, 2014; 81 FR 3293, Jan. 21, 2016; 81 FR 10063, Feb. 29, 2016; 86 FR 43391, Aug. 9, 2021]

§ 764.102 General limitations.

(a) Limitations specific to each loan program are contained in subparts D through I of this part.

(b) The total principal balance owed to the Agency at any one time by the applicant, or any one who will sign the promissory note, cannot exceed the limits established in § 761.8 of this chapter.

(c) The funds from the FLP loan must be used for farming operations located in the United States.

(d) The Agency will not make a loan if the proceeds will be used:

(1) For any purpose that contributes to excessive erosion of highly erodible land, or to the conversion of wetlands;

(2) To drain, dredge, fill, level, or otherwise manipulate a wetland; or

(3) To engage in any activity that results in impairing or reducing the flow, circulation, or reach of water, except in the case of activity related to the maintenance of previously converted wetlands as defined in the Food Security Act of 1985.

(e) Any construction financed by the Agency must comply with the standards established in § 761.10 of this chapter.

(f) Loan funds will not be used to establish or support a non-eligible enterprise, even if the non-eligible enterprise contributes to the farm. Notwithstanding this limitation, an EM loan may cover qualified equine losses as specified in subpart I of this part.

[72 FR 63298, Nov. 8, 2007, as amended at 75 FR 54015, Sept. 3, 2010; 76 FR 75434, Dec. 2, 2011]

§ 764.103 General security requirements.

(a) Security requirements specific to each loan program are outlined in subparts D through I of this part.

(b) All loans must be secured by assets having a security value of at least 100 percent of the loan amount, except for EM loans as provided in subpart I of this part. If the applicant's assets do not provide adequate security, the Agency may accept:

(1) A pledge of security from a third party; or

(2) Interests in property not owned by the applicant (such as leases that provide a mortgageable value, water rights, easements, mineral rights, and royalties).

(c) An additional amount of security up to 150 percent of the loan amount will be taken when available, except for downpayment loans, MLs made for purposes other than annual operating, and youth loans.

(d) The Agency will choose the best security available when there are several alternatives that meet the Agency's security requirements.

(e) The Agency will take a lien on all assets that are not essential to the farming operation and are not being converted to cash to reduce the loan amount when each such asset, or aggregate value of like assets (such as stocks), has a value in excess of $15,000. The value of this security is not included in the Agency's additional security requirement stated in paragraph (c) of this section. This requirement does not apply to downpayment loans, CL, ML, or youth loans.

[72 FR 63298, Nov. 8, 2007, as amended at 73 FR 74345, Dec. 8, 2008; 75 FR 54015, Sept. 3, 2010; 78 FR 3835, Jan. 17, 2013; 86 FR 43391, Aug. 9, 2021]

§764.104 General real estate security requirements.

(a) *Agency lien position requirements.* If real estate is pledged as security for a loan, the Agency must obtain a first lien, if available. When a first lien is not available, the Agency may take a junior lien under the following conditions:

(1) The prior lien does not contain any provisions that may jeopardize the Agency's interest or the applicant's ability to repay the FLP loan;

(2) Prior lienholders agree to notify the Agency prior to foreclosure;

(3) The applicant must agree not to increase an existing prior lien without the written consent of the Agency; and

(4) Equity in the collateral exists.

(b) *Real estate held under a purchase contract.* If the real estate offered as security is held under a recorded purchase contract:

(1) The applicant must provide a security interest in the real estate;

(2) The applicant and the purchase contract holder must agree in writing that any insurance proceeds received for real estate losses will be used only for one or more of the following purposes:

(i) To replace or repair the damaged real estate improvements which are essential to the farming operation;

(ii) To make other essential real estate improvements; or

(iii) To pay any prior real estate lien, including the purchase contract.

(3) The purchase contract must provide the applicant with possession, control and beneficial use of the property, and entitle the applicant to marketable title upon fulfillment of the contract terms.

(4) The purchase contract must not:

(i) Be subject to summary cancellation upon default;

(ii) Contain provisions which jeopardize the Agency's security position or the applicant's ability to repay the loan.

(5) The purchase contract holder must agree in writing to:

(i) Not sell or voluntarily transfer their interest without prior written consent of the Agency;

(ii) Not encumber or cause any liens to be levied against the property;

(iii) Not take any action to accelerate, forfeit, or foreclose the applicant's interest in the security property until a specified period of time after notifying the Agency of the intent to do so;

(iv) Consent to the Agency making the loan and taking a security interest in the applicant's interest under the purchase contract as security for the FLP loan;

(v) Not take any action to foreclose or forfeit the interest of the applicant under the purchase contract because the Agency has acquired the applicant's interest by foreclosure or voluntary conveyance, or because the Agency has subsequently sold or assigned the applicant's interest to a third party who will assume the applicant's obligations under the purchase contract;

(vi) Notify the Agency in writing of any breach by the applicant; and

(vii) Give the Agency the option to rectify the conditions that amount to a breach within 30 days after the date the Agency receives written notice of the breach.

(6) If the Agency acquires the applicant's interest under the purchase contract by foreclosure or voluntary conveyance, the Agency will not be deemed to have assumed any of the applicant's obligations under the contract, provided that if the Agency fails to perform the applicant's obligations while it holds the applicant's interest is grounds for terminating the purchase contract.

(c) *Tribal lands held in trust or restricted.* The Agency may take a lien on Indian Trust lands as security provided the applicant requests the Bureau of Indian Affairs to furnish Title Status Reports to the agency and the Bureau of Indian Affairs provides the reports and approves the lien.

(d) *Security for more than one loan.* The same real estate may be pledged as security for more than one direct or guaranteed loan.

(e) *Loans secured by leaseholds.* A loan may be secured by a mortgage on a leasehold, if the leasehold has negotiable value and can be mortgaged.

§ 764.105 General chattel security requirements.

The same chattel may be pledged as security for more than one direct or guaranteed loan.

§ 764.106 Exceptions to security requirements.

Notwithstanding any other provision of this part, the Agency will not take a security interest:

(a) When adequate security is otherwise available and the lien will prevent the applicant from obtaining credit from other sources;

(b) When the property could have significant environmental problems or costs as described in part 799 of this chapter;

(c) When the Agency cannot obtain a valid lien;

(d) When the property is the applicant's personal residence and appurtenances and:

(1) They are located on a separate parcel; and

(2) The real estate that serves as security for the FLP loan plus crops and chattels are greater than or equal to 150 percent of the unpaid balance due on the loan;

(e) When the property is subsistence livestock, cash, working capital accounts the applicant uses for the farming operation, retirement accounts, personal vehicles necessary for family living, household contents, or small equipment such as hand tools and lawn mowers; or

(f) On marginal land and timber that secures an outstanding ST loan.

[72 FR 63298, Nov. 8, 2007, as amended at 81 FR 51284, Aug. 3, 2016]

§ 764.107 General appraisal requirements.

(a) *Establishing value for real estate.* The value of real estate will be established by an appraisal completed in accordance with § 761.7 of this chapter, except that for MLs made for FO purposes, the appraisal requirement may be satisfied by an evaluation by an authorized agency official that establishes the value of the real estate.

(b) *Establishing value for chattels.* The value of chattels will be established as follows:

(1) *Annual production.* The security value of annual livestock and crop production is presumed to be 100 percent of the amount loaned for annual operating and family living expenses, as outlined in the approved farm operating plan.

(2) *Livestock and equipment.* The value of livestock and equipment will be established by an appraisal completed in accordance with § 761.7 of this chapter.

[72 FR 63298, Nov. 8, 2007, as amended at 81 FR 3293, Jan. 21, 2016]

§ 764.108 General insurance requirements.

The applicant must obtain and maintain insurance, equal to the lesser of the value of the security at the time of loan closing or the principal of all FLP and non-FLP loans secured by the property, subject to the following:

(a) All security, except growing crops, must be covered by hazard insurance if it is readily available (sold by insurance agents in the applicant's normal trade area) and insurance premiums do not exceed the benefit. The Agency must be listed as loss payee for the insurance indemnity payment or as a beneficiary in the mortgagee loss payable clause.

(b) Real estate security located in flood or mudslide prone areas must be covered by flood or mudslide insurance. The Agency must be listed as a beneficiary in the mortgagee loss payable clause.

(c) Growing crops used to provide adequate security must be covered by crop insurance if such insurance is available. The Agency must be listed as loss payee for the insurance indemnity payment.

(d) Prior to closing the loan, the applicant must have obtained at least the catastrophic risk protection level of crop insurance coverage for each crop which is a basic part of the applicant's total operation, if such insurance is available, unless the applicant executes a written waiver of any emergency crop loss assistance with respect to such crop. The applicant must execute an assignment of indemnity in favor of the Agency for this coverage.

§§764.109–764.150 [Reserved]

Subpart D—Farm Ownership Loan Program

§764.151 Farm Ownership loan uses.

FO loan funds may only be used to:

(a) Acquire or enlarge a farm or make a down payment on a farm;

(b) Make capital improvements to a farm owned by the applicant, for construction, purchase or improvement of farm dwellings, service buildings or other facilities and improvements essential to the farming operation. In the case of leased property, the applicant must have a lease to ensure use of the improvement over its useful life or to ensure that the applicant receives compensation for any remaining economic life upon termination of the lease;

(c) Promote soil and water conservation and protection;

(d) Pay loan closing costs;

(e) Refinance a bridge loan if the following conditions are met:

(1) The applicant obtained the loan to be refinanced to purchase a farm after a direct FO was approved;

(2) Direct FO funds were not available to fund the loan at the time of approval;

(3) The loan to be refinanced is temporary financing; and

(4) The loan was made by a commercial or cooperative lender.

§764.152 Eligibility requirements.

The applicant:

(a) Must comply with the general eligibility requirements established at §764.101;

(b) And anyone who will sign the promissory note, must not have received debt forgiveness from the Agency on any direct or guaranteed loan;

(c) Must be the owner-operator of the farm financed with Agency funds after the loan is closed. Ownership of the family farm operation and farm real estate may be held either directly in the individual's name or indirectly through interest in a legal entity. In the case of an entity:

(1) The entity is controlled by farmers engaged primarily and directly in farming in the United States, after the loan is made;

(2) An ownership entity must be authorized to own a farm in the state or states in which the farm is located. An operating entity must be authorized to operate a farm in the state or states in which the farm is located.

(3) If the entity members holding majority interest are:

(i) Related by blood or marriage, at least one member of the entity must operate the family farm and at least one member of the entity or the entity must own the farm; or,

(ii) Not related by blood or marriage, the entity members holding at least 50 percent interest must operate the family farm and the entity members holding at least 50 percent interest or the entity must own the farm.

(4) If the entity is an operator only entity, the individuals that own the farm (real estate) must own at least 50 percent of the family farm (operating entity).

(d) And in the case of an entity, one or more members constituting a majority interest, must have participated in the business operations of a farm for at least 3 years out of the 10 years prior to the date the application is submitted. One of these three years can be substituted with the following experience:

(1) Postsecondary education in agriculture business, horticulture, animal science, agronomy, or other agricultural related fields,

(2) Significant business management experience, or

(3) Leadership or management experience while serving in any branch of the military.

(e) For an ML made for FO purposes, if an ML applicant has successfully repaid an FSA financed youth loan, the term of that loan may be used toward the 3 years of management experience required for a FO direct loan.

(f) And anyone who will sign the promissory note, must satisfy at least one of the following conditions:

(1) Meet the definition of a beginning farmer;

(2) Have not had a direct FO loan outstanding for more than a total of 10 years prior to the date the new FO loan is closed;

(3) Have never received a direct FO loan.

[72 FR 63298, Nov. 8, 2007, as amended at 79 FR 60744, Oct. 8, 2014; 81 FR 3293, Jan. 21, 2016; 86 FR 43391, Aug. 9, 2021]

§ 764.153　Limitations.

The applicant must:

(a) Comply with the general limitations established at § 764.102;

(b) Have dwellings and other buildings necessary for the planned operation of the farm available for use after the loan is made.

§ 764.154　Rates and terms.

(a) *Rates.* (1) The interest rate is the Agency's Direct Farm Ownership rate, available in each Agency office.

(2) The limited resource Farm Ownership interest rate is available to applicants who are unable to develop a feasible plan at regular interest rates.

(3) If the FO loan is part of a joint financing arrangement and the amount of the Agency's loan does not exceed 50 percent of the total amount financed, the interest rate charged will be the greater of the following:

(i) The Agency's Direct Farm Ownership rate, available in each Agency office, minus 2 percent; or

(ii) 2.5 percent.

(4) The interest rate charged will be the lower of the rate in effect at the time of loan approval or loan closing.

(b) *Terms.* Except for MLs made for FO purposes, the Agency schedules repayment of an FO loan based on the applicant's ability to repay and the useful life of the security. In no event will the term be more than 40 years from the date of the note.

(1) For MLs made for FO purposes the Agency schedules repayment of an FO based on the applicant's ability to repay and the useful life of the security. In no event will the term be more than 25 years from the date of the note.

(2) [Reserved]

[72 FR 63298, Nov. 8, 2007, as amended at 79 FR 78693, Dec. 31, 2014; 81 FR 3293, Jan. 21, 2016]

§ 764.155　Security requirements.

An FO loan must be secured:

(a) In accordance with §§ 764.103 through 764.106;

(b) At a minimum, by the real estate being purchased or improved.

(1) An ML made for FO purposes, may be secured only by the real estate being purchased or improved, as long as its value is at least 100 percent of the loan amount.

(2) [Reserved]

[72 FR 63298, Nov. 8, 2007, as amended at 81 FR 3293, Jan. 21, 2016]

§§ 764.156–764.200　[Reserved]

Subpart E—Downpayment Loan Program

§ 764.201　Down payment loan uses.

Down payment loan funds may be used to partially finance the purchase of a family farm by an eligible beginning farmer or socially disadvantaged farmer.

[72 FR 63298, Nov. 8, 2007, as amended at 73 FR 74345, Dec. 8, 2008; 86 FR 43392, Aug. 9, 2021]

§ 764.202　Eligibility requirements.

The applicant must:

(a) Comply with the general eligibility requirements established at § 764.101 and the FO eligibility requirements of § 764.152; and

(b) Be a beginning farmer or socially disadvantaged farmer.

[72 FR 63298, Nov. 8, 2007, as amended at 73 FR 74345, Dec. 8, 2008]

§ 764.203　Limitations.

(a) The applicant must:

(1) Comply with the general limitations established at § 764.102; and

(2) Provide a minimum down payment of 5 percent of the purchase price of the farm.

(b) Down payment loans will not exceed 45 percent of the lesser of:

(1) The purchase price,

(2) The appraised value of the farm to be acquired, or

(3) $667,000; subject to the direct FO dollar limit specified in 7 CFR 761.8(a)(1)(i).

(c) Down payment loans made as an ML for FO purposes may not exceed $50,000.

(d) Financing provided by the Agency and all other creditors must not exceed 95 percent of the purchase price. Financing provided by eligible lenders may be guaranteed by the Agency under part 762 of this chapter.

[72 FR 63298, Nov. 8, 2007, as amended at 73 FR 74345, Dec. 8, 2008; 79 FR 78693, Dec. 31, 2014; 81 FR 3293, Jan. 21, 2016; 86 FR 43392, Aug. 9, 2021]

§764.204 Rates and terms.

(a) *Rates.* The interest rate for Down payment loans will be the regular direct FO rate minus 4 percent, but in no case less than 1.5 percent.

(b) *Terms.* (1) The Agency schedules repayment of Down payment loans in equal, annual installments over a term not to exceed 20 years.

(2) The non-Agency financing must have an amortization period of at least 30 years and cannot have a balloon payment due within the first 20 years of the loan.

[72 FR 63298, Nov. 8, 2007, as amended at 73 FR 74345, Dec. 8, 2008; 86 FR 43392, Aug. 9, 2021]

§764.205 Security requirements.

A Down payment loan must:

(a) Be secured in accordance with §§764.103 through 764.106;

(b) Be secured by a lien on the property being acquired with the loan funds and junior only to the party financing the balance of the purchase price.

[72 FR 63298, Nov. 8, 2007, as amended at 73 FR 74345, Dec. 8, 2008; 86 FR 43392, Aug. 9, 2021]

§§764.206–764.230 [Reserved]

Subpart F—Conservation Loan Program

SOURCE: 75 FR 54015, Sept. 3, 2010, unless otherwise noted.

§764.231 Conservation loan uses.

(a) CL funds may be used for any conservation activities included in a conservation or Forestry Service Stewardship Management Plan, including but not limited to:

(1) The installation of conservation structures to address soil, water, and related resources;

(2) The establishment of forest cover for sustained yield timber management, erosion control, or shelter belt purposes;

(3) The installation of water conservation measures;

(4) The installation of waste management systems;

(5) The establishment or improvement of permanent pasture; and

(6) Other purposes including the adoption of any other emerging or existing conservation practices, techniques, or technologies.

(b) [Reserved]

[75 FR 54015, Sept. 3, 2010, as amended at 77 FR 15938, Mar. 19, 2012]

§764.232 Eligibility requirements.

(a) The applicant:

(1) Must comply with general eligibility requirements specified in §764.101 except paragraphs (e) and (k) of that section;

(2) And anyone who will sign the promissory note, must not have received debt forgiveness from the Agency on any direct or guaranteed loan; and

(3) Must be the owner-operator or tenant-operator of a farm and be engaged in agricultural production after the time of loan is closed. In the case of an entity:

(i) The entity is controlled by farmers engaged primarily and directly in farming in the United States;

(ii) The entity must be authorized to operate a farm in the State in which the farm is located.

(b) [Reserved]

§764.233 Limitations.

(a) The applicant must comply with the general limitations specified in §764.102 except §764.102(f), which does not apply to applicants for the CL Program.

(b) The applicant must agree to repay any duplicative financial benefits or assistance to CL.

§764.234 Rates and terms.

(a) *Rates.* The interest rate:

(1) Will be the Agency's Direct Farm Ownership rate, available in each Agency office.

(2) Charged will be the lower rate in effect either at the time of loan approval or loan closing.

(b) *Terms.* The following terms apply to CLs:

(1) The Agency schedules repayment of a CL based on the useful life of the security.

(2) The maximum term for loans secured by chattels only will not exceed 7 years from the date of the note.

(3) In no event will the term of the loan exceed 20 years from the date of the note.

§ 764.235 Security requirements.

(a) The loan must be secured in accordance with requirements established in §§ 764.103 through 764.106.

(b) Loans to purchase chattels will be secured by a first lien on chattels purchased with loan funds. Real estate may be taken as additional security if needed.

(c) Loans of $25,000 of less for real estate purposes will be secured in the following order of priority:

(1) By a lien on chattels determined acceptable by the Agency, and then

(2) By a lien on real estate, if available and necessary. When real estate is taken as security a certification of ownership in real estate is required. Certification of ownership may be in the form of an affidavit that is signed by the applicant, names all of the record owners of the real estate in question and lists the balances due on all known debts against the real estate. Whenever the Agency is uncertain of the record owner or debts against the real estate security, a tile search is required.

(d) Loans greater than $25,000 for real estate purposes will be secured in the following order of priority:

(1) By a lien on real estate, if available, and then

(2) By a lien on chattels, if needed and determined acceptable by the Agency.

(e) For loans greater than $25,000 title clearance is required when real estate is taken as security.

[77 FR 15938, Mar. 19, 2012]

§§ 764.236-764.250 [Reserved]

Subpart G—Operating Loan Program

SOURCE: 72 FR 63298, Nov. 8, 2007, unless otherwise noted. Redesignated at 75 FR 54015, Sept. 3, 2010.

§ 764.251 Operating loan uses.

(a) Except as provided in paragraph (b), OL and ML used for OL purposes loan funds may only be used for:

(1) Costs associated with reorganizing a farm to improve its profitability;

(2) Purchase of livestock, including poultry, farm equipment or fixtures, quotas and bases, and cooperative stock for credit, production, processing or marketing purposes;

(3) Farm operating expenses, including, but not limited to, feed, seed, fertilizer, pesticides, farm supplies, repairs and improvements which are to be expensed, cash rent and family living expenses;

(4) Scheduled principal and interest payments on term debt provided the debt is for authorized FO or OL purposes;

(5) Other farm needs;

(6) Costs associated with land and water development, use, or conservation;

(7) Loan closing costs;

(8) Costs associated with Federal or State-approved standards under the Occupational Safety and Health Act of 1970 (29 U.S.C. 655 and 667) if the applicant can show that compliance or noncompliance with the standards will cause substantial economic injury;

(9) Borrower training costs required or recommended by the Agency;

(10) Refinancing farm-related debts other than real estate to improve the farm's profitability provided the applicant has refinanced direct or guaranteed OL loans four times or fewer and one of the following conditions is met:

(i) A designated or declared disaster caused the need for refinancing; or

(ii) The debts to be refinanced are owed to a creditor other than the USDA;

(11) Costs for minor real estate repairs or improvements, provided the loan can be repaid within 7 years.

(b) [Reserved]

[72 FR 63298, Nov. 8, 2007, as amended at 78 FR 3835, Jan. 17, 2013; 81 FR 3293, Jan. 21, 2016]

§ 764.252 Eligibility requirements.

(a) The applicant must comply with the general eligibility requirements established in § 764.101.

(b) The applicant and anyone who will sign the promissory note, except as provided in paragraph (c) of this section, must not have received debt forgiveness from the Agency on any direct or guaranteed loan.

(c) The applicant and anyone who will sign the promissory note, may receive direct OL loans to pay annual farm operating and family living expenses, provided that the applicant meets all other applicable requirements under this part, if the applicant:

(1) Received a write-down under section 353 of the Act;

(2) Is current on payments under a confirmed reorganization plan under Chapter 11, 12, or 13 of Title 11 of the United States Code; or

(3) Received debt forgiveness on not more than one occasion after April 4, 1996, resulting directly and primarily from a Presidentially-designated emergency for the county or contiguous county in which the applicant operates. Only applicants who were current on all existing direct and guaranteed FLP loans prior to the beginning date of the incidence period of a Presidentially-designated emergency and received debt forgiveness on that debt within 3 years after the designation of such emergency meet this exception.

(d) In the case of an entity applicant, the entity must be:

(1) Controlled by farmers engaged primarily and directly in farming in the United States; and

(2) Authorized to operate the farm in the State in which the farm is located.

(e) The applicant and anyone who will sign the promissory note, may close an OL in no more than 7 calendar years, either as an individual or as a member of an entity, except as provided in paragraphs (e)(1) through (4) of this section. The years may be consecutive or nonconsecutive, and there is no limit on the number of OLs closed in a year. Microloans made to a begin-ning farmer or a veteran farmer are not counted toward this limitation. Youth loans are not counted toward this limitation. The following exceptions apply:

(1) This limitation does not apply if the applicant and anyone who will sign the promissory note is a beginning farmer.

(2) This limitation does not apply if the applicant's land is subject to the jurisdiction of an Indian tribe, the loan is secured by one or more security instruments subject to the jurisdiction of an Indian tribe, and commercial credit is generally not available to such farm operations.

(3) If the applicant, and anyone who will sign the promissory note, has closed direct OL loans in 4 or more previous calendar years as of April 4, 1996, the applicant is eligible to close OL loans in any 3 additional years after that date.

(4) On a case-by-case basis, may be granted a one-time waiver of OL term limits for a period of 2 years, not subject to administrative appeal, if the applicant:

(i) Has a financially viable operation;

(ii) And in the case of an entity, the members holding the majority interest, applied for commercial credit from at least two lenders and were unable to obtain a commercial loan, including an Agency-guaranteed loan; and

(iii) Has successfully completed, or will complete within one year, borrower training. Previous waivers to the borrower training requirements are not applicable under this paragraph.

[79 FR 78693, Dec. 31, 2014]

§ 764.253 Limitations.

The applicant must comply with the general limitations established at § 764.102.

§ 764.254 Rates and terms.

(a) *Rates.* (1) The interest rate is the Agency's Direct Operating Loan rate, available in each Agency office.

(2) The limited resource Operating Loan interest rate is available to applicants who are unable to develop a feasible plan at regular interest rates.

(3) The interest rate charged will be the lower rate in effect at the time of loan approval or loan closing.

(4) The Agency's Direct ML OL interest rate on an ML to a beginning farmer or veteran farmer is available in each Agency office. ML borrowers in these groups have the option of choosing the ML OL interest rate or the Direct OL interest rate in effect at the time of approval, or if lower, the rate in effect at the time of closing.

(b) *Terms.* (1) The Agency schedules repayment of annual OL loans made for family living and farm operating expenses when planned income is projected to be available.

(i) The term of the loan may not exceed 24 months from the date of the note.

(ii) The term of the loan may exceed 24 months in unusual situations such as establishing a new enterprise, developing a farm, purchasing feed while crops are being established, marketing plans, or recovery from a disaster or economic reverse. In no event will the term of the loan exceed 7 years from the date of the note. Crops and livestock produced for sale will not be considered adequate security for such loans.

(2) The Agency schedules the repayment of all other OL loans based on the applicant's ability to repay and the useful life of the security. In no event will the term of the loan exceed 7 years from the date of the note. Repayment schedules may include equal, unequal, or balloon installments if needed to establish a new enterprise, develop a farm, or recover from a disaster or economic reversal. Loans with balloon installments:

(i) Must have adequate security at the time the balloon installment comes due. Crops, livestock other than breeding stock, or livestock products produced are not adequate collateral for such loans;

(ii) Are only authorized when the applicant can project the ability to refinance the remaining debt at the time the balloon payment comes due based on the expected financial condition of the operation, the depreciated value of the collateral, and the principal balance on the loan;

(iii) Are not authorized when loan funds are used for real estate repairs or improvements.

[72 FR 63298, Nov. 8, 2007, as amended at 79 FR 78694, Dec. 31, 2014; 86 FR 43392, Aug. 9, 2021]

§ 764.255   Security requirements.

An OL loan must be secured:
(a) In accordance with §§ 764.103 through 764.106.
(b) Except for MLs, by a:
(1) First lien on all property or products acquired or produced with loan funds;
(2) Lien of equal or higher position of that held by the creditor being refinanced with loan funds.
(c) For MLs used for OL purposes:
(1) For annual operating purposes, loans must be secured by a first lien on farm property or products having a security value of at least 100 percent of the loan amount, and up to 150 percent, when available.
(2) For loans made for purposes other than annual operating purposes, loans must be secured by a first lien on farm property or products purchased with loan funds and having a security value of at least 100 percent of the loan amount.
(3) A lien on real estate is not required unless the value of the farm products, farm property, and other assets available to secure the loan is not at least equal to 100 percent of the loan amount.
(4) Notwithstanding the provisions of paragraphs (c)(1), (c)(2), and (c)(3) of this section, FSA will not require a lien on a personal residence.

[72 FR 63298, Nov. 8, 2007, as amended at 78 FR 3835, Jan. 17, 2013; 81 FR 3293, Jan. 21, 2016]

§§ 764.256–764.300   [Reserved]

Subpart H—Youth Loan Program

SOURCE: 72 FR 63298, Nov. 8, 2007, unless otherwise noted. Redesignated at 75 FR 54015, Sept. 3, 2010.

§ 764.301   Youth loan uses.

Youth loan funds may only be used to finance a modest, income-producing, agriculture-related, educational project while participating in 4–H, FFA, or a similar organization.

§ 764.302 Eligibility requirements.

The applicant:

(a) Must comply with the general eligibility requirements established at § 764.101(a) through (g);

(b) And anyone who will sign the promissory note, must not have received debt forgiveness from the Agency on any direct or guaranteed loan;

(c) Must be at least 10 but not yet 21 years of age at the time the loan is closed;

(d) Must be recommended and continuously supervised by a project advisor, such as a 4–H Club advisor, a vocational teacher, a county extension agent, or other agriculture-related organizational sponsor; and

(e) Must obtain a written recommendation and consent from a parent or guardian if the applicant has not reached the age of majority under state law.

[72 FR 63298, Nov. 8, 2007. Redesignated at 75 FR 54015, Sept. 3, 2010, as amended at 79 FR 78694, Dec. 31, 2014]

§ 764.303 Limitations.

(a) The applicant must comply with the general limitations established at § 764.102.

(b) The total principal balance owed by the applicant to the Agency on all Youth loans at any one time cannot exceed $5,000.

§ 764.304 Rates and terms.

(a) *Rates.* (1) The interest rate is the Agency's Direct Operating Loan rate, available in each Agency office.

(2) The limited resource Operating Loan interest rate is not available for Youth loans.

(3) The interest rate charged will be the lower rate in effect at the time of loan approval or loan closing.

(b) *Terms.* Youth loan terms are the same as for an OL established at § 764.254(b).

§ 764.305 Security requirements.

A first lien will be obtained on property or products acquired or produced with loan funds.

§§ 764.306–764.350 [Reserved]

Subpart I—Emergency Loan Program

SOURCE: 72 FR 63298, Nov. 8, 2007, unless otherwise noted. Redesignated at 75 FR 54015, Sept. 3, 2010.

§ 764.351 Emergency loan uses.

(a) *Physical losses*—(1) *Real estate losses.* EM loan funds for real estate physical losses may only be used to repair or replace essential property damaged or destroyed as a result of a disaster as follows:

(i) For any FO purpose, as specified in § 764.151, except subparagraph (e) of that section;

(ii) To establish a new site for farm dwelling and service buildings outside of a flood or mudslide area; and

(iii) To replace land from the farm that was sold or conveyed, if such land is necessary for the farming operation to be effective.

(2) *Chattel losses.* EM loan funds for chattel physical losses may only be used to repair or replace essential property damaged or destroyed as a result of a disaster as follows:

(i) Purchase livestock, farm equipment, quotas and bases, and cooperative stock for credit, production, processing, or marketing purposes;

(ii) Pay customary costs associated with obtaining and closing a loan that an applicant cannot pay from other sources (e.g., fees for legal, architectural, and other technical services, but not fees for agricultural management consultation, or preparation of Agency forms);

(iii) Repair or replace household contents damaged in the disaster;

(iv) Pay the costs to restore perennials, which produce an agricultural commodity, to the stage of development the damaged perennials had obtained prior to the disaster;

(v) Pay essential family living and farm operating expenses, in the case of an operation that has suffered livestock losses not from breeding stock, or losses to stored crops held for sale; and

(vi) Refinance farm-related debts other than real estate to improve farm profitability, if the applicant has refinanced direct or guaranteed loans four times or fewer and one of the following conditions is met:

(A) A designated or declared disaster caused the need for refinancing; or

(B) The debts to be refinanced are owed to a creditor other than the USDA.

(b) *Production losses.* EM loan funds for production losses to agricultural commodities (except the losses associated with the loss of livestock) may be used to:

(1) Pay costs associated with reorganizing the farm to improve its profitability except that such costs must not include the payment of bankruptcy expenses;

(2) Pay annual operating expenses, which include, but are not limited to, feed, seed, fertilizer, pesticides, farm supplies, and cash rent;

(3) Pay costs associated with Federal or State-approved standards under the Occupational Safety and Health Act of 1970 (29 U.S.C. 655 and 667) if the applicant can show that compliance or noncompliance with the standards will cause substantial economic injury;

(4) Pay borrower training costs required or recommended by the Agency;

(5) Pay essential family living expenses;

(6) Refinance farm-related debts other than real estate to improve farm profitability, if the applicant has refinanced direct or guaranteed loans four times or fewer and one of the following conditions is met:

(i) A designated or declared disaster caused the need for refinancing; or

(ii) The debts to be refinanced are owed to a creditor other than the USDA; and

(7) Replace lost working capital.

§ 764.352 Eligibility requirements.

The applicant:

(a) Must comply with the general eligibility requirements established at § 764.101;

(b) Must be an established farmer;

(c) Must be the owner-operator or tenant-operator as follows:

(1) For a loan made under § 764.351(a)(1), must have been:

(i) The owner-operator of the farm at the time of the disaster; or

(ii) The tenant-operator of the farm at the time of the disaster whose lease on the affected real estate exceeds the term of the loan. The operator will provide prior notification to the Agency if the lease is proposed to terminate during the term of the loan. The lessor will provide the Agency a mortgage on the real estate as security for the loan;

(2) For a loan made under § 764.351(a)(2) or (b), must have been the operator of the farm at the time of the disaster; and

(3) In the case of an entity, the entity must be:

(i) Engaged primarily and directly in farming in the United States;

(ii) Authorized to operate and own the farm, if the funds are used for farm ownership loan purposes, in the State in which the farm is located;

(d) Must demonstrate the intent to continue the farming operation after the designated or declared disaster;

(e) And all entity members must be unable to obtain sufficient credit elsewhere at reasonable rates and terms. To establish this, the applicant must obtain written declinations of credit, specifying the reasons for declination, from legally organized commercial lending institutions within reasonable proximity of the applicant as follows:

(1) In the case of a loan in excess of $300,000, two written declinations of credit are required;

(2) In the case of a loan of $300,000 or less, one written declination of credit is required; and

(3) In the case of a loan of $100,000 or less, the Agency may waive the requirement for obtaining a written declination of credit, if the Agency determines that it would pose an undue burden on the applicant, the applicant certifies that they cannot get credit elsewhere, and based on the applicant's circumstances credit is not likely to be available;

(4) Notwithstanding the applicant's submission of the required written declinations of credit, the Agency may contact other commercial lending institutions within reasonable proximity of the applicant and make an independent determination of the applicant's ability to obtain credit elsewhere;

(f) And all entity members in the case of an entity, must not have received debt forgiveness from the Agency on more than one occasion on or before April 4, 1996, or any time after April 4, 1996.

(g) Must submit an application to be received by the Agency no later than 8 months after the date the disaster is declared or designated in the county of the applicant's operation.

(h) For production loss loans, must have a disaster yield that is at least 30 percent below the normal production yield of the crop, as determined by the Agency, which comprises a basic part of an applicant's total farming operation.

(i) For physical loss loans, must have suffered disaster-related damage to chattel or real estate essential to the farming operation, or to household contents that must be repaired or replaced, to harvested or stored crops, or to perennial crops.

(j) Must meet all of the following requirements if the ownership structure of the family farm changes between the time of a qualifying loss and the time an EM loan is closed:

(1) The applicant, including all owners must meet all of the eligibility requirements;

(2) The individual applicant, or all owners of a entity applicant, must have had an ownership interest in the farming operation at the time of the disaster; and

(3) The amount of the loan will be based on the percentage of the former farming operation transferred to the applicant and in no event will the individual portions aggregated equal more than would have been authorized for the former farming operation.

(k) Must agree to repay any duplicative Federal assistance to the agency providing such assistance. An applicant receiving Federal assistance for a major disaster or emergency is liable to the United States to the extent that the assistance duplicates benefits available to the applicant for the same purpose from another source.

(1) Whose primary enterprise is to breed, raise, and sell horses may be eligible under this part.

[72 FR 63298, Nov. 8, 2007, as amended at 76 FR 75434, Dec. 2, 2011]

§ 764.353 Limitations.

(a) EM loans must comply with the general limitations established at § 764.102.

(b) EM loans may not exceed the lesser of:

(1) The amount of credit necessary to restore the farming operation to its pre-disaster condition;

(2) In the case of a physical loss loan, the total eligible physical losses caused by the disaster; or

(3) In the case of a production loss loan, 100 percent of the total actual production loss sustained by the applicant as calculated in paragraph (c) of this section.

(c) For production loss loans, the applicant's actual crop production loss will be calculated as follows:

(1) Subtract the disaster yield from the normal yield to determine the per acre production loss;

(2) Multiply the per acre production loss by the number of acres of the farming operation devoted to the crop to determine the volume of the production loss;

(3) Multiply the volume of the production loss by the market price for such crop as determined by the Agency to determine the dollar value for the production loss; and

(4) Subtract any other disaster related compensation or insurance indemnities received or to be received by the applicant for the production loss.

(d) For a physical loss loan, the applicant's total eligible physical losses will be calculated as follows:

(1) Add the allowable costs associated with replacing or repairing chattel covered by hazard insurance (excluding labor, machinery, equipment, or materials contributed by the applicant to repair or replace chattel);

(2) Add the allowable costs associated with repairing or replacing real estate, covered by hazard insurance;

(3) Add the value of replacement livestock and livestock products for which the applicant provided:

(i) Written documentation of inventory on hand immediately preceding the loss;

(ii) Records of livestock product sales sufficient to allow the Agency to establish a value;

(4) Add the allowable costs to restore perennials to the stage of development the damaged perennials had obtained prior to the disaster;

(5) Add, in the case of an individual applicant, the allowable costs associated with repairing or replacing household contents, not to exceed $20,000; and

(6) Subtract any other disaster related compensation or insurance indemnities received or to be received by the applicant for the loss or damage to the chattel or real estate.

(e) EM loan funds may not be used for physical loss purposes unless:

(1) The physical property was covered by general hazard insurance at the time that the damage caused by the natural disaster occurred. The level of the coverage in effect at the time of the disaster must have been the tax or cost depreciated value, whichever is less. Chattel property must have been covered at the tax or cost depreciated value, whichever is less, when such insurance was readily available and the benefit of the coverage was greater than the cost of the insurance; or

(2) The loan is to a poultry farmer to cover the loss of a chicken house for which the applicant did not have hazard insurance at the time of the loss and the applicant:

(i) Applied for, but was unable to obtain hazard insurance for the chicken house;

(ii) Uses the loan to rebuild the chicken house in accordance with industry standards in effect on the date the applicant submits an application for the loan;

(iii) Obtains, for the term of the loan, hazard insurance for the full market value of the chicken house; and

(iv) Meets all other requirements for the loan.

(f) EM loan funds may not be used to refinance consumer debt, such as automobile loans, or credit card debt unless such credit card debt is directly attributable to the farming operation.

(g) Losses associated with horses used for racing, showing, recreation, or pleasure or loss of income derived from racing, showing, recreation, boarding, or pleasure are not considered qualified losses under this section.

[72 FR 63298, Nov. 8, 2007, as amended at 76 FR 75434, Dec. 2, 2011]

§ 764.354 Rates and terms.

(a) *Rates.* (1) The interest rate is the Agency's Emergency Loan Actual Loss rate, available in each Agency office.

(2) The interest rate charged will be the lower rate in effect at the time of loan approval or loan closing.

(b) *Terms.* (1) The Agency schedules repayment of EM loans based on the useful life of the security, the applicant's repayment ability, and the type of loss.

(2) The repayment schedule must include at least one payment every year.

(3) EM loans for annual operating expenses, except expenses associated with establishing a perennial crop that are subject to paragraph (b)(4), must be repaid within 12 months. The Agency may extend this term to not more than 24 months to accommodate the production cycle of the agricultural commodities.

(4) EM loans for production losses or physical losses to chattel (including, but not limited to, assets with an expected life between one and 7 years) may not exceed 7 years. The Agency may extend this term up to a total length not to exceed 20 years, if necessary to improve the applicant's repayment ability and real estate security is available.

(5) The repayment schedule for EM loans for physical losses to real estate is based on the applicant's repayment ability and the useful life of the security, but in no case will the term exceed 40 years.

[72 FR 63298, Nov. 8, 2007, as amended at 86 FR 43392, Aug. 9, 2021]

§ 764.355 Security requirements.

(a) EM loans made under § 764.351(a)(1) must comply with the general security requirements established at §§ 764.103, 764.104 and 764.155(b).

(b) EM loans made as specified in § 764.351(a)(2) and (b) must generally comply with the general security requirements established in §§ 764.103, 764.104, and 764.255(b). These general security requirements, however, do not apply to equine loss loans to the extent that a lien is not obtainable or obtaining a lien may prevent the applicant from carrying on the normal course of business. Other security may be considered for an equine loss loan in the order of priority as follows:

(1) Real estate,

(2) Chattels and crops, other than horses,

(3) Other assets owned by the applicant,

(4) Third party pledges of property not owned by the applicant,

(5) Repayment ability under paragraph (c) of this section.

(c) Notwithstanding the requirements of paragraphs (a) and (b) of this section, when adequate security is not available because of the disaster, the loan may be approved if the Agency determines, based on an otherwise feasible plan, there is a reasonable assurance that the applicant has the ability to repay the loan provided:

(1) The applicant has pledged as security for the loan all available personal and business security, except as provided in § 764.106;

(2) The farm operating plan, approved by the Agency, indicates the loan will be repaid based upon the applicant's production and income history; addresses applicable pricing risks through the use of marketing contracts, hedging, options, or other revenue protection mechanisms, and includes a marketing plan or similar risk management practice;

(3) The applicant has had positive net cash farm income in at least 3 of the past 5 years; and

(4) The applicant has provided the Agency an assignment on any USDA program payments to be received.

(d) For loans over $25,000, title clearance is required when real estate is taken as security.

(e) For loans of $25,000 or less, when real estate is taken as security, a certification of ownership in real estate is required. Certification of ownership may be in the form of an affidavit which is signed by the applicant, names the record owner of the real estate in question and lists the balances due on all known debts against the real estate. Whenever the Agency is uncertain of the record owner or debts

against the real estate security, a title search is required.

[72 FR 63298, Nov. 8, 2007, as amended at 76 FR 75434, Dec. 2, 2011]

§ 764.356 Appraisal and valuation requirements.

(a) In the case of physical losses associated with livestock, the applicant must have written documentation of the inventory of livestock and records of livestock product sales sufficient to allow the Agency to value such livestock or livestock products just prior to the loss.

(b) In the case of farm assets damaged by the disaster, the value of such security shall be established as of the day before the disaster occurred.

(c) In the case of an equine loss loan:

(1) The applicant's Federal income tax and business records will be the primary source of financial information. Sales receipts, invoices, or other official sales records will document the sales price of individual animals.

(2) If the applicant does not have 3 complete years of business records, the Agency will obtain the most reliable and reasonable information available from sources such as the Cooperative Extension Service, universities, and breed associations to document production for those years for which the applicant does not have a complete year of business records.

[72 FR 63298, Nov. 8, 2007, as amended at 76 FR 75435, Dec. 2, 2011]

§§ 764.357–764.400 [Reserved]

Subpart J—Loan Decision and Closing

SOURCE: 72 FR 63298, Nov. 8, 2007, unless otherwise noted. Redesignated at 75 FR 54015, Sept. 3, 2010.

§ 764.401 Loan decision.

(a) *Loan approval.* (1) The Agency will approve a loan only if it determines that:

(i) The applicant's farm operating plan reflects a feasible plan, which includes repayment of the proposed loan and demonstrates that all other credit needs can be met;

289

(ii) The proposed use of loan funds is authorized for the type of loan requested;

(iii) The applicant has been determined eligible for the type of loan requested;

(iv) All security requirements for the type of loan requested have been, or will be met before the loan is closed;

(v) The applicant's total indebtedness to the Agency, including the proposed loan, will not exceed the maximum limits established in § 761.8 of this chapter;

(vi) There have been no significant changes in the farm operating plan or the applicant's financial condition since the time the Agency received a complete application; and

(vii) All other pertinent requirements have been, or will be met before the loan is closed.

(2) The Agency will place conditions upon loan approval it determines necessary to protect its interest and maximize the applicant's potential for success.

(b) *Loan denial.* The Agency will not approve a loan if it determines that:

(1) The applicant's farm operating plan does not reflect a feasible plan;

(2) The proposed use of loan funds is not authorized for the type of loan requested;

(3) The applicant does not meet the eligibility requirements for the type of loan requested;

(4) There is inadequate security for the type of loan requested;

(5) Approval of the loan would cause the applicant's total indebtedness to the Agency to exceed the maximum limits established in § 761.8 of this chapter;

(6) The applicant's circumstances may not permit continuous operation and management of the farm; or

(7) The applicant, the farming operation, or other circumstances surrounding the loan are inconsistent with the authorizing statutes, other Federal laws, or Federal credit policies.

(c) *Overturn of an Agency decision by appeal.* If an FLP loan denial is overturned on administrative appeal, the Agency will not automatically approve the loan. Unless prohibited by the final appeal determination or otherwise advised by the Office of General Counsel, the Agency will:

(1) Request current financial information from the applicant as necessary to determine whether any changes in the applicant's financial condition or agricultural conditions which occurred after the Agency's adverse decision was made will adversely affect the applicant's farming operation;

(2) Approve a loan for crop production:

(i) Only if the Agency can determine that the applicant will be able to produce a crop in the production cycle for which the loan is requested; or

(ii) For the next production cycle, upon review of current financial data and a farm operating plan for the next production cycle, if the Agency determines the loan can be repaid. The new farm operating plan must reflect any financial issues resolved in the appeal.

(3) Determine whether the applicant's farm operating plan, as modified based on the appeal decision, reflects a feasible plan, which includes repayment of the proposed loan and demonstrates that all other credit needs can be met.

§ 764.402   Loan closing.

(a) *Signature requirements.* Signatures on loan documents are required as follows:

(1) For individual applicants, only the applicant is required to sign the promissory note.

(2) For entity applicants, the promissory note will be executed to evidence the liability of the entity, any embedded entities, and the individual liability of all entity members.

(3) Despite minority status, a youth executing a promissory note for a Youth loan will incur full personal liability for the debt.

(4) A cosigner will be required to sign the promissory note if they assist the applicant in meeting the repayment requirements for the loan requested.

(5) All signatures needed for the Agency to acquire the required security interests will be obtained according to State law.

(b) *Payment of fees.* The applicant, or in the case of a real estate purchase, the applicant and seller, must pay all filing, recording, notary, lien search, and any other fees necessary to process and close a loan.

(c) *Chattel-secured loans.* The following requirements apply to loans secured by chattel:

(1) The Agency will close a chattel loan only when it determines the Agency requirements for the loan have been satisfied;

(2) A financing statement is required for every loan except when a filed financing statement covering the applicant's property is still effective, covers all types of chattel property that will serve as security for the loan, describes the land on which crops and fixtures are or will be located, and complies with the law of the jurisdiction where filed;

(3) A new security agreement is required for new loans, as necessary to secure the loan under State law, prior to the disbursement of loan funds.

(d) *Real estate-secured loans.* (1) The Agency will close a real estate loan only when it determines that the Agency requirements for the loan have been satisfied and the closing agent can issue a policy of title insurance or final title opinion as of the date of closing. The title insurance or final title opinion requirement may be waived:

(i) For loans of $10,000 or less;

(ii) As provided in §764.235 for CLs and §764.355 for EMs;

(iii) When the real estate is considered additional security by the Agency; or

(iv) When the real estate is a non-essential asset.

(2) The title insurance or final title opinion must show title vested as required by the Agency, the lien of the Agency's security instrument in the priority required by the Agency, and title to the security property, subject only to those exceptions approved in writing by the Agency.

(3) The Agency must approve agents who will close FLP loans. Closing agents must meet all of the following requirements to the Agency's satisfaction:

(i) Be licensed in the state where the loan will be closed;

(ii) Not be debarred or suspended from participating in any Federal programs;

(iii) Maintain liability insurance;

(iv) Have a fidelity bond that covers all employees with access to loan funds;

(v) Have current knowledge of the requirements of State law in connection with the loan closing and title clearance;

(vi) Not represent both the buyer and seller in the transaction;

(vii) Not be related as a family member or business associate with the applicant; and

(viii) Act promptly to provide required services.

(e) *Disbursement of funds.* (1) Loan funds will be made available to the applicant within 15 days of loan approval, subject to the availability of funding.

(2) If the loan is not closed within 90 days of loan approval or if the applicant's financial condition changes significantly, the Agency must reconfirm the requirements for loan approval prior to loan closing. The applicant may be required to provide updated information for the Agency to reconfirm approval and proceed with loan closing.

(3) The Agency or closing agent will be responsible for disbursing loan funds. The electronic funds transfer process, followed by Treasury checks, are the Agency's preferred methods of loan funds disbursement. The Agency will use these processes on behalf of borrowers to disburse loan proceeds directly to creditors being refinanced with loan funds or to sellers of chattel property that is being acquired with loan funds. A supervised bank account will be used according to subpart B of part 761 of this chapter when these processes are not practicable.

[72 FR 63298, Nov. 8, 2007. Redesignated at 75 FR 54015, Sept. 3, 2010, as amended at 77 FR 15939, Mar. 19, 2012; 79 FR 60745, Oct. 8, 2014]

§§ 764.403–764.450 [Reserved]

Subpart K—Borrower Training and Training Vendor Requirements

SOURCE: 72 FR 63298, Nov. 8, 2007, unless otherwise noted. Redesignated at 75 FR 54015, Sept. 3, 2010.

§ 764.451 Purpose.

The purpose of production and financial management training is to help an applicant develop and improve skills necessary to:

(a) Successfully operate a farm;

(b) Build equity in the operation; and

(c) Become financially successful and prepared to graduate from Agency financing to commercial sources of credit.

§ 764.452 Borrower training requirements.

(a) The applicant must agree to complete production and financial management training, unless the Agency provides a waiver in accordance with § 764.453, or the applicant has previously satisfied the training requirements. In the case of an entity:

(1) Any individual member holding a majority interest in the entity or who is operating the farm must complete training on behalf of the entity, except as provided in paragraph (a)(2) of this section;

(2) If one entity member is solely responsible for production or financial management, then only that member will be required to complete training.

(b) When the Agency determines that production training is required, the applicant must agree to complete course work covering production management in each crop or livestock enterprise the Agency determines necessary.

(c) When the Agency determines that financial management training is required, the applicant must agree to complete course work covering all aspects of farm accounting and integrating accounting elements into a financial management system.

(d) An applicant who applies for a loan to finance a new enterprise, such as a new crop or a new type of livestock, must agree to complete production training with regard to that enterprise, even if production training requirements were waived or satisfied under a previous loan request, unless the Agency provides a waiver in accordance with § 764.453.

(e) Even if a waiver is granted, the borrower must complete borrower training as a condition for future loans if and when Agency supervision provided in 7 CFR part 761 subpart C reflects that such training is needed.

(f) The Agency cannot reject a request for a direct loan based solely on an applicant's need for training.

(g) The Agency will provide written notification of required training or waiver of training.

§ 764.453 Agency waiver of training requirements.

(a) The applicant must request the waiver in writing.

(b) The Agency will grant a waiver for training in production, financial management, or both, under the following conditions:

(1) The applicant submits evidence of successful completion of a course similar to a course approved under section § 764.457 and the Agency determines that additional training is not needed; or

(2) The applicant submits evidence which demonstrates to the Agency's satisfaction the applicant's experience and training necessary for a successful and efficient operation.

(c) If the production and financial functions of the operation are shared among individual entity members, the Agency will consider the collective knowledge and skills of those individuals when determining whether to waive training requirements.

(d) When considering subsequent loan actions, previous training requirements that have not yet been satisfied may be waived by the Agency should the borrower submit satisfactory evidence in accordance with paragraph (b) of this section.

[72 FR 63298, Nov. 8, 2007, as amended at 86 FR 43392, Aug. 9, 2021]

§ 764.454 Actions that an applicant must take when training is required.

(a) *Deadline for completion of training.* (1) If the Agency requires an applicant to complete training, at loan closing the applicant must agree in writing to complete all required training within 2 years.

(2) The Agency will grant a one-year extension to complete training if the applicant is unable to complete training within the 2-year period due to circumstances beyond the applicant's control.

(3) The Agency will grant an extension longer than one year for extraordinary circumstances as determined by the Agency.

(4) An applicant who does not complete the required training within the specified time-period will be ineligible for additional direct FLP loans until the training is completed.

(b) *Arranging training with a vendor.* The applicant must select and contact an Agency approved vendor and make all arrangements to begin training.

(c) *Payment of training fees.* (1) The applicant is responsible for the cost of training and must include training fees in the farm operating plan as a farm operating expense.

(2) The payment of training fees is an authorized use of OL funds.

(3) The Agency is not a party to fee or other agreements between the applicant and the vendor.

(d) *Evaluation of a vendor.* Upon completion of the required training, the applicant will complete an evaluation of the course and submit it to the vendor. The vendor will forward the completed evaluation forms to the Agency.

§764.455 **Potential training vendors.**

The Agency will contract for training services with State or private providers of production and financial management training services.

§764.456 **Applying to be a vendor.**

(a) A vendor for borrower training services must apply to the Agency for approval.

(b) The vendor application must include:

(1) A sample of the course materials and a description of the vendor's training methods;

(2) Specific training objectives for each section of the course;

(3) A detailed course agenda specifying the topics to be covered, the time devoted to each topic, and the number of sessions to be attended;

(4) A list of instructors and their qualifications;

(5) The criteria by which additional instructors will be selected;

(6) The proposed locations where training will take place;

(7) The cost per participant, including cost for additional members of a farming operation;

(8) The minimum and maximum class size;

(9) The vendor's experience in developing and administering training to farmers;

(10) The monitoring and quality control methods the vendor will use;

(11) The policy on allowing Agency employees to attend the course for monitoring purposes;

(12) A plan of how the needs of applicants with physical, mental, or learning disabilities will be met; and

(13) A plan of how the needs of applicants who do not speak English as their primary language will be met.

§764.457 **Vendor requirements.**

(a) *Minimum experience.* The vendor must demonstrate a minimum of 3 years of experience in conducting training courses or teaching the subject matter.

(b) *Training objectives.* The courses provided by a vendor must enable the applicant to accomplish one or more of the following objectives:

(1) Describe the specific goals of the farming operation, any changes required to attain the goals, and outline how these changes will occur using present and projected cash flow budgets;

(2) Maintain and use a financial management information system to make financial decisions;

(3) Understand and use an income statement;

(4) Understand and use a balance sheet;

(5) Understand and use a cash flow budget; and

(6) Use production records and other production information to identify problems, evaluate alternatives, and correct current production practices to improve efficiency and profitability.

(c) *Curriculum.* At least one of the following subjects must be covered:

(1) Business planning courses, covering general goal setting, risk management, and planning.

(2) Financial management courses, covering all aspects of farm accounting and focusing on integrating accounting

elements into a financial management system.

(3) Crop and livestock production courses focusing on improving the profitability of the farm.

(d) *Instructor qualifications.* All instructors must have:

(1) Sufficient knowledge of the material and experience in adult education;

(2) A bachelor's degree or comparable experience in the subject area to be taught; and

(3) A minimum of 3 years experience in conducting training courses or teaching.

§ 764.458   Vendor approval.

(a) *Agreement to conduct training.* (1) Upon approval, the vendor must sign an agreement to conduct training for the Agency's borrowers.

(2) The agreement to conduct training is valid for 3 years.

(3) Any changes in curriculum, instructor, or cost require prior approval by the Agency.

(4) The vendor may revoke the agreement by giving the Agency a written 30-day notice.

(5) The Agency may revoke the agreement if the vendor does not comply with the responsibilities listed in the agreement by giving the vendor a written 30-day notice.

(b) *Renewal of agreement to conduct training.* (1) To renew the agreement to conduct training, the vendor must submit in writing to the Agency:

(i) A request to renew the agreement;

(ii) Any changes in curricula, instructor, or cost; and

(iii) Documentation that the vendor is providing effective training.

(2) The Agency will review renewal requests in accordance with § 764.457.

§ 764.459   Evaluation of borrower progress.

(a) The vendor must provide the Agency with a periodic progress report for each borrower enrolled in training in accordance with the agreement to complete training. The reports will indicate whether the borrower is attending sessions, completing the training program, and demonstrating an understanding of the course material.

(b) Upon borrower completion of the training, the vendor must provide the Agency with an evaluation of the borrower's knowledge of the course material and assign a score. The following table lists the possible scores, the criteria used to assign each score, and Agency consideration of each score:

Score	Criteria used to determine score	Agency consideration
1	If the borrower: • Attended sessions as agreed, • Satisfactorily completed all assignments, and • Demonstrated an understanding of the course material..	Training requirement associated with course is complete.
2	If the borrower: • Attended sessions as agreed, and • Attempted to complete all assignments, but • Does not demonstrate an understanding of the course material..	Training requirement associated with couse is complete. Additional Agency supervision may be necessary.
3	If the borrower did not: • Attend sessions as agreed, or • Attempt to complete assignments, or • Otherwise make a good faith effort to complete the training..	Training requirement associated with course is not complete. The borrower is ineligible for future direct loans until the training is completed.

PART 765—DIRECT LOAN SERVICING—REGULAR

Sec.

Subpart A—Overview

765.1   Introduction.
765.2   Abbreviations and definitions.
765.3–765.50   [Reserved]

Subpart B—Borrowers with Limited Resource Interest Rate Loans

765.51   Required review.
765.52–765.100   [Reserved]

Subpart C—Borrower Graduation

765.101   Borrower graduation requirements.
765.102   Borrower noncompliance with graduation requirements.

AUTHORITY: 5 U.S.C. 301 and 7 U.S.C. 1989.

SOURCE: 72 FR 63309, Nov. 8, 2007, unless otherwise noted.

Subpart A—Overview

§765.1 Introduction.

(a) *Purpose.* This part describes the policies for servicing direct FLP loans, except for borrowers who are delinquent, financially distressed, or otherwise in default on their loan.

(b) *Servicing actions.* Servicing actions described in this part include:

(1) Limited resource reviews;

(2) Graduation to commercial credit;

(3) Application of payments;

(4) Maintaining and disposing of security;

(5) Transfer of security and assumption of debt; and

(6) Servicing accounts of deceased borrowers.

(c) *Loans covered.* The Agency services direct FLP loans under the policies contained in this part. This part is not applicable to Non-program loans, except where noted.

§765.2 Abbreviations and definitions.

Abbreviations and definitions for terms used in this part are provided in §761.2 of this chapter.

§§765.3–765.50 [Reserved]

Subpart B—Borrowers With Limited Resource Interest Rate Loans

§765.51 Required review.

(a) At least every 2 years, a borrower with limited resource interest rate loans is required to provide the operation's financial information to the Agency; for the Agency to determine if the borrower can afford to pay a higher interest rate on the loan. The Agency will review the information provided in accordance with § 761.105 of this chapter.

(b) If the borrower's farm operating plan shows that the debt service margin exceeds 110 percent, the Agency will increase the interest rate on the loans with a limited resource interest rate until:

(1) A further increase in the interest rate results in a debt service margin of less than 110 percent; or

(2) The interest rate is equal to the interest rate currently in effect for the type of loan.

(c) Except as provided in paragraph (d) of this section, the Agency will increase the limited resource interest rate to the current interest rate for the type of loan, if the borrower:

(1) Purchases items not planned during the term of the loan;

(2) Refuses to submit information the Agency requests for use in reviewing the borrower's financial condition;

(3) Ceases farming, as described in § 765.253; or

(4) Is ineligible due to disqualification resulting from Federal crop insurance violation according to 7 CFR part 718.

(d) If the borrower has limited resource interest rate loans that are deferred, the Agency will not change the interest rate during the deferral period.

[72 FR 63309, Nov. 8, 2007, as amended at 86 FR 43392, Aug. 9, 2021]

§§ 765.52–765.100 [Reserved]

Subpart C—Borrower Graduation

§ 765.101 Borrower graduation requirements.

(a) In accordance with the promissory note and security instruments, the borrower must graduate to another source of credit if the Agency determines that:

(1) The borrower has the ability to obtain credit from other sources; and

(2) Adequate credit is available from other sources at reasonable rates and terms.

(b) The Agency may require partial or full graduation.

(1) In a partial graduation, all FLP loans of one type (i.e. all chattel loans or all real estate loans) must be paid in full by refinancing with other credit with or without an Agency guarantee.

(2) In a full graduation, all FLP loans are paid in full by refinancing with other credit with or without an Agency guarantee.

(3) A loan made for chattel and real estate purposes will be categorized according to how the majority of the loan's funds are expended.

(c) The borrower must submit all information that the Agency requests in conjunction with the review of the borrower's financial condition.

(d) The Agency may provide a borrower's prospectus to lenders in an attempt to identify sources of non-Agency credit and assess the lenders' interest in refinancing the borrower's loan. The Agency will notify the borrower when the borrower's prospectus is provided to one or more lenders.

(e) If a lender expresses an interest in refinancing the borrower's FLP loan, the borrower must:

(1) Apply for a loan from the interested lender within 30 days of notice; or

(2) Seek guaranteed loan assistance under the market placement program in accordance with § 762.110(g) of this chapter.

(f) The borrower will be responsible for any application fees or purchase of stock in conjunction with graduation.

(g) CLs are not subject to graduation requirements under this part.

[72 FR 63309, Nov. 8, 2007, as amended at 75 FR 54016, Sept. 3, 2010]

§ 765.102 Borrower noncompliance with graduation requirements.

Borrower failure to fulfill all graduation requirements within the time-period specified by the Agency constitutes default on the loan. The Agency will accelerate the borrower's loan without offering servicing options provided in 7 CFR part 766.

§ 765.103 Transfer and assignment of Agency liens.

The Agency may assign its lien to the new lender when the borrower is graduating and all FLP debt will be paid in full.

§§ 765.104–765.150 [Reserved]

Subpart D—Borrower Payments

§ 765.151 Handling payments.

(a) *Borrower payments.* Borrowers must submit their loan payments in a form acceptable to the Agency, such as checks, cash, and money orders. Forms of payment not acceptable to the Agency include, but are not limited to, foreign currency, foreign checks, and sight drafts.

(b) *Crediting account.* The Agency credits the borrower's account as of the date the Agency receives payment.

§ 765.152 Types of payments.

(a) *Regular payments.* Regular payments are derived from, but are not limited to:

(1) The sale of normal income security;

(2) The sale of farm products;

(3) Lease income, including mineral lease signing bonus;

(4) Program or disaster-related disbursements from USDA or crop insurance entities; and

(5) Non-farm income.

(b) *Extra payments.* Extra payments are derived from any of the following:

(1) Sale of chattel security other than normal income security;

(2) Sale of real estate security;

(3) Refinancing of FLP debt;

(4) Cash proceeds of insurance claims received on Agency security, if not being used to repair or replace the security;

(5) Any transaction that results in a loss in the value of any Agency basic security;

(6) Refunds of duplicate program benefits or assistance to be applied on CL or EM loans; or

(7) Refunds of unused loan funds.

(c) *Payments from sale of real estate.* Notwithstanding any other provision of this section, payments derived from the sale of real estate security will be treated as regular payments at the Agency's discretion, if the FLP loans will be adequately secured after the transaction.

[72 FR 63309, Nov. 8, 2007, as amended at 75 FR 54016, Sept. 3, 2010]

§ 765.153 Application of payments.

(a) *Regular payments.* A regular payment is credited to a scheduled installment on program and non-program loans. Regular payments are applied to loans in the following order:

(1) Annual operating loan;

(2) Delinquent FLP installments, paying least secured loans first;

(3) Non-delinquent FLP installments due in the current production cycle in order of security priority, paying least secured loans first;

(4) Any future installments due.

(b) *Extra payments.* An extra payment is not credited to a scheduled installment and does not relieve the borrower's responsibility to make scheduled loan installments, but will reduce the borrower's FLP indebtedness. Extra payments are applied to FLP loans in order of lien priority except for refunds of unused loan funds, which shall be applied to the loan for which the funds were advanced.

§ 765.154 Distribution of payments.

The Agency applies both regular and extra payments to each loan in the following order, as applicable:

(a) Recoverable costs and protective advances plus interest;

(b) Deferred non-capitalized interest;

(c) Accrued deferred interest;

(d) Interest accrued to date of payment; and

(e) Loan principal.

§ 765.155 Final loan payments.

(a) *General.* (1) Unless the Agency has reservations regarding the validity of the payment, the Agency may release the borrower's security instruments at the time payment is made, if the borrower makes a final payment by one of the following methods:

(i) Cash;

(ii) U.S. Treasury check;

(iii) Cashier's check; or

(iv) Certified check.

(2) Security instruments will only be released when all loans secured by the instruments have been paid in full or otherwise satisfied.

(3) The Agency will return the paid note and satisfied security instruments to the borrower after the Agency processes the final payment and determines

that the total indebtedness is paid in full.

(b) *Borrower refunds.* If the borrower refunds the entire loan after the loan is closed, the borrower must pay interest from the date of the note to the date the Agency received the funds.

(c) *Overpayments.* If an Agency miscalculation of a final payment results in an overpayment by the borrower of less than $10, the borrower must request a refund from the Agency in writing. Overpayments of $10 or more automatically will be refunded by the Agency.

(d) *Underpayments.* If an Agency miscalculation of a final payment amount results in an underpayment, the Agency may collect all account balances resulting from its error. If the Agency cannot collect an underpayment from the borrower, the Agency will service the debt in accordance with part 761, subpart F of this chapter.

[72 FR 63309, Nov. 8, 2007, as amended at 85 FR 36693, June 17, 2020]

§§ 765.156–765.160 [Reserved]

§ 765.161 Borrowers entering the Armed Forces.

(a) *Protections for borrowers on active duty.* The Servicemembers Civil Relief Act (Pub. L. 108–189) and the Ronald W. Reagan National Defense Authorization Act for Fiscal Year (FY) 2005 (Pub. L. 108–375) provide certain loan servicing protections for military borrowers. The Agency will apply those loan servicing protections to applicable Farm Loan borrowers.

(1) The benefits and protections of the Servicemembers Civil Relief Act apply to borrowers on active duty at all times.

(2) The requirements of the Ronald W. Reagan National Defense Authorization Act for Fiscal Year (FY) 2005 apply during a time of a war or national emergency as declared by the President or Congress.

(b) *Eligibility for National Guard members and military reservists.* Borrowers who are National Guard members or military reservists will be eligible for the protections covered by this section, as specified in paragraphs (b)(1) and (2) of this section:

(1) National Guard members must be on duty for at least 30 consecutive calendar days.

(2) Military reservists are eligible from the date orders are received to report for active duty.

(c) *Entity eligibility.* National Guard members and military reservists on active duty and any operating entity owned solely by the active duty borrower may be considered for protections specified in paragraph (a) of this section.

[86 FR 43392, Aug. 9, 2021]

§§ 765.162–765.200 [Reserved]

Subpart E—Protecting the Agency's Security Interest

§ 765.201 General policy.

All Agency servicing actions regarding preservation and protection of Agency security will be consistent with the covenants and agreements contained in all loan agreements and security instruments.

§ 765.202 Borrower responsibilities.

The borrower must:

(a) Comply with all provisions of the loan agreements;

(1) Non-compliance with the provisions of loan agreements and documents, other than failure to meet scheduled loan repayment installments contained in the promissory note, constitutes non-monetary default on FLP loans by the borrower;

(2) Borrower non-compliance will be considered by the Agency when making eligibility determinations for future requests for assistance and may adversely impact such requests;

(b) Maintain, protect, and account for all security;

(c) Pay the following, unless State law requires the Agency to pay:

(1) Fees for executing, filing or recording financing statements, continuation statements or other security instruments; and

(2) The cost of lien search reports;

(d) Pay taxes on property securing FLP loans when they become due;

(e) Maintain insurance coverage in an amount specified by the Agency;

(f) Protect the interests of the Agency when a third party brings suit or takes other action that could affect Agency security.

§ 765.203 Protective advances.

When necessary to protect the Agency's security interest, costs incurred for the following actions will be charged to the borrower's account:

(a) Maintain abandoned security property;

(b) Preserve inadequately maintained security;

(c) Pay real estate taxes and assessments;

(d) Pay property, hazard, or flood insurance;

(e) Pay harvesting costs;

(f) Maintain Agency security instruments;

(g) Pay ground rents;

(h) Pay expenses for emergency measures to protect the Agency's collateral; and

(i) Protect the Agency from actions by third parties.

§ 765.204 Notifying potential purchasers.

(a) *States with Central Filing System (CFS).* The Agency participates and complies with central filing systems in States where CFS has been organized. In a State with a CFS, the Agency is not required to additionally notify potential purchasers that the Agency has a lien on a borrower's chattel security, unless specifically required by State law.

(b) *States without CFS.* In a State without CFS, the Agency follows the filing requirements specified for perfecting a lien on a borrower's chattel security under State law. The Agency will distribute the list of chattel and crop borrowers to sale barns, warehouses, and other businesses that buy or sell chattels or crops. In addition, the Agency may provide the list of borrowers to potential purchasers upon request.

§ 765.205 Subordination of liens.

(a) *Borrower application requirements.* The borrower must submit the following, unless it already exists in the Agency's file and is still current as determined by the Agency:

(1) Completed Agency application for subordination form;

(2) A current financial statement, including, in the case of an entity, financial statements from all entity members;

(3) Documentation of compliance with the Agency's environmental regulations contained in part 799 of this chapter;

(4) Verification of all non-farm income;

(5) The farm's operating plan, including a projected cash flow budget reflecting production, income, expenses, and debt repayment plan; and

(6) Verification of all debts.

(b) *Subordination of real estate security.* For loans secured by real estate, the Agency will approve a request for subordination subject to the following conditions:

(1) If a lender requires that the Agency subordinate its lien position on the borrower's existing property in order for the borrower to acquire new property and the request meets the requirements in paragraph (b)(3) of this section, the request may be approved. The Agency will obtain a valid mortgage and the required lien position on the new property. The Agency will require title clearance and loan closing for the property in accordance with § 764.402 of this chapter.

(2) If the borrower is an entity and the Agency has taken real estate as additional security on property owned by a member, a subordination for any authorized loan purpose may be approved when it meets the requirements in paragraph (b)(3) of this section and it is needed for the entity member to finance a separate farming operation. The subordination must not cause the unpaid principal and interest on the FLP loan to exceed the value of loan security or otherwise adversely affect the security.

(3) The Agency will approve a request for subordination of real estate to a creditor if:

(i) The loan will be used for an authorized loan purpose or is to refinance a loan made for an authorized loan purpose by the Agency or another creditor;

(ii) The credit is essential to the farming operation, and the borrower cannot obtain the credit without a subordination;

(iii) The FLP loan is still adequately secured after the subordination, or the value of the loan security will be increased by an amount at least equal to the advance to be made under the subordination;

(iv) Except as authorized by paragraph (c)(2) of this section, there is no other subordination outstanding with another lender in connection with the same security;

(v) The subordination is limited to a specific amount;

(vi) The loan made in conjunction with the subordination will be closed within a reasonable time and has a definite maturity date;

(vii) If the loan is made in conjunction with a guaranteed loan, the guaranteed loan meets the requirements of § 762.142(c) of this chapter;

(viii) The borrower is not in default or will not be in default on FLP loans by the time the subordination closing is complete;

(ix) The borrower can demonstrate, through a current farm operating plan, the ability to repay all debt payments scheduled, and to be scheduled, during the production cycle;

(x) Except for CL, the borrower is unable to partially or fully graduate;

(xi) The borrower must not be ineligible as a result of a conviction for controlled substances according to part 718 of this chapter;

(xii) The borrower must not be ineligible due to disqualification resulting from Federal crop insurance violation according to part 718 of this chapter;

(xiii) The borrower will not use loan funds in a way that will contribute to erosion of highly erodible land or conversion of wetlands as described in part 799 of this chapter;

(xiv) Any planned development of real estate security will be performed as directed by the lessor or creditor, as approved by the Agency, and will comply with the terms and conditions of § 761.10 of this chapter;

(xv) If a borrower with an SAA mortgage is refinancing a loan held by a lender, subordination of the SAA mortgage may only be approved when the refinanced loan does not increase the amount of debt; and

(xvi) In the case of a subordination of non-program loan security, the non-program loan security also secures a program loan with the same borrower.

(4) The Agency will approve a request for subordination of real estate to a lessee if the conditions in paragraphs (b)(3)(viii) through (xvi) of this section are met.

(c) *Chattel security.* The requirements for chattel subordinations are as follows:

(1) For loans secured by chattel, the subordination must meet the conditions contained in paragraphs (b)(3)(i) through (xiii) of this section.

(2) The Agency will approve a request for a second subordination to enable a borrower to obtain crop insurance, if the following conditions are met:

(i) The creditor to whom the first subordination was given did not provide for payment of the current year's crop insurance premium, and consents in writing to the provisions of the second subordination to pay insurance premiums from the crop or insurance proceeds;

(ii) The borrower assigns the insurance proceeds to the Agency or names the Agency in the loss payable clause of the policy; and

(iii) The subordination meets the conditions under paragraphs (b)(1) through (12) of this section.

(d) *Appraisals.* An appraisal of the property that secures the FLP loan will be required when the Agency determines it necessary to protect its interest. Appraisals will be obtained in accordance with § 761.7 of this chapter.

[72 FR 63309, Nov. 8, 2007, as amended at 75 FR 54016, Sept. 3, 2010; 78 FR 65530, Nov. 1, 2013; 81 FR 51284, Aug. 3, 2016]

§ 765.206   Junior liens.

(a) *General policy.* The borrower will not give a lien on Agency security without the consent of the Agency. Failure to obtain Agency consent will be considered by the Agency when making eligibility determinations for future requests for assistance and may adversely impact such requests.

(b) *Conditions for consent.* The Agency will consent to the terms of a junior lien if all of the following conditions are met:

(1) The borrower's ability to make scheduled loan payments is not jeopardized;

(2) The borrower provides the Agency a copy of the farm operating plan submitted to the junior lienholder, and the plan is consistent with the Agency operating plan;

(3) The total debt against the security does not exceed the security's market value;

(4) The junior lienholder agrees in writing not to foreclose the security instrument unless written notice is provided to the Agency;

(5) The borrower is unable to graduate on any program except for CL; and

(6) The junior lien will not otherwise adversely impact the Agency's financial interests.

[72 FR 63309, Nov. 8, 2007, as amended at 75 FR 54016, Sept. 3, 2010]

§765.207 Conditions for severance agreements.

For loans secured by real estate, a borrower may request Agency consent to a severance agreement or similar instrument so that future chattel acquired by the borrower will not become part of the real estate securing the FLP debt. The Agency will consent to severance agreements if all of the following conditions are met:

(a) The financing arrangements are in the financial interest of the Agency and the borrower;

(b) The transaction will not adversely affect the Agency's security position;

(c) The borrower is unable to graduate on any program except for CL;

(d) The transaction will not jeopardize the borrower's ability to pay all outstanding debts to the Agency and other creditors; and

(e) The property acquired is consistent with authorized loan purposes.

[72 FR 63309, Nov. 8, 2007, as amended at 75 FR 54016, Sept. 3, 2010]

§§765.208–765.250 [Reserved]

Subpart F—Required Use and Operation of Agency Security

§765.251 General.

(a) A borrower is required to be the operator of Agency security in accordance with loan purposes, loan agreements, and security instruments.

(b) A borrower who fails to operate the security without Agency consent is in violation of loan agreements and security instruments.

(c) The Agency will consider a borrower's request to lease or cease to operate the security as provided in §§765.252 and 765.253.

§765.252 Lease of security.

(a) *Real estate surface leases.* The borrower must request prior approval to lease the surface of real estate security. The Agency will approve requests provided the following conditions are met:

(1) The lease will not adversely affect the Agency's security interest;

(2) The term of consecutive leases for agricultural purposes does not exceed 3 years, or 5 years if the borrower and the lessee are related by blood or marriage. The term of surface leases for farm property no longer in use, such as old barns, or for nonfarm purposes, such as wind turbines, communication towers, or similar installations can be for any term;

(3) The lease does not contain an option to purchase; and

(4) The lease does not hinder the future operation or success of the farm, or, if the borrower has ceased to operate the farm, the requirements specified in §765.253 are met. Leases for nonfarm enterprises, such as solar farms, which take significant acreage of the operation out of agriculture production are not authorized. Non-productive land may be considered for this type of lease; and.

(5) The lease and any contracts or agreements in connection with the lease must be reviewed and approved by the Government.

(b) *Mineral leases.* The borrower must request Agency consent to lease any mineral rights used as security for FLP loans.

(1) For FO loans made from December 23, 1985, to February 7, 2014, and loans other than FO loans secured by real estate and made from December 23, 1985, to November 1, 2013, the value of the mineral rights must have been included in the original appraisal in

order for the Agency to obtain a security interest in any oil, gas, and other mineral associated with the real estate security.

(2) For all other loans not covered by paragraph (b)(1) of this section, the Agency will obtain a security interest in any oil, gas, and other mineral on or under the real estate pledged as collateral in accordance with the applicable security agreement, regardless of whether such minerals were included in the original appraisal.

(3) The Agency may consent to a mineral lease if the proposed use of the leased rights will not adversely affect either:

(i) The Agency's security interest; or

(ii) Compliance with any applicable environmental requirements of part 799 of this chapter.

(4) The term of the mineral lease is not limited.

(c) *Lease of chattel security.* Lease of chattel security is not authorized.

(d) *Lease proceeds.* Lease proceeds are considered normal income security and may be used in accordance with § 765.303.

(e) *Lease of allotments.* (1) The Agency will not approve any crop allotment lease that will adversely affect its security interest in the allotment.

(2) The borrower must assign all rental proceeds from an allotment lease to the Agency.

[72 FR 63309, Nov. 8, 2007, as amended at 78 FR 65531, Nov. 1, 2013; 79 FR 78694, Dec. 31, 2014; 81 FR 51284, Aug. 3, 2016; 86 FR 43392, Aug. 9, 2021]

§ 765.253 Ceasing to operate security.

If the borrower requests Agency consent to cease operating the security or if the Agency discovers that the borrower is failing to operate the security, the Agency will give consent if:

(a) Such action is in the Agency's best interests;

(b) The borrower is unable to graduate on any program except for CL;

(c) The borrower is not ineligible as a result of disqualification for Federal crop insurance violation according to 7 CFR part 718;

(d) Any one of the following conditions is met:

(1) The borrower is involved in the day-to-day operational activities, management decisions, costs and returns of the farming operation, and will continue to reside in the immediate farming community for reasonable management and operation involvement;

(2) The borrower's failure to operate the security is due to age or poor health, and the borrower continues to reside in the immediate farming community for reasonable management and operation involvement; or

(3) The borrower's failure to operate the security is beyond the borrower's control, and the borrower will resume the farming operation within 3 years.

[72 FR 63309, Nov. 8, 2007, as amended at 75 FR 54016, Sept. 3, 2010; 78 FR 65531, Nov. 1, 2013]

§§ 765.254–765.300 [Reserved]

Subpart G—Disposal of Chattel Security

§ 765.301 General.

(a) The borrower must account for all chattel security, and maintain records of dispositions of chattel security and the actual use of proceeds. The borrower must make these records available to the Agency upon request.

(b) The borrower may not dispose of chattel security for an amount less than its market value. All proceeds, including any amount in excess of the market value, must be distributed to lienholders for application to the borrower's account in the order of lien priority.

(1) The Agency considers the market value of normal income security to be the prevailing market price of the commodity in the area in which the farm is located.

(2) The market value for basic security is determined by an appraisal obtained in accordance with § 761.7 of this chapter.

(c) When the borrower sells chattel security, the property and proceeds remain subject to the Agency lien until the lien is released by the Agency.

(d) The Agency and all other lienholders must provide written consent before a borrower may use proceeds for a purpose other than payment of lienholders in the order of lien priority.

(e) The transaction must not interfere with the borrower's farming operation or jeopardize the borrower's ability to repay the FLP loan.

(f) The disposition must enhance the program objectives of the FLP loan.

(g) When the borrower exchanges security property for other property or purchases new property with sale proceeds, the acquisition must be essential to the farming operation as well as meet the program objectives, purposes, and limitations for the type of loan.

(h) All checks, drafts, or money orders which the borrower receives from the sale of Agency security must be payable to the borrower and the Agency. If all FLP loan installments and any past due installments, for the period of the agreement for the use of proceeds have been paid, however, these payments from the sale of normal income security may be payable solely to the borrower.

[72 FR 63309, Nov. 8, 2007, as amended at 78 FR 65531, Nov. 1, 2013]

§ 765.302 Use and maintenance of the agreement for the use of proceeds.

(a) The borrower and the Agency will execute an agreement for the use of proceeds.

(b) The borrower must report any disposition of basic or normal income security to the Agency as specified in the agreement for the use of proceeds.

(c) If a borrower wants to dispose of normal income security in a way different than provided by the agreement for the use of proceeds, the borrower must obtain the Agency's consent before the disposition unless all FLP payments planned on the agreement have been paid.

(d) If the borrower sells normal income security to a purchaser not listed in the agreement for the use of proceeds, the borrower must immediately notify the Agency of what property has been sold and of the name and business address of the purchaser.

(e) The borrower must provide the Agency with the necessary information to update the agreement for the use of proceeds.

(f) Changes to the agreement on the use of proceeds will be recorded, dated and initialed by the borrower and the Agency.

[72 FR 63309, Nov. 8, 2007, as amended at 78 FR 65531, Nov. 1, 2013]

§ 765.303 Use of proceeds from chattel security.

(a) *General.* (1) Proceeds from the sale of basic security and normal income security must be remitted to lienholders in order of lien priority.

(2) Proceeds remitted to the Agency may be used as follows:

(i) Applied to the FLP loan;

(ii) Pay customary costs appropriate to the transaction.

(3) With the concurrence of all lienholders, proceeds may be used to preserve the security because of a natural disaster or other severe catastrophe, when funds cannot be obtained by other means in time to prevent the borrower and the Agency from suffering a substantial loss.

(4) Security may be consumed as follows:

(i) Livestock may be used by the borrower's family for subsistence;

(ii) If crops serve as security and usually would be marketed, the Agency may allow such crops to be fed to the borrower's livestock, if this is preferable to marketing, provided the Agency obtains a lien or assignment on the livestock, and livestock products, at least equal to the lien on the crops.

(b) *Proceeds from the sale of normal income security.* In addition to the uses specified in paragraph (a) of this section, the agreement for the use of proceeds will allow for release of proceeds from the sale of normal income security to be used to pay essential family living and farm operating expenses. Such releases will be terminated when an account is accelerated.

(c) *Proceeds from the sale of basic security.* In addition to the uses specified in paragraph (a) of this section:

(1) Proceeds from the sale of basic security may not be used for any family living and farm operating expenses.

(2) Security may be exchanged for chattel property better suited to the borrower's needs if the Agency will acquire a lien on the new property at least equal in value to the lien held on the property exchanged.

(3) Proceeds may be used to purchase chattel property better suited to the borrower's needs if the Agency will acquire a lien on the purchased property. The value of the purchased property, together with any proceeds applied to the FLP loan, must at least equal the value of the Agency lien on the old security.

§ 765.304 Unapproved disposition.

(a) If a borrower disposes of chattel security without Agency approval, or misuses proceeds, the borrower must:

(1) Make restitution to the Agency within 30 days of Agency notification; or

(2) Provide disposition or use information to enable the Agency to consider post-approval within 30 days of Agency notification.

(b) Failure to cure the first unauthorized disposition in accordance with paragraph (a) of this section, or a second unauthorized disposition, whether or not cured, constitutes a non-monetary default, will be considered by the Agency when making eligibility determinations for future requests for assistance, may adversely impact such requests, and may result in civil or criminal action.

§ 765.305 Release of security interest.

(a) When Agency security is sold, exchanged, or consumed in accordance with the agreement for the use of proceeds, the Agency will release its security interest to the extent of the value of the security disposed.

(b) Security interests on wool and mohair may be released when the security is marketed by consignment, provided all of the following conditions are met:

(1) The borrower assigns to the Agency the proceeds of any advances made, or to be made, on the wool or mohair by the broker, less shipping, handling, processing, and marketing costs;

(2) The borrower assigns to the Agency the proceeds of the sale of the wool or mohair, less any remaining costs in shipping, handling, processing, and marketing, and less the amount of any advance (including any interest which may have accrued on the advance) made by the broker against the wool or mohair; and

(3) The borrower and broker agree that the net proceeds of any advances on, or sale of, the wool or mohair will be paid by checks made payable jointly to the borrower and the Agency.

(c) The Agency will release its lien on chattel security without compensation, upon borrower request provided:

(1) The borrower has not received primary loan servicing or Disaster Set-Aside within the last 3 years;

(2) The borrower will retain the security and use it as collateral for other credit, including partial graduation as specified in § 765.101;

(3) The security margin on each FLP direct loan will be 150 percent or more after the release. The value of the retained and released security will normally be based on appraisals obtained as specified in § 761.7 of this chapter; however, well documented recent sales of similar properties can be used if the Agency determines a supportable decision can be made without current appraisals;

(4) The release is approved by the FSA State Executive Director; and

(5) Except for CL, the borrower is unable to fully graduate as specified in § 765.101.

[72 FR 63309, Nov. 8, 2007, as amended at 78 FR 65531, Nov. 1, 2013]

§§ 765.306–765.350 [Reserved]

Subpart H—Partial Release of Real Estate Security

§ 765.351 Requirements to obtain Agency consent.

The borrower must obtain prior consent from the Agency for any transactions affecting the real estate security, including, but not limited to, sale or exchange of security, a right-of-way across security, and a partial release. The Agency may consent to such transactions provided the conditions in this section are met.

(a) *General.* The following conditions apply to all transactions affecting real estate:

(1) The transaction will enhance the objectives for which the FLP loan or loans were made;

(2) The transaction will not jeopardize the borrower's ability to repay the FLP loan, or is necessary to place

the borrower's farming operation on a sound basis;

(3) Except for releases in paragraph (f) of this section, the amount received by the borrower for the security being disposed of, or the rights being granted, is not less than the market value and will be remitted to the lienholders in the order of lien priority;

(4) The transaction must not interfere with the borrower's farming operation;

(5) The market value of the remaining security is adequate to secure the FLP loans, or if the market value of the security before the transaction was inadequate to fully secure the FLP loans, the Agency's equity in the security is not diminished;

(6) The environmental requirements of part 799 of this chapter must be met;

(7) The borrower cannot graduate to other credit on any program except for CL;

(8) The borrower must not be ineligible due to disqualification resulting from Federal crop insurance violation according to 7 CFR part 718; and

(9) The disposition of real estate security for an outstanding ST loan will only be authorized if the transaction will result in full repayment of the loan.

(b) *Sale of timber, gravel, oil, gas, coal, or other minerals.* (1) Agency security instruments require that the borrower request and receive written consent from the Agency prior to certain transactions, including, but not limited to, cutting, removal, or lease of timber, gravel, oil, gas, coal, or other minerals, except small amounts used by the borrower for ordinary household purposes.

(i) The sale of timber from real estate that secures an FLP loan will be considered a disposition of a portion of the security.

(ii) When the Agency has a security interest in oil, gas, or other minerals as provided by §765.252(b), the sale of such products will be considered a disposition of a portion of the security by the Agency.

(2) Any compensation the borrower may receive for damages to the surface of the real estate security resulting from exploration for, or recovery of, minerals must be assigned to the Agency. Such proceeds will be used to repair the damage, and any remaining funds must be remitted to lienholders in the order of lien priority or, with all lienholders' consent, used for an authorized loan purpose.

(c) *Exchange of security property.* (1) When an exchange of security results in a balance owing to the borrower, the proceeds must be used in accordance with §765.352.

(2) Property acquired by the borrower must meet program objectives, purposes and limitations relating to the type of loan involved as well as applicable requirements for appraisal, title clearance and security.

(d) *Sale under contract for deed.* A borrower may sell a portion of the security for not less than its market value under a contract for deed subject to the following:

(1) Not less than 10 percent of the purchase price will be paid as a down payment and remitted to lienholders in the order of lien priority;

(2) Payments will not exceed 10 annual installments of principal plus interest or the remaining term of the FLP loan, whichever is less. The interest rate will be the current rate being charged on a regular FO loan plus 1 percent or the rate on the borrower's notes, whichever is greater. Payments may be in equal or unequal installments with a balloon final installment;

(3) The Agency's security rights, including the right to foreclose on either the portion being sold or retained, will not be impaired;

(4) Any subsequent payments must be assigned to the lienholders and remitted in order of lien priority, or with lienholder's approval, used in accordance with §765.352;

(5) The mortgage on the property sold will not be released prior to either full payment of the borrower's account or receipt of the full amount of sale proceeds;

(6) The sale proceeds applied to the borrower's loan accounts will not relieve the borrower from obligations under the terms of the note or other agreements approved by the Agency;

(7) All other requirements of this section are met.

(e) *Transfer of allotments.* (1) The Agency will not approve any crop allotment lease that will adversely affect its security interest.

(2) The sale of an allotment must comply with all conditions of this subpart.

(3) The borrower may transfer crop allotments to another farm owned or controlled by the borrower. Such transfer will be treated as a lease under § 765.252.

(f) *Release without compensation.* Real estate security may be released by FSA without compensation when the requirements of paragraph (a) of this section, except paragraph (a)(3) of this section, are met, and:

(1) The borrower has not received primary loan servicing or Disaster Set-Aside within the last 3 years;

(2) The security is:

(i) To be retained by the borrower and used as collateral for other credit, including partial graduation as specified in § 765.101; or

(ii) No more than 10 acres, or the minimum size that meets all State and local requirements for a division into a separate legal lot, whichever is greater, and is transferred without compensation to a person who is related to the borrower by blood or marriage.

(3) The property released will not interfere with access to or operation of the remaining farm;

(4) Essential buildings and facilities will not be released if they reduce the utility or marketability of the remaining property;

(5) Any issues arising due to legal descriptions, surveys, environmental concerns, utilities are the borrower's responsibility and no costs or fees will be paid by FSA;

(6) The security margin on each FLP direct loan will be above 150 percent after the release. The value of the retained security will normally be based on appraisals obtained as specified in § 761.7 of this chapter; however, well documented recent sales of similar properties can be used if the Agency determines the criteria have been met and a sound decision can be made without current appraisals;

(7) The release is approved by the FSA State Executive Director; and

(8) Except for CL, the borrower is unable to fully graduate as specified in § 765.101.

[72 FR 63309, Nov. 8, 2007, as amended at 75 FR 54016, Sept. 3, 2010; 78 FR 65531, Nov. 1, 2013; 81 FR 51284, Aug. 3, 2016; 86 FR 43392, Aug. 9, 2021]

§ 765.352   Use of proceeds.

(a) Proceeds from transactions affecting the real estate security may only be used as follows:

(1) Applied on liens in order of priority;

(2) To pay customary costs appropriate to the transaction, which meet the following conditions:

(i) Are reasonable in amount;

(ii) Cannot be paid by the borrower;

(iii) Will not be paid by the purchaser;

(iv) Must be paid to consummate the transaction; and

(v) May include postage and insurance when it is necessary for the Agency to present the promissory note to the recorder to obtain a release of a portion of the real estate from the mortgage.

(3) For development or enlargement of real estate owned by the borrower as follows:

(i) Development or enlargement must be necessary to improve the borrower's debt repayment ability, place the borrower's farming operation on a sound basis, or otherwise enhance the objectives of the loan;

(ii) Such use will not conflict with the loan purposes, restrictions or requirements of the type of loan involved;

(iii) Funds will be deposited in a supervised bank account in accordance with subpart B of part 761 of this chapter;

(iv) The Agency has, or will obtain, a lien on the real estate developed or enlarged;

(v) Construction and development will be completed in accordance with § 761.10 of this chapter.

(b) After acceleration, the Agency may approve transactions only when all the proceeds will be applied to the liens against the security in the order of their priority, after deducting customary costs appropriate to the transaction. Such approval will not cancel or delay liquidation, unless all loan defaults are otherwise cured.

§765.353 Determining market value.

(a) *Security proposed for disposition.* (1) The Agency will obtain an appraisal of the security proposed for disposition.

(2) The Agency may waive the appraisal requirement when the estimated value is less than $50,000.

(b) *Security remaining after disposition.* The Agency will obtain an appraisal of the remaining security if it determines that the transaction will reduce the value of the remaining security.

(c) *Appraisal requirements.* Appraisals, when required, will be conducted in accordance with §761.7 of this chapter.

[72 FR 63309, Nov. 8, 2007, as amended at 86 FR 43392, Aug. 9, 2021]

§§765.354–765.400 [Reserved]

Subpart I—Transfer of Security and Assumption of Debt

§765.401 Conditions for transfer of real estate and chattel security.

(a) *General conditions.* (1) Approval of a security transfer and corresponding loan assumption obligates a new borrower to repay an existing FLP debt.

(2) All transferees will become personally liable for the debt and assume the full responsibilities and obligations of the debt transferred when the transfer and assumption is complete. If the transferee is an entity, the entity and each entity member must assume personal liability for the loan.

(3) A transfer and assumption will only be approved if the Agency determines it is in the Agency's financial interest.

(b) *Agency consent.* A borrower must request and obtain written Agency consent prior to selling or transferring security to another party.

[72 FR 63309, Nov. 8, 2007, as amended at 79 FR 60745, Oct. 8, 2014]

§765.402 Transfer of security and loan assumption on same rates and terms.

An eligible applicant may assume an FLP loan on the same rates and terms as the original note if:

(a) The original borrower has died and the spouse, other relative, or joint tenant who is not obligated on the note inherits the security property;

(b) A family member of the borrower or an entity comprised solely of family members of the borrower assumes the debt along with the original borrower;

(c) An individual with an ownership interest in the borrower entity buys the entire ownership interest of the other members and continues to operate the farm in accordance with loan requirements. The new owner must assume personal liability for the loan;

(d) A new entity buys the borrower entity and continues to operate the farm in accordance with loan requirements; or

(e) The original loan is an EM loan for physical or production losses and persons who were directly involved in the farm's operation at the time of the loss will assume the loan. If the original loan was made to:

(1) An individual borrower, the transferee must be a family member of the original borrower or an entity in which the entity members are comprised solely of family members of the original borrower.

(2) A trust, partnership or joint operation, the transferee must have been a member, partner or joint operator when the Agency made the original loan or remain an entity comprised solely of people who were original entity members, partners or joint operators when the entity received the original loan.

(3) A corporation, limited liability company, cooperative, or other legal business organization, the transferee must:

(i) Have been a corporate stockholder, cooperative member or other member of a legal business organization, when the Agency made the original loan or will be an entity comprised solely of entity members who were entity members when the entity received the loan; and

(ii) Assume only the portion of the physical or production loss loan equal to the transferee's percentage of ownership. In the case of entity transferees, the transferee must assume that portion of the loan equal to the combined percentages of ownership of the individual stockholders or entity members in the transferee.

[72 FR 63309, Nov. 8, 2007, as amended at 79 FR 60745, Oct. 8, 2014]

§ 765.403 Transfer of security to and assumption of debt by eligible applicants.

(a) *Transfer of real estate and chattel security.* The Agency may approve transfers of security with assumption of FLP debt, other than EM loans for physical or production losses, by transferees eligible for the type of loan being assumed if:

(1) The transferee meets all loan and security requirements in part 764 of this chapter for the type of loan being assumed; and

(2) The outstanding loan balance (principal and interest) does not exceed the maximum loan limit for the type of loan as contained in § 761.8 of this chapter.

(b) *Assumption of Non-program loans.* Applicants eligible for FO loans under part 764 of this chapter may assume Non-program loans made for real estate purposes if the Agency determines the property meets program requirements. In such case, the Agency will reclassify the Non-program loan as an FO loan.

(c) *Loan types that the Agency no longer makes.* Real estate loan types the Agency no longer makes (*i.e.* EE, RL, RHF) may be assumed and reclassified as FO loans if the transferee is eligible for an FO loan under part 764 of this chapter and the property proposed for transfer meets program requirements.

(d) *Amount of assumption.* The transferee must assume the lesser of:

(1) The outstanding balance of the transferor's loan; or

(2) The market value of the security, less prior liens and authorized costs, if the outstanding loan balance exceeds the market value of the property.

(e) *Rates and terms.* The interest rate and loan term will be determined according to rates and terms established in part 764 of this chapter for the type of loan being assumed.

§ 765.404 Transfer of security to and assumption of debt by ineligible applicants.

(a) *General.* (1) The Agency will allow the transfer of real estate and chattel security property to applicants who are ineligible for the type of loan being assumed only on Non-program loan rates and terms.

(2) The Agency will reclassify the assumed loan as a Non-program loan.

(b) *Eligibility.* Transferees must:

(1) Provide written documentation verifying their credit worthiness and debt repayment ability;

(2) Not have received debt forgiveness from the Agency;

(3) Not be ineligible for loans as a result of a conviction for controlled substances according to 7 CFR part 718; and

(4) Not be ineligible due to disqualification resulting from Federal crop insurance violation according to 7 CFR part 718.

(c) *Assumption amount.* The transferee must assume the total outstanding FLP debt or if the value of the property is less than the entire amount of debt, an amount equal to the market value of the security less any prior liens. The total outstanding FLP debt will include any unpaid deferred interest that accrued on the loan to the extent that the debt does not exceed the security's market value.

(d) *Downpayment.* Non-program transferees must make a downpayment to the Agency of not less than 10 percent of the lesser of the market value or unpaid debt.

(e) *Interest rate.* The interest rate will be the Non-program interest rate in effect at the time of loan approval.

(f) *Loan terms.* (1) For a Non-program loan secured by real estate, the Agency schedules repayment in 25 years or less, based on the applicant's repayment ability.

(2) For a Non-program loan secured by chattel property only, the Agency schedules repayment in 5 years or less, based on the applicant's repayment ability.

§ 765.405 Payment of costs associated with transfers.

The transferor and transferee are responsible for paying transfer costs such as real estate taxes, title examination, attorney's fees, surveys, and title insurance. When the transferor is unable to pay its portion of the transfer costs, the transferee, with Agency approval, may pay these costs provided:

(a) Any cash equity due the transferor is applied first to payment of costs and the transferor does not receive any cash payment above these costs;

(b) The transferee's payoff of any junior liens does not exceed $5,000;

(c) Fees are customary and reasonable;

(d) The transferee can verify that personal funds are available to pay transferor and transferee fees; and

(e) Any equity due the transferor is held in escrow by an Agency designated closing agent and is disbursed at closing.

§ 765.406 Release of transferor from liability.

(a) *General.* Agency approval of an assumption does not automatically release the transferor from liability.

(b) *Requirements for release.* (1) The Agency may release the transferor from liability when all of the security is transferred and the total outstanding debt is assumed.

(2) If an outstanding debt balance will remain and only part of the transferor's Agency security is transferred, the written request for release of liability will not be approved, unless the deficiency is otherwise resolved to the Agency's satisfaction.

(3) If an outstanding balance will remain and all of the transferor's security has been transferred, the transferor may pay the remaining balance or request debt settlement in accordance with part 761 subpart F of this chapter. If the transferor does not resolve the debt by paying the remaining balance or submitting a debt settlement offer that is acceptable to the Agency, the Agency will service the debt in accordance with part 3 of this title using all applicable collection tools including, but not limited to, administrative offset, AWG, cross-servicing, Federal salary offset, and TOP.

(4) Except for loans in default being serviced under 7 CFR part 766, if an individual who is jointly liable for repayment of an FLP loan withdraws from the farming operation and conveys all of their interest in the security to the remaining borrower, the withdrawing party may be released from liability under the following conditions:

(i) A divorce decree or property settlement states that the withdrawing party is no longer responsible for repaying the loan;

(ii) All of the withdrawing party's interests in the security are conveyed to the persons with whom the loan will be continued; and

(iii) The persons with whom the loan will be continued can demonstrate the ability to repay all of the existing and proposed debt obligations.

[72 FR 63309, Nov. 8, 2007, as amended at 85 FR 36693, June 17, 2020]

§§ 765.407–765.450 [Reserved]

Subpart J—Deceased Borrowers

§ 765.451 Continuation of FLP debt and transfer of security.

(a) *Individuals who are liable.* Following the death of a borrower, the Agency will continue the loan with any individual who is liable for the indebtedness provided that the individual complies with the obligations of the loan and security instruments.

(b) *Individuals who are not liable.* The Agency will continue the loan with a person who is not liable for the indebtedness in accordance with subpart I of this part.

§ 765.452 Borrowers with Non-program loans.

(a) *Loan continuation.* (1) The Agency will continue the loan with a jointly liable borrower if the remaining borrower continues to pay the deceased borrower's loan in accordance with the loan and security instruments.

(2) The Agency may continue the loan with an individual who inherits title to the property and is not liable for the indebtedness provided the individual makes payments as scheduled and fulfills all other responsibilities of the borrower according to the loan and security instruments.

(b) *Loan assumption.* A deceased borrower's loan may be assumed by an individual not liable for the indebtedness in accordance with subpart I of this part.

(c) *Loan discontinuation.* (1) The Agency will not continue a loan for any subsequent transfer of title by the heirs, or sale of interests between heirs to consolidate title; and

(2) The Agency treats any subsequent transfer of title as a sale subject to requirements listed in subpart I of this part.

§§ 765.453–765.500   [Reserved]

Subpart K—Exception Authority

§ 765.501   Agency exception authority.

On an individual case basis, the Agency may consider granting an exception to any regulatory requirement or policy of this part if:

(a) The exception is not inconsistent with the authorizing statute or other applicable law; and

(b) The Agency's financial interest would be adversely affected by acting in accordance with published regulations or policies and granting the exception would resolve or eliminate the adverse effect upon the Agency's financial interest.

PART 766—DIRECT LOAN SERVICING—SPECIAL

Subpart A—Overview

Subpart B—Disaster Set-Aside

Subpart C—Loan Servicing Programs

Subpart D—Homestead Protection Program

Subpart E—Servicing Shared Appreciation Agreements and Net Recovery Buyout Agreements

Subpart F—Unauthorized Assistance

766.253 Unauthorized assistance resulting from submission of inaccurate information by borrower or Agency error.
766.254–766.300 [Reserved]

Subpart G—Loan Servicing For Borrowers in Bankruptcy

766.301 Notifying borrower in bankruptcy of loan servicing.
766.302 Loan servicing application requirements for borrowers in bankruptcy.
766.303 Processing loan servicing requests from borrowers in bankruptcy.
766.304–766.350 [Reserved]

Subpart H—Loan Liquidation

766.351 Liquidation.
766.352 Voluntary sale of real property and chattel.
766.353 Voluntary conveyance of real property.
766.354 Voluntary conveyance of chattel.
766.355 Acceleration of loans.
766.336 Acceleration of loans to American Indian borrowers.
766.357 Involuntary liquidation of real property and chattel.
766.358 Acceleration and foreclosure moratorium.
766.359–766.400 [Reserved]

Subpart I—Exception Authority

766.401 Agency exception authority.

AUTHORITY: 5 U.S.C. 301, 7 U.S.C. 1989, and 1981d(c).

SOURCE: 72 FR 63316, Nov. 8, 2007, unless otherwise noted.

Subpart A—Overview

§ 766.1 Introduction.

(a) This part describes the Agency's servicing policies for direct loan borrowers who:

(1) Are financially distressed;

(2) Are delinquent in paying direct loans or otherwise in default;

(3) Have received unauthorized assistance;

(4) Have filed bankruptcy or are involved in other civil or criminal cases affecting the Agency; or

(5) Have loan security being liquidated voluntarily or involuntarily.

(b) The Agency services direct FLP loans under the policies contained in this part.

(1) Youth loans:

(i) May not receive Disaster Set-Aside under subpart B of this part;

(ii) Will only be considered for rescheduling according to § 766.107 and deferral according to § 766.109.

(2) The Agency does not service Non-program loans under this part except where noted.

(c) The Agency requires the borrower to make every reasonable attempt to make payments and comply with loan agreements before the Agency considers special servicing.

§ 766.2 Abbreviations and definitions.

Abbreviations and definitions for terms used in this part are provided in § 761.2 of this chapter.

§§ 766.3–766.50 [Reserved]

Subpart B—Disaster Set-Aside

§ 766.51 General.

(a) DSA is available to borrowers with program loans who suffered losses as a result of a natural disaster.

(b) DSA is not intended to circumvent other servicing available under this part.

(c) Non-program loans may be serviced under this subpart for borrowers who also have program loans.

§ 766.52 Eligibility.

(a) *Borrower eligibility.* The borrower must meet all of the following requirements to be eligible for a DSA:

(1) The borrower must have operated the farm in a county designated or declared a disaster area or a contiguous county at the time of the disaster. Farmers who have rented out their land base for cash are not operating the farm:

(2) The borrower must have acted in good faith, and the borrower's inability to make the upcoming scheduled loan payments must be for reasons not within the borrower's control.

(3) The borrower cannot have more than one installment set aside on each loan.

(4) As a direct result of the natural disaster, the borrower does not have sufficient income available to pay all family living and farm operating expenses, other creditors, and debts to the Agency. This determination will be based on:

(i) The borrower's actual production, income and expense records for the year the natural disaster occurred;

(ii) Any other records required by the Agency;

(iii) Compensation received for losses; and

(iv) Increased expenses incurred because of the natural disaster.

(5) For the next production cycle, the borrower must develop a feasible plan showing that the borrower will at least be able to pay all operating expenses and taxes due during the year, essential family living expenses, and meet scheduled payments on all debts, including FLP debts. The borrower must provide any documentation required to support the farm operating plan.

(6) The borrower must not be in nonmonetary default.

(7) The borrower must not be ineligible due to disqualification resulting from Federal crop insurance violation according to 7 CFR part 718.

(8) The borrower must not become 165 days past due before the appropriate Agency DSA documents are executed.

(b) *Loan eligibility.* (1) Any FLP loan to be considered for DSA must have been outstanding at the time the natural disaster occurred.

(2) All of the borrower's program and non-program loans must be current after the Agency completes a DSA of the scheduled installment.

(3) All FLP loans must be current or less than 90 days past due at the time the application for DSA is complete.

(4) The Agency has not accelerated or applied any special servicing action under this part to the loan since the natural disaster occurred.

(5) For any loan that will receive a DSA, the remaining term of the loan must equal or exceed 2 years from the due date of the installment set-aside.

(6) The loan must not have a DSA in place.

§ 766.53 Disaster Set-Aside amount limitations.

(a) The DSA amount is limited to the lesser of:

(1) The first or second scheduled annual installment on the FLP loans due after the disaster occurred; or

(2) The amount the borrower is unable to pay the Agency due to the dis-

aster. Borrowers are required to pay any portion of an installment they are able to pay.

(b) The amount set aside will be the unpaid balance remaining on the installment at the time the DSA is complete. This amount will include the unpaid interest and any principal that would be credited to the account as if the installment were paid on the due date, taking into consideration any payments applied to principal and interest since the due date.

(c) Recoverable cost items may not be set aside.

§ 766.54 Borrower application requirements.

(a) *Requests for DSA.* (1) A borrower must submit a request for DSA in writing within eight months from the date the natural disaster was designated.

(2) All borrowers must sign the DSA request.

(b) *Required financial information.* (1) The borrower must submit actual production, income, and expense records for the production cycle in which the disaster occurred unless the Agency already has this information.

(2) The Agency may request other information needed to make an eligibility determination.

§ 766.55 Eligibility determination.

Within 30 days of a complete DSA application, the Agency will determine if the borrower meets the eligibility requirements for DSA.

§ 766.56 Security requirements.

If, prior to executing the appropriate DSA Agency documents, the borrower is not current on all FLP loans, the borrower must execute and provide to the Agency a best lien obtainable on all of their assets except those listed under § 766.112(b).

§ 766.57 Borrower acceptance of Disaster Set-Aside.

The borrower must execute the appropriate Agency documents within 45 days after the borrower receives notification of Agency approval of DSA.

§766.58 Installment to be set aside.

(a) The Agency will set-aside the first installment due immediately after the disaster occurred.

(b) If the borrower has already paid the installment due immediately after the disaster occurred, the Agency will set-aside the next annual installment.

§766.59 Payments toward set-aside installments.

(a) *Interest accrual.* (1) Interest will accrue on any principal portion of the set-aside installment at the same rate charged on the balance of the loan.

(2) If the borrower's set-aside installment is for a loan with a limited resource rate and the Agency modifies that limited resource rate, the interest rate on the set-aside portion will be modified concurrently.

(b) *Due date.* The amount set-aside, including interest accrued on the principal portion of the set-aside, is due on or before the final due date of the loan.

(c) *Applying payments.* The Agency will apply borrower payments toward set-aside installments first to interest and then to principal.

§766.60 Canceling a Disaster Set-Aside.

The Agency will cancel a DSA if:

(a) The Agency takes any primary loan servicing action on the loan;

(b) The borrower pays the current market value buyout in accordance with §766.113; or

(c) The borrower pays the set-aside installment.

§766.61 Reversal of a Disaster Set-Aside.

If the Agency determines that the borrower received an unauthorized DSA, the Agency will reverse the DSA after all appeals are concluded.

§§766.62–766.100 [Reserved]

Subpart C—Loan Servicing Programs

§766.101 Initial Agency notification to borrower of loan servicing programs.

(a) *Borrowers notified.* The Agency will provide servicing information under this section to borrowers who:

(1) Have a current farm operating plan that demonstrates the borrower is financially distressed;

(2) Are 90 days or more past due on loan payments, even if the borrower has submitted an application for loan servicing as a financially distressed borrower;

(3) Are in non-monetary default on any loan agreements;

(4) Have filed bankruptcy;

(5) Request this information;

(6) Request voluntary conveyance of security;

(7) Have only delinquent SA; or

(8) Are subject to any other collection action, except when such action is a result of failure to graduate. Borrowers who fail to graduate when required and are able to do so, will be accelerated without providing notification of loan servicing options.

(b) *Form of notification.* The Agency will notify borrowers of the availability of primary loan servicing programs, conservation contract, current market value buyout, debt settlement programs, and homestead protection as follows:

(1) A borrower who is financially distressed, or current and requesting servicing will be provided FSA–2512;

(2) A borrower who is 90 days past due will be sent FSA–2510 (Appendix A to this subpart) or FSA–2510–IA (Appendix B to this subpart);

(3) A borrower who is in non-monetary or both monetary and non-monetary default will receive FSA–2514;

(4) A borrower who has only delinquent SA will be notified of available loan servicing;

(5) Notification to a borrower who files bankruptcy will be provided in accordance with subpart G of this part.

(c) *Mailing.* Notices to delinquent borrowers or borrowers in non-monetary default will be sent by certified mail to the last known address of the borrower. If the certified mail is not accepted, the notice will be sent immediately by first class mail to the last known address. The appropriate response time will begin three days following the date of the first class mailing. For all other borrowers requesting the notices, the notices will be sent by regular mail or hand-delivered.

(d) *Borrower response timeframes.* To be considered for loan servicing, a borrower who is:

(1) Current or financially distressed may submit a complete application any time prior to becoming 90 days past due;

(2) Ninety (90) days past due must submit a complete application within 60 days from receipt of or FSA–2510–IA;

(3) In non-monetary default with or without monetary default must submit a complete application within 60 days from receipt of FSA–2514.

[72 FR 63316, Nov. 8, 2007, as amended at 85 FR 36693, June 17, 2020]

§ 766.102 Borrower application requirements.

(a) Except as provided in paragraph (e) of this section, an application for primary loan servicing, conservation contract, current market value buyout, homestead protection, or some combination of these options, must include the following to be considered complete:

(1) Completed acknowledgment form provided with the Agency notification and signed by all borrowers;

(2) Completed Agency application form;

(3) Financial records for the 3 most recent years, including income tax returns;

(4) The farming operation's production records for the 3 most recent years or the years the borrower has been farming, whichever is less;

(5) Documentation of compliance with the Agency's environmental regulations contained in subpart G of 7 CFR part 1940;

(6) Verification of all non-farm income;

(7) A current financial statement and the operation's farm operating plan, including the projected cash flow budget reflecting production, income, expenses, and debt repayment plan. In the case of an entity, the entity and all entity members must provide current financial statements; and

(8) Verification of all debts and collateral.

(b) In addition to the requirements contained in paragraph (a) of this section, the borrower must submit an aerial photo delineating any land to be considered for a conservation contract.

(c) To be considered for debt settlement, the borrower must provide the appropriate Agency form, and any additional information required under part 761, subpart F of this chapter.

(d) If a borrower who submitted a complete application while current or financially distressed is renotified as a result of becoming 90 days past due, the borrower must only submit a request for servicing in accordance with paragraph (a)(1) of this section, provided all other information is less than 90 days old and is based on the current production cycle. Any information 90 or more days old or not based on the current production cycle must be updated.

(e) The borrower need not submit any information under this section that already exists in the Agency's file and is still current as determined by the Agency.

(f) When jointly liable borrowers have been divorced and one has withdrawn from the farming operation, the Agency may release the withdrawing individual from liability, provided:

(1) The remaining individual submits a complete application in accordance with this section;

(2) Both parties have agreed in a divorce decree or property settlement that only the remaining individual will be responsible for all FLP loan payments;

(3) The withdrawing individual has conveyed all ownership interest in the security to the remaining individual; and

(4) The withdrawing individual does not have repayment ability and does not own any non-essential assets.

[72 FR 63316, Nov. 8, 2007, as amended at 85 FR 36693, June 17, 2020]

EDITORIAL NOTE: At 81 FR 51285, Aug. 3, 2016, § 766.102, in paragraph (b)(3)(ii), the words "subpart G of 7 CFR part 1940" were removed and the words "part 799 of this chapter" were added in their place. However, paragraph (b)(3)(ii) does not exist, and this amendment could not be incorporated.

§766.103 Borrower does not respond or does not submit a complete application.

(a) If a borrower, who is financially distressed or current, requested loan servicing and received FSA–2512, but fails to respond timely and subsequently becomes 90 days past due, the Agency will notify the borrower in accordance with §766.101(a)(2).

(b) If a borrower who is 90 days past due and received FSA–2510 or FSA–2510–IA, or is in non-monetary, or both monetary and non-monetary default and received FSA–2514, and fails to timely respond or does not submit a complete application within the 60-day timeframe, the Agency will notify the borrower by certified mail of the following:

(1) The Agency's intent to accelerate the loan; and

(2) The borrower's right to request reconsideration, mediation and appeal in accordance with 7 CFR parts 11 and 780.

[72 FR 63316, Nov. 8, 2007, as amended at 85 FR 36693, June 17, 2020]

§766.104 Borrower eligibility requirements.

(a) A borrower must meet the following eligibility requirements to be considered for primary loan servicing:

(1) The delinquency or financial distress is the result of reduced repayment ability due to one of the following circumstances beyond the borrower's control:

(i) Illness, injury, or death of a borrower or other individual who operates the farm;

(ii) Natural disaster, adverse weather, disease, or insect damage which caused severe loss of agricultural production;

(iii) Widespread economic conditions such as low commodity prices;

(iv) Damage or destruction of property essential to the farming operation; or

(v) Loss of, or reduction in, the borrower or spouse's essential non-farm income.

(2) The borrower does not have non-essential assets for which the net recovery value is sufficient to resolve the financial distress or pay the delinquent portion of the loan.

(3) If the borrower is in non-monetary default, the borrower will resolve the non-monetary default prior to closing the servicing action.

(4) The borrower has acted in good faith.

(5) Financially distressed or current borrowers requesting servicing must pay a portion of the interest due on the loans.

(6) The borrower must not be ineligible due to disqualification resulting from Federal crop insurance violation according to 7 CFR part 718.

(b) Debtors with SA only must:

(1) Be delinquent due to circumstances beyond their control;

(2) Have acted in good faith.

§766.105 Agency consideration of servicing requests.

(a) *Order in which Agency considers servicing options.* The Agency will consider loan servicing options and combinations of options to maximize loan repayment and minimize losses to the Agency. The Agency will consider loan servicing options in the following order for each eligible borrower who requests servicing:

(1) Conservation Contract, if requested;

(2) Consolidation and rescheduling or reamortization;

(3) Deferral;

(4) Writedown; and

(5) Current market value buyout.

(b) *Debt service margin.* (1) The Agency will attempt to achieve a 110 percent debt service margin for the servicing options listed in paragraphs (a)(2) through (4) of this section.

(2) If the borrower cannot develop a feasible plan with the 110 percent debt service margin, the Agency will reduce the debt service margin by one percent and reconsider all available servicing authorities. This process will be repeated until a feasible plan has been developed or it has been determined that a feasible plan is not possible with a 100 percent margin.

(3) The borrower must be able to develop a feasible plan with at least a 100 percent debt service margin to be considered for the servicing options listed in paragraphs (a)(1) through (4) of this section.

(c) *Appraisal of borrower's assets.* The Agency will obtain an appraisal on:

(1) All Agency security, non-essential assets, and real property unencumbered by the Agency that does not meet the criteria established in §766.112(b), when:

(i) A writedown is required to develop a feasible plan;

(ii) The borrower will be offered current market value buyout.

(2) The borrower's non-essential assets when their net recovery value may be adequate to bring the delinquent loans current.

§766.106 Agency notification of decision regarding a complete application.

The Agency will send the borrower notification of the Agency's decision within 60 calendar days after receiving a complete application for loan servicing. Except that when a real estate appraisal is involved, the Agency will send the borrower notification of the Agency's decision within 90 calendar days after receiving a complete application.

(a) *Notification to financially distressed or current borrowers.* (1) If the borrower can develop a feasible plan and is eligible for primary loan servicing, the Agency will offer to service the account.

(i) The borrower will have 45 days to accept the offer of servicing. After accepting the Agency's offer, the borrower must execute loan agreements and security instruments, as appropriate.

(ii) If the borrower does not accept the offer, the Agency will send the borrower another notification of the availability of loan servicing if the borrower becomes 90 days past due in accordance with §766.101(a)(2).

(2) If the borrower cannot develop a feasible plan, or is not eligible for loan servicing, the Agency will send the borrower the calculations used and the reasons for the adverse decision.

(i) The borrower may request reconsideration, mediation and appeal in accordance with 7 CFR parts 11 and 780 of this title.

(ii) The Agency will send the borrower another notification of the availability of loan servicing if the borrower

becomes 90 days past due in accordance with §766.101(a)(2).

(b) *Notification to borrowers 90 days past due or in non-monetary default.* (1) If the borrower can develop a feasible plan and is eligible for primary loan servicing, the Agency will offer to service the account.

(i) The borrower will have 45 days to accept the offer of servicing. After accepting the Agency's offer, the borrower must execute loan agreements and security instruments, as appropriate.

(ii) If the borrower does not timely accept the offer, or fails to respond, the Agency will notify the borrower of its intent to accelerate the account.

(2) If the borrower cannot develop a feasible plan, or is not eligible for loan servicing, the Agency will send the borrower notification within 15 days, including the calculations used and reasons for the adverse decision, of its intent to accelerate the account in accordance with subpart H of this part, unless the account is resolved through any of the following options:

(i) The borrower may request reconsideration, mediation or voluntary meeting of creditors, or appeal in accordance with 7 CFR parts 11 and 780.

(ii) The borrower may request negotiation of appraisal within 30 days in accordance with §766.115.

(iii) If the net recovery value of non-essential assets is sufficient to pay the account current, the borrower has 90 days to pay the account current.

(iv) The borrower, if eligible in accordance with §766.113, may buy out the loans at the current market value within 90 days.

(v) The borrower may request homestead protection if the borrower's primary residence was pledged as security by providing the information required under §766.151.

[72 FR 63316, Nov. 8, 2007, as amended at 86 FR 43392, Aug. 9, 2021]

§766.107 Consolidation and rescheduling.

(a) *Loans eligible for consolidation.* The Agency may consolidate OL loans if:

(1) The borrower meets the loan servicing eligibility requirements in §766.104;

(2) The Agency determines that consolidation will assist the borrower to repay the loans;

(3) Consolidating the loans will bring the borrower's account current or prevent the borrower from becoming delinquent;

(4) The Agency has not referred the borrower's account to OGC or the U.S. Attorney, and the Agency does not plan to refer the account to either of these two offices in the near future;

(5) The borrower is in compliance with the Highly Erodible Land and Wetland Conservation requirements of 7 CFR part 12, if applicable;

(6) The loans are not secured by real estate;

(7) The Agency holds the same lien position on each loan;

(8) The Agency has not serviced the loans for unauthorized assistance under subpart F of this part; and

(9) The loan is not currently deferred, as described in §766.109, or set-aside, as described in subpart B of this part. The Agency may consolidate loans upon cancellation of the deferral or DSA.

(b) *Loans eligible for rescheduling.* The Agency may reschedule loans made for chattel purposes, including OL, CL, SW, RL, EE, or EM if:

(1) The borrower meets the loan servicing eligibility requirements in §766.104;

(2) Rescheduling the loans will bring the borrower's account current or prevent the borrower from becoming delinquent;

(3) The Agency determines that rescheduling will assist the borrower to repay the loans;

(4) The Agency has not referred the borrower's account to OGC or the U.S. Attorney, and the Agency does not plan to refer the account to either of these two offices in the near future;

(5) The borrower is in compliance with the Highly Erodible Land and Wetland Conservation requirements of 7 CFR part 12, if applicable; and

(6) The loan is not currently deferred, as described in §766.109, or set-aside, as described in subpart B of this part. The Agency may reschedule loans upon cancellation of the deferral or DSA.

(c) *Consolidated and rescheduled loan terms.* (1) The Agency determines the repayment schedule for consolidated and rescheduled loans according to the borrower's repayment ability.

(2) Except for CL and RL loans, the repayment period cannot exceed 15 years from the date of the consolidation and rescheduling.

(3) The repayment schedule for RL loans may not exceed 7 years from the date of rescheduling.

(4) The repayment schedule for CLs may not exceed 20 years from the date of the original note or assumption agreement.

(d) *Consolidated and rescheduled loan interest rate.* The interest rate of consolidated and rescheduled loans will be as follows:

(1) The interest rate for loans made at the regular interest rate will be the lesser of:

(i) The interest rate for that type of loan on the date a complete servicing application was received;

(ii) The interest rate for that type of loan on the date of restructure; or

(iii) The lowest original loan note rate on any of the original notes being consolidated and rescheduled.

(2) The interest rate for loans made at the limited resource interest rate will be the lesser of:

(i) The limited resource interest rate for that type of loan on the date a complete servicing application was received;

(ii) The limited resource interest rate for that type of loan on the date of restructure; or

(iii) The lowest original loan note rate on any of the original notes being consolidated and rescheduled.

(3) At the time of consolidation and rescheduling, the Agency may reduce the interest rate to a limited resource rate, if available, if:

(i) The borrower meets the requirements for the limited resource interest rate; and

(ii) A feasible plan cannot be developed at the regular interest rate and maximum terms permitted in this section.

(4) Loans consolidated and rescheduled at the limited resource interest rate will be subject to annual limited resource review in accordance with §765.51 of this chapter.

(e) *Capitalizing accrued interest and adding protective advances to the loan*

317

principal. (1) The Agency capitalizes the amount of outstanding accrued interest on the loan at the time of consolidation and rescheduling.

(2) The Agency adds protective advances for the payment of real estate taxes to the principal balance at the time of consolidation and rescheduling.

(3) The borrower must resolve all other protective advances not capitalized prior to closing the servicing actions.

(f) *Installments.* If there are no deferred installments, the first installment payment under the consolidation and rescheduling will be at least equal to the interest amount which will accrue on the new principal between the date the promissory note is executed and the next installment due date.

[72 FR 63316, Nov. 8, 2007, as amended at 75 FR 54016, Sept. 3, 2010]

§ 766.108 Reamortization.

(a) *Loans eligible for reamortization.* The Agency may reamortize loans made for real estate purposes, including FO, SW, RL, SA, EE, RHF, CL, and EM if:

(1) The borrower meets the loan servicing eligibility requirements in § 766.104;

(2) Reamortization will bring the borrower's account current or prevent the borrower from becoming delinquent;

(3) The Agency determines that reamortization will assist the borrower to repay the loan;

(4) The Agency has not referred the borrower's account to OGC or the U.S. Attorney, and the Agency does not plan to refer the account to either of these two offices in the near future;

(5) The borrower is in compliance with the Highly Erodible Land and Wetland Conservation requirements of 7 CFR part 12, if applicable; and

(6) The loan is not currently deferred, as described in § 766.109, or set-aside, as described in subpart B of this part. The Agency may reamortize loans upon cancellation of the deferral or DSA.

(b) *Reamortized loan terms.* (1) Except as provided in paragraph (b)(2), the Agency will reamortize loans within the remaining term of the original loan or assumption agreement unless a feasible plan cannot be developed or debt

forgiveness will be required to develop a feasible plan.

(2) If the Agency extends the loan term, the repayment period from the original loan date may not exceed the maximum number of years for the type of loan being reamortized in paragraphs (2)(i) through (iv), or the useful life of the security, whichever is less.

(i) FO, SW, RL, EE real estate-type, and EM loans made for real estate purposes may not exceed 40 years from the date of the original note or assumption agreement.

(ii) EE real estate-type loans secured by chattels only may not exceed 20 years from the date of the original note or assumption agreement.

(iii) RHF loans may not exceed 33 years from the date of the original note or assumption agreement.

(iv) SA loans may not exceed 25 years from the date of the original Shared Appreciation note.

(v) CLs may not exceed 20 years from the date of the original note or assumption agreement.

(c) *Reamortized loan interest rate.* The interest rate will be as follows:

(1) The interest rate for loans made at the regular interest rate will be the lesser of:

(i) The interest rate for that type of loan on the date a complete servicing application was received;

(ii) The interest rate for that type of loan on the date of restructure; or

(iii) The original loan note rate of the note being reamortized.

(2) The interest rate for loans made at the limited resource interest rate will be the lesser of:

(i) The limited resource interest rate for that type of loan on the date a complete servicing application was received;

(ii) The limited resource interest rate for that type of loan on the date of restructure; or

(iii) The original loan note rate of the note being reamortized.

(3) At the time of reamortization, the Agency may reduce the interest rate to a limited resource rate, if available, if:

(i) The borrower meets the requirements for the limited resource interest rate; and

(ii) A feasible plan cannot be developed at the regular interest rate and maximum terms permitted in this section.

(4) Loans reamortized at the limited resource interest rate will be subject to annual limited resource review in accordance with § 765.51 of this chapter.

(5) SA payment agreements will be reamortized at the current SA amortization rate in effect on the date of approval or the rate on the original payment agreement, whichever is less.

(d) *Capitalizing accrued interest and adding protective advances to the loan principal.* (1) The Agency capitalizes the amount of outstanding accrued interest on the loan at the time of reamortization.

(2) The Agency adds protective advances for the payment of real estate taxes to the principal balance at the time of reamortization.

(3) The borrower must resolve all other protective advances not capitalized prior to closing the reamortization.

(e) *Installments.* If there are no deferred installments, the first installment payment under the reamortization will be at least equal to the interest amount which will accrue on the new principal between the date the promissory note is executed and the next installment due date.

[72 FR 63316, Nov. 8, 2007, as amended at 75 FR 54016, Sept. 3, 2010]

§ 766.109 Deferral.

(a) *Conditions for approving deferrals.* The Agency will only consider deferral of loan payments if:

(1) The borrower meets the loan servicing eligibility requirements in § 766.104;

(2) Rescheduling, consolidation, and reamortization of all the borrower's loans, will not result in a feasible plan with 110 percent debt service margin;

(3) The need for deferral is temporary; and

(4) The borrower develops feasible first-year deferral and post-deferral farm operating plans subject to the following:

(i) The deferral will not create excessive net cash reserves beyond that necessary to develop a feasible plan.

(ii) The Agency will consider a partial deferral if deferral of the total Agency payment would result in the borrower developing more cash availability than necessary to meet debt repayment obligations.

(b) *Deferral period.* (1) The deferral term will not exceed 5 years and will be determined based on the post-deferral plan that results in the:

(i) Greatest improvement over the first year cash available to service FLP debt;

(ii) The shortest possible deferral period.

(2) The Agency will distribute interest accrued on the deferred principal portion of the loan equally to payments over the remaining loan term after the deferral period ends.

(c) *Agency actions when borrower's repayment ability improves.* (1) If during the deferral period the borrower's repayment ability has increased to allow the borrower to make payments on the deferred loans, the borrower must make supplemental payments, as determined by the Agency. If the borrower agrees to make supplemental payments, but does not do so, the borrower will be considered to be in nonmonetary default.

(2) If the Agency determines that the borrower's improved repayment ability will allow graduation, the Agency will require the borrower to graduate in accordance with part 765, subpart C of this chapter.

(d) *Associated loan servicing.* (1) The Agency must cancel an existing deferral if the Agency approves any new primary loan servicing action.

(2) Loans deferred will also be serviced in accordance with §§ 766.107, 766.108 and 766.111, as appropriate.

§ 766.110 Conservation Contract.

(a) *General.* (1) A debtor with only SA or Non-program loans is not eligible for a Conservation Contract. However, an SA or Non-program loan may be considered for a Conservation Contract if the borrower also has program loans.

(2) A current or financially distressed borrower may request a Conservation Contract at any time prior to becoming 90 days past due.

(3) A delinquent borrower may request a Conservation Contract during the same 60-day time period in which the borrower may apply for primary

loan servicing. The borrower eligibility requirements in §766.104 will apply.

(4) A Conservation Contract may be established for conservation, recreation, and wildlife purposes.

(5) The land under a Conservation Contract cannot be used for the production of agricultural commodities during the term of the contract.

(6) Only loans secured by the real estate that will be subject to the Conservation Contract may be considered for debt reduction under this section.

(b) *Eligible lands.* The following types of lands are eligible to be considered for a Conservation Contract by the Conservation Contract review team:

(1) Wetlands or highly erodible lands; and

(2) Uplands that meet any one of the following criteria:

(i) Land containing aquatic life, endangered species, or wildlife habitat of local, State, tribal, or national importance;

(ii) Land in 100-year floodplains;

(iii) Areas of high water quality or scenic value;

(iv) Historic or cultural properties listed in or eligible for the National Register of Historic Places;

(v) Aquifer recharge areas of local, regional, State, or tribal importance;

(vi) Buffer areas necessary for the adequate protection of proposed Conservation Contract areas, or other areas enrolled in other conservation programs;

(vii) Areas that contain soils generally not suited for cultivation; or

(viii) Areas within or adjacent to Federal, State, tribal, or locally administered conservation areas.

(c) *Unsuitable acreage.* Notwithstanding paragraph (b) of this section, acreage is unsuitable for a Conservation Contract if:

(1) It is not suited or eligible for the program due to legal restrictions;

(2) It has on-site or off-site conditions that prohibit the use of the land for conservation, wildlife, or recreational purposes; or

(3) The Conservation Contract review team determines that the land is not suitable for conservation, wildlife, or recreational purposes.

(3) The Conservation Contract review team determines that the land does not provide measurable conservation, wildlife, or recreational benefits;

(4) There would be a duplication of benefits as determined by the Conservation Contract review team because the acreage is encumbered under another Federal, State, or local government program for which the borrower has been or is being compensated for conservation, wildlife, or recreation benefits;

(5) The acreage subject to the proposed Conservation Contract is encumbered under a Federal, State, or local government cost share program that is inconsistent with the purposes of the proposed Conservation Contract, or the required practices of the cost share program are not identified in the conservation management plan;

(6) The tract does not contain a legal right of way or other permanent access for the term of the contract that can be used by the Agency or its designee to carry out the contract; or

(7) The tract, including any buffer areas, to be included in a Conservation Contract is less than 10 acres.

(d) *Conservation Contract terms.* The borrower selects the term of the contract, which may be 10, 30, or 50 years.

(e) *Conservation management plan.* The Agency, with the recommendations of the Conservation Contract review team, is responsible for developing a conservation management plan. The conservation management plan will address the following:

(1) The acres of eligible land and the approximate boundaries, and

(2) A description of the conservation, wildlife, or recreation benefits to be realized.

(f) *Management authority.* The Agency has enforcement authority over the Conservation Contract. The Agency, however, may delegate contract management to another entity if doing so is in the Agency's best interest.

(g) *Limitations.* The Conservation Contract must meet the following conditions:

(1) Result in a feasible plan for current borrowers; or

(2) Result in a feasible plan with or without primary loan servicing for financially distressed or delinquent borrowers; and

(3) Improve the borrower's ability to repay the remaining balance of the loan.

(h) *Maximum debt reduction for a financially distressed or current borrower.* The amount of debt reduction by a Conservation Contract is calculated as follows:

(1) Divide the contract acres by the total acres that secure the borrower's FLP loans to determine the contract acres percentage.

$$\frac{}{\text{Contract acres}} \text{divided by} \frac{}{\text{Total acres}} = \frac{}{\text{Percent of contract acres to total acres}}$$

(2) Multiply the borrower's total unpaid FLP loan balance (principal, interest, and recoverable costs already paid by the Agency) by the percentage calculated under paragraph (h)(1) of this section to determine the amount of FLP debt that is secured by the contract acreage.

$$\frac{}{\text{Total FLP debt}} \times \frac{}{\text{Percent calculated under (h)(1)}} = \frac{}{\text{FLP debt secured by contract acres}}$$

(3) Multiply the borrower's total unpaid FLP loan balance (principal, interest, and recoverable costs already paid by the Agency) by 33 percent.

$$\frac{}{\text{Total FLP debt}} \times 33\% = \frac{}{}$$

(4) The lesser of the amounts calculated in paragraphs (h)(2) and (h)(3) of this section is the maximum amount of debt reduction for a 50-year contract.

(5) The borrower will receive 60 percent of the amount calculated in paragraph (h)(4) of this section for a 30-year contract.

$$\frac{}{\text{Result from (h)(4)}} \times 60\% = \frac{}{\text{Maximum debt reduction for a 30-year contract}}$$

(6) The borrower will receive 20 percent of the amount calculated in paragraph (h)(4) of this section for a 10-year contract.

$$\frac{}{\text{Result from (h)(4)}} \times 20\% = \frac{}{\text{Maximum debt reduction for a 10-year contract}}$$

(i) *Maximum debt reduction for a delinquent borrower.* The amount of debt reduction by a Conservation Contract is calculated as follows:

(1) Divide the contract acres by the total acres that secure the borrower's FLP loans to determine the contract acres percentage.

$$\frac{}{\text{Contract acres}} \text{ divided by } \frac{}{\text{Total acres}} = \frac{}{\text{Percent of contract acres to total acres}}$$

(2) Multiply the borrower's total unpaid FLP loan balance (principal, interest, and recoverable costs already paid by the Agency) by the percentage calculated in paragraph (i)(1) of this section to determine the amount of FLP debt that is secured by the contract acreage.

$$\frac{}{\text{Total FLP debt}} \times \frac{}{\text{Percent calculated in (i)(1)}} = \frac{}{\text{FLP debt secured by contract acres}}$$

(3) Multiply the market value of the total acres, less contributory value of any structural improvements, that secure the borrower's FLP loans by the percent calculated in paragraph (i)(1) of this section to determine the current value of the acres in the contract.

$$\frac{}{\substack{\text{Market value of total acres} \\ \text{less contributory value of} \\ \text{structural improvements}}} \times \frac{}{\text{Percent calculated in (i)(1)}} = \frac{}{\substack{\text{Market value of} \\ \text{acres in the contract}}}$$

(4) Subtract the market value of the contract acres calculated in paragraph (i)(3) of this section from the FLP debt secured by the contract acres as calculated in paragraph (i)(2) of this section.

$$\frac{}{\text{Result from (i)(2)}} - \frac{}{\text{Result from (i)(3)}} = \frac{}{\text{Difference}}$$

(5) Select the greater of the amounts calculated in either paragraphs (i)(3) and (i)(4) of this section.

(6) The lesser of the amounts calculated in paragraphs (i)(2) and (i)(5) of this section will be the maximum amount of debt reduction for a 50-year contract term.

(7) The borrower will receive 60 percent of the amount calculated in paragraph (i)(6) of this section for a 30-year contract term.

$$\frac{}{\text{Result from (i)(6)}} \times 60\% = \frac{}{\text{Maximum debt cancellation for a 30-year term}}$$

(8) The borrower will receive 20 percent of the amount calculated in paragraph (i)(6) of this section for a 10-year contract term.

$$\frac{}{\text{Result from (i)(6)}} \times 20\% = \frac{}{\text{Maximum debt cancellation for a 10-year term}}$$

(j) *Conservation Contract Agreement.* The borrower must sign the Conservation Contract Agreement establishing the contract's terms and conditions.

(k) *Transferring title to land under Conservation Contract.* If the borrower or any subsequent landowner transfers title to the property, the Conservation Contract will remain in effect for the duration of the contract term.

(1) *Borrower appeals of technical decisions.* Borrower appeals of the Natural Resources Conservation Service's (NRCS) technical decisions made in connection with a Conservation Contract, will be handled in accordance with applicable NRCS regulations. Other aspects of the denial of a conservation contract may be appealed in accordance with 7 CFR parts 11 and 780.

(m) *Subordination.* For real estate with a Conservation Contract:

(1) Subordination will be required for all liens that are in a prior lien position to the Conservation Contract.

(2) The Agency will not subordinate Conservation Contracts to liens of other lenders or other Governmental entities.

(n) *Breach of Conservation Contract.* If the borrower or a subsequent owner of the land under the Conservation Contract fails to comply with any of its provisions, the Agency will declare the Conservation Contract breached. If the Conservation Contract is breached, the borrower or subsequent owner of the land must restore the land to be in compliance with the Conservation Contract and all terms of the conservation management plan within 90 days. If this cure is not completed, the Agency will take the following actions:

(1) For borrowers who have or had a loan in which debt was exchanged for the Conservation Contract and breach the Conservation Contract, the Agency may reinstate the debt that was cancelled, plus interest to the date of payment at the rate of interest in the promissory note, and assess liquidated damages in the amount of 25 percent of the debt cancelled, plus any actual expenses incurred by the Agency in enforcing the terms of the Conservation Contract. The borrower's account will be considered in non-monetary default; and

(2) Subsequent landowners who breach the Conservation Contract must pay the Agency the amount of the debt cancelled when the contract was executed, plus interest at the non-program interest rate to the date of payment, plus liquidated damages in the amount of 25 percent of the cancelled debt, plus any actual expenses incurred by the Agency in enforcing the terms of the Conservation Contract.

[72 FR 63316, Nov. 8, 2007, as amended at 78 FR 65532, Nov. 1, 2013]

§ 766.111 **Writedown.**

(a) *Borrower eligibility.* The Agency will only consider a writedown if the borrower:

(1) Meets the eligibility criteria in §766.104;

(2) Is delinquent;

(3) Has not previously received debt forgiveness on any FLP direct loan; and

(4) Complies with the Highly Erodible Land and Wetland Conservation requirements of 7 CFR part 12.

(b) *Conditions.* The conditions required for approval of writedown are:

(1) Rescheduling, consolidation, reamortization, deferral or some combination of these options on all of the borrower's loans would not result in a feasible plan with a 110 percent debt service margin. If a feasible plan is achieved with a debt service margin of 101 percent or more, the Agency will permit a borrower to accept a non-writedown servicing offer and waive the right to a writedown offer when the

writedown offer will require additional time and appraisals to fully develop. If after obtaining an appraisal a feasible plan is achieved with and without a writedown and the borrower meets all the eligibility requirements, both options will be offered and the borrower may choose one option.

(2) The present value of the restructured loan must be greater than or equal to the net recovery value of Agency security and any non-essential assets.

(3) The writedown amount, excluding debt reduction received through Conservation Contract, does not exceed $300,000.

(4) A borrower who owns real estate must execute an SAA in accordance with § 766.201.

(c) *Associated loan servicing.* Loans written down will also be serviced in accordance with §§ 766.107 and 766.108, as appropriate.

[72 FR 63316, Nov. 8, 2007, as amended at 86 FR 43392, Aug. 9, 2021]

§ 766.112 Additional security for restructured loans.

(a) If the borrower is delinquent prior to restructuring, the borrower, and all entity members in the case of an entity, must execute and provide to the Agency a lien on all of their assets, except as provided in paragraph (b) of this section, when the Agency is servicing a loan.

(b) The Agency will take the best lien obtainable on all assets the borrower owns, except:

(1) When taking a lien on such property will prevent the borrower from obtaining credit from other sources;

(2) When the property could have significant environmental problems or costs as described in subpart G of 7 CFR part 1940;

(3) When the Agency cannot obtain a valid lien;

(4) When the property is subsistence livestock, cash, special collateral accounts the borrower uses for the farming operation, retirement accounts, personal vehicles necessary for family living, household contents, or small equipment such as hand tools and lawn mowers; or

(5) When a contractor holds title to a livestock or crop enterprise, or the borrower manages the enterprise under a share lease or share agreement.

EDITORIAL NOTE: At 81 FR 51285, Aug. 3, 2016, § 766.112, in paragraph (a)(6), the words "subpart G of 7 CFR part 1940" were removed and the words "part 799 of this chapter" were added in their place. However, paragraph (b)(3)(ii) does not exist, and this amendment could not be incorporated.

§ 766.113 Buyout of loan at current market value.

(a) *Borrower eligibility.* A delinquent borrower may buy out the borrower's FLP loans at the current market value of the loan security, including security not in the borrower's possession, and all non-essential assets if:

(1) The borrower has not previously received debt forgiveness on any other FLP direct loan;

(2) The borrower has acted in good faith;

(3) The borrower does not have non-essential assets for which the net recovery value is sufficient to pay the account current;

(4) The borrower is unable to develop a feasible plan through primary loan servicing programs or a Conservation Contract, if requested;

(5) The present value of the restructured loans is less than the net recovery value of Agency security;

(6) The borrower pays the amount required in a lump sum without guaranteed or direct credit from the Agency; and

(7) The amount of debt forgiveness does not exceed $300,000.

(b) *Buyout time frame.* After the Agency offers current market value buyout of the loan, the borrower has 90 days from the date of Agency notification to pay that amount.

§ 766.114 State-certified mediation or voluntary meeting of creditors.

(a) A borrower who is unable to develop a feasible plan but is otherwise eligible for primary loan servicing may request:

(1) State-certified mediation; or

(2) Voluntary meeting of creditors when a State does not have a certified mediation program.

(b) Any negotiation of the Agency's appraisal must be completed before State-certified mediation or voluntary meeting of creditors.

§ 766.115 Challenging the Agency appraisal.

(a) A borrower considered for primary loan servicing who does not agree with the Agency's appraisal of the borrower's assets may:

(1) Obtain a USPAP compliant technical appraisal review prepared by a State Certified General Appraiser of the Agency's appraisal and provide it to the Agency prior to reconsideration or the appeal hearing;

(2) Obtain an independent appraisal completed in accordance with § 761.7 as part of the appeals process. The borrower must:

(i) Pay for this appraisal;

(ii) Choose which appraisal will be used in Agency calculations, if the difference between the two appraisals is five percent or less.

(3) Negotiate the Agency's appraisal by obtaining a second appraisal.

(i) If the difference between the two appraisals is five percent or less, the borrower will choose the appraisal to be used in Agency calculations.

(ii) If the difference between the two appraisals is greater than five percent, the borrower may request a third appraisal. The Agency and the borrower will share the cost of the third appraisal equally. The average of the two appraisals closest in value will serve as the final value.

(iii) A borrower may request a negotiated appraisal only once in connection with an application for primary loan servicing.

(iv) The borrower may not appeal a negotiated appraisal.

(b) If the appraised value of the borrower's assets change as a result of the challenge, the Agency will reconsider its previous primary loan servicing decision using the new appraisal value.

(c) If the appeal process results in a determination that the borrower is eligible for primary loan servicing, the Agency will use the information utilized to make the appeal decision, unless stated otherwise in the appeal decision letter.

[72 FR 63316, Nov. 8, 2007, as amended at 78 FR 65532, Nov. 1, 2013]

§§ 766.116–766.150   [Reserved]

APPENDIX A TO SUBPART C OF PART 766—FSA–2510, NOTICE OF AVAILABILITY OF
LOAN SERVICING TO BORROWERS WHO ARE 90 DAYS PAST DUE

This appendix contains the notification (form letter) that the Farm Service Agency will send to borrowers who are at least 90 days past due on their loan payments. It provides information about the loan servicing that is available to the borrower. As stated below on the notification, the borrower is to respond within 60 days from receiving the notification (see §766.101(b)(2) and (d)(2) for the requirements). The notification is provided here as required by 7 U.S.C. 1981d.

This form is available electronically.

FSA-2510   U.S. DEPARTMENT OF AGRICULTURE   Position 4
(April 2020)   Farm Service Agency

NOTICE OF AVAILABILITY OF LOAN SERVICING
TO BORROWERS WHO ARE 90 DAYS PAST DUE

Date

[Borrower's Name]
[Borrower Name/Address]   [MAILING INSTRUCTIONS]
[Borrower Address]
[City, State, Zip Code]

This notice informs you that you are seriously delinquent with your Farm Loan Programs (FLP) loan payment and notifies you of options that may be available to you. The Agency's primary loan servicing programs, Conservation Contract Program, current market value buyout, Homestead Protection Program, and debt settlement programs may help you repay your loan or retain your farm property and settle your FLP debt.

How to apply

To apply, you must complete, where applicable, and provide all items required in paragraph (f), within 60 days of the date you receive this notice.

Help in responding to this notice

The servicing options available to you may become complicated. You may need help to understand them and their impact on your operation. You may want to ask an attorney to help you or there are organizations that give free or low-cost advice to farmers. You may contact your State Department of Agriculture or the U.S. Department of Agriculture (USDA) Extension Service for available services in your State.

Note: Agency employees cannot recommend a particular attorney or organization.

Who will decide if you qualify?

After you submit a complete application, the Agency will determine if you meet all eligibility requirements and can develop a farm operating plan that shows that you can pay all debts and expenses.

What happens if you do not bring the account current or apply within 60 days?

The Agency will accelerate your loans if you do not bring your account current or timely apply for loan servicing. This means the Agency will take legal action to collect all the money you owe to the Agency under FLP. After acceleration of your loan accounts, the Agency will start foreclosure proceedings. The Agency will repossess or take legal action to sell your real estate, personal property, crops, livestock, equipment, or any other assets in which the Agency has a security interest. The Agency will also obtain and file judgments against you and your property or refer your account to the Department of the Treasury for collection.

Included with this notice you will find information on:

- (a) Primary loan servicing programs;
- (b) Conservation Contract Program;
- (c) Current market value buyout;
- (d) Homestead Protection Program;
- (e) Debt settlement programs;
- (f) Forms, documentation, and information needed to apply;
- (g) How to get copies of Agency handbooks and forms;
- (h) Reconsideration, mediation, and appeal to NAD;
- (i) Challenging the Agency appraisal;
- (j) Acceleration and foreclosure;
- (k) The right not to be discriminated against.

(a) Primary Loan Servicing Programs

Eligibility

You must meet the following eligibility requirements to obtain primary loan servicing:

(1) You cannot repay your FLP debt due to one of the following circumstances beyond your control:

    (i) Illness, injury, or death of a borrower or other individual who operates the farm;

    (ii) Natural disaster, adverse weather, disease, or insect damage which caused severe loss of agricultural production;

    (iii) Widespread economic conditions such as low commodity prices;

    (iv) Damage or destruction of property essential to the farming operation; or

    (v) Loss of, or reduction in, your or your spouse's essential non-farm income.

(2) You do not have non-essential assets for which the net recovery value is sufficient to pay the delinquent portion of the loan. The Agency cannot write down or write off debt that you could pay with the value of your equity in these assets.

(3) If you are in non-monetary default as a result of non-compliance with the Agency's loan agreements, you must resolve the non-monetary default prior to closing the servicing action.

(4) You must have acted in good faith in all past dealings with the Agency and in accordance with your loan agreements.

Time limits

If the Agency determines that you can develop a feasible plan and are eligible for primary loan servicing, you will have 45 days from the date you receive the Agency's offer to accept loan servicing.

Lien requirements

If you are offered loan servicing and accept the offer, you must agree to give the Agency a lien on your other assets and you must provide this lien at closing.

Youth Loans

If you have a Youth Loan, it is not eligible for debt writedown, current market value buyout, or limited resource interest rates, but can be rescheduled or deferred. This has no effect on any other loans you may have with the Agency.

Loan consolidation, rescheduling, and reamortization

In loan consolidation, the unpaid principal and interest of two or more operating loans can be combined into one larger operating loan.

In loan rescheduling, the repayment schedule may be changed to cure the delinquency and give you new terms to repay loans made for equipment, livestock, or annual operating purposes.

In loan reamortization, the repayment schedule may be changed to cure the delinquency and give you a new schedule of repayment on loans made for real estate purposes.

When loans are consolidated, rescheduled or reamortized, accrued interest becomes principal and interest is charged on the new principal balance. The interest rate will be the lesser of:

(1) The interest rate for that type of loan on the date a complete servicing application was received;
(2) The interest rate for that type of loan on the date of restructure; or
(3) The lowest original loan note rate on any of the original notes being restructured.

In addition, the Agency will consider the maximum loan terms. This means that operating loans, including carry over annual operating and family living expenses may have repayment terms of up to 15 years.

Limited resource interest rate

Limited resource interest rates are available for certain types of loans. If you have existing loans which are not at the limited resource rate, and a limited resource rate is available, the Agency will consider reducing the rate of the loans. The limited resource interest rate can be as low as five percent, however, this rate may change depending on what it costs the Government to borrow money.

For information about current interest rates, contact this office.

Loan deferral

Partial or full payments of principal and interest may be temporarily delayed for up to 5 years. You will only be considered for loan deferral if the loan servicing programs discussed above will not allow you to pay all essential family living and farm operating expenses, maintain your property, and pay your debts.

You must be able to show through a farm operating plan that you are unable to pay all essential family living and farm operating expenses, maintain your property, and pay your debts. The farm operating plan must also show that you will be able to pay your full installment at the end of the deferral period.

The interest that accrues during the deferral period must be paid in yearly payments for the rest of the loan term after the deferral period ends.

Debt writedown

Debt writedown can reduce the principal and interest on your loan. The Agency offers a writedown only when the loan servicing programs discussed above and the Conservation Contract Program, if requested, will not result in a feasible plan. To receive debt writedown, the value of your restructured loan must be equal to or greater than the recovery value to the Agency from foreclosure and repossession of your security property.

The recovery value is the market value of:

(1) The collateral pledged as security for FLP loans minus expenses (such as the sale costs, attorneys' fees, management costs, taxes, and payment of prior liens) on the collateral that the Agency would have to pay if it foreclosed, or repossessed, and sold the collateral;

(2) Any collateral that is not in your possession and has not been released for sale by the Agency in writing; and

(3) Any other non-essential assets you may own.

A qualified appraiser determines the value of the collateral and any other assets you own. You may receive a writedown only if you have not previously received any form of debt forgiveness on any other FLP direct loan. The maximum amount of debt that can be written down on all direct loans is $300,000.

Shared Appreciation Agreement

If you own real estate and receive a debt writedown, you must sign a Shared Appreciation Agreement. The term of the agreement is 5 years. Under the terms of the agreement you must repay all or a part of the amount written down at the maturity of your Shared Appreciation Agreement if your real estate collateral increased in value. Payment of shared appreciation will be required prior to the maturity of your Shared Appreciation Agreement if you:

(1) Sell or convey the real estate;

(2) Stop farming;

(3) Pay off your entire FLP debt; or

(4) Have your FLP accounts accelerated by the Agency.

If any of these events occur within the first 4 years of the agreement, you will have to pay 75 percent of the increase in value of the real estate. If any of these events occur after the fourth anniversary of the agreement, or if the Shared Appreciation Agreement matures without having previously been fully triggered, you will have to pay only 50 percent of the increase in value. You will not have to pay more than the amount of the debt written down.

Time limits

To buyout your FLP debt at the current market value, you must pay the Agency within 90 days of the date you receive the offer.

Method of payment

To buyout your FLP debt at the current market value, you must pay by cashier's check or U.S. Treasury check. The Agency will not make or guarantee a loan for this purpose.

(b) Conservation Contract Program

You may request a Conservation Contract to protect highly erodible land, wetlands, or wildlife habitats located on your real estate property that serves as security for your FLP debt. In exchange for such contract, the Agency would reduce your FLP debt. The amount of land left after the contract must be sufficient to continue your farming operation.

(c) Current Market Value Buyout

If the analysis of your debt shows that you cannot achieve a feasible plan even if the present value of your FLP debt is reduced to the value of the security, the Agency may offer you buyout of your FLP debt. You would pay the market value of all FLP security and non-essential assets, minus any prior liens. The market value is determined by a current appraisal completed by a qualified appraiser. In exchange, your loans would be satisfied.

Limits

To receive a current market value buyout offer:

(1) You must not have previously received any form of debt forgiveness from the Agency on any other direct FLP loan;
(2) The maximum debt to be written off with buyout does not exceed $300,000; and
(3) You must not have non-essential assets with a net recovery value sufficient to pay your account current.

Eligibility

To qualify, you must prove that:

(1) You cannot repay your delinquent FLP debt due to circumstances beyond your control; and
(2) You have acted in good faith in all past dealings with the Agency and in accordance with your loan agreements.

(d) Homestead Protection Program

Under the Homestead Protection Program, you may repurchase your primary residence, certain outbuildings, and up to 10 acres of land. If you cannot pay cashier's check or U.S. Treasury check or Agency financing is not available, you may lease your primary residence. The lease will include an option for you to purchase the property you lease.

This program may apply when primary loan servicing, the Conservation Contract Program, or current market value buyout is not available or not accepted.

You must agree to give the Agency title to your land at the time the Agency signs the Homestead Protection Agreement with you. The Agency will compute the costs of taking title including the cost of paying other creditors with outstanding liens on the property. The Agency will take title only if it can obtain a positive recovery.

Farm Service Agency, USDA   **Pt. 766, Subpt. C, App. A**

FSA-2510 (April 2020)   Page 6 of 10

Eligibility requirements

(1) Your gross annual income from the farming operation must have been similar to other comparable operations in your area in at least two of the last 6 years.
(2) Sixty percent (60%) of your gross annual income in at least two of the last 6 years must have come from the farming operation.
(3) You must have lived in your homestead property for 6 years immediately before your application. If you had to leave for less than 12 months during the 6-year period and you had no control over the circumstances, you may still qualify.
(4) You must be the owner of the property immediately prior to the Agency obtaining title.

Property restrictions and easements

The Agency may place restrictions or easements on your property which restrict your use if the property is located in a special area or has special characteristics. These restrictions and easements will be placed in leases and in deeds on properties containing wetlands, floodplains, endangered species, wild and scenic rivers, historic and cultural properties, coastal barriers, and highly erodible lands.

Leasing the homestead property

(1) You must pay rent to the Agency to lease the property determined eligible for homestead protection. The rent the Agency charges will be similar to comparable property in your area.
(2) You must maintain the property in good condition during the term of the lease.
(3) You may lease the property for up to 5 years but no less than 3 years.
(4) You cannot sublease the property.
(5) If you do not make the rental payments to the Agency, the Agency will cancel the lease and take legal action to force you to leave.
(6) Lease payments are not applied toward the final purchase price of the property.

Purchasing the homestead protection property

You can repurchase your homestead property at market value at any time during the lease. The market value of the property will be decided by a qualified appraiser and will reflect the value of the land after any placement of a restriction or easement such as a wetland conservation easement.

(e) Debt Settlement Programs

You can apply for debt settlement at any time; however, these programs are usually used only after it has been determined that primary loan servicing programs and the Conservation Contract Program cannot help you. Under the debt settlement programs, the debt you owe the Agency under FLP may be settled for less than the amount you owe. These programs are subject to the discretion of the Agency and are not a matter of entitlement or right. If you do not have any Agency security, you may apply for debt settlement only. If you do not apply, or do not receive approval of a debt settlement request, any FLP loan account balance remaining after liquidation of loan collateral will be forwarded to the Department of Treasury for cross-servicing and administrative wage garnishment.

Settlement alternatives

Settlement alternatives include:

(1) Compromise: A lump-sum payment of less than the total FLP debt owed;
(2) Adjustment: Two or more payments of less than the total amount owed to the Agency. Payments can be spread out over a maximum of 5 years if the Agency determines you will be able to make the payments as they become due; and
(3) Cancellation: Satisfaction of Agency debt without payment.

Note: The Agency will not finance these alternatives.

Processing and requirements

If you sell loan collateral, you must apply the proceeds from the sale to your FLP loans before you can be considered for debt settlement. In the case of compromise or adjustment you may keep your collateral, if you pay the Agency the market value of your collateral along with any additional amount the Agency determines you are able to pay.

Debt amounts which are collectible through administrative offset, judgment, or by the Department of the Treasury will not be settled through debt settlement procedures. You must certify that you do not have assets or income in addition to what you stated in your application. If you qualify, your application must also be approved by the State Executive Director or the Administrator, depending on the amount of the debt to be settled.

(f) Forms, documentation, and information needed to apply

A complete application for primary loan servicing must include items (1) through (10). Additional information is required as noted if you want to be considered for the Conservation Contract Program or debt settlement programs. If you need help to complete the required forms, you may request an Agency official to assist you. The forms for requirements (1) through (8) and (11) are included with this package.

(1) FSA-2511, "Borrower Response to Notice of the Availability of Loan Servicing-For Borrowers who Received FSA-2510," signed by all borrowers.
(2) FSA-2001, "Request for Direct Loan Assistance."
(3) FSA-2002, "Three Year Financial History," or other financial records, including copies of your income tax returns and any supporting documents, for each of the 3 years immediately preceding the year of application or the years you have been farming, whichever is less and if not already in the Agency case file. If your copies of tax returns are not readily available, you can obtain copies from the Internal Revenue Service.
(4) FSA-2003, "Three Year Production History," or any other format that provides production and expense history for crops, livestock, livestock products, etc., for each of the 3 years immediately preceding the year of application or the years you have been farming, whichever is less and if not already in the Agency case file. You must be able to support this information with farm records.
(5) FSA-2004, "Authorization to Release Information." The Agency will use this form to verify your debts and assets, as well as your non-farm income.
(6) FSA-2005, "Creditor List." The Agency will use this form to verify your debts. Any debts less than $1,000 can be verified by a credit report. If debts of $1,000 or more appear on your credit report and the creditor is not listed on FSA 2005, the application cannot be considered complete.
(7) FSA-2037, "Farm Business Plan Worksheet - Balance Sheet." In the case of an entity, the entity and all entity members must provide current financial statements.
(8) FSA-2038, "Farm Business Plan Worksheet -Projected/Actual Income and Expenses," or other acceptable farm operating plan.

Farm Service Agency, USDA   **Pt. 766, Subpt. C, App. A**

FSA-2510 (April 2020)   Page 8 of 10

(9) AD-1026, "Highly Erodible Land Conservation (HELC) and Wetland Conservation (WC) Certification." You will be required to complete this form if the one you have on file does not reflect all the land you own and lease.

(10) SCS-CPA-026, "Highly Erodible Land and Wetland Conservation Determination." This form must be obtained from and completed by the Natural Resources Conservation Service office, if not already on file with the Agency.

(11) FSA-2732, "Debt Settlement Application." Complete this form only if you wish to apply for debt settlement. You must also comply with any Agency request for additional information needed to process a debt settlement request.

(12) If you are applying for a Conservation Contract, a map or aerial photo of your farm identifying the portion of the land and approximate number of acres to be considered.

<u>Divorced spouses</u>

If you are an FLP borrower who has left the farming operation due to divorce, you may request release of liability. To be released of liability after a divorce, you must present the Agency with the following within 60 days of receiving this notice:

(1) A divorce decree or property settlement document which states the remaining party will be responsible for all repayment to the Agency;

(2) Evidence that you have conveyed your ownership interest in FLP security to the remaining party; and

(3) Evidence that you do not have any repayment ability for the FLP loan through cash, income, or other non-essential assets.

The Agency will make a determination on your request and will inform you of the decision within 60 days of receiving your request.

If you are not released of liability, you will need to include all of your relevant financial information if applying for primary loan servicing, homestead protection, or debt settlement programs.

(g) How to get copies of Agency handbooks and forms

Copies of the forms for requirements (f)(1) through (f)(8) and (f)(11) have been included in this package. You may obtain copies of Agency handbooks, which include the pertinent regulations, describing available programs or additional copies of forms from this office.

(h) Reconsideration, mediation, and appeal to NAD

Reconsideration, mediation, and appeal rights pursuant to 7 CFR parts 780 and 11, respectively, will be provided to you if the Agency makes an adverse decision on your request for loan servicing or prior to acceleration of your account.

<u>Reconsideration</u> – according to FSA's appeal procedures in 7 CFR part 780.

<u>Mediation</u> – according to FSA's appeal procedures in 7 CFR part 780.

<u>Appeal to NAD</u> – according to the NAD appeal procedures in 7 CFR part 11.

(i) Challenging the Agency appraisal

If you timely submit a complete application for primary loan servicing, but disagree with the appraisal used by the Agency for processing your request, you may 1.) obtain a USPAP compliant technical appraisal review by a State Certified General Appraiser of the Agency appraisal and submit it to the Agency prior to reconsideration or an appeal hearing, 2.) obtain an independent appraisal, and 3.) possibly negotiate the appraised value based on the specifics of the two appraisals.

If this applies to you, the Agency will provide additional information in the notification letter advising you of the Agency's decision concerning your loan servicing application.

(j) Acceleration and foreclosure

If you do not appeal an adverse determination, if you appeal, but are denied relief on appeal, or if you do not otherwise resolve your delinquency, the Agency will accelerate your loan accounts and demand payment of the entire debt. You may prevent Agency foreclosure on the loan collateral if, with prior Agency approval, you:

(1) Sell all loan collateral for not less than its market value and apply all proceeds to your creditors in order of lien priority.
(2) Transfer the collateral to someone else and have that person assume all or part of your FLP debt.
(3) Transfer the collateral to the Agency.

If any of these options result in payment of less than you owe, you may apply for debt settlement, even if you applied before and were denied. However, applications for debt settlement filed after the 60-day time period provided in this notice will not delay acceleration, administrative offset, and foreclosure.

If the Agency determines that you cannot qualify for debt settlement, you can:

(1) Pay your FLP loan accounts current;
(2) Pay your FLP loan accounts in full;
(3) Request reconsideration, mediation or appeal.

If your real estate security contains your primary residence and becomes inventory property of the Agency, homestead protection rights will be provided.

(k) The right not to be discriminated against

The Federal Equal Credit Opportunity Act prohibits creditors from discriminating against credit applicants on the basis of race, color, religion, national origin, sex, marital status, age (provided the applicant has the capacity to enter into a binding contract); because all or part of the applicant's income derives from any public assistance program; or because the applicant has in good faith exercised any right under the Consumer Credit Protection Act. The Federal agency that administers compliance with this law is the Federal Trade Commission, Equal Credit Opportunity, Washington, D.C. 20580.

The servicing programs described by this Notice are subject to applicable Agency regulations published at 7 CFR Part 766.

For more information or if you have any questions, please contact [this office or the specific office name]at [County Office Address] or telephone [phone number].

1A. Authorized Agency Official Name	1B. Signature	1C. Title

[85 FR 36693, June 17, 2020]

APPENDIX B TO SUBPART C OF PART 766—FSA–2510–IA, NOTICE OF AVAILABILITY OF LOAN SERVICING TO BORROWERS WHO ARE 90 DAYS PAST DUE (FOR USE IN IOWA ONLY)

This appendix contains the notification (form letter) that the Farm Service Agency will send to borrowers with loans in Iowa who are at least 90 days past due on their loan payments. It provides information about the loan servicing that is available to the borrower. As stated below on the notification, the borrower is to respond within 60 days from receiving the notification (see § 766.101(b)(2) and (d)(2) for the requirements). The notification is provided here as required by 7 U.S.C. 1981d.

This form is available electronically.

FSA-2510-IA   U.S. DEPARTMENT OF AGRICULTURE   Position 4
(April 2020)   Farm Service Agency

**NOTICE OF AVAILABILITY OF LOAN SERVICING
TO BORROWERS WHO ARE 90 DAYS PAST DUE
(For Use in Iowa Only)**

Date

[Borrower's Name]
[Borrower Name/Address]   [MAILING INSTRUCTIONS]
[Borrower Address]
[City, State, Zip Code]

This notice informs you that you are seriously delinquent with your Farm Loan Programs (FLP) loan payment and notifies you of options that may be available to you. The Agency's primary loan servicing programs, Conservation Contract Program, current market value buyout, Homestead Protection Program, and debt settlement programs may help you repay your loan or retain your farm property and settle your FLP debt.

How to apply

To apply, you must complete, where applicable, and provide all items required in paragraph (f), within 60 days of the date you receive this notice.

Help in responding to this notice

The servicing options available to you may become complicated. You may need help to understand them and their impact on your operation. You may want to ask an attorney to help you or there are organizations that give free or low-cost advice to farmers. You may contact your State Department of Agriculture or the U.S. Department of Agriculture (USDA) Extension Service for available services in your State.

Note: Agency employees cannot recommend a particular attorney or organization.

Who will decide if you qualify?

After you submit a complete application, the Agency will determine if you meet all eligibility requirements and can develop a farm operating plan that shows that you can pay all debts and expenses.

What happens if you do not bring the account current or apply within 60 days?

The Agency will accelerate your loans if you do not bring your account current or timely apply for loan servicing. This means the Agency will take legal action to collect all the money you owe to the Agency under FLP. After acceleration of your loan accounts, the Agency will start foreclosure proceedings. The Agency will repossess or take legal action to sell your real estate, personal property, crops, livestock, equipment, or any other assets in which the Agency has a security interest. The Agency will also obtain and file judgments against you and your property or refer your account to the Department of the Treasury for collection.

In accordance with Federal civil rights law and U.S. Department of Agriculture (USDA) civil rights regulations and policies, the USDA, its Agencies, offices, and employees, and institutions participating in or administering USDA programs are prohibited from discriminating based on race, color, national origin, religion, sex, gender identity (including gender expression), sexual orientation, disability, age, marital status, family/parental status, income derived from a public assistance program, political beliefs, or reprisal or retaliation for prior civil rights activity, in any program or activity conducted or funded by USDA (not all bases apply to all programs). Remedies and complaint filing deadlines vary by program or incident.

Persons with disabilities who require alternative means of communication for program information (e.g., Braille, large print, audiotape, American Sign Language, etc.) should contact the responsible Agency or USDA's TARGET Center at (202) 720-2600 (voice and TTY) or contact USDA through the Federal Relay Service at (800) 877-8339. Additionally, program information may be made available in languages other than English.

To file a program discrimination complaint, complete the USDA Program Discrimination Complaint Form, AD-3027, found online at http://www.ascr.usda.gov/complaint_filing_cust.html and at any USDA office or write a letter addressed to USDA and provide in the letter all of the information requested in the form. To request a copy of the complaint form, call (866) 632-9992. Submit your completed form or letter to USDA by: (1) mail: U.S. Department of Agriculture Office of the Assistant Secretary for Civil Rights 1400 Independence Avenue, SW Washington, D.C. 20250-9410; (2) fax: (202) 690-7442; or (3) email: program.intake@usda.gov. USDA is an equal opportunity provider, employer, and lender.

Paperwork Reduction Act: This information collection is exempted from the Paperwork Reduction Act as specified in 5 CFR 1320.4(a)(2) because the form is used when FSA conducts administrative action against individuals or debtors.

Included with this notice you will find information on:

- (a) Primary loan servicing programs;
- (b) Conservation Contract Program;
- (c) Current market value buyout;
- (d) Homestead Protection Program;
- (e) Debt settlement programs;
- (f) Forms, documentation, and information needed to apply;
- (g) How to get copies of Agency handbooks and forms;
- (h) Reconsideration, mediation, and appeal to NAD;
- (i) Challenging the Agency appraisal;
- (j) Acceleration and foreclosure;
- (k) The right not to be discriminated against.

(a) **Primary Loan Servicing Programs**

Eligibility

You must meet the following eligibility requirements to obtain primary loan servicing:

(1) You cannot repay your FLP debt due to one of the following circumstances beyond your control:

- (i) Illness, injury, or death of a borrower or other individual who operates the farm;
- (ii) Natural disaster, adverse weather, disease, or insect damage which caused severe loss of agricultural production;
- (iii) Widespread economic conditions such as low commodity prices;
- (iv) Damage or destruction of property essential to the farming operation; or
- (v) Loss of, or reduction in, your or your spouse's essential non-farm income.

(2) You do not have non-essential assets for which the net recovery value is sufficient to pay the delinquent portion of the loan. The Agency cannot write down or write off debt that you could pay with the value of your equity in these assets.

(3) If you are in non-monetary default as a result of non-compliance with the Agency's loan agreements, you must resolve the non-monetary default prior to closing the servicing action.

(4) You must have acted in good faith in all past dealings with the Agency and in accordance with your loan agreements.

Time limits

If the Agency determines that you can develop a feasible plan and are eligible for primary loan servicing, you will have 45 days from the date you receive the Agency's offer to accept loan servicing.

Lien requirements

If you are offered loan servicing and accept the offer, you must agree to give the Agency a lien on your other assets and you must provide this lien at closing.

Youth Loans

If you have a Youth Loan, it is not eligible for debt writedown, current market value buyout, or limited resource interest rates, but can be rescheduled or deferred. This has no effect on any other loans you may have with the Agency.

Farm Service Agency, USDA   **Pt. 766, Subpt. C, App. B**

FSA-2510-IA (April 2020)   Page 3 of 10

Loan consolidation, rescheduling, and reamortization

In loan consolidation, the unpaid principal and interest of two or more operating loans can be combined into one larger operating loan.

In loan rescheduling, the repayment schedule may be changed to cure the delinquency and give you new terms to repay loans made for equipment, livestock, or annual operating purposes.

In loan reamortization, the repayment schedule may be changed to cure the delinquency and give you a new schedule of repayment on loans made for real estate purposes.

When loans are consolidated, rescheduled or reamortized, accrued interest becomes principal and interest is charged on the new principal balance. The interest rate will be the lesser of:

(1) The interest rate for that type of loan on the date a complete servicing application was received;
(2) The interest rate for that type of loan on the date of restructure; or
(3) The lowest original loan note rate on any of the original notes being restructured.

In addition, the Agency will consider the maximum loan terms. This means that operating loans, including carry over annual operating and family living expenses may have repayment terms of up to 15 years.

Limited resource interest rate

Limited resource interest rates are available for certain types of loans. If you have existing loans which are not at the limited resource rate, and a limited resource rate is available, the Agency will consider reducing the rate of the loans. The limited resource interest rate can be as low as five percent, however, this rate may change depending on what it costs the Government to borrow money.

For information about current interest rates, contact this office.

Loan deferral

Partial or full payments of principal and interest may be temporarily delayed for up to 5 years. You will only be considered for loan deferral if the loan servicing programs discussed above will not allow you to pay all essential family living and farm operating expenses, maintain your property, and pay your debts.

You must be able to show through a farm operating plan that you are unable to pay all essential family living and farm operating expenses, maintain your property, and pay your debts. The farm operating plan must also show that you will be able to pay your full installment at the end of the deferral period.

The interest that accrues during the deferral period must be paid in yearly payments for the rest of the loan term after the deferral period ends.

Debt writedown

Debt writedown can reduce the principal and interest on your loan. The Agency offers a writedown only when the loan servicing programs discussed above and the Conservation Contract Program, if requested, will not result in a feasible plan. To receive debt writedown, the value of your restructured loan must be equal to or greater than the recovery value to the Agency from foreclosure and repossession of your security property.

The recovery value is the market value of:

(1) The collateral pledged as security for FLP loans minus expenses (such as the sale costs, attorneys' fees, management costs, taxes, and payment of prior liens) on the collateral that the Agency would have to pay if it foreclosed, or repossessed, and sold the collateral;
(2) Any collateral that is not in your possession and has not been released for sale by the Agency in writing; and
(3) Any other non-essential assets you may own.

A qualified appraiser determines the value of the collateral and any other assets you own. You may receive a writedown only if you have not previously received any form of debt forgiveness on any other FLP direct loan. The maximum amount of debt that can be written down on all direct loans is $300,000.

Shared Appreciation Agreement

If you own real estate and receive a debt writedown, you must sign a Shared Appreciation Agreement. The term of the agreement is 5 years. Under the terms of the agreement you must repay all or a part of the amount written down at the maturity of your Shared Appreciation Agreement if your real estate collateral increased in value. Payment of shared appreciation will be required prior to the maturity of your Shared Appreciation Agreement if you:

(1) Sell or convey the real estate;
(2) Stop farming;
(3) Pay off your entire FLP debt; or
(4) Have your FLP accounts accelerated by the Agency.

If any of these events occur within the first 4 years of the agreement, you will have to pay 75 percent of the increase in value of the real estate. If any of these events occur after the fourth anniversary of the agreement, or if the Shared Appreciation Agreement matures without having previously been fully triggered, you will have to pay only 50 percent of the increase in value. You will not have to pay more than the amount of the debt written down.

Time limits

To buyout your FLP debt at the current market value, you must pay the Agency within 90 days of the date you receive the offer.

Method of payment

To buyout your FLP debt at the current market value, you must pay by cashier's check or U.S. Treasury check. The Agency will not make or guarantee a loan for this purpose.

Farm Service Agency, USDA   **Pt. 766, Subpt. C, App. B**

FSA-2510-IA (April 2020)   Page 5 of 10

(b) Conservation Contract Program

You may request a Conservation Contract to protect highly erodible land, wetlands, or wildlife habitats located on your real estate property that serves as security for your FLP debt. In exchange for such contract, the Agency would reduce your FLP debt. The amount of land left after the contract must be sufficient to continue your farming operation.

(c) Current Market Value Buyout

If the analysis of your debt shows that you cannot achieve a feasible plan even if the present value of your FLP debt is reduced to the value of the security, the Agency may offer you buyout of your FLP debt. You would pay the market value of all FLP security and non-essential assets, minus any prior liens. The market value is determined by a current appraisal completed by a qualified appraiser. In exchange, your loans would be satisfied.

Limits

To receive a current market value buyout offer:

(1) You must not have previously received any form of debt forgiveness from the Agency on any other direct FLP loan;
(2) The maximum debt to be written off with buyout does not exceed $300,000; and
(3) You must not have non-essential assets with a net recovery value sufficient to pay your account current.

Eligibility

To qualify, you must prove that:

(1) You cannot repay your delinquent FLP debt due to circumstances beyond your control; and
(2) You have acted in good faith in all past dealings with the Agency and in accordance with your loan agreements.

(d) Homestead Protection Program

Under the Homestead Protection Program, you may repurchase your primary residence, certain outbuildings, and up to 40 acres of land. If you cannot pay cashier's check or U.S. Treasury check or Agency financing is not available, you may lease your primary residence. The lease will include an option for you to purchase the property you lease.

This program may apply when primary loan servicing, the Conservation Contract Program, or current market value buyout is not available or not accepted.

You must agree to give the Agency title to your land at the time the Agency signs the Homestead Protection Agreement with you. The Agency will compute the costs of taking title including the cost of paying other creditors with outstanding liens on the property. The Agency will take title only if it can obtain a positive recovery.

Eligibility requirements

(1) Your gross annual income from the farming operation must have been similar to other comparable operations in your area in at least two of the last 6 years.
(2) Sixty percent (60%) of your gross annual income in at least two of the last 6 years must have come from the farming operation.
(3) You must have lived in your homestead property for 6 years immediately before your application. If you had to leave for less than 12 months during the 6-year period and you had no control over the circumstances, you may still qualify.
(4) You must be the owner of the property immediately prior to the Agency obtaining title.

Property restrictions and easements

The Agency may place restrictions or easements on your property which restrict your use if the property is located in a special area or has special characteristics. These restrictions and easements will be placed in leases and in deeds on properties containing wetlands, floodplains, endangered species, wild and scenic rivers, historic and cultural properties, coastal barriers, and highly erodible lands.

Leasing the homestead property

(1) You must pay rent to the Agency to lease the property determined eligible for homestead protection. The rent the Agency charges will be similar to comparable property in your area.
(2) You must maintain the property in good condition during the term of the lease.
(3) You may lease the property for up to 5 years but no less than 3 years.
(4) You cannot sublease the property.
(5) If you do not make the rental payments to the Agency, the Agency will cancel the lease and take legal action to force you to leave.
(6) Lease payments are not applied toward the final purchase price of the property.

Purchasing the homestead protection property

You can repurchase your homestead property at market value at any time during the lease. The market value of the property will be decided by a qualified appraiser and will reflect the value of the land after any placement of a restriction or easement such as a wetland conservation easement.

(e) Debt Settlement Programs

You can apply for debt settlement at any time; however, these programs are usually used only after it has been determined that primary loan servicing programs and the Conservation Contract Program cannot help you. Under the debt settlement programs, the debt you owe the Agency under FLP may be settled for less than the amount you owe. These programs are subject to the discretion of the Agency and are not a matter of entitlement or right. If you do not have any Agency security, you may apply for debt settlement only. If you do not apply, or do not receive approval of a debt settlement request, any FLP loan account balance remaining after liquidation of loan collateral will be forwarded to the Department of Treasury for cross-servicing and administrative wage garnishment.

Settlement alternatives

Settlement alternatives include:

(1) Compromise: A lump-sum payment of less than the total FLP debt owed;

(2) Adjustment: Two or more payments of less than the total amount owed to the Agency. Payments can be spread out over a maximum of 5 years if the Agency determines you will be able to make the payments as they become due; and

(3) Cancellation: Satisfaction of Agency debt without payment.

Note: The Agency will not finance these alternatives.

Processing and requirements

If you sell loan collateral, you must apply the proceeds from the sale to your FLP loans before you can be considered for debt settlement. In the case of compromise or adjustment you may keep your collateral, if you pay the Agency the market value of your collateral along with any additional amount the Agency determines you are able to pay.

Debt amounts which are collectible through administrative offset, judgment, or by the Department of the Treasury will not be settled through debt settlement procedures. You must certify that you do not have assets or income in addition to what you stated in your application. If you qualify, your application must also be approved by the State Executive Director or the Administrator, depending on the amount of the debt to be settled.

(f) Forms, documentation, and information needed to apply

A complete application for primary loan servicing must include items (1) through (10). Additional information is required as noted if you want to be considered for the Conservation Contract Program or debt settlement programs. If you need help to complete the required forms, you may request an Agency official to assist you. The forms for requirements (1) through (8) and (11) are included with this package.

(1) FSA-2511, "Borrower Response to Notice of the Availability of Loan Servicing-For Borrowers who Received FSA-2510-IA," signed by all borrowers.

(2) FSA-2001, "Request for Direct Loan Assistance."

(3) FSA-2002, "Three Year Financial History," or other financial records, including copies of your income tax returns and any supporting documents, for each of the 3 years immediately preceding the year of application or the years you have been farming, whichever is less and if not already in the Agency case file. If your copies of tax returns are not readily available, you can obtain copies from the Internal Revenue Service.

(4) FSA-2003, "Three Year Production History," or any other format that provides production and expense history for crops, livestock, livestock products, etc., for each of the 3 years immediately preceding the year of application or the years you have been farming, whichever is less and if not already in the Agency case file. You must be able to support this information with farm records.

(5) FSA-2004, "Authorization to Release Information." The Agency will use this form to verify your debts and assets, as well as your non-farm income.

(6) FSA-2005, "Creditor List." The Agency will use this form to verify your debts. Any debts less than $1,000 can be verified by a credit report. If debts of $1,000 or more appear on your credit report and the creditor is not listed on FSA 2005, the application cannot be considered complete.

(7) FSA-2037, "Farm Business Plan Worksheet - Balance Sheet." In the case of an entity, the entity and all entity members must provide current financial statements.

(8) FSA-2038, "Farm Business Plan Worksheet -Projected/Actual Income and Expenses," or other acceptable farm operating plan.

(9) AD-1026, "Highly Erodible Land Conservation (HELC) and Wetland Conservation (WC) Certification." You will be required to complete this form if the one you have on file does not reflect all the land you own and lease.

(10) SCS-CPA-026, "Highly Erodible Land and Wetland Conservation Determination." This form must be obtained from and completed by the Natural Resources Conservation Service office, if not already on file with the Agency.

(11) FSA-2732, "Debt Settlement Application." Complete this form only if you wish to apply for debt settlement. You must also comply with any Agency request for additional information needed to process a debt settlement request.

(12) If you are applying for a Conservation Contract, a map or aerial photo of your farm identifying the portion of the land and approximate number of acres to be considered.

Divorced spouses

If you are an FLP borrower who has left the farming operation due to divorce, you may request release of liability. To be released of liability after a divorce, you must present the Agency with the following within 60 days of receiving this notice:

(1) A divorce decree or property settlement document which states the remaining party will be responsible for all repayment to the Agency;

(2) Evidence that you have conveyed your ownership interest in FLP security to the remaining party; and

(3) Evidence that you do not have any repayment ability for the FLP loan through cash, income, or other non-essential assets.

The Agency will make a determination on your request and will inform you of the decision within 60 days of receiving your request.

If you are not released of liability, you will need to include all of your relevant financial information if applying for primary loan servicing, homestead protection, or debt settlement programs.

(g) How to get copies of Agency handbooks and forms

Copies of the forms for requirements (f)(1) through (f)(8) and (f)(11) have been included in this package. You may obtain copies of Agency handbooks, which include the pertinent regulations, describing available programs or additional copies of forms from this office.

(h) Reconsideration, mediation, and appeal to NAD

Reconsideration, mediation, and appeal rights pursuant to 7 CFR parts 780 and 11, respectively, will be provided to you if the Agency makes an adverse decision on your request for loan servicing or prior to acceleration of your account.

Reconsideration – according to FSA's appeal procedures in 7 CFR part 780.

Mediation – according to FSA's appeal procedures in 7 CFR part 780.

Appeal to NAD – according to the NAD appeal procedures in 7 CFR part 11.

(i) Challenging the Agency appraisal

If you timely submit a complete application for primary loan servicing, but disagree with the appraisal used by the Agency for processing your request, you may 1.) obtain a USPAP compliant technical appraisal review by a State Certified General Appraiser of the Agency appraisal and submit it to the Agency prior to reconsideration or an appeal hearing, 2.) obtain an independent appraisal, and 3.) possibly negotiate the appraised value based on the specifics of the two appraisals.

If this applies to you, the Agency will provide additional information in the notification letter advising you of the Agency's decision concerning your loan servicing application.

(j) Acceleration and foreclosure

If you do not appeal an adverse determination, if you appeal, but are denied relief on appeal, or if you do not otherwise resolve your delinquency, the Agency will accelerate your loan accounts and demand payment of the entire debt. You may prevent Agency foreclosure on the loan collateral if, with prior Agency approval, you:

(1). Sell all loan collateral for not less than its market value and apply all proceeds to your creditors in order of lien priority.
(2) Transfer the collateral to someone else and have that person assume all or part of your FLP debt.
(3) Transfer the collateral to the Agency.

If any of these options result in payment of less than you owe, you may apply for debt settlement, even if you applied before and were denied. However, applications for debt settlement filed after the 60-day time period provided in this notice will not delay acceleration, administrative offset, and foreclosure.

If the Agency determines that you cannot qualify for debt settlement, you can:

(1) Pay your FLP loan accounts current;
(2) Pay your FLP loan accounts in full;
(3) Request reconsideration, mediation or appeal.

If your real estate security contains your primary residence and becomes inventory property of the Agency, homestead protection rights will be provided.

(k) The right not to be discriminated against

The Federal Equal Credit Opportunity Act prohibits creditors from discriminating against credit applicants on the basis of race, color, religion, national origin, sex, marital status, age (provided the applicant has the capacity to enter into a binding contract); because all or part of the applicant's income derives from any public assistance program; or because the applicant has in good faith exercised any right under the Consumer Credit Protection Act. The Federal agency that administers compliance with this law is the Federal Trade Commission, Equal Credit Opportunity, Washington, D.C. 20580.

The servicing programs described by this Notice are subject to applicable Agency regulations published at 7 CFR Part 766.

For more information or if you have any questions, please contact [this office or the specific office name]at [County Office Address] or telephone [phone number].

1A. Authorized Agency Official Name	1B. Signature	1C. Title

[85 FR 36703, June 17, 2020]

Subpart D—Homestead Protection Program

§ 766.151 Applying for Homestead Protection.

(a) *Pre-acquisition*—(1) *Notification.* If the borrower requested primary loan servicing but cannot develop a feasible plan, the Agency will notify the borrower of any additional information needed to process the homestead protection request. The borrower must provide this information within 30 days of Agency notification.

(2) *Borrower does not respond.* If the borrower does not timely provide the information requested, the Agency will deny the homestead protection request and provide appeal rights.

(3) *Application requirements.* A complete application for homestead protection will include:

(i) Updates to items required under § 766.102;

(ii) Information required under § 766.353; and

(iii) Identification of land and buildings to be considered.

(b) *Post-acquisition*—(1) *Notification.* After the Agency acquires title to the real estate property, the Agency will notify the borrower of the availability of homestead protection. The borrower must submit a complete application within 30 days of Agency notification.

(2) *Borrower does not respond.* If the borrower does not respond to the Agency notice, the Agency will dispose of the property in accordance with 7 CFR part 767.

(3) *Application requirements.* A complete application for homestead protection will include:

(i) Updates to items required under § 766.102; and

(ii) Identification of land and buildings to be considered.

§ 766.152 Eligibility.

(a) *Property.* (1) The principal residence and the adjoining land of up to 10 acres, must have served as real estate security for the FLP loan and may include existing farm service buildings. Homestead protection does not apply if the FLP loans were secured only by chattels.

(2) The applicant may propose a homestead protection site. Any proposed site is subject to Agency approval.

(3) The proposed homestead protection site must meet all State and local requirements for division into a separate legal lot.

(4) Where voluntary conveyance of the property to the Agency is required to process the homestead protection request, the Agency will process any request for voluntary conveyance according to § 766.353.

(b) *Applicant.* To be eligible for homestead protection, the applicant:

(1) Must be the owner, or former owner from whom the Agency acquired title of the property pledged as security for an FLP loan. For homestead protection purposes, an owner or former owner includes:

(i) A member of an entity who is or was personally liable for the FLP loan secured by the homestead protection property when the applicant or entity held fee title to the property; or

(ii) A member of an entity who is or was personally liable for the FLP loan that possessed and occupied a separate dwelling on the security property;

(2) Must have earned gross farm income commensurate with:

(i) The size and location of the farm; and

(ii) The local agricultural conditions in at least 2 calendar years during the 6-year period immediately preceding the calendar year in which the applicant applied for homestead protection;

(3) Must have received 60 percent of gross income from farming in at least two of the 6 years immediately preceding the year in which the applicant applied for homestead protection;

(4) Must have lived in the home during the 6-year period immediately preceding the year in which the applicant applied for homestead protection. The applicant may have left the home for not more than 12 months if it was due to circumstances beyond their control;

(5) Must demonstrate sufficient income to make rental payments on the homestead property for the term of the lease, and maintain the property in good condition. The lessee will be responsible for any normal maintenance; and

(6) Must not be ineligible due to disqualification resulting from Federal

crop insurance violation according to 7 CFR part 718.

§766.153 Homestead Protection transferability.

Homestead protection rights are not transferable or assignable, unless the eligible party dies or becomes legally incompetent, in which case the homestead protection rights may be transferred to the spouse only, upon the spouse's agreement to comply with the terms and conditions of the lease.

§766.154 Homestead Protection leases.

(a) *General.* (1) The Agency may approve a lease-purchase agreement on the appropriate Agency form subject to obtaining title to the property.

(2) If a third party obtains title to the property:

(i) The applicant and the property are no longer eligible for homestead protection;

(ii) The Agency will not implement any outstanding lease-purchase agreement.

(3) The borrower may request homestead protection for property subject to third party redemption rights. In such case, homestead protection will not begin until the Agency obtains title to the property.

(b) *Lease terms and conditions.* (1) The amount of rent will be based on equivalent rents charged for similar residential properties in the area in which the dwelling is located.

(2) All leases will include an option to purchase the homestead protection property as described in paragraph (c) of this section.

(3) The lease term will not be less than 3 years and will not exceed 5 years.

(4) The lessee must agree to make lease payments on time and maintain the property.

(5) The lessee must cooperate with Agency efforts to sell the remaining portion of the farm.

(c) *Lease-purchase options.* (1) The lessee may exercise in writing the purchase option and complete the homestead protection purchase at any time prior to the expiration of the lease provided all lease payments are current.

(2) If the lessee is a member of a socially disadvantaged group, the lessee may designate a member of the lessee's immediate family (that is, parent, sibling, or child) (designee) as having the right to exercise the option to purchase.

(3) The purchase price is the market value of the property when the option is exercised as determined by a current appraisal obtained by the Agency.

(4) The lessee or designee may purchase homestead protection property with cash or other credit source.

(5) The lessee or designee may receive Agency program or non-program financing provided:

(i) The lessee or designee has not received previous debt forgiveness;

(ii) The Agency has funds available to finance the purchase of homestead protection property;

(iii) The lessee or designee demonstrates an ability to repay such an FLP loan; and

(iv) The lessee or designee is otherwise eligible for the FLP loan.

(d) *Lease terminations.* The Agency may terminate the lease if the lessee does not cure any lease defaults within 30 days of Agency notification.

(e) *Appraisal of homestead protection property.* The Agency will use an appraisal obtained within six months from the date of the application for considering homestead protection. If a current appraisal does not exist, the applicant will select an independent real estate appraiser from a list of appraisers approved by the Agency.

[72 FR 63316, Nov. 8, 2007, as amended at 76 FR 5058, Jan. 28, 2011]

§766.155 Conflict with State law.

If there is a conflict between a borrower's homestead protection rights and any provisions of State law relating to redemption rights, the State law prevails.

§§766.156–766.200 [Reserved]

Subpart E—Servicing Shared Appreciation Agreements and Net Recovery Buyout Agreements

§766.201 Shared Appreciation Agreement.

(a) *When a SAA is required.* The Agency requires a borrower to enter into a

SAA with the Agency covering all real estate security when the borrower:

(1) Owns any real estate that serves or will serve as loan security; and

(2) Accepts a writedown in accordance with § 766.111.

(b) *When SAA is due.* The borrower must repay the calculated amount of shared appreciation after a term of 5 years from the date of the writedown, or earlier if:

(1) The borrower sells or conveys all or a portion of the Agency's real estate security, unless real estate is conveyed upon the death of a borrower to a spouse who will continue farming;

(2) The borrower repays or satisfies all FLP loans;

(3) The borrower ceases farming; or

(4) The Agency accelerates the borrower's loans.

§ 766.202 Determining the shared appreciation due.

(a) The value of the real estate security at the time of maturity of the SAA (market value) will be the appraised value of the security at the highest and best use, less the increase in the value of the security resulting from capital improvements added during the term of the SAA (contributory value). The market value of the real estate security property will be determined based on a current appraisal completed within the previous 18 months in accordance with § 761.7 of this chapter, and subject to the following:

(1) Prior to completion of the appraisal, the borrower will identify any capital improvements that have been added to the real estate security since the execution of the SAA.

(2) The appraisal must specifically identify the contributory value of capital improvements made to the real estate security during the term of the SAA to make deductions for that value.

(3) For calculation of shared appreciation recapture, the contributory value of capital improvements added during the term of the SAA will be deducted from the market value of the property. Such capital improvements must also meet at least one of the following criteria:

(i) It is the borrower's primary residence. If the new residence is affixed to the real estate security as a replacement for a residence which existed on the security property when the SAA was originally executed, or, the living area square footage of the original residence was expanded, only the value added to the real property by the new or expanded portion of the original residence (if it added value) will be deducted from the market value.

(ii) It is an improvement to the real estate with a useful life of over one year and is affixed to the property, the following conditions must be met:

(A) The item must have been capitalized and not taken as an annual operating expense on the borrower's Federal income tax returns. The borrower must provide copies of appropriate tax returns to verify that capital improvements claimed for shared appreciation recapture reduction are capitalized.

(B) If the new item is affixed to the real estate as a replacement for an item that existed on the real estate at the time the SAA was originally executed, only the value added by the new item will be deducted from the market value.

(b) In the event of a partial sale, an appraisal of the property being sold may be required to determine the market value at the time the SAA was signed if such value cannot be obtained through another method.

[72 FR 63316, Nov. 8, 2007, as amended at 86 FR 43392, Aug. 9, 2021]

§ 766.203 Payment of recapture.

(a) The borrower must pay on the due date or 30 days from Agency notification, whichever is later:

(1) Seventy-five percent of the appreciation in the real estate security if the agreement is triggered within 4 years or less from the date of the writedown; or

(2) Fifty percent of such appreciation if the agreement is triggered more than 4 years from the date of the writedown or when the agreement matures.

(b) If the borrower sells a portion of the security, the borrower must pay shared appreciation only on the portion sold. Shared appreciation on the remaining portion will be due in accordance with paragraph (a) of this section.

(c) The amount of recapture cannot exceed the amount of the debt written off through debt writedown.

§766.204 Amortization of recapture.

(a) The Agency will amortize the recapture into a Shared Appreciation Payment Agreement provided the borrower:

(1) Has not ceased farming and the borrower's account has not been accelerated;

(2) Provides a complete application in accordance with §764.51(b), by the recapture due date or within 60 days of Agency notification of the amount of recapture due, whichever is later;

(3) Is unable to pay the recapture and cannot obtain funds from any other source;

(4) Develops a feasible plan that includes repayment of the shared appreciation amount;

(5) Provides a lien on all assets, except those listed in §766.112(b); and

(6) Signs loan agreements and security instruments as required.

(b) If the borrower later becomes delinquent or financially distressed, reamortization of the Shared Appreciation Payment Agreement can be considered under subpart C of this part.

§766.205 Shared Appreciation Payment Agreement rates and terms.

(a) The interest rate for Shared Appreciation Payment Agreements is the Agency's SA amortization rate.

(b) The term of the Shared Appreciation Payment Agreement is based on the borrower's repayment ability and the useful life of the security. The term will not exceed 25 years.

§766.206 Net Recovery Buyout Recapture Agreement.

(a) *Servicing existing Net Recovery Buyout Recapture Agreements.* Prior to July 3, 1996, the Agency was authorized to offer borrowers buy out their loans at the net recovery value. A Net Recovery Buyout Agreement was required for borrowers who bought out their loans at the net recovery value. The Agency services existing Net Recovery Buyout Recapture Agreements as described in this section.

(b) *Requirements and terms.* (1) The term of a Net Recovery Buyout Recapture Agreement is 10 years. Net Recovery Buyout Recapture Agreements are secured by a lien on the former borrower's real estate.

(2) If the former borrower sells or conveys real estate within the 10-year term, the former borrower must repay the Agency the lesser of:

(i) The market value of the real estate parcel at the time of sale or conveyance, as determined by an Agency appraisal, minus the portion of the recovery value of the real estate paid to the Agency in the buyout;

(ii) The market value of the real estate parcel at the time of the sale or conveyance, as determined by an Agency appraisal, minus:

(A) The unpaid balance of prior liens at the time of the sale or conveyance; and

(B) The net recovery value of the real estate the borrower paid to the Agency in the buyout if this amount has not been accounted for as a prior lien;

(iii) The total amount of the FLP debt the Agency wrote off for loans secured by real estate.

(3) If the former borrower does not pay the amount due, the Agency will liquidate the Net Recovery Buyout account in accordance with subpart H of this part.

(4) If the former borrower does not sell or convey the real estate within the 10-year term, no recapture is due.

§§766.207–766.250 [Reserved]

Subpart F—Unauthorized Assistance

§766.251 Repayment of unauthorized assistance.

(a) Except where otherwise specified, the borrower is responsible for repaying any unauthorized assistance in full within 90 days of Agency notice. The Agency may reverse any unauthorized loan servicing actions, when possible.

(b) The borrower has the opportunity to meet with the Agency to discuss or refute the Agency's findings.

§ 766.252 Unauthorized assistance resulting from submission of false information.

A borrower is ineligible for continued Agency assistance if the borrower, or a third party on the borrower's behalf, submits information to the Agency that the borrower knows to be false.

§ 766.253 Unauthorized assistance resulting from submission of inaccurate information by borrower or Agency error.

(a) *Borrower options.* (1) The borrower may repay the amount of the unauthorized assistance in a lump sum within 90 days of Agency notice.

(2) If the borrower is unable to repay the entire amount in a lump sum, the Agency will accept partial repayment of the unauthorized assistance within 90 days of Agency notice to the extent of the borrower's ability to repay.

(3) If the borrower is unable to repay all or part of the unauthorized amount, the loan will be converted to a Non-program loan under the following conditions:

(i) The borrower did not provide false information;

(ii) It is in the interest of the Agency;

(iii) The debt will be subject to the interest rate for Non-program loans;

(iv) The debt will be serviced as a Non-program loan;

(v) The term of the Non-program loan will be as short as feasible, but in no case will exceed:

(A) The remaining term of the FLP loan;

(B) Twenty-five (25) years for real estate loans; or

(C) The life of the security for chattel loans.

(b) *Borrower refusal to pay.* If the borrower is able to pay the unauthorized assistance amount but refuses to do so, the Agency will notify the borrower of the availability of loan servicing in accordance with subpart C of this part.

§§ 766.254-766.300   [Reserved]

Subpart G—Loan Servicing For Borrowers in Bankruptcy

§ 766.301 Notifying borrower in bankruptcy of loan servicing.

If a borrower files for bankruptcy, the Agency will provide written notification to the borrower's attorney with a copy to the borrower as follows:

(a) *Borrower not previously notified.* The Agency will provide notice of all loan servicing options available under subpart C of this part, if the borrower has not been previously notified of these options.

(b) *Borrower with prior notification.* If the borrower received notice of all loan servicing options available under subpart C of this part prior to the time of bankruptcy filing but all loan servicing was not completed, the Agency will provide notice of any remaining loan servicing options available.

§ 766.302 Loan servicing application requirements for borrowers in bankruptcy.

(a) *Borrower not previously notified.* To be considered for loan servicing, the borrower or borrower's attorney must sign and return the appropriate response form and any forms or information requested by the Agency within 60 days of the date of receipt of Agency notice on loan servicing options.

(b) *Borrower previously notified.* To be considered for continued loan servicing, the borrower or borrower's attorney must sign and return the appropriate response form and any forms or information requested by the Agency within the greater of:

(1) Sixty days after the borrower's attorney received the notification of any remaining loan servicing options; or

(2) The remaining time from the Agency's previous notification of all servicing options that the Agency suspended when the borrower filed bankruptcy.

(c) *Court approval.* The borrower is responsible for obtaining court approval prior to exercising any available servicing rights.

§766.303 Processing loan servicing requests from borrowers in bankruptcy.

(a) *Considering borrower requests for servicing.* Any request for servicing is the borrower's acknowledgment that the Agency will not interfere with any rights or protections under the Bankruptcy Code and its automatic stay provisions.

(b) *Borrowers with confirmed bankruptcy plans.* If a plan is confirmed before servicing and any appeal is completed under 7 CFR part 11, the Agency will complete the servicing or appeals process and may consent to a post-confirmation modification of the plan if it is consistent with the Bankruptcy Code and subpart C of this part, as appropriate.

(c) *Chapter 7 borrowers.* A borrower filing for bankruptcy under chapter 7 of the Bankruptcy Code may not receive primary loan servicing unless the borrower reaffirms the entire FLP debt. A borrower who filed chapter 7 does not have to reaffirm the debt in order to be considered for homestead protection.

§§766.304–766.350 [Reserved]

Subpart H—Loan Liquidation

§766.351 Liquidation.

(a) *General.* (1) When a borrower cannot or will not meet a loan obligation, the Agency will consider liquidating the borrower's account in accordance with this subpart.

(2) The Agency will charge protective advances against the borrower's account as necessary to protect the Agency's interests during liquidation in accordance with §765.203 of this chapter.

(3) When no surviving family member or third party assumes or repays a deceased borrower's loan in accordance with part 765, subpart J, of this chapter, or when the estate does not otherwise fully repay or sell loan security to repay a deceased borrower's FLP loans, the Agency will liquidate the security as quickly as possible in accordance with State and local requirements.

(b) *Liquidation for Program borrowers.* (1) If the borrower does not apply, does not accept, or is not eligible for pri-

mary loan servicing, conservation contract, market value buyout or homestead protection, and all administrative appeals are concluded, the Agency will accelerate the borrower's account in accordance with §§766.355 and 766.356, as appropriate.

(2) Borrowers may voluntarily liquidate their security in accordance with §§766.352, 766.353 and 766.354. In such case, the Agency will:

(i) Not delay involuntary liquidation action.

(ii) Notify the borrower in accordance with subpart C of this part, prior to acting on the request for voluntary liquidation, if the conditions of paragraph (b)(1) of this section have not been met.

(c) *Liquidation for Non-program borrowers.* If a borrower has both program and Non-program loans, the borrower's account will be handled in accordance with paragraph (b) of this section. If a borrower with only Non-program loans is in default, the borrower may liquidate voluntarily, subject to the following:

(1) The Agency may delay involuntary liquidation actions when in the Agency's financial interest for a period not to exceed 60 days.

(2) The borrower must obtain the Agency's consent prior to the sale of the property.

(3) If the borrower will not pay the Agency in full, the minimum sales price must be the market value of the property as determined by the Agency.

(4) The Agency will accept a conveyance offer only when it is in the Agency's financial interest.

(5) If a Non-program borrower does not cure the default, or cannot or will not voluntarily liquidate, the Agency will accelerate the loan.

§766.352 Voluntary sale of real property and chattel.

(a) *General.* A borrower may voluntarily sell real property or chattel security to repay FLP debt in lieu of involuntary liquidation if all applicable requirements of this section are met. Partial dispositions are handled in accordance with part 765, subparts G and H, of this chapter.

(1) The borrower must sell all real property and chattel that secure FLP

debt until the debt is paid in full or until all security has been liquidated.

(2) The Agency must approve the sale and approve the use of proceeds.

(3) The sale proceeds are applied in order of lien priority, except that proceeds may be used to pay customary costs appropriate to the transaction provided:

(i) The costs are reasonable in amount;

(ii) The borrower is unable to pay the costs from personal funds or have the purchaser pay;

(iii) The costs must be paid to complete the sale;

(iv) Costs are not for postage and insurance of the note while in transit when required for the Agency to present the promissory note to the recorder to obtain a release of a portion of the real property from the mortgage.

(4) The Agency will approve the sale of property when the proceeds do not cover the borrower's full debt only if:

(i) The sales price must be equal to or greater than the market value of the property; and

(ii) The sale is in the Agency's financial interest.

(5) If an unpaid loan balance remains after the sale, the Agency will continue to service the loan in accordance with part 761, subpart F of this chapter and part 3 of this title.

(b) *Voluntary sale of chattel.* If the borrower complies with paragraph (a) of this section, the borrower may sell chattel security by:

(1) Public sale if the borrower obtains the agreement of lienholders as necessary to complete the public sale; or

(2) Private sale if the borrower:

(i) Sells all of the security for not less than the market value;

(ii) Obtains the agreement of lienholders as necessary to complete the sale;

(iii) Has a buyer who is ready and able to purchase the property; and

(iv) Obtains the Agency's agreement for the sale.

[72 FR 63316, Nov. 8, 2007, as amended at 85 FR 36713, June 17, 2020]

§ 766.353 **Voluntary conveyance of real property.**

(a) *Requirements for conveying real property.* The borrower must supply the Agency with the following:

(1) An Agency application form;

(2) A current financial statement. If the borrower is an entity, all entity members must provide current financial statements;

(3) Information on present and future income and potential earning ability;

(4) A warranty deed or other deed acceptable to the Agency;

(5) A resolution approved by the governing body that authorizes the conveyance in the case of an entity;

(6) Assignment of all leases to the Agency. The borrower must put all oral leases in writing;

(7) Title insurance or title record for the security, if available;

(8) Complete debt settlement application in accordance with subpart B of part 761, subpart F of this chapter before, or in conjunction with, the voluntary conveyance offer if the value of the property to be conveyed is less than the FLP debt; and

(9) Any other documentation required by the Agency to evaluate the request.

(b) *Conditions for conveying real property.* The Agency will accept voluntary conveyance of real property by a borrower if:

(1) Conveyance is in the Agency's financial interest;

(2) The borrower conveys all real property securing the FLP loan; and

(3) The borrower has received prior notification of the availability of loan servicing in accordance with subpart C of this part.

(c) *Prior and junior liens.* (1) The Agency will pay prior liens to the extent consistent with the Agency's financial interest.

(2) Before conveyance, the borrower must pay or obtain releases of all junior liens, real estate taxes, judgments, and other assessments. If the borrower is unable to pay or obtain a release of the liens, the Agency may attempt to negotiate a settlement with the lienholder if it is in the Agency's financial interest.

(d) *Charging and crediting the borrower's account.* (1) The Agency will charge the borrower's account for all recoverable costs incurred in connection with a conveyance.

(2) The Agency will credit the borrower's account for the amount of the market value of the property less any prior liens, or the debt, whichever is less. In the case of an American Indian borrower whose loans are secured by real estate located within the boundaries of a Federally recognized Indian reservation, however, the Agency will credit the borrower's account with the greater of the market value of the security or the borrower's FLP debt.

(e) *Right of possession.* After voluntary conveyance, the borrower or former owner retains no statutory, implied, or inherent right of possession to the property beyond those rights under an approved lease-purchase agreement executed according to §766.154 or required by State law.

[72 FR 63316, Nov. 8, 2007, as amended at 85 FR 36713, June 17, 2020]

§766.354 Voluntary conveyance of chattel.

(a) *Requirements for conveying chattel.* The borrower must supply the Agency with the following:

(1) An Agency application form;

(2) A current financial statement. If the borrower is an entity, all entity members must provide current financial statements;

(3) Information on present and future income and potential earning ability;

(4) A bill of sale including each item and titles to all vehicles and equipment, as applicable;

(5) A resolution approved by the governing body that authorizes the conveyance in the case of an entity borrower;

(6) Complete debt settlement application in accordance with part 761, subpart F of this chapter before, or in conjunction with, the voluntary conveyance offer if the value of the property to be conveyed is less than the FLP debt.

(b) *Conditions for conveying chattel.* The Agency will accept conveyance of chattel only if:

(1) The borrower has made every possible effort to sell the property voluntarily;

(2) The borrower can convey the chattel free of other liens;

(3) The conveyance is in the Agency's financial interest;

(4) The borrower conveys all chattel securing the FLP loan; and

(5) The borrower has received prior notification of the availability of loan servicing in accordance with subpart C of this part.

(c) *Charging and crediting the borrower's account.* (1) The Agency will charge the borrower's account for all recoverable costs incurred in connection with the conveyance.

(2) The Agency will credit the borrower's account in the amount of the market value of the chattel.

[72 FR 63316, Nov. 8, 2007, as amended at 85 FR 36713, June 17, 2020]

§766.355 Acceleration of loans.

(a) *General.* (1) The Agency accelerates loans in accordance with this section, unless:

(i) State law imposes separate restrictions on accelerations;

(ii) The borrower is American Indian, whose real estate is located on an Indian reservation.

(2) The Agency accelerates all of the borrower's loans at the same time, regardless of whether each individual loan is delinquent or not.

(3) All borrowers must receive prior notification in accordance with subpart C of this part, except for borrowers who fail to graduate in accordance with §766.101(a)(8).

(b) *Time limitations.* The borrower has 30 days from the date of the Agency acceleration notice to pay the Agency in full.

(c) *Borrower options.* The borrower may:

(1) Pay cash;

(2) Transfer the security to a third party in accordance with part 765, subpart I of this chapter;

(3) Sell the security property in accordance with §766.352; or

(4) Voluntarily convey the security to the Agency in accordance with §§766.353 and 766.354, as appropriate.

(d) *Partial payments.* The Agency may accept a payment that does not cover the unpaid balance of the accelerated loan if the borrower is in the process of selling security, unless acceptance of the payment would reverse the acceleration.

(e) *Failure to satisfy the debt.* The Agency will liquidate the borrower's account in accordance with §766.357 if the borrower does not pay the account in full within the time period specified in the acceleration notice.

§766.356 Acceleration of loans to American Indian borrowers.

(a) *General.* (1) The Agency accelerates loans to American Indian borrowers whose real estate is located on an Indian reservation in accordance with this section, unless State law imposes separate restrictions on accelerations.

(2) The Agency accelerates all of the borrower's loans at the same time, regardless of whether each individual loan is delinquent or not.

(3) All borrowers must receive prior notification in accordance with subpart C of this part, except for borrowers who fail to graduate in accordance with §766.101(a)(8).

(4) At the time of acceleration, the Agency will notify the borrower and the Tribe that has jurisdiction over the Indian reservation of:

(i) The possible outcomes of a foreclosure sale and the potential impacts of those outcomes on rights established under paragraphs (a)(4)(ii) and (iii) of this section;

(ii) The priority for purchase of the property acquired by the Agency through voluntary conveyance or foreclosure;

(iii) Transfer of acquired property to the Secretary of the Interior if the priority of purchase of the property established under paragraph (a)(4)(ii) of this section is not exercised.

(b) *Borrower options.* The Agency will notify an American Indian borrower of the right to:

(1) Request the Tribe, having jurisdiction over the Indian reservation in which the real property is located, be assigned the loan;

(i) The Tribe will have 30 calendar days after the Agency notification of such request to accept the assignment of the loan.

(ii) The Tribe must pay the Agency the lesser of the outstanding Agency indebtedness secured by the real estate or the market value of the property.

(iii) The Tribe may pay the amount in a lump sum or according to the rates, terms and requirements established in part 770 of this chapter, subject to the following:

(A) The Tribe must execute the promissory note and loan documents within 90 calendar days of receipt from the Agency;

(B) Such loan may not be considered for debt writedown under 7 CFR part 770.

(iv) The Tribe's failure to respond to the request for assignment of the loan or to finalize the assignment transaction within the time provided, shall be treated as the Tribe's denial of the request.

(2) Request the loan be assigned to the Secretary of the Interior. The Secretary of the Interior's failure to respond to the request for assignment of the loan or to finalize the assignment transaction, shall be treated as denial of the request;

(3) Voluntarily convey the real estate property to the Agency;

(i) The Agency will conduct a environmental review before accepting voluntary conveyance.

(ii) The Agency will credit the account with the greater of the market value of the real estate or the amount of the debt.

(4) Sell the real estate;

(i) The buyer must have the financial ability to buy the property.

(ii) The sale of the property must be completed within 90 calendar days of the Agency's notification.

(iii) The loan can be transferred and assumed by an eligible buyer.

(5) Pay the FLP debt in full.

(6) Consult with the Tribe that has jurisdiction over the Indian reservation to determine if State or Tribal law provides rights and protections that are more beneficial than those provided under this section.

(c) *Tribe notification.* At the time of acceleration, the Agency will notify the Tribe that has jurisdiction over the Indian reservation in which the property is located, of the:

(1) Sale of the American Indian borrower's property;

(2) Market value of the property;

(3) Amount the Tribe would be required to pay the Agency for assignment of the loan.

(d) *Partial payments*. The Agency may accept a payment that does not cover the unpaid balance of the accelerated loan if the borrower is in the process of selling security, unless acceptance of the payment would reverse the acceleration.

(e) *Failure to satisfy the debt*. The Agency will liquidate the borrower's account in accordance with §766.357 if:

(1) The borrower does not pay the account in full within the time period specified in the acceleration notice;

(2) The borrower does not voluntarily convey the property to the Agency;

(3) Neither the Tribe nor the Secretary of the Interior accepts assignment of the borrower's loan.

§766.357 Involuntary liquidation of real property and chattel.

(a) *General*. The Agency will liquidate the borrower's security if:

(1) The borrower does not satisfy the account in accordance with §§766.355 and 766.356, as appropriate;

(2) The involuntary liquidation is in the Agency's financial interest.

(b) *Foreclosure on loans secured by real property*. (1) The Agency will charge the borrower's account for all recoverable costs incurred in connection with the foreclosure and sale of the property.

(2) If the Agency acquires the foreclosed property, the Agency will credit the borrower's account in the amount of the Agency's bid except when incremental bidding was used, in which case the amount of credit will be the maximum bid that was authorized. If the Agency does not acquire the foreclosed property, the Agency will credit the borrower's account in accordance with State law and guidance from the Regional OGC.

(3) Notwithstanding paragraph (b)(2), for an American Indian borrower whose real property secures an FLP loan and is located within the confines of a Federally-recognized Indian reservation, the Agency will credit the borrower's account in the amount that is the greater of:

(i) The market value of the security; or

(ii) The amount of the FLP debt against the property.

(4) After the date of foreclosure, the borrower or former owner retains no statutory, implied, or inherent right of possession to the property beyond those rights granted by State law.

(5) If an unpaid balance on the FLP loan remains after the foreclosure sale of the property, the Agency will service the account in accordance with part 761, subpart F of this chapter and part 3 of this title.

(c) *Foreclosure of loans secured by chattel*. (1) The Agency will charge the borrower's account for all recoverable costs incurred by the Agency as a result of the repossession and sale of the property.

(2) The Agency will apply the proceeds from the repossession sale to the borrower's account less prior liens and all authorized liquidation costs.

(3) If an unpaid balance on the FLP loan remains after the sale of the repossessed property, the Agency will service the account in accordance with part 761, subpart F of this chapter and part 3 of this title.

[72 FR 63316, Nov. 8, 2007, as amended at 85 FR 36713, June 17, 2020]

§766.358 Acceleration and foreclosure moratorium.

(a) Notwithstanding any other provisions of this subpart, borrowers who file or have filed a program discrimination complaint that is accepted by USDA Office of Adjudication or successor office (USDA), and have been serviced to the point of acceleration or foreclosure on or after May 22, 2008, will not have their account accelerated or liquidated until such complaint has been resolved by USDA or closed by a court of competent jurisdiction. This moratorium applies only to program loans made under subtitle A, B, or C of the Act (for example, CL, FO, OL, EM, SW, or RL). Interest will not accrue and no offsets will be taken on these loans during the moratorium. Interest accrual and offsets will continue on all other loans, including, but not limited to, non-program loans.

(1) If the Agency prevails on the program discrimination complaint, the interest that would have accrued during the moratorium will be reinstated on

353

the account when the moratorium terminates, and all offsets and servicing actions will resume.

(2) If the borrower prevails on the program discrimination complaint, the interest that would have accrued during the moratorium will not be reinstated on the account unless specifically required by the settlement agreement or court order.

(b) The moratorium will begin on:

(1) May 22, 2008, if the borrower had a pending program discrimination claim that was accepted by USDA as valid and the account was at the point of acceleration or foreclosure on or before that date; or

(2) The date after May 22, 2008, when the borrower has a program discrimination claim accepted by USDA as valid and the borrower's account is at the point of acceleration or foreclosure.

(c) The point of acceleration under this section is the earliest of the following:

(1) The day after all rights offered on the Agency notice of intent to accelerate expire if the borrower does not appeal;

(2) The day after all appeals resulting from an Agency notice of intent to accelerate are concluded if the borrower appeals and the Agency prevails on the appeal;

(3) The day after all appeal rights have been concluded relating to a failure to graduate and the Agency prevails on any appeal;

(4) Any other time when, because of litigation, third party action, or other unforeseen circumstance, acceleration is the next step for the Agency in servicing and liquidating the account.

(d) A borrower is considered to be in foreclosure status under this section anytime after acceleration of the account.

(e) The moratorium will end on the earlier of:

(1) The date the program discrimination claim is resolved by USDA or

(2) The date that a court of competent jurisdiction renders a final decision on the program discrimination claim if the borrower appeals the decision of USDA.

[76 FR 5058, Jan. 28, 2011]

§§ 766.359–766.400　　[Reserved]

Subpart I—Exception Authority

§ 766.401　Agency exception authority.

On an individual case basis, the Agency may consider granting an exception to any regulatory requirement or policy of this part if:

(a) The exception is not inconsistent with the authorizing statute or other applicable law; and

(b) The Agency's financial interest would be adversely affected by acting in accordance with published regulations or policies and granting the exception would resolve or eliminate the adverse effect upon its financial interest.

PART 767—INVENTORY PROPERTY MANAGEMENT

Subpart A—Overview

767.156–767.200 [Reserved]

Subpart E—Real Estate Property with Important Resources or Located in Special Hazard Areas

767.201 Real estate inventory property with important resources.
767.202 Real estate inventory property located in special hazard areas.
767.203–767.250 [Reserved]

Subpart F—Exception Authority

767.251 Agency exception authority.

AUTHORITY: 5 U.S.C. 301 and 7 U.S.C. 1989.

SOURCE: 72 FR 63358, Nov. 8, 2007, unless otherwise noted.

Subpart A—Overview

§ 767.1 Introduction.

(a) *Purpose.* This part describes the Agency's policies for:

(1) Managing inventory property;

(2) Selling inventory property;

(3) Leasing inventory property;

(4) Managing real and chattel property the Agency takes into custody after abandonment by the borrower;

(5) Selling or leasing inventory property with important resources, or located in special hazard areas; and

(6) Conveying interest in real property for conservation purposes.

(b) *Basic policy.* The Agency maintains, manages and sells inventory property as necessary to protect the Agency's financial interest.

§ 767.2 Abbreviations and definitions.

Abbreviations and definitions for terms used in this part are provided in § 761.2 of this chapter.

§§ 767.3–767.50 [Reserved]

Subpart B—Property Abandonment and Personal Property Removal

§ 767.51 Property abandonment.

The Agency will take actions necessary to secure, maintain, preserve, manage, and operate the abandoned security property, including marketing perishable security property on behalf of the borrower when such action is in the Agency's financial interest. If the security is in jeopardy, the Agency will take the above actions prior to completing servicing actions contained in 7 CFR part 766.

§ 767.52 Disposition of personal property from real estate inventory property.

(a) *Preparing to dispose of personal property.* If, at the time of acquisition, personal property has been left on the real estate inventory property, the Agency will notify the former real estate owner and any known lienholders that the Agency will dispose of the personal property. Property of value may be sold at a public sale.

(b) *Reclaiming personal property.* The owner or lienholder may reclaim personal property at any time prior to the property's sale or disposal by paying all expenses incurred by the Agency in connection with the personal property.

(c) *Use of proceeds from sale of personal property.* Proceeds from the public sale of personal property will be distributed as follows:

(1) To lienholders in order of lien priority less a pro rata share of the sale expenses;

(2) To the inventory account up to the amount of expenses incurred by the Agency in connection with the sale of personal property;

(3) To the outstanding balance on the FLP loan; and

(4) To the borrower, if the borrower's whereabouts are known.

§§ 767.53–767.100 [Reserved]

Subpart C—Lease of Real Estate Inventory Property

§ 767.101 Leasing real estate inventory property.

(a) The Agency may lease real estate inventory property:

(1) To the former owner under the Homestead Protection Program;

(2) To a beginning farmer or socially disadvantaged farmer selected to purchase the property but who was unable to purchase it because of a lack of Agency direct or guaranteed loan funds;

(3) When the Agency is unable to sell the property because of lengthy litigation or appeal processes.

(b) The Agency will lease real estate inventory property in an "as is" condition.

(c) The Agency will lease property for:

(1) Homestead protection in accordance with part 766, subpart D, of this chapter.

(2) A maximum of 18 months to a beginning farmer or socially disadvantaged farmer the Agency selected as purchaser when no Agency loan funds are available; or

(3) The shortest possible duration for all other cases subject to the following:

(i) The maximum lease term for such a lease is 12 months.

(ii) The lease is not subject to renewal or extension.

(d) The lessee may pay:

(1) A lump sum;

(2) On an annual installment basis; or

(3) On a crop-share basis, if the lessee is a beginning farmer or socially disadvantaged farmer under paragraph (a) of this section.

(e) The Agency leases real estate inventory property for a market rent amount charged for similar properties in the area.

(f) The Agency may require the lessee to provide a security deposit.

(g) Only leases to a beginning farmer or socially disadvantaged farmer or Homestead Protection Program participant will contain an option to purchase the property.

[72 FR 63358, Nov. 8, 2007, as amended at 73 FR 74345, Dec. 8, 2008]

§ 767.102 Leasing non-real estate inventory property.

The Agency does not lease non-real estate property unless it is attached as a fixture to real estate inventory property that is being leased and it is essential to the farming operation.

§ 767.103 Managing leased real estate inventory property.

(a) The Agency will pay for repairs to leased real estate inventory property only when necessary to protect the Agency's interest.

(b) If the lessee purchases the real estate inventory property, the Agency will not credit lease payments to the purchase price of the property.

§§ 767.104–767.150 [Reserved]

Subpart D—Disposal of Inventory Property

§ 767.151 General requirements.

Subject to § 767.152, the Agency will attempt to sell its inventory property as follows:

(a) The Agency will combine or divide inventory property, as appropriate, to maximize the opportunity for beginning farmers or socially disadvantaged farmersto purchase real property.

(b) The Agency will advertise all real estate inventory property that can be used for any authorized FO loan purpose for sale to beginning farmers or socially disadvantaged farmers no later than 15 days after the Agency obtains title to the property.

(c) If more than one eligible beginning farmer or socially disadvantaged farmer applies, the Agency will select a purchaser by a random selection process open to the public.

(1) All applicants will be advised of the time and place of the selection.

(2) All drawn offers will be numbered.

(3) Offers drawn after the first will be held in suspense pending sale to the successful applicant.

(4) Random selection is final and not subject to administrative appeal.

(d) If there are no offers from beginning farmers or socially disadvantaged farmers, the Agency will sell inventory property by auction or sealed bid to the general public no later than 165 days after the Agency obtains title to the property. All bidders will be required to submit a 10 percent deposit with their bid.

(e) If the Agency receives no acceptable bid through an auction or sealed bid, the Agency will attempt to sell the property through a negotiated sale at the best obtainable price.

(f) If the Agency is not able to sell the property through negotiated sale, the Agency may list the property with a real estate broker. The broker must be properly licensed in the State in which the property is located.

[72 FR 63358, Nov. 8, 2007, as amended at 73 FR 74345, Dec. 8, 2008]

§767.152 Exceptions.

The Agency's disposition procedure under §767.151 is subject to the following:

(a) If the Agency leases real estate inventory property to a beginning farmer or socially disadvantaged farmer in accordance with §767.101(a)(2), and the lease expires, the Agency will not advertise the property if the Agency has direct or guaranteed loan funds available to finance the transaction.

(b) The Agency will not advertise a property for sale until the homestead protection rights have terminated in accordance with part 766, subpart D of this chapter.

(c) The Agency may allow an additional 60 days if needed for conservation easements or environmental reviews.

(d) If the property was owned by an American Indian borrower and is located on an Indian reservation, the Agency will:

(1) No later than 90 days after acquiring the property, offer the opportunity to purchase or lease the property in accordance with:

(i) The priorities established by the Indian Tribe having jurisdiction over the Indian reservation;

(ii) In cases where priorities have not been established, the following order:

(A) A member of the Indian Tribe that has jurisdiction over the Indian reservation;

(B) An Indian entity;

(C) The Indian Tribe.

(2) Transfer the property to the Secretary of the Interior if the property is not purchased or leased under paragraph (1) of this section.

(e) If Agency analysis of farm real estate market conditions indicates the sale of the Agency's inventory property will have a negative effect on the value of farms in the area, the Agency may withhold inventory farm properties in the affected area from the market until further analysis indicates otherwise.

[72 FR 63358, Nov. 8, 2007, as amended at 73 FR 74345, Dec. 8, 2008]

§767.153 Sale of real estate inventory property.

(a) *Pricing.* (1) The Agency will advertise property for sale at its market value, as established by an appraisal obtained in accordance with §761.7.

(2) Property sold by auction or sealed bid will be sold for the best obtainable price. The Agency reserves the right to reject any and all bids.

(b) *Agency-financed sales.* The Agency may finance sales to purchasers if:

(1) The Agency has direct or guaranteed FO loan funds available;

(2) All applicable loan making requirements are met; and

(3) All purchasers who are not beginning farmers or socially disadvantaged farmers make a 10 percent down payment.

(c) *Taxes and assessments.* (1) Property taxes and assessments will be prorated between the Agency and the purchaser based on the date the Agency conveys title to the purchaser.

(2) The purchaser is responsible for paying all taxes and assessments after the Agency conveys title to the purchaser.

(d) *Loss or damage to property.* If, through no fault of either party, the property is lost or damaged as a result of fire, vandalism, or act of God before the Agency conveys the property, the Agency may reappraise the property and set the sale price accordingly.

(e) *Termination of contract.* Either party may terminate the sales contract. If the contract is terminated by the Agency, the Agency returns any deposit to the bidder. If the contract is terminated by the purchaser, any deposit will be retained by the Agency as full liquidated damages, except where failure to close is due to Agency nonapproval of credit.

(f) *Warranty on title.* The Agency will not provide any warranty on the title or on the condition of the property.

[72 FR 63358, Nov. 8, 2007, as amended at 73 FR 74345, Dec. 8, 2008]

§767.154 Conveying easements, rights-of-way, and other interests in inventory property.

(a) *Appraisal of real property and real property interests.* The Agency will determine the value of real property and real property interests being transferred in accordance with §761.7 of this chapter.

(b) *Easements and rights-of-way on inventory property.* (1) The Agency may

357

grant or sell an easement or right-of-way for roads, utilities, and other appurtenances if the conveyance is in the public interest and does not adversely affect the value of the real property.

(2) The Agency may sell an easement or right-of-way by negotiation for market value to any purchaser for cash without giving public notice if:

(i) The sale would not prevent the Agency from selling the property; and

(ii) The sale would not decrease the value of the property by an amount greater than the price received.

(3) In the case of condemnation proceedings by a State or political subdivision, the transfer of title will not be completed until adequate compensation and damages have been determined and paid.

(c) *Disposal of other interests in inventory property.* (1) If applicable, the Agency will sell mineral and water rights, mineral lease interests, mineral royalty interests, air rights, and agricultural and other lease interests with the surface land except as provided in paragraph (b) of this section.

(2) If the Agency sells the land in separate parcels, any rights or interests that apply to each parcel are included with the sale.

(3) The Agency will assign lease or royalty interests not passing by deed to the purchaser at the time of sale.

(4) Appraisals of property will reflect the value of such rights, interests, or leases.

§ 767.155   Selling chattel property.

(a) *Method of sale.* (1) The Agency will use sealed bid or established public auctions for selling chattel. The Agency does not require public notice of sale in addition to the notice commonly used by the auction facility.

(2) The Agency may sell chattel inventory property, including fixtures, concurrently with real estate inventory property if, by doing so, the Agency can obtain a higher aggregate price. The Agency may accept an offer for chattel based upon the combined final sales price of both the chattel and real estate.

(b) *Agency-financed sales.* The Agency may finance the purchase of chattel inventory property if the Agency has direct or guaranteed OL loan funds available and all applicable loan making requirements are met.

§§ 767.156–767.200   [Reserved]

Subpart E—Real Estate Property With Important Resources or Located in Special Hazard Areas

§ 767.201   Real estate inventory property with important resources.

In addition to the requirements established in part 799 of this chapter, the following apply to inventory property with important resources:

(a) *Wetland conservation easements.* The Agency will establish permanent wetland conservation easements to protect and restore certain wetlands that exist on inventory property prior to the sale of such property, regardless of whether the sale is cash or credit.

(1) The Agency establishes conservation easements on all wetlands or converted wetlands located on real estate inventory property that:

(i) Were not considered cropland on the date the property was acquired by the Agency; and

(ii) Were not used for farming at any time during the 5 years prior to the date of acquisition by the Agency.

(A) The Agency will consider property to have been used for farming if it was used for agricultural purposes including, but not limited to, cropland, pastures, hayland, orchards, vineyards, and tree farming.

(B) In the case of cropland, hayland, orchards, vineyards, or tree farms, the Agency must be able to demonstrate that the property was harvested for crops.

(C) In the case of pastures, the Agency must be able to demonstrate that the property was actively managed for grazing by documenting practices such as fencing, fertilization, and weed control.

(2) The wetland conservation easement will provide for access to other portions of the property as necessary for farming or other uses.

(b) *Mandatory conservation easements.* The Agency will establish conservation easements to protect 100-year floodplains and other Federally-designated important resources. Federally-designated important resources include, but are not limited to:

(1) Listed or proposed endangered or threatened species;

(2) Listed or proposed critical habitats for endangered or threatened species;

(3) Designated or proposed wilderness areas;

(4) Designated or proposed wild or scenic rivers;

(5) Historic or archeological sites listed or eligible for listing on the National Register of Historic Places;

(6) Coastal barriers included in Coastal Barrier Resource Systems;

(7) Natural landmarks listed on National Registry of Natural Landmarks; and

(8) Sole source aquifer recharge areas as designated by EPA.

(c) *Discretionary easements.* The Agency may grant or sell an easement, restriction, development right, or similar legal right to real property for conservation purposes to a State government, a political subdivision of a State government, or a private non-profit organization.

(1) The Agency may grant or sell discretionary easements separate from the underlying fee or property rights.

(2) The Agency may convey property interests under this paragraph by negotiation to any eligible recipient without giving public notice if the conveyance does not change the intended use of the property.

(d) *Conservation transfers.* The Agency may transfer real estate inventory property to a Federal or State agency provided the following conditions are met:

(1) The transfer of title must serve a conservation purpose;

(2) A predominance of the property must:

(i) Have marginal value for agricultural production;

(ii) Be environmentally sensitive; or

(iii) Have special management importance;

(3) The homestead protection rights of the previous owner have been exhausted;

(4) The Agency will notify the public of the proposed transfer; and

(5) The transfer is in the Agency's financial interest.

(e) *Use restrictions on real estate inventory property with important resources.*

(1) Lessees and purchasers receiving Agency credit must follow a conservation plan developed with assistance from NRCS.

(2) Lessees and purchasers of property with important resources or real property interests must allow the Agency or its representative to periodically inspect the property to determine if it is being used for conservation purposes.

[72 FR 63358, Nov. 8, 2007, as amended at 81 FR 51285, Aug. 3, 2016]

§ 767.202 **Real estate inventory property located in special hazard areas.**

(a) The Agency considers the following to be special hazard areas:

(1) Mudslide hazard areas;

(2) Special flood areas; and

(3) Earthquake areas.

(b) The Agency will use deed restrictions to prohibit residential use of properties determined to be unsafe in special hazard areas.

(c) The Agency will incorporate use restrictions in its leases of property in special hazard areas.

§§ 767.203–767.250 **[Reserved]**

Subpart F—Exception Authority

§ 767.251 **Agency exception authority.**

On an individual case basis, the Agency may consider granting an exception to any regulatory requirement or policy of this part if:

(a) The exception is not inconsistent with the authorizing statute or other applicable law; and

(b) The Agency's financial interest would be adversely affected by acting in accordance with published regulations or policies and granting the exception would reduce or eliminate the adverse effect upon the its financial interest.

PART 768 [RESERVED]

PART 769—FARM LOAN PROGRAMS RELENDING PROGRAMS

Subpart A—Highly Fractionated Indian Land Loan Program

Subpart B—Heirs' Property Relending Program

AUTHORITY: 5 U.S.C. 301, 7 U.S.C. 1989, and 25 U.S.C. 488.

SOURCE: 80 FR 74970, Dec. 1, 2015, unless otherwise noted.

Subpart A—Highly Fractionated Indian Land Loan Program

§ 769.101 Purpose.

(a) This part contains regulations for loans made by the Agency to eligible intermediary lenders and applies to intermediary lenders and ultimate recipient involved in making and servicing Highly Fractionated Indian Land (HFIL) loans.

(b) The purpose of the HFIL Loan Program is to establish policies and procedures for a revolving loan fund through intermediary lenders for the purchase of HFIL by a Native American tribe, tribal entity, or member of either.

§ 769.102 Abbreviations and definitions

(a) *Abbreviations.* The following abbreviations are used in this part:

BIA—The Department of the Interior's Bureau of Indian Affairs (BIA).

HFIL—Highly Fractionated Indian Land.

(b) *Definitions.* The following definitions are used in this part:

Administrator means the head of the Farm Service Agency or designee.

Highly Fractionated Indian Land (HFIL) means for the purpose of this part only, Highly Fractionated Indian Land is undivided interests held by four or more individuals as a result of ownership or original allotments passing by state laws of intestate succession for multiple generations.

Indian Country land, communities, and allotments means the following:

(1) All land within the limits of any Indian reservation under the jurisdiction of the U.S. Government, notwithstanding the issuance of any patent, and, including rights-of-way running through the reservation,

(2) All dependent Indian communities within the borders of the United States whether within the original or subsequently acquired territory thereof, and whether within or without the limits of a state, and

(3) All Indian allotments, the Indian titles to which have not been extinguished, including rights-of-way running through the same; or

(4) All land, communities, and allotments that meet the definition of 18 U.S.C. 1151.

Intermediary lender means the entity requesting or receiving HFIL loan funds for establishing a revolving fund and relending to ultimate recipients.

Intermediary relending agreement means the signed agreement between

FSA and the intermediary that specifies the terms and conditions of the HFIL loan.

Native American tribe means the following:

(1) An Indian tribe recognized by the U.S. Department of the Interior; or

(2) A community in Alaska incorporated by the U.S. Department of the Interior pursuant to the Indian Reorganization Act.

Revolving funds means a fund that has two types of deposit accounts, one of which will be HFIL funds from FSA and the other will be comprised of repayments of loans from the ultimate recipients, interest earned on funds in the account and cash, or other short-term marketable assets that the intermediary lender chooses to deposit. Revolving funds are not considered Federal funds.

Tribal entity means an eligible entity established pursuant to the Indian Reorganization Act.

Ultimate recipient means Native American tribe, tribal entity, or member of either that receives a loan from an intermediary lender's HFIL revolving fund.

Undivided interest means a common interest in the whole parcel of land that is owned by two or more people. Owners of undivided interest do not own a specific piece of a parcel of land; rather they own a percentage interest in the whole.

§769.103 Eligibility requirements of the intermediary lender.

(a) *Eligible entity types.* The types of entities that may become an intermediary lender are:

(1) Private and Tribal operated non-profit corporations;

(2) Public agencies—Any State or local government, or any branch or agency of such government having authority to act on behalf of that government, borrow funds, and engage in activities eligible for funding under this part;

(3) Indian tribes or tribal corporations; or

(4) Lenders who are subject to credit examination and supervision by an acceptable State or Federal regulatory agency.

(b) *Intermediary lender requirements.* The intermediary lender must:

(1) Have the legal authority necessary for carrying out the proposed loan purposes and for obtaining, giving security for, and repaying the proposed loan;

(2) Have a record of successful lending in Indian Country and knowledge and experience working with the BIA. The Agency will assess the applicant staff's training and experience in lending in Indian Country based on recent experience in loan making and servicing with loans that are similar in nature to the HFIL program. If consultants will be used, FSA will assess the staff's experience in choosing and supervising consultants; and

(3) Have an adequate assurance of repayment of the loan based on the fiscal and managerial capabilities of the proposed intermediary lender.

(c) *The Intermediary Relending Agreement.* The intermediary lender and the Agency will enter into an Intermediary Relending Agreement, satisfactory to the Agency based on:

(1) Loan documentation requirements including planned application forms, security instruments, and loan closing documents;

(2) List of proposed fees and other charges it will assess the ultimate recipients;

(3) The plan for relending the loan funds. The plan must have sufficient detail to provide the Agency with a complete understanding of the complete mechanics of how the funds will get from the intermediary lender to the ultimate recipient. Included in the plan are the service area, eligibility criteria, loan purposes, rates, terms, collateral requirements, a process for addressing environmental issues on property to be purchased, limits, priorities, application process, analysis of new loan requests, and method of disbursement of the funds to the ultimate recipient;

(4) Loan review plans that specify how the intermediary lender will review the loan request from the ultimate recipient and make an eligibility determination;

(5) An explanation of the intermediary lender's established internal credit review process; and

(6) An explanation of how the intermediary lender will monitor the loans to the ultimate recipients.

§769.104 Requirements of the ultimate recipient.

(a) Ultimate recipients must be individual Tribal members, Tribes or eligible Tribal entities, with authority to incur the debt and carry out the purpose of the loan.

(b) The intermediary lender will make this determination in accordance with the Intermediary Relending Agreement.

§769.105 Authorized loan purposes.

(a) *Intermediary lender.* Agency HFIL loan funds must be placed in the intermediary's HFIL revolving fund and used by the intermediary to provide direct loans to eligible ultimate recipients.

(b) *Ultimate recipient.* Loans from the intermediary lender to the ultimate recipient using the HFIL revolving fund:

(1) Must be used to acquire and consolidate at least 50 percent of the highly fractionated Indian land parcel and interests in the land. The interests include rights-of-way, water rights, easements, and other appurtenances that would normally pass with the land or are necessary for the proposed operation of the land located within the tribe's reservation;

(2) Must finance land that will be used for agricultural purposes during the term of the loan;

(3) May be used to pay costs incidental to land acquisition, including, but not limited to, title clearance, legal services, archeological or land surveys, and loan closing; and

(4) May be used to pay for the costs of any appraisal conducted in accordance with this part.

§769.106 Limitations.

(a) Loan funds may not be used for any land improvement or development purposes, acquisition or repair of buildings or personal property, payment of operating costs, payment of finders' fees, or similar costs, or for any purpose that will contribute to excessive erosion of highly erodible land or to the conversion of wetlands to produce an agricultural commodity as specified in 7 CFR part 12.

(b) The amount of loan funds used to acquire land may not exceed the current market value of the land as determined by a current appraisal that meets the requirements as specified in 7 CFR 761.7(b)(1).

(c) Agency HFIL loan funds may not be used for payment of the intermediary's administrative costs or expenses. The amount removed from the HFIL revolving fund for administrative costs in any year must be reasonable, must not exceed the actual cost of operating the HFIL revolving fund and must not exceed the amount approved by the Agency in the intermediary lender's annual loan monitoring report.

(d) No loan to an intermediary lender may exceed the maximum amount the intermediary can reasonably expect to lend to eligible ultimate recipients, based on anticipated demand for loans to consolidate fractioned interests and capacity of the intermediary to effectively carry out the terms of the loan.

§769.107 Rates and terms.

(a) Loans made by the Agency to the intermediary lender will bear interest at a fixed rate as determined by the Administrator, but not less than 1 percent per year over the term of the loan.

(1) Interest rates charged by intermediary lender to ultimate recipients on loans from the HFIL revolving fund will be negotiated between the intermediary lender and ultimate recipient, but the rate must be within limits established by the Intermediary Relending Agreement.

(2) The rate should normally be the lowest rate sufficient to cover the loan's proportional share of the revolving fund's debt service costs and administrative costs.

(b) No loan to an intermediary lender will be extended for a period exceeding 30 years. Interest will be due annually but principal payments may be deferred by the Agency.

(1) Loans made by an intermediary lender to an ultimate recipient from the HFIL revolving fund will be scheduled for repayment over a term negotiated by the intermediary lender and ultimate recipient but will not exceed

30 years or the date of the end of the term of the HFIL loan, whichever is sooner.

(2) The term of an HFIL loan must be reasonable and prudent considering the purpose of the loan, expected repayment ability of the ultimate recipient, and the useful life of collateral, and must be within any limits established by the intermediary lender's Intermediary Relending Agreement.

§ 769.108 Security requirements for HFIL loans and the ultimate recipients.

(a) *HFIL loans.* Security for all loans to intermediaries must be such that the repayment of the loan is reasonably assured, taking into consideration the intermediary's financial condition, Intermediary Relending Agreement, and management ability. The intermediary is responsible to make loans to ultimate recipients in such a manner that will fully protect the interest of the intermediary and the Government. The Agency will require adequate security, as determined by the Agency, to fully secure the loan, including but not limited to the following:

(1) Assignments of assessments, taxes, levies, or other sources of revenue as authorized by law;

(2) Investments and deposits of the intermediary; and

(3) Capital assets or other property of the intermediary and its members.

(b) *Liens.* In addition to normal security documents, a first lien interest in the intermediary's revolving fund account will be accomplished by a control agreement satisfactory to the Agency. The control agreement does not require the Agency's signature for withdrawals. The depository bank must waive its offset and recoupment rights against the depository account to the Agency and subordinate any liens it may have against the HFIL depository bank account.

(c) *Ultimate recipient.* Security for a loan from an intermediary lender's HFIL revolving fund to an ultimate recipient will be adequate to fully secure the loan as specified in the relending agreement.

(1) The Agency will only require concurrence in the intermediary lender's security requirement for a specific loan when security for the loan from the intermediary lender to the ultimate recipient will also serve as security for an Agency loan.

(2) The ultimate recipient will take appropriate action to obtain and provide security for the loan.

§ 769.109 Intermediary lender's application.

(a) The application will consist of:

(1) An application form provided by the Agency;

(2) A draft Intermediary Relending Agreement and other evidence the Agency requires to show the feasibility of the intermediary lender's program to meet the objectives of the HFIL Loan Program; and

(3) Applications from intermediary lenders that already have an active HFIL loan may be streamlined by filing a new application and a statement that the new loan would be operated in accordance with the Intermediary Relending Agreement on file for the previous loan. This statement may be submitted at the time of application in lieu of a new Intermediary Relending Agreement.

(4) Documentation of the intermediary lender's ability to administer HFIL in accordance with this part;

(5) Submission of a completed Agency application form;

(6) Prior to approval of a loan or advance of funds, certification of whether or not the intermediary lender is delinquent on any Federal debt, including, but not limited to, Federal income tax obligations or a loan or loan guarantee or from another Federal agency. If delinquent, the intermediate lender must explain the reasons for the delinquency, and the Agency will take such written explanation into consideration in deciding whether to approve the loan or advance of funds;

(7) Prior to approval of a loan or advance of funds, certification as to whether the intermediary lender has been convicted of a felony criminal violation under Federal law in the 24 months preceding the date of application.

(8) Certification of compliance with the restrictions and requirements in 31

U.S.C. 1352, and 2 CFR 200.450 and part 418.

(9) Certification to having been informed of the collection options the Federal government may use to collect delinquent debt.

(b) An intermediary lender that has received one or more HFIL loans may apply for and be considered for subsequent HFIL loans provided:

(1) The intermediary lender is relending all collections from loans made from its revolving fund in excess of what is needed for required debt service, approved administration costs, and a reserve for debt service;

(2) The outstanding loans of the intermediary lender's HFIL revolving fund are performing; and

(3) The intermediary lender is in compliance with all regulations and its loan agreements with the Agency.

§ 769.110   Letter of conditions.

(a) The Agency will provide the intermediary lender a letter listing all requirements for the loan. After reviewing the conditions and requirements in the letter of conditions, the intermediary lender must complete, sign, and return the form provided by the Agency indicating the intermediary lender's intent to meet the conditions. If certain conditions cannot be met, the intermediary lender may propose alternate conditions in writing to the Agency. The Agency loan approval official must concur with any changes made to the initially issued or proposed letter of conditions prior to acceptance. The loan request will be withdrawn if the intermediary lender does not respond within 15 days.

(b) At loan closing, the intermediary lender must certify that:

(1) No major changes have been made in the Intermediary Relending Agreement except those approved in the interim by the Agency;

(2) All requirements of the letter of conditions have been met; and

(3) There has been no material change in the intermediary lender or its financial condition since the issuance of the letter of conditions. If there have been changes, the intermediary lender must explain the changes to the Agency. The changes

may be waived, at the sole discretion of the Agency.

§ 769.111   Loan approval and obligating funds.

(a) Loan requests will be processed based on the date the Agency receives the application. Loan approval is subject to the availability of funds.

(b) The loan will be considered approved for the intermediary lender on the date the signed copy of the obligation of funds document is mailed to the intermediary lender.

§ 769.120   Loan closing.

(a) *Loan agreement.* A loan agreement or supplement to a previous loan agreement must be executed by the intermediary lender and the Agency at loan closing for each loan setting forth, at a minimum,

(1) The amount of the loan, the interest rate, the term and repayment schedule,

(2) The requirement to maintain a separate ledger and segregated account for the HFIL revolving fund; and

(3) It agrees to comply with Agency reporting requirements.

(b) *Loan closing.* Intermediary lenders receiving HFIL loans will be governed by this part, the loan agreement, the approved Intermediary Relending Agreement, security instruments, and any other conditions that the Agency requires on loans made from the "HFIL revolving fund." The requirement applies to all loans made by an intermediary lender to an ultimate recipient from the intermediary lender's HFIL revolving fund for as long as any portion of the intermediary lender's HFIL loan from the Agency remains unpaid.

(c) *Intermediary lender certification.* The intermediary lender must include in their file a certification that:

(1) The proposed ultimate recipient is eligible for the loan;

(2) The proposed loan is for eligible purposes; and

(3) The proposed loan complies with all applicable laws and regulations.

§ 769.121   Maintenance and monitoring of HFIL revolving fund.

(a) *Maintenance of revolving fund.* The intermediary lender must maintain the

HFIL revolving fund until all of its HFIL obligations have been paid in full. All HFIL loan funds received by an intermediary lender must be deposited into an HFIL revolving fund account. Such accounts must be fully covered by Federal deposit insurance or fully collateralized with U.S. Government obligations. All cash of the HFIL revolving fund must be deposited in a separate bank account or accounts so as not to be commingled with other financial assets of the intermediary lender. All money deposited in such bank account or accounts must be security assets of the HFIL revolving fund. Loans to ultimate recipients must be from the HFIL revolving fund.

(1) The portion of the HFIL revolving fund that consists of Agency HFIL loan funds may only be used for making loans in accordance with § 769.105. The portion of the HFIL revolving fund that consists of repayments from ultimate recipients may be used for debt service, reasonable administrative costs, or for making additional loans;

(2) An intermediary lender may use revolving funds and HFIL loan funds to make loans to ultimate recipients without obtaining prior Agency concurrence in accordance with the Intermediary Relending Agreement;

(3) Any funds in the HFIL revolving fund from any source that is not needed for debt service, approved administrative costs, or reasonable reserves must be available for additional loans to ultimate recipients;

(4) All reserves and other funds in the HFIL revolving loan fund not immediately needed for loans to ultimate recipients or other authorized uses must be deposited in accounts in banks or other financial institutions. Such accounts must be fully covered by Federal deposit insurance or fully collateralized with U.S. Government obligations, and will be interest bearing. Any interest earned thereon remains a part of the HFIL revolving fund;

(5) If an intermediary lender receives more than one HFIL loan, it does not need to establish and maintain a separate HFIL revolving loan fund for each loan; it may combine them and maintain only one HFIL revolving fund, unless the Agency requires separate HFIL revolving funds because there are significant differences in the loan purposes, Intermediary Relending Agreement, loan agreements, or requirements for the loans; and

(6) A reasonable amount of revolved funds must be used to create a reserve for bad debts. Reserves should be accumulated over a period of years. The total amount should not exceed maximum expected losses, considering the quality of the intermediary lender's portfolio of loans. Unless the intermediary lender provides loss and delinquency records that, in the opinion of the Agency, justifies different amounts, a reserve for bad debts of 6 percent of outstanding loans must be accumulated over 5 years and then maintained.

(b) *Loan monitoring reviews.* The intermediary lender must complete loan monitoring reviews, including annual and periodic reviews, and performance monitoring.

(1) At least annually, the intermediary lender must provide the Agency documents for the purpose of reviewing the financial status of the intermediary Lender, assessing the progress of utilizing loan funds, and identifying any potential problems or concerns. Non-regulated intermediary lenders must furnish audited financial statements at least annually.

(2) At any time the Agency determines it is necessary, the intermediary lender must allow the Agency or its representative to review the operations and financial condition of the intermediary lender. Upon the Agency requests, the Intermediary must submit financial or other information within 14 days unless the data requested is not available within that time frame.

(c) *Progress reports.* Each intermediary lender will be monitored by the Agency based on progress reports submitted by the intermediary lender, audit findings, disbursement transactions, visitations, and other contact with the intermediary lender as necessary.

§ 769.122 Loan servicing.

(a) *Payments.* Payments will be made to the Agency as specified in loan agreements and debt instruments. The funds from any extra payments will be applied entirely to loan principal.

(b) *Restructuring.* The Agency may restructure the intermediary lender's loan debt, if:

(1) The Government's interest will be protected;

(2) The restructuring will be performed within the Agency's budget authority; and

(3) The loan objectives cannot be met unless the HFIL loan is restructured.

(c) *Default.* In the event of monetary or non-monetary default, the Agency will take all appropriate actions to protect its interest, including, but not limited to, declaring the debt fully due and payable and may proceed to enforce its rights under the loan agreement or any other loan instruments relating to the loan under applicable law and regulations, and commencement of legal action to protect the Agency's interest. The Agency will work with the intermediary lender to correct any default, subject to the requirements of paragraph (b) of this section. Violation of any agreement with the Agency or failure to comply with reporting or other program requirements will be considered non-monetary default.

§ 769.123　Transfer and assumption.

(a) All transfers and assumptions must be approved in advance in writing by the Agency. The assuming entity must meet all eligibility criteria for the HFIL Loan Program.

(b) Available transfer and assumption options to eligible intermediary lenders include the following:

(1) The total indebtedness may be transferred to another eligible intermediary lender on the same terms; or

(2) The total indebtedness may be transferred to another eligible intermediary lender on different terms not to exceed the term for which an initial loan can be made. The assuming entity must meet all eligibility criteria for the HFIL Loan Program.

(c) The transferor must prepare the transfer document for the Agency review prior to the transfer and assumption.

(d) The transferee must provide the Agency with information required in the application as specified in § 769.109.

(e) The Agency prepared assumption agreement will contain the Agency case number of the transferor and transferee.

(f) The transferee must complete an application as specified in § 769.109(a).

(g) When the transferee makes a cash down-payment in connection with the transfer and assumption, any proceeds received by the transferor will be credited on the transferor's loan debt in order of maturity date.

(h) The Administrator or designee will approve or decline all transfers and assumptions.

§ 769.124　Appeals.

Any appealable adverse decision made by the Agency may be appealed upon written request of the intermediary as specified in 7 CFR part 11.

§ 769.125　Exceptions.

The Agency may grant an exception to any of the requirements of this part if the proposed change is in the best financial interest of the Government and not inconsistent with the authorizing law or any other applicable law.

Subpart B—Heirs' Property Relending Program

SOURCE: 86 FR 43393, Aug. 9, 2021, unless otherwise noted.

§ 769.150　Purpose.

(a) This subpart contains regulations for loans made by the Agency to eligible intermediaries that will make and service loans to ultimate recipients pursuant to requirements in this subpart. This subpart applies to intermediaries, ultimate recipients, and other parties involved in making such loans.

(b) The purpose of HPRP is to assist heirs with undivided ownership interests resolve ownership and succession issues on a farm that is owned by multiple owners. This purpose is achieved by providing loan funds to eligible intermediaries who will re-lend to individuals and entities for the purpose of developing and implementing a succession plan and to resolve title issues.

(c) Intermediaries receiving HPRP loans must comply with this subpart, the HPRP loan agreement, the intermediary's relending plan approved by the Agency, the HPRP loan documents and security instruments and any other conditions that the Agency may impose in making a loan.

§ 769.151 Abbreviations and definitions.

Abbreviations and definitions used in this subpart are found in §761.2 of this chapter.

§ 769.152 Eligibility requirements of the intermediary.

(a) *Eligible entity types.* Cooperatives, credit unions, and nonprofit organizations are eligible to participate as intermediaries.

(b) *Certification.* The intermediary must be certified as a community development financial institution under 12 CFR 1805.201 to operate as a lender.

(c) *Citizenship.* The applicant and the members of the intermediary must be a U.S. citizen or qualified alien (see 8 U.S.C. 1641). Each intermediary must certify to the citizenship requirement in the HPRP loan application.

(d) *Experience.* The intermediary must have:

(1) The requisite experience and capability in making and servicing agricultural and commercial loans that are similar in nature to HPRP. If consultants will be used in the making and servicing of HPRP loans, the Agency will assess the intermediary's experience in choosing and supervising consultants based on information intermediaries include in their application describing the particular lending functions they typically rely on agents to fulfill and also describe their policies and procedures for monitoring these agents;

(2) The legal authority necessary to carry out the proposed loan purposes and to obtain, provide security for, and repay the proposed loan; and

(3) Demonstrated ability and willingness to repay the loan based on the intermediary's financial condition, managerial capabilities, and other resources.

§ 769.153 Eligibility requirements of the ultimate recipient.

(a) The eligibility requirements for the ultimate recipient are:

(1) Ultimate recipients must be individuals or legal entities, with authority to incur the debt and to resolve ownership and succession of a farm owned by multiple owners;

(2) Individual ultimate recipients or members of entity ultimate recipients must be a family member or heir-at-law related by blood or marriage to the previous owner of the real property; and

(3) The ultimate recipient must agree to complete a succession plan.

(b) The intermediary will determine the eligibility of the applicant to become the ultimate recipient in accordance with the rules provided in this subpart and in accordance with the intermediary's relending plan as approved by the Agency in the HPRP loan agreement.

§ 769.154 Authorized loan purposes.

(a) *Loans to the intermediary.* HPRP loan funds must be used by the intermediary to provide direct loans to eligible ultimate recipients according to the rules provided in this subpart and pursuant to the HPRP loan agreement approved by the Agency.

(b) *Loans to the ultimate recipients.* HPRP loan funds:

(1) Must be used to assist heirs with undivided ownership interests to resolve ownership and succession of a farm owned by multiple owners;

(2) Must be sufficient to cover costs and fees associated with development and implementation of the succession plan, including closing costs (such as costs for preparing documents, appraisals, surveys, and title reports) and other associated legal services (such as fees incurred for mediation); and

(3) May be used to purchase and consolidate fractional interests held by other heirs in jointly-owned property, and to purchase rights-of-way, water rights, easements, and other appurtenances that would normally pass with the property and are necessary for the proposed operation of the farm.

§ 769.155 Loan limitations.

(a) For each application period:

(1) Loans to intermediaries will not exceed $5,000,000 to any intermediary;

(2) Loans to ultimate recipients will not exceed the loan limit for a Direct Farm Ownership loan as specified in § 761.8(a)(1)(i) of this chapter to any ultimate recipient.

(b) Loans to the ultimate recipient may not be used:

(1) For any land improvement, development purpose, acquisition or repair of buildings, acquisition of personal property, payment of operating costs, payment of finders' fees, or similar costs;

(2) For any purpose that will contribute to excessive erosion of highly erodible land or for the conversion of wetlands to produce an agricultural commodity as specified in 7 CFR part 12; or

(3) To resolve heirs' property issues on property that will not be used, or has traditionally not been used, for production agricultural purposes.

(c) The HPRP loan amount may not exceed the current market value of the land determined by an appraisal that meets the requirements specified in § 761.7(b)(1) of this chapter; and

(d) Intermediaries who receive HPRP funding are not permitted to charge the ultimate recipients for mediation services provided through grants received under the Agency's State Agriculture Mediation Program (part 785 of this chapter).

§ 769.156 Rates and terms.

(a) For loans to intermediaries:

(1) The rate of interest for an HPRP loan will bear a fixed rate over the term of the loan of 1 percent or less as determined by the Administrator;

(2) The repayment term for an HPRP loan will not exceed 30 years; and

(3) Annual payments will be established. Interest will be due annually; however, principal payments may be deferred by the Agency.

(b) Loans to the ultimate recipient from the HPRP revolving loan fund are required to have rates and terms clearly and publicly disclosed to qualified ultimate recipients.

(1) The interest rate for loans to ultimate recipients will be set by the intermediary within the limits established by the intermediary's relending plan approved by the Agency. The rate should normally be the lowest rate sufficient to cover the loan's proportional share of the HPRP revolving loan fund's debt service costs, reserve for bad debts, and administrative costs.

(2) Loans made by an intermediary to an ultimate recipient will be scheduled for repayment over a term negotiated by the intermediary and ultimate recipient; but in no case will the loan term exceed 30 years, unless otherwise specified by the Agency.

§ 769.157 Intermediary's relending plan.

(a) The intermediary must submit a proposed relending plan which, once approved by the Agency, will be incorporated by reference as an attachment to the HPRP loan agreement. The relending plan will explain in sufficient detail the mechanics of how the funds will be distributed from the intermediary to the ultimate recipient.

(b) The intermediary's relending plan must include copies of the intermediary's proposed application forms, loan documents and security instruments, and should include information regarding:

(1) The service area;

(2) The proposed fees and other charges the intermediary will assess the ultimate recipients;

(3) Eligibility criteria for the ultimate recipient;

(4) Authorized loan purposes;

(5) Loan limitations;

(6) Loan underwriting methods and criteria;

(7) Loan rates and terms;

(8) Security requirements;

(9) The method of disbursement of the funds to the ultimate recipient;

(10) The process for addressing environmental issues on property to be purchased;

(11) The proposed process for reviewing loan requests from ultimate recipients and making eligibility determinations;

(12) A description of the established internal credit review process;

(13) The monitoring and servicing of loans distributed to the ultimate recipients;

(14) The amount that will be set aside to maintain a reserve for bad debts; and

(15) A description of the requirements for maintaining adequate hazard insurance, life insurance (for principals and key employees of the ultimate recipient), workmen's compensation insurance on ultimate recipients, flood insurance, and fidelity bond coverage.

§769.158 Intermediary's loan application.

(a) The intermediary's loan application will consist of:

(1) An application form provided by the Agency;

(2) A relending plan addressing the items in §769.157;

(3) A copy of the intermediary's certification as a community development financial institution;

(4) A signed form, to be provided by the Agency, assuring the intermediary's compliance and continued compliance with Title IX of the Education Amendments of 1972 (20 U.S.C. 1681–1688) and Title VI of the Civil Rights Act of 1964 (42 U.S.C. 2000d–1–2000d–7);

(5) Other evidence the Agency requires to determine that the intermediary satisfies the eligibility requirements in §769.152, and that the intermediary's proposed relending plan is feasible and meets the objectives of HPRP;

(6) Documentation of the intermediary's ability to administer the HPRP loan funds in accordance with this subpart; and

(7) The name(s) of attorneys or any third parties involved with the application process.

(b) Prior to loan approval and advancing funds, the intermediary must certify that:

(1) The intermediary and its officers, or agents are not delinquent on any Federal debt, including, but not limited to, federal income tax obligations, federal loan or loan guarantee, or obligation from another Federal agency. If delinquent, the intermediary must provide in writing the reasons for the delinquency, and the Agency will take this into consideration in deciding whether to approve the loan or advance of funds;

(2) The intermediary and its officers have not been convicted of a felony criminal violation under Federal law in the 24 months preceding the date of the loan application;

(3) The intermediary is in compliance with the restrictions and requirements in 31 U.S.C. 1352, limitation on use of appropriated funds to influence certain Federal contracting and financial transactions;

(4) The intermediary has been informed of the options by the Federal Government to collect delinquent debt; and

(5) The intermediary, its officers, or agents are not debarred or suspended from participation in Government contracts or programs.

(c) An intermediary that has received one or more HPRP loans may apply for and be considered for subsequent HPRP loans provided:

(1) The intermediary is relending all collections from loans made from its revolving fund in excess of what is needed for the required debt service reserve and approved administrative costs;

(2) The outstanding loans of the intermediary's HPRP revolving loan fund are performing; and

(3) The intermediary is following all regulatory requirements and is complying with the terms and conditions of its HPRP loan agreement(s) and the intermediary's relending plan(s) approved by the Agency.

(d) The Agency may require the intermediary to provide information relating to applications from ultimate recipients the intermediary has in process.

§769.159 Processing loan applications.

(a) *Application dates.* The opening and closing dates for the HPRP applications submission will be announced in FEDERAL REGISTER.

(b) *Intermediary loan application review.* After the closing date, the Agency will review applications from intermediaries for compliance with the provisions of this subpart.

(c) *Loan approval.* Loan approval is subject to the availability of funds. The loan will be considered approved for the intermediary on the date the

Agency signs the obligation of funds confirmation.

(d) *Preferences for loan funding.* The Agency will fund eligible applications from intermediaries:

(1) First, to those with not less than 10 years' experience serving socially disadvantaged farmers and ranchers that are located in states that have adopted a statute consisting of an enactment or adoption of the Uniform Partition of Heirs Property Act, as approved and recommended for enactment in all States by the National Conference of Commissioners on Uniform State Laws in 2010, that relend to owners of heirs property (as defined by the Uniform Partition of Heirs Property Act);

(2) Second, to those that have applications from ultimate recipients already in process, or that have a history of successfully relending previous HPRP funds; and

(3) Multiple applications in the same priority tier, will be processed based by date of application received; and

(4) Any remaining applications, after priority tiers 1 and 2 have be funded, will be funded in order of the date the application was received.

(e) *Current information required.* Information supplied by the intermediary in the loan application must be updated by the intermediary if the information is more than 90 days old at the time of loan closing.

§ 769.160 Letter of conditions.

(a) If the Agency approves a loan application, the Agency will provide the intermediary with a letter of conditions listing all requirements for the loan.

(b) Immediately after reviewing the conditions and requirements in the letter of conditions, the intermediary should complete, sign, and return the form provided by the Agency indicating the intermediary's intent to meet the conditions.

(1) If certain conditions cannot be met, the intermediary may propose alternative conditions to the Agency.

(2) The Agency loan approval official must concur with any changes made to the initially issued or proposed letter of conditions prior to loan approval.

(c) The loan request will be considered withdrawn if the intermediary does not respond within 15 calendar days from the date the letter of conditions was sent.

§ 769.161 Loan agreements.

(a) The HPRP loan agreement will specify the terms of each loan, such as:

(1) The amount of the loan;

(2) The interest rate;

(3) The term and repayment schedule;

(4) Any provisions for late charges;

(5) The disbursement procedure;

(6) Provisions regarding default; and

(7) Fidelity insurance.

(b) As a condition of receiving HPRP loan funds, the intermediary will agree:

(1) To obtain written approval from the Agency prior to making any changes in the intermediary's articles of incorporation, charter, or by-laws;

(2) To maintain a separate ledger and segregated account for the HPRP revolving loan fund;

(3) To comply with the Agency's annual reporting requirements in § 769.164(g);

(4) To obtain prior written approval from the Agency regarding all forms to be used for relending purposes, as well as the intermediary's policy with regard to the amount and security to be required;

(5) To obtain written approval from the Agency prior to making any significant changes in the proposed forms, security policy, or the intermediary's relending plan;

(6) To maintain the collateral pledged as security for the HPRP loan; and

(7) To request demographics data from ultimate recipients on race, ethnicity, and gender. The response to the data request will be voluntary. The intermediary will maintain the information when voluntarily submitted by the ultimate recipient. The intermediary agrees to make this information available when requested by FSA.

§ 769.162 Security.

(a) *Loans to intermediaries.* Security pledged to the Agency by intermediaries must be sufficient to reasonably assure repayment of the loan, while taking into consideration the intermediary's financial condition, the intermediary's relending plan, and the intermediary's management ability. The Agency will require adequate security, as determined by the Agency, to fully secure the loan:

(1) Primary security for HPRP loan will be in the form of a first lien upon the intermediary's revolving loan fund and such accounts must be fully covered by Federal deposit insurance or fully collateralized with other securities in accordance with normal banking practices and all applicable State laws. The form of the control agreement with the depository bank that will be used to perfect the Agency's security interest in the depository accounts used by the intermediary to maintain HPRP funds must be approved by the Agency. The control agreement will not require the Agency's signature for withdrawals. Among other things, the intermediary must use a depository bank that agrees to waive its offset and recoupment rights against the depository account and subordinate any liens it may have against the HPRP depository account in favor of the Agency;

(2) Additional security as needed, which includes, but is not limited to:

(i) Assignments of assessments, taxes, levies, or other sources of revenue as authorized by law;

(ii) Financial assets of the intermediary and its members; and

(ii) Capital assets or other property of the intermediary and its members.

(b) *Loans to the ultimate recipient.* The intermediary is responsible for obtaining adequate security for all loans made to ultimate recipients from the HPRP revolving loan funds as specified in the HPRP loan agreement and intermediary's relending plan. The Agency will only require concurrence with the intermediary's proposed security for a loan to an ultimate recipient from the HPRP revolving loan fund if the proposed security will also serve as security for an unrelated Agency loan.

§769.163 Loan closing.

(a) *HPRP loan documents and security instruments.* At loan closing, the intermediary will execute the HPRP loan agreement or supplemental loan agreement, HPRP promissory note, the HPRP security agreement, the control agreement, and any other security instruments required by the Agency.

(b) *Intermediary certification.* At loan closing, the intermediary must certify that:

(1) No changes have been made in the intermediary's relending plan except those approved in the interim by the Agency;

(2) All requirements in the letter of conditions have been met; and

(3) There has been no material change in the intermediary or its financial condition since the issuance of the letter of conditions. If there have been changes, the intermediary must explain the changes to the Agency. The Agency will review the changes and respond in writing prior to loan closing.

§769.164 Post award requirements.

(a) *Applicability.* Whenever this subpart imposes a requirement on loan funds from the HPRP revolving loan fund, the requirement will apply to all loans made by an intermediary to an ultimate recipient from the intermediary's HPRP revolving loan fund for as long as any portion of the intermediary's HPRP loan remains unpaid.

(b) *Applicability for HPRP loan funds.* Whenever this subpart imposes a requirement on loans made by intermediaries from HPRP loan funds, without specific reference to the HPRP revolving loan fund, such requirement only applies to loans made by an intermediary using HPRP loan funds, and will not apply to loans made from revolved funds.

(c) *File maintenance.* In addition to information normally maintained by lenders in each loan file associated with a relending loan to an ultimate recipient, the intermediary must include a certification and supporting documentation in its file demonstrating that:

(1) The ultimate recipient is eligible for the loan;

(2) The loan is for eligible purposes; and

(3) The loan complies with all applicable laws, regulations, and the intermediary's HPRP loan agreement.

(d) *Maintenance of HPRP revolving loan fund.* For as long as any part of an

HPRP loan remains unpaid, the intermediary must maintain the HPRP revolving loan fund in accordance with the requirements in paragraphs (d)(1) through (11) of this section:

(1) All HPRP loan funds received by an intermediary must be deposited into the HPRP revolving loan fund. The intermediary may transfer additional assets into the HPRP revolving loan fund;

(2) All cash of the HPRP revolving loan fund must be deposited in a separate bank account or accounts;

(3) The HPRP revolving loan fund must be segregated from other financial assets of the intermediary, and no other funds of the intermediary will be commingled with the HPRP revolving loan fund;

(4) All moneys deposited in the HPRP revolving loan fund account or accounts will be money from the HPRP revolving loan fund;

(5) Loans to ultimate recipients are advanced from the HPRP revolving loan fund;

(6) The receivables created by making loans to ultimate recipients, the intermediary's security interest in collateral pledged by ultimate recipients, collections on the receivables, interest, fees, and any other income or assets derived from the operation of the HPRP revolving loan fund are a part of the HPRP revolving loan fund;

(7) The portion of the HPRP revolving loan fund consisting of HPRP loan funds may only be used for making loans in accordance with § 769.154. The portion of the HPRP revolving loan fund that consists of revolved funds may be used for debt service reserve, approved administrative costs, or for making additional loans;

(8) A reasonable amount of revolved funds must be maintained as a reserve for bad debts. The total amount should not exceed maximum expected losses, considering the credit quality of the intermediary's portfolio of loans. The amount of reserved funds proposed by the intermediary requires written concurrence from the Agency. Unless the intermediary provides loss and delinquency records that, in the opinion of the Agency, justifies different amounts, a reserve for bad debts of 6 percent of outstanding loans must be

accumulated over 5 years and then maintained; and

(9) Any funds in the HPRP revolving loan fund from any source that is not needed for debt service reserve, approved administrative costs, or reasonable reserves must be available for additional loans to ultimate recipients.

(i) Funds may not be used for any investments in securities or certificates of deposit of over 30-day duration without the Agency's concurrence.

(ii) The intermediary must make one or more loans to ultimate recipients within 6 months of any disbursement it receives from the Agency. If funds have been unused to make loans to ultimate recipients for 6 months or more, those funds will be returned to the agency unless the Agency provides a written exception based on evidence satisfactory to the Agency that every effort is being made by the intermediary to utilize the HPRP funding in conformance with HPRP objectives;

(10) All reserves and other cash in the HPRP revolving loan fund that are not immediately needed for loans to ultimate recipients or other authorized uses must be deposited in accounts in banks or other financial institutions. Such accounts must be fully covered by Federal deposit insurance or fully collateralized with other securities in accordance with normal banking practices and all applicable State laws. Any interest earned on the account remains a part of the HPRP revolving loan fund; and

(11) If an intermediary receives more than one HPRP loan, it does not need to establish and maintain a separate HPRP revolving loan fund for each loan; it may combine them and maintain only one HPRP revolving loan fund.

(e) *Budgets and administrative costs.* The intermediary must submit an annual budget of proposed administrative costs for Agency approval. The annual budget should itemize cash income and cash out-flow. Projected cash income should consist of, but is not limited to, collection of principal repayment, interest repayment, interest earnings on deposits, fees, and other income. Projected cash out-flow should consist of, but is not limited to, principal and interest payments, reserve for bad debt, and an itemization of administrative costs to operate the HPRP revolving loan fund.

(1) Proceeds received from the collection of principal repayment cannot be used for administrative expenses.

(2) The amount removed from the HPRP revolving loan fund for administrative costs in any year must be reasonable, must not exceed the actual cost of operating the HPRP revolving loan fund, including loan servicing and providing technical assistance, and must not exceed the amount approved by the Agency in the intermediary's annual budget.

(f) *Loan monitoring reviews.* The Agency may conduct loan monitoring reviews, including annual and periodic reviews, and performance monitoring.

(1) At least annually, the intermediary must provide the Agency documents for reviewing the financial status of the intermediary, assessing the progress of using loan funds, and identifying any potential problems or concerns. Non-regulated intermediaries must furnish audited financial statements at least annually.

(2) The intermediary must allow the Agency or its representative to review the operations and financial condition of the intermediary upon the Agency's request. The intermediary and its agents must provide access to all pertinent information to allow the Agency, or any party authorized by the Agency, to conduct such reviews. The intermediary must submit financial or other information within 14 calendar days upon receipt of the Agency's request, unless the data requested is not available within that time frame. Failure to supply the requested information to the satisfaction of the Agency will constitute non-monetary default. The Agency may conduct reviews, including on-site reviews, of the intermediary's operations and the operations of any agent of the intermediary, for the purpose of verifying compliance with Agency regulations and guidelines. These reviews may include, but are not limited to, audits of case files; interviews with owners, managers, and staff; audits of collateral; and inspections of the intermediary's and its agents under-

writing, servicing, and liquidation guidelines.

(g) *Annual monitoring reports.* Each intermediary will be monitored by the Agency through annual monitoring reports submitted by the intermediary. Annual monitoring reports must include a description of the use of loan funds, information regarding the acreage, the number of heirs both before and after loan was made, audit findings, disbursement transactions, and any other information required by the Agency, as necessary.

(h) *Unused loan funds.* If any part of the HPRP loan has not been used in accordance with the intermediary's relending plan within 3 years from the date of the HPRP loan agreement, the Agency may cancel the approval of any funds not delivered to the intermediary. The Agency may also direct the intermediary to return any funds delivered to the intermediary that have not been used by that intermediary in accordance with the intermediary's relending plan. The Agency may, at its sole discretion, allow the intermediary additional time to use the HPRP loan funds.

§769.165 Loan servicing.

(a) *Payments.* The intermediary will make payments to the Agency as specified in the HPRP loan documents. All payments will be applied to interest first, any additional amount will be applied to principal.

(b) *Restructuring.* The Agency may restructure the intermediary's loan debt, if:

(1) The loan objectives cannot be met unless the HPRP loan is restructured;

(2) The Agency's interest will be protected; and

(3) The restructuring will be within the Agency's budget authority.

(c) *Default.* The Agency will work with the intermediary to correct any default, subject to the requirements of paragraph (b) of this section. In the event of monetary or non-monetary default, the Agency will take all appropriate actions to protect its interest, including, but not limited to, declaring the debt fully due and payable and may proceed to enforce its rights under the HPRP loan agreement, and any other loan instruments relating to the loan under applicable law and regulations, and commencement of legal action to protect the Agency's interest. Violation of any agreement with the Agency or failure to comply with reporting or other HPRP requirements will be considered non-monetary default.

§ 769.166　Transfers and assumptions.

(a) All transfers and assumptions must be approved in advance by the Agency. The assuming entity must meet all eligibility criteria for HPRP.

(b) Available transfer and assumption options to eligible intermediaries include:

(1) The total indebtedness may be transferred to another eligible intermediary on the same rates and terms; or

(2) The total indebtedness may be transferred to another eligible intermediary on different terms not to exceed the term for which an initial loan can be made.

(c) The transferor must prepare the transfer document for the Agency's review prior to the transfer and assumption.

(d) The transferee must provide the Agency with information required in the application as specified in § 769.158.

(e) The Agency's approved form of the assumption agreement will formally authorize the transfer and assumption and will contain the Agency case number of the transferor and transferee.

(f) When the transferee makes a cash down-payment in connection with the transfer and assumption, any proceeds received by the transferor will be credited on the transferor's loan debt in order of maturity date.

§ 769.167　Appeals.

Any appealable adverse decision made by the Agency may be appealed upon written request of the intermediary as specified in 7 CFR part 11.

§ 769.168　Exceptions.

The Agency may grant an exception to any of the requirements of this subpart if the proposed change is in the best financial interest of the Government and not inconsistent with the authorizing law or any other applicable law.

PART 770—INDIAN TRIBAL LAND ACQUISITION LOANS

AUTHORITY: 5 U.S.C. 301 and 25 U.S.C. 488.

SOURCE: 66 FR 1567, Jan. 9, 2001, unless otherwise noted.

§ 770.1　Purpose.

This part contains the Agency's policies and procedures for making and servicing loans to assist a Native American tribe or tribal corporation with the acquisition of land interests within the tribal reservation or Alaskan community.

§ 770.2　Abbreviations and definitions.

(a) *Abbreviations.*

FSA Farm Service Agency, an Agency of the United States Department of Agriculture, including its personnel and any successor Agency.

ITLAP Indian Tribal Land Acquisition Program.

USPAP Uniform Standards of Professional Appraisal Practice.

(b) *Definitions.*

Administrator is the head of the Farm Service Agency.

Agency is Farm Service Agency (FSA).

Appraisal is an appraisal for the purposes of determining the market value of land (less value of any existing improvements that pass with the land) that meets the requirements of part 761 of this chapter.

Applicant is a Native American tribe or tribal corporation established pursuant to the Indian Reorganization Act seeking a loan under this part.

Loan funds refers to money loaned under this part.

Native American tribe is:

(1) An Indian tribe recognized by the Department of the Interior; or

(2) A community in Alaska incorporated by the Department of the Interior pursuant to the Indian Reorganization Act.

Rental value for the purpose of rental value write-downs, equals the average actual rental proceeds received from the lease of land acquired under ITLAP. If there are no rental proceeds, then rental value will be based on market data according to §770.10(e)(4).

Reservation is lands or interests in land within:

(1) The Native American tribe's reservation as determined by the Department of the Interior; or

(2) A community in Alaska incorporated by the Department of the Interior pursuant to the Indian Reorganization Act.

Reserve is an account established for loans approved in accordance with regulations in effect prior to February 8, 2001 which required that an amount equal to 10 percent of the annual payment be set aside each year until at least one full payment is available.

Tribal corporation is a corporation established pursuant to the Indian Reorganization Act.

[66 FR 1567, Jan. 9, 2001, as amended at 70 FR 7167, Feb. 11, 2005; 72 FR 51990, Sept. 12, 2007]

§770.3 Eligibility requirements.

An applicant must:

(a) Submit a completed Agency application form;

(b) Except for refinancing activities authorized in §770.4(c), obtain an option or other acceptable purchase agreement for land to be purchased with loan funds;

(c) Be a Native American tribe or a tribal corporation of a Native American tribe without adequate uncommitted funds, based on Generally Accepted Accounting Principles, or another financial accounting method acceptable to Secretary of Interior to acquire lands or interests therein within the Native American tribe's reservation for the use of the Native American tribe or tribal corporation or the members of either;

(d) Be unable to obtain sufficient credit elsewhere at reasonable rates and terms for purposes established in §770.4;

(e) Demonstrate reasonable prospects of success in the proposed operation of the land to be purchased with funds provided under this part by providing:

(1) A feasibility plan for the use of the Native American tribe's land and other enterprises and funds from any other source from which payment will be made;

(2) A satisfactory management and repayment plan; and

(3) A satisfactory record for paying obligations.

(f) Unless waived by the FSA Administrator, not have any outstanding debt with any Federal Agency (other than debt under the Internal Revenue Code of 1986) which is in a delinquent status.

(g) Not be subject to a judgment lien against the tribe's property arising out of a debt to the United States.

(h) Have not received a write-down as provided in §770.10(e) within the preceding 5 years.

[66 FR 1567, Jan. 9, 2001, as amended at 70 FR 7167, Feb. 11, 2005]

§770.4 Authorized loan uses.

Loan funds may only be used to:

(a) Acquire land and interests therein (including fractional interests, rights-of-way, water rights, easements, and other appurtenances (excluding improvements) that would normally pass with the land or are necessary for the proposed operation of the land) located within the Native American tribe's reservation which will be used for the benefit of the tribe or its members.

(b) Pay costs incidental to land acquisition, including but not limited to, title clearance, legal services, land surveys, and loan closing.

(c) Refinance non-United States Department of Agriculture preexisting debts the applicant incurred to purchase the land provided the following conditions exist:

(1) Prior to the acquisition of such land, the applicant filed a loan application regarding the purchase of such land and received the Agency's approval for the land purchase;

(2) The applicant could not acquire an option on such land;

(3) The debt for such land is a short term debt with a balloon payment that cannot be paid by the applicant and that cannot be extended or modified to

enable the applicant to satisfy the obligation; and

(4) The purchase of such land is consistent with all other applicable requirements of this part.

(d) Pay for the costs of any appraisal conducted pursuant to this part.

§ 770.5   Loan limitations.

(a) Loan funds may not be used for any land improvement or development purposes, acquisition or repair of buildings or personal property, payment of operating costs, payment of finder's fees, or similar costs, or for any purpose that will contribute to excessive erosion of highly erodible land or to the conversion of wetlands to produce an agriculture commodity as further established in part 799 of this chapter.

(b) The amount of loan funds used to acquire land may not exceed the market value of the land (excluding the value of any improvements) as determined by a current appraisal.

(c) Loan funds for a land purchase must be disbursed over a period not to exceed 24 months from the date of loan approval.

(d) The sale of assets that are not renewable within the life of the loan will require a reduction in loan principal equal to the value of the assets sold.

[66 FR 1567, Jan. 9, 2001, as amended at 81 FR 51285, Aug. 3, 2016]

§ 770.6   Rates and terms.

(a) *Term.* Each loan will be scheduled for repayment over a period not to exceed 40 years from the date of the note.

(b) *Interest rate.* The interest rate charged by the Agency will be the lower of the interest rate in effect at the time of the loan approval or loan closing, which is the current rate available in any FSA office. Except as provided in § 770.10(b) the interest rate will be fixed for the life of the loan.

§ 770.7   Security requirements.

(a) The applicant will take appropriate action to obtain and provide security for the loan.

(b) A mortgage or deed of trust on the land to be purchased by the applicant will be taken as security for a loan, except as provided in paragraph (c) of this section.

(1) If a mortgage or deed of trust is to be obtained on trust or restricted land and the applicant's constitution or charter does not specifically authorize mortgage of such land, the mortgage must be authorized by tribal referendum.

(2) All mortgages or deeds of trust on trust or restricted land must be approved by the Department of the Interior.

(c) The Agency may take an assignment of income in lieu of a mortgage or deed of trust provided:

(1) The Agency determines that an assignment of income provides as good or better security; and

(2) Prior approval of the Administrator has been obtained.

§ 770.8   Use of acquired land.

(a) *General.* Subject to § 770.5(d) land acquired with loan funds, or other property serving as the security for a loan under this part, may be leased, sold, exchanged, or subject to a subordination of the Agency's interests, provided:

(1) The Agency provides prior written approval of the action;

(2) The Agency determines that the borrower's loan obligations to the Agency are adequately secured; and

(3) The borrower's ability to repay the loan is not impaired.

(b) *Title.* Title to land acquired with a loan made under this part may, with the approval of the Secretary of the Interior, be taken by the United States in trust for the tribe or tribal corporation.

§ 770.9   Appraisals.

(a) The applicant or the borrower, as appropriate, will pay the cost of any appraisal required under this part.

(b) Appraisals must be completed in accordance with § 761.7 of this chapter.

§ 770.10   Servicing.

(a) *Reamortization*—(1) *Eligibility.* The Agency may consider reamortization of a loan provided:

(i) The borrower submits a completed Agency application form; and

(ii) The account is delinquent due to circumstances beyond the borrower's control and cannot be brought current within 1 year; or

(iii) The account is current, but due to circumstances beyond the borrower's control, the borrower will be unable to meet the annual loan payments.

(2) *Terms.* The term of a loan may not be extended beyond 40 years from the date of the original note.

(i) Reamortization within the remaining term of the loan will be predicated on a projection of the tribe's operating expenses indicating the ability to meet the new payment schedule; and

(ii) No intervening lien exists on the security for the loan which would jeopardize the Government's security priority.

(3) *Consolidation of notes.* If one or more notes are to be reamortized, consolidation of the notes is authorized.

(b) *Interest rate reduction.* The Agency may consider a reduction of the interest rate for an existing loan to the current interest rate as available from any Agency office provided:

(1) The borrower submits a completed Agency application form;

(2) The loan was made more than 5 years prior to the application for the interest reduction; and

(3) The Department of the Interior and the borrower certify that the borrower meets at least one of the criteria contained in paragraph (e)(2) of this section.

(c) *Deferral.* The Agency may consider a full or partial deferral for a period not to exceed 5 years provided:

(1) The borrower submits a completed Agency application form;

(2) The borrower presents a plan which demonstrates that due to circumstances beyond their control, they will be unable to meet all financial commitments unless the Agency payment is deferred; and

(3) The borrower will be able to meet all financial commitments, including the Agency payments, after the deferral period has ended.

(d) *Land exchanges.* In the cases where a borrower proposes to exchange any portion of land securing a loan for other land, title clearance and a new mortgage on the land received by the borrower in exchange, which adequately secures the unpaid principal balance of the loan, will be required unless the Agency determines any remaining land or other loan security is adequate security for the loan.

(e) *Debt write-down*—(1) *Application.* The Agency will consider debt write-down under either the land value option or rental value option, as requested by the borrower.

(i) The borrower must submit a completed Agency application form;

(ii) If the borrower applies and is determined eligible for a land value and a rental value write-down, the borrower will receive a write-down based on the write-down option that provides the greatest debt reduction.

(2) *Eligibility.* To be eligible for debt write-down, the borrower (in the case of a tribal corporation, the Native American tribe of the borrower) must:

(i) Be located in a county which is identified as a persistent poverty county by the United States Department of Agriculture, Economic Research Service pursuant to the most recent data from the Bureau of the Census; and

(ii) Have a socio-economic condition over the immediately preceding 5 year period that meets the following two factors as certified by the Native American tribe and the Department of the Interior:

(A) The Native American tribe has a per capita income for individual enrolled tribal members which is less than 50 percent of the Federally established poverty income rate established by the Department of Health and Human Services;

(B) The tribal unemployment rate exceeds 50 percent;

(3) *Land value write-down.* The Agency may reduce the unpaid principal and interest balance on any loan made to the current market value of the land that was purchased with loan funds provided:

(i) The market value of such land has declined by at least 25 percent since the land was purchased as established by a current appraisal;

(ii) Land value decrease is not attributed to the depletion of resources contained on or under the land;

(iii) The loan was made more than 5 years prior to the application for land value write-down;

(iv) The loan has not previously been written down under paragraph (e)(4) of this section and has not been written

377

down within the last 5 years under this paragraph, and

(v) The borrower must meet the eligibility requirements of paragraphs (a)(1)(ii) or (iii) of this section.

(4) *Rental value write-down.* The Agency may reduce the unpaid principal and interest on any loan, so the annual loan payment for the remaining term of each loan equals the average of annual rental value of the land purchased by each such loan for the immediately preceding 5-year period provided:

(i) The loan was made more than 5 years prior to the rental value writedown;

(ii) The description of the land purchased with the loan funds and the rental values used to calculate the 5 year average annual rental value of the land have been certified by the Department of the Interior;

(iii) The borrower provides a record of any actual rents received for the land for the preceding 5 years, which will be used to calculate the average rental value. This record must be certified by the Department of the Interior. For land that has not been leased or has not received any rental income, the borrower must provide a market value rent study report for the preceding 5 years, which identifies the average annual rental value based on the market data. The market value rent study report must be prepared by a certified general appraiser and meet the requirements of USPAP.

(iv) The borrower has not previously received a write-down under this paragraph and has not had a loan written down within the last 5 years under paragraph (e)(3) of this section, and

(v) The borrower must meet the eligibility requirements of paragraph (a)(1)(ii) or (iii) of this section.

(f) *Release of reserve.* Existing reserve accounts may be released for the purpose of making ITLAP loan payments or to purchase additional lands, subject to the following:

(1) A written request is received providing details of the use of the funds;

(2) The loan is not delinquent;

(3) The loan adequately secured by a general assignment of tribal income.

[66 FR 1567, Jan. 9, 2001; 66 FR 47877, Sept. 14, 2001, as amended at 70 FR 7167, Feb. 11, 2005; 72 FR 51990, Sept. 12, 2007]

PART 771—BOLL WEEVIL ERADICATION LOAN PROGRAM

AUTHORITY: 5 U.S.C. 301; 7 U.S.C. 1989; and Pub. L. 104-180, 110 Stat. 1569.

SOURCE: 67 FR 59771, Sept. 24, 2002, unless otherwise noted.

§ 771.1   Introduction.

The regulations in this part set forth the terms and conditions under which loans are made through the Boll Weevil Eradication Loan Program. The regulations in this part are applicable to applicants, borrowers, and other parties involved in the making, servicing, and liquidation of these loans. The program's objective is to assist producers and state government agencies in the eradication of boll weevils from cotton producing areas.

§ 771.2   Abbreviations and definitions.

The following abbreviations and definitions apply to this part:

(a) Abbreviations:

APHIS means the Animal and Plant Health Inspection Service of the United States Department of Agriculture, or any successor Agency.

FSA means the Farm Service Agency, its employees, and any successor agency.

(b) Definitions:

Extra payment means a payment derived from the sale of property serving as security for a loan, such as real estate or vehicles. Proceeds from program assessments and other normal operating income, when remitted for payment on a loan, will not be considered as an extra payment.

Non-profit corporation means a private domestic corporation created and organized under the laws of the State(s) in which the entity will operate whose net earnings are not distributable to any private shareholder or individual, and which qualifies under the Internal Revenue Service code.

Restructure means to modify the terms of a loan. This may include a modification of the interest rate and/or repayment terms of the loan.

Security means assets pledged as collateral to assure repayment of a loan in the event of default on the loan.

State organization means a quasi-state run public operation exclusively established and managed by state and/or non-state employees, with all employees currently dedicated to the specific task of eliminating the boll weevil from the cotton growing area of the state.

§ 771.3 [Reserved]

§ 771.4 Eligibility requirements.

(a) An eligible applicant must:

(1) Meet all requirements prescribed by APHIS to qualify for cost-share grant funds as determined by APHIS, (FSA will accept the determination by APHIS as to an organization's qualification);

(2) Have the appropriate charter and/ or legal authority as a non-profit corporation or as a State organization specifically organized to operate the boll weevil eradication program in any State, biological, or geographic region of any State in which it operates;

(3) Possess the legal authority to enter into contracts, including debt instruments;

(4) Operate in an area in which producers have approved a referendum authorizing producer assessments and in which an active eradication or post-eradication program is underway or scheduled to begin no later than the fiscal year following the fiscal year in which the application is submitted;

(5) Have the legal authority to pledge producer assessments as security for loans from FSA.

(b) Individual producers are not eligible for loans.

§ 771.5 Loan purposes.

(a) Loan funds may be used for any purpose directly related to boll weevil eradication activities, including, but not limited to:

(1) Purchase or lease of supplies and equipment;

(2) Operating expenses, including but not limited to, travel and office operations;

(3) Salaries and benefits.

(b) Loan funds may not be used to pay expenses incurred for lobbying, public relations, or related activities, or to pay interest on loans from the Agency.

§ 771.6 Environmental requirements.

No loan will be made until all Federal and state statutory and regulatory environmental requirements have been complied with.

§ 771.7 Equal opportunity and non-discrimination requirements.

No recipient of a boll weevil eradication loan shall directly, or through contractual or other arrangement, subject any person or cause any person to be subjected to discrimination on the basis of race, religion, color, national origin, gender, or other prohibited basis. Borrowers must comply with all applicable Federal laws and regulations regarding equal opportunity in hiring, procurement, and related matters.

§ 771.8 Other Federal, State, and local requirements.

(a) In addition to the specific requirements in this subpart, loan applications will be coordinated with all appropriate Federal, State, and local agencies.

(b) Borrowers are required to comply with all applicable:

(1) Federal, State, or local laws;

(2) Regulatory commission rules; and

(3) Regulations which are presently in existence, or which may be later adopted including, but not limited to, those governing the following:

(i) Borrowing money, pledging security, and raising revenues for repayment of debt;

(ii) Accounting and financial reporting; and

(iii) Protection of the environment.

§ 771.9 Interest rates, terms, security requirements, and repayment.

(a) *Interest rate.* The interest rate will be fixed for the term of the loan. The rate will be established by FSA, based upon the cost of Government borrowing for instruments on terms similar to that of the loan requested.

(b) *Term.* The loan term will be based upon the needs of the applicant to accomplish the objectives of the loan program as determined by FSA, but may not exceed 10 years.

(c) *Security requirements.* (1) Loans must be adequately secured as determined by FSA. FSA may require certain security, including but not limited to the following:

(i) Assignments of assessments, taxes, levies, or other sources of revenue as authorized by State law;

(ii) Investments and deposits of the applicant; and

(iii) Capital assets or other property of the applicant or its members.

(2) In those cases in which FSA and another lender will hold assignments of the same revenue as collateral, the other lender must agree to a prorated distribution of the assigned revenue. The distribution will be based upon the proportionate share of the applicant's debt the lender holds for the eradication zone from which the revenue is derived at the time of loan closing.

(d) *Repayment.* The applicant must demonstrate that income sources will be sufficient to meet the repayment requirements of the loan and pay operating expenses.

§ 771.10 [Reserved]

§ 771.11 Application.

A complete application will consist of the following:

(a) An application for Federal assistance (available in any FSA office);

(b) Applicant's financial projections including a cash flow statement showing the plan for loan repayment;

(c) Copies of the applicant's authorizing State legislation and organizational documents;

(d) List of all directors and officers of the applicant;

(e) Copy of the most recent audited financial statements along with updates through the most recent quarter;

(f) Copy of the referendum used to establish the assessments and a certification from the Board of Directors that the referendum passed;

(g) Evidence that the officers and employees authorized to disburse funds are covered by an acceptable fidelity bond;

(h) Evidence of acceptable liability insurance policies;

(i) Statement from the applicant addressing any current or pending litigation against the applicant as well as any existing judgments;

(j) A copy of a resolution passed by the Board of Directors authorizing the officers to incur debt on behalf of the borrower;

(k) Any other information deemed to be necessary by FSA to render a decision.

§ 771.12 Funding applications.

Loan requests will be processed based on the date FSA receives the application. Loan approval is subject to the availability of funds. However, when multiple applications are received on the same date and available funds will not cover all applications received, applications from active eradication areas, which FSA determines to be most critical for the accomplishment of program objectives, will be funded first.

§ 771.13 Loan closing.

(a) *Conditions.* The applicant must meet all conditions specified by the loan approval official in the notification of loan approval prior to closing.

(b) *Loan instruments and legal documents.* The borrower, through its authorized representatives will execute all loan instruments and legal documents required by FSA to evidence the debt, perfect the required security interest in property and assets securing the loan, and protect the Government's interest, in accordance with applicable State and Federal laws.

(c) *Loan agreement.* A loan agreement between the borrower and FSA will be required. The agreement will set forth performance criteria and other loan requirements necessary to protect the Government's financial and programmatic interest and accomplish the objectives of the loan. Specific provisions of the agreement will be developed on a case-by-case basis to address the particular situation associated with the loan being made. However, all loan agreements will include at least the following provisions:

(1) The borrower must submit audited financial statements to FSA at least annually;

(2) The borrower will immediately notify FSA of any adverse actions such as:

(i) Anticipated default on FSA debt;

(ii) Potential recall vote of an assessment referendum; or

(iii) Being named as a defendant in litigation;

(3) Submission of other specific financial reports for the borrower;

(4) The right of deferral under 7 U.S.C. 1981a; and

(5) Applicable liquidation procedures upon default.

(d) *Fees.* The borrower will pay all fees for recording any legal instruments determined to be necessary and all notary, lien search, and similar fees incident to loan transactions. No fees will be assessed for work performed by FSA employees.

§ 771.14 Loan monitoring.

(a) *Annual and periodic reviews.* At least annually, the borrower will meet with FSA representatives to review the financial status of the borrower, assess the progress of the eradication program utilizing loan funds, and identify any potential problems or concerns.

(b) *Performance monitoring.* At any time FSA determines it necessary, the borrower must allow FSA or its representative to review the operations and financial condition of the borrower. This may include, but is not limited to, field visits, and attendance at Foundation Board meetings. Upon FSA request, a borrower must submit any financial or other information within 14 days unless the data requested is not available within that time frame.

§ 771.15 Loan servicing.

(a) *Advances.* FSA may make advances to protect its financial interests and charge the borrower's account for the amount of any such advances.

(b) *Payments.* Payments will be made to FSA as set forth in loan agreements and debt instruments. The funds from extra payments will be applied entirely to loan principal.

(c) *Restructuring.* The provisions of 7 CFR part 766 are not applicable to loans made under this section. However, FSA may restructure loan debts; provided:

(1) The Government's interest will be protected;

(2) The restructuring will be performed within FSA budgetary restrictions; and

(3) The loan objectives cannot be met unless the loan is restructured.

(d) *Default.* In the event of default, FSA will take all appropriate actions to protect its interest.

[67 FR 59771, Sept. 24, 2002, as amended at 72 FR 64121, Nov. 15, 2007]

PART 772—SERVICING MINOR PROGRAM LOANS

AUTHORITY: 5 U.S.C. 301, 7 U.S.C. 1989, and 25 U.S.C. 490.

SOURCE: 68 FR 69949, Dec. 16, 2003, unless otherwise noted.

§ 772.1 Policy.

(a) *Purpose.* This part contains the Agency's policies and procedures for servicing Minor Program loans which include: Grazing Association loans, Irrigation and Drainage Association loans, and Non-Farm Enterprise and Recreation loans to individuals.

(b) *Appeals.* The regulations at 7 CFR parts 11 and 780 apply to decisions made under this part.

§ 772.2 Abbreviations and Definitions.

(a) *Abbreviations.*

AMP Association-Type Minor Program loan;
CFR Code of Federal Regulations;
FO Farm Ownership Loan;
FSA Farm Service Agency;
IMP Individual-Type Minor Program loan;
OL Operating Loan;
USDA United States Department of Agriculture.

(b) *Definitions.*

Association-Type Minor Program loans (AMP): Loans to Grazing Associations and Irrigation and Drainage Associations.

Entity: Cooperative, corporation, partnership, joint operation, trust, or limited liability company.

Graduation: The requirement contained in loan documents that borrowers pay their FSA loan in full with funds received from a commercial lending source as a result of improvement in their financial condition.

Individual-type Minor Program loans (IMP): Non-Farm Enterprise or Recreation loans to individuals.

Member: Any individual who has an ownership interest in the entity which has received the Minor Program loan.

Minor Program: Non-Farm Enterprise, Individual Recreation, Grazing Association, or Irrigation and Drainage loan programs administered or to be administered by FSA

Review official: An agency employee, contractor or designee who is authorized to conduct a compliance review of a Minor Program borrower under this part.

§ 772.3 Compliance.

(a) *Requirements.* No Minor Program borrower shall directly, or through contractual or other arrangement, subject any person or cause any person to be subjected to discrimination on the basis of race, color, national origin, or disability. Borrowers must comply with all applicable Federal laws and regulations regarding equal opportunity in hiring, procurement, and related matters. AMP borrowers are subject to the nondiscrimination provisions applicable to Federally assisted programs contained in 7 CFR part 15, subparts A and C, and part 15b. IMP loans are subject to the nondiscrimination provisions applicable to federally conducted programs contained in 7 CFR parts 15d and 15e.

(b) *Reviews.* In accordance with Title VI of the Civil Rights Act of 1964, the Agency will conduct a compliance review of all Minor Program borrowers, to determine if a borrower has directly, or through contractual or other arrangement, subjected any person or caused any person to be subjected to discrimination on the basis of race, color, or national origin. The borrower must allow the review official access to their premises and all records necessary to carry out the compliance review as determined by the review official.

(c) *Frequency and timing.* Compliance reviews will be conducted no later than October 31 of every third year until the Minor Program loan is paid in full or otherwise satisfied.

(d) *Violations.* If a borrower refuses to provide information or access to their premises as requested by a review official during a compliance review, or is determined by the Agency to be not in compliance in accordance with this section or Departmental regulations and procedures, the Agency will service the loan in accordance with the provisions of § 772.16 of this part.

§ 772.4 Environmental requirements.

Servicing activities such as transfers, assumptions, subordinations, sale or exchange of security property, and leasing of security will be reviewed for compliance with 7 CFR part 799.

[68 FR 69949, Dec. 16, 2003, as amended at 81 FR 51285, Aug. 3, 2016]

§ 772.5 Security maintenance.

(a) *General.* Borrowers are responsible for maintaining the collateral that is serving as security for their Minor Program loan in accordance with their lien instruments, security agreement and promissory note.

(b) *Security inspection.* The Agency will inspect real estate that is security for a Minor Program loan at least once every 3 years, and chattel security at least annually. More frequent security inspections may be made as determined necessary by the Agency. Borrowers will allow representatives of the Agency, or any agency of the U.S. Government, in accordance with statutes and regulations, such access to the security property as the agency determines is necessary to document compliance with the requirements of this section.

(c) *Violations.* If the Agency determines that the borrower has failed to adequately maintain security, made unapproved dispositions of security, or otherwise has placed the repayment of the Minor Program loan in jeopardy, the Agency will:

(1) For chattel security, service the account according to part 765 of this chapter. If any normal income security as defined in that subpart secures a Minor Program loan, the reporting, approval and release provisions in that subpart shall apply.

(2) For real estate security for AMP loans, contact the Regional Office of General Counsel for advice on the appropriate servicing including liquidation if warranted.

(3) For real estate security for IMP loans, service the account according to part 765 of this chapter.

[68 FR 69949, Dec. 16, 2003, as amended at 78 FR 65541, Nov. 1, 2013]

§772.6 Subordination of security.

(a) *Eligibility.* The Agency shall grant a subordination of Minor Program loan security when the transaction will further the purposes for which the loan was made, and all of the following are met:

(1) The loan will still be adequately secured after the subordination, or the value of the loan security will be increased by the amount of advances to be made under the terms of the subordination.

(2) The borrower can document the ability to pay all debts including the new loan.

(3) The action does not change the nature of the borrower's activities to the extent that they would no longer be eligible for a Minor Program loan.

(4) The subordination is for a specific amount.

(5) The borrower is unable, as determined by the Agency, to refinance its loan and graduate in accordance with this subpart.

(6) The loan funds will not be used in such a way that will contribute to erosion of highly erodible land or conversion of wetlands for the production of an agricultural commodity according to part 799 of this chapter.

(7) The borrower has not been convicted of planting, cultivating, growing, producing, harvesting or storing a controlled substance under Federal or state law. "Borrower," for purposes of this subparagraph, specifically includes an individual or entity borrower and any member of an entity borrower. "Controlled substance," for the purpose of this subparagraph, is defined at 21 CFR part 1308. The borrower will be ineligible for a subordination for the crop year in which the conviction occurred and the four succeeding crop years. An applicant must attest on the Agency application form that it, and its members if an entity, have not been convicted of such a crime.

(b) *Application.* To request a subordination, a Minor Program borrower must make the request in writing and provide the following:

(1) The specific amount of debt for which a subordination is needed;

(2) An appraisal prepared in accordance with §761.7 of this chapter, if the request is for a subordination of more than $10,000, unless a sufficient appraisal report, as determined by the Agency, that is less than one year old, is on file with the Agency; and

(3) Consent and subordination, as necessary, of all other creditors' security interests.

[68 FR 69949, Dec. 16, 2003, as amended at 81 FR 51285, Aug. 3, 2016]

§ 772.7 Leasing minor program loan security.

(a) *Eligibility.* The Agency may consent to the borrower leasing all or a portion of security property for Minor Program loans to a third party when:

(1) Leasing is the only feasible way to continue to operate the enterprise and is a customary practice;

(2) The lease will not interfere with the purpose for which the loan was made;

(3) The borrower retains ultimate responsibility for the operation, maintenance and management of the facility or service for its continued availability and use at reasonable rates and terms;

(4) The lease prohibits amendments to the lease or subleasing arrangements without prior written approval from the Agency;

(5) The lease terms provide that the Agency is a lienholder on the subject property and, as such, the lease is subordinate to the rights and claims of the Agency as lienholder; and

(6) The lease is for less than 3 years and does not constitute a lease/purchase arrangement, unless the transfer and assumption provisions of this subpart are met.

(b) *Application.* The borrower must submit a written request for Agency consent to lease the property.

§ 772.8 Sale or exchange of security property.

(a) For AMP loans.

(1) Sale of all or a portion of the security property may be approved when all of the following conditions are met:

(i) The property is sold for market value based on a current appraisal prepared in accordance with § 761.7 of this chapter.

(ii) The sale will not prevent carrying out the original purpose of the loan. The borrower must execute an Assurance Agreement as prescribed by the Agency. The covenant involved will remain in effect as long as the property continues to be used for the same or similar purposes for which the loan was made. The instrument of conveyance will contain the following non-discrimination covenant:

The property described herein was obtained or improved with Federal financial assistance and is subject to the non-discrimination provisions of title VI of the Civil Rights Act of 1964, title IX of the Education Amendments of 1972, section 504 of the Rehabilitation Act of 1973, and other similarly worded Federal statutes, and the regulations issued pursuant thereto that prohibit discrimination on the basis of race, color, national origin, handicap, religion, age, or sex in programs or activities receiving Federal financial assistance. Such provisions apply for as long as the property continues to be used for the same or similar purposes for which the Federal assistance was extended, or for so long as the purchaser owns it, whichever is later.

(iii) The remaining security for the loan is adequate or will not change after the transaction.

(iv) Sale proceeds remaining after paying any reasonable and necessary selling expenses are applied to the Minor Program loan according to lien priority.

(2) Exchange of all or a portion of security property for an AMP loan may be approved when:

(i) The Agency will obtain a lien on the property acquired in the exchange;

(ii) Property more suited to the borrower's needs related to the purposes of the loan is to be acquired in the exchange;

(iii) The AMP loan will be as adequately secured after the transaction as before; and

(iv) It is necessary to develop or enlarge the facility, improve the borrower's debt-paying ability, place the operation on a more sound financial basis or otherwise further the loan objectives and purposes, as determined by the Agency.

(b) For IMP loans, a sale or exchange of real estate or chattel that is serving as security must be done as specified in part 765 of this chapter.

[68 FR 69949, Dec. 16, 2003, as amended at 69 FR 18741, Apr. 8, 2004; 78 FR 65533, Nov. 1, 2013]

§ 772.9 Releases.

(a) *Security.* Minor Program liens may be released when:

(1) The debt is paid in full;

(2) Security property is sold for market value and sale proceeds are received and applied to the borrower's creditors according to lien priority; or

(3) An exchange in accordance with § 772.8 has been concluded.

Farm Service Agency, USDA §772.12

(b) *Borrower liability.* The Agency may release a borrower from liability when the Minor Program loan, plus all administrative collection costs and charges are paid in full. IMP borrowers who have had previous debt forgiveness on a farm loan program loan as defined in 7 CFR part 761, however, cannot be released from liability by FSA until the previous loss to the Agency has been repaid with interest from the date of debt forgiveness. An AMP borrower may also be released in accordance with §772.10 in conjunction with a transfer and assumption.

(c) *Servicing of debt not satisfied through liquidation.* Balances remaining after the sale or liquidation of the security will be serviced in accordance with part 761, subpart F of this chapter and part 3 of this title.

[68 FR 69949, Dec. 16, 2003, as amended at 69 FR 7679, Feb. 19, 2004; 72 FR 64121, Nov. 15, 2007; 85 FR 36713, June 17, 2020]

§772.10 Transfer and assumption—AMP loans.

(a) *Eligibility.* The Agency may approve transfers and assumptions of AMP loans when:

(1) The present borrower is unable or unwilling to accomplish the objectives of the loan;

(2) The transfer will not harm the Government or adversely affect the Agency's security position;

(3) The transferee will continue with the original purpose of the loan;

(4) The transferee will assume an amount at least equal to the present market value of the loan security;

(5) The transferee documents the ability to pay the AMP loan debt as provided in the assumption agreement and has the legal capacity to enter into the contract;

(6) If there is a lien or judgment against the Agency security being transferred, the transferee is subject to such claims. The transferee must document the ability to repay the claims against the land; and

(7) If the transfer is to one or more members of the borrower's organization and there is no new member, there must not be a loss to the Government.

(b) *Withdrawal.* Withdrawal of a member and transfer of the withdrawing member's interest in the Association to a new eligible member may be approved by the Agency if all of the following conditions are met:

(1) The entire unpaid balance of the withdrawing member's share of the AMP loan must be assumed by the new member;

(2) In accordance with the Association's governing articles, the required number of remaining members must agree to accept any new member; and

(3) The transfer will not adversely affect collection of the AMP loan.

(c) *Requesting a transfer and assumption.* The transferor/borrower and transferee/applicant must submit:

(1) The written consent of any other lienholder, if applicable.

(2) A current balance sheet and cash flow statement.

(d) *Terms.* The interest rate and term of the assumed AMP loan will not be changed. Any delinquent principal and interest of the AMP loan must be paid current before the transfer and assumption will be approved by the Agency.

(e) *Release of liability.* Transferors may be released from liability with respect to an AMP loan by the Agency when:

(1) The full amount of the loan is assumed; or

(2) Less than the full amount of the debt is assumed, and the balance remaining will be serviced in accordance with §772.9(c).

§772.11 Transfer and assumption—IMP loans.

Transfers and assumptions for IMP loans are processed in accordance with 7 CFR part 765. Any remaining transferor liability will be serviced in accordance with §772.9(c) of this subpart.

[68 FR 69949, Dec. 16, 2003, as amended at 72 FR 64121, Nov. 15, 2007]

§772.12 Graduation.

(a) *General.* This section only applies to Minor Program borrowers with promissory notes which contain provisions requiring graduation.

(b) *Graduation reviews.* Borrowers shall provide current financial information when requested by the Agency or its representatives to conduct graduation reviews.

(1) AMP loans shall be reviewed at least every two years. In the year to be reviewed, each borrower must submit, at a minimum, a year-end balance sheet and cash flow projection for the current year.

(2) All IMP borrowers classified as "commercial" or "standard" by the agency must be reviewed at least every 2 years. In the year to be reviewed, each borrower must submit a year-end balance sheet, actual financial performance for the most recent year, and a projected budget for the current year.

(c) *Criteria.* Borrowers must graduate from the Minor Programs as follows:

(1) Borrowers with IMP loans that are classified as "commercial" or "standard" must apply for private financing within 30 days from the date the borrower is notified of lender interest, if an application is required by the lender. For good cause, the Agency may grant the borrower a reasonable amount of additional time to apply for refinancing.

(2) Borrowers with AMP loans will be considered for graduation at least every two years or more frequently if the Agency determines that the borrower's financial condition has significantly improved.

[68 FR 69949, Dec. 16, 2003, as amended at 72 FR 64121, Nov. 15, 2007]

§ 772.13　Delinquent account servicing.

(a) *AMP loans.* If the borrower does not make arrangements to cure the default after notice by the Agency and is not eligible for reamortization in accordance with § 772.14, the Agency will liquidate the account in accordance with § 772.16. Delinquent AMP loans will be serviced in accordance with part 761, subpart F of this chapter and part 3 of this title.

(b) *IMP loans.* Delinquent IMP loans will be serviced in accordance with part 761, subpart F of this chapter and part 3 of this title.

[85 FR 36713, June 17, 2020]

§ 772.14　Reamortization of AMP loans.

The Agency may approve reamortization of AMP loans provided:

(a) There is no extension of the final maturity date of the loan;

(b) No intervening lien exists on the security for the loan which would jeopardize the Government's security position;

(c) If the account is delinquent, it cannot be brought current within one year and the borrower has presented a cash flow budget which demonstrates the ability to meet the proposed new payment schedule; and

(d) If the account is current, the borrower will be unable to meet the annual loan payments due to circumstances beyond the borrower's control.

§ 772.15　Protective advances.

(a) The Agency may approve, without regard to any loan or total indebtedness limitation, vouchers to pay costs, including insurance and real estate taxes, to preserve and protect the security, the lien, or the priority of the lien securing the debt owed to the Agency if the debt instrument provides that the Agency may voucher the account to protect its lien or security.

(b) The Agency may pay protective advances only when it determines it to be in the Government's best financial interest.

(c) Protective advances are immediately due and payable.

§ 772.16　Liquidation.

When the Agency determines that continued servicing will not accomplish the objectives of the loan and the delinquency or financial distress cannot be cured by the options in § 772.13, or the loan is in non-monetary default, the borrower will be encouraged to dispose of the Agency security voluntarily through sale or transfer and assumption in accordance with this part. If such a transfer or voluntary sale is not carried out, the loan will be liquidated according to 7 CFR part 766. For AMP loans, appeal rights under 7 CFR part 11 are provided in the notice of acceleration. For IMP loans, appeal rights must be exhausted before acceleration, and the notice of acceleration is not appealable.

[68 FR 69949, Dec. 16, 2003, as amended at 72 FR 64121, Nov. 15, 2007]

§772.17 Equal opportunity and non-discrimination requirements.

With respect to any aspect of a credit transaction, the Agency will comply with the requirements of the Equal Credit Opportunity Act and the Department's civil rights policy in 7 CFR part 15d.

[72 FR 64121, Nov. 15, 2007]

§772.18 Exception authority.

Exceptions to any requirement in this subpart can be approved in individual cases by the Administrator if application of any requirement or failure to take action would adversely affect the Government's financial interest. Any exception must be consistent with the authorizing statute and other applicable laws.

PART 773—SPECIAL APPLE LOAN PROGRAM

AUTHORITY: Pub. L. 106–224.

SOURCE: 65 FR 76117, Dec. 6, 2000, unless otherwise noted.

§773.1 Introduction.

This part contains the terms and conditions for loans made under the Special Apple Loan Program. These regulations are applicable to applicants, borrowers, and other parties involved in making, servicing, and liquidating these loans. The program objective is to assist producers of apples suffering from economic loss as a result of low apple prices.

§773.2 Definitions.

As used in this part, the following definitions apply:

Agency is the Farm Service Agency, its employees, and any successor agency.

Apple producer is a farmer in the United States or its territories that produced apples, on not less than 10 acres, for sale in 1999 or 2000.

Applicant is the individual or business entity applying for the loan.

Business entity is a corporation, partnership, joint operation, trust, limited liability company, or cooperative.

Cash flow budget is a projection listing all anticipated cash inflows (including all farm income, nonfarm income and all loan advances) and all cash outflows (including all farm and nonfarm debt service and other expenses) to be incurred by the borrower during the period of the budget. A cash flow budget may be completed either for a 12 month period, a typical production cycle or the life of the loan, as appropriate.

Domestically owned enterprise is an entity organized in the United States under the law of the state or states in which the entity operates and a majority of the entity is owned by members meeting the citizenship test.

False information is information provided by an applicant, borrower, or other source to the Agency which information is known by the provider to be incorrect, and was given to the Agency in order to obtain benefits for which the applicant or borrower would not otherwise have been eligible.

Feasible plan is a plan that demonstrates that the loan will be repaid as agreed, as determined by the Agency.

Security is real or personal property pledged as collateral to assure repayment of a loan in the event there is a default on the loan.

USPAP is Uniform Standards of Professional Appraisal Practice.

§773.3 Appeals.

A loan applicant or borrower may request an appeal or review of an adverse decision made by the Agency in accordance with 7 CFR part 11.

§§ 773.4–773.5 [Reserved]

§ 773.6 Eligibility requirements.

Loan applicants must meet all of the following requirements to be eligible for a Special Apple Program Loan:

(a) The loan applicant must be an apple producer;

(b) The loan applicant must be a citizen of the United States or an alien lawfully admitted to the United States for permanent residence under the Immigration and Nationalization Act. For a business entity applicant, the majority of the business entity must be owned by members meeting the citizenship test or, other entities that are domestically owned. Aliens must provide the appropriate Immigration and Naturalization Service forms to document their permanent residency;

(c) The loan applicant and anyone who will execute the promissory note must possess the legal capacity to enter into contracts, including debt instruments;

(d) At loan closing the loan applicant and anyone who will execute the promissory note must not be delinquent on any Federal debt, other than a debt under the Internal Revenue Code of 1986;

(e) At loan closing the loan applicant and anyone who will execute the promissory note must not have any outstanding unpaid judgments obtained by the United States in any court. Such judgments do not include those filed as a result of action in the United States Tax Courts;

(f) The loan applicant, in past or present dealings with the Agency, must not have provided the Agency with false information; and

(g) The individual or business entity loan applicant and all entity members must have acceptable credit history demonstrated by debt repayment. A history of failure to repay past debts as they came due (including debts to the Internal Revenue Service) when the ability to repay was within their control will demonstrate unacceptable credit history. Unacceptable credit history will not include isolated instances of late payments which do not represent a pattern and were clearly beyond the applicant's control or lack of credit history.

§ 773.7 Loan uses.

Loan funds may be used for any of the following purposes related to the production or marketing of apples:

(a) Payment of costs associated with reorganizing a farm to improve its profitability;

(b) Payment of annual farm operating expenses;

(c) Purchase of farm equipment or fixtures;

(d) Acquiring, enlarging, or leasing a farm;

(e) Making capital improvements to a farm;

(f) Refinancing indebtedness;

(g) Purchase of cooperative stock for credit, production, processing or marketing purposes; or

(h) Payment of loan closing costs.

§ 773.8 Limitations.

(a) The maximum loan amount any individual or business entity may receive under the Special Apple Loan Program is limited to $500,000.

(b) The maximum loan is further limited to $300 per acre of apple trees in production in 1999 or 2000, whichever is greater.

(c) Loan funds may not be used to pay expenses incurred for lobbying or related activities.

(d) Loans may not be made for any purpose which contributes to excessive erosion of highly erodible land or to the conversion of wetlands to produce an agricultural commodity.

§ 773.10 Other Federal, State, and local requirements.

Borrowers are required to comply with all applicable:

(a) Federal, State, or local laws;

(b) Regulatory commission rules; and

(c) Regulations which are presently in existence, or which may be later adopted including, but not limited to, those governing the following:

(1) Borrowing money, pledging security, and raising revenues for repayment of debt;

(2) Accounting and financial reporting; and

(3) Protection of the environment.

§§ 773.11–773.17 [Reserved]

§ 773.18 Loan application.

(a) A complete application will consist of the following:

(1) A completed Agency application form;

(2) If the applicant is a business entity, any legal documents evidencing the organization and any State recognition of the entity;

(3) Documentation of compliance with the Agency's environmental regulations contained in part 799 of this chapter;

(4) A balance sheet on the applicant;

(5) The farm's operating plan, including the projected cash flow budget reflecting production, income, expenses, and loan repayment plan;

(6) The last 3 years of production and income and expense information;

(7) Payment to the Agency for ordering a credit report; and

(8) Any additional information required by the Agency to determine the eligibility of the applicant, the feasibility of the operation, or the adequacy and availability of security.

(b) Except as required in § 773.19(e), the Agency will waive requirements for a complete application, listed in paragraphs (a)(5) and (a)(6) of this section, for requests of $30,000 or less.

[65 FR 76117, Dec. 6, 2000, as amended at 81 FR 51285, Aug. 3, 2016]

§ 773.19 Interest rate, terms, security requirements, and repayment.

(a) *Interest rate.* The interest rate will be fixed for the term of the loan. The rate will be established by the Agency and available in each Agency Office, based upon the cost of Government borrowing for loans of similar maturities.

(b) *Terms.* The loan term will be for up to 3 years, based upon the useful life of the security offered.

(c) *Security requirements.* The Agency will take a lien on the following security, if available, as necessary to adequately secure the loan:

(1) Real estate;

(2) Chattels;

(3) Crops;

(4) Other assets owned by the applicant; and

(5) Assets owned and pledged by a third party.

(d) *Documentation of security value.* (1) For loans that are for $30,000 or less, collateral value will be based on the best available, verifiable information.

(2) For loans of greater than $30,000 where the applicant's balance sheet shows a net worth of three times the loan amount or greater, collateral value will be based on tax assessment of real estate and depreciation schedules of chattels, as applicable, less any existing liens.

(3) For loans of greater than $30,000 where the applicant's balance sheet shows a net worth of less than three times the loan amount, collateral value will be based on an appraisal. Such appraisals must be obtained by the applicant, at the applicant's expense and acceptable to the Agency. Appraisals of real estate must be completed in accordance with USPAP.

(e) *Repayment.* (1) All loan applicants must demonstrate that the loan can be repaid.

(2) For loans that are for $30,000 or less where the applicant's balance sheet shows a net worth of three times the loan amount or greater, repayment ability will be considered adequate without further documentation.

(3) For loans that are for $30,000 or less where the applicant's balance sheet shows a net worth of less than three times the loan amount, repayment ability must be demonstrated using the farm's operating plan, including a projected cash flow budget based on historical performance. Such operating plan is required notwithstanding § 773.18 of this part.

(4) For loans that are for more than $30,000, repayment ability must be demonstrated using the farm's operating plan, including a projected cash flow budget based on historical performance.

(f) *Creditworthiness.* All loan applicants must have an acceptable credit history demonstrated by debt repayment. A history of failure to repay past debts as they came due (including debts to the Internal Revenue Service) when the ability to repay was within their control will demonstrate unacceptable credit history. Unacceptable credit history will not include isolated instances of late payments which do not represent a pattern and were clearly beyond the applicant's control or lack of credit history.

§ 773.20 Funding applications.

Loan requests will be funded based on the date the Agency approves the application. Loan approval is subject to the availability of funds.

§ 773.21 Loan decision, closing, and fees.

(a) *Loan decision.* (1) The Agency will approve a loan if it determines that:

(i) The loan can be repaid;

(ii) The proposed use of loan funds is authorized;

(iii) The applicant has been determined eligible;

(iv) All security requirements have been, or will be met at closing;

(vi) All other pertinent requirements have been, or will be met at closing.

(2) The Agency will place conditions upon loan approval as necessary to protect its interest.

(b) *Loan closing.* (1) The applicant must meet all conditions specified by the loan approval official in the notification of loan approval prior to loan closing;

(2) There must have been no significant changes in the plan of operation or the applicant's financial condition since the loan was approved; and

(2) The applicant will execute all loan instruments and legal documents required by the Agency to evidence the debt, perfect the required security interest in property securing the loan, and protect the Government's interests, in accordance with applicable State and Federal laws. In the case of an entity applicant, all officers or partners and any board members also will be required to execute the promissory notes as individuals.

(c) *Fees.* The applicant will pay all loan closing fees including credit report fees, fees for appraisals, fees for recording any legal instruments determined to be necessary, and all notary, lien search, and similar fees incident to loan transactions. No fees will be assessed for work performed by Agency employees.

§ 773.22 Loan servicing.

Loans will be serviced as a Non-program loan in accordance with 7 CFR part 766 during the term of the loan. If the loan is not paid in full during this term, servicing will proceed in accordance with 7 CFR part 766, subpart H.

[72 FR 64121, Nov. 15, 2007]

§ 773.23 Exception.

The Agency may grant an exception to the security requirements of this section, if the proposed change is in the best financial interest of the Government and not inconsistent with the authorizing statute or other applicable law.

PART 774—EMERGENCY LOAN FOR SEED PRODUCERS PROGRAM

AUTHORITY: Pub. L. 106–224

SOURCE: 65 FR 76119, Dec. 6, 2000, unless otherwise noted.

§ 774.1 Introduction.

The regulations of this part contain the terms and conditions under which loans are made under the Emergency Loan for Seed Producers Program. These regulations are applicable to applicants, borrowers, and other parties involved in making, servicing, and liquidating these loans. The program objective is to assist certain seed producers adversely affected by the bankruptcy filing of AgriBiotech.

§774.2 Definitions.

As used in this part, the following definitions apply:

Agency is the Farm Service Agency, its employees, and any successor agency.

Applicant is the individual or business entity applying for the loan.

Business entity is a corporation, partnership, joint operation, trust, limited liability company, or cooperative.

Domestically owned enterprise is an entity organized in the United States under the law of the state or states in which the entity operates and a majority of the entity is owned by members meeting the citizenship test.

False information is information provided by an applicant, borrower or other source to the Agency that the borrower knows to be incorrect, and that the borrower or other source provided in order to obtain benefits for which the borrower would not otherwise have been eligible.

Seed producer is a farmer that produced a 1999 crop of grass, forage, vegetable, or sorghum seed for sale to AgriBiotech under contract.

§774.3 Appeals.

A loan applicant or borrower may request an appeal or review of an adverse decision made by the Agency in accordance with 7 CFR part 11.

§§774.4–774.5 [Reserved]

§774.6 Eligibility requirements.

Loan applicants must meet all of the following requirements to be eligible under the Emergency Loan for Seed Producers Program;

(a) The loan applicant must be a seed producer;

(b) The individual or entity loan applicant must have a timely filed proof of claim in the Chapter XI bankruptcy proceedings involving AgriBiotech and the claim must have arisen from acontract to grow seeds in the United States;

(c) The loan applicant must be a citizen of the United States or an alien lawfully admitted to the United States for permanent residence under the Immigration and Nationalization Act. For a business entity applicant, the majority of the business entity must be owned by members meeting the citizenship test or, other entities that are domestically owned. Aliens must provide the appropriate Immigration and Naturalization Service forms to document their permanent residency;

(d) The loan applicant and anyone who will execute the promissory note must possess the legal capacity to enter into contracts, including debt instruments;

(e) At loan closing, the applicant and anyone who will execute the promissory note must not be delinquent on any Federal debt, other than a debt under the Internal Revenue Code of 1986;

(f) At loan closing, the applicant and anyone who will execute the promissory note must not have any outstanding unpaid judgments obtained by the United States in any court. Such judgments do not include those filed as a result of action in the United States Tax Courts;

(g) The loan applicant, in past and current dealings with the Agency, must not have provided the Agency with false information.

§774.7 [Reserved]

§774.8 Limitations.

(a) The maximum loan amount any individual or business entity may receive will be 65% of the value of the timely filed proof of claim against AgriBiotech in the bankruptcy proceeding as determined by the Agency.

(b) Loan funds may not be used to pay expenses incurred for lobbying or related activities.

(c) Loans may not be made for any purpose which contributes to excessive erosion of highly erodible land or to the conversion of wetlands to produce an agricultural commodity.

§774.10 Other Federal, State, and local requirements.

Borrowers are required to comply with all applicable:

(a) Federal, State, or local laws;

(b) Regulatory commission rules; and

(c) Regulations which are presently in existence, or which may be later adopted including, but not limited to, those governing the following:

(1) Borrowing money, pledging security, and raising revenues for repayment of debt;

(2) Accounting and financial reporting; and

(3) Protection of the environment.

§§ 774.11–774.16 [Reserved]

§ 774.17 Loan application.

A complete application will consist of the following:

(a) A completed Agency application form;

(b) Proof of a bankruptcy claim in the AgriBiotech bankruptcy proceedings;

(c) If the applicant is a business entity, any legal documents evidencing the organization and any State recognition of the entity;

(d) Documentation of compliance with the Agency's environmental regulations contained in part 799 of this chapter;

(e) A balance sheet on the applicant; and

(f) Any other additional information the Agency needs to determine the eligibility of the applicant and the application of any Federal, State or local laws.

[65 FR 76119, Dec. 6, 2000, as amended at 81 FR 51285, Aug. 3, 2016]

§ 774.18 Interest rate, terms and security requirements.

(a) *Interest rate.* (1) The interest rate on the loan will be zero percent for 36 months or until the date of settlement of, completion of, or final distribution of assets in the bankruptcy proceeding involving AgriBiotech, whichever comes first.

(2) Thereafter interest will begin to accrue at the regular rate for an Agency Farm operating-direct loan (available in any Agency office).

(b) *Terms.* (1) Loans shall be due and payable upon the earlier of the settlement of the bankruptcy claim or 36 months from the date of the note.

(2) However, any principal remaining thereafter will be amortized over a term of 7 years at the Farm operating-direct loan interest rate (available in any Agency office). If the loan is not paid in full during this time and default occurs, servicing will proceed in accordance with 7 CFR part 766, subpart H.

(c) *Security requirements.* (1) The Agency will require a first position pledge and assignment of the applicant's monetary claim in the AgriBiotech bankruptcy estate to secure the loan.

(2) If the applicant has seed remaining in their possession that was produced under contract to AgriBiotech, the applicant also will provide the Agency with a first lien position on this seed. It is the responsibility of the applicant to negotiate with any existing lienholders to secure the Agency's first lien position.

[65 FR 76119, Dec. 6, 2000, as amended at 68 FR 7696, Feb. 18, 2003; 72 FR 64121, Nov. 15, 2007]

§ 774.19 Processing applications.

Applications will be processed until such time that funds are exhausted, or all claims have been paid and the bankruptcy involving AgriBiotech has been discharged. When all loan funds have been exhausted or the bankruptcy is discharged, no further applications will be accepted and any pending applications will be considered withdrawn.

§ 774.20 Funding applications.

Loan requests will be funded based on the date the Agency approves an application. Loan approval is subject to the availability of funds.

§ 774.21 [Reserved]

§ 774.22 Loan closing.

(a) *Conditions.* The applicant must meet all conditions specified by the loan approval official in the notification of loan approval prior to closing.

(b) *Loan instruments and legal documents.* The applicant will execute all loan instruments and legal documents required by the Agency to evidence the debt, perfect the required security interest in the bankruptcy claim, and protect the Government's interest, in accordance with applicable State and Federal laws. In the case of an entity applicant, all officers or partners and any board members also will be required to execute the promissory notes as individuals.

(c) *Fees.* The applicant will pay all loan closing fees for recording any legal instruments determined to be necessary and all notary, lien search, and similar fees incident to loan transactions. No fees will be assessed for work performed by Agency employees.

§ 774.23 Loan servicing.

Loans will be serviced as a Non-program loan in accordance with 7 CFR part 766. If the loan is not repaid as agreed and default occurs, servicing will proceed in accordance with 7 CFR part 766, subpart H.

[72 FR 64121, Nov. 15, 2007]

§ 774.24 Exception.

The Agency may grant an exception to any of the requirements of this section, if the proposed change is in the best financial interest of the Government and not inconsistent with the authorizing statute or other applicable law.

PART 780—APPEAL REGULATIONS

AUTHORITY: 5 U.S.C. 301 and 574; 7 U.S.C. 6995; 15 U.S.C. 714b and 714c; 16 U.S.C. 590h.

SOURCE: 70 FR 43266, July 27, 2005, unless otherwise noted.

§ 780.1 General.

This part sets forth rules applicable to appealability reviews, reconsiderations, appeals and alternative dispute resolution procedures comprising in aggregate the informal appeals process of FSA. FSA will apply these rules to facilitate and expedite participants' submissions and FSA reviews of documentary and other evidence material to resolution of disputes arising under agency program regulations.

§ 780.2 Definitions.

For purposes of this part:

1994 Act means the Federal Crop Insurance Reform and Department of Agriculture Reorganization Act of 1994 (Pub. L. 103–354).

Adverse decision means a program decision by an employee, officer, or committee of FSA that is adverse to the participant. The term includes any denial of program participation, benefits, written agreements, eligibility, etc., that results in a participant receiving less funds than the participant believes should have been paid or not receiving a program benefit to which the participant believes the participant was entitled.

Agency means FSA and its county and State committees and their personnel, CCC, NRCS, and any other agency or office of the Department which the Secretary may designate, or any successor agency.

Agency record means all documents and materials maintained by FSA that are related to the adverse decision under review that are compiled and reviewed by the decision-maker or that are compiled in the record provided to the next level reviewing authority.

Appeal means a written request by a participant asking the next level reviewing authority within FSA to review a decision. However, depending on the context, the term may also refer to a request for review by NAD.

Appealability review means review of a decision-maker's determination that a decision is not appealable under this part. That decision is, however, subject to review according to § 780.5 or 7 CFR part 11 to determine whether the decision involves a factual dispute that is appealable or is, instead, an attempt to challenge generally applicable program policies, provisions, regulations, or statutes that were not appealable.

Appellant means any participant who appeals or requests reconsideration or mediation of an adverse decision in accordance with this part or 7 CFR part 11.

Authorized representative means a person who has obtained a Privacy Act waiver and is authorized in writing by a participant in a reconsideration, mediation, or appeal.

CCC means the Commodity Credit Corporation, a wholly owned Government corporation within USDA.

Certified State means, in connection with mediation, a State with a mediation program, approved by the Secretary, that meets the requirements of 7 CFR part 785.

Confidential mediation means a mediation process in which neither the mediator nor parties participating in mediation will disclose to any person oral or written communications provided to the mediator in confidence, except as allowed by 5 U.S.C. 574 or 7 CFR part 785.

County committee means an FSA county or area committee established in accordance with section 8(b) of the Soil Conservation and Domestic Allotment Act (16 U.S.C. 590h(b)).

Determination of NRCS means a decision by NRCS made pursuant to Title XII of the Food Security Act of 1985 (16 U.S.C. 3801 *et seq.*), as amended.

FSA means the Farm Service Agency, an agency within USDA.

Final decision means a program decision rendered by an employee or officer of FSA pursuant to delegated authority, or by the county or State committee upon written request of a participant. A decision that is otherwise final shall remain final unless the decision is timely appealed to the State committee or NAD. A decision of FSA made by personnel subordinate to the county committee is considered "final" for the purpose of appeal to NAD only after that decision has been appealed to the county committee under the provisions of this part.

Hearing means an informal proceeding on an appeal to afford a participant opportunity to present testimony, documentary evidence, or both to show why an adverse decision is in error and why the adverse decision should be reversed or modified.

Implement means the taking of action by FSA, NRCS, or CCC that is necessary to effectuate fully and promptly a final decision.

Mediation means a technique for resolution of disputes in which a mediator assists disputing parties in voluntarily reaching mutually agreeable settlement of issues within the laws, regulations, and the agency's generally applicable program policies and procedures, but in which the mediator has no authoritative decision making power.

Mediator means a neutral individual who functions specifically to aid the parties in a dispute during a mediation process.

NAD means the USDA National Appeals Division established pursuant to the 1994 Act.

NAD rules means the NAD rules of procedure published at 7 CFR part 11, implementing title II, subtitle H of the 1994 Act.

Non-certified State means a State that is not approved to participate in the certified mediation program under 7 CFR part 785, or any successor regulation.

NRCS means the Natural Resources Conservation Service of USDA.

Participant means any individual or entity who has applied for, or whose right to participate in or receive, a payment, loan, loan guarantee, or other benefit in accordance with any program of FSA to which the regulations in this part apply is affected by a decision of FSA. The term includes anyone meeting this definition regardless of whether, in the particular proceeding, the participant is an appellant or a third party respondent. The term does not include individuals or entities whose claim(s) arise under the programs excluded in the definition of "participant" published at 7 CFR 11.1.

Qualified mediator means a mediator who meets the training requirements established by State law in the State in which mediation services will be provided or, where a State has no law prescribing mediator qualifications, an individual who has attended a minimum of 40 hours of core mediator knowledge and skills training and, to remain in a qualified mediator status, completes a minimum of 20 hours of additional training or education during each 2-year period. Such training or education must be approved by USDA, by an accredited college or university, or by one of the following organizations: State Bar of a qualifying State, a State mediation association, a State approved mediation program, or a society of dispute resolution professionals.

Reconsideration means a subsequent consideration of a program decision by the same level of decision-maker or reviewing authority.

Reviewing authority means a person or committee assigned the responsibility of making a decision on reconsideration or an appeal filed by a participant in accordance with this part.

State committee means an FSA State committee established in accordance with Section 8(b) of the Soil Conservation and Domestic Allotment Act (16 U.S.C. 590h(b)) including, where appropriate, the Director of the Caribbean Area FSA office for Puerto Rico and the Virgin Islands.

State Conservationist means the NRCS official in charge of NRCS operations within a State, as set forth in part 600 of this title.

State Executive Director means the executive director of an FSA State office with administrative responsibility for a FSA State office as established under the Reorganization Act.

USDA means the U.S. Department of Agriculture.

Verbatim transcript means an official, written record of proceedings in an appeal hearing or reconsideration of an adverse decision appealable under this part.

§780.3 Reservations of authority.

(a) Representatives of FSA and CCC may correct all errors in data entered on program contracts, loan agreements, and other program documents and the results of the computations or calculations made pursuant to the contract or agreement. FSA and CCC will furnish appropriate notice of such corrections when corrections are deemed necessary.

(b) Nothing contained in this part shall preclude the Secretary, or the Administrator of FSA, Executive Vice President of CCC, the Chief of NRCS, if applicable, or a designee, from determining at any time any question aris-

ing under the programs within their respective authority or from reversing or modifying any decision made by a subordinate employee of FSA or its county and State committees, or CCC.

§780.4 Applicability.

(a)(1) Except as provided in other regulations, this part applies to decisions made under programs and by agencies, as set forth herein:

(i) Decisions in programs administered by FSA to make, guarantee or service farm loans set forth in chapters VII and XVIII of this title relating to farm loan programs;

(ii) Decisions in those domestic programs administered by FSA on behalf of CCC through State and county committees, or itself, which are generally set forth in chapters VII and XIV of this title, or in part VII relating to conservation or commodities;

(iii) Appeals from adverse decisions, including technical determinations, made by NRCS under title XII of the Food Security Act of 1985, as amended;

(iv) Penalties assessed by FSA under the Agricultural Foreign Investment Disclosure Act of 1978, 5 U.S.C. 501 *et seq.*;

(v) Decisions on equitable relief made by a State Executive Director or State Conservationist pursuant to section 1613 of the Farm Security and Rural Investment Act of 2002, Pub. L. 107–171; and

(vi) Other programs to which this part is made applicable by specific program regulations or notices in the FEDERAL REGISTER.

(2) The procedures contained in this part may not be used to seek review of statutes or regulations issued under Federal law or review of FSA's generally applicable interpretations of such laws and regulations.

(3) For covered programs, this part is applicable to any decision made by an employee of FSA or of its State and county committees, CCC, the personnel of FSA, or CCC, and by the officials of NRCS to the extent otherwise provided in this part, and as otherwise may be provided in individual program requirements or by the Secretary.

(b) With respect to matters identified in paragraph (a) of this section, participants may request appealability review, reconsideration, mediation, or appeal under the provisions of this part, of decisions made with respect to:

(1) Denial of participation in a program;

(2) Compliance with program requirements;

(3) Issuance of payments or other program benefits to a participant in a program; and

(4) Determinations under Title XII of the Food Security Act of 1985, as amended, made by NRCS.

(c) Only a participant directly affected by a decision may seek administrative review under § 780.5(c).

§ 780.5 Decisions that are not appealable.

(a) Decisions that are not appealable under this part shall include the following:

(1) Any general program provision or program policy or any statutory or regulatory requirement that is applicable to similarly situated participants;

(2) Mathematical formulas established under a statute or program regulation and decisions based solely on the application of those formulas;

(3) Decisions made pursuant to statutory provisions that expressly make agency decisions final or their implementing regulations;

(4) Decisions on equitable relief made by a State Executive Director or State Conservationist pursuant to Section 1613 of the Farm Security and Rural Investment Act of 2002, Pub. L. 107–171;

(5) Decisions of other Federal or State agencies;

(6) Requirements and conditions designated by law to be developed by agencies other than FSA.

(7) Disapprovals or denials because of a lack of funding.

(8) Decisions made by the Administrator or a Deputy Administrator.

(b) A participant directly affected by an adverse decision that is determined not to be subject to appeal under this part may request an appealability review of the determination by the State Executive Director of the State from which the underlying decision arose in accordance with § 780.15.

(c) Decisions that FSA renders under this part may be reviewed by NAD under part 11 of this title to the extent otherwise allowed by NAD under its rules and procedures. An appealability determination of the State Executive Director in an administrative review is considered by FSA to be a new decision.

§ 780.6 Appeal procedures available when a decision is appealable.

(a) For covered programs administered by FSA for CCC, the following procedures are available:

(1) Appeal to the county committee of decisions of county committee subordinates;

(2) Reconsideration by the county committee;

(3) Appeal to the State committee;

(4) Reconsideration by the State committee;

(5) Appeal to NAD;

(6) Mediation under guidelines specified in § 780.9.

(b) For decisions in agricultural credit programs administered by FSA, the following procedures are available:

(1) Reconsideration under § 780.7;

(2) Mediation under § 780.9;

(3) Appeal to NAD.

(c) For programs and regulatory requirements under Title XII of the Food Security Act of 1985, as amended, to the extent not covered by paragraph (a) of this section, the following procedures are available:

(1) Appeal to the county committee;

(2) Appeal to the State committee;

(3) Mediation under § 780.9;

(4) Appeal to NAD.

§ 780.7 Reconsideration.

(a) A request for reconsideration must be submitted in writing by a participant or by a participant's authorized representative and addressed to the FSA decision maker as will be instructed in the adverse decision notification.

(b) A participant's right to request reconsideration is waived if, before requesting reconsideration, a participant:

(1) Has requested and begun mediation of the adverse decision;

(2) Has appealed the adverse decision to a higher reviewing authority in FSA; or

(3) Has appealed to NAD.

(c) Provided a participant has not waived the right to request reconsideration, FSA will consider a request for reconsideration of an adverse decision under these rules except when a request concerns a determination of NRCS appealable under the procedures in § 780.11, the decision has been mediated, the decision has previously been reconsidered, or the decision-maker is the Administrator, Deputy Administrator, or other FSA official outside FSA's informal appeals process.

(d) A request for reconsideration will be deemed withdrawn if a participant requests mediation or appeals to a higher reviewing authority within FSA or requests an appeal by NAD before a request for reconsideration has been acted upon.

(e) The Federal Rules of Evidence do not apply to reconsiderations. Proceedings may be confined to presentations of evidence to material facts, and evidence or questions that are irrelevant, unduly repetitious, or otherwise inappropriate may be excluded.

(f) The official decision on reconsideration will be the decision letter that is issued following disposition of the reconsideration request.

(g) A decision on reconsideration is a new decision that restarts applicable time limitations periods under § 780.15 and part 11 of this title.

[70 FR 43266, July 27, 2005, as amended at 71 FR 30573, May 30, 2006]

§ 780.8 County committee appeals.

(a) A request for appeal to a county committee concerning a decision of a subordinate of the county committee must be submitted by a participant or by a participant's authorized representative in writing and must be addressed to the office in which the subordinate is employed.

(b) The Federal Rules of Evidence do not apply to appeals to a county committee. However, a county committee may confine presentations of evidence to material facts and may exclude evidence or questions that are irrelevant, unduly repetitious, or otherwise inappropriate.

(c) The official county committee decision on an appeal will be the decision letter that is issued following disposition of the appeal.

(d) Deliberations shall be in confidence except to the extent that a county committee may request the assistance of county committee or FSA employees during deliberations.

§ 780.9 Mediation.

(a) Any request for mediation must be submitted after issuance of an adverse decision but before any hearing in an appeal of the adverse decision to NAD.

(b) An adverse decision and any particular issues of fact material to an adverse decision may be mediated only once:

(1) If resolution of an adverse decision is not achieved in mediation, a participant may exercise any remaining appeal rights under this part or appeal to NAD in accordance with part 11 of this title and NAD procedures.

(2) If an adverse decision is modified as a result of mediation, a participant may exercise any remaining appeal rights as to the modified decision under this part or appeal to NAD, unless such appeal rights have been waived pursuant to agreement in the mediation.

(c) Any agreement reached during, or as a result of, the mediation process shall conform to the statutory and regulatory provisions governing the program and FSA's generally applicable interpretation of those statutes and regulatory provisions.

(d) FSA will participate in mediation in good faith and to do so will take steps that include the following:

(1) Designating a representative in the mediation;

(2) Instructing the representative that any agreement reached during, or as a result of, the mediation process must conform to the statutes, regulations, and FSA's generally applicable interpretations of statutes and regulations governing the program;

(3) Assisting as necessary in making pertinent records available for review and discussion during the mediation; and

(4) Directing the representative to forward any written agreement proposed in mediation to the appropriate FSA official for approval.

(e) Mediations will be treated in a confidential manner consistent with the purposes of the mediation.

(f) For requests for mediation in a Certified State, if the factual issues implicated in an adverse decision have not previously been mediated, notice to a participant of an adverse decision will include notice of the opportunity for mediation, including a mailing address and facsimile number, if available, that the participant may use to submit a written request for mediation.

(1) If the participant desires mediation, the participant must request mediation in writing by contacting the certified mediation program or such other contact as may be designated by FSA in an adverse decision letter. The request for mediation must include a copy of the adverse decision to be mediated.

(2) Participants in mediation may be required to pay fees established by the mediation program.

(3) A listing of certified State mediation programs and means for contact may be found on the FSA Web site at *http://www.usda.gov/fsa/ disputemediation.htm.*

(g) For requests for mediation in a Non-certified State, if the factual issues implicated in an adverse decision have not previously been mediated, notice to a participant of an adverse decision will, as appropriate, include notice of the opportunity for mediation, including the mailing address of the State Executive Director and a facsimile number, if available, that the participant may use to submit a written request for mediation.

(1) It is the duty of the participant to contact the State Executive Director in writing to request mediation. The request for mediation must include a copy of the adverse decision to be mediated.

(2) If resources are available for mediation, the State Executive Director will select a qualified mediator and provide written notice to the participant that mediation is available and

the fees that the participant will incur for mediation.

(3) If the participant accepts such mediation, FSA may give notice of the mediation to interested parties and third parties whose interests are known to FSA.

(h) Mediation will be considered to be at an end on that date set out in writing by the mediator or mediation program, as applicable, or when the participant receives written notice from the State Executive Director that the State Executive Director believes the mediation is at an impasse, whichever is earlier.

(i) To provide for mediator impartiality:

(1) No person shall be designated as mediator in an adverse program dispute who has previously served as an advocate or representative for any party in the mediation.

(2) As a condition of retention to mediate in an adverse program dispute under this part, the mediator shall agree not to serve thereafter as an advocate or representative for a participant or party in any other proceeding arising from or related to the mediated dispute, including, without limitation, representation of a mediation participant before an administrative appeals entity of USDA, or any other Federal Government department.

[70 FR 43266, July 27, 2005, as amended at 71 FR 30573, May 30, 2006]

§ 780.10　State committee appeals.

(a) A request for appeal to the State committee from a decision of a county committee must be submitted by a participant or by a participant's authorized representative in writing and addressed to the State Executive Director.

(b) A participant's right to appeal a decision to a State committee is waived if a participant has appealed the adverse decision to NAD before requesting an appeal to the State Committee.

(c) If a participant requests mediation or requests an appeal to NAD before a request for an appeal to the State Committee has been acted upon, the appeal to the State Committee will be deemed withdrawn. The deemed withdrawal of a participant's appeal to the State Committee will not preclude a subsequent request for a State Committee hearing on appealable matters not resolved in mediation.

(d) The Federal Rules of Evidence do not apply in appeals to a State committee. Notwithstanding, a State committee may confine presentations of evidence to material facts and exclude evidence or questions as irrelevant, unduly repetitious, or otherwise inappropriate.

(e) The official record of a State committee decision on an appeal will be the decision letter that is issued following disposition of the appeal.

(f) Deliberations shall be in confidence except to the extent that a State committee may request the assistance of FSA employees during deliberations.

[70 FR 43266, July 27, 2005, as amended at 71 FR 30573, May 30, 2006]

§780.11 Appeals of NRCS determinations.

(a) Notwithstanding any other provision of this part, a determination of NRCS issued to a participant pursuant to Title XII of the Food Security Act of 1985, as amended, including a wetland determination, may be appealed to the county committee in accordance with the procedures in this part.

(b) If the county committee hears the appeal and believes that the challenge to the NRCS determination is not frivolous, the county committee shall refer the case with its findings on other issues to the NRCS State Conservationist to review the determination, or may make such a referral in advance of resolving other issues.

(c) A decision of the county committee not to refer the case with its findings to the NRCS State Conservationist may be appealed to the State Committee.

(d) The county or State committee decision must incorporate, and be based upon, the results of the NRCS State Conservationist's review and subsequent determination.

§780.12 Appeals of penalties assessed under the Agricultural Foreign Investment Disclosure Act of 1978.

(a) Requests for appeals of penalties assessed under the Agricultural For-

eign Investment Disclosure Act of 1978 must be addressed to: Administrator, Farm Service Agency, Stop 0572, 1400 Independence Avenue, SW., Washington, DC 20250–0572.

(b) Decisions in appeals under this section are not subject to reconsideration and are administratively final.

§780.13 Verbatim transcripts.

(a) Appellants and their representatives are precluded from making any electronic recording of any portion of a hearing or other proceeding conducted in accordance with this part. Appellants interested in obtaining an official recording of a hearing or other proceeding may request a verbatim transcript in accordance with paragraph (b) of this section.

(b) Any party to an appeal or request for reconsideration under this part may request that a verbatim transcript be made of the hearing proceedings and that such transcript be made the official record of the hearing. The party requesting a verbatim transcript shall pay for the transcription service, provide a copy of the transcript to FSA free of charge, and allow any other party in the proceeding desiring to purchase a copy of the transcript to order it from the transcription service.

§780.14 [Reserved]

§780.15 Time limitations.

(a) To the extent practicable, no later than 10 business days after an agency decision maker renders an adverse decision that affects a participant, FSA will provide the participant written notice of the adverse decision and available appeal rights.

(b) A participant requesting an appealability review by the State Executive Director of an agency decision made at the county, area, district or State level that is otherwise determined by FSA not to be appealable must submit a written request for an appealability review to the State Executive Director that is received no later than 30 calendar days from the date a participant receives written notice of the decision.

(c) A participant requesting reconsideration, mediation or appeal must submit a written request as instructed in the notice of decision that is received no later than 30 calendar days from the date a participant receives written notice of the decision. A participant that receives a determination made under part 1400 of this title will be deemed to have consented to an extension of the time limitation for a final determination as provided in part 1400 of this title if the participant requests mediation.

(d) Notwithstanding the time limits in paragraphs (b) and (c) of this section, a request for an appealability review, reconsideration, or appeal may be accepted if, in the judgment of the reviewing authority with whom such request is filed, exceptional circumstances warrant such action. A participant does not have the right to seek an exception under this paragraph. FSA's refusal to accept an untimely request is not appealable.

(e) Decisions appealable under this part are final unless review options available under this part or part 11 are timely exercised.

(1) Whenever the final date for any requirement of this part falls on a Saturday, Sunday, Federal holiday, or other day on which the pertinent FSA office is not open for the transaction of business during normal working hours, the time for submission of a request will be extended to the close of business on the next working day.

(2) The date when an adverse decision or other notice pursuant to these rules is deemed received is the earlier of physical delivery by hand, by facsimile with electronic confirmation of receipt, actual stamped record of receipt on a transmitted document, or 7 calendar days following deposit for delivery by regular mail.

[70 FR 43266, July 27, 2005, as amended at 71 FR 30574, May 30, 2006]

§ 780.16 Implementation of final agency decisions.

To the extent practicable, no later than 30 calendar days after an agency decision becomes a final administrative decision of USDA, FSA will implement the decision.

§ 780.17 Judicial review.

(a) Decisions of the Administrator in appeals under this part from Agriculture Foreign Investment Disclosure Act penalties are administratively final decisions of USDA.

(b) The decision of a State Executive Director or State Conservationist on equitable relief made under § 718.307 of this title is administratively final and also not subject to judicial review.

PART 781—DISCLOSURE OF FOREIGN INVESTMENT IN AGRICULTURAL LAND

Sec.
781.1 General.
781.2 Definitions.
781.3 Reporting requirements.
781.4 Assessment of penalties.
781.5 Penalty review procedure.
781.6 Paperwork Reduction Act assigned number.

AUTHORITY: Sec. 1–10, 92 Stat. 1266 (7 U.S.C. 3501 et seq.).

SOURCE: 49 FR 35074, Sept. 6, 1984, unless otherwise noted.

§ 781.1 General.

The purpose of these regulations is to set forth the requirements designed to implement the Agricultural Foreign Investment Disclosure Act of 1978. The regulations require that a foreign person who acquires, disposes of, or holds an interest in United States agricultural land shall disclose such transactions and holdings to the Secretary of Agriculture. In particular, the regulations establish a system for the collection of information by the Agricultural Stablization and Conservation Service (FSA) pertaining to foreign investment in United States agricultural land. The information collected will be utilized in the preparation of periodic reports to Congress and the President by the Economic Research Service (ERS) concerning the effect of such holdings upon family farms and rural communities.

§ 781.2 Definitions.

In determining the meaning of the provisions of this part, unless the context indicates otherwise, words importing the singular include and apply to several persons or things, words importing the plural include the singular, and words used in the present tense include the future as well as the present.

The following terms shall have the following meanings:

(a) *AFIDA.* AFIDA means the Agricultural Foreign Investment Disclosure Act of 1978.

(b) *Agricultural land.* Agricultural land means land in the United States used for forestry production and land in the United States currently used for, or, if currently idle, land last used within the past five years, for farming, ranching, or timber production, except land not exceeding ten acres in the aggregate, if the annual gross receipts from the sale of the farm, ranch, or timber products produced thereon do not exceed $1,000. Farming, ranching, or timber production includes, but is not limited to, activities set forth in the Standard Industrial Classification Manual (1987), Division A, exclusive of industry numbers 0711–0783, 0851, and 0912–0919 which cover animal trapping, game management, hunting carried on as a business enterprise, trapping carried on as a business enterprise, and wildlife management. Land used for forestry production means, land exceeding 10 acres in which 10 percent is stocked by trees of any size, including land that formerly had such tree cover and that will be naturally or artificially regenerated.

(c) *Any interest.* Any interest means all interest acquired, transferred or held in agricultural lands by a foreign person, except:

(1) Security interests;

(2) Leaseholds of less than 10 years;

(3) Contingent future interests;

(4) Noncontingent future interests which do not become possessory upon the termination of the present possessory estate;

(5) Surface or subsurface easements and rights of way used for a purpose unrelated to agricultural production; and

(6) An interest solely in mineral rights.

(d) *County.* County means a political subdivision of a State identified as a County or parish. In Alaska, the term means an area so designated by the State Agricultural Stabilization and Conservation committee.

(e) *Foreign government.* Foreign government means any government other than the United States government, the government of a State, or a political subdivision of a State.

(f) *Foreign individual.* Foreign individual means foreign person as defined in paragraph (g)(1) of this section.

(g) *Foreign person.* Foreign person means:

(1) Any individual:

(i) Who is not a citizen or national of the United States; or

(ii) Who is not a citizen of the Northern Mariana Islands or the Trust Territory of the Pacific Islands; or

(iii) Who is not lawfully admitted to the United States for permanent residence or paroled into the United States under the Immigration and Nationality Act;

(2) Any person, other than an individual or a government, which is created or organized under the laws of a foreign government or which has its principal place of business located outside of all the States;

(3) Any foreign government;

(4) Any person, other than an individual or a government:

(i) Which is created or organized under the laws of any State; and

(ii) In which a significant interest or substantial control is directly or indirectly held:

(A) By any individual referred to in paragraph (g)(1) of this section; or

(B) By any person referred to in paragraph (g)(2) of this section; or

(C) By any foreign government referred to in paragraph (g)(3) of this section; or

(D) By any numerical combination of such individuals, persons, or governments, which combination need not have a common objective.

(h) *Person.* Person means any individual, corporation, company, association, partnership, society, joint stock company, trust, estate, or any other legal entity.

(i) *Secretary.* Secretary means the Secretary of Agriculture.

(j) *Security interest.* Security interest means a mortgage or other debt securing instrument.

(k) *Significant interest of substantial control.* Significant interest or substantial control means:

(1) An interest of 10 percent or more held by a person referred to in paragraph (g)(4) of this section, by a single

individual referred to in paragraph (g)(1) of this section, by a single person referred to in paragraph (g)(2) of this section, by a single government referred to in paragraph (g)(3) of this section; or

(2) An interest of 10 percent or more held by persons referred to in paragraph (g)(4) of this section, by individuals referred to in paragraph (g)(1) of this section, by persons referred to in paragraph (g)(2) of this section, or by governments referred to in paragraph (g)(3) of this section, whenever such persons, individuals, or governments are acting in concert with respect to such interest even though no single individual, person, or government holds an interest of 10 percent or more; or

(3) An interest of 50 percent or more, in the aggregate, held by persons referred to in paragraph (g)(4) of this section, by individuals referred to in paragraph (g)(1) of this section, by persons referred to in paragraph (g)(2) of this section, or by governments referred to in paragraph (g)(3) of this section, even though such individuals, persons, or governments may not be acting in concert.

(1) *State.* State means any of the several States, the District of Columbia, the Commonwealth of Puerto Rico, the Northern Mariana Islands, Guam, the Virgin Islands, American Samoa, the Trust Territory of the Pacific Islands or any other territory or possession of the United States.

[49 FR 35074, Sept. 6, 1984, as amended at 58 FR 48274, Sept. 15, 1993]

§ 781.3 Reporting requirements.

(a) All reports required to be filed pursuant to this part shall be filed with the FSA County office in the county where the land with respect to which such report must be filed is located or where the FSA County office administering programs carried out on such land is located; Provided, that the FSA office in Washington, DC, may grant permission to foreign persons to file reports directly with its Washington office when complex filings are involved, such as where the land being reported is located in more than one county.

(b) Any foreign person who held, holds, acquires, or transfers any interest in United States agricultural land is subject to the requirement of filing a report on form FSA-153 by the following dates:

(1) August 1, 1979, if the interest in the agricultural land was held on the day before February 2, 1979, or

(2) Ninety days after the date of acquisition or transfer of the interest in the agricultural land, if the interest was acquired or transferred on or after February 2, 1979.

(c) Any person who holds or acquires any interest in United States agricultural land at a time when such person is not a foreign person and who subsequently becomes a foreign person must submit, not later than 90 days after the date on which such person becomes a foreign person, a report containing the information required to be submitted under paragraph (e) of this section.

(d) Any foreign person who holds or acquires any interest in United States land at a time when such land is not agricultural land and such land subsequently becomes agricultural land must submit, not later than 90 days after the date on which such land becomes agricultural, a report containing the information required to be submitted under paragraph (e) of this section.

(e) Any foreign person required to submit a report under this regulation, except under paragraph (g) of this section, shall file an FSA-153 report containing the following information:

(1) The legal name and the address of such foreign person;

(2) In any case in which such foreign person is an individual, the citizenship of such foreign person;

(3) In any case in which such foreign person is not an individual or a government, the nature and name of the person holding the interest, the country in which such foreign person is created or organized, and the principal place of business of such foreign person;

(4) The type of interest held by a foreign person who acquired or transferred an interest in agricultural land;

(5) The legal description and acreage of such agricultural land;

(6) The purchase price paid for, or any other consideration given for, such interest; the amount of the purchase price or the value of the consideration

yet to be given; the current estimated value of the land reported;

(7) In any case in which such foreign person transfers such interest, the legal name and the address of the person to whom such interest is transferred; and

(i) In any case in which such transferee is an individual, the citizenship of such transferee; and

(ii) In any case in which such transferee is not an individual, or a government, the nature of the person holding the interest, the country in which such transferee is created or organized, and the principal place of business;

(8) The agricultural purposes for which such foreign person intends, on the date on which such report is submitted, to use such agricultural land;

(9) When applicable, the name, address and relationship of the representative of the foreign person who is completing the FSA–153 form for the foreign person;

(10) How the tract of land was acquired or transferred, the relationship of the foreign person to the previous owner, producer, manager, tenant or sharecropper, and the rental agreement; and

(11) The date the interest in the land was acquired or transferred.

(f)(1) Any foreign person, other than an individual or government, required to submit a report under paragraphs (b), (c), and (d) of this section, must submit, in addition to the report required under paragraph (e) of this section, a report containing the following information:

(i) The legal name and the address of each foreign individual or government holding significant interest or substantial control in such foreign person;

(ii) In any case in which the holder of such interest is an individual, the citizenship of such holder; and

(iii) In any case in which the holder of significant interest or substantial control in such foreign person is not an individual or a government, the nature and name of the foreign person holding such interest, the country in which such holder is created or organized, and the principal place of business of such holder.

(2) In addition, any such foreign person required to submit a report under paragraph (f)(1) of this section may also be required, upon request, to submit a report containing:

(i) The legal name and the address of each individual or government whose legal name and address did not appear on the report required to be submitted under paragraph (f)(1) of this section, if such individual or government holds any interest in such foreign person:

(ii) In any case in which the holder of such interest is an individual, the citizenship of such holder; and

(iii) In any case in which the holder of such interest is not an individual or a government, the nature and name of the person holding the interest, the country in which such holder is created or organized, and the principal place of business of such holder.

(g) Any foreign person, other than an individual or a government, whose legal name is contained on any report submitted in satisfaction of paragraph (f) of this section may also be required, upon request, to:

(1) Submit a report containing:

(i) The legal name and the address of each foreign individual or government holding significant interest or substantial control in such foreign person;

(ii) In any case in which the holder of such interest is an individual, the citizenship of such holder; and

(iii) In any case in which the holder of such interest in such foreign person is not an individual or a government, the nature and name of the foreign person holding such interest, the country in which each holder is created or organized, and the principal place of business of such holder.

(2) Submit a report containing:

(i) The legal name and address of each individual or government whose legal name and address did not appear on the report required to be submitted under paragraph (g)(1) of this section if such individual or government holds any interest in such foreign person and, except in the case of a request which involves a foreign person, a report was required to be submitted pursuant to paragraph (f)(2) of this section, disclosing information relating to nonforeign interest holders;

(ii) In any case in which the holder of such interest is an individual, the citizenship of such holder; and

(iii) In any case in which the holder of such interest is not an individual or government and, except in a situation where the information is requested from a foreign person, a report was required to be submitted pursuant to paragraph (f)(2) of this section disclosing information relating to nonforeign interest holders, the nature and name of the person holding the interest, the country in which such holder is created or organized, and the principal place of business of such holder.

(h)(1) Any person which has issued fewer than 100,000 shares of common and preferred stock and instruments convertible into equivalents thereof shall be considered to have satisfactorily determined that it has no obligation to file a report pursuant to § 781.3 if, in addition to information within its knowledge, a quarterly examination of its business records fails to reveal that persons with foreign mailing addresses hold significant interest or substantial control in such person.

(2) Any person which has issued 100,000 or more shares of common and preferred stock and instruments convertible into equivalents thereof shall be considerd to have satisfactorily determined that it has no obligation to file a report pursuant to § 781.3 if, in addition to information within its knowledge, a quarterly examination of its business records fails to reveal that the percentage of shares held in such person both by persons with foreign mailing addresses and investment institutions which manage shares does not equal or exceed significant interest or substantial control in such person.

(3) If the person in paragraph (h)(2) of this section determines that the percentage of shares, which is held in it both by persons with foreign mailing addresses and investment institutions which manage shares, equals or exceeds significant interest or substantial control in such persons, then such person shall be considered to have satisfactorily attempted to determine whether it has an obligation to file a report pursuant to § 781.3 if it sends questionnaires to each such investment institution holding an interest in it inquiring as to whether the persons for which they are investing are foreign persons and the percentage of shares reflected by the affirmative responses from each such investment institution plus the percentage of shares held by persons listed on the business records with foreign mailing addresses does not reveal that foreign persons hold significant interest or substantial control in such person.

(i) Any foreign person, who submitted a report under paragraph (b), (c), or (d) of this section at a time when such land was agricultural, and such agricultural land later ceases to be agricultural, must submit, not later than 90 days after the date on which such land ceases being agricultural, a revised report from FSA–153 or a written notification of the change of status of the land to the FSA office where the report form was originally filed. The report form and notification must contain the following information:

(1) The legal name and the address of such foreign person;

(2) The legal description, which includes the State and county where the land is located, and the acreage of such land;

(3) The date the land ceases to be agricultural;

(4) The use of the land while agricultural.

(j) If any foreign person who submitted a report under paragraph (b), (c), or (d) of this section ceases to be a foreign person, such person must submit, not later than 90 days after the date such person ceases being a foreign person, a written notification of the change of status of the person to the FSA office where the report form FSA–153 was originally filed. The notification must contain the following information:

(1) The legal name of such person;

(2) The legal description and acreage of such land;

(3) The date such person ceases to be foreign.

(k) Any foreign person who submitted a report under paragraph (b), (c), or (d) of this section must submit, not later than 90 days after the change of information contained on the report, a written notification of the change to the FSA office where the report form FSA–153 was originally filed. The following information must be kept current on the report:

(1) The legal address of such foreign person;

(2) The legal name and the address required to be submitted under (f)(1) of this section;

(3) The legal name and the address required to be submitted under (g)(1) of this section.

[49 FR 35074, Sept. 6, 1984, as amended at 51 FR 25993, July 18, 1986]

§781.4 Assessment of penalties.

(a) Violation of the reporting obligations will consist of:

(1) Failure to submit any report in accordance with §781.3;

(2) Failure to maintain any submitted report with accurate information; or

(3) Submission of a report which the foreign person knows:

(i) Does not contain, initially or within thirty days from the date of a letter returning for completion such incomplete report, all the information required to be in such report; or

(ii) Contains misleading or false information.

(b) Any foreign person who violates the reporting obligation as described in paragraph (a) of this section shall be subject to the following penalties:

(1) Late-filed reports: One-tenth of one percent of the fair market value, as determined by the Farm Service Agency, of the foreign person's interest in the agricultural land, with respect to which such violation occurred, for each week or portion thereof that such violation continues, but the total penalty imposed shall not exceed 25 percent of the fair market value of the foreign person's interest in such land.

(2) Submission of an incomplete report or a report containing misleading or false information, failure to submit a report or failure to maintain a submitted report with accurate information: 25 percent of the fair market value, as determined by the Farm Service Agency, of the foreign person's interest in the agricultural land with respect to which such violation occurred.

(3) Penalties prescribed above are subject to downward adjustments based on factors including:

(i) Total time the violation existed.

(ii) Method of discovery of the violation.

(iii) Extenuating circumstances concerning the violation.

(iv) Nature of the information misstated or not reported.

(c) The fair market value for the land, with respect to which such violation occurred, shall be such value on the date the penalty is assessed, or if the land is no longer agricultural, on the date it was last used as agricultural land. The price or current estimated value reported by the foreign person, as verified and/or adjusted by the County Agricultural Stabilization and Conservation Committee for the County where the land is located, will be considered to be the fair market value.

§781.5 Penalty review procedure.

(a) Whenever it appears that a foreign person has violated the reporting obligation as described in paragraph (a) of §781.4, a written notice of apparent liability will be sent to the foreign person's last known address by the Farm Service Agency. This notice will set forth the facts which indicate apparent liability, identify the type of violation listed in paragraph (a) of §781.4 which is involved, state the amount of the penalty to be imposed, include a statement of fair market value of the foreign person's interest in the subject land, and summarize the courses of action available to the foreign person.

(b) The foreign person involved shall respond to a notice of apparent liability within 60 days after the notice is mailed. If a foreign person fails to respond to the notice of apparent liability, the proposed penalty shall become final. Any of the following actions by the foreign person shall constitute a response meeting the requirements of this paragraph.

(1) Payment of the proposed penalty in the amount specified in the notice of apparent liability and filing of a report, if required, in compliance with §781.3. The amount shall be paid by check or money order drawn to the Treasurer of the United States and shall be mailed to the U.S. Department of Agriculture, P.O. Box 2415, Washington, DC 20013. The Department is not responsible for the loss of currency sent through the mails.

(2) Submission of a written statement denying liability for the penalty in whole or in part. Allegations made in any such statement must be supported by detailed factual data. The statement should be mailed to the Administrator, Farm Service Agency, U.S. Department of Agriculture, P.O. Box 2415, Washington, DC 20013.

(3) A request for a hearing on the proposed penalty may be filed in accordance with part 780 of this title.

(c) After a final decision is issued pursuant to an appeal under part 780 of this title, the Administrator or Administrator's designee shall mail the foreign person a notice of the determination on appeal, stating whether a report must be filed or amended in compliance with §781.3, the amount of the penalty (if any), and the date by which it must be paid. The foreign person shall file or amend the report as required by the Administrator. The penalty in the amount stated shall be paid by check or money order drawn to the Treasurer of the United States and shall be mailed to the United States Department of Agriculture, P.O. Box 2415, Washington, DC 20013. The Department is not responsible for the loss of currency sent through the mails.

(d) If the foreign person contests the notice of apparent liability by submitting a written statement or a request for a hearing thereon, the foreign person may elect either to pay the penalty or decline to pay the penalty pending resolution of the matter by the Administrator. If the Administrator determines that the foreign person is not liable for the penalty or is liable for less than the amount paid, the payment will be wholly or proportionally refunded. If the Administrator ultimately determines that the foreign person is liable, the penalty finally imposed shall not exceed the amount imposed in the notice of apparent liability.

(e) If a foreign person fails to respond to the notice of apparent liability as required by paragraph (b) of this section, or fails to pay the penalty imposed by the Administrator under paragraph (d) of this section, the case will, without further notice, be referred by the Department to the Department of Justice for prosecution in the appropriate District Court to recover the amount of the penalty.

(f) Any amounts approved by the U.S. Department of Agriculture for disbursement to a foreign person under the programs administered by the Department may be setoff against penalties assessed hereunder against such person, in accordance with the provisions of 7 CFR part 13.

[49 FR 35074, Sept. 6, 1984, as amended at 60 FR 67318, Dec. 29, 1995]

§781.6 Paperwork Reduction Act assigned number.

The information collection requirements contained in these regulations (7 CFR part 781) have been approved by the Office of Management and Budget (OMB) under the provisions of 44 U.S.C. Chapter 35 and have been assigned OMB control number 0560–0097.

PART 782—END-USE CERTIFICATE PROGRAM

Subpart A—General

Sec.
782.1 Basis and purpose.
782.2 Definitions.
782.3 Administration.
782.4 OMB control numbers assigned pursuant to the Paperwork Reduction Act.

Subpart B—Implementation of the End-Use Certificate Program

782.10 Identification of commodities subject to end-use certificate regulations.
782.11 Extent to which commodities are subject to end-use certificate regulations.
782.12 Filing FSA–750, End-Use Certificate for Wheat.
782.13 Importer responsibilities.
782.14 Identity preservation.
782.15 Filing FSA–751, Wheat Consumption and Resale Report.
782.16 Designating end use on form FSA–751.
782.17 Wheat purchased for resale.
782.18 Wheat purchased for export.
782.19 Penalty for noncompliance.

Subpart C—Records and Reports

782.20 Importer records and reports.
782.21 End-user and exporter records and reports.
782.22 Subsequent buyer records and reports.
782.23 Failure to file end-use certificates or consumption and resale reports.

782.24 Recordkeeping and examination of records.

782.25 Length of time records are to be kept.

AUTHORITY: 19 U.S.C. 3391(f).

SOURCE: 60 FR 5089, Jan. 26, 1995, unless otherwise noted.

EDITORIAL NOTE: Nomenclature changes to part 782 appear at 61 FR 32643, June 25, 1996.

EFFECTIVE DATE NOTE: At 77 FR 51459, Aug. 24, 2012, part 782 was suspended, effective Aug. 31, 2012.

Subpart A—General

§782.1 Basis and purpose.

The regulations contained in this part are issued pursuant to and in accordance with Section 321(f) of the North American Free Trade Agreement Implementation Act. These regulations govern the establishment of the end-use certificate program, the completion of end-use certificates, the identification of commodities requiring end-use certificates, the submission of reports, and the keeping of records and making of reports incident thereto.

§782.2 Definitions.

As used in this part and in all instructions, forms, and documents in connection therewith, the words and phrases defined in this section shall have the meanings herein assigned to them unless the context or subject matter requires otherwise. References contained herein to other parts of this chapter or title shall be construed as references to such parts and amendments now in effect or later issued.

Date of entry means the effective time of entry of the merchandise, as defined in 19 CFR part 101.

End Use means the actual manner in which Canadian-produced wheat was used, including, among other uses, milling, brewing, malting, distilling, manufacturing, or export.

End user means the entity that uses Canadian-produced wheat for, among other uses, milling, brewing, malting, distilling, manufacturing, or other use, except resale.

Entity means a legal entity including, but not limited to, an individual, joint stock company, corporation, association, partnership, cooperative, trust, and estate.

Entry means that documentation required by 19 CFR part 142 to be filed with the appropriate U.S. Customs officer to secure the release of imported merchandise from U.S. Customs custody, or the act of filing that documentation.

Grain handler means an entity other than the importer, exporter, subsequent buyer, or end user that handles wheat on behalf of an importer, exporter, subsequent buyer, or end user.

Importer means a party qualifying as an Importer of Record pursuant to 19 U.S.C. 1484(a).

Metric ton means a unit of measure that equals 2,204.6 pounds.

Subsequent buyer means an entity other than the end user or importer which owns wheat originating in Canada.

Workdays means days that the Federal government normally conducts business, which excludes Saturdays, Sundays, and Federal holidays.

[60 FR 5089, Jan. 26, 1995, as amended at 61 FR 32643, June 25, 1996; 64 FR 12885, Mar. 16, 1999]

§782.3 Administration.

The end-use certificate program will be administered under the general supervision and direction of the Administrator, Farm Service Agency (FSA), U.S. Department of Agriculture (USDA), through the Office of the Deputy Administrator for Commodity Operations (DACO), FSA, Washington, D.C., and the Kansas City Commodity Office (KCCO), FSA, Kansas City, MO, in coordination with the Commissioner of Customs pursuant to a Memorandum of Understanding.

§782.4 OMB control numbers assigned pursuant to the Paperwork Reduction Act.

The information collection requirements in this part have been approved by the Office of Management and Budget and assigned OMB control number 0560–0151.

[61 FR 32643, June 25, 1996]

Subpart B—Implementation of the End-Use Certificate Program

§ 782.10 Identification of commodities subject to end-use certificate regulations.

(a) The regulations in this part are applicable to wheat and barley, respectively, imported into the U.S. from any foreign country, as defined in 19 CFR 134.1, or instrumentality of such foreign country that, as of April 8, 1994, required end-use certificates for imports of U.S.-produced wheat or barley.

(b) Because Canada is the only country with such requirements on wheat, and no country has an end-use certificate requirement for barley, only wheat originating in Canada is affected by the regulations in this part.

§ 782.11 Extent to which commodities are subject to end-use certificate regulations.

(a) In the event that Canada eliminates the requirement for end-use certificates on imports from the U.S., the provisions of the regulations in this part shall be suspended 30 calendar days following the date Canada eliminates its end-use certificate requirement, as determined by the Secretary.

(b) The provisions of the regulations in this part may be suspended if the Secretary, after consulting with domestic producers, determines that the program has directly resulted in the:

(1) Reduction of income to U.S. producers of agricultural commodities, or

(2) Reduction of the competitiveness of U.S. agricultural commodities in world export markets.

§ 782.12 Filing FSA–750, End-Use Certificate for Wheat.

(a) Each entity that imports wheat originating in Canada shall, for each entry into the U.S., obtain form FSA–750, End-Use Certificate for Wheat, from Kansas City Commodity Office, Warehouse Contract Division, P.O. Box 419205, Kansas City, MO 64141–6205, and submit the completed original form FSA–750 to KCCO within 10 workdays following the date of entry or release. Each form FSA–750 shall set forth, among other things, the:

(1) Name, address, and telephone number of the importer,

(2) Customs entry number,

(3) Date of entry,

(4) Importer number,

(5) Class of wheat being imported,

(6) Grade, protein content, moisture content, and dockage level of wheat being imported,

(7) If imported as a result of a contract for sale, the date of such contract.

(8) Quantity imported, in net metric tons, rounded to the nearest hundredth of a metric ton, per conveyance,

(9) Storage location of the wheat,

(10) Mode of transportation and the name of the transportation company used to import the wheat, and

(11) A certification that the identity of the Canadian-produced wheat will be preserved until such time as the wheat is either delivered to a subsequent buyer or end-user, or loaded onto a conveyance for direct delivery to an end user.

(b) Importers may provide computer generated form FSA–750, provided such computer generated forms:

(1) Are approved in advance by KCCO,

(2) Contain a KCCO-assigned serial number, and

(3) Contain all of the information required in paragraphs (a)(1) through (a)(9).

(c) KCCO will accept form FSA–750 submitted through the following methods:

(1) Mail service, including express mail,

(2) Facsimile machine, and

(3) Other electronic transmissions, provided such transmissions are approved in advance by KCCO. The importer remains responsible for ensuring that electronically transmitted forms are received in accordance with paragraph (a).

(d) The original form FSA–750 and one copy of form FSA–750 shall be signed and dated by the importer.

(e) Distribution of form FSA–750 will be as follows:

(1) If form FSA–750 is submitted to KCCO in accordance with paragraph (c)(1);

(i) The original shall be forwarded to Kansas City Commodity Office, Warehouse License and Contract Division, P.O. Box 419205, Kansas City, MO 64141–6205, by the importer,

(ii) One copy shall be retained by the importer.

(2) If form FSA–750 is submitted to KCCO in accordance with paragraphs (c)(2) or (c)(3), the original form FSA–750 that is signed and dated by the importer in accordance with paragraph (d) shall be maintained by the importer,

(3) The importer shall provide a photocopy to the end user or, if the wheat is purchased for purposes of resale, the subsequent buyer(s).

(f) The completion and filing of an end-use certificate does not relieve the importer of other legal requirements, such as those imposed by other U.S. agencies, pertaining to the importation.

[60 FR 5089, Jan. 26, 1995, as amended at 61 FR 32643, June 25, 1996; 64 FR 12885, Mar. 16, 1999]

§ 782.13 Importer responsibilities.

The importer shall:

(a) File form FSA–750 in accordance with § 782.12.

(b) Immediately notify each subsequent buyer, grain handler, or end user that the wheat being purchased or handled originated in Canada and may only be commingled with U.S.-produced wheat by the end user or when loaded onto a conveyance for direct delivery to the end user or a foreign country.

(c) Provide each subsequent buyer or end user with a copy of form FSA–750 that was filed when the Canadian wheat entered the U.S.

(d) Submit to KCCO, within 15 workdays following the date of sale, form FSA–751, Wheat Consumption and Resale Report, in accordance with § 782.15.

[60 FR 5089, Jan. 26, 1995, as amended at 61 FR 32643, June 25, 1996]

§ 782.14 Identity preservation.

(a) The importer and all subsequent buyers of the imported wheat shall preserve the identity of the Canadian-produced wheat.

(b) Canadian-produced wheat may only be commingled with U.S.-produced wheat by the end user, or when loaded onto a conveyance for direct delivery to the end user or foreign country.

(c) Failure to meet the requirements in paragraphs (a) and (b) of this section shall constitute noncompliance by the importer or subsequent buyer for the purposes of this part.

§ 782.15 Filing FSA–751, Wheat Consumption and Resale Report.

(a) For purposes of providing information relating to the consumption and resale of Canadian-produced wheat, form FSA–751, Wheat Consumption and Resale Report, shall be filed with KCCO by each:

(1) Importer and subsequent buyer, for each sale to a subsequent buyer or end user, within 15 workdays following the date of sale.

(2) End user and exporter, for full and partial consumption or export, within 15 workdays following:

(i) March 31,

(ii) June 30,

(iii) September 30, and

(iv) December 31.

(b) Each form FSA–751 shall set forth, among other things, the:

(1) Name, address, and telephone number of the filer,

(2) Storage location of the wheat,

(3) Name and address of the importer,

(4) Form FSA–750, End-Use Certificate for Wheat, serial number,

(5) Class of wheat,

(6) Date the wheat was received at the filer's facility,

(7) Quantity of wheat received, in net metric tons, rounded to the nearest hundredth of a metric ton,

(8) Certification to be completed by end users and exporters that requires the end user or exporter to provide, among other things:

(i) A certification of compliance with these regulations,

(ii) The quantity consumed or exported,

(iii) The quantity remaining,

(iv) The manner in which the commodity was used.

(v) The signature of an authorized representative of the end user or exporter.

(9) Certification to be completed by subsequent buyers and importers that requires the subsequent buyer or importer to provide, among other things:

(i) A certification of compliance with the regulations in this part,

(ii) The quantity resold,

(iii) The name, address, and telephone number of the buyer, and

(iv) The signature of an authorized representative of the subsequent buyer or importer.

(c) End user and exporter shall submit form FSA–751 to KCCO quarterly until the wheat has been fully utilized or exported in accordance with the regulations in this part.

(d) Importers and subsequent buyers shall, for each individual sale, submit form FSA–751 to KCCO until the imported wheat has been fully resold.

(e) Filers may provide computer generated form FSA–751, provided such computer generated forms:

(1) Are approved in advance by KCCO, and

(2) Contain the information required in paragraphs (b)(1) through (b)(9) of this section.

(f) KCCO will accept form FSA–751 submitted through the following methods:

(1) Mail service, including express mail,

(2) Facsimile machine, and

(3) Other electronic transmissions, provided such transmissions are approved in advance by KCCO. The importer, end user, exporter, or subsequent buyer remains responsible for ensuring that electronically transmitted forms are received in accordance with this section.

(g) Distribution of form FSA–751 will be as follows:

(1) If form FSA–751 is submitted to KCCO in accordance with paragraph (f)(1) of this section:

(i) The original shall be forwarded to Kansas City Commodity Office, Warehouse License and Contract Division, P.O. Box 419205, Kansas City, MO 64141–6205, by the importer, end user, exporter, or subsequent buyer.

(ii) One copy shall be retained by the importer, end user, exporter, or subsequent buyer.

(2) If form FSA–751 is submitted to KCCO in accordance with paragraphs (f)(2) or (f)(3) of this section, the original form FSA–751 shall be maintained by the importer, end user, exporter, or subsequent buyer.

[60 FR 5089, Jan. 26, 1995, as amended at 61 FR 32643, June 25, 1996]

§ 782.16 Designating end use on form FSA–751.

(a) If the end use specified on the applicable form FSA–751, Wheat Consumption and Resale Report, is "export," the exporter must specify the final destination, by country, on form FSA–751.

(b) If the end user utilizes the wheat for purposes other than milling, brewing, malting, distilling, export, or manufacturing, such use must be specifically designated on form FSA–751.

§ 782.17 Wheat purchased for resale.

(a) This section applies to an importer or subsequent buyer who imports or purchases Canadian-produced wheat for the purpose of reselling the wheat.

(b) The importer or subsequent buyer shall immediately notify each subsequent buyer, grain handler, exporter, or end user that the wheat being purchased or handled originated in Canada and may only be commingled with U.S.-produced wheat by the end user or when loaded onto a conveyance for direct delivery to the end user or a foreign country.

(c) The importer or subsequent buyer shall provide all purchasers of Canadian-produced wheat with a photocopy of the form FSA–750 submitted to KCCO by the importer in accordance with § 782.12(a).

[60 FR 5089, Jan. 26, 1995, as amended at 61 FR 32643, June 25, 1996]

§ 782.18 Wheat purchased for export.

(a) This section applies to an importer or subsequent buyer who imports or purchases Canadian-produced wheat for the purpose of export to a foreign country or instrumentality.

(b) Wheat that is purchased for the purpose of export must be stored identity preserved while the importer or subsequent buyer maintains control of the wheat, except that such wheat may be commingled when loaded onto a conveyance for delivery to the foreign country or instrumentality.

(c) Importers or subsequent buyers that purchase wheat for export to a foreign country or instrumentality must complete form FSA–751 quarterly, in accordance with § 782.15.

§782.19 Penalty for noncompliance.

It shall be a violation of 18 U.S.C. 1001 for any entity to engage in fraud with respect to, or to knowingly violate, the provisions set forth in this part.

Subpart C—Records and Reports

§782.20 Importer records and reports.

(a) The importer shall retain a copy of each form:

(1) FSA–750, End-Use Certificate for Wheat, that is submitted to KCCO in accordance with §782.12(a); and

(2) FSA–751, Wheat Consumption and Resale Report, that is submitted to KCCO in accordance with §782.15(a)(1).

(b) The importer shall maintain records to verify that the wheat was identity preserved until such time as the wheat was:

(1) Loaded onto the conveyance for direct delivery to an end user, or

(2) Delivered to an end user, or

(3) Delivered to a subsequent buyer.

(c) Copies of the documents, information, and records required in paragraphs (a) and (b) of this section shall be kept on file at the importer's headquarters office or other location designated by the importer for the period specified in §782.25.

§782.21 End-user and exporter records and reports.

(a) The end user or exporter shall retain a copy of each form FSA–751, Wheat Consumption and Resale Report, that is filed with KCCO in accordance with §782.15(a)(2).

(b) The end user or exporter shall retain a copy of each form FSA–750, End-Use Certificate for Wheat, provided to the end-user or exporter in accordance with §782.17(b).

(c) The exporter shall maintain records to verify that wheat purchased for the purpose of export was stored identity preserved until such time as the wheat was loaded onto a conveyance for delivery to the foreign country or instrumentality.

(d) Copies of the documents required in paragraphs (a), (b), and (c) of this section shall be kept on file at the end-user's or exporter's headquarters office or other location designated by the end user or exporter for the period specified in §782.25.

§782.22 Subsequent buyer records and reports.

(a) The subsequent buyer shall retain a copy of each form FSA–751, Wheat Consumption and Resale Report, that is filed with KCCO in accordance with §782.15(a)(1).

(b) The subsequent buyer shall retain a copy of each form FSA–750, End-Use Certificate for Wheat, provided to the subsequent buyer in accordance with §782.17(b).

(c) The subsequent buyer shall maintain records to verify that the wheat specified on the end-use certificate was identity preserved during the time that the subsequent buyer maintained control of the wheat, or until the wheat was loaded onto a conveyance for direct delivery to an end user.

(d) Copies of the documents and records required in paragraphs (a) through (c) of this section shall be kept on file at the subsequent buyer's headquarters office or other location designated by the subsequent buyer for the period specified in §782.25.

§782.23 Failure to file end-use certificates or consumption and resale reports.

Failure by importers, end users, exporters, and subsequent buyers to file form FSA–750, End-Use Certificate for Wheat, and form FSA–751, Wheat Consumption and Resale Report, as applicable, and retain or maintain related copies and records shall constitute noncompliance for the purposes of §782.19.

§782.24 Recordkeeping and examination of records.

(a) *Examination.* For the purpose of verifying compliance with the requirements of this part, each importer, end-user, exporter, and subsequent buyer shall make available at one place at all reasonable times for examination by representatives of USDA, all books, papers, records, contracts, scale tickets, settlement sheets, invoices, written price quotations, or other documents related to the importation of the Canadian-produced wheat that is within the control of such entity.

(b) *Orderly retention of records.* To facilitate examination and verification of the records and reports required by this part, copies of form FSA-750, End-Use Certificate for Wheat, and form FSA-751, Wheat Consumption and Resale Report, shall be filed in an orderly manner, and must be made available for inspection by representatives of USDA.

§ 782.25 Length of time records are to be kept.

The records required to be kept under this part shall be retained for 3 years following the filing date of the applicable record. Records shall be kept for such longer period of time as may be requested in writing by USDA representatives.

PART 784—2004 EWE LAMB REPLACEMENT AND RETENTION PAYMENT PROGRAM

Sec.
784.1 Applicability.
784.2 Administration.
784.3 Definitions.
784.4 Time and method of application.
784.5 Payment eligibility.
784.6 Rate of payment and limitations on funding.
784.7 Availability of funds.
784.8 Appeals.
784.9 Misrepresentation and scheme or device.
784.10 Estates, trusts, and minors.
784.11 Death, incompetence, or disappearance.
784.12 Maintaining records.
784.13 Refunds; joint and several liability.
784.14 Offsets and withholdings.
784.15 Assignments.
784.16 Termination of program.

AUTHORITY: Clause (3) of section 32 of the Act of August 24, 1935, as amended; 7 U.S.C. 612c.

SOURCE: 69 FR 76837, Dec. 23, 2004, unless otherwise noted.

§ 784.1 Applicability.

(a) Subject to the availability of funds, this part establishes terms and conditions under which the 2004 Ewe Lamb Replacement and Retention Payment Program will be administered.

(b) Unless otherwise determined by the Farm Service Agency (FSA) in accordance with the provisions of this part, the amount that may be expended under this part for program payments shall not exceed $18 million. Claims that exceed that amount will be pro-rated in accordance with § 784.7.

(c) To be eligible for payments, producers must comply with all provisions of this part and with any other conditions imposed by FSA.

§ 784.2 Administration.

(a) This part shall be administered by FSA under the general direction and supervision of the Deputy Administrator for Farm Programs, FSA. The program shall be carried out in the field by FSA State and county committees (State and county committees) in accordance with their assigned duties and the regulations of this part.

(b) The Deputy Administrator for Farm Programs, FSA, or a designee, may reverse or modify a determination made by a State or county committee.

(c) The Deputy Administrator for Farm Programs, FSA, may waive or modify deadlines and other program requirements in cases where timeliness or failure to meet such other requirements does not adversely affect the operation of the program.

(d) The program described under this part is a one-time program to be administered with respect to eligibility and qualifying factors occurring during or related to the base period of August 1, 2003 through July 31, 2004, as specified in this part.

§ 784.3 Definitions.

The definitions in this section shall apply to the 2004 Ewe Lamb Replacement and Retention Payment Program and this part.

Agricultural Marketing Service or AMS means the Agricultural Marketing Service of the Department.

Application means the Ewe Lamb Replacement and Retention Payment Program Application.

Application period means the date established by the Deputy Administrator for producers to apply for program benefits. Unless otherwise announced, that period will end January 13, 2005.

Base period means the period from August 1, 2003, through July 31, 2004, during and after which ewe lambs must meet all qualifying eligibility criteria.

Ewe lamb means a female lamb no more than 18 months of age that has not produced an offspring.

Farm Service Agency or FSA means the Farm Service Agency of the Department.

Foot rot means an infectious, contagious disease of sheep that causes severe lameness and economic loss from decreased flock production.

Lambing cycle means the period of time from birth to weaning.

Parrot mouth means a genetic defect resulting in the failure of the incisor teeth to meet the dental pad correctly.

Person means any individual, group of individuals, partnership, corporation, estate, trust, association, cooperative, or other business enterprise or other legal entity who is, or whose members are, a citizen or citizens of, or legal resident alien or aliens in the United States.

Sheep and lamb operation means any self-contained, separate enterprise operated as an independent unit exclusively within the United States in which a person or group of persons raise sheep and/or lambs.

United States means the 50 States of the United States of America, the District of Columbia, and the Commonwealth of Puerto Rico.

§784.4 Time and method of application.

(a) A request for benefits under this part must be submitted on the Ewe Lamb Replacement and Retention Program Application. The application form may be obtained in person, by mail, by telephone, or by facsimile from any county FSA office. In addition, applicants may download a copy of the form at *http://www.usda.gov/dafp/psd/*.

(b) The form may be obtained from and must be submitted to the FSA county office serving the county where the sheep and lamb operation is located. The completed form must be received by the FSA county office by the date established by FSA. Applications not received by that date will be disapproved and returned as not having been timely filed and the sheep and lamb operation filing the application will not be eligible for benefits under this program.

(c) The sheep and lamb operation requesting benefits under this part must certify to the accuracy of the information provided in their application for benefits. All information provided is subject to verification and spot checks by FSA. Refusal to allow FSA or any other agency of the Department of Agriculture to verify any information provided will result in a determination of ineligibility. Data furnished by the applicant will be used to determine eligibility for program benefits. Providing a false certification will lead to ineligibility for payments and may be subject to additional civil and criminal sanctions.

(d) The sheep and lamb operation requesting benefits under this part must maintain accurate records that document that they meet all eligibility requirements specified herein, as may be requested by FSA. Acceptable forms of supporting documentation include, but are not limited to: Sales receipts, farm management records, veterinarian certifications, scrapie program forms and identification numbers, as well as, other types of documents that prove the eligibility of the qualifying ewe lambs and the sheep and lamb operation. The supporting documentation provided must, at a minimum, include: date of lamb purchase or date of birth, date of lamb death (if applicable), lamb identification and control information, number of ewe lambs purchased or retained, and scrapie program identification numbers.

§784.5 Payment eligibility.

(a) Payments can be made, as agreed to by FSA and subject to the availability of funds, for eligible ewe lambs considered by FSA, as determined by FSA only, to have been acquired or held during the base period by eligible sheep and lamb operations for breeding purposes. Payments may be made for eligible ewe lambs held continuously by the operation, through the end of the compliance period, from the time of the first possession of the ewe lamb in the base period. The payment rate cannot exceed the rate provided for in §784.6 and may be prorated pursuant to §784.7. For purposes of this section, the "base period" is the period from August 1, 2003, through July 31, 2004. A purchase in the base period without possession in the base period will not be considered an acquisition in the base period for purposes of this section unless otherwise allowed by FSA.

(b) For the ewe lamb to be eligible, a sheep and lamb operation must certify that the ewe lamb:

(1) During at least part of the base period was a ewe lamb which was both, at the same time, not older than 18 months of age and had not produced an offspring; and

(2) At the time of certification, does not possess any of the following characteristics:

(i) Parrot mouth; or

(ii) Foot rot.

(c) The sheep and lamb operation must certify and agree to:

(1) Maintain the qualifying ewe lambs in the herd for at least one complete, normal offspring lambing cycle, the end of which shall constitute the end of the compliance period for the purposes of paragraph (a) of this section, and actually maintain the lambs for that period in accord with that certification. The "offspring" lambing cycle refers to the time in which the qualifying ewe lamb's own offspring would be weaned, in a normal course, from that qualifying ewe if the ewe were to have offspring, irrespective of whether the ewe actually produces offspring.

(2) Upon request by an AMS agent or FSA representative, allow the AMS agent or FSA Representative to verify that the ewe lambs meet qualifying characteristics. Spot checks will be conducted by FSA within 30 days of the end of the sign-up period. Any animal showing evidence of parrot mouth, foot rot, or scrapie in such spot checks will be considered to have had those conditions at the time of certification. Other spot checks may be conducted as needed.

(3) Maintain documentation of any death loss of qualifying ewe lambs.

(4) Agree to refund any payments made with respect to any ewe lamb or offspring that has died before completing the full program requirements where said deaths for the operation exceed 10 percent.

(5) Be in compliance with all requirements relating to scrapie, as described in 9 CFR parts 54 and 79.

(d) To be eligible for any payments addressed under this section, a sheep and lamb operation must be engaged in the business of producing and marketing agricultural products at the time of filing the application.

(e) In addition, to be eligible for payment, a sheep and lamb operation must submit a timely application during the application period for benefits and comply with all other terms and conditions of this part or that are contained in the application for such benefits, and such other conditions as may be imposed by FSA.

(f) Proof that each lamb was held during and through the end of the base period as required by paragraph (a) of this section, as must be determined individually for each lamb, shall be provided in such manner, and with such access to the operation and the documents and information related to the operation, as FSA may request.

§ 784.6 Rate of payment and limitations on funding.

(a) Subject to the availability of funds and to the proration provisions of § 784.7, payments for qualifying operations shall be $18 for each qualifying ewe lamb retained or purchased for breeding purposes.

§ 784.7 Availability of funds.

Total payments under this part, unless otherwise determined by the FSA, cannot exceed $18 million. In the event that approval of all eligible applications would result in expenditures in excess of the amount available, FSA shall prorate the available funds by a national factor to reduce the expected payments to be made to the amount available. The payment shall be made based on the national factored rate as determined by FSA. FSA shall prorate the payments in such manner as it, in its sole discretion, finds appropriate and reasonable.

§ 784.8 Appeals.

The appeal regulations set forth at parts 11 and 780 of this title apply to determinations made pursuant to this part.

§784.9 Misrepresentation and scheme or device.

(a) A sheep and lamb operation shall be ineligible to receive assistance under this program if it is determined by the State committee or the county committee to have:

(1) Adopted any scheme or device that tends to defeat the purpose of this program;

(2) Made any fraudulent representation; or

(3) Misrepresented any fact affecting a program determination.

(b) Any funds disbursed pursuant to this part to any person or operation engaged in a misrepresentation, scheme, or device, shall be refunded with interest together with such other sums as may become due. Any sheep and lamb operation or person engaged in acts prohibited by this section and any sheep and lamb operation or person receiving payment under this part shall be jointly and severally liable with other persons or operations involved in such claim for benefits for any refund due under this section and for related charges. The remedies provided in this part shall be in addition to other civil, criminal, or administrative remedies that may apply.

§784.10 Estates, trusts, and minors.

(a) Program documents executed by persons legally authorized to represent estates or trusts will be accepted only if such person furnishes evidence of the authority to execute such documents.

(b) A minor who is otherwise eligible for assistance under this part must, also:

(1) Establish that the right of majority has been conferred on the minor by court proceedings or by statute;

(2) Show a guardian has been appointed to manage the minor's property and the applicable program documents are executed by the guardian; or

(3) Furnish a bond under which the surety guarantees any loss incurred for which the minor would be liable had the minor been an adult.

§784.11 Death, incompetence, or disappearance.

In the case of death, incompetence, disappearance or dissolution of a person that is eligible to receive benefits in accordance with this part, such person or persons specified in 7 CFR part 707 may receive such benefits, as determined appropriate by FSA.

§784.12 Maintaining records.

Persons making application for benefits under this program must maintain accurate records and accounts that will document that they meet all eligibility requirements specified herein. Such records and accounts must be retained for 3 years after the date of payment to the sheep and lamb operations under this program. Destruction of the records after such date shall be at the risk of the party undertaking the destruction.

§784.13 Refunds; joint and several liability.

(a) In the event there is an inaccurate certification or a failure to comply with any term, requirement, or condition for payment arising under the application, or this part, and if any refund of a payment to FSA shall otherwise become due in connection with the application, or this part, all related payments made under this part to any sheep and lamb operation shall be refunded to FSA together with interest as determined in accordance with paragraph (c) of this section and late payment charges as provided in part 1403 of this title.

(b) All persons signing a sheep and lamb operation's application for payment as having an interest in the operation shall be jointly and severally liable for any refund, including related charges, that is determined to be due for any reason under the terms and conditions of the application or this part with respect to such operation.

(c) Interest shall be charged on refunds required of any person under this part if FSA determines that payments or other assistance was provided to a person who was not eligible for such assistance. Such interest shall be charged at the rate of interest that the United States Treasury charges the Commodity Credit Corporation for funds, from the date FSA made such benefits available to the date of repayment or the date interest increases as determined in accordance with applicable regulations. FSA may waive the accrual of interest if FSA determines that the cause of the erroneous determination was not due to any action of the person.

(d) Interest determined in accordance with paragraph (c) of this section may be waived at the discretion of FSA alone for refunds resulting from those violations determined by FSA to have been beyond the control of the person committing the violation.

(e) Late payment interest shall be assessed on all refunds in accordance with the provisions of, and subject to the rates prescribed in 7 CFR part 792.

(f) Any excess payments made by FSA with respect to any application under this part must be refunded.

(g) In the event that a benefit under this subpart was provided as the result of erroneous information provided by any person, the benefit must be repaid with any applicable interest.

§ 784.14 Offsets and withholdings.

FSA may offset or withhold any amounts due FSA under this subpart in accordance with the provisions of 7 CFR part 792, or successor regulations, as designated by the Department.

§ 784.15 Assignments.

Any person who may be entitled to a payment may assign his rights to such payment in accordance with 7 CFR part 1404, or successor regulations, as designated by the Department.

§ 784.16 Termination of program.

This program will be terminated after payment has been made to those applications certified as eligible pursuant to the application period established in § 784.4.

PART 785—CERTIFIED STATE MEDIATION PROGRAM

AUTHORITY: 5 U.S.C. 301; 7 U.S.C. 1989; and 7 U.S.C. 5101–5104.

SOURCE: 67 FR 57315, Sept. 10, 2002, unless otherwise noted.

§ 785.1 General.

(a) States meeting conditions specified in this part may have their mediation programs certified by the Farm Service Agency (FSA) and receive Federal grant funds for the operation and administration of agricultural mediation programs.

(b) USDA agencies participate in mediations pursuant to agency rules governing their informal appeals processes. Where mediation of an agency decision by a certified State mediation program is available to participants in an agency program as part of the agency's informal appeal process, the agency will offer a participant receiving notice of an agency decision the opportunity to mediate the decision under the State's certified mediation program, in accordance with the agency's informal appeals regulations.

(c) USDA agencies making mediation available as part of the agency informal appeals process may execute memoranda of understanding with a certified mediation program concerning procedures and policies for mediations during agency informal appeals that are not inconsistent with this part or other applicable regulations. Each such memorandum of understanding will be deemed part of the grant agreement governing the operation and administration of a State certified mediation program receiving Federal grant funds under this part.

(d) A mediator in a program certified under this part has no authority to make decisions that are binding on parties to a dispute.

(e) No person may be compelled to participate in mediation provided through a mediation program certified under this part. This provision shall not affect a State law requiring mediation before foreclosure on agricultural land or property.

§785.2 Definitions.

Administrator means the Administrator, FSA, or authorized designee.

Certified State mediation program means a program providing mediation services that has been certified in accordance with §785.3.

Confidential mediation means a mediation process in which the mediator will not disclose to any person oral or written communications provided to the mediator in confidence, except as allowed by 5 U.S.C. 574 or §785.9.

Covered persons means producers, their creditors (as applicable), and other persons directly affected by actions of the USDA involving one or more of the following issues:

(1) Wetlands determinations;

(2) Compliance with farm programs, including conservation programs;

(3) Agricultural loans (regardless of whether the loans are made or guaranteed by the USDA or are made by a third party);

(4) Rural water loan programs;

(5) Grazing on National Forest System lands;

(6) Pesticides; or

(7) Such other issues as the Secretary may consider appropriate.

Fiscal year means the period of time beginning October 1 of one year and ending September 30 of the next year and designated by the year in which it ends.

FSA means the Farm Service Agency of the U.S. Department of Agriculture, or a successor agency.

Mediation services means all activities relating to the intake and scheduling of mediations; the provision of background and selected information regarding the mediation process; financial advisory and counseling services (as reasonable and necessary to prepare parties for mediation) performed by a person other than a State mediation program mediator; and mediation sessions in which a mediator assists disputing parties in voluntarily reaching mutually agreeable settlement of issues within the laws, regulations, and the agency's generally applicable program policies and procedures, but has

no authoritative decision making power.

Mediator means a neutral individual who functions specifically to aid the parties in a dispute during a mediation process.

Qualified mediator means a mediator who meets the training requirements established by State law in the State in which mediation services will be provided or, where a State has no law prescribing mediator qualifications, an individual who has attended a minimum of 40 hours of core mediator knowledge and skills training and, to remain in a qualified mediator status, completes a minimum of 20 hours of additional training or education during each 2-year period. Such training or education must be approved by the USDA, by an accredited college or university, or by one of the following organizations: State Bar of a qualifying State, a State mediation association, a State approved mediation program, or a society of professionals in dispute resolution.

Qualifying State means a State with a State mediation program currently certified by FSA.

§785.3 Annual certification of State mediation programs.

To obtain FSA certification of the State's mediation program, the State must meet the requirements of this section.

(a) *New request for certification.* A new request for certification of a State mediation program must include descriptive and supporting information regarding the mediation program and a certification that the mediation program meets certain requirements as prescribed in this subsection. If a State is also qualifying its mediation program to request a grant of Federal funds under the certified State mediation program, the State must submit with its request for certification additional information in accordance with §785.4.

(1) *Description of mediation program.* The State must submit a narrative describing the following with supporting documentation:

(i) A summary of the program;

(ii) An identification of issues available for mediation under the program;

417

(iii) Management of the program;

(iv) Mediation services offered by the program;

(v) Program staffing and staffing levels;

(vi) Uses of contract mediation services in the program describing both services provided by contractors and costs of such services;

(vii) State statutes and regulations in effect that are applicable to the State's mediation program; and

(viii) A description of the State program's education and training requirements for mediators including:

(A) Training in mediation skills and in USDA programs;

(B) Identification and compliance with any State law requirements; and

(C) Other steps by the State's program to recruit and deploy qualified mediators.

(ix) Any other information requested by FSA;

(2) *Certification.* The Governor, or head of a State agency designated by the Governor, must certify in writing to the Administrator that the State's mediation program meets the following program requirements:

(i) That the State's mediation program provides mediation services to covered persons with the aim of reaching mutually agreeable decisions between the parties under the program;

(ii) That the State's mediation program is authorized or administered by an agency of the State government or by the Governor of the State;

(iii) That the State's mediation program provides for training of mediators in mediation skills and in all issues covered by the State's mediation program;

(iv) That the State's mediation program shall provide confidential mediation as defined in·§ 785.2;

(v) That the State's mediation program ensures, in the case of agricultural loans, that all lenders and borrowers of agricultural loans receive adequate notification of the mediation program;

(vi) That the State's mediation program ensures, in the case of other issues covered by the mediation program, that persons directly affected by actions of the USDA receive adequate notification of the mediation program; and

(vii) That the State's mediation program prohibits discrimination in its programs on the basis of race, color, national origin, sex, religion, age, disability, political beliefs, and marital or familial status.

(b) *Request for re-certification by qualifying State.* If a State is a qualifying State at the time its request is made, the written request need only describe the changes made in the program since the previous year's request, together with such documents and information as are necessary concerning such changes, and a written certification that the remaining elements of the program will continue as described in the previous request.

§ 785.4 **Grants to certified State mediation programs.**

(a) *Eligibility.* To be eligible to receive a grant, a State mediation program must:

(1) Be certified as described in § 785.3; and

(2) Submit an application for a grant with its certification or re-certification request as set forth in this section.

(b) *Application for grant.* A State requesting a grant will submit the following to the Administrator:

(1) Application for Federal Assistance, Standard Form 424 (available in any FSA office and on the Internet, *http://www.whitehouse.gov/omb/grants/*);

(2) A budget with supporting details providing estimates of the cost of operation and administration of the program. Proposed direct expenditures will be grouped in the categories of allowable direct costs under the program as set forth in paragraph (c)(1) of this section;

(3) Other information pertinent to the funding criteria specified in § 785.7(b); and

(4) Any additional supporting information requested by FSA in connection with its review of the grant request.

(c) *Grant purposes.* Grants made under this part will be used only to pay the allowable costs of operation and administration of the components of a qualifying State's mediation program that have been certified as set forth in §785.3(b)(2). Costs of services other than mediation services to covered persons within the State are not considered part of the cost of operation and administration of the mediation program for the purpose of determining the amount of a grant award.

(1) *Allowable costs.* Subject to applicable cost principles in 2 CFR part 200, subpart E, allowable costs for operations and administration are limited to those that are reasonable and necessary to carry out the State's certified mediation program in providing mediation services for covered persons within the State. Specific categories of costs allowable under the certified State mediation program include, and are limited to:

(i) Staff salaries and fringe benefits;

(ii) Reasonable fees and costs of mediators;

(iii) Office rent and expenses, such as utilities and equipment rental;

(iv) Office supplies;

(v) Administrative costs, such as workers' compensation, liability insurance, employer's share of Social Security, and travel that is necessary to provide mediation services;

(vi) Education and training of participants and mediators involved in mediation;

(vii) Security systems necessary to assure confidentiality of mediation sessions and records of mediation sessions;

(viii) Costs associated with publicity and promotion of the program; and

(ix) Financial advisory and counseling services for parties requesting mediation (as reasonable and necessary to prepare parties for mediation) that are performed by a person other than a state mediation program mediator and as approved under guidelines established by the state mediation program and reported to FSA.

(2) *Prohibited expenditures.* Expenditures of grant funds are not allowed for:

(i) Purchase of capital assets, real estate, or vehicles and repair, or maintenance of privately-owned property;

(ii) Political activities;

(iii) Routine administrative activities not allowable under 2 CFR part 200, subpart E; and

(iv) Services provided by a State mediation program that are not consistent with the features of the mediation program certified by the State, including advocacy services on behalf of a mediation participant, such as representation of a mediation client before an administrative appeals entity of the USDA or other Federal Government department or Federal or State Court proceeding.

[67 FR 57315, Sept. 10, 2002, as amended at 79 FR 75996, Dec. 19, 2014]

§785.5 Fees for mediation services.

A requirement that non-USDA parties who elect to participate in mediation pay a fee for mediation services will not preclude certification of a certified State mediation program or its eligibility for a grant; however, if participation in mediation is mandatory for a USDA agency, a certified State mediation program may not require the USDA agency to pay a fee to participate in a mediation.

§785.6 Deadlines and address.

(a) *Deadlines.* (1) To be a qualifying State as of the beginning of a fiscal year and to be eligible for grant funding as of the beginning of the fiscal year, the Governor of a State or head of a State agency designated by the Governor of a State must submit a request for certification and application for grant on or before August 1 of the calendar year in which the fiscal year begins.

(2) *Requests received after August 1.* FSA will accept requests for re-certifications and for new certifications of State mediation programs after August 1 in each calendar year; however, such requests will not be considered for grant funding under §785.7(c) until after March 1.

(3) *Requests for additional grant funds during a fiscal year.* Any request by a State mediation program that is eligible for grant funding as of the beginning of the fiscal year for additional grant funds during that fiscal year for additional, unbudgeted demands for mediation services must be submitted on or before March 1 of the fiscal year.

(b) *Address.* The request for certification or re-certification and any grant request must be mailed or delivered to:

Administrator, Farm Service Agency, U.S. Department of Agriculture, Stop 0501, 1400 Independence Avenue, SW., Washington, DC 20250–0501.

§ 785.7 Distribution of Federal grant funds.

(a) *Maximum grant award.* A grant award shall not exceed 70 percent of the budgeted allowable costs of operation and administration of the certified State mediation program. In no case will the sum granted to a State exceed $500,000 per fiscal year.

(b) *Funding criteria.* FSA will consider the following in determining the grant award to a qualifying State:

(1) Demand for and use of mediation services (historical and projected);

(2) Scope of mediation services;

(3) Service record of the State program, as evidenced by:

(i) Number of inquiries;

(ii) Number of requests for and use of mediation services, historical and projected, as applicable;

(iii) Number of mediations resulting in signed mediation agreements;

(iv) Timeliness of mediation services; and

(v) Activities promoting awareness and use of mediation;

(4) Historic use of program funds (budgeted versus actual); and

(5) Material changes in the State program.

(c) *Disbursements of grant funds.* (1) Grant funds will be paid in advance, in installments throughout the Federal fiscal year as requested by a certified State mediation program and approved by FSA. The initial payment to a program in a qualifying State eligible for grant funding as of the beginning of a fiscal year shall represent at least one-fourth of the State's annual grant award. The initial payment will be made as soon as practicable after certification, or re-certification, after grant funds are appropriated and available.

(2) Payment of grant funds will be by electronic funds transfer to the designated account of each certified State mediation program, as approved by FSA.

(d) *Administrative reserve fund.* After funds are appropriated, FSA will set aside 5 percent of the annual appropria-

tion for use as an administrative reserve.

(1) Subject to paragraph (a) of this section and the availability of funds, the Administrator will allocate and disburse sums from the administrative reserve in the following priority order:

(i) Disbursements to cover additional, unbudgeted demands for mediation services in qualifying States eligible for grant funding as of the beginning of the fiscal year;

(ii) Grants to qualifying States whose requests for new certification or re-certification were received between August 2 and March 1. A previously qualifying State that submits a request for re-certification received after August 1 may receive a grant award effective as of the beginning of the fiscal year. A newly qualifying State that submits a request for certification received after August 1 may receive a grant award effective March 31 of the fiscal year.

(iii) Any balance remaining in the administrative reserve will be allocated pro rata to certified State mediation programs based on their initial fiscal year grant awards.

(2) All funds from the administrative reserve will be made available on or before March 31 of the fiscal year.

(e) *Period of availability of funds.* (1) Certified State mediation programs receiving grant funds are encouraged to obligate award funds within the Federal fiscal year of the award. A State may, however, carry forward any funds disbursed to its certified State mediation program that remain unobligated at the end of the fiscal year of award for use in the next fiscal year for costs resulting from obligations in the subsequent funding period. Any carryover balances plus any additional obligated fiscal year grant will not exceed the lesser of 70 percent of the State's budgeted allowable costs of operation and administration of the certified State mediation program for the subsequent fiscal year, or $500,000.

(2) Grant funds not spent in accordance with this part will be subject to de-obligation and must be returned to the USDA.

§785.8 Reports by qualifying States receiving mediation grant funds.

(a) *Annual report by certified State mediation program.* No later than 30 days following the end of a fiscal year during which a qualifying State received a grant award under this part, the State must submit to the Administrator an annual report on its certified State mediation program. The annual report must include the following:

(1) A review of mediation services provided by the certified State mediation program during the preceding Federal fiscal year providing information concerning the following matters:

(i) A narrative review of the goals and accomplishments of the certified State mediation program in providing intake and scheduling of cases; the provision of background and selected information regarding the mediation process; financial advisory and counseling services, training, notification, public education, increasing resolution rates, and obtaining program funding from sources other than the grant under this part.

(ii) A quantitative summary for the preceding fiscal year, and for prior fiscal years, as appropriate, for comparisons of program activities and outcomes of the cases opened and closed during the reporting period; mediation services provided to clients grouped by program and subdivided by issue, USDA agency, types of covered persons and other participants; and the resolution rate for each category of issue reported for cases closed during the year;

(2) An assessment of the performance and effectiveness of the State's certified mediation program considering:

(i) Estimated average costs of mediation services per client with estimates furnished in terms of the allowable costs set forth in §785.4(b)(1).

(ii) Estimated savings to the State as a result of having the State mediation program certified including:

(A) Projected costs of avoided USDA administrative appeals based on projections of the average costs of such appeals furnished to the State by FSA, with the assistance of the USDA National Appeals Division and other agencies as appropriate;

(B) In agricultural credit mediations that do not result from a USDA ad-verse program decision, projected cost savings to the various parties as a result of resolution of their dispute in mediation. Projected cost savings will be based on such reliable statistical data as may be obtained from State statistical sources including the certified State's bar association, State Department of Agriculture, State court system or Better Business Bureau, or other reliable State or Federal sources;

(iii) Recommendations for improving the delivery of mediation services to covered persons, including:

(A) Increasing responsiveness to needs for mediation services.

(B) Promoting increases in dispute resolution rates.

(C) Improving assessments of training needs.

(D) Improving delivery of training.

(E) Reducing costs per mediation.

(3) Such other matters relating to the program as the State may elect to include, or as the Administrator may require.

(b) *Audits.* Any qualifying State receiving a grant under this part is required to submit an audit report in compliance with 2 CFR part 200, subpart F.

[67 FR 57315, Sept. 10, 2002, as amended at 79 FR 75996, Dec. 19, 2014]

§785.9 Access to program records.

The regulations in 2 CFR 200.333 through 200.337 provide general record retention and access requirements for records pertaining to grants. In addition, the State must maintain and provide the Government access to pertinent records regarding services delivered by the certified State mediation program for purposes of evaluation, audit and monitoring of the certified State mediation program as follows:

(a) For purposes of this section, pertinent records consist of: the names and addresses of applicants for mediation services; dates mediations opened and closed; issues mediated; dates of sessions with mediators; names of mediators; mediation services furnished to participants by the program; the sums charged to parties for each mediation service; records of delivery of services to prepare parties for mediation (including financial advisory and counseling services); and the outcome of the mediation services including formal settlement results and supporting documentation.

(b) State mediators will notify all participants in writing at the beginning of the mediation session that the USDA, including the USDA Inspector General, the Comptroller General of the United States, the Administrator, and any of their representatives will have access to pertinent records as necessary to monitor and to conduct audits, investigations, or evaluations of mediation services funded in whole or in part by the USDA.

(c) All participants in a mediation must sign and date an acknowledgment of receipt of such notice from the mediator. The certified State mediation program shall maintain originals of such acknowledgments in its mediation files for at least 5 years.

[67 FR 57315, Sept. 10, 2002, as amended at 79 FR 75996, Dec. 19, 2014]

§ 785.10　Penalty for non-compliance.

(a) The Administrator is authorized to withdraw certification of a State mediation program, terminate or suspend the grant to such program, require a return of unspent grant funds, a reimbursement of grant funds on account of expenditures that are not allowed, and may impose any other penalties or sanctions authorized by law if the Administrator determines that:

(1) The State's mediation program, at any time, does not meet the requirements for certification;

(2) The mediation program is not being operated in a manner consistent with the features of the program certified by the State, with applicable regulations, or the grant agreement;

(3) Costs that are not allowed under § 785.4(b) are being paid out of grant funds;

(4) The mediation program fails to grant access to mediation records for purposes specified in § 785.8; or

(5) Reports submitted by the State pursuant to § 785.7 are false, contain misrepresentations or material omissions, or are otherwise misleading.

(b) In the event that FSA gives notice to the State of its intent to enforce any withdrawal of certification or other penalty for non-compliance, USDA agencies will cease to partici-

pate in any mediation conducted by the State's mediation program immediately upon delivery of such notice to the State.

§ 785.11　Reconsideration by the Administrator.

(a) A State mediation program may request that the Administrator reconsider any determination that a State is not a qualifying State under § 785.3 and any penalty decision made under § 785.10. The decision of the Administrator upon reconsideration shall be the final administrative decision of FSA.

(b) Nothing in this part shall preclude action to suspend or debar a State mediation program or administering entity under 2 CFR parts 180 and 417 following a withdrawal of certification of the State mediation program.

[67 FR 57315, Sept. 10, 2002, as amended at 79 FR 75996, Dec. 19, 2014]

§ 785.12　Nondiscrimination.

The provisions of parts 15, 15b and 1901, subpart E, of this title and part 90 of title 45 apply to activities financed by grants made under this part.

§ 785.13　OMB Control Number.

The information collection requirements in this regulation have been approved by the Office of Management and Budget and assigned OMB control number 0560-0165.

PART 786—DAIRY DISASTER ASSISTANCE PAYMENT PROGRAM (DDAP-III)

AUTHORITY: Sec. 9007, Pub. L. 110–28, 121 Stat. 112; and Sec. 743, Pub. L. 110–161.

SOURCE: 73 FR 11522, Mar. 4, 2008, unless otherwise noted.

§ 786.100   Applicability.

(a) Subject to the availability of funds, this part specifies the terms and conditions applicable to the Dairy Disaster Assistance Payment Program (DDAP–III) authorized by section 9007 of Public Law 110–28 (extended by Pub. L. 110–161). Benefits are available to eligible United States producers who have suffered dairy production losses in eligible counties as a result of a natural disaster declared during the period between January 1, 2005, and December 31, 2007, (that is, after January 1, 2005, and before December 31, 2007).

(b) To be eligible for this program, a producer must have been a milk producer anytime during the period of January 2, 2005, through December 30, 2007, in a county declared a natural disaster by the Secretary of Agriculture, declared a major disaster or emergency designated by the President of the United States. For a county for which there was a timely Presidential declaration, but the declaration did not cover the loss, the county may still be eligible if the county is one for which an appropriate determination of a Farm Service Agency (FSA) Administrator's Physical Loss Notice applies. Counties contiguous to a county that is directly eligible by way of a natural disaster declaration are also eligible. Only losses occurring in eligible counties are eligible for payment in this program.

(c) Subject to the availability of funds, FSA will provide benefits to eligible dairy producers. Additional terms and conditions may be specified in the payment application that must be completed and submitted by producers to receive a disaster assistance payment for dairy production losses.

(d) To be eligible for payments, producers must meet the provisions of, and their losses must meet the conditions of, this part and any other conditions imposed by FSA.

§ 786.101   Administration.

(a) DDAP–III will be administered under the general supervision of the Administrator, FSA, or a designee, and be carried out in the field by FSA State and county committees (State and county committees) and FSA employees.

(b) State and county committees, and representatives and employees thereof, do not have the authority to modify or waive any of the provisions of the regulations of this part.

(c) The State committee will take any action required by the regulations of this part that has not been taken by the county committee. The State committee will also:

(1) Correct, or require the county committee to correct, any action taken by such county committee that is not in accordance with the regulations of this part; and

(2) Require a county committee to withhold taking any action that is not in accordance with the regulations of this part.

(d) No provision of delegation in this part to a State or county committee will preclude the Administrator, FSA, or a designee, from determining any question arising under the program or from reversing or modifying any determination made by the State or county committee.

(e) The Deputy Administrator, Farm Programs, FSA, may authorize State and county committees to waive or modify deadlines in cases where lateness or failure to meet such requirements do not adversely affect the operation of the DDAP–III and does not violate statutory limitations of the program.

(f) Data furnished by the applicants is used to determine eligibility for program benefits. Although participation in DDAP–III is voluntary, program benefits will not be provided unless the producer furnishes all requested data.

§ 786.102   Definitions.

The definitions in 7 CFR part 718 apply to this part except to the extent they are inconsistent with the provisions of this part. In addition, for the purpose of this part, the following definitions apply.

Administrator means the FSA Administrator, or a designee.

Application means DDAP-III application.

Application period means the time period established by the Deputy Administrator for producers to apply for program benefits.

Base annual production means the pounds of production determined by multiplying the average annual production per cow calculated from base period information times the average number of cows in the dairy herd during each applicable disaster year.

County committee means the FSA county committee.

County office means the FSA office responsible for administering FSA programs for farms located in a specific area in a State.

Dairy operation means any person or group of persons who, as a single unit, as determined by FSA, produces and markets milk commercially from cows and whose production facilities are located in the United States.

Department or USDA means the United States Department of Agriculture.

Deputy Administrator means the Deputy Administrator for Farm Programs (DAFP), FSA, or a designee.

Disaster claim period means the calendar year(s) applicable to the disaster declaration during the eligible period in which the production losses occurred.

Disaster county means a county included in the geographic area covered by a natural disaster declaration, and any county contiguous to a county that qualifies by a natural disaster declaration.

Farm Service Agency or FSA means the Farm Service Agency of the Department.

Hundredweight or cwt. means 100 pounds.

Milk handler or cooperative means the marketing agency to, or through, which the producer commercially markets whole milk.

Milk marketings means a marketing of milk for which there is a verifiable sale or delivery record of milk marketed for commercial use.

Natural disaster declaration means a natural disaster declaration issued by the Secretary of Agriculture after January 1, 2005, but before December 31, 2007, under section 321(a) of the Consolidated Farm and Rural Development Act (7 U.S.C. 1961(a)), a major disaster or emergency designation by the President of the United States in that period under the Robert T. Stafford Disaster Relief and Emergency Assistance Act, or a determination of a Farm Service Agency Administrator's Physical Loss Notice for a county covered in an otherwise eligible Presidential declaration.

Payment pounds means the pounds of milk production from a dairy operation for which the dairy producer is eligible to be paid under this part.

Producer means any individual, group of individuals, partnership, corporation, estate, trust association, cooperative, or other business enterprise or other legal entity who is, or whose members are, a citizen of, or a legal resident alien in, the United States, and who directly or indirectly, as determined by the Secretary, have a share entitlement or ownership interest in a commercial dairy's milk production and who share in the risk of producing milk, and make contributions (including land, labor, management, equipment, or capital) to the dairy farming operation of the individual or entity.

Reliable production evidence means records provided by the producer subject to a determination of acceptability by the county committee that are used to substantiate the amount of production reported when verifiable records are not available; the records may include copies of receipts, ledgers of income, income statements of deposit slips, register tapes, and records to verify production costs, contemporaneous measurements, and contemporaneous diaries.

Verifiable production records means evidence that is used to substantiate the amount of production marketed, including any dumped production, and that can be verified by FSA through an independent source.

§786.103 Time and method of application.

(a) Dairy producers may obtain an application, in person, by mail, by telephone, or by facsimile from any FSA county office. In addition, applicants may download a copy of the application at *http://www.sc.egov.usda.gov.*

(b) A request for benefits under this part must be submitted on a completed DDAP–III application. Applications and any other supporting documentation must be submitted to the FSA county office serving the county where the dairy operation is located, but, in any case, must be received by the FSA county office by the close of business on the date established by the Deputy Administrator. Applications not received by the close of business on such date will be disapproved as not having been timely filed and the dairy producer will not be eligible for benefits under this program.

(c) All persons who share in the milk production of the dairy operation and risk of the dairy operation's total production must certify to the information on the application before the application will be considered complete.

(d) Each dairy producer requesting benefits under this part must certify to the accuracy and truthfulness of the information provided in their application and any supporting documentation. Any information entered on the application will be considered information from the applicant regardless of who entered the information on the application. All information provided is subject to verification by FSA. Refusal to allow FSA or any other agency of the Department of Agriculture to verify any information provided may result in a denial of eligibility. Furnishing the information is voluntary; however, without it program benefits will not be approved. Providing a false certification to the Government may be punishable by imprisonment, fines, and other penalties or sanctions.

§786.104 Eligibility.

(a) Producers in the United States will be eligible to receive dairy disaster benefits under this part only if they have suffered dairy production losses, previously uncompensated by disaster payments including any previous dairy disaster payment program, during the claim period applicable to a natural disaster declaration in a disaster county. To be eligible to receive payments under this part, producers in a dairy operation must:

(1) Have produced and commercially marketed milk in the United States and commercially marketed the milk produced anytime during the period of January 2, 2005 through December 30, 2007;

(2) Be a producer on a dairy farm operation physically located in an eligible county where dairy production losses were incurred as a result of a disaster for which an applicable natural disaster declaration was issued between January 1, 2005 and December 31, 2007, and limit their claims to losses that occurred in those counties, specific to conditions resulting from the declared disaster as described in the natural disaster declaration;

(3) Provide adequate proof, to the satisfaction of the FSA county committee, of the average number of cows in the dairy herd and annual milk production commercially marketed by all persons in the eligible dairy operation during the years of the base period (2003 and 2004 calendar years) and applicable disaster year that corresponds with the issuance date of the applicable natural disaster declaration, or other period as determined by FSA, to determine the total pounds of eligible losses that will be used for payment; and

(4) Apply for payments during the application period established by the Deputy Administrator.

(b) Payments may be made for losses suffered by an otherwise eligible producer who is now deceased or is a dissolved entity if a representative who currently has authority to enter into a contract for the producer or the producer's estate signs the application for payment. Proof of authority to sign for the deceased producer's estate or a dissolved entity must be provided. If a producer is now a dissolved general partnership or joint venture, all members of the general partnership or joint venture at the time of dissolution or their duly-authorized representatives must sign the application for payment.

(c) Producers associated with a dairy operation must submit a timely application and satisfy the terms and conditions of this part, instructions issued by FSA, and instructions contained in the application to be eligible for benefits under this part.

(d) As a condition to receive benefits under this part, a producer must have been in compliance with the Highly Erodible Land Conservation and Wetland Conservation provisions of 7 CFR part 12 for the calendar year applicable to the natural disaster declaration and loss claim period, and must not otherwise be barred from receiving benefits under 7 CFR part 12 or any other law or regulation.

(e) Payments are limited to losses in eligible counties, in eligible disaster years.

(f) All payments under this part are subject to the availability of funds.

(g) Eligible losses are determined from the applicable base annual production, as defined in § 786.102, that corresponds to the natural disaster declaration and must have occurred during that same period as follows:

(1) For disaster declarations for disasters during a calendar year (2005, 2006, or 2007), the disaster claim period is the full calendar year and

(2) For disaster declarations issued during one calendar year that ends in another calendar year, the producer will be eligible for both disaster years.

(h) Deductions in eligibility will be made for any disaster payments previously received for the loss including any made under a previous dairy disaster assistance payment program for 2005.

§ 786.105   Proof of production.

(a) Evidence of production is required to establish the commercial marketing and production history of the dairy operation so that dairy production losses can be computed in accordance with § 786.106.

(b) A dairy producer must, based on the instructions issued by the Deputy Administrator, provide adequate proof of the dairy operation's commercial production, including any dairy herd inventory records available for the operation, for the years of the base period (2003 and 2004 calendar years) and disaster claim period that corresponds with the issuance date of the applicable natural disaster declaration.

(1) A producer must certify and provide such proof as requested that losses for which compensation is claimed were related to the disaster declaration issued and occurred in an eligible county during the eligible claim period.

(2) A producer must certify to the average number of cows in the dairy herd during the base period and applicable disaster claim period when there is insufficient documentation available for verification.

(3) Additional supporting documentation may be requested by FSA as necessary to verify production losses to the satisfaction of FSA.

(c) Adequate proof of production history of the dairy operation under paragraph (b) of this section must be based on milk marketing statements obtained from the dairy operation's milk handler or marketing cooperative. Supporting documents may include, but are not limited to: Tank records, milk handler records, daily milk marketings, copies of any payments received from other sources for production losses, or any other documents available to confirm or adjust the production history losses incurred by the dairy operation. All information provided is subject to verification, spot check, and audit by FSA.

(d) As specified in § 786.106, loss calculations will be based on comparing the expected base annual production consistent with this part and the actual production during the applicable disaster claim year. Such calculations are subject to adjustments as may be appropriate such as a correction for losses not due to the disaster. If adequate proof of normally marketed production and any other production for relevant periods is not presented to the satisfaction of FSA, the request for benefits will be rejected. Special adjustments for new producers may be made as determined necessary by the Administrator.

§ 786.106   Determination of losses incurred.

(a) Eligible payable losses are calculated on a dairy operation by dairy operation basis and are limited to those occurring during the applicable disaster claim period, as provided by §786.104(g), that corresponds with the applicable natural disaster declaration. Specifically, dairy production losses incurred by producers under this part are determined on the established history of the dairy operation's average number of cows in the dairy herd and actual commercial production marketed during the base period and applicable disaster claim period that corresponds with the applicable natural disaster declaration, as provided by the dairy operation consistent with §786.105. Except as otherwise provided in this part, the base annual production, as defined in §786.102 and established in §786.104(g) is determined for each applicable disaster year based on the average annual production per cow determined according to the following:

(1) The average of annual marketed production during the base period calendar years of 2003 and 2004, divided by;

(2) The average number of cows in the dairy operation's herd during the base period calendar years of 2003 and 2004.

(b) If relevant information to calculate the average annual production per cow for one or both of the base period calendar years of 2003 and 2004, is not available, an alternative method of determining the average annual production per cow may be established by the FSA Administrator. For example, for new dairies not in operation during 2003 and 2004, information from three similar farms may be obtained by FSA to estimate base annual production.

(c) The average annual production per cow, as determined according to paragraphs (a)(1) and (a)(2) of this section, is multiplied by the average number of cows in the dairy operation's herd during the applicable disaster year (excluding cow losses resulting from the disaster occurrence), to determine base annual production for the dairy operation for each applicable disaster claim period year.

(d) The eligible dairy production losses for a dairy operation for each of the authorized disaster claim period years will be:

(1) The relevant period's base annual production for the dairy operation calculated under paragraph (c) of this section less,

(2) For each such disaster claim period for each dairy operation the actual commercially-marketed production relevant to that period.

(e) Spoiled or dumped milk, disposed of for reasons unrelated to the disaster occurrence, must be counted as production for the relevant disaster claim period. Actual production losses may be adjusted to the extent the reduction in production is not certified by the producer to be the result of the disaster identified in the natural disaster declaration or is determined by FSA not to be related to the natural disaster identified in the natural disaster declaration. FSA county committees will determine production losses that are not caused by the disaster associated with the natural disaster declaration. The calculated production loss determined in §786.106(d) will be adjusted to account for pounds of production losses determined by the FSA county committee to not have been associated with the declared natural disaster for an eligible disaster county. The FSA county committee may convert cow numbers to actual pounds of production used in the adjustment, by multiplying the average annual production per cow determined from base period information, by the applicable number of cows determined to be ineligible to generate claims for benefits. Other appropriate adjustments will be made on such basis as the Deputy Administrator finds to be consistent with the objectives of the program.

(f) Actual production, as adjusted, that exceeds the base annual production will mean that the dairy operation incurred no eligible production losses for the corresponding claim period as a result of the natural disaster.

(g) Eligible production losses as otherwise determined under paragraphs (a) through (f) of this section for each authorized year of the program are added together to determine total eligible losses incurred by the dairy operation under DDAP–III subject to all other eligibility requirements as may be included in this part or elsewhere, including the deduction for previous payments including those made under a previous DDAP program.

(h) Payment on eligible dairy operation losses will be calculated using whole pounds of milk. No double counting is permitted, and only one payment will be made for each pound of milk calculated as an eligible loss after the distribution of the dairy operation's eligible production loss among the producers of the dairy operation according to § 786.107(b). Payments under this part will not be affected by any payments for dumped or spoiled milk that the dairy operation may have received from its milk handler, marketing cooperative, or any other private party; however, produced milk that was dumped or spoiled for reasons unrelated to the disaster occurrence will still count as production.

§ 786.107 Rate of payment and limitations on funding.

(a) Subject to the availability of funds, the payment rate for eligible production losses determined according to § 786.106 is, depending on the State, the annual average Mailbox milk price for the Marketing Order, applicable to the State where the eligible disaster county is located, as reported by the Agricultural Marketing Service during the relevant period. States not regulated under a Marketing Order will be assigned a payment rate based on contiguous or nearby State's Mailbox milk price. Maximum per pound payment rates for eligible losses for dairy operations located in specific states during the relevant period are as follows:

State	Mailbox price 2005	Mailbox price 2006	Mailbox price 2007*
Alabama	0.1596	0.1443	
Alaska	0.2040	0.2010	
Arizona	0.1388	0.1128	
Arkansas	0.1596	0.1443	
California	0.1388	0.1128	
Colorado	0.1403	0.1214	
Connecticut	0.1539	0.1344	
Delaware	0.1539	0.1344	
Florida	0.1758	0.1603	
Georgia	0.1596	0.1443	
Hawaii	0.2700	0.2600	
Idaho	0.1402	0.1215	
Illinois	0.1514	0.1283	
Indiana	0.1503	0.1294	
Iowa	0.1507	0.1285	
Kansas	0.1403	0.1214	
Kentucky	0.1527	0.1349	
Louisiana	0.1596	0.1443	
Maine	0.1539	0.1344	
Maryland	0.1539	0.1344	
Massachusetts	0.1539	0.1344	
Michigan	0.1478	0.1264	

State	Mailbox price 2005	Mailbox price 2006	Mailbox price 2007*
Minnesota	0.1512	0.1277	
Mississippi	0.1596	0.1443	
Missouri (Northern)	0.1403	0.1214	
Missouri (Southern)	0.1467	0.1254	
Montana	0.1512	0.1277	
Nebraska	0.1403	0.1214	
Nevada	0.1388	0.1128	
New Hampshire	0.1539	0.1344	
New Jersey	0.1539	0.1344	
New Mexico	0.1323	0.1108	
New York	0.1539	0.1303	
North Carolina	0.1527	0.1349	
North Dakota	0.1512	0.1277	
Ohio	0.1506	0.1302	
Oklahoma	0.1596	0.1443	
Oregon	0.1402	0.1215	
Pennsylvania (Eastern)	0.1539	0.1340	
Pennsylvania (Western)	0.1539	0.1302	
Puerto Rico	0.2550	0.2570	
Rhode Island	0.1539	0.1344	
South Carolina	0.1527	0.1349	
South Dakota	0.1512	0.1277	
Tennessee	0.1527	0.1349	
Texas	0.1405	0.1194	
Vermont	0.1539	0.1344	
Virginia	0.1527	0.1349	
Washington	0.1402	0.1215	
West Virginia	0.1506	0.1302	
Wisconsin	0.1535	0.1305	
Wyoming	0.1403	0.1214	

NOTE: Calculations are rounded to 7 decimal places.
*Payment rates for 2007 are currently unavailable, but will be based on the annual average Mailbox milk price for the Marketing Order, applicable to the State where the eligible disaster county is located, as reported by the Agricultural Marketing Service, consistent with payment rates provided for 2005 and 2006.

(b) Subject to the availability of funds, each eligible dairy operation's payment is calculated by multiplying the applicable payment rate under paragraph (a) of this section by the operation's total eligible losses as adjusted pursuant to this part. Where there are multiple producers in the dairy operation, individual producers' payments are disbursed according to each producer's share of the dairy operation's production as specified in the application.

(c) If the total value of losses claimed nationwide under paragraph (b) of this section exceeds the $16 million available for the DDAP-III, less any reserve that may be created under paragraph (e) of this section, total eligible losses of individual dairy operations that, as calculated as an overall percentage for each full disaster claim period applicable to the disaster declaration, are greater than 20 percent of the total base annual production will be paid at the maximum rate under paragraph (a) of this section to the extent available funding allows. A loss of over 20 percent in only one or two months during the applicable disaster claim period does not of itself qualify for the maximum per-pound payment. Rather, the priority level must be reached as an average over the whole disaster claim period for the relevant calendar year. Total eligible losses for a producer, as calculated under §786.106, of less than or equal to 20 percent during the eligible claim period will then be paid at a rate, not to exceed the rate allowed in paragraph (a) of this section, determined by dividing the eligible losses of less than 20 percent by the funds remaining after making payments for all eligible losses above the 20-percent threshold.

(d) In no event will the payment exceed the value determined by multiplying the producer's total eligible loss times the average price received for commercial milk production in the producer's area as defined in paragraph (a) of this section.

(e) No participant will receive disaster benefits under this part that in combination with the value of production not lost would result in an amount that exceeds 95 percent of the value of the expected production for the relevant period as estimated by the Secretary. Unless otherwise program funds would not be fully expended, the sum of the value of the production not lost, if any; and the disaster payment received under this part, cannot exceed 95 percent of what the production's value would have been if there had been no loss. In no case, however, may the value of production and the payment exceed the value the milk would have without the loss.

(f) A reserve may be created to handle pending or disputed claims, but claims will not be payable once the available funding is expended.

§786.108 Availability of funds.

The total available program funds are $16 million as provided by section 9007 of Title IX of Public Law 110–28.

§786.109 Appeals.

Provisions of the appeal regulations set forth at 7 CFR parts 11 and 780 apply to this part. Appeals of deter-

minations of ineligibility or payment amounts are subject to the limitations in §§786.107 and 786.108 and other limitations that may apply.

§786.110 Misrepresentation, scheme, or device.

(a) In addition to other penalties, sanctions, or remedies that may apply, a dairy producer is ineligible to receive assistance under this program if the producer is determined by FSA to have:

(1) Adopted any scheme or device that tends to defeat the purpose of this program,

(2) Made any fraudulent representation,

(3) Misrepresented any fact affecting a program determination, or

(4) Violated 7 CFR 795.17 and thus be ineligible for the year(s) of violation and the subsequent year.

(b) Any funds disbursed pursuant to this part to any person or dairy operation engaged in a misrepresentation, scheme, or device, must be refunded with interest together with such other sums as may become due. Interest will run from the date of the disbursement to the producer or other recipient of the payment from FSA. Any person or dairy operation engaged in acts prohibited by this section and any person or dairy operation receiving payment under this part is jointly and severally liable with other persons or dairy operations involved in such claim for benefits for any refund due under this section and for related charges. The remedies provided in this part are in addition to other civil, criminal, or administrative remedies that may apply.

§786.111 Death, incompetence, or disappearance.

In the case of death, incompetency, disappearance, or dissolution of an individual or entity that is eligible to receive benefits in accordance with this part, such alternate person or persons specified in 7 CFR part 707 may receive such benefits, as determined appropriate by FSA.

§786.112 Maintaining records.

Persons applying for benefits under this program must maintain records and accounts to document all eligibility requirements specified herein and must keep such records and accounts for 3 years after the date of payment to their dairy operations under this program. Destruction of the records after such date is at the risk of the party required, by this part, to keep the records.

§786.113 Refunds; joint and several liability.

(a) Excess payments, payments provided as the result of erroneous information provided by any person, or payments resulting from a failure to meet any requirement or condition for payment under the application or this part, must be refunded to FSA.

(b) A refund required under this section is due with interest determined in accordance with paragraph (d) of this section and late payment charges as provided in 7 CFR part 792. Notwithstanding any other regulation, interest will be due from the date of the disbursement to the producer or other recipient of the funds.

(c) Persons signing a dairy operation's application as having an interest in the operation will be jointly and severally liable for any refund and related charges found to be due under this section.

(d) In the event FSA determines a participant owes a refund under this part, FSA will charge program interest from the date of disbursement of the erroneous payment. Such interest will accrue at the rate that the United States Department of the Treasury charges FSA for funds plus additional charges as deemed appropriate by the Administrator or provided for by regulation or statute.

(e) The debt collection provisions of part 792 of this chapter applies to this part except as is otherwise provided in this part.

§786.114 Miscellaneous provisions.

(a) Payments or any portion thereof due under this part must be made without regard to questions of title under State law and without regard to any claim or lien against the livestock, or proceeds thereof, in favor of the owner or any other creditor except agencies and instrumentalities of the U.S. Government.

(b) Any producer entitled to any payment under this part may assign any payments in accordance with the provisions of 7 CFR part 1404.

§786.115 Termination of program.

This program will be terminated after payment has been made to those applicants certified as eligible pursuant to the application period established in §786.104. All eligibility determinations will be final except as otherwise determined by the Deputy Administrator. Any claim for payment may be denied once the allowed funds are expended, irrespective of any other provision of this part.

PART 789—AGRICULTURE PRIORITIES AND ALLOCATIONS SYSTEM (APAS)

Subpart A—General

AUTHORITY: 50 U.S.C. App. 2061–2170, 2171, and 2172; 42 U.S.C. 5195–5197h.

SOURCE: 80 FR 63898, Oct. 22, 2015, unless otherwise noted.

Subpart A—General

§789.1 Purpose.

This part provides guidance and procedures for use of the Defense Production Act priorities and allocations authority by the United States Department of Agriculture (USDA) with respect to food resources, food resource facilities, livestock resources, veterinary resources, plant health resources, and the domestic distribution of farm equipment and commercial fertilizer in this part. (The guidance and procedures in this part are consistent with the guidance and procedures provided in other regulations issued under Executive Order 13603. Guidance and procedures for use of the Defense Production

Act priorities and allocations authority with respect to other types of resources are as follows: For all forms of energy, refer to the Department of Energy's Energy Priorities and Allocations System (EPAS) regulation in 10 CFR part 217; for all forms of civil transportation, refer to the Department of Transportation's Transportation Priorities and Allocations System (TPAS) regulation in 49 CFR part 33; for water resources, refer to the Department of Defense; for health resources, refer to the Department of Health and Human Services' Health Resources Priorities and Allocations System in 45 CFR part 101; and for all other materials, services, and facilities, including construction materials, refer to the Department of Commerce's Defense Priorities and Allocations System (DPAS) regulation in 15 CFR part 700.)

§789.2 Priorities and allocations authority.

(a) Section 201 of Executive Order 13603 (3 CFR, 2012 Comp., p. 225) delegates the President's authority under section 101 of the Defense Production Act to require acceptance and priority performance of contracts and orders (other than contracts of employment) to promote the national defense over performance of any other contracts or orders, and to allocate materials, services, and facilities as deemed necessary or appropriate to promote the national defense to the following agencies. Essentially, this allows the following agencies to place priority on the performance of contracts for items and materials under their jurisdiction as required for national defense initiatives including emergency preparedness activities:

(1) The Secretary of Agriculture with respect to food resources, food resource facilities, livestock resources, veterinary resources, plant health resources, and the domestic distribution of farm equipment and commercial fertilizer;

(2) The Secretary of Energy with respect to all forms of energy;

(3) The Secretary of Health and Human Services with respect to health resources;

(4) The Secretary of Transportation with respect to all forms of civil transportation;

(5) The Secretary of Defense with respect to water resources; and

(6) The Secretary of Commerce with respect to all other materials, services, and facilities, including construction materials.

(b) Section 202 of Executive Order 13603 specifies that the priorities and allocations authority may be used only to support programs that have been determined in writing as necessary or appropriate to promote the national defense by:

(1) The Secretary of Defense with respect to military production and construction, military assistance to foreign nations, military use of civil transportation, stockpiles managed by the Department of Defense, space, and directly related activities;

(2) The Secretary of Energy with respect to energy production and construction, distribution and use, and directly related activities; or

(3) The Secretary of Homeland Security with respect to all other national defense programs, including civil defense and continuity of Government.

§ 789.3   Program eligibility.

Certain programs that promote the national defense are eligible for priorities and allocations support. These include programs for military and energy production or construction, military or critical infrastructure assistance to any foreign nation, homeland security, stockpiling, space, and any directly related activity. Other eligible programs include emergency preparedness activities conducted pursuant to Title VI of the Stafford Act and critical infrastructure protection and restoration.

Subpart B—Definitions

§ 789.8   Definitions.

Allocations means the control of the distribution of materials, services, or facilities for a purpose deemed necessary or appropriate to promote the national defense.

Allocations order means an official action to control the distribution of materials, services, or facilities for a purpose deemed necessary or appropriate to promote the national defense.

Allotment means an official action that specifies the maximum quantity for a specific use of a material, service, or facility authorized to promote the national defense.

Animal means any member of the animal kingdom (except a human).

APAS means the Agriculture Priorities and Allocations System established by this part.

Applicant means the person applying for assistance under APAS. (See definition of "person.")

Approved program means a program determined by the Secretary of Defense, the Secretary of Energy, or the Secretary of Homeland Security to be necessary or appropriate to promote the national defense, as specified in section 202 of Executive Order 13603.

Civil transportation includes movement of persons and property by all modes of transportation in interstate, intrastate, or foreign commerce within the United States, its territories and possessions, and the District of Columbia, and related public storage and warehousing, ports, services, equipment and facilities, such as transportation carrier shop and repair facilities. "Civil transportation" also includes direction, control, and coordination of civil transportation capacity regardless of ownership. "Civil transportation" does not include transportation owned or controlled by the Department of Defense, use of petroleum and gas pipelines, and coal slurry pipelines used only to supply energy production facilities directly.

Construction means the erection, addition, extension, or alteration of any building, structure, or project, using materials or products that are to be an integral and permanent part of the building, structure, or project. Construction does not include maintenance and repair.

Critical infrastructure means any systems and assets, whether physical or cyber-based, so vital to the United States that the degradation or destruction of such systems and assets would have a debilitating impact on national security, including, but not limited to, national economic security and national public health or safety.

Defense Production Act means the Defense Production Act of 1950, as amended (50 U.S.C. App. 2061 to 2170, 2171, and 2172).

Delegate agency means a government agency authorized by delegation from USDA to place priority ratings on contracts or orders needed to support approved programs.

Directive means an official action that requires a person to take or refrain from taking certain actions in accordance with the provisions.

Emergency preparedness means all those activities and measures designed or undertaken to prepare for or minimize the effects of a hazard upon the civilian population, to deal with the immediate emergency conditions that would be created by the hazard, and to make emergency repairs to, or the emergency restoration of, vital utilities and facilities destroyed or damaged by the hazard. Emergency preparedness includes the following:

(1) Measures to be undertaken in preparation for anticipated hazards (including the establishment of appropriate organizations, operational plans, and supporting agreements, the recruitment and training of personnel, the conduct of research, the procurement and stockpiling of necessary materials and supplies, the provision of suitable warning systems, the construction or preparation of shelters, shelter areas, and control centers, and, when appropriate, the non-military evacuation of the civilian population).

(2) Measures to be undertaken during a hazard (including the enforcement of passive defense regulations prescribed by duly established military or civil authorities, the evacuation of personnel to shelter areas, the control of traffic and panic, and the control and use of lighting and civil communications).

(3) Measures to be undertaken following a hazard (including activities for fire fighting, rescue, emergency medical, health and sanitation services, monitoring for specific dangers of special weapons, unexploded bomb reconnaissance, essential debris clearance, emergency welfare measures, and immediately essential emergency repair or restoration of damaged vital facilities).

Energy means all forms of energy including petroleum, gas (both natural and manufactured), electricity, solid fuels (including all forms of coal, coke, coal chemicals, coal liquefaction and coal gasification), solar, wind, other types of renewable energy, atomic energy, and the production, conservation, use, control, and distribution (including pipelines) of all of these forms of energy.

Facilities includes all types of buildings, structures, or other improvements to real property (but excluding farms, churches or other places of worship, and private dwelling houses), and services relating to the use of any such building, structure, or other improvement.

Farm equipment means equipment, machinery, and repair parts manufactured for use on farms in connection with the production or preparation for market use of food resources.

Feed is a nutritionally adequate manufactured food for animals (livestock and poultry raised for agriculture production); and by specific formula is compounded to be fed as the sole ration and is capable of maintaining life and promoting production without any additional substance being consumed except water.

Fertilizer means any product or combination of products that contain one or more of the elements—nitrogen, phosphorus, and potassium—for use as a plant nutrient.

Food resources means all commodities and products (simple, mixed, or compound), or complements to such commodities or products, that are capable of being ingested by either human beings or animals, irrespective of other uses to which such commodities or products may be put, at all stages of processing from the raw commodity to the products suitable for sale for human or animal consumption. Food resources also means potable water packaged in commercially marketable containers, all starches, sugars, vegetable and animal or marine fats and oils, seed, cotton, hemp, and flax fiber, but does not mean any such material after it loses its identity as an agricultural commodity or agricultural product.

Food resource facilities means plants, machinery, vehicles (including on-farm), and other facilities required for the production, processing, distribution, and storage (including cold storage) of food resources, and for the domestic distribution of farm equipment and fertilizer (excluding transportation for that distribution).

Hazard means an emergency or disaster resulting from a natural disaster; or from an accidental or man-caused event.

Health resources means drugs, biological products, medical devices, materials, facilities, health supplies, services, and equipment required to diagnose, mitigate, or prevent the impairment of, improve, treat, cure, or restore the physical or mental health conditions of the population.

Homeland security includes efforts:

(1) To prevent terrorist attacks within the United States;

(2) To reduce the vulnerability of the United States to terrorism;

(3) To minimize damage from a terrorist attack in the United States; and

(4) To recover from a terrorist attack in the United States.

Industrial resources means all materials, services, and facilities, including construction materials, but not including: Food resources, food resource facilities, livestock resources, veterinary resources, plant health resources, and the domestic distribution of farm equipment and commercial fertilizer; all forms of energy; health resources; all forms of civil transportation; and water resources.

Item means any raw, in process, or manufactured material, article, commodity, supply, equipment, component, accessory, part, assembly, or product of any kind, technical information, process, or service.

Letter of understanding means an official action that may be issued in resolving special priorities assistance cases to reflect an agreement reached by all parties (USDA, the Department of Commerce (if applicable), a delegate agency (if applicable), the supplier, and the customer).

Livestock means all farm-raised animals.

Livestock resources means materials, facilities, vehicles, health supplies, services, and equipment required for the production and distribution of livestock.

Maintenance and repair and operating supplies (MRO) means:

(1) *Maintenance* is the upkeep necessary to continue any plant, facility, or equipment in working condition.

(2) *Repair* is the restoration of any plant, facility, or equipment to working condition when it has been rendered unsafe or unfit for service by wear and tear, damage, or failure of parts.

(3) *Operating supplies* are any resources carried as operating supplies according to a person's established accounting practice. Operating supplies may include hand tools and expendable tools, jigs, dies, fixtures used on production equipment, lubricants, cleaners, chemicals, and other expendable items.

(4) *MRO* does not include items produced or obtained for sale to other persons or for installation upon or attachment to the property of another person, or items required for the production of such items; items needed for the replacement of any plant, facility, or equipment; or items for the improvement of any plant, facility, or equipment by replacing items that are still in working condition with items of a new or different kind, quality, or design.

Materials includes:

(1) Any raw materials (including minerals, metals, and advanced processed materials), commodities, articles, components (including critical components), products, and items of supply; and

(2) Any technical information or services ancillary to the use of any such materials, commodities, articles, components, products, or items.

National defense means programs for military and energy production or construction, military or critical infrastructure assistance to any foreign nation, homeland security, stockpiling, space, and any directly related activity. Such term includes emergency preparedness activities conducted pursuant to Title VI of the Stafford Act and critical infrastructure protection and restoration.

Official action means an action taken by USDA or another resource agency under the authority of the Defense Production Act, Executive Order 13603, or this part. Such actions include the issuance of rating authorizations, directives, set-asides, allotments, letters of understanding, demands for information, inspection authorizations, and administrative subpoenas.

Person includes an individual, corporation, partnership, association, or any other organized group of persons, or legal successor or representative thereof, or any State or local government or agency thereof, or any Federal agency.

Plant health resources means biological products, materials, facilities, vehicles, supplies, services, and equipment required to prevent the impairment of, improve, or restore plant health conditions.

Rated order means a prime contract, a subcontract, or a purchase order in support of an approved program issued as specified in the provisions of this part. Persons may request an order (contract) be rated in response to a need that is defined in this part. However, an order does not become rated until the request is approved by USDA. USDA will assign a rating priority for each rating request approved that designates the priority of that order over other orders that have similar order specifics.

Resource agency means any agency that is delegated priorities and allocations authority as specified in §789.2.

Secretary means the Secretary of Agriculture.

Seed is used with its commonly understood meaning and includes all seed grown for and customarily sold to users for planting for the production of agriculture crops.

Services includes any effort that is needed for or incidental to:

(1) The development, production, processing, distribution, delivery, or use of an industrial resource or a critical technology item;

(2) The construction of facilities;

(3) The movement of individuals and property by all modes of civil transportation; or

(4) Other national defense programs and activities.

Set-aside means an official action that requires a person to reserve materials, services, or facilities capacity in anticipation of the receipt of rated orders.

Stafford Act means the Robert T. Stafford Disaster Relief and Emergency Assistance Act, as amended (42 U.S.C. 5195–5197h).

USDA means the U.S. Department of Agriculture.

Veterinary resources means drugs, biological products, medical devices, materials, facilities, vehicles, health supplies, services, and equipment required to diagnose, mitigate or prevent the impairment of, improve, treat, cure, or restore the health conditions of the animal population.

Water resources means all usable water, from all sources, within the jurisdiction of the United States, that can be managed, controlled, and allocated to meet emergency requirements, except water resources does not include usable water that qualifies as food resources.

Subpart C—Placement of Rated Orders

§789.10 Delegations of authority.

(a) [Reserved]

(b) Within USDA, authority to administer APAS has been delegated to the Administrator, Farm Service Agency, through the Under Secretary for Farm and Foreign Agricultural Services. (See §§2.16(a)(6) and 2.42(a)(5) of this title.) The Farm Service Agency Administrator will coordinate APAS implementation and administration through the Director, USDA Office of Homeland Security and Emergency Coordination, as delegated by the Assistant Secretary for Administration. (See §§2.24(a)(8)(ii)(A) and 2.24(a)(8)(v); 2.95(b)(1)(i) and 2.95(b)(4) of this title.)

§789.11 Priority ratings.

(a) *Levels of priority.* Priority levels designate differences between orders based on national defense including emergency preparedness requirements.

(1) There are two levels of priority established by APAS, identified by the rating symbols "DO" and "DX."

(2) All DO-rated orders have equal priority with each other and take precedence over unrated orders. All DX-rated orders have equal priority with each other and take precedence over DO-rated orders and unrated orders. (For resolution of conflicts among rated orders of equal priority, see § 789.14(c).)

(3) In addition, a directive regarding priority treatment for a given item issued by the resource agency with priorities jurisdiction for that item takes precedence over any DX-rated order, DO-rated order, or unrated order, as stipulated in the directive. (For more information on directives, see § 789.42.)

(b) *Program identification symbols.* Program identification symbols indicate which approved program is being supported by a rated order. The list of currently approved programs and their identification symbols are listed in Schedule I. For example, P1 identifies a program involving food and food resources processing and storage. Program identification symbols, in themselves, do not connote any priority. Additional programs may be approved under the procedures of Executive Order 13603 at any time.

(c) *Priority ratings.* A priority rating consists of the rating symbol DO or DX followed by the program identification symbol, such as P1 or P2. Thus, a contract for the supply of livestock feed will contain a DO–P1 or DX–P1 priority rating.

§ 789.12 Elements of a rated order.

(a) Each rated order must include:

(1) The appropriate priority rating (for example, DO–P1 for food and food resources processing and storage);

(2) A required delivery date or dates. The words "immediately" or "as soon as possible" do not constitute a delivery date. Some purchase orders, such as a "requirements contract," "basic ordering agreement," "prime vendor contract," or similar procurement document, bearing a priority rating may contain no specific delivery date or dates if it provides for the furnishing of items or services from time-to-time or within a stated period against specific purchase orders, such as calls, requisitions, and delivery orders. Specific purchase orders must specify a required delivery date or dates and are to be considered as rated as of the date of their receipt by the supplier and not as of the date of the original procurement document;

(3) The written signature on a manually placed order, or the digital signature or name on an electronically placed order, of an individual authorized to sign rated orders for the person placing the order. The signature or use of the name certifies that the rated order is authorized under this part and that the requirements of this part are being followed; and

(4) A statement as follows:

(i) A statement that reads:

This is a rated order certified for national defense use, and you are required to follow all the provisions of the Agriculture Priorities and Allocations System regulation in 7 CFR part 789.

(ii) If the rated order is placed in support of emergency preparedness requirements and expedited action is necessary and appropriate to meet these requirements, the following sentences should be added following the statement specified in paragraph (a)(4)(i) of this section:

This rated order is placed for the purpose of emergency preparedness. It must be accepted or rejected within six (6) hours after receipt of the order if the order is issued in response to a hazard that has occurred; or within the greater of twelve (12) hours or the time specified in the order, if the order is issued to prepare for an imminent hazard, in accordance with 7 CFR 789.13(e).

(b) [Reserved]

§ 789.13 Acceptance and rejection of rated orders.

(a) *Mandatory acceptance.* A person must accept a rated order in accordance with the following requirements:

(1) Except as otherwise specified in this section, a person must accept every rated order received and must fill such orders regardless of any other rated or unrated orders that have been accepted.

(2) A person must not discriminate against rated orders in any manner such as by charging higher prices or by imposing different terms and conditions than for comparable unrated orders.

(b) *Mandatory rejection.* Unless otherwise directed by USDA for a rated order involving food resources, food resource facilities, livestock resources, veterinary resources, plant health resources, or the domestic distribution of farm equipment and commercial fertilizer:

(1) A person must not accept a rated order for delivery on a specific date if unable to fill the order by that date. However, the person must inform the customer of the earliest date on which delivery can be made and offer to accept the order on the basis of that date. Scheduling conflicts with previously accepted lower rated or unrated orders are not sufficient reason for rejection in this section.

(2) A person must not accept a DO-rated order for delivery on a date that would interfere with delivery of any previously accepted DO- or DX-rated orders. However, the person must offer to accept the order based on the earliest delivery date otherwise possible.

(3) A person must not accept a DX-rated order for delivery on a date that would interfere with delivery of any previously accepted DX-rated orders, but must offer to accept the order based on the earliest delivery date otherwise possible.

(4) If a person is unable to fill all of the rated orders of equal priority status received on the same day, the person must accept, based upon the earliest delivery dates, only those orders that can be filled, and reject the other orders. For example, a person must accept order A requiring delivery on December 15 before accepting order B requiring delivery on December 31. However, the person must offer to accept the rejected orders based on the earliest delivery dates otherwise possible.

(5) A person must reject the rated order if the person is prohibited by Federal law from meeting the terms of the order.

(c) *Optional rejection.* Unless otherwise directed by USDA for a rated order involving food resources, food resource facilities, livestock resources, veterinary resources, plant health resources, or the domestic distribution of farm equipment and commercial fertilizer, rated orders may be rejected in any of the following cases as long as a supplier does not discriminate among customers:

(1) If the person placing the order is unwilling or unable to meet regularly established terms of sale or payment;

(2) If the order is for an item not supplied or for a service not capable of being performed;

(3) If the order is for an item or service produced, acquired, or provided only for the supplier's own use for which no orders have been filled for 2 years prior to the date of receipt of the rated order. If, however, a supplier has sold some of these items or provided similar services, the supplier is obligated to accept rated orders up to that quantity or portion of production or service, whichever is greater, sold or provided within the past 2 years;

(4) If the person placing the rated order, other than the Federal Government, makes the item or performs the service being ordered;

(5) If acceptance of a rated order or performance against a rated order would violate any other regulation, official action, or order of USDA, issued under the authority of the Defense Production Act or another relevant law.

(d) *Customer notification requirements.* A person in receipt of a rated order is required to provide to the customer placing the order written or electronic notification of acceptance or rejection of the order.

(1) Except as provided in paragraph (e) of this section, a person must accept or reject a rated order in writing or electronically within fifteen (15) working days after receipt of a DO-rated order and within ten (10) working days after receipt of a DX-rated order. If the order is rejected, the person must give reasons in writing or electronically for the rejection.

(2) If a person has accepted a rated order and subsequently finds that shipment or performance will be delayed, the person must notify the customer immediately, give the reasons for the delay, and advise of a new shipment or performance date. If notification is given verbally, written or electronic confirmation must be provided within 5 working days.

(e) *Exception for emergency preparedness conditions.* If the rated order is placed for the purpose of emergency preparedness and includes the additional statement as specified in § 789.12(a)(4)(ii), a person must accept or reject a rated order and send the acceptance or rejection in writing or in an electronic format:

(1) Within 6 hours after receipt of the order if the order is issued in response to a hazard that has occurred; or

(2) Within the greater of 12 hours or the time specified in the order, if the order is issued to prepare for an imminent hazard.

§ 789.14 Preferential scheduling.

(a) A person must schedule operations, including the acquisition of all needed production items or services, in a timely manner to satisfy the delivery requirements of each rated order. Modifying production or delivery schedules is necessary only when required delivery dates for rated orders cannot otherwise be met.

(b) DO-rated orders must be given production preference over unrated orders, if necessary to meet required delivery dates, even if this requires the diversion of items being processed or ready for delivery or services being performed against unrated orders. Similarly, DX-rated orders must be given preference over DO-rated orders and unrated orders. (Examples: If a person receives a DO-rated order with a delivery date of June 3 and if meeting that date would mean delaying production or delivery of an item for an unrated order, the unrated order must be delayed. If a DX-rated order is received calling for delivery on July 15 and a person has a DO-rated order requiring delivery on June 2 and operations can be scheduled to meet both deliveries, there is no need to alter production schedules to give any additional preference to the DX-rated order.)

(c) For conflicting rated orders:

(1) If a person finds that delivery or performance against any accepted rated orders conflicts with the delivery or performance against other accepted rated orders of equal priority status, the person must give precedence to the conflicting orders in the sequence in which they are to be delivered or performed (not to the receipt dates). If the conflicting orders are scheduled to be delivered or performed on the same day, the person must give precedence to those orders that have the earliest receipt dates.

(2) If a person is unable to resolve rated order delivery or performance conflicts as specified in this section, the person should promptly seek special priorities assistance as provided in §§ 789.20 through 789.24. If the person's customer objects to the rescheduling of delivery or performance of a rated order, the customer should promptly seek special priorities assistance as specified in §§ 789.20 through 789.24. For any rated order against which delivery or performance will be delayed, the person must notify the customer as provided in § 789.13(d)(2).

(d) If a person is unable to purchase needed production items in time to fill a rated order by its required delivery date, the person must fill the rated order by using inventoried production items. A person who uses inventoried items to fill a rated order may replace those items with the use of a rated order as provided in § 789.17(b).

§ 789.15 Extension of priority ratings.

(a) A person must use rated orders as necessary with suppliers to obtain items or services needed to fill a rated order. The person must use the priority rating indicated on the customer's rated order, except as otherwise provided in this part or as directed by USDA.

(b) The priority rating must be included as necessary on each successive order placed to obtain items or services needed to fill a customer's rated order. This continues from contractor to subcontractor to supplier throughout the entire procurement chain.

§ 789.16 Changes or cancellations of priority ratings and rated orders.

(a) The priority rating on a rated order may be changed or canceled by:

(1) An official action of USDA; or

(2) Written notification from the person who placed the rated order.

(b) If an unrated order is amended so as to make it a rated order, or a DO rating is changed to a DX rating, the supplier must give the appropriate preferential treatment to the order as

438

of the date the change is received by the supplier.

(c) An amendment to a rated order that significantly alters a supplier's original production or delivery schedule constitutes a new rated order as of the date of its receipt. The supplier must accept or reject the amended order according to the provisions of §789.13.

(d) The following amendments do not constitute a new rated order:

(1) A change in shipping destination;

(2) A reduction in the total amount of the order;

(3) An increase in the total amount of the order that has a negligible impact upon deliveries;

(4) A minor variation in size or design; or

(5) A change that is agreed upon between the supplier and the customer.

(e) If a person no longer needs items or services to fill a rated order, any rated orders placed with suppliers for the items or services, or the priority rating on those orders, must be canceled.

(f) When a priority rating is added to an unrated order, or is changed or canceled, all suppliers must be promptly notified in writing.

§789.17 Use of rated orders.

(a) A person must use rated orders as necessary to obtain:

(1) Items that will be physically incorporated into other items to fill rated orders, including that portion of such items normally consumed or converted into scrap or by-products in the course of processing;

(2) Containers or other packaging materials required to make delivery of the finished items against rated orders;

(3) Services, other than contracts of employment, needed to fill rated orders; and

(4) MRO needed to produce the finished items to fill rated orders.

(b) A person may use a rated order to replace inventoried items (including finished items) if such items were used to fill rated orders, as follows:

(1) The order must be placed within 90 days of the date of use of the inventory.

(2) A DO rating and the program identification symbol indicated on the customer's rated order must be used on the order. A DX rating must not be used even if the inventory was used to fill a DX-rated order.

(3) If the priority ratings on rated orders from one customer or several customers contain different program identification symbols, the rated orders may be combined. In this case, the program identification symbol P4 must be used (that is DO–P4).

(c) A person may combine DX- and DO-rated orders from one customer or several customers if the items or services covered by each level of priority are identified separately and clearly. If different program identification symbols are indicated on those rated orders of equal priority, the person must use the program identification symbol P4 (that is DO–P4 or DX–P4).

(d) For combining rated and unrated orders:

(1) A person may combine rated and unrated order quantities on one purchase order provided that:

(i) The rated quantities are separately and clearly identified; and

(ii) The four elements of a rated order, as required by §789.12, are included on the order with the statement required in §789.12(a)(4)(i) modified to read:

This purchase order contains rated order quantities certified for national defense use, and you are required to follow all the provisions of the Agriculture Priorities and Allocations System regulation in 7 CFR part 789 only as it pertains to the rated quantities.

(2) A supplier must accept or reject the rated portion of the purchase order as provided in §789.13 and give preferential treatment only to the rated quantities as required by this part. This part must not be used to require preferential treatment for the unrated portion of the order.

(3) Any supplier who believes that rated and unrated orders are being combined in a manner contrary to the intent of this part or in a fashion that causes undue or exceptional hardship may submit a request for adjustment or exception as specified in §789.60.

(e) A person may place a rated order for the minimum commercially procurable quantity even if the quantity needed to fill a rated order is less than that minimum. However, a person must combine rated orders as provided in paragraph (c) of this section, if possible, to obtain minimum procurable quantities.

(f) A person is not required to place a priority rating on an order for less than $75,000 or one-half of the Simplified Acquisition Threshold (as established in the Federal Acquisition Regulation (FAR) (see 48 CFR 2.101) or in other authorized acquisition regulatory or management systems) whichever amount is greater, provided that delivery can be obtained in a timely fashion without the use of the priority rating.

§ 789.18   Limitations on placing rated orders.

(a) *General limitations.* Rated orders may be placed only by persons with the proper authority for items and services that are needed to support approved programs.

(1) A person must not place a DO- or DX-rated order unless authorized by USDA to do so under this part.

(2) Rated orders must not be used to obtain:

(i) Delivery on a date earlier than needed;

(ii) A greater quantity of the item or services than needed, except to obtain a minimum procurable quantity. Separate rated orders must not be placed solely for the purpose of obtaining minimum procurable quantities on each order;

(iii) Items or services in advance of the receipt of a rated order, except as specifically authorized by USDA (see § 789.21(c) for information on obtaining authorization for a priority rating in advance of a rated order);

(iv) Items that are not needed to fill a rated order, except as specifically authorized by USDA or as otherwise permitted by this part;

(v) Any of the following items unless specific priority rating authority has been obtained from USDA or the Department of Commerce, as appropriate:

(A) Items for plant improvement, expansion, or construction, unless they will be physically incorporated into a construction project covered by a rated order; and

(B) Production or construction equipment or items to be used for the manufacture of production equipment. For information on requesting priority rating authority, see § 789.21; or

(vi) Any items related to the development of chemical or biological warfare capabilities or the production of chemical or biological weapons, unless such development or production has been authorized by the President or the Secretary of Defense.

(b) *Jurisdictional limitations.* (1) Unless authorized by the resource agency with jurisdiction (see § 789.10), the provisions of this part are not applicable to the following resources:

(i) All forms of energy (Resource agency with jurisdiction—Department of Energy);

(ii) Health resources (Resource agency with jurisdiction—Department of Health and Human Services);

(iii) All forms of civil transportation (Resource agency with jurisdiction—Department of Transportation);

(iv) Water resources (Resource agency with jurisdiction—Department of Defense, U.S. Army Corps of Engineers);

(v) All materials, services, and facilities, including construction materials for which the authority has not been delegated to other agencies under Executive Order 13603 (Resource agency with jurisdiction—Department of Commerce); and

(2) The priorities and allocations authority in this part may not be applied to communications services subject to Executive Order 13618 of July 6, 2012 (3 CFR, 2012 Comp., p. 273).

Subpart D—Special Priorities Assistance

§ 789.20   General provisions.

(a) APAS is designed to be largely self-executing. However, if production or delivery problems arise, a person should immediately contact the Farm Service Agency Administrator for special priorities assistance pursuant to §§ 789.20 through 789.24 and as directed by § 789.73. If the Farm Service Agency is unable to resolve the problem or to authorize the use of a priority rating and believes additional assistance is warranted, USDA may forward the request to another resource agency, as appropriate, for action. Special priorities assistance is a service provided to alleviate problems.

(b) Special priorities assistance is available for any reason consistent with this part. Generally, special priorities assistance is provided to expedite deliveries, resolve delivery conflicts, place rated orders, locate suppliers, or verify information supplied by customers and vendors. Special priorities assistance may also be used to request rating authority for items that are not normally eligible for priority treatment.

(c) A request for special priorities assistance or priority rating authority must be submitted on Form AD–2102 (OMB Control Number 0560–0280) to the Farm Service Agency as provided in paragraph (a) of this section. Form AD–2102 may be obtained from USDA by downloading the form and instructions from *http://forms.sc.egov.usda.gov/ eForms/welcomeAction.do?Home* or by contacting the Administrator of the Farm Service Agency as specified in § 789.73. Either mail or fax the form to USDA, using the address or fax number shown on the form.

§ 789.21 Requests for priority rating authority.

(a) *Rating authority for items or services not normally rated.* If a rated order is likely to be delayed because a person is unable to obtain items or services not normally rated under this part, the person may request the authority to use a priority rating in ordering the needed items or services.

(b) *Rating authority for production or construction equipment.* For a rated order for production or construction equipment not under the resource jurisdiction of USDA, follow the regulation in 15 CFR part 700.

(1) A request for priority rating authority for production or construction equipment must be submitted to the U.S. Department of Commerce on Form BIS–999 (see 15 CFR 700.51). Form BIS–999 may be obtained from USDA as specified in § 789.20(c) or from the Department of Commerce as specified in 15 CFR 700.50.

(2) When the use of a priority rating is authorized for the procurement of production or construction equipment,

a rated order may be used either to purchase or to lease such equipment. However, in the latter case, the equipment may be leased only from a person engaged in the business of leasing such equipment or from a person willing to lease rather than sell.

(c) For rating authority in advance of a rated prime contract:

(1) In certain cases and upon specific request, USDA, in order to promote the national defense, may authorize a person to place a priority rating on an order to a supplier in advance of the issuance of a rated prime contract. In these instances, the person requesting advance rating authority must obtain sponsorship of the request from USDA. The person assumes any business risk associated with the placing of a rated order if the order has to be canceled in the event the rated prime contract is not issued.

(2) The person must state the following in the request:

It is understood that the authorization of a priority rating in advance of our receiving a rated prime contract from USDA and our use of that priority rating with our suppliers in no way commits USDA or any other government agency to enter into a contract or order or to expend funds. Further, we understand that the Federal Government will not be liable for any cancellation charges, termination costs, or other damages that may accrue if a rated prime contract is not eventually placed and, as a result, we must subsequently cancel orders placed with the use of the priority rating authorized as a result of this request.

(3) In reviewing requests for rating authority in advance of a rated prime contract, USDA will consider, among other things, the following criteria:

(i) The probability that the prime contract will be awarded;

(ii) The impact of the resulting rated orders on suppliers and on other authorized programs;

(iii) Whether the contractor is the sole source;

(iv) Whether the item being produced has a long lead time; and

(v) The time period for which the rating is being requested.

(4) USDA may require periodic reports on the use of the rating authority granted through paragraph (c) of this section.

(5) If a rated prime contract is not issued, the person will promptly notify each supplier who has received any rated order related to the advanced rating authority that the priority rating on the order is canceled.

§ 789.22 Examples of assistance.

(a) While special priorities assistance may be provided for any reason in support of this part, it is usually provided in situations in which:

(1) A person is experiencing difficulty in obtaining delivery against a rated order by the required delivery date; or

(2) A person cannot locate a supplier for an item or service needed to fill a rated order.

(b) Other examples of special priorities assistance include:

(1) Ensuring that rated orders receive preferential treatment by suppliers;

(2) Resolving production or delivery conflicts between various rated orders;

(3) Assisting in placing rated orders with suppliers;

(4) Verifying the urgency of rated orders; and

(5) Determining the validity of rated orders.

§ 789.23 Criteria for assistance.

(a) Requests for special priorities assistance should be timely (for example, the request has been submitted promptly and enough time exists for USDA to meaningfully resolve the problem), and must establish that:

(1) There is an urgent need for the item; and

(2) The applicant has made a reasonable effort to resolve the problem.

(b) [Reserved]

§ 789.24 Instances in which assistance must not be provided.

(a) Special priorities assistance is provided at the discretion of USDA when it is determined that such assistance is warranted to meet the objectives of this part. Examples in which assistance must not be provided include situations in which a person is attempting to:

(1) Secure a price advantage;

(2) Obtain delivery prior to the time required to fill a rated order;

(3) Gain competitive advantage;

(4) Disrupt an industry apportionment program in a manner designed to provide a person with an unwarranted share of scarce items; or

(5) Overcome a supplier's regularly established terms of sale or conditions of doing business.

(b) [Reserved]

Subpart E—Allocations Actions

§ 789.30 Policy.

(a) It is the policy of the Federal Government that the allocations authority under Title I of the Defense Production Act may:

(1) Only be used when there is insufficient supply of a material, service, or facility to satisfy national defense supply requirements through the use of the priorities authority or when the use of the priorities authority would cause a severe and prolonged disruption in the supply of materials, services, or facilities available to support normal U.S. economic activities; and

(2) Not be used to ration materials or services at the retail level.

(b) Allocations orders, when used, will be distributed equitably among the suppliers of the materials, services, or facilities being allocated and not require any person to relinquish a disproportionate share of the civilian market.

§ 789.31 General procedures.

(a) When USDA plans to execute its allocations authority to address a supply problem within its resource jurisdiction, USDA will develop a plan that includes the following information:

(1) A copy of the written determination made in accordance with section 202 of Executive Order 13603, that the program or programs that would be supported by the allocations action are necessary or appropriate to promote the national defense;

(2) A detailed description of the situation to include any unusual events or circumstances that have created the requirement for an allocations action;

(3) A statement of the specific objective(s) of the allocations action;

(4) A list of the materials, services, or facilities to be allocated;

(5) A list of the sources of the materials, services, or facilities that will be subject to the allocations action;

(6) A detailed description of the provisions that will be included in the allocations orders, including the type(s) of allocations orders, the percentages or quantity of capacity or output to be allocated for each purpose, and the duration of the allocations action (for example, anticipated start and end dates);

(7) An evaluation of the impact of the proposed allocations action on the civilian market; and

(8) Proposed actions, if any, to mitigate disruptions to civilian market operations.

(b) [Reserved]

§ 789.32 Precedence over priority rated orders.

If a conflict occurs between an allocations order and an unrelated rated order or priorities directive, the allocations order takes precedence.

§ 789.33 Controlling the general distribution of a material in the civilian market.

(a) No allocations by USDA may be used to control the general distribution of a material in the civilian market, unless the Secretary has:

(1) Made a written finding that:

(i) Such material is a scarce and critical material essential to the national defense; and

(ii) The requirements of the national defense for such material cannot otherwise be met without creating a significant dislocation of the normal distribution of such material in the civilian market to such a degree as to create appreciable hardship;

(2) Submitted the finding for the President's approval through the Assistant to the President and National Security Advisor and the Assistant to the President for Homeland Security and Counterterrorism; and

(3) The President has approved the finding.

(b) [Reserved]

§ 789.34 Types of allocations orders.

(a) The three types of allocations orders that may be used for allocations actions are:

(1) Set-asides;

(2) Directives; and

(3) Allotments.

(b) [Reserved]

§ 789.35 Elements of an allocations order.

(a) Each allocations order will include:

(1) A detailed description of the required allocations action(s);

(2) Specific start and end calendar dates for each required allocations action;

(3) The Secretary's written signature on a manually placed order, or the digital signature or name on an electronically placed order, of the Secretary. The signature or use of the name certifies that the order is authorized as specified in this part and that the requirements of this part are being followed;

(4) A statement that reads: "This is an allocations order certified for national defense use. [Insert the legal name of the person receiving the order] is required to comply with this order, in accordance with the provisions of 7 CFR part 789;" and

(5) A current copy of the APAS regulation (7 CFR part 789).

(b) [Reserved]

§ 789.36 Mandatory acceptance of allocations orders.

(a) A person must accept every allocations order received that the person is capable of fulfilling, and must comply with such orders regardless of any rated order that the person may be in receipt of or other commitments involving the resource(s) covered by the allocations order.

(b) A person must not discriminate against an allocations order in any manner such as by charging higher prices for resources covered by the order or by imposing terms and conditions for contracts and orders involving allocated resources(s) that differ from the person's terms and conditions for contracts and orders for the resource(s) prior to receiving the allocations order.

(c) If circumstances prevent a person from being able to accept an allocations order, the person must comply with the provisions specified in §789.60

upon realization of the inability to accept the order.

§ 789.37 Changes or cancellations of allocations orders.

An allocations order may be changed or canceled by an official action of USDA.

Subpart F—Official Actions

§ 789.40 General provisions.

(a) USDA may take specific official actions to implement the provisions of this part.

(b) Several of these official actions (rating authorizations, directives, and letters of understanding) are discussed in this subpart. Other official actions that pertain to compliance (administrative subpoenas, demands for information, and inspection authorizations) are discussed in § 789.51(c).

§ 789.41 Rating authorizations.

(a) A rating authorization is an official action granting specific priority rating authority that:

(1) Permits a person to place a priority rating on an order for an item or service not normally ratable under this part; or

(2) Authorizes a person to modify a priority rating on a specific order or series of contracts or orders.

(b) To request priority rating authority, see section § 789.21.

§ 789.42 Directives.

(a) A directive is an official action that requires a person to take or refrain from taking certain actions in accordance with the provisions of the directive.

(b) A person must comply with each directive issued. However, a person may not use or extend a directive to obtain any items from a supplier, unless expressly authorized to do so in the directive.

(c) A priorities directive takes precedence over all DX-rated orders, DO-rated orders, and unrated orders previously or subsequently received, unless a contrary instruction appears in the directive.

(d) An allocations directive takes precedence over all priorities directives, DX-rated orders, DO-rated orders, and unrated orders previously or subsequently received, unless a contrary instruction appears in the directive.

§ 789.43 Letters of understanding.

(a) A letter of understanding is an official action that may be issued in resolving special priorities assistance cases to reflect an agreement reached by all parties (USDA, the Department of Commerce (if applicable), a delegate agency (if applicable), the supplier, and the customer).

(b) A letter of understanding is not used to alter scheduling between rated orders, to authorize the use of priority ratings, to impose restrictions under this part, or to take other official actions. Rather, letters of understanding are used to confirm production or shipping schedules that do not require modifications to other rated orders.

Subpart G—Compliance

§ 789.50 General provisions.

(a) USDA may take specific official actions for any reason necessary or appropriate to the enforcement or the administration of the Defense Production Act and other applicable statutes, this part, or an official action. Such actions include administrative subpoenas, demands for information, and inspection authorizations.

(b) Any person who places or receives a rated order or an allocations order must comply with the provisions of this part.

(c) Willful violation of the provisions of Title I or section 705 of the Defense Production Act and other applicable statutes, this part, or an official action of USDA, is a criminal act, punishable as provided in the Defense Production Act and other applicable statutes, and as specified in § 789.54.

§ 789.51 Audits and investigations.

(a) Audits and investigations are official examinations of books, records, documents, other writings, and information to ensure that the provisions of the Defense Production Act and other applicable statutes, this part, and official actions have been properly followed. An audit or investigation may also include interviews and a systems evaluation to detect problems or failures in the implementation of this part.

(b) When undertaking an audit, investigation, or other inquiry, USDA will:

(1) *Scope and purpose.* Define the scope and purpose in the official action given to the person under investigation; and

(2) *Information not available.* Have ascertained that the information sought or other adequate and authoritative data are not available from any Federal or other responsible agency.

(c) In administering this part, USDA may issue the following documents that constitute official actions:

(1) *Administrative subpoenas.* An administrative subpoena requires a person to appear as a witness before an official designated by USDA to testify under oath on matters of which that person has knowledge relating to the enforcement or the administration of the Defense Production Act and other applicable laws, this part, or official actions. An administrative subpoena may also require the production of books, papers, records, documents, and physical objects or property.

(2) *Demands for information.* A demand for information requires a person to furnish to a duly authorized representative of USDA any information necessary or appropriate to the enforcement or the administration of the Defense Production Act and other applicable statutes, this part, or official actions.

(3) *Inspection authorizations.* An inspection authorization requires a person to permit a duly authorized representative of USDA to interview the person's employees or agents, to inspect books, records, documents, other writings, and information, including electronically-stored information, in the person's possession or control at the place where that person usually keeps them or otherwise, and to inspect a person's property when such interviews and inspections are necessary or appropriate to the enforcement or the administration of the Defense Production Act and other related laws, this part, or official actions.

(d) The production of books, records, documents, other writings, and information will not be required at any place other than where they are usually kept if, prior to the return date specified in the administrative subpoena or demand for information, a duly authorized official of USDA is furnished with copies of such material that are certified under oath to be true copies. As an alternative, a person may enter into a stipulation with a duly authorized official of USDA as to the content of the material.

(e) An administrative subpoena, demand for information, or inspection authorization will include the name, title, or official position of the person to be served, the evidence sought, and its general relevance to the scope and purpose of the audit, investigation, or other inquiry. If employees or agents are to be interviewed; if books, records, documents, other writings, or information are to be produced; or if property is to be inspected; the administrative subpoena, demand for information, or inspection authorization will describe the requirements.

(f) Service of documents will be made in the following manner:

(1) *In person.* Service of a demand for information or inspection authorization will be made personally, or by certified mail-return receipt requested at the person's last known address. Service of an administrative subpoena will be made personally. Personal service may also be made by leaving a copy of the document with someone at least 18 years old at the person's last known dwelling or place of business.

(2) *Other than to the named individual.* Service upon other than an individual may be made by serving a partner, corporate officer, or a managing or general agent authorized by appointment or by law to accept service of process. If an agent is served, a copy of the document will be mailed to the person named in the document.

(3) *Delivering individual and documentation.* Any individual 18 years of age or over may serve an administrative subpoena, demand for information, or inspection authorization. When personal service is made, the individual making the service must prepare an affidavit specifying the manner in which service was made and the identity of

the person served, and return the affidavit, and in the case of subpoenas, the original document, to the issuing officer. In case of failure to make service, the reasons for the failure will be stated on the original document.

§ 789.52 Compulsory process.

(a) If a person refuses to permit a duly authorized representative of USDA to have access to any premises or source of information necessary to the administration or the enforcement of the Defense Production Act and other applicable laws, this part, or official actions, the USDA representative may seek compulsory process. Compulsory process is the institution of appropriate legal action, including ex parte application for an inspection warrant or its equivalent, in any forum of appropriate jurisdiction.

(b) Compulsory process may be sought in advance of an audit, investigation, or other inquiry, if, in the judgment of USDA, there is reason to believe that a person will refuse to permit an audit, investigation, or other inquiry, or that other circumstances exist that make such process desirable or necessary.

§ 789.53 Notification of failure to comply.

(a) At the conclusion of an audit, investigation, or other inquiry, or at any other time, USDA may inform the person in writing when compliance with the requirements of the Defense Production Act and other applicable laws, this part, or an official action was not met.

(b) In cases in which USDA determines that failure to comply with the provisions of the Defense Production Act and other applicable laws, this part, or an official action was inadvertent, the person may be informed in writing of the particulars involved and the corrective action to be taken. Failure to take corrective action may then be construed as a willful violation of the Defense Production Act and other applicable laws, this part, or an official action.

§ 789.54 Violations, penalties, and remedies.

(a) Willful violation of the Defense Production Act, the priorities provisions of the Military Selective Service Act (50 U.S.C. App. 468), this part, or an official action, is a crime and upon conviction, a person may be punished by fine or imprisonment, or both. The maximum penalty provided by the Defense Production Act is a $10,000 fine, or 1 year in prison, or both. The maximum penalty provided by the Military Selective Service Act is a $50,000 fine, or 3 years in prison, or both.

(b) The Government may also seek an injunction from a court of appropriate jurisdiction to prohibit the continuance of any violation of, or to enforce compliance with, the Defense Production Act, this part, or an official action.

(c) In order to secure the effective enforcement of the Defense Production Act and other applicable laws, this part, and official actions, certain actions as follows are prohibited:

(1) Soliciting, influencing, or permitting another person to perform any act prohibited by, or to omit any act required by, the Defense Production Act and other applicable laws, this part, or an official action.

(2) Conspiring or acting in concert with any other person to perform any act prohibited by, or to omit any act required by, the Defense Production Act and other applicable laws, this part, or an official action.

(3) Delivering any item if the person knows or has reason to believe that the item will be accepted, redelivered, held, or used in violation of the Defense Production Act and other applicable laws, this part, or an official action. In such instances, the person must immediately notify USDA that, in accordance with this provision, delivery has not been made.

§ 789.55 Compliance conflicts.

If compliance with any provision of the Defense Production Act and other applicable laws, this part, or an official action would prevent a person from filling a rated order or from complying with another provision of the Defense Production Act and other applicable laws, this part, or an official action, the person must immediately notify USDA for resolution of the conflict.

Subpart H—Adjustments, Exceptions, and Appeals

§789.60 Adjustments or exceptions.

(a) A person may submit a request to the Farm Service Agency Deputy Administrator for Management, as directed in §789.73, for an adjustment or exception on the ground that:

(1) A provision of this part or an official action results in an undue or exceptional hardship on that person not suffered generally by others in similar situations and circumstances; or

(2) The consequences of following a provision of this part or an official action is contrary to the intent of the Defense Production Act and other applicable laws, or this part.

(b) Each request for adjustment or exception must be in writing and contain a complete statement of all the facts and circumstances related to the provision of this part or official action from which adjustment is sought and a full and precise statement of the reasons why relief should be provided.

(c) The submission of a request for adjustment or exception will not relieve any person from the obligation of complying with the provision of this part or official action in question while the request is being considered unless such interim relief is granted in writing by the Farm Service Agency Deputy Administrator for Management.

(d) A decision of the Farm Service Agency Deputy Administrator for Management under this section may be appealed to the Farm Service Agency Administrator. (For information on the appeal procedure, see §789.61.)

§789.61 Appeals.

(a) Any person whose request for adjustment or exception has been denied by the Farm Service Agency Deputy Administrator for Management as specified in §789.60, may appeal to the Farm Service Agency Administrator who will review and reconsider the denial.

(b) A person must submit the appeal in writing to the Farm Service Agency Administrator as follows:

(1) Except as provided in paragraph (b)(2) of this section, an appeal must be received by the Farm Service Agency Administrator no later than 45 days after receipt of a written notice of denial from the Farm Service Agency Deputy Administrator for Management. After the 45-day period, an appeal may be accepted at the discretion of the Farm Service Agency Administrator if the person shows good cause.

(2) For requests for adjustment or exception involving rated orders placed for the purpose of emergency preparedness (see §789.13(e)), an appeal must be received by the Farm Service Agency Administrator no later than 15 days after receipt of a written notice of denial from the Farm Service Agency Deputy Administrator for Management.

(c) Contract performance under the order may not be stayed pending resolution of the appeal.

(d) Each appeal must be in writing and contain a complete statement of all the facts and circumstances related to the appealed action and a full and precise statement of the reasons the decision should be modified or reversed.

(e) In addition to the written materials submitted in support of an appeal, an appellant may request, in writing, an opportunity for an informal hearing. This request may be granted or denied at the discretion of the Farm Service Agency Administrator.

(f) When a hearing is granted, the Farm Service Agency Administrator may designate an employee of the Farm Service Agency to conduct the hearing and to prepare a report. The hearing officer will determine all procedural questions and impose such time or other limitations deemed reasonable. If the hearing officer decides that a printed transcript is necessary, the transcript expenses must be paid by the appellant.

(g) When determining an appeal, the Farm Service Agency Administrator may consider all information submitted during the appeal as well as any recommendations, reports, or other relevant information and documents available to USDA, or consult with any other person or group.

(h) The submission of an appeal under this section will not relieve any person from the obligation of complying with the provision of this part or official action in question while the appeal is being considered unless such relief is granted in writing by the Farm Service Agency Administrator.

(i) The decision of the Farm Service Agency Administrator will be made within 5 days after receipt of the appeal, or within 1 day for appeals pertaining to emergency preparedness, and will be the final administrative action. The Administrator will issue a written statement of the reasons for the decision to the appellant.

Subpart I—Miscellaneous Provisions

§ 789.70   Protection against claims.

A person will not be held liable for damages or penalties for any act or failure to act resulting directly or indirectly from compliance with any provision of this part, or an official action, even if such provision or action is subsequently declared invalid by judicial or other competent authority.

§ 789.71   Records and reports.

(a) Persons are required to make and preserve for at least 3 years, accurate and complete records of any transaction covered by this part or an official action.

(b) Records must be maintained in sufficient detail to permit the determination, upon examination, of whether each transaction complies with the provisions of this part or any official action. However, this part does not specify any particular method or system to be used.

(c) Records required to be maintained by this part must be made available for examination on demand by duly authorized representatives of USDA as provided in § 789.51.

(d) In addition, persons must develop, maintain, and submit any other records and reports to USDA that may be required for the administration of the Defense Production Act and other applicable statutes, and this part..

(e) Section 705(d) of the Defense Production Act, as implemented by Executive Order 13603, provides that information obtained under that section which the Secretary deems confidential, or with reference to which a request for confidential treatment is made by the person furnishing such information, will not be published or disclosed unless the Secretary determines that the withholding of this information is contrary to the interest of the national defense. Information required to be submitted to USDA in connection with the enforcement or administration of the Defense Production Act, this part, or an official action, is deemed to be confidential under section 705(d) of the Defense Production Act and will be handled in accordance with applicable Federal law.

§ 789.72   Applicability of this part and official actions.

(a) This part and all official actions, unless specifically stated otherwise, apply to transactions in any State, territory, or possession of the United States and the District of Columbia.

(b) This part and all official actions apply not only to deliveries to other persons but also include deliveries to affiliates and subsidiaries of a person and deliveries from one branch, division, or section of a single entity to another branch, division, or section under common ownership or control.

(c) This part and its schedules will not be construed to affect any administrative actions taken by USDA, or any outstanding contracts or orders placed based on any of the regulations, orders, schedules, or delegations of authority previously issued by USDA based on authority granted to the President in the Defense Production Act. Such actions, contracts, or orders will continue in full force and effect under this part unless modified or terminated by proper authority.

§ 789.73   Communications.

Except as otherwise provided, all communications concerning this part, including requests for copies of this part and explanatory information, requests for guidance or clarification, and submission of appeals as specified in § 789.61 will be addressed to the Administrator, Farm Service Agency, Room 4752, Mail Stop 0512, USDA, 1400 Independence Ave. SW., Washington,

DC 20250–0512 or email: *FSA.EPD@wdc.usda.gov.* This address is also to be used for requests for adjustments or exceptions to the Farm Service Agency Deputy Administrator for Management as specified in § 789.60.

SCHEDULE I TO PART 789—APPROVED PROGRAMS AND DELEGATE AGENCIES

The programs listed in this schedule have been approved for priorities and allocations support under this part by the Department of Defense, Department of Energy, or Department of Homeland Security as required by section 202 of Executive Order 13603. They have equal preferential status. USDA has authorized the delegate agencies to use the authorities in this part in support of those programs assigned to them, as indicated below.

Program identification symbol	Approved program	Authorized delegate agency
Agriculture programs:		
P1	Food and food resources (civilian)	USDA, Department of Homeland Security, Federal Emergency Management Agency
P2	Agriculture and food critical infrastructure protection and restoration.	USDA
P3	Food resources (combat rations)	Department of Defense [1]
P4	Certain combined orders (see § 789.17)	USDA

[1] Department of Defense includes: The Office of the Secretary of Defense, the Military Departments, the Joint Staff, the Combatant Commands, the Defense Agencies, the Defense Field Activities, all other organizational entities in the Department of Defense, and for purpose of this part, the Central Intelligence Agency, and the National Aeronautics and Space Administration as Associated Agencies.

SUBCHAPTER E—PROVISIONS COMMON TO MORE THAN ONE PROGRAM

PART 795—PAYMENT LIMITATION

AUTHORITY: Sec. 1001 of the Food Security Act of 1985, as amended, 99 Stat, 1444, as amended, 7 U.S.C 1308; Pub. L. 99–500 and Pub. L. 99–591.

SOURCE: 43 FR 9784, Mar. 10, 1978, unless otherwise noted.

GENERAL

§ 795.1 [Reserved]

§ 795.2 Applicability.

(a) The provisions of this part are applicable to payments when so provided by the individual program regulations under which the payments are made. The amount of the limitation shall be as specified in the individual program regulations.

(b) The limitation shall be applied to the payments for a commodity for a crop year.

(c) The limitation shall not be applicable to payments made to States, political subdivisions, or agencies thereof for participation in the programs on lands owned by such States, political subdivisions, or agencies thereof so long as such lands are farmed primarily in the direct furtherance of a public function. However, the limitation is applicable to persons who rent or lease land owned by States, political subdivisions, or agencies thereof.

(d) The limitation shall not be applicable to payments made to Indian tribal ventures participating in the programs where a responsible official of the Bureau of Indian Affairs or the Indian Tribal Council certifies that no more than the program payment limitation shall accrue directly or indirectly to any individual Indian and the State committee reviews and approves the exemption.

(e) Except as provided in part 1497 of this title, this part shall not be applicable to contracts entered into on or after August 1, 1988 in accordance with part 704 of this chapter.

[49 FR 14719, Apr. 13, 1984, as amended at 51 FR 8453, Mar. 11, 1986; 51 FR 36905, Oct. 16, 1986; 53 FR 29570, Aug. 5, 1988]

§ 795.3 Definitions.

(a) The terms defined in part 719 of this chapter, governing reconstitutions of farms, shall be applicable to this part and all documents issued in accordance with this part, except as otherwise provided in this section.

(b)(1) Subject to the provisions of this part, the term "person" shall mean an individual, joint stock company, corporation, association, trust, estate, or other legal entity. In order to be considered to be a separate person for the purposes of this part with respect to any crop, in addition to any other provision of this part, an individual or other legal; entity must:

(i) Have a separate and distinct interest in the crop or the land on which the crop is produced;

(ii) Exercise separate responsibility for such interest; and

(iii) Be responsible for payment of the cost of farming related to such interest from a fund or account separate from that of any other individual or entity.

(2) The term "person" shall not include any cooperative association of producers that markets commodities for producers with respect to the commodities so marketed for producers.

(c) The term "family member" shall mean the individual, the great-grand-parent, grand-parent, child, grandchild, and great-grandchild of such individual and the spouses of all such individuals.

(d) The term "separate unit" shall mean an individual who, prior to December 31, 1985: (1) Had been engaged in a separate farming operation and (2) in accordance with the provisions of this part, had been determined to be a separate person or could have so determined under the circumstances existing at such time.

[52 FR 26295, July 14, 1987]

§ 795.4 Family members.

Effective for the 1987 through 1990 crops, an individual shall not be denied a determination that such individual was a "person" solely on the basis that:

(a) A family member cosigns for, or makes a loan to, such individual and leases, loans or gives equipment, land or labor to such an individual; and

(b) Such family members were organized as separate units prior to December 31, 1985.

[52 FR 26295, July 14, 1987]

§ 795.5 Timing for determining status of persons.

Except as otherwise set forth in this part, the status of individuals or entities as of March 1, or such other date as may be determined and announced by the Administrator shall be the basis on which determinations are made in accordance with this part for the year for which the determination is made.

[51 FR 21836, June 16, 1986; 51 FR 36905, Oct. 16, 1986]

§ 795.6 Multiple individuals or other entities.

The rules in §§ 795.5 through 795.16 shall be used to determine whether certain multiple individuals or legal entities are to be treated as one person or as separate persons for the purpose of applying the limitation. In cases in which more than one rule would appear to be applicable, the rule which is most restrictive on the number of persons shall apply.

§ 795.7 Entities or joint operations not considered as a person.

A partnership, joint venture, tenants-in-common, or joint tenants shall not be considered as a person but, notwithstanding the provisions of §795.3, each individual or other legal entity who shares in the proceeds derived from farming by such joint operations shall be considered a separate person, except as otherwise provided in this part, and shall be listed as a producer for payment purposes on program documents. The payment shares listed on the program documents for each individual or other legal entity shall be the same as each individual or other legal entity shares in the proceeds derived from farming by such joint operation. Notwithstanding the foregoing, each individual or other legal entity who shares in the proceeds derived from farming by such joint operation shall not be considered as a separate person unless the individual or other legal entity is actively engaged in the farming operations of the partnership or other joint operation. An individual or other legal entity shall be considered as actively engaged in the farming operation only if its contribution to the joint operation is commensurate with its share in the proceeds derived from farming by such joint operation. Members of the partnership or joint venture must furnish satisfactory evidence that their contributions of land, labor, management, equipment, or capital to the joint operation are commensurate with their claimed shares of the proceeds. A capital contribution may be a direct out-of-pocket input of a specified sum or an amount borrowed by the individual. If the contribution consists substantially of capital, such capital must have been contributed directly to the joint operation by the individual or other legal entity and not acquired as a result of (a) a loan made to the joint operation, (b) a loan which was made to such individual or other legal entity by the joint operation or any of its members or related entities, or (c) a loan made to such individual or other legal entity which was guaranteed by the joint operation or any of its members or related entities.

§ 795.8 Corporations and stockholders.

(a) A corporation (including a limited partnership) shall be considered as one person, and an individual stockholder of the corporation may be considered as a separate person to the extent that such stockholder is engaged in the production of the crop as a separate producer and otherwise meets the requirements of § 795.3, except that a corporation in which more than 50 percent of the stock is owned by an individual (including the stock owned by the individual's spouse, minor children, and trusts for the benefit of such minor children), or by a legal entity, shall not be considered as a separate person from such individual or legal entity.

(b) Where the same two or more individuals or other legal entities own more than 50 percent of the stock in each of two or more corporations, all such corporations shall be considered as one person.

(c) The percentage share of the value of the stock owned by an individual or other legal entity shall be determined as of March 1 of the crop year, except that where a stockholder voluntarily acquires stock after March 1 and before the harvest of the crop, the amount of any stock so acquired shall be included in determining the percentage share of the value of the stock owned by the stockholder. Where there is only one class of stock, a stockholder's percentage share of the value of the outstanding stock shall be equal to the percentage of the outstanding stock owned by the stockholder. If the corporation has more than one class of stock the percentage share of the value of the stock owned by a stockholder shall be determined by the Deputy Administrator on the basis of market quotations, and if market quotations are lacking or too scarce to be recognized the percentage share of the value of the stock shall be determined by the Deputy Administrator on the basis of all relevant factors affecting the fair market value, including the rights and privileges of the various stock issues.

(Title I, Agricultural Act of 1970, as amended by the Agriculture and Consumer Protection Act of 1973, Pub. L. 93–86, 87 Stat. 221 (7 U.S.C. 1307) and under Title I, Rice Production Act of 1975, Pub. L. 94–214, 90 Stat. 181 (7 U.S.C. 428c note), and Pub. L. 95–156, 91 Stat. 1264 (7 U.S.C. 1307 note, 7 U.S.C. 1307, 7 U.S.C. 1441))

[43 FR 9784, Mar. 10, 1978, as amended at 45 FR 10311, Feb. 15, 1980; 45 FR 11795, Feb. 22, 1980]

§ 795.9 Estate or trust.

(a) An estate or irrevocable trust shall be considered as one person except that, where two or more estates or irrevocable trusts have common beneficiaries or heirs (including spouses and minor children) with more than a 50–percent interest, all such estates or irrevocable trusts shall be considered as one person.

(b) An individual heir of an estate or beneficiary of a trust may be considered as a separate person to the extent that such heir or beneficiary is engaged in the production of crops as a separate producer and otherwise meets the requirements of § 795.3, except that an estate or irrevocable trust which has a sole heir or beneficiary shall not be considered as a separate person from such heir or beneficiary.

(c) Where an irrevocable trust or an estate is a producer on a farm and one or more of the beneficiaries or heirs of such trust or estate are minor children, the minor children's pro rata share of the program payments to the trust or estate shall be attributed to the parent of the minor children except as otherwise provided in § 795.12.

(d) A revocable trust shall not be considered as a separate person from the grantor.

§ 795.10 Club, society, fraternal or religious organization.

Each individual club, society, fraternal or religious organization may be considered as a separate person to the extent that each such club, society, fraternal or religious organization is engaged in the production of crops as a separate producer and otherwise meets the requirements of § 795.3.

§795.11 Husband and wife.

With respect to the 1988 crop year, a husband and wife shall be considered to be one person except that such individuals who, prior to their marriage, were separately engaged in unrelated farming operations will be determined to be separate persons with respect to such farming operations so long as the operations remain separate and distinct from any farming operation conducted by the other spouse if such individuals have executed a Contract to Participate in the 1988 Price Support and Production Adjustment Programs by April 15, 1988. Such individuals must file a form FSA-561 with the county committee for each such farming operation by July 8, 1988, if they desire to be considered as separate persons under this section.

[53 FR 21410, June 8, 1988]

§795.12 Minor children.

(a) A minor child and his parents or guardian (or other person responsible for him) shall be considered as one person, except that the minor child may be considered as a separate person if such minor child is a producer on a farm in which the parents or guardian or other person responsible for him (including any entity in which the parents or guardian or other person responsible for him has a substantial interest, i.e., more than a 20-percent interest) takes no part in the operation of the farm (including any activities as a custom farmer) and owns no interest in the farm or allotment or in any portion of the production on the farm, and if such minor child:

(1) Is represented by a court-appointed guardian who is required by law to make a separate accounting for the minor and ownership of the farm is vested in the minor, or

(2) Has established and maintains a different household from his parents or guardian and personally carries out the actual farming operations on the farm for which there is a separate accounting, or

(3) Has a farming operation resulting from his being the beneficiary of an irrevocable trust and ownership of the property is vested in the trust or the minor.

(b) A person shall be considered a minor until he reaches 18 years of age. Court proceedings conferring majority on a person under 18 years of age will not change such person's status as a minor for purposes of applying the regulations.

§795.13 Other cases.

Where the county committee is unable to determine whether certain individuals or legal entities involved in the production of a commodity are to be treated as one person or separate persons, all the facts regarding the arrangement under which the commodity is produced shall be submitted to the State committee for decision. Where the State committee is unable to determine whether such individuals or legal entities are to be treated as one person or separate persons, all the facts regarding the arrangement under which the farming operation is conducted shall be submitted to the Deputy Administrator for decision.

§795.14 Changes in farming operations.

(a) Subject to the provisions of this part, a person may exercise his or her right heretofore existing under law, to divide, sell, transfer, rent, or lease his or her property if such division, sale, transfer, rental arrangement, or lease is legally binding as between the parties thereto. However, any document representing a division, sale, transfer, rental arrangement, or lease which is fictitious or not legally binding as between the parties thereto shall be considered to be for the purpose of evading the payment limitation and shall be disregarded for the purpose of applying the payment limitation. Any change in farming operations that would otherwise serve to increase the number of persons for application of the payment limitation must be bona fide and substantive.

(b) A substantive change includes, for example, a substantial increase or decrease in the size of the farm by purchase, sale, or lease; a substantial increase or decrease in the size of allotment by purchase, sale, or lease; a change from a cash lease to a share lease or vice versa; and dissolution of an entity such as a corporation or partnership.

(c) Examples of the types of changes that would not be considered as substantive are the following:

Example 1. A corporation is owned equally by four shareholders. The corporation owns land, buildings, and equipment and in the prior year carried out substantial farming operations. Three of the shareholders propose forming a partnership which they would own equally. The partnership would cash lease land and equipment from the corporation with the objective of having the three partners considered as separate persons for purposes of applying the payment limitation under the provisions of § 795.7 of the regulations.

The formation of such a partnership and the leasing of land from a corporation in which they hold a major interest would not constitute a substantive and bona fide change in operations. Therefore, the corporation and the partners would be limited to a single payment limitation.

Example 2. Three individuals each have individual farming operations which, if continued unchanged, would permit them to have a total of three payment limitations.

The three individuals propose forming a corporation which they would own equally. The corporation would then cash lease a portion of the farmland owned and previously operated by the individuals with the objective of having the corporation considered as a separate person for purposes of applying the payment limitation under the provisions of § 795.8 of the regulations. The formation of such a corporation and the leasing of land from the stockholders would not constitute a substantive and bona fide change in operations. Therefore, the corporation and the three individuals would be limited to three payment limitations.

§ 795.15 Determining whether agreement is a share lease or a cash lease.

(a) *Cash lease.* If a rental agreement contains provisions for a guaranteed minimum rental with respect to the amount of rent to be paid to the landlord by a tenant, such agreement shall be considered to be a cash rental agreement. In addition, the rental agreement must be customary and reasonable for the area.

(b) *Share lease.* If a rental agreement contains provisions that require the payment of rent on the basis of the amount of the crop produced or the proceeds derived from the crop, such agreement shall be considered to be a

share rental agreement. In addition, the rental agreement must be customary and reasonable for the area.

[51 FR 8454, Mar. 11, 1986]

§ 795.16 Custom farming.

(a) Custom farming is the performance of services on a farm such as land preparation, seeding, cultivating, applying pesticides, and harvesting for hire with remuneration on a unit of work basis, except that, for the purpose of applying the provisions of this section, the harvesting of crops and the application of agricultural chemicals by firms regularly engaged in such businesses shall not be regarded as custom farming. A person performing custom farming shall be considered as being separate from the person for whom the custom farming is performed only if:

(1) The compensation for the custom farming is paid at a unit of work rate customary in the area and is in no way dependent upon the amount of the crop produced, and (2) the person performing the custom farming (and any other entity in which such person has more than a 20-percent interest) has no interest, directly or indirectly, (i) in the crop on the farm by taking any risk in the production of the crop, sharing in the proceeds of the crop, granting or guaranteeing the financing of the crop, (ii) in the allotment on the farm, or (iii) in the farm as landowner, landlord, mortgage holder, trustee, lienholder, guarantor, agent, manager, tenant, sharecropper, or any other similar capacity.

(b) A person having more than a 20-percent interest in any legal entity performing custom farming shall be considered as being separate from the person for whom the custom farming is performed only if:

(1) The compensation for the custom farming service is paid at a unit of work rate customary in the area and is in no way dependent upon the amount of the crop produced, and (2) the person having such interest in the legal entity performing the custom farming has no interest, directly or indirectly, (i) in the crop on the farm by taking any risk in the production of the crop, sharing in the proceeds of the corp, granting or guaranteeing the financing

of the crop, (ii) in the allotment on the farm, or (iii) in the farm as landowner, landlord, mortgage holder, trustee, lienholder, guarantor, agent, manager, tenant, sharecropper, or in any other similar capacity.

§ 795.17 Scheme or device.

All or any part of the payments otherwise due a person under the upland cotton, wheat, feed grain and rice programs on all farms in which the person has an interest may be withheld or required to be refunded if the person adopts or participates in adopting any scheme or device designed to evade or which has the effect of evading the rules of this part. Such acts shall include, but are not limited to, concealing from the county committee any information having a bearing on the application of the rules of this part or submitting false information to the county committee (for example, a set-aside agreement which is entered into that differs from information furnished to the county committee concerning the manner in which program payments are actually shared, concerning the actual facts of a sale, or concerning the transfer of property) or creating fictitious entities for the purpose of concealing the interest of a person in a farming operation.

§ 795.20 Joint and several liability.

Where two or more individuals or legal entities, who are treated as one person hereunder, receive payments which in the aggregate exceed the limitation, such individuals or legal entities shall be liable, jointly and severally, for any liability arising therefrom. The provisions of this part requiring the refund of payments shall be applicable in addition to any liability under criminal and civil fraud statutes.

§ 795.21 Appeals.

Any person may obtain reconsideration and review of determinations made under this part in accordance with the appeal regulations, part 780 of this chapter, as amended.

§ 795.22 Interpretations.

In interpretations previously issued pursuant to the payment limitation regulations and published at 36 FR 16569, 37 FR 3049, 39 FR 15021 and 41 FR 17527 shall be applicable in construing the provisions of this part.

§ 795.23 Paperwork Reduction Act assigned number.

The information collection requirements contained in these regulations (7 CFR part 795) have been approved by the Office of Management and Budget under the provisions of 44 U.S.C. Chapter 35 and have been assigned OMB control number 0560–0096.

[49 FR 14719, Apr. 13, 1984]

§ 795.24 Relief.

If a producer relied on a county committee and/or State committee "person" determination for a crop year and higher reviewing authority makes a more restrictive determination, the Deputy Administrator may grant relief only for such crop year if the producer was not afforded an opportunity to exercise other alternatives with respect to the producer's farming operation and the program provisions and the county committee has determined that the producers acted in good faith based upon the original "person" determination.

[51 FR 8454, Mar. 11, 1986; 51 FR 36905, Oct. 16, 1986]

SUBCHAPTER F—PUBLIC RECORDS

PART 798—AVAILABILITY OF INFORMATION TO THE PUBLIC

AUTHORITY: 5 U.S.C. 301, 552; 7 CFR 1.1 through 1.16.

SOURCE: 44 FR 10353, Feb. 20, 1979, unless otherwise noted.

§ 798.1 General statement.

This part is issued in accordance with the regulations of the Secretary of Agriculture at 7 CFR 1.1 through 1.16, and appendix A, implementing the Freedom of Information Act (5 U.S.C. 552). The Secretary's regulations as implemented by the regulations in this part, govern the availability of records of the FSA and Commodity Credit Corporation (CCC) to the public.

§ 798.2 Public inspection and copying.

5 U.S.C. 552(a)(2) requires that certain materials be made available for public inspection and copying. Members of the public may request access to such materials maintained by FSA and/or CCC at the Office of the Director, Information Division, Farm Service Agency, Room 3608 South Building, P.O. Box 2415, Washington, DC 20013, between the hours of 8:15 and 4:45 p.m., Monday through Friday.

[50 FR 53259, Dec. 31, 1985]

§ 798.3 Index.

5 U.S.C. 552(a)(2) requires that each agency publish or otherwise make available a current index of all materials required to be made available for public inspection and copying. FSA maintains an index of FSA National Handbooks, CCC Board Dockets, decisions of the Board of Contract Appeals of the Department of Agriculture affecting FSA or CCC, and Marketing Quota Review Committee determinations. In view of the small number of public requests for such index, publica-tion of the index is unnecessary and impractical. The index is maintained and available to the public at the office shown in § 798.2 and copies of the index are available upon request in person or by mail to that office.

§ 798.4 Request for records.

Request for records under 5 U.S.C. 552(a)(3) shall be made in accordance with 7 CFR 1.3. Reasonable requests for material not in existence may also be honored where their compilation will not unduly interfere with FSA operations and programs. Each FSA office in the field and each FSA office and division in Washington (see statement of Organization and Functions of FSA, 40 FR 18815, and of CCC, 35 FR 14951, and any amendments thereto) is designated as an "information center" and shall make space available to inspect and copy records in their custody not exempted from disclosure. Copies of records shall also be made available upon request. The head of each office or division is authorized to receive requests for records and to make determinations regarding requests for records in the office's custody in accordance with 7 CFR 1.4(c). Requests to Washington divisions and offices shall be addressed to USDA, FSA, P.O. Box 2415, Washington, D.C. 20013. The heads of FSA field offices shall be addressed as listed in the local telephone directory under "U.S. Government, Department of Agriculture, FSA". Names and addresses of heads of field offices may also be obtained from the office indicated in § 798.2.

§ 798.5 Appeals.

Any person whose request under § 798.4 of this part is denied shall have the right to appeal such denial. This appeal shall be submitted in accordance with 7 CFR 1.3(e) and addressed to the Administrator, FSA (Executive Vice-President, CCC), USDA, FSA, P.O. Box 2415, Washington, D.C. 20013.

§ 798.6 Fees.

This schedule supplements the fee schedule in 7 CFR, part 1, subpart A, appendix A and sets forth the fees to be charged by FSA for providing copies of records, materials, and services not covered in appendix A:

(a) Records, materials and services furnished without cost.

(1) One copy each of related directives, or blank forms required by FSA for program participation, if requester is a program participant.

(2) List of names and addresses of county and/or community committee members, and names of county employees in the county.

(3) One copy of an investigation report furnished to an appellant for a program appeal.

(b) Records, materials and services for which fees are charged.

(1) *National handbooks.* Three dollars for the first copy. One dollar for each additional copy. (The term "copy" includes all national amendments to date. They will be furnished separately for the requester to assemble).

(2) *Field supplementation to national handbooks.* Five cents per page, not to exceed $3, for each supplement.

(3) *Computerized records.* The requester shall furnish the necessary reels when computerized records are furnished on magnetic tape.

SUBCHAPTER G—ENVIRONMENTAL PROTECTION

PART 799—COMPLIANCE WITH THE NATIONAL ENVIRONMENTAL POLICY ACT

AUTHORITY: 42 U.S.C. 4321–4370.

SOURCE: 81 FR 51285, Aug. 3, 2016, unless otherwise noted.

Subpart A—General FSA Implementing Regulations for NEPA

§ 799.1 Purpose.

(a) This part:

(1) Explains major U.S. Department of Agriculture (USDA) Farm Service Agency (FSA) environmental policies.

(2) Establishes FSA procedures to implement the:

(i) National Environmental Policy Act (NEPA) of 1969, as amended (42 U.S.C. 4321 through 4370);

(ii) Council on Environmental Quality (CEQ) regulations (40 CFR parts 1500 through 1518); and

(iii) USDA NEPA regulations (§§ 1b.1 through 1b.4 of this title).

(3) Establishes procedures to ensure that FSA complies with other applicable laws, regulations, and Executive Orders, including, but not limited to, the following:

(i) American Indian Religious Freedom Act (42 U.S.C. 1996);

(ii) Archaeological and Historic Preservation Act (16 U.S.C. 469 through 469c);

(iii) Archaeological Resources Protection Act of 1979 (16 U.S.C. 470aa through 470mm);

(iv) Clean Air Act (42 U.S.C. 7401 through 7671q);

(v) Clean Water Act (33 U.S.C. 1251 through 1387);

(vi) Coastal Barrier Resources Act (16 U.S.C. 3501 through 3510);

(vii) Coastal Zone Management Act of 1972 (CZMA) (16 U.S.C. 1451 through 1466);

(viii) Comprehensive Environmental Response, Compensation, and Liability Act (42 U.S.C. 9601 through 9675);

(ix) Endangered Species Act (ESA) (16 U.S.C. 1531 through 1544);

(x) Farmland Protection Policy Act (7 U.S.C. 4201 through 4209);

(xi) Migratory Bird Treaty Act (16 U.S.C. 703 through 712);

(xii) National Historic Preservation Act (NHPA) of 1966, as amended (54 U.S.C. 300101 through 307101),

(xiii) Native American Graves Protection and Repatriation Act (25 U.S.C. 3001 through 3013);

(xiv) Resource Conservation and Recovery Act (42 U.S.C. 6901 through 6992k);

(xv) Safe Drinking Water Act (42 U.S.C. 300h through 300h.8);

(xvi) Wild and Scenic Rivers Act (16 U.S.C. 1271 through 1287);

(xvii) Wilderness Act (16 U.S.C. 1131 through 1136);

(xviii) Advisory Council on Historic Preservation regulations in 36 CFR part 800 "Protection of Historic Properties;"

(xix) USDA, Office of Environmental Quality regulations in part 3100 of this title, "Cultural and Environmental Quality" (see part 190, subpart F, of this title, "Procedures for the Protection of Historic and Archaeological Properties," for more specific implementation procedures);

(xx) USDA, Natural Resources Conservation Service regulations in part 658 of this title, "Farmland Protection Policy Act;"

(xxi) USDA regulations in part 12 of this title, "Highly Erodible Land and Wetland Conservation;"

(xxii) U.S. Department of the Interior, National Park Service regulations in 36 CFR part 60, "National Register of Historic Places;"

(xxiii) U.S. Department of the Interior, National Park Service regulations in 36 CFR part 63, "Determinations of Eligibility for Inclusion in the National Register of Historic Places;"

(xxiv) USDA, Departmental Regulation 9500-3, "Land Use Policy;"

(xxv) USDA, Departmental Regulation 9500-4, "Fish and Wildlife Policy;"

(xxvi) Executive Order 11514, "Protection and Enhancement of Environmental Quality;"

(xxvii) Executive Order 11593, "Protection and Enhancement of the Cultural Environment;"

(xxviii) Executive Order 11988, "Floodplain Management;"

(xxix) Executive Order 11990, "Protection of Wetlands;"

(xxx) Executive Order 11991, "Relating to Protection and Enhancement of Environmental Quality;"

(xxxi) Executive Order 12898, "Federal Actions to Address Environmental Justice in Minority Populations and Low-Income Populations;"

(xxxii) Executive Order 13007, "Indian Sacred Sites;"

(xxxiii) Executive Order 13175, "Consultation and Coordination with Indian Tribal Governments;"

(xxxiv) Executive Order 13186, "Responsibilities of Federal Agencies to Protect Migratory Birds;"

(xxxv) Executive Order 13287, "Preserve America;" and

(xxxvi) Executive Order 13690, "Establishing a Federal Flood Risk Management Standard and a Process for Further Soliciting and Considering Stakeholder Input."

(b) The procedures and requirements in this part supplement CEQ and USDA regulations; they do not replace or supersede them.

§799.2 FSA environmental policy.

(a) FSA will:

(1) Use all practical means to protect and, where possible, improve the quality of the human environment and avoid or minimize any adverse environmental effects of FSA actions;

(2) Ensure protection of basic resources, including important farmlands and forestlands, prime rangelands, wetlands, floodplains, and other protected resources. Consistent with Departmental Regulations and related Executive Orders, it is FSA policy not to approve or fund proposed actions that, as a result of their identifiable impacts, direct, indirect, or cumulative, would lead to or accommodate either the conversion of these land uses or encroachment upon them.

(3) Ensure that the requirements of NEPA and other State and national environmental policies designed to protect and manage impacts on the human environment are addressed:

(i) As required by 40 CFR 1501.2, at the earliest feasible stage in the planning of any FSA action,

(ii) Concurrently and in a coordinated manner,

(iii) During all stages of the decision making process,

(iv) Using professional and scientific integrity in their discussions and analyses, identifying applicable methodologies, and explaining the use of the best available information, and

(v) In consultation with all interested parties, including Federal, State, and Tribal governments;

(4) As appropriate, make environmental review available to the public through various means, which can include, but are not limited to: Posting on the National FSA Web site or a State FSA Web site, publishing in the *Federal Register,* or publishing in a newspaper in the area of interest; and

(5) Ensure that, if an FSA proposed action represents one of several phases of a larger action, the entire action is the subject of an environmental review independent of the phases of funding. If the FSA proposed action is one segment of a larger action, the entire action will be used in determining the appropriate level of FSA environmental review.

(b) A proposed action that consists of more than one categorically excluded proposed action may be categorically excluded only if all components of the proposed action are included within one or more categorical exclusions and trigger no extraordinary circumstances. The component of a proposed action that requires the highest level of NEPA review will be used to determine the required level of the NEPA review.

§799.3　Applicability.

(a) Except as provided for in paragraph (b) of this section, this part applies to:

(1) The development or revision of FSA rules, regulations, plans, policies, or procedures;

(2) New or continuing FSA proposed actions and programs, including, on behalf of the Commodity Credit Corporation (CCC), CCC programs, Farm Loan Programs, and Farm Programs; and

(3) FSA legislative proposals, not including appropriations requests, developed by FSA or with significant FSA cooperation and support.

(b) This part does not apply to FSA programs specifically exempted from environmental review by the authorizing legislation for those programs.

§799.4　Abbreviations and definitions.

(a) The following abbreviations apply to this part:

CAAP Concentrated Aquatic Animal Production Facilities.
CAFO Concentrated Animal Feeding Operation.
CCC Commodity Credit Corporation.
CEQ Council on Environmental Quality.
EA Environmental Assessment.
EIS Environmental Impact Statement.
ESA Endangered Species Act.
ESW Environmental Screening Worksheet.
FONSI Finding of No Significant Impact.
FPO Federal Preservation Officer.
FSA Farm Service Agency.
MOA Memorandum of Agreement.
MOU Memorandum of Understanding.
NECM National Environmental Compliance Manager.
NEPA National Environmental Policy Act.
NHPA National Historic Preservation Act.
NOA Notice of Availability.
NOI Notice of Intent.
PEA Programmatic Environmental Assessment.
PEIS Programmatic Environmental Impact Statement.
RAO Responsible Approving Official.
RFO Responsible Federal Officer
ROD Record of Decision.
SEC State Environmental Coordinator.
SED State Executive Director for FSA.
SEIS Supplemental Environmental Impact Statement.
SHPO State Historic Preservation Officer.
THPO Tribal Historic Preservation Officer.
USDA United States Department of Agriculture.

(b) The definitions in 40 CFR part 1508 apply and are supplemented by parts 718 and 1400 of this title; in the event of a conflict the definitions in this section will be controlling. In addition, the following definitions apply to this part:

Administrator means the Administrator, Farm Service Agency, including designees.

Application means the formal process of requesting FSA assistance.

Construction means actions that include building, rehabilitation, modification, repair, and demolition of facilities, and earthmoving.

Consultation means the process of soliciting, discussing, and considering the views of other participants in the environmental review process and working toward agreement where feasible.

Environmental screening worksheet, or ESW, means the FSA screening procedure used to record the use of categorical exclusions, review if a proposed action that can be categorically excluded involves extraordinary circumstances, and evaluate the appropriate level and extent of environmental review needed in an EA or EIS when a categorical exclusion is not available or not appropriate. For the purposes of this part, the ESW may be represented by alternate documentation comparable to the ESW, and that has been approved in advance by the NECM, such as related environmental documentation, including, but not limited to, the related documentation from another agency.

Financial assistance means any form of loan, loan guarantee, grant, guaranty, insurance, payment, rebate, subsidy, or any other form of direct or indirect Federal monetary assistance.

Floodplains means the lowland and relatively flat areas adjoining inland and coastal waters, including floodprone areas of offshore islands, including, at a minimum, those that are subject to a 1-percent or greater chance of flooding in any given year.

Historic property means any prehistoric or historic district, site, building, structure, or object included in, or eligible for inclusion in, the National Register of Historic Places maintained by the Secretary of the Interior as defined in 36 CFR 800.16.

Memorandum of Agreement means a document that records the terms and conditions agreed upon to resolve the potential effects of a Federal agency proposed action or program. Often used interchangeably with Memorandum of Understanding.

Plow zone means the depth of previous tillage or disturbance.

Programmatic Environmental Assessment (PEA) means an assessment prepared when the significance of impacts of a program are uncertain to assist in making this determination.

Programmatic Environmental Impact Statement (PEIS) means an analysis of the potential impacts that could be associated with various components of a program or proposed action that may not yet be clearly defined or even known, to determine if the program or its various components have the potential to significantly affect the quality of the human environment.

Program participant means any person, agency, or other entity that applies for or receives FSA program benefits or assistance.

Protected resources means environmentally sensitive resources that are protected by laws, regulations, or Executive Orders for which FSA proposed actions may pose potentially significant environmental effects.

State Historic Preservation Officer (SHPO) means the state official appointed or designated under the NHPA to administer a State historic preservation program, or a representative to act for the SHPO.

Tribal Historic Preservation Officer (THPO) means the Tribal official appointed by a Tribe's chief governing authority or designated by a Tribal ordinance or preservation program, who has assumed the responsibilities of the SHPO on Tribal lands under the NHPA.

Wetlands means areas that are inundated by surface or ground water with a frequency sufficient to support and, under normal circumstances, do support or would support a prevalence of vegetative or aquatic life that requires saturated or seasonally saturated soil conditions for growth and reproduction. Wetlands generally include swamps, marshes, bogs, and similar areas, such as sloughs, prairie potholes, wet meadows, river overflows, mudflats, and natural ponds.

Subpart B—FSA and Program Participant Responsibilities

§ 799.5 National office environmental responsibilities.

(a) The FSA Administrator or designee:

(1) Is the Responsible Federal Officer (RFO) for FSA compliance with applicable environmental laws, regulations, and Executive Orders, including NEPA, and unless otherwise specified, will make all determinations under this part;

(2) Will ensure responsibilities for complying with NEPA are adequately delegated to FSA personnel within their areas of responsibility at the Federal, State, and county levels;

(3) Will appoint a National Environmental Compliance Manager (NECM), as required by 40 CFR 1507.2(a), who reports directly to the FSA Administrator; and

(4) Will appoint a qualified Federal Preservation Officer (FPO), as required by Executive Order 13287 "Preserve America" section 3(e) and by section 110 of NHPA (54 U.S.C. 306101). This individual must meet the National Park Service professional qualification standards requirements referenced in 36 CFR part 61 and will report directly to the NECM.

(b) The NECM or designee coordinates FSA environmental policies and reviews under this part on a national basis and is responsible for:

(1) Ensuring FSA legislative proposals and multistate and national programs are in compliance with NEPA and other applicable environmental and cultural resource laws, regulations, and Executive Orders;

(2) Providing education and training on implementing NEPA and other environmental compliance requirements to appropriate FSA personnel;

(3) Serving as the principal FSA advisor to the FSA Administrator on NEPA and other environmental compliance requirements;

(4) Representing FSA, and serving as an intra- and inter-agency liaison, on NEPA- and environmental compliance-related matters on a national basis;

(5) Maintaining a record of FSA environmental compliance actions; and

(6) Ensuring State and county office compliance with NEPA and other applicable environmental laws, regulations, and Executive Orders.

(c) The FPO or designee coordinates NHPA compliance under this part and is responsible for:

(1) Serving as the principal FSA advisor to the NECM on NHPA requirements;

(2) Representing FSA, and serving as FSA intra- and inter-agency liaison, on all NHPA-related matters on a national basis;

(3) Maintaining current FSA program guidance on NHPA requirements;

(4) Maintaining a record of FSA environmental actions related to the NHPA; and

(5) Ensuring State and county office compliance with the NHPA and other cultural resource-related requirements.

§ 799.6 FSA State office environmental responsibilities.

(a) FSA State Executive Directors (SEDs) or designees are the responsible approving officials (RAOs) in their respective States and are responsible for:

(1) Ensuring FSA proposed actions within their State comply with applicable environmental laws, regulations, and Executive Orders, including NEPA; and

(2) Appointing two or more collateral duty State Environmental Coordinators (SECs) or at least one full time SEC.

(b) An SED will not appoint more than one SEC for Farm Programs and one SEC for Farm Loan Programs in a State unless approved in writing by the NECM.

(c) SECs or designees are responsible for:

(1) Serving as the environmental compliance coordinators on all environmental-related matters within their respective State;

(2) Advising SEDs on environmental issues;

(3) Providing training, in coordination with the NECM, on NEPA and other environmental compliance requirements to appropriate FSA State and county office personnel;

(4) Providing assistance on environmental-related matters on a proposed

action-by-action basis to State and county office personnel, as needed;

(5) When feasible, developing controls for avoiding or mitigating adverse environmental impacts and monitoring the implementation of those controls;

(6) Reviewing FSA proposed actions that are not categorically excluded from documentation in an environmental assessment or environmental impact statement, or that otherwise require State office approval or clearance, and making appropriate recommendations to the approving official;

(7) Providing assistance to resolve post-approval environmental issues at the State office level;

(8) Maintaining decision records for State office environmental compliance matters;

(9) Monitoring their respective State's compliance with environmental laws, regulations, and Executive Orders;

(10) Acting as a liaison on FSA State office environmental compliance matters with the public and other Federal, State, and Tribal governments;

(11) Representing the SED on environmental issues, as requested;

(12) Delegating duties under this section with the approval of both the SED and NECM; and

(13) Other NEPA and environmental compliance-related duties as assigned.

(d) County Executive Directors, District Directors, and Farm Loan Programs loan approval officers or designees are responsible for compliance with this part within their geographical areas.

§799.7 FSA program participant responsibilities.

(a) Potential FSA program participants requesting FSA assistance must do all of the following:

(1) Consult with FSA early in the process about potential environmental concerns associated with program participation. The program participation information required to start participation in an FSA program varies by FSA program and may be in the form of an offer, enrollment, sign-up, contract, note and security agreement, or other as is required by the relevant FSA program.

(2) Submit applications for all Federal, regional, State, and local approvals and permits early in the planning process.

(3) Coordinate the submission of program participation information to FSA and other agencies (for example, if a conservation plan is required, then the program participation information is also submitted to USDA's Natural Resources Conservation Service).

(4) Work with other appropriate Federal, State, and Tribal governments to ensure all environmental factors are identified and impacts addressed and, to the extent possible, mitigated, consistent with how mitigation is defined in 40 CFR 1508.20.

(5) Inform FSA of other Federal, State, and Tribal government environmental reviews that have previously been completed or required of the program participant.

(6) Provide FSA with a list of all parties affected by or interested in the proposed action.

(7) If requested by FSA, provide information necessary for FSA to evaluate a proposed action's potential environmental impacts and alternatives.

(8) Ensure that all compliance documentation provided is current, sufficiently detailed, complete, and submitted in a timely fashion.

(9) Be in compliance with all relevant laws, regulations, and policies regarding environmental management and protection.

(10) Not implement any component of the proposed action prior to the completion of FSA's environmental review and final decision, or FSA's approval for that proposed action, consistent with 40 CFR 1506.1.

(b) When FSA receives program participation information for assistance or notification that program participation information will be filed, FSA will contact the potential program participant about the environmental information the program participant must provide as part of the process. This required information may include:

(1) Design specifications;

(2) Topographical, aerial, and location maps;

(3) Surveys and assessments necessary for determining the impact on

protected resources listed in § 799.33(a)(2);

(4) Nutrient management plans; and

(5) Applications, plans, and permits for all Federal, regional, State and local approvals including construction permits, storm water run-off and operational plans and permits, and engineering designs and plans.

§ 799.8 Significant environmental effect.

(a) In determining whether a proposed action will have a significant effect on the quality of the human environment, FSA will consider the proposed action's potential effects in the context of society as a whole, the affected region and interests, the locality, and the intensity of the potential impact as specified in 40 CFR 1508.27.

(b) [Reserved]

§ 799.9 Environmental review documents.

(a) FSA may prepare the following documents during the environmental review process:

(1) ESW;

(2) Programmatic Environmental Assessment (PEA);

(3) Environmental Assessment (EA);

(4) Supplemental Environmental Assessment;

(4) Programmatic Environmental Impact Statement (PEIS);

(5) Environmental Impact Statement (EIS);

(6) Supplemental Environmental Impact Statement (SEIS);

(7) Finding of No Significant Impact (FONSI);

(8) Record of Decision (ROD);

(9) Notice of Intent (NOI) to prepare any type of EIS;

(10) Notice of Availability (NOA) of environmental documents;

(11) Notice of public scoping meetings;

(12) Other notices, including those required under Executive Order 11988, "Floodplain Management," Executive Order 13690, "Establishing a Federal Flood Risk Management Standard and a Process for Further Soliciting and Considering Stakeholder Input," and Executive Order 11990, "Protection of Wetlands;"

(13) Memorandums of Agreement or Understanding (MOA or MOU), such as those for mitigation of adverse effects on historic properties as specified in 36 CFR part 800, "Protection of Historic Properties;" and

(14) Environmental studies, as indicated and appropriate.

(b) [Reserved]

§ 799.10 Administrative records.

(a) FSA will maintain an administrative record of documents and materials that FSA created or considered during its NEPA decision making process for a proposed action and referenced as such in the NEPA documentation, which can include any or all the following:

(1) Any NEPA environmental review documents listed in § 799.9, as applicable;

(2) Technical information, permits, plans, sampling results, survey information, engineering reports, and studies, including environmental impact studies and assessments;

(3) Policies, guidelines, directives, and manuals;

(4) Internal memorandums or informational papers;

(5) Contracts or agreements;

(6) Notes of professional telephone conversations and meetings;

(7) Meeting minutes;

(8) Correspondence with agencies and stakeholders;

(9) Communications to and from the public;

(10) Documents and materials that contain any information that supports or conflicts with the FSA decision;

(11) Maps, drawings, charts, and displays; and

(12) All public comments received during the NEPA comment periods.

(b) The administrative record may be used, among other purposes, to facilitate better decision making, as determined by FSA.

§ 799.11 Actions during NEPA reviews.

(a) Except as specified in paragraphs (b) and (c) of this section, FSA or a program participant must not take any action, implement any component of a proposed action, or make any final decision during FSA's NEPA and environmental compliance review process that could have an adverse environmental impact or limit the range of alternatives until FSA completes its environmental review by doing one of the following:

(1) Determines that the proposed action is categorically excluded under NEPA under subpart D of this part and does not trigger any extraordinary circumstances; or

(2) Issues a FONSI or ROD under subpart E or F of this part.

(b) FSA may approve interim actions related to proposed actions provided the:

(1) Interim actions will not have an adverse environmental impact;

(2) Expenditure is necessary to maintain a schedule for the proposed action;

(3) Interim actions and expenditures will not compromise FSA's environmental compliance review and decision making process for the larger action;

(4) Interim actions and expenditures will not segment otherwise connected actions; and

(5) NEPA and associated environmental compliance review has been completed for the interim action or expenditure.

(c) FSA and program participants may develop preliminary plans or designs, or perform work necessary to support an application for Federal, State, or local permits or assistance, during the NEPA review process, provided all requirements in paragraphs (a) and (b) of this section are met.

§799.12 Emergency circumstances.

(a) If emergency circumstances exist that make it necessary to take action to mitigate harm to life, property, or important natural, cultural, or historic resources, FSA may take an action with significant environmental impact without complying with the requirements of this part.

(b) If emergency circumstances exist, the NECM will consult with CEQ as soon as feasible about alternative NEPA arrangements for controlling the immediate impact of the emergency, as specified in 40 CFR 1506.11.

(c) If emergency circumstances exist, the FPO will follow the emergency procedures specified in 36 CFR 800.12 regarding preservation of historic properties, if applicable.

(d) FSA assistance provided in response to a Presidentially-declared disaster under the Robert T. Stafford Disaster Relief and Emergency Assistance Act, as amended, 42 U.S.C. 5121—5207, is exempt from NEPA requirements, as specified in 42 U.S.C. 5159. Under a Presidentially-declared disaster, the following actions to specifically address immediate post-emergency health or safety hazards are exempt from environmental compliance requirements:

(1) Clearing roads and constructing temporary bridges necessary for performing emergency tasks and essential community services;

(2) Emergency debris removal in support of performing emergency tasks and essential community services;

(3) Demolishing unsafe structures that endanger the public or could create a public health hazard if not demolished;

(4) Disseminating public information and assistance for health and safety measures;

(5) Providing technical assistance to State, regional, local, or Tribal governments on disaster management control;

(6) Reducing immediate threats to life, property, and public health and safety; and

(7) Warning of further risks and hazards.

(c) Proposed actions other than those specified in paragraph (d) of this section that are not specifically to address immediate post-emergency health or safety hazards require the full suite of environmental compliance requirements and are not exempt.

§799.13 FSA as lead agency.

(a) When FSA acts as the lead agency in a NEPA review as specified in 40 CFR 1501.5, FSA will:

(1) Coordinate its review with other appropriate Federal, State, and Tribal governments; and

(2) Request other agencies to act as cooperating agencies as specified in 40 CFR 1501.6, and defined in 40 CFR 1508.5, as early in the review process as possible.

(b) If FSA acts as a lead agency for a proposed action that affects more than one State, the NECM will designate one SEC to act as RAO.

(c) If the role of lead agency is disputed, the NECM will refer the matter to the FSA Administrator, who will attempt to resolve the matter with the other agency. If the Federal agencies cannot agree which will serve as the lead agency, the FSA Administrator will follow the procedures specified in 40 CFR 1501.5(e) to request that CEQ determine the lead agency.

§ 799.14 FSA as cooperating agency.

(a) FSA will act as a cooperating agency if requested by another agency, as specified in 40 CFR 1501.6 and defined in 40 CFR 1508.5. However, FSA may decline another agency's request if FSA determines the proposed action does not fall within FSA's area of expertise or FSA does not have jurisdiction by law. If FSA declines such a request to cooperate, that will be documented in writing to the requesting agency and a copy will be provided to CEQ.

(b) FSA may request to be designated as a cooperating agency if another agency's proposed action falls within FSA's area of expertise.

§ 799.15 Public involvement in environmental review.

(a) FSA will involve the public in the environmental review process as early as possible and in a manner consistent with 40 CFR 1506.6. To determine the appropriate level of public participation, FSA will consider:

(1) The scale of the proposed action and its probable effects;

(2) The likely level of public interest and controversy; and

(3) Advice received from knowledgeable parties and experts.

(b) Depending upon the scale of the proposed action, FSA will:

(1) Coordinate public notices and consultation with the U.S. Fish and Wildlife Service, USDA's Natural Resources Conservation Service, Federal Emergency Management Agency, the National Marine Fisheries Service, the U.S. Army Corps of Engineers, and other agencies, as appropriate, if wetlands, floodplains, ESA-listed species, or other protected resources have the potential to be impacted;

(2) Make appropriate environmental documents available to interested partiesby request;

(3) Publish a Notice of Intent (NOI) to prepare an EIS, as specified in subpart F of this part; and

(4) Publish a Notice of Availability (NOA) of draft and final EISs and RODs, as specified in subpart F of this part.

(c) If the effects of a proposed action are local in nature and the scale of the proposed action is likely to generate interest and controversy at the local level, then in addition to the proposed actions specified in paragraphs (a) and (b) of this section, FSA will:

(1) Notify appropriate State, local, regional, and Tribal governments and clearinghouses, and parties and organizations, including the State Historic Preservation Officer (SHPO) and Tribal Historic Preservation Officer (THPO), known to have environmental, cultural, and economic interests in the locality affected by the proposed action; and

(2) Publish notice of the proposed action in the local media.

(d) Public review for 30 days for a FONSI is necessary if any of the limited circumstances specified in 40 CFR 1501.4(e)(2)(i) or (ii) applies.

§ 799.16 Scoping.

(a) FSA will determine the appropriate scoping process for the environmental review of a proposed action based on the nature, complexity, potential significance of effects, and level of controversy of the proposed action.

(b) As part of its scoping process, FSA will:

(1) Invite appropriate Federal, State, and Tribal governments, and other interested parties to participate in the process, if determined necessary by FSA;

(2) Identify the significant issues to be analyzed;

(3) Identify and eliminate from further review issues that were determined not significant or have been adequately addressed in any prior environmental reviews;

(4) Determine the roles of lead and cooperating agencies, if appropriate;

(5) Identify any related EAs or EISs;

(6) Identify other environmental reviews and consultation requirements, including NHPA requirements and State, local, regional, and Tribal requirements, so they are integrated into the NEPA process;

(7) Identify the relationship between the timing of the environmental review process and FSA's decision making process;

(8) Determine points of contact within FSA; and

(9) Establish time limits for the environmental review process.

(c) FSA may hold public meetings as part of the scoping process, if appropriate and as time permits. The process that FSA will use to determine if a public scoping meeting is needed, and how such meetings will be announced, is specified in §799.17.

§799.17 Public meetings.

(a) In consultation with the NECM, the SEC will determine if public meetings will be held on a proposed action to:

(1) Inform the public about the details of a proposed action and its possible environmental effects;

(2) Gather information about the public concerns; and

(3) Resolve, address, or respond to issues raised by the public.

(b) In determining whether to hold a public meeting, FSA will consider and determine whether:

(1) There is substantial controversy concerning the environmental impact of the proposed action;

(2) There is substantial interest in holding a public meeting;

(3) Another Federal agency or Tribal government has requested a public scoping meeting and their request is warranted; or

(4) The FSA Administrator has determined that a public meeting is needed.

(c) FSA will publish notice of a public meeting, including the time, date and location of the meeting, in the local media or *Federal Register,* as appropriate, at least 15 days before the first meeting. A notice of a public scoping meeting may be included in a Notice of Intent to prepare an EIS.

(d) If a NEPA document is to be considered at a public meeting, FSA will make the appropriate documentation available to the public at least 15 days before the meeting.

§799.18 Overview of FSA NEPA process.

If the proposed action:	FSA:
Is an emergency action	Follows the procedures in §799.12
Is exempt from section 102(2)(C) of NEPA (42 U.S.C. 4332(2)(C)) by authorizing legislation for the program.	Implements the action.
Is categorically excluded under §799.31(b) or §1b.3 of this title	Implements the action after recording the specific categorical exclusion on the ESW (no review needed).
Is a proposed action that has the potential to impact historic properties as specified in §799.33(e) and therefore requires the completion of an ESW.	Completes an ESW to determine if there will be an impact on historic properties. FSA will prepare an EA or EIS, as indicated, before implementing the action.
Is a categorically excluded proposed action listed in §799.32 that requires the completion of an ESW.	Completes an ESW to determine whether extraordinary circumstances are present, as defined in §799.33. This review includes a determination of whether the proposed action will potentially impact protected resources. If there are no extraordinary circumstances, FSA implements the action; if there are extraordinary circumstances, FSA will prepare an EA or EIS, as indicated, before implementing the action.
Involves a category of proposed actions requiring an EA listed in §799.41.	Prepares an EA.
Involves a category of proposed actions requiring an EIS listed in §799.51.	Prepares an EIS.

Subpart C—Environmental Screening Worksheet

§799.20 Purpose of the ESW.

(a) FSA uses the ESW as an initial screening tool to evaluate record the use of a categorical exclusion for a pro-

posed action and to determine the required type of environmental review.

(b) Review with the ESW is not required for proposed actions that are categorically excluded as specified in §799.31(b) or §1b.3 of this title, or for proposed actions where FSA determines at an early stage that there is a need to prepare an EA or EIS.

Subpart D—Categorical Exclusions

§ 799.30 Purpose of categorical exclusion process.

(a) FSA has determined that the categories of proposed actions listed in §§ 799.31 and 799.32 do not normally individually or cumulatively have a significant effect on the human environment and do not threaten a violation of applicable statutory, regulatory, or permit requirements for environment, safety, and health, including requirements of Executive Orders and other USDA regulations in this chapter. Based on FSA's previous experience implementing these actions and similar actions through the completion of EAs, these proposed actions are categorically excluded.

(b) If a proposed action falls within one of the categories of proposed actions listed in § 1b.3 of this title, § 799.31, or § 799.32, and there are no extraordinary circumstances present as specified in § 799.33, then the proposed action is categorically excluded from the requirements to prepare an EA or an EIS.

(c) Those proposed actions in categories in § 799.31 or § 799.32 will be considered categorical exclusions unless it is determined there are extraordinary circumstances, as specified in § 799.33.

§ 799.31 Categorical exclusions to be recorded on an ESW.

(a) Proposed actions listed in this section involve no new ground disturbance below the existing plow zone (does not exceed the depth of previous tillage or disturbance) and therefore only need to be recorded on the ESW; no further review will be required. Unless otherwise noted, the proposed actions in this section also do not have the potential to cause effects to historic properties, and will therefore not be reviewed for compliance with section 106 of NHPA (54 U.S.C. 306108) or its implementing regulations, 36 CFR part 800. However, some proposed actions may require other Federal consultation to determine if there are extraordinary circumstances as specified in § 799.33.

(b) The following proposed actions are categorically excluded. These proposed actions are grouped into broader categories of similar types of proposed actions. Those proposed actions that are similar in scope (purpose, intent, and breadth) and the potential significance of impacts to those listed in this section, but not specifically listed in § 799.31 or § 799.32, will be considered categorical exclusions in this category, unless it is determined that extraordinary circumstances exist, as specified in § 799.33:

(1) *Loan actions.* The following list includes categorical exclusions for proposed actions related to FSA loans:

(i) Closing cost payments;

(ii) Commodity loans;

(iii) Debt set asides;

(iv) Deferral of loan payments;

(v) Youth loans;

(vi) Loan consolidation;

(vii) Loans for annual operating expenses, except livestock;

(viii) Loans for equipment;

(ix) Loans for family living expenses;

(x) Loan subordination, with no or minimal construction below the depth of previous tillage or ground disturbance, and no change in operations, including, but not limited to, an increase in animal numbers to exceed the current CAFO designation (as defined by the U.S. Environmental Protection Agency in 40 CFR 122.23);

(xi) Loans to pay for labor costs;

(xii) Loan (debt) transfers and assumptions with no new ground disturbance;

(xiii) Partial or complete release of loan collateral;

(xiv) Re-amortization of loans;

(xv) Refinancing of debt;

(xvi) Rescheduling loans;

(xvii) Restructuring of loans; and

(xvii) Writing down of debt;

(2) *Repair, improvement, or minor modification actions.* The following list includes categorical exclusions for repair, improvement, or minor modification proposed actions:

(i) Existing fence repair;

(ii) Improvement or repair of farm-related structures under 50 years of age; and

(iii) Minor amendments or revisions to previously approved projects, provided such proposed actions do not substantively alter the purpose, operation, location, impacts, or design of the project as originally approved;

(3) *Administrative actions.* The following list includes categorically excluded administrative proposed actions:

(i) Issuing minor technical corrections to regulations, handbooks, and internal guidance, as well as amendments to them;

(ii) Personnel actions, reduction-in-force, or employee transfers; and

(iii) Procurement actions for goods and services conducted in accordance with Executive Orders;

(4) *Planting actions.* The following list includes categorical exclusions for planting proposed actions:

(i) Bareland planting or planting without site preparation;

(ii) Bedding site establishment for wildlife;

(iii) Chiseling and subsoiling;

(iv) Clean tilling firebreaks;

(v) Conservation crop rotation;

(vi) Contour farming;

(vii) Contour grass strip establishment;

(viii) Cover crop and green manure crop planting;

(ix) Critical area planting;

(x) Firebreak installation;

(xi) Grass, forbs, or legume planting;

(xii) Heavy use area protection;

(xiii) Installation and maintenance of field borders or field strips;

(xiv) Pasture, range, and hayland planting;

(xv) Seeding of shrubs;

(xvi) Seedling shrub planting;

(xvii) Site preparation;

(xviii) Strip cropping;

(xix) Wildlife food plot planting; and

(xx) Windbreak and shelterbelt establishment;

(5) *Management actions.* The following list includes categorical exclusions of land and resource management proposed actions:

(i) Forage harvest management;

(ii) Integrated crop management;

(iii) Mulching, including plastic mulch;

(iv) Netting for hard woods;

(v) Obstruction removal;

(vi) Pest management (consistent with all labelling and use requirements);

(vii) Plant grafting;

(viii) Plugging artesian wells;

(ix) Residue management including seasonal management;

(x) Roof runoff management;

(xi) Thinning and pruning of plants;

(xii) Toxic salt reduction; and

(xiii) Water spreading; and

(6) *Other FSA actions.* The following list includes categorical exclusions for other FSA proposed actions:

(i) Conservation easement purchases with no construction planned;

(ii) Emergency program proposed actions (including Emergency Conservation Program and Emergency Forest Restoration Program) that have a total cost share of less than $5,000;

(iii) Financial assistance to supplement income, manage the supply of agricultural commodities, or influence the cost and supply of such commodities or programs of a similar nature or intent (that is, price support programs);

(iv) Individual farm participation in FSA programs where no ground disturbance or change in land use occurs as a result of the proposed action or participation;

(v) Inventory property disposal or lease with protective easements or covenants;

(vi) Safety net programs administered by FSA;

(vii) Site characterization, environmental testing, and monitoring where no significant alteration of existing ambient conditions would occur, including air, surface water, groundwater, wind, soil, or rock core sampling; installation of monitoring wells; installation of small scale air, water, or weather monitoring equipment;

(viii) Stand analysis for forest management planning;

(ix) Tree protection including plastic tubes; and

(x) Proposed actions involving another agency that are fully covered by one or more of that agency's categorical exclusions (on the ESW, to record the categorical exclusion, FSA will name the other agency and list the specific categorical exclusion(s) that applies).

§ 799.32 Categorical exclusions requiring review with an ESW.

(a) Proposed actions listed in this section may be categorically excluded after completion of a review with an ESW to document that a proposed action does not involve extraordinary circumstances as specified in § 799.33.

(b) This section has two types of categorical exclusions, one without construction and ground disturbance and one with construction and ground disturbance that will require additional environmental review and consultation in most cases.

(c) Consultations under NHPA, ESA, and other relevant environmental mandates, may be required to document that no extraordinary circumstances exist.

(d) The following proposed actions are grouped into broader categories of similar types of proposed actions without ground disturbance. Those proposed actions that are similar in scope (purpose, intent, and breadth) and the potential significance of impacts to those listed in this section, but not specifically listed in this section, will be considered categorical exclusions in this category, unless it is determined that extraordinary circumstances exist, as specified in § 799.33:

(1) *Loan actions.* The following list includes categorical exclusions for proposed actions related to FSA loans:

(i) Farm storage and drying facility loans for added capacity;

(ii) Loans for livestock purchases;

(iii) Release of loan security for forestry purposes;

(iv) Reorganizing farm operations; and

(v) Replacement building loans;

(2) *Minor management, construction, or repair actions.* The following list includes categorical exclusions for minor construction or repair proposed actions:

(i) Minor construction, such as a small addition;

(ii) Drain tile replacement;

(iii) Erosion control measures;

(iv) Grading, leveling, shaping, and filling;

(v) Grassed waterway establishment;

(vi) Hillside ditches;

(vii) Land-clearing operations of no more than 15 acres, provided any amount of land involved in tree harvesting (without stump removal) is to be conducted on a sustainable basis and according to a Federal, State, Tribal, or other governmental unit approved forestry management plan;

(viii) Nutrient management;

(ix) Permanent establishment of a water source for wildlife (not livestock);

(x) Restoring and replacing property;

(xi) Soil and water development;

(xii) Spring development;

(xiii) Trough or tank installation; and

(xiv) Water harvesting catchment; and

(3) *Other FSA actions.* The following list includes categorical exclusions for other FSA proposed actions:

(i) Fence installation and replacement;

(ii) Fish stream improvement;

(iii) Grazing land mechanical treatment; and

(iv) Inventory property disposal or lease without protective easements or covenants (this proposed action, in particular, has the potential to cause effects to historic properties and therefore requires analysis under section 106 of NHPA (54 U.S.C. 306108), as well as under the ESA and wetland protection requirements).

(e) The following proposed actions are grouped into broader categories of similar types of proposed actions with ground disturbance, each of the listed proposed actions has the potential for extraordinary circumstances because they include construction or ground disturbance. Therefore, additional environmental review and consultation will be necessary in most cases. Those proposed actions that are similar in scope (purpose, intent, and breadth) and the potential significance of impacts to those listed in this section, but not specifically listed in this section, will be considered categorical exclusions in this category, unless it is determined that extraordinary circumstances exist, as specified in § 799.33:

(1) *Loan actions.* The following list includes categorical exclusions for proposed actions related to FSA loans:

(i) Loans and loan subordination with construction, demolition, or ground disturbance planned;

(ii) Real estate purchase loans with new ground disturbance planned; and

(iii) Term operating loans with construction or demolition planned;

(2) *Construction or ground disturbance actions.* The following list includes categorical exclusions for construction or ground disturbance proposed actions:

(i) Bridges;

(ii) Chiseling and subsoiling in areas not previously tilled;

(iii) Construction of a new farm storage facility;

(iv) Dams;

(v) Dikes and levees;

(vi) Diversions;

(vii) Drop spillways;

(viii) Dugouts;

(ix) Excavation;

(x) Grade stabilization structures;

(xi) Grading, leveling, shaping and filling in areas or to depths not previously disturbed;

(xii) Installation of structures designed to regulate water flow such as pipes, flashboard risers, gates, chutes, and outlets;

(xiii) Irrigation systems;

(xiv) Land smoothing;

(xv) Line waterways or outlets;

(xvi) Lining;

(xvii) Livestock crossing facilities;

(xviii) Pesticide containment facility;

(xix) Pipe drop;

(xx) Pipeline for watering facility;

(xxi) Ponds, including sealing and lining;

(xxii) Precision land farming with ground disturbance;

(xxiii) Riparian buffer establishment;

(xxiv) Roads, including access roads;

(xxv) Rock barriers;

(xxvi) Rock filled infiltration trenches;

(xxvii) Sediment basin;

(xxviii) Sediment structures;

(xxix) Site preparation for planting or seeding in areas not previously tilled;

(xxx) Soil and water conservation structures;

(xxxi) Stream bank and shoreline protection;

(xxxii) Structures for water control;

(xxxiii) Subsurface drains;

(xxxiv) Surface roughening;

(xxxv) Terracing;

(xxxvi) Underground outlets;

(xxxvii) Watering tank or trough installation, if in areas not previously disturbed;

(xxxviii) Wells; and

(xxxix) Wetland restoration.

(3) *Management and planting type actions.* The following list includes categorical exclusions for resource management and planting proposed actions:

(i) Establishing or maintaining wildlife plots in areas not previously tilled or disturbed;

(ii) Prescribed burning;

(iii) Tree planting when trees have root balls of one gallon container size or larger; and

(iv) Wildlife upland habitat management.

§799.33 Extraordinary circumstances.

(a) As specified in 40 CFR 1508.4, in the definition of categorical exclusion, procedures are required to provide for extraordinary circumstances in which a normally categorically excluded action may have a significant environmental effect. The presence and impacts of extraordinary circumstances require heightened review of proposed actions that would otherwise be categorically excluded. Extraordinary circumstances include, but are not limited to:

(1) Scientific controversy about environmental effects of the proposed action;

(2) Impacts that are potentially adverse, significant, uncertain, or involve unique or unknown risks, including, but not limited to, impacts to protected resources. Protected resources include, but are not limited to:

(i) Property (for example, sites, buildings, structures, and objects) of historic, archeological, or architectural significance, as designated by Federal, Tribal, State, or local governments, or property eligible for listing on the National Register of Historic Places;

(ii) Federally-listed threatened or endangered species or their habitat (including critical habitat), or Federally-proposed or candidate species or their habitat;

(iii) Important or prime agricultural, forest, or range lands, as specified in part 657 of this chapter and in USDA Departmental Regulation 9500-3;

(iv) Wetlands, waters of the United States, as regulated under the Clean Water Act (33 U.S.C. 1344), highly erodible land, or floodplains;

(v) Areas having a special designation, such as Federally- and State-designated wilderness areas, national parks, national natural landmarks, wild and scenic rivers, State and Federal wildlife refuges, and marine sanctuaries; and

(vi) Special sources of water, such as sole-source aquifers, wellhead protection areas, or other water sources that are vital in a region;

(3) A proposed action that is also "connected" (as specified in 40 CFR 1508.25(a)(1)) to other actions with potential impacts;

(4) A proposed action that is related to other proposed actions with cumulative impacts (40 CFR 1508.25(a)(2));

(5) A proposed action that does not comply with 40 CFR 1506.1, "Limitations on actions during NEPA process;" and

(6) A proposed action that violates any existing Federal, State, or local government law, policy, or requirements (for example, wetland laws, Clean Water Act-related requirements, water rights).

(b) FSA will use the ESW to review proposed actions that are eligible for categorical exclusion to determine if extraordinary circumstances exist that could impact protected resources. If an extraordinary circumstance exists, and cannot be avoided or appropriately mitigated, an EA or EIS will be prepared, as specified in this part. Specifically, FSA will complete a review with the ESW for proposed actions that fall within the list of categorical exclusions specified in § 799.32 to determine whether extraordinary circumstances are present.

(c) For any proposed actions that have the potential to cause effects to historic properties, endangered species, waters of the United States, wetlands, and other protected resources, FSA will ensure appropriate analyses is completed to comply with the following mandates:

(1) For section 106 of the NHPA (54 U.S.C. 306108), the regulations in 36 CFR part 800, "Protection of Historic Properties;" if an authorized technical representative from another Federal agency assists with compliance with 36 CFR part 800, FSA will remain responsible for any consultation with SHPO, THPO, or Tribal governments;

(2) For section 7 of the ESA that governs the protection of Federally proposed, threatened and endangered species and their designated and proposed critical habitats; and

(3) For the Clean Water Act and related Executive Order provisions for avoiding impacts to wetlands and waters of the United States, including impaired waters listed under Section 303(d) of the Clean Water Act.

(d) If technical assistance is provided by another Federal agency, FSA will ensure that the environmental documentation provided is commensurate to or exceeds the requirements of the FSA ESW. If it is not, a review with an ESW is needed to determine if an EA or EIS is warranted.

§ 799.34   Establishing and revising categorical exclusions.

(a) As part of the process to establish a new categorical exclusion, FSA will consider all relevant information, including the following:

(1) Completed FSA NEPA documents;

(2) Other Federal agency NEPA documents on proposed actions that could be considered similar to the categorical exclusion being considered;

(3) Results of impact demonstration or pilot projects;

(4) Information from professional staff, expert opinions, and scientific analyses; and

(5) The experiences of FSA, private, and public parties that have taken similar actions.

(b) FSA will consult with CEQ and appropriate Federal agencies while developing or modifying a categorical exclusion.

(c) Before establishing a new final categorical exclusion, FSA will follow the CEQ specified process for establishing Categorical Exclusions, including consultation with CEQ and an opportunity for public review and comment as required by 40 CFR 1507.3.

(d) FSA will maintain an administrative record that includes the supporting information and findings used in establishing a categorical exclusion.

(e) FSA will periodically review its categorical exclusions to identify and revise exclusions that no longer effectively reflect environmental circumstances or current FSA program scope.

(f) FSA will use the same process specified in this section and the results of its periodic reviews to revise a categorical exclusion or remove a categorical exclusion.

Subpart E—Environmental Assessments

§ 799.40 Purpose of an EA.

(a) FSA prepares an EA to determine whether a proposed action would significantly affect the environment, and to consider the potential impacts of reasonable alternatives and the potential mitigation measures to the alternatives and proposed action.

(b) FSA will prepare a PEA to determine if proposed actions that are broad in scope or similar in nature have cumulative significant environmental impacts, although the impacts of the proposed actions may be individually insignificant.

(c) The result of the EA process will be either a FONSI or a determination that an EIS is required. FSA may also determine that a proposed action will significantly affect the environment without first preparing an EA; in that case, an EIS is required.

§ 799.41 When an EA is required.

(a) Proposed actions that require the preparation of an EA include the following:

(1) New Conservation Reserve Enhancement Program (CREP) agreements;

(2) Development of farm ponds or lakes greater than or equal to 20 acres;

(3) Restoration of wetlands greater than or equal to 100 acres aggregate;

(4) Installation or enlargement of irrigation facilities, including storage reservoirs, diversions, dams, wells, pumping plants, canals, pipelines, and sprinklers designed to irrigate greater than 320 acres aggregate;

(5) Land clearing operations (for example, vegetation removal, including tree stumps; grading) involving greater than or equal to 40 acres aggregate;

(6) Clear cutting operations for timber involving greater than or equal to 100 acres aggregate;

(7) Construction or major enlargement of a Concentrated Aquatic Animal Production Facility (CAAP), as defined by the U.S. Environmental Protection Agency in 40 CFR 122.24;

(8) Construction of commercial facilities or structures for processing or handling of farm production or for public sales;

(9) Construction or major expansion of a large CAFO, as defined by the U.S. Environmental Protection Agency in 40 CFR 122.23, regardless of the type of manure handling system or water system;

(10) Refinancing of a newly constructed large CAFO, as defined by the U.S. Environmental Protection Agency in 40 CFR 122.23, or CAAPs as defined by the U.S. Environmental Protection Agency in 40 CFR 122.24 through 122.25, that has been in operation for 24 months or less;

(11) Issuance of substantively discretionary FSA regulations, FEDERAL REGISTER notices, or amendments to existing programs that authorize FSA or CCC funding for proposed actions that have the potential to significantly affect the human environment;

(12) Newly authorized programs that involve substantively discretionary proposed actions and are specified in § 799.32(d);

(13) Any FSA proposed action that has been determined to trigger extraordinary circumstances specified in § 799.33(c); and

(14) Any proposed action that will involve the planting of a potentially invasive species, unless exempted by Federal law.

(b) Proposed actions that do not reach the thresholds defined in paragraph (a) of this section, unless otherwise identified under § 799.31(b) or § 799.32(c), require a review using the ESW to determine if an EA is warranted.

§ 799.42 Contents of an EA.

(a) The EA should include at least the following:

(1) FSA cover sheet;

(2) Executive summary;

(3) Table of contents;

(4) List of acronyms;

(5) A discussion of the purpose of and need for the proposed action;

(6) A discussion of alternatives, if the proposed action involves unresolved conflicts concerning the uses of available resources;

(7) A discussion of the existing pre-project environment and the potential environmental impacts of the proposed action, with reference to the significance of the impact as specified in § 799.8 and 40 CFR 1508.27;

(8) Likelihood of any significant impact and potential mitigation measures that FSA will require, if needed, to support a FONSI;

(9) A list of preparers and contributors;

(10) A list of agencies, tribes, groups, and persons solicited for feedback and the process used to solicit that feedback;

(11) References; and

(12) Appendixes, if appropriate.

(b) FSA will prepare a Supplemental EA, and place the supplements in the administrative record of the original EA, if:

(1) Substantial changes occur in the proposed action that are relevant to environmental concerns previously presented, or

(2) Significant new circumstances or information arise that are relevant to environmental concerns and to the proposed action or its impacts.

(c) FSA may request that a program participant prepare or provide information for FSA to use in the EA and may use the program participant's information in the EA or Supplemental EA, provided that FSA also:

(1) Independently evaluates the environmental issues;

(2) Takes responsibility for the scope and content of the EA and the process utilized, including any required public involvement; and

(3) Prepares the FONSI or NOI to prepare an EIS.

§ 799.43 Tiering.

(a) As specified in 40 CFR 1508.28, tiering is a process of covering general environmental review in a broad PEA, followed by subsequent narrower scope analysis to address specific proposed actions, action stages, or sites. FSA will use tiering when FSA prepares a broad PEA and subsequently prepares a site-specific ESW, EA, or PEA for a proposed action included within the program addressed in the original, broad PEA.

(b) When FSA uses tiering in a broad PEA, the subsequent ESW, EA, or PEA will:

(1) Summarize the issues discussed in the broader statement;

(2) Incorporate by reference the discussions from the broader statement and the conclusions carried forward into the subsequent tiered analysis and documentation; and

(3) State where the PEA document is available.

§ 799.44 Adoption of an EA prepared by another entity.

(a) FSA may adopt an EA prepared by another Federal agency, State, or Tribal government if the EA meets the requirements of this subpart.

(b) If FSA adopts another agency's EA and issues a FONSI, FSA will follow the procedures specified in § 799.44.

§ 799.45 Finding of No Significant Impact (FONSI).

(a) If after completing the EA, FSA determines that the proposed action will not have a significant effect on the quality of the human environment, FSA will issue a FONSI.

(b) The FONSI will include the reasons FSA determined that the proposed action will have no significant environmental impacts.

(c) If the decision to issue the FONSI is conditioned upon the implementation of measures (mitigation actions) to ensure that impacts will be held to a nonsignificant level, the FONSI must include an enforceable commitment to implement such measures on the part of FSA, and any applicant or other party responsible for implementing the measures will be responsible for the commitments outlined in the FONSI.

Subpart E—Environmental Impact Statements

§799.50 Purpose of an Environmental Impact Statement (EIS).

(a) FSA will prepare an EIS for proposed actions that are expected to have a significant effect on the human environment. The purpose of the EIS is to ensure that all significant environmental impacts and reasonable alternatives are fully considered in connection with the proposed action.

(b) FSA will prepare a PEIS for proposed actions that are broad in scope or similar in nature and may cumulatively have significant environmental impacts, although the impact of the individual proposed actions may be insignificant.

§799.51 When an EIS is required.

(a) The following FSA proposed actions normally require preparation of an EIS:

(1) Legislative proposals, not including appropriations requests, with the potential for significant environmental impact that are drafted and submitted to Congress by FSA;

(2) Broad Federal assistance programs administered by FSA, involving significant financial assistance or payments to program participants, that may have significant cumulative impacts on the human environment; and

(3) Ongoing programs that have been found through previous environmental analyses to have major environmental concerns.

(b) [Reserved]

§799.52 Notice of intent to prepare an EIS.

(a) FSA will publish a Notice of Intent to prepare an EIS in the FEDERAL REGISTER and, depending on the scope of the proposed action, may publish a notice in other media.

(b) The notice will include the following:

(1) A description of the proposed action and possible alternatives;

(2) A description of FSA's proposed scoping process, including information about any public meetings; and

(3) The name of an FSA point of contact who can receive input and answer questions about the proposed action and the preparation of the EIS.

§799.53 Contents of an EIS.

(a) FSA will prepare the EIS as specified in 40 CFR part 1502 and in section 102 of NEPA (42 U.S.C. 4332).

(b) The EIS should include at least the following:

(1) An FSA cover sheet;

(2) An executive summary explaining the major conclusions, areas of controversy, and the issues to be resolved;

(3) A table of contents;

(4) List of acronyms and abbreviations;

(5) A brief statement explaining the purpose and need of the proposed action;

(6) A detailed discussion of the environmental impacts of the proposed action and reasonable alternatives to the proposed action, a description and brief analysis of the alternatives considered but eliminated from further consideration, the no-action alternative, FSA's preferred alternative(s), and discussion of appropriate mitigation measures;

(7) A discussion of the affected environment;

(8) A detailed discussion of:

(i) The direct and indirect environmental consequences, including any cumulative impacts, of the proposed action and of the alternatives;

(ii) Unavoidable adverse environmental effects;

(iii) The relationship between local short-term uses of the environment and long-term ecosystem productivity;

(iv) Any irreversible and irretrievable commitments of resources;

(vi) Possible conflicts with the objectives of Federal, regional, State, local, regional, and Tribal land use plans, policies, and controls for the area concerned;

(vii) Energy and natural depletable resource requirements, including, but not limited to natural gas and oil, and conservation potential of the alternatives and mitigation measures; and

(viii) Urban quality, historic, and cultural resources and the design of the built environment, including the reuse and conservation potential of the alternatives and mitigation measures;

(9) In the draft EIS, a list of all Federal permits, licenses, and other entitlements that must be obtained for implementation of the proposed action;

(10) A list of preparers;

(11) Persons and agencies contacted;

(12) References, if appropriate;

(13) Glossary, if appropriate;

(14) Index;

(15) Appendixes, if any;

(16) A list of agencies, organizations, and persons to whom copies of the EIS are sent; and

(17) In the final EIS, a response to substantive comments on environmental issues.

(c) FSA may have a contractor prepare an EIS as specified in 40 CFR 1506.5(c). If FSA has a contractor prepare an EIS, FSA will:

(1) Require the contractor to sign a disclosure statement specifying it has no financial or other interest in the outcome of the proposed action, which will be included in the administrative record; and

(2) Furnish guidance and participate in the preparation of the EIS, and independently evaluate the EIS before its approval.

§ 799.54 Draft EIS.

(a) FSA will prepare the draft EIS addressing the information specified in § 799.53.

(b) FSA will circulate the draft EIS as specified in 40 CFR 1502.19.

(c) In addition to the requirements of 40 CFR 1502.19, FSA will request comments on the draft EIS from:

(1) Appropriate State and local agencies authorized to develop and enforce environmental standards relevant to the scope of the EIS;

(2) Tribal governments that have interests that could be impacted; and

(3) If the proposed action affects historic properties, the appropriate SHPO, THPO, and the Advisory Council on Historic Preservation.

(d) FSA will file the draft EIS with the U.S. Environmental Protection Agency as specified in 40 CFR 1506.9 and in accordance with U.S. Environmental Protection Agency filing requirements (available at *http://www.epa.gov/compliance/nepa/submiteis/index.html*).

(e) The draft EIS will include a cover sheet with the information specified in 40 CFR 1502.11.

(f) FSA will provide for a minimum 45-day comment period calculated from the date the U.S. Environmental Protection Agency publishes the NOA of the draft EIS.

§ 799.55 Final EIS.

(a) FSA will prepare the final EIS addressing the information specified in § 799.53.

(b) FSA will evaluate the comments received on the draft EIS and respond in the final EIS as specified in 40 CFR 1503.4. FSA will discuss in the final EIS any issues raised by commenters that were not discussed in the draft EIS and provide a response to those comments.

(c) FSA will attach substantive comments, or summaries of lengthy comments, to the final EIS and will include all comments in the administrative record.

(d) FSA will circulate the final EIS as specified in 40 CFR 1502.19.

(e) FSA will file the final EIS with the U.S. Environmental Protection Agency as specified in 40 CFR 1506.9.

(f) The final EIS will include a cover sheet with the information specified in 40 CFR 1502.11.

§ 799.56 Supplemental EIS.

(a) FSA will prepare supplements to a draft or final EIS if:

(1) Substantial changes occur in the proposed action that are relevant to environmental concerns; or

(2) Significant new circumstances or information arise that are relevant to environmental concerns and bearing on the proposed action or its impacts.

(b) The requirements of this subpart for completing the original EIS apply to the supplemental EIS, with the exception of the scoping process, which is optional.

§ 799.57 Tiering.

(a) As specified in 40 CFR 1508.28, tiering is a process of covering general environmental review in a broad PEIS, followed by subsequent narrower scope analysis to address specific proposed actions, action stages, or sites. FSA will use tiering when FSA prepares a broad PEIS and subsequently prepares a site-specific ESW, EA, or PEA for a proposed action included within the program addressed in the original, broad PEIS.

(b) When FSA uses tiering in a broad PEIS, the subsequent ESW, EA, or PEA will:

(1) Summarize the issues discussed in the broader statement;

(2) Incorporate by reference the discussions from the broader statement and the conclusions carried forward into the subsequent tiered analysis and documentation; and

(3) State where the PEIS document is available.

§799.58 Adoption of an EIS prepared by another entity.

(a) FSA may elect to adopt an EIS prepared by another Federal agency, State, or Tribal government if:

(1) The NECM determines that the EIS and the analyses and procedures by which they were developed meet the requirements of this part; and

(2) The agency responsible for preparing the EIS concurs.

(b) For the adoption of another Federal agency EIS, FSA will follow the procedures specified in the CEQ regulations in 40 CFR 1506.3.

(c) For the adoption of an EIS from a state or tribe that has an established state or tribal procedural equivalent to the NEPA process (generally referred to as "mini-NEPA"), FSA will follow the procedures specified in the CEQ regulations in 40 CFR 1506.3.

§799.59 Record of Decision.

(a) FSA will issue a Record of Decision (ROD) within the time periods specified in 40 CFR 1506.10(b) but no sooner than 30 days after the U.S. Environmental Protection Agency's publication of the NOA of the final EIS. The ROD will:

(1) State the decision reached;

(2) Identify all alternatives considered by FSA in reaching its decision, specifying the alternative or alternatives considered to be environmentally preferable;

(3) Identify and discuss all factors, including any essential considerations of national policy, which were considered by FSA in making its decision, and state how those considerations entered into its decision; and

(4) State whether all practicable means to avoid or minimize environmental harm from the alternative selected have been adopted and, if not, explain why these mitigation measures were not adopted. A monitoring and enforcement program will be adopted and summarized where applicable for any mitigation.

(b) FSA will distribute the ROD to all parties who request it.

(c) FSA will publish the ROD or a notice of availability of the ROD in the FEDERAL REGISTER.

CHAPTER VIII—AGRICULTURAL MARKETING SERVICE (FEDERAL GRAIN INSPECTION SERVICE, FAIR TRADE PRACTICES PROGRAM), DEPARTMENT OF AGRICULTURE

SUBCHAPTER A—FEDERAL GRAIN INSPECTION

AUTHORITY: 7 U.S.C. 71–87K, 1621–1627.

EDITORIAL NOTE: Nomenclature changes to subchapter A appear at 84 FR 45645, Aug. 30, 2019.

PART 800—GENERAL REGULATIONS

AUTHORITY: 7 U.S.C. 71–87k.

SOURCE: 45 FR 15810, Mar. 11, 1980, unless otherwise noted.

DEFINITIONS

§ 800.0 Meaning of terms.

(a) *Construction.* Words used in the singular form shall be considered to imply the plural and vice versa, as appropriate. When a section; e.g., § 800.2, is cited, it refers to the indicated section in these regulations.

(b) *Definitions.* For the purpose of these regulations, unless the context requires otherwise, the following terms shall have the meanings given for them below. The terms defined in the Act have been incorporated herein for easy reference.

Act. The United States Grain Standards Act, as amended (39 Stat. 482–485, as amended 7 U.S.C. 71 *et seq.*).

Additives. Materials approved by the Food and Drug Administration or the Environmental Protection Agency and added to grain for purposes of insect and fungi control, dust suppression, or identification.

Administrator. The Administrator of the Agricultural Marketing Service or any person to whom authority has been delegated.

Agency. A delegated State or an official agency designated by the Administrator, as appropriate.

Appeal inspection service. An official review by a field office of the results of an original inspection service or a reinspection service.

Applicant. An interested person who requests an official inspection or a Class X or Class Y weighing service.

Approved scale testing organization. A State or local governmental agency, or person, approved by the Service to perform official equipment testing services with respect to weighing equipment.

Approved weigher. A person employed by or at an approved weighing facility and approved by the Service to physically perform Class X or Class Y weighing services, and certify the results of Class Y weighing.

Approved weighing equipment. Any weighing device or related equipment approved by the Service for the performance of Class X or Class Y weighing services.

Approved weighing facility. An elevator that is approved by the Service to receive Class X or Class Y weighing services.

Assigned area of responsibility. A geographical area assigned to an agency or to a field office for the performance of official inspection or Class X or Class Y weighing services.

Average grade. Multiple carrier units or sublots that are graded individually then averaged to form a single lot inspection.

Board appeal inspection service. An official review by the Board of Appeals and Review of the results on an appeal inspection service.

Board of Appeals and Review. The Board of Appeals and Review of the Service.

Business day. The established field office working hours, any Monday through Friday that is not a holiday, or the working hours and days established by an agency.

Cargo shipment. Bulk or sacked grain that is loaded directly aboard waterborne carrier for shipment. Grain loaded aboard a land carrier for shipment aboard a waterborne carrier shall not be considered to be a cargo shipment.

Carrier. A truck, trailer, truck/trailer(s) combination, railroad car, barge, ship, or other container used to transport bulk or sacked grain.

Chapter. Chapter VIII of the Code of Federal Regulations (7 CFR chapter VIII).

Circuit. A geographical area assigned to a field office.

Class X or Class Y weighing equipment testing. Any operation or procedure performed by official personnel to determine the accuracy of the equipment used, or to be used, in the performance of Class X or Class Y weighing services.

Combined lot. Grain loaded aboard, or being loaded aboard, or discharged from two or more carriers as one lot.

Compliance. Conformance with all requirements and procedures established by statute, regulation, instruction, or directive so that managerial, administrative, and technical functions are accomplished effectively. Compliance functions include: evaluating alleged violations, initiating preliminary investigations; initiating implementation of all necessary corrective actions; conducting management and technical reviews; administering the designation of agencies and the delegation of State agencies to perform official functions; identifying and, where appropriate, waiving and monitoring conflicts of interest; licensing agency personnel; responding to audits of FGIS programs; and reviewing and, when appropriate, approving agency fee schedules.

Composite grade. Multiple samples obtained from the same type of carriers (*e.g.,* trucklots, containers) that are combined into one sample for grade to form a single lot inspection.

Container. A carrier, or a bin, other storage space, bag, box, or other receptacle for grain.

Contract grade. The official grade, official factors, or official criteria specified in a contract for sale or confirmation of sale; or in the absence of a contract the official grade, official factors, or official criteria specified by the applicant for official service.

Contract service. An inspection or weighing service performed under a contract between an applicant and the Service.

Contractor. A person who enters into a contract with the Service for the performance of specified official inspection or official monitoring services.

Date of official inspection service or Class X or Class Y weighing services. The day on which an official inspection, or a Class X or Class Y weighing service is completed. For certification purposes, a day shall be considered to end at midnight, local time.

Deceptive loading, handling, weighing, or sampling. Any manner of loading, handling, weighing, or sampling that knowingly deceives or attempts to deceive official personnel.

Delegated State. A State agency delegated authority under the Act to provide official inspection service, or Class X or Class Y weighing services, or both, at one or more export port locations in the State.

Department of Agriculture and Department. The United States Department of Agriculture (USDA).

Designated agency. A State or local governmental agency, or person, designated under the Act to provide either official inspection service, or Class X or Class Y weighing services, or both, at locations other than export port locations.

Door-probe sample. A sample taken with a probe from a lot of bulk grain that is loaded so close to the top of the carrier that it is possible to insert the probe in the grain only in the vicinity of the tailgate of the truck or trailer, the door of the railroad boxcar, or in a similarly restricted opening or area in the carrier in which the grain is located or is loaded in hopper cars or barges in such a manner that a representative sample cannot be obtained.

Elevator. Any warehouse, storage, or handling facility used primarily for receiving, storing, or shipping grain. In a facility that is used primarily for receiving, storing, and shipping grain, all parts of the main facility, as well as annexes, shall be considered to be part of the elevator. A warehouse, storage, and handling facility that is located adjacent to and is operated primarily as an adjunct of a grain processing facility shall not be considered to be an elevator.

Elevator areas and facilities. All operational areas, including the automated data processing facilities that are an integral part of the inspection or weighing operations of an elevator; loading and unloading docks; the headhouse and control rooms; all storage areas, including the bins, the interstices, the bin floor, and the basement; and all handling facilities, including the belts, other conveyors, distributor scales, spouting, mechanical samplers, and electronic controls.

Emergency. A situation that is outside the control of the applicant that prevents official inspection or weighing services within 24 hours of the scheduled service time.

Employed. An individual is employed if the individual is actually employed or the employment is being withheld pending issuance of a license under the Act.

Exporter. Any person who ships or causes to be shipped any bulk or sacked grain in a final carrier or container in which the grain is transported from the United States to any place outside the United States.

Export elevator. Any grain elevator, warehouse, or other storage or handling facility in the United States (i) from which bulk or sacked export grain is loaded (A) aboard a carrier in which the grain is shipped from the United States to any place outside thereof, or (B) into a container for shipment to an export port location where the grain and the container will be loaded aboard a carrier in which it will be shipped from the United States to any place outside thereof; and (ii) which has been approved by the Service as a facility where Class X or Class Y weighing of grain may be obtained.

Export grain. Grain for shipment from the United States to any place outside thereof.

Export port location. A commonly recognized port of export in the United States or Canada, as determined by the Administrator, from which grain produced in the United States is shipped to any place outside the United States. Such locations include any coastal or border location or site in the United States which contains one or more export elevators, and is identified by the Service as an export port location.

False, incorrect, and misleading. Respectively false, incorrect, and misleading in any particular.[2]

Federal Register. An official U.S. Government publication issued under the Federal Register Act of July 26, 1935, as amended (44 U.S.C. 301 *et seq.*).

Field Office. An office of the Service designated to perform or supervise official inspection services and Class X and Class Y weighing services.

Field Office administrative costs. The costs of management, support, and maintenance of a Field Office, including, but not limited to, the management and administrative support personnel, rent, and utilities. This does not include any costs directly related to providing original or review inspection or weighing services.

Grain. Corn, wheat, rye, oats, barley, flaxseed, sorghum, soybeans, triticale, mixed grain, sunflower seed, canola, and any other food grains, feed grains, and oilseeds for which standards are established under section 4 of the Act.

Handling. Loading, unloading, elevating, storing, binning, mixing, blending, drying, aerating, screening, cleaning, washing, treating, or fumigating grain.

High quality specialty grain. Grain sold under contract terms that specify all factors exceed the grade limits for U.S. No. 1 grain, except for the factor test weight, or specify "organic" as defined by 7 CFR part 205.

Holiday. The legal public holidays specified in paragraph (a) of section 6103, Title 5, of the United States Code (5 U.S.C. 6103(a)) and any other day declared to be a holiday by Federal statute or Executive Order. Under section 610 and Executive Order No. 10357, as amended, if the specified legal public holiday falls on a Saturday, the preceding Friday shall be considered to be the holiday, or if the specified legal public holiday falls on a Sunday, the following Monday shall be considered to be the holiday.

"IN" movement. A movement of grain into an elevator, or into or through a city, town, port, or other location without a loss of identity.

Instructions. The Notices, Instructions, Handbooks, and other directives issued by the Service.

Interested person. Any person having a contract or other financial interest in grain as the owner, seller, purchaser, warehouseman, or carrier, or otherwise.

Interstate or foreign commerce. Commerce from any State to or through any other State, or to or through any foreign country.

Licensee. Any person licensed by the Service.

Loading. Placing grain in or aboard any carrier or container.

("LOCAL" movement. A bin run or other inhouse movement, or grain in bins, tanks, or similar containers which are not in transit or designed to transport grain

Lot. A specific quantity of grain identified as such.

Material error. An error in the results of an official inspection service that exceeds the official tolerance, or any error in the results of a Class X or Class Y weighing service

Material portion. A subsample, component, or sublot which is determined to be inferior to the contract or declared grade. A subsample is a material portion when it has sour, musty, or commercially objectionable foreign odors,

---

[1] [Reserved]

[2] A definition taken from the U.S. Grain Standards Act, as amended, with certain modifications which do not change the meanings.

when it is heating; or when it is of distinctly low quality. A component is a material portion when it is infested or when it is determined to be inferior in quality by more than one numerical grade to the contract or declared grade. A sublot is a material portion when a factor result causes a breakpoint to be exceeded or when a factor result exceeds specific sublot contract requirements. A sublot designated a material portion shall include only one sublot.

Merchandiser. Any person, other than a producer, who buys and sells grain and takes title to the grain. A person who operates as a broker or commission agent and does not take title to the grain shall not be considered to be a merchandiser.

Monitoring. Observing or reviewing activities performed under or subject to the Act for adherence to the Act, the regulations, standards, and instructions and preparing reports thereon.

National program administrative costs. The costs of national management and support of official grain inspection and/or weighing. This does not include the Field Office administrative costs and any costs directly related to providing service.

Nonregular workday. Any Sunday or holiday.

Official agency. Any State or local government agency, or any person, designated by the Administrator pursuant to subsection (f) of section 7 of the Act for the conduct of official inspection (other than appeal inspection), or subsection (c) of section 7A of the Act for the conduct of Class X or Class Y weighing (other than review of weighing).

Official certificate. Those certificates which show the results of official services performed under the Act as provided in the instructions, and any other official certificates which may be approved by the Service in accordance with the instructions.

Official criteria. A quantified physical or chemical property of grain that is approved by the Service to determine the quality or condition of grain or other facts relating to grain.

Official factor. A quantified physical or chemical property of grain as identified in the Official U.S. Standards for Grain.

Official forms. License, authorizations, and approvals; official certificates; official pan tickets; official inspection or weighing logs; weight sheets; shipping bin weight loading logs; official equipment testing reports; official certificates of registration; and any other forms which may be issued or approved by the Service that show the name of the Service or an agency and a form number.

Official grade designation. A numerical or sample grade designation, specified in the standards relating to kind, class, quality, and condition of grain provided for in the Act.

Official inspection. The determination (by original inspection, and when requested, reinspection and appeal inspection) and the certification, by official personnel, of the kind, class, quality, or condition of grain, under standards provided for in the Act; or the condition of vessels and other carriers or receptacles for the transportation of grain insofar as it may affect the quality of such grain; or other facts relating to grain under other criteria approved by the Administrator (the term "officially inspected" shall be construed accordingly).

Official inspection equipment testing. Any operation or procedure by official personnel to determine the accuracy of equipment used, or to be used, in the performance of official inspection services.

Official inspection technician. Any official personnel who perform or supervise the performance of specified official inspection services and certify the results thereof, other than certifying the grade of the grain.

Official inspector. Any official personnel who perform or supervise the performance of official inspection services and certify the results thereof including the grade of the grain.

Official marks. The symbols or terms "official certificate," "official grade," "officially sampled," "officially inspected," "official inspection," "U.S. inspected," "loaded under continuous official inspection," "official weighing," "officially weighed," "official

485

weight," "official supervision of weighing," "supervision of weighing," "officially supervised weight," "loaded under continiuous official weighing," "loaded under continuous official inspection and weighing," "officially tested," "Class X weight," "official Class X weighing," "Class X weighing," "official Class Y weighing," "Class Y weighing," and "Class Y weight."

Official personnel. Persons licensed or otherwise authorized by the Administrator pursuant to Section 8 of the Act to perform all or specified functions involved in official inspection, Class X or Class Y weighing, or in the supervision of official inspection, or Class X or Class Y weighing.

Official sample. A sample obtained from a lot of grain by, and submitted for official inspection by, official personnel (the term "official sampling" shall be construed accordingly).

Official sampler. Any official personnel who perform or supervise the performance of official sampling services and certify the results thereof.

Official stowage examination. Any examining operation or procedure performed by official personnel to determine the suitability of a carrier or container to receive or store grain.

Official tolerance. A statistical allowance prescribed by the Service, on the basis of expected variation, for use in performing or supervising the performance of official inspection services, official equipment testing services, and, when determined under an established loading plan, reinspection services and appeal inspection services.

Official U.S. Standards for Grain. The Official U.S. Standards for Grain established under the Act describe the physical and biological condition of grain at the time of inspection.

Official weigher. Any official personnel who perform or supervise the performance of Class X or Class Y weighing services and certify the results thereof, including the weight of the grain.

Official weighing. (Referred to as Class X weighing.) The determination and certification by official personnel of the quantity of a lot of grain under standards provided for in the Act, based on the actual performance of weighing or the physical supervision

thereof, including the physical inspection and testing for accuracy of the weights and scales and the physical inspection of the premises at which weighing is performed and the monitoring of the discharge of grain into the elevator or conveyance. (The terms "officially weigh" and "officially weighed" shall be construed accordingly.)

Official weighing technician. Any personnel who perform or supervise specified weighing services and certify the results thereof other than certifying the weight of grain.

Official weight sample. Sacks of grain obtained at random by, or under the complete supervision of, official personnel from a lot of sacked grain for the purpose of computing the weight of the grain in the lot.

Operating expenses. The total costs to the Service to provide official grain inspection and/or weighing services.

Operating reserve. The amount of funds the Service has available to provide official grain inspection and/or weighing services.

Original inspection. An initial official inspection of grain.

"Out" movement. A movement of grain out of an elevator or out of a city, town, port, or other location.

Person. Any individual, partnership, corporation, association, or other business entity.

Quantity. Pounds or kilograms, tons or metric tons, or bushels.

Reasonably continuous operation. A loading or unloading operation in one specific location which does not include inactive intervals in excess of 88 consecutive hours.

Regular workday. Any Monday, Tuesday, Wednesday, Thursday, Friday, or Saturday that is not a holiday.

Regulations. The regulations in parts 800, 801, and 802 of this chapter.

Reinspection service. An official review of the results of an original inspection service by the agency or field office that performed the original inspection service.

Respondent. The party proceeded against.

Review of weighing service. An official review of the results of a Class X or Class Y weighing service.

Secretary. The Secretary of Agriculture of the United States or any person to whom authority has been delegated.

Service. The Federal Grain Inspection Service of the Agricultural Marketing Service of the United States Department of Agriculture.

Service representative. An authorized salaried employee of the Service; or a person licensed by the Administrator under a contract with the Service.

Shallow-probe sample. A sample taken with a probe from a lot of bulk grain that is loaded so close to the top of the carrier that it is possible to insert the probe in the grain at the prescribed locations, but only at an angle greater or more obtuse from the vertical than the angle prescribed in the instructions.

Ship. The verb "ship" with respect to grain means transfer physical possession of the grain to another person for the purpose of transportation by any means of conveyance, or transport one's own grain by any means of conveyance.

Shiplot grain. Grain loaded aboard, or being loaded aboard, or discharged from an ocean-going vessel including a barge, lake vessel, or other vessel of similar capacity.

Shipper's Export Declaration. The Shipper's Export Declaration certificate filed with the U.S. Department of Commerce, Bureau of Census.

Specified service point. A city, town, or other location specified by an agency for the performance of official inspection or Class X or Class Y weighing services and within which the agency or one or more of its inspectors or weighers is located.

Standardization. The act, process, or result of standardizing methodology and measurement of quality and quantity. Standardization functions include: compiling and evaluating data to develop and to update grading and weighing standards, developing or evaluating new methodology for determining grain quality and quantity, providing reference standards for official grading methods, and reviewing official results through the use of a quality control and weight monitoring program.

State. Any one of the States (including Puerto Rico) or territories or possessions of the United States (including the District of Columbia).

Submitted sample. A sample submitted by or for an interested person for official inspection, other than an official sample.

Supervision. The effective guidance of agencies, official personnel and others who perform activities under the Act, so as to reasonably assure the integrity and accuracy of the program activities. Supervision includes overseeing, directing, and coordinating the performance of activities under the Act, reviewing the performance of these activities; and effecting appropriate action. FGIS supervisory personnel supervise agencies, official personnel and others who perform activities under the Act. Agency supervisors are responsible for the direct supervision of their own official personnel and employees. FGIS provides oversight, guidance, and assistance to agencies as they carry out their responsibilities.

Supervision of weighing (Referred to as Class Y weighing.) Such supervision by official personnel of the grain-weighing process as is determined by the Administrator to be adequate to reasonably assure the integrity and accuracy of the weighing and of certificates which set forth the weight of the grain and such physical inspection by such personnel of the premises at which the grain weighing is performed as will reasonably assure that all the grain intended to be weighed has been weighed and discharged into the elevator or conveyance.

United States. The States (including Puerto Rico) and the territories and possessions of the United States (including the District of Columbia).

Use of official inspection service. The use of the services provided under a delegation or designation or provided by the Service.

Uniform in quality. A lot of grain in which there are no material portions.

Warehouseman's sampler. An elevator employee licensed by the Service to obtain samples of grain for a warehouseman's sample-lot inspection service. Warehouseman's samplers are not considered official personnel, but they are licensed under authority of section 11 of the Act.

[45 FR 15810, Mar. 11, 1980, as amended at 49 FR 36068, Sept. 14, 1984; 49 FR 37055, Sept. 21, 1984; 49 FR 49586, Dec. 21, 1984; 52 FR 6495, Mar. 4, 1987; 55 FR 24041, June 13, 1990; 57 FR 3273, Jan. 29, 1992; 60 FR 5835, Jan. 31, 1995; 70 FR 21923, Apr. 28, 2005; 70 FR 73558, Dec. 13, 2005; 75 FR 41695, July 19, 2010; 76 FR 45399, July 29, 2011; 78 FR 43755, July 22, 2013; 81 FR 49859, July 29, 2016]

ADMINISTRATION

§ 800.1 Mission.

The mission of the Federal Grain Inspection Service is to facilitate the marketing of grain, oilseeds, pulses, rice, and related commodities by:

(a) Establishing descriptive standards and terms,

(b) Accurately and consistently certifying quality,

(c) Providing for uniform official inspection and weighing,

(d) Carrying out assigned regulatory and service responsibilities, and

(e) Providing the framework for commodity quality improvement incentives to both domestic and foreign buyers.

[54 FR 9197, Mar. 6, 1989]

§ 800.2 Administrator.

The Administrator is delegated, from the Secretary, responsibility for administration of the United States Grain Standards Act and responsibilities under the Agricultural Marketing Act of 1946 (7 U.S.C. 1621 et seq.). The Administrator is responsible for the establishment of policies, guidelines, and regulations by which the Service is to carry out the provisions of the Act and the Agricultural Marketing Act of 1946. The regulations promulgated under the Agricultural Marketing Act of 1946 appear at part 68 of this title (7 CFR part 68). The Administrator is authorized by the Secretary to take any action required by law or considered to be necessary and proper to the discharge of the functions and services under the Act. The Administrator may delegate authority to the Deputy Administrator and other appropriate officers and employees. The Administrator may, in emergencies or other circumstances which would not impair the objectives

of the Act, suspend for period determined by the Administrator any provision of the regulations or official grain standards. The Administrator may authorize research; experimentation; and testing of new procedures, equipment, and handling techniques to improve the inspection and weighing of grain. The Administrator may waive the official inspection and official weighing requirements pursuant to Section 5 of the Act.

[60 FR 5835, Jan. 31, 1995]

§ 800.3 Nondiscrimination—policy and provisions.

In implementing, administering, and enforcing the Act and the regulations, standards, and instructions, it is the policy of the Service to promote adherence to the provisions of the Civil Rights Act of 1964 (42 U.S.C. 2000a et seq.), (Pub. L. 88–352).

§ 800.4 Procedures for establishing regulations, official standards, and official criteria.

Notice of proposals to prescribe, amend, or revoke regulations, official standards, and official criteria under the Act shall be published in accordance with applicable provisions of the Administrative Procedure Act (5 U.S.C. 551, et seq.). Proposals to establish, amend, or revoke grain standards will be made effective not less than 1 calendar year after promulgation unless, for good cause, the Service determines that the public health, interest, or safety require that they become effective sooner. Any interested person desiring to file a petition for the issuance, amendment, or revocation of regulations, Official U.S. Standards for Grain, or official criteria may do so in accordance with § 1.28 of the regulations of the Office of the Secretary of Agriculture (7 CFR 1.28).

§ 800.5 Complaints and reports of alleged violations.

(a) *General.* Except as provided in paragraphs (b) and (c) of this section, complaints and reports of violations involving the Act or the regulations, standards, and instructions issued under the Act should be filed with the Service in accordance with § 1.133 of the regulations of the Office of the Secretary of Agriculture (7 CFR 1.133) and with the regulations and the instructions.

(b) *Reinspection, review of weighing, and appeal services.* Complaints involving the results of official inspection or Class X or Class Y weighing services shall, to the extent practicable, be submitted as requests for a reinspection service, a review of weighing service, an appeal inspection service, or a Board appeal inspection service as set forth in these regulations.

(c) *Foreign buyer complaints.* Inquiries or complaints from importers or other purchasers in foreign countries involving alleged discrepancies in the quality or weight of officially inspected or Class X weighed export grain shall, to the extent possible, be submitted by the importers or purchasers to the appropriate U.S. Agricultural Attache in accordance with § 2.68(a)(14) of the regulations of the Office of the Secretary of Agriculture (7 CFR 2.68(a)(14)) and the instructions issued by the Foreign Agricultural Service of the Department.

[45 FR 15810, Mar. 11, 1980, as amended at 54 FR 5924, Feb. 7, 1989]

§ 800.6 Provisions for hearings.

Opportunities will be provided for hearings prescribed or authorized by sections 7(g)(3), 7A(c)(2), 9, 10(d), and 17A(d) of the Act, and the hearings shall be conducted in accordance with the Rules of Practice Governing Formal Adjudicatory Administrative Proceedings Instituted by the Secretary under Various Statutes (7 CFR, part 1, subpart H).

§ 800.7 Information about the Service, Act, and regulations.

Information about the Agricultural Marketing Service, Service, Act, regulations, official standards, official criteria, rules of practice, instructions, and other matters related to the official inspection or Class X or Class Y weighing of grain may be obtained by telephoning or writing the Service at its headquarters or any one of its field offices at the numbers and addresses listed on the Service's website.

[84 FR 45646, Aug. 30, 2019]

§ 800.8 Public information.

(a) *General.* This section is issued in accordance with §§ 1.1 through 1.23 of the regulations of the Secretary of Agriculture in part 1, subpart A, of subtitle A of title 7 (7 CFR 1.1 through 1.23), and appendix A thereto, implementing the Freedom of Information Act (5 U.S.C. 552). The Secretary's regulations, as implemented by this section, govern the availability of records of the Service to the public.

(b) *Public inspection and copying.* Materials maintained by the Service, including those described in 7 CFR 1.5, will be made available, upon a request which has not been denied, for public inspection and copying at the U.S. Department of Agriculture, Agricultural Marketing Service, at 14th Street and Independence Avenue, SW., Washington, D.C. 20250. The public may request access to these materials during regular working hours, 8:00 a.m. to 4:30 p.m., est, Monday through Friday except for holidays.

(c) *Indexes.* FGIS shall maintain an index of all material required to be made available in 7 CFR 1.5. Copies of these indexes will be maintained at the location given in paragraph (b) of this section. Notice is hereby given that quarterly publication of these indexes is unnecessary and impracticable, because the material is voluminous and does not change often enough to justify the expense of quarterly publication. However, upon specific request, copies of any index will be provided at a cost not to exceed the direct cost of duplication.

(d) *Requests for records.* Requests for records under 5 U.S.C. 552(a)(3) shall be made in accordance with 7 CFR 1.6 and shall be addressed as follows: AMS FOIA Officer, Agricultural Marketing Service, FOIA Request, 1400 Independence Avenue SW, Room 2095–S, Stop 0203, Washington, DC 20250–0203.

(e) *Appeals.* Any person whose request under paragraph (d) of this section is denied shall have the right to appeal such denial in accordance with 7 CFR 1.13. Appeals shall be addressed to the Administrator, Agricultural Marketing

Service, FOIA Appeal, 1400 Independence Avenue SW, Room 3071–S, Stop 0201, Washington, DC 20250–0201.

(Secs. 5, 18, Pub. L. 94–582, 90 Stat. 2869, 2884; (7 U.S.C. 76, 87e))

[48 FR 57467, Dec. 30, 1983, as amended at 54 FR 5924, Feb. 7, 1989; 60 FR 5836, Jan. 31, 1995; 84 FR 45646, Aug. 30, 2019]

OFFICIAL INSPECTION AND CLASS X OR CLASS Y WEIGHING REQUIREMENTS

§ 800.15 Services.

(a) *General.* These regulations implement requirements for a national inspection and weighing system. This system promotes the uniform and accurate application of the official grain standards and provides inspection and weighing services required by the Act and as requested by applicants for official services. The types and kinds of services available under the Act and regulations can be obtained at all specified service points in the United States and on U.S. grain in Canadian ports.

(b) *Responsibilities for complying with the official inspection, aflatoxin testing, and weighing requirements—*(1) *Export grain.* Exporters are responsible for (i) complying with all inspection, Class X weighing, and other certification provisions and requirements of section 5(a)(1) of the Act and the regulations applicable to export grain and (ii) having all corn, as defined in § 810.401, exported from the United States tested for aflatoxin contamination unless the buyer and seller agree not to have the corn tested. The Service shall perform the aflatoxin testing service unless the buyer and seller agree to have the corn tested by an entity other than the Service.

(2) *Grain in marked containers.* When grain is in a container that bears an official grade designation or mark, the person who places the designation or mark on the container or the person who places the grain in a container that bears the designation or mark shall be responsible for determining that the grain has been inspected or weighed by official personnel and qualifies for the official grade designation or mark.

(3) *Grain for which representations have been made.* Any person who makes a representation that (i) grain has been officially inspected or weighed; or (ii) grain has been officially inspected or weighed and found to be of a particular kind, class, quality, condition, or weight; or (iii) particular facts have been established with respect to the grain by official inspection or weighing, shall be responsible for determining that the representation is true and is not in violation of the Act and regulations.

[50 FR 49668, Dec. 4, 1985, as amended at 57 FR 2439, Jan. 22, 1992; 81 FR 49860, July 29, 2016]

§ 800.16 Certification requirements for export grain.

(a) *General.* Official Export Grain Inspection and Weight Certificates, Official Export Grain Inspection Certificates, and Official Export Grain Weight Certificates for bulk or sacked grain shall be issued according to § 800.162 for export grain loaded by an export elevator. Only these types of export certificates showing the official grade, official aflatoxin test results if required under the Act and the regulations, and/or the Class X weight of the grain shall be considered to be in compliance with inspection and weighing requirements under the Act for export grain.

(b) *Promptly furnished.* Export certificates shall be considered promptly furnished if they are forwarded by the shipper or the shipper's agent to the consignee not later than 10 business days after issuance.

[50 FR 49668, Dec. 4, 1985, as amended at 57 FR 2439, Jan. 22, 1992]

§ 800.17 Special inspection and weighing requirements for sacked export grain.

(a) *General.* Subject to the provisions of § 800.18, sacked export grain shall be (1) officially inspected on the basis of official samples obtained with an approved sampling device and operated in accordance with instructions, (2) Class X weighed or checkweighed, and (3) officially checkloaded by official personnel at the time the grain is loaded aboard the export carrier, in accordance with the provisions of paragraphs (b) and (c) of this section.

(b) *Services at time of loading.* When official sampling, official inspection,

Class X weighing or checkweighing, and checkloading of sacked export grain loaded aboard an export carrier is performed at one location and time, official export inspection and weight certificate(s) which identify the export carrier shall be issued.

(c) *Services prior to loading.* When official sampling, official inspection, and Class X weighing or checkweighing of sacked export grain is performed prior to the date of loading aboard an export carrier, official "OUT" certificates shall be issued. An examination by official personnel for condition and checkloading of the grain shall be made as the grain is loaded aboard the export carrier. If the examination for condition and the checkloading shows that the identity or quantity of the grain has not changed or the condition of the grain has not changed beyond expected variations prescribed in the instruction, official export inspection and weight certificates shall be issued on the basis of the official "OUT" certificates and the checkloading. If the identity, quantity, or the condition has changed, official export inspection and weight certificates shall be issued on the basis of the most representative samples, including weight samples, obtained at the time the grain is loaded aboard the export carrier.

[50 FR 49668, Dec. 4, 1985]

§800.18 Waivers of the official inspection and Class X weighing requirements.

(a) *General.* Waivers from the official inspection and Class X weighing requirements for export grain under section 5 of the Act shall be provided in accordance with this section and the Act.

(b) *Waivers*—(1) *15,000 metric-ton waiver.* Official inspection and Class X weighing requirements apply only to exporters and individual elevator operators who (i) exported 15,000 metric tons or more of grain during the preceding calendar year, or (ii) have exported 15,000 metric tons or more of grain during the current calendar year. Exporters and elevator operators who are granted a waiver by reason of this paragraph shall, as a condition of the waiver, keep such accounts, records, and memorandum to fully and cor-

rectly disclose all transactions concerning lots of all export grain shipments. In addition, the exporters or elevator operators shall notify the Service in writing of the intention to export grain under this waiver. In the case of lots waived under this provision, if such lots are required by contract to be inspected or weighed, or if the lots are represented by official inspection or weight certificates, then such certificates shall meet the requirements of section 5 of the Act.

(2) *Grain exported for seeding purposes.* Official inspection and Class X weighing requirements do not apply to grain exported for seeding purposes, provided that (i) the grain is (A) sold or consigned for sale and invoiced as seed; and (B) identified as seed for seeding purposes on the Shipper's Export Declaration; and (ii) records pertaining to these shipments are made available, upon request by the Service, for review or copying purposes.

(3) *Grain shipped in bond.* Official inspection and weighing requirements do not apply to grain that is shipped from a foreign country to a foreign country through the United States in bond in accordance with applicable regulations of the United States Customs Service (19 CFR part 18).

(4) *Grain exported by rail or truck to Canada or Mexico.* Inspection and weighing requirements do not apply to grain exported by rail or truck from the United States to Canada or Mexico.

(5) *Grain not sold by grade.* Official inspection requirements may be waived by the Service on a shipment-by-shipment basis for export grain not sold, offered for sale, or consigned for sale by official grade if (i) the contract and any amendments clearly show that the buyer and seller mutually agree to ship the grain without official inspection and (ii) a copy of the contract and any amendments is furnished in advance of loading, along with a completed application on a form prescribed by the Service.

(6) *Service not available.* Upon request, any required official inspection or Class X weighing of grain may be waived on a shipment-by-shipment basis if (i) official personnel are not and will not be available within a 24-

hour period to perform needed inspection or weighing services and (ii) both the buyer and seller of the grain are made aware that the grain has not been officially inspected or Class X weighed.

(7) *Emergency waiver.* (i) Upon request, the requirements for official inspection or Class X weighing will be waived whenever the Service determines that an emergency exists that precludes official inspection or Class X weighing;

(ii) To qualify for an emergency waiver, the exporter or elevator operator must submit a timely written request to the Service for the emergency waiver and also comply with all conditions that the Service may require.

(8) *High quality specialty grain shipped in containers.* Official inspection and weighing requirements do not apply to high quality specialty grain exported in containers. Records generated during the normal course of business that pertain to these shipments must be made available to the Service upon request, for review or copying. These records must be maintained for a period of 3 years.

(Approved by the Office of Management and Budget under control number 0580-0011)

[50 FR 49669, Dec. 4, 1985, as amended at 70 FR 21923, Apr. 28, 2005; 70 FR 73559, Dec. 13, 2005; 75 FR 41695, July 19, 2010; 76 FR 45399, July 29, 2011; 81 FR 49860, July 29, 2016]

RECORDKEEPING AND ACCESS TO
FACILITIES

§ 800.25 Required elevator and merchandising records.

(a) *Elevator and merchandiser recordkeeping.* Every person and every State or political subdivision of a State that owns or operates an elevator and every merchandiser that has obtained or obtains official inspection or official weighing services other than (1) submitted sample inspection service, or (2) official sampling service, or (3) official stowage examination service shall keep such accounts, records, and memoranda that fully and correctly disclose all transactions concerning the lots of grain for which the elevator or merchandiser received official services, except as provided under § 800.18.

(b) *Retention period.* Records specified in this section may be disposed of after a period of 3 years from the date of the official service; provided, the 3-year period may be extended if the elevator owner or operator, or merchandiser is notified in writing by the Administrator that specific records should be retained for a longer period for effective administration and enforcement of the Act. This requirement does not restrict or modify the requirements of any other Federal, State, or local statute concerning recordkeeping.

(Approved by the Office of Management and Budget under control number 0580-0011)

[51 FR 1768, Jan. 15, 1986]

§ 800.26 Access to records and facilities.

(a) *Inspection of records and facilities.* Prior to the examination of records or inspection of facilities by an authorized representative of the Secretary or the Administrator, the authorized representative shall contact or otherwise notify the elevator manager or manager's representative of their presence and furnish proof of identity and authority. While in the elevator, the authorized representative shall abide by the safety regulations in effect at the elevator. Every elevator owner and operator and every merchandiser shall permit authorized representatives of the Secretary or Administrator to enter its place of business during normal business hours and have access to the facilities and to inspect any books, documents, papers, and records that are maintained by such persons. Such access and inspection will be to effectuate the purpose, provisions, and objectives of the Act and to assure the integrity of official services under the Act or of any official transaction with which the Act is concerned. All copies of such records will be made at the Service's expense. Reasonable accommodations shall be made available to the duly authorized representative by elevator owners and operators, and merchandisers for such examination of records.

(b) *Disclosure of business information.* FGIS employees or persons acting for FGIS under the Act shall not, without the consent of the elevator operator or merchandiser concerned, divulge or make known in any manner, any facts or information acquired pursuant to the Act and regulations except as authorized by the Administrator, by a court of competent jurisdiction, or otherwise by law.

[51 FR 1768, Jan. 15, 1986]

REGISTRATION

§800.30 Foreign commerce grain business.

"Foreign commerce grain business" is defined as the business of buying grain for sale in foreign commerce or the business of handling, weighing, or transporting grain for sale in foreign commerce. This provision shall not include:

(a) Any person who only incidentally or occasionally buys for sale, or handles, weighs, or transports grain for sale and is not engaged in the regular business of buying grain for sale, or handling, weighing, or transporting grain for sale;

(b) Any producer of grain who only incidentally or occasionally sells or transports grain which the producer has purchased;

(c) Any person who transports grain for hire and does not own a financial interest in such grain; or

(d) Any person who buys grain for feeding or processing and not for the purpose of reselling and only incidentally or occasionally sells such grain as grain.

[48 FR 44455, Sept. 29, 1983]

§800.31 Who must register.

Each person who has engaged in foreign commerce grain business totaling 15,000 or more metric tons during the preceding or current calendar year must register with the Service and shall be deemed to be regularly engaged in foreign commerce grain business. This includes foreign-based firms operating in the United States but does not include foreign governments or their agents. The Service will, upon request, register persons not required to register under this section if they comply with the requirements of §§800.33 and 800.34.

(Approved by the Office of Management and Budget under control number 0580–0012)

[48 FR 44453 and 44455, Sept. 29, 1983, as amended at 54 FR 5924, Feb. 7, 1989]

§800.32 When to register.

A person shall submit an application for registration to the Service at least 30 calendar days before regularly engaging in foreign commerce grain business according to §800.31. For good cause shown, the Service may waive this 30-day requirement.

(Approved by the Office of Management and Budget under control number 0580–0012)

[48 FR 44453 and 44455, Sept. 29, 1983, as amended at 54 FR 5924, Feb. 7, 1989]

§800.33 How to register.

Any person who is required or desires to register must submit an application for registration to the Service. Application forms can be obtained from the Service. Each application shall: (a) Be typewritten or legibly written in English; (b) include all information required by the application form; and (c) be signed by the applicant. The information required by this paragraph may be submitted to the Service via telephone, subject to written confirmation. An applicant shall furnish any additional information requested by the Service for consideration of the application.

(Approved by the Office of Management and Budget under control number 0580–0012)

[48 FR 44453 and 44456, Sept. 29, 1983, as amended at 54 FR 5924, Feb. 7, 1989]

§800.34 Registration fee.

An applicant shall submit the registration fee prescribed in §800.71 with the completed application. If an application is dismissed, the fee shall be refunded by the Service. No fee or portion of a fee shall be refunded if a person is registered and the registration is subsequently suspended or revoked under §800.39.

[48 FR 44456, Sept. 29, 1983]

§800.35 Review of applications.

(a) The Service shall review each application to determine if it complies with §§ 800.32, 800.33, and 800.34. If the application complies and the fee has been paid, the applicant shall be registered.

(b) If the application does not comply with §§ 800.32, 800.33, and 800.34 and the omitted information prevents a satisfactory review by the Service, the applicant shall be provided an opportunity to submit the needed information. If the needed information is not submitted within a reasonable time, the application may be dismissed. The Service shall promptly notify the applicant, in writing, of the reasons for the dismissal.

[48 FR 44456, Sept. 29, 1983]

§ 800.36 Certificates of registration.

The Service shall furnish the applicant with an original and three copies of the registration certificate. The registration shall be effective on the issue date shown on the certificate. Each certificate of registration is issued on the condition that the registrant will comply with all provisions of the Act, regulations, and instructions. The Service shall charge a fee, in accordance with § 800.71, for each additional copy of a certificate of registration requested by a registrant.

[48 FR 44456, Sept. 29, 1983]

§ 800.37 Notice of change in information.

Each registrant shall notify the Service within 30 days of any change in the information contained in the application for registration. If the notice is submitted orally, it shall be promptly confirmed in writing.

(Approved by the Office of Management and Budget under control number 0580–0012)

[48 FR 44453 and 44456, Sept. 29, 1983, as amended at 54 FR 5924, Feb. 7, 1989]

§ 800.38 Termination and renewal of registration.

Each certificate of registration shall terminate on December 31 of the calendar year for which it is issued. The Service shall send a letter to each registrant notifying the registrant of the impending termination of the registration and providing instructions for requesting renewal. The registration may be renewed in accordance with §§ 800.33

and 800.34. Failure to receive the letter shall not exempt registrants from the responsibility of renewing their registration if required by § 800.31.

[48 FR 44456, Sept. 29, 1983]

§ 800.39 Suspension or revocation of registration for cause.

(a) *General.* Registration is subject to suspension or revocation whenever the Administrator determines that the registrant has violated any provision of the Act or regulations, or has been convicted of any violation involving the handling, weighing, or inspection of grain under Title 18 of the United States Code.

(b) *Procedure.* Before the Service suspends or revokes a registration, the registrant (hereinafter the "respondent"): (1) Shall be notified of the proposed action and the reasons therefor and (2) shall be afforded opportunity for a hearing in accordance with the Rules of Practice Governing Formal Adjudicatory Proceedings Instituted by the Secretary under Various Statutes (7 CFR, 1.130 through 1.151). Prior to formal adjudicatory proceedings, the Service may allow the respondent to express views on the action proposed by the Service in an informal conference before the Administrator. If the Service and the respondent enter into a consent agreement, no formal adjudicatory proceedings shall be initiated.

[48 FR 44456, Sept. 29, 1983]

CONDITIONS FOR OBTAINING OR
WITHHOLDING OFFICIAL SERVICES

§ 800.45 Availability of official services.

(a) *Original inspection and weighing services.* Original inspection and weighing services on grain are available according to this section and §§ 800.115 through 800.118 when requested by an interested person.

(b) *Reinspection, review of weighing, and appeal inspection services.* Reinspection, review of weighing, appeal inspection, and Board appeal inspection services are available when requested by an interested person, according to §§ 800.125 through 800.129 and §§ 800.135 through 800.139.

(c) *Proof of authorization.* If an application for official services is filed by a person representing the applicant, the agency or the field office receiving the application may require written proof of the authority to file the application.

(Approved by the Office of Management and Budget under control number 0580–0012)

(Secs. 8, 9, 10, 13 and 18, Pub. L. 94–582, 90 Stat. 2870, 2875, 2877, 2880, and 2884, 7 U.S.C. 79, 79a, 79b, 84, 87, and 87e)

[49 FR 30913, Aug. 2, 1984, as amended at 50 FR 45392, Oct. 31, 1985; 54 FR 5924, Feb. 7, 1989]

§800.46 Requirements for obtaining official services.

(a) *Consent and agreement by applicant.* In submitting a request for official services, the applicant and the owner of the grain consent to the special and general requirements specified in paragraphs (b) and (c) of this section. These requirements are essential to carry out the purposes or provisions of the Act.

(b) *General requirements*—(1) *Access to grain.* Grain on which official services are to be performed shall, except as provided in §§800.85, 800.86, 800.98, and 800.99, be made accessible by the applicant for the performance of the requested official service and related monitoring and supervision activities. For the purposes of this section, grain is not "accessible" if it is offered for official services (i) in containers or carriers that are closed and cannot, with reasonable effort, be opened by or for official personnel; (ii) when any portion is located so as to prohibit the securing or a representative sample; or (iii) under conditions prescribed in the instructions as being hazardous to the health or safety of official personnel.

(2) *Working space.* When official services are performed at an elevator, adequate and separate space must be provided by the applicant for the performance of the requested service and related monitoring and supervision activities. Space will be "adequate" if it meets the space, location, and safety requirements specified in the instructions.

(3) *Notice of changes.* The operator of each facility at which official services are performed must notify the appropriate agency or field office promptly, in full detail, of changes in the grain handling and weighing facilities, equip-

ment, or procedures at the elevator that could or would affect the proper performance of official services.

(4) *Loading and unloading conditions.* As applicable, each applicant for official services must provide or arrange for suitable conditions in the (i) loading and unloading areas and the truck and railroad holding areas; (ii) gallery and other grain-conveying areas; (iii) elevator legs, distributor, and spout areas; (iv) pier or dock areas; (v) deck and stowage areas in the carrier; and (vi) equipment used in loading or unloading and handling the grain. Suitable conditions are those which will facilitate accurate inspection and weighing, maintain the quantity and the quality of the grain that is to be officially inspected or weighed, and not be hazardous to the health and safety of official personnel, as prescribed in the instructions.

(5) *Timely arrangements.* Requests for official service shall be made in a timely manner; otherwise, official personnel may not be available to provide the requested service. For the purpose of this paragraph, "timely manner" shall mean not later than 2 p.m., local time, of the preceding business day.

(6) *Observation of activities.* Each applicant for official services must provide any interested person, or their agent, an opportunity to observe sampling, inspection, weighing, and loading or unloading of grain. Appropriate observation areas shall be mutually defined by the Service and facility operator. The areas shall be safe and shall afford a clear and unobstructed view of the performance of the activity, but shall not permit a close over-the-shoulder type of observation by the interested person.

(7) *Payment of bills.* Each applicant, for services under the Act, must pay bills for the services according to §§800.70 through 800.73.

(8) *Written confirmations.* When requested by the agency or field office, verbal requests for official services shall be confirmed in writing. Each written request shall be signed by the applicant, or the applicant's agent, and shall show or be accompanied by the following information:

(i) The identification, quantity, and specific location of the grain;

495

(ii) The name and mailing address of the applicant;

(iii) The kind and scope of services desired; and

(iv) Any other information requested by the agency or field office.

(9) *Names and addresses of interested persons.* When requested, each applicant for official services shall show on the application form the name and address of each known interested person.

(10) *Surrender of superseded certificates.* When a request for official service results in a certificate being superseded, the superseded certificate must be promptly surrendered.

(11) *Recordkeeping and access.* Each applicant for official services must comply with applicable recordkeeping and access-to-facility provisions in §§ 800.25 and 800.26.

(12) *Monitoring equipment.* Owners and operators of elevators shall, upon a finding of need by the Administrator, provide equipment necessary for the monitoring by official personnel of grain loading, unloading, handling, sampling, weighing, inspection, and related activities. The finding of need will be based primarily on a consideration of manpower and efficiency.

(c) *Special requirements for official Class X and Class Y weighing services—* (1) *General.* Weighing services shall be provided only at weighing facilities which have met the conditions, duties, and responsibilities specified in section 7A(f) of the Act and this section of the regulations. Weighing services will be available only in accordance with the requirements of § 800.115. Facilities desiring weighing services should contact the Service in advance to allow the Service time to determine if the facility complies with the provisions of the Act and regulations.

(2) *Conditions.* The facility shall provide the following information annually to the Service:

(i) The facility owner's name and address;

(ii) The facility operator's name and address;

(iii) The name of each individual employed by the facility as a weigher and a statement that each individual:

(A) Has a technical ability to operate grain weighing equipment and

(B) Has a reputation for honesty and integrity;

(iv) A blueprint or similar drawing of the facility showing the location of:

(A) The loading, unloading, and grain handling systems;

(B) The scale systems used or to be used in weighing grain; and

(C) The bins and other storage areas;

(v) The identification of each scale in the facility that is to be used for weighing grain under the Act;

(vi) The following information regarding automated data processing systems:

(A) Overall system intent, design, and layout;

(B) Make, model, and technical specifications of all hardware;

(C) Description of software, language used, and flow charts of all programs, subprograms, routines, and subroutines; and

(D) Complete operating instructions; and

(vii) Any other information deemed necessary to carry out the provisions of the Act.

If a facility has, or plans to have, an automated data processing system which is used in conjunction with any portion of the scale system, grain handling system, or the preparing or printing of official weight certificates, the facility shall make available to the Service sufficient documentation to ensure that the system cannot be used deceptively or otherwise provide inaccurate information. The Service or approved scale testing and certification organization shall conduct an onsite review to evaluate the performance and accuracy of each scale that will be used for weighing grain under the Act, and the performance of the grain loading, unloading, and related grain handling equipment and systems.

(3) *Duties and responsibilities of weighing facilities requesting official services—* (i) *Providing official services.* Upon request, each weighing facility shall permit official weighing services to be performed promptly.

(ii) *Supervision.* Each weighing facility shall supervise its employees and shall take action necessary to assure that employees are performing their duties according to the Act, regulations, and instructions and are not performing prohibited functions or are not involved in any action prohibited by the Act, regulations, and instructions.

(iii) *Facilities and equipment*—(A) *General.* Each weighing facility shall obtain and maintain facilities and equipment which the Service determines are needed for weighing services performed at the facility. Each facility shall operate and shall maintain each scale system and related grain handling system used in weighing according to instructions issued by the manufacturer and by the Service. A scale log book for each approved scale used for official weighing services shall be maintained according to instructions at each weighing facility.

(B) *Malfunction of scales.* Scales or scale systems that are operating in other than a correct and approved manner shall not be used for weighing grain under the Act. Before the malfunctioning scale or scale system can be used again for weighing grain under the Act, it shall be repaired and determined to be operating properly by the Service or approved scale testing and certification organization.

(iv) *Oral directives.* FGIS oral directives issued to elevator personnel shall be confirmed in writing upon request by elevator management. Whenever practicable, the Service shall issue oral directives through elevator management officials.

(Approved by the Office of Management and Budget under control number 0580–0012)

(Secs. 8, 9, 10, 13 and 18, Pub. L. 94–582, 90 Stat. 2870, 2875, 2877, 2880, and 2884, 7 U.S.C. 79, 79a, 79b, 84, 87, and 87e)

[49 FR 30915, Aug. 2, 1984, as amended at 49 FR 49587, Dec. 21, 1984; 50 FR 45392, Oct. 31, 1985; 54 FR 5924, Feb. 7, 1989]

§ 800.47 Withdrawal of request for official services.

An applicant may withdraw a request for official services any time before official personnel release results, either verbally or in writing. See § 800.51 for reimbursement of expenses, if any.

(Secs. 8, 9, 10, 13 and 18, Pub. L. 94–582, 90 Stat. 2870, 2875, 2877, 2880, and 2884, 7 U.S.C. 79, 79a, 79b, 84, 87, and 87e)

[49 FR 30915, Aug. 2, 1984]

§ 800.48 Dismissal of request for official services.

(a) *Conditions for dismissal*—(1) *General.* An agency or the Service shall dismiss requests for official services when (i) § 800.76 prohibits the requested service; (ii) performing the requested service is not practicable; (iii) the agency or the Service lacks authority under the Act or regulations; or (iv) sufficient information is not available to make an accurate determination.

(2) *Original services.* A request for original services shall be dismissed if a reinspection, review of weighing, appeal inspection, or Board appeal inspection has been performed on the same lot at the same specified service point within 5 business days.

(3) *Reinspection, appeal inspection, or Board appeal inspection services.* A request for a reinspection, appeal inspection, or Board appeal inspection service shall be dismissed when:

(i) The kind and scope are different from the kind and scope of the last inspection service;

(ii) The condition of the grain has undergone a material change;

(iii) The request specifies a representative file sample and a representative file sample is not available,

(iv) The applicant requests that a new sample be obtained and a new sample cannot be obtained; or

(v) The service cannot be performed within 5 business days of the date of the last inspection date.

(4) *Review of weighing services.* A request for review of weighing services shall be dismissed when the request (i) is filed before the weighing results have been released, or (ii) is filed more than 90 calendar days after the date of the original service.

(b) *Procedure for dismissal.* When an agency or the Service proposes to dismiss a request for official services, the applicant shall be notified of the proposed action. The applicant will then be afforded reasonable time to take corrective action or to demonstrate there is no basis for the dismissal. If the agency or the Service determines that corrective action has not been adequate, the applicant will be notified again of the decision to dismiss the request for service, and any results of official services shall not be released.

(Secs. 8, 9, 10, 13 and 18, Pub. L. 94–582, 90 Stat. 2870, 2875, 2877, 2880, and 2884, 7 U.S.C. 79, 79a, 79b, 84, 87, and 87e)

[49 FR 30915, Aug. 2, 1984, as amended at 50 FR 45392, Oct. 31, 1985]

§ 800.49   Conditional withholding of official services.

(a) *Conditional withholding.* An agency or the Service shall conditionally withhold requests for official services when an applicant fails to meet any requirement prescribed in § 800.46.

(b) *Procedure and withholding.* When an agency or the Service proposes to conditionally withhold official services, the applicant shall be notified of the reason for the proposed action. The applicant will then be afforded reasonable time to take corrective action or to show that there is no basis for withholding services. If the agency or the Service determines that corrective action has not been adequate, the applicant will be notified. Any results of official services shall not be released when a request for service is withheld.

(Secs. 8, 9, 10, 13 and 18, Pub. L. 94–582, 90 Stat. 2870, 2875, 2877, 2880, and 2884, 7 U.S.C. 79, 79a, 79b, 84, 87, and 87e)

[49 FR 30915, Aug. 2, 1984]

§ 800.50   Refusal of official services and civil penalties.

(a) *Grounds for refusal.* Any or all services available to an applicant under the Act may be refused, either temporarily or indefinitely, by the Service for causes prescribed in section 10(a) of the Act. Such refusal by the Service may be restricted to the particular facility or applicant (if not a facility) found in violation or to a particular type of service, as the facts may warrant. Such action may be in addition to, or in lieu of, criminal penalties or other remedial action authorized by the Act.

(b) *Provision and procedure for summary refusal.* The Service may, without first affording the applicant (hereafter in this section "respondent") a hearing, refuse to provide official inspection and Class X or Y weighing services pending final determination of the proceeding whenever the Service has reason to believe there is cause, as prescribed in section 10 of the Act, for refusing such official services and considers such action to be in the best interest of the official services system under the Act: *Provided that* within 7 days after refusal of such service, the Service shall afford the respondent an opportunity for a hearing as provided under paragraph (c)(2) of this section. Pending final determination, the Service may terminate the temporary refusal if alternative managerial, staffing, financial, or operational arrangements satisfactory to the Service can be and are made by the respondent.

(c) *Procedure for other than summary refusal.* Except as provided in paragraph (b) of this section, before the Service refuses to provide official services the respondent shall be (1) notified of the services that are to be refused, the locations at which are and the time period for which service will be refused, and the reasons for the refusal; and (2) afforded an opportunity for a hearing in accordance with the provisions of the Rules of Practice Governing Formal Adjudicatory Proceedings Instituted by the Secretary Under Various Statutes (7 CFR 1.130 *et seq.*). At the discretion of the Service, prior to initiation of formal adjudicatory proceedings, the respondent may be given an opportunity to express his or her views on the action proposed by the Service in an informal conference before the Administrator of the Service. If, as a result of such an informal conference, the Service and the respondent enter into a consent agreement, no formal adjudicatory proceedings shall be initiated.

(d) *Assessment of civil penalties.* Any person who has knowingly committed any violation of section 13 of the Act or has been convicted of any violation of other Federal law with respect to the handling, weighing, or official inspection of grain may be assessed a civil penalty not to exceed the amount specified at § 3.91(b)(6)(viii) of this title for each such violation as the Administrator determines is appropriate to effect compliance with the Act. Such action may be in addition to, or in lieu of, criminal penalties under section 14 of the Act, or in addition to, or in lieu of, the refusal of official services authorized by the Act.

(e) *Provisions for civil penalty hearings.* Before a civil penalty is assessed against any person, such person shall be afforded an opportunity for a hearing as provided under paragraph (c)(2) of this section.

(f) *Collection of civil penalties.* Upon failure to pay the civil penalty, the Service may request the Attorney General to file civil action to collect the penalty in a court of appropriate jurisdiction.

[45 FR 15810, Mar. 11, 1980, as amended at 51 FR 12830, Apr. 16, 1986, 75 FR 17560, Apr. 7, 2010]]

§ 800.51 Expenses of agency, field office, or Board of Appeals and Review.

For any request that has been dismissed or withdrawn under § 800.47, § 800.48, or § 800.49, respectively, each applicant shall pay expenses incurred by the agency or the Service.

(Secs. 8, 9, 10, 13 and 18, Pub. L. 94–582, 90 Stat. 2870, 2875, 2877, 2880, and 2884, 7 U.S.C. 79, 79a, 79b, 84, 87, and 87e)

[49 FR 30915, Aug. 2, 1984]

§ 800.52 Official services not to be denied.

Subject to the provisions of §§ 800.48, 800.49, and 800.50, no person entitled to official services under the Act shall be denied or deprived of the right thereto by reason of any rule, regulation, bylaw, or custom of any market, board of trade, chamber of commerce, exchange, inspection department, or similar organization; or by any contract, agreement, or other understanding.

DESCRIPTIONS

§ 800.55 Descriptions by grade.

(a) *General.* In any sale, offer for sale, or consignment for sale, which involves the shipment of grain in interstate or foreign commerce, the description of grain, as being of a grade in any advertising, price quotation, other negotiation of sale, contract of sale, invoice, bill of lading, other document, or description on bags or other containers of the grain, is prohibited if such description is other than by an official grade designation, with or without additional information as to specified factors. An official grade designation contains any of the following: The term "U.S.," the numerals 1 through 5, the term "Sample grade," or the name of a subclass or a special grade of grain specified in the Official United States Standards for Grain.

(b) *Proprietary brand names or trademarks.* A description of grain by a proprietary brand name or a trademark that does not resemble an official grade designation will not be considered to be a description by grade; but a description by a proprietary brand name or trademark that contains singly or in combination any of the terms referenced in paragraph (a) of this section shall be considered to resemble an official grade designation.

(c) *Use of one or more factor designations.* In interstate commerce, a description of grain by the use of one or more grade factor designations which appear in the Official United States Standards for Grain or by other criteria will not be considered to be a description by grade.

(d) *False or misleading descriptions.* In any sale, offer for sale, or consignment for sale of any grain which involves the shipment of grain from the United States to any place outside thereof, knowingly using a false or misleading description of grain by official grade designation, or other description is prohibited.

[50 FR 9982, Mar. 13, 1985]

§ 800.56 Requirements on descriptions.

Section 13 of the Act contains certain prohibitions with respect to the use of official grade designations, official marks, and other representations with respect to grain.

(a) The use of an official grade designation, with or without factor information, or of official criteria information, or of the term "official grain standards," shall not, without additional information, be considered to be a representation that the grain was officially inspected.

(b) The use of any symbol or term listed as an official mark, at § 800.0(b)(68), with respect to grain shall be considered to be a representation of official service under the Act: Provided

499

however, that the use of the official marks "official certificate;" "officially inspected;" "official inspection;" "officially weighed;" "official weight;" and "official weighing" shall not be considered to be a representation of official service under the Act if it is clearly shown that the activity occurred under the U.S. Warehouse Act (7 U.S.C. 241 *et seq.*): Provided further, that the use of the official mark "officially tested" with respect to grain inspection and weighing equipment shall not be considered to be a representation of testing under the Act if it is clearly shown that the equipment was tested under a State statute.

[50 FR 9982, Mar. 13, 1985]

GRAIN HANDLING PRACTICES

§ 800.60 Deceptive actions and practices.

In the absence of prior adequate notice to appropriate official personnel, any action or practice, including the loading, weighing, handling, or sampling of grain that knowingly causes or is an attempt to cause the issuance by official personnel of a false or incorrect official certificate or other official form, is deemed to be deceptive and, as such, is a violation of section 13(a)(3) of the Act. For the purposes of this paragraph, adequate notice is written or oral notice given to an agency or the Service, as applicable, before official personnel begin to perform official inspection or weighing services. If oral notice is given, it must be confirmed in writing within 2 business days. To be adequate, the notice must explain the nature and extent of the action or practice in question and must identify the grain, stowage container, equipment, facility, and the official personnel actually or potentially involved.

(Approved by the Office of Management and Budget under control number 0580–0011)

[48 FR 17330, Apr. 22, 1983, as amended at 48 FR 44453, Sept. 29, 1983; 54 FR 5924, Feb. 7, 1989]

§ 800.61 Prohibited grain handling practices.

(a) *Definitions.* For the purpose of this section, dockage and foreign material in grain shall be:

(1) Defined for export elevators at export port locations as set forth in 7 CFR part 810 and as dust removed from grain and collected in a bin/container and as dust settling on floors, equipment, and other areas, commonly referred to as dust sweepings; and

(2) Defined for other than export elevators as set forth in 7 CFR part 810.

(b) *Prohibited practices.* Except as permitted in paragraphs (c) and (d) of this section, no person shall:

(1) Recombine or add dockage or foreign material to any grain, or

(2) Blend different kinds of grain except when such blending will result in grain being designated as Mixed grain in accordance with subpart E of the Official United States Standards for Grain.

(3) Add water to grain for purposes other than milling, malting, or similar processing operations.

(c) *Exemption.* (1) The Administrator may grant exemptions from paragraph (b) of this section for grain shipments sent directly to a domestic end-user or processor. Requests for exemptions shall be submitted by grain handlers to the Service through the domestic end-users or processors or their representatives.

(2) Grain sold under an exemption shall be consumed or processed into a product(s) by the purchaser and not resold into the grain market.

(3) Products or byproducts from grain sold under an exemption shall not be blended with or added to grain in commercial channels, except for vegetable oil which may be used as a dust suppressant in accordance with (d)(4) of this section.

(d) *Exceptions.* Paragraph (b) shall not be construed as prohibiting the following grain handling practices. Compliance with paragraphs (d)(1) through (d)(6) of this section does not excuse compliance with applicable Federal, State, and local laws.

(1) *Blending.* Grain of the same kind, as defined by the Official United States Standards for Grain, may be blended to adjust quality. Broken corn or broken kernels may be recombined or added to whole grain of the same kind provided that no foreign material or dockage has been added to the broken corn or broken kernels.

(2) *Insect and fungi control.* Grain may be treated to control insects and fungi. Elevators, other grain handlers, and their agents are responsible for the proper use and applications of insecticides and fungicides. Sections 800.88 and 800.96 include additional requirements for grain that is officially inspected and weighed.

(3) *Marketing dockage and foreign material.* Dockage and foreign material may be marketed separately.

(4) *Dust suppressants.* Grain may be treated with an additive, other than water, to suppress dust during handling. Elevators, other grain handlers, and their agents are responsible for the proper use and application of dust suppressants. Sections 800.88 and 800.96 include additional requirements for grain that is officially inspected and weighed.

(5) *Identification.* Confetti or similar material may be added to grain for identification purposes. Elevators, other grain handlers, and their agents are responsible for the proper use and application of such materials. Sections 800.88 and 800.96 include additional requirements for grain that is officially inspected or weighed.

(6) *Export loading facilities.* Between May 1, 1987, and December 31, 1987, export elevators at export port locations may recombine dockage and foreign material, but not dust, with grain provided such recombination occurs during the loading of a vessel with the intended purpose of ensuring uniformity of dockage and foreign material in the cargo.

(Approved by the Office of Management and Budget under control number 0580–0011)

[52 FR 24437, June 30, 1987, as amended at 59 FR 52077, Oct. 14, 1994]

FEES

§ 800.70 Fees for official services performed by agencies.

(a) *Assessment and use of fees.* (1) Fees assessed by an agency for official inspection and Class X or Class Y weighing services or testing of inspection equipment shall be reasonable and nondiscriminatory.

(2) In the case of a State or local governmental agency, fees shall not be used for any purpose other than to finance the cost of the official inspection and Class X or Class Y weighing service and inspection equipment testing service performed by the agency or the cost of other closely related programs administered by the agency.

(b) *Approval required*—(1) *Restriction.* Only fees that meet the requirements stated in this section and are approved by the Service as reasonable and nondiscriminatory may be charged by an agency.

(2) *Exceptions.* For good cause shown by an agency, the Administrator may grant case-by-case exceptions to the requirements in this section, provided that a determination is made that the agency fees would be reasonable and nondiscriminatory.

(c) *Reasonable fees.* In determining if an agency's fees are reasonable, the Service will consider whether the fees:

(1) Cover the estimated total cost to the agency of

(i) Official inspection services,

(ii) Class X or Class Y weighing services,

(iii) Inspection equipment testing services, and

(iv) Related supervision and monitoring activities performed by the agency;

(2) Are reasonably consistent with fees assessed by adjacent agencies for similar services;

(3) Are assessed on the basis of the average cost of performing the same or similar services at all locations served by the agency; and

(4) Are supported by sufficient information which shows how the fees were developed.

(d) *Nondiscriminatory fees.* In determining if fees are nondiscriminatory, the Service will consider whether the fees are collected from all applicants for official service in accordance with the approved fee schedule. Charges for time and travel incurred in providing service at a location away from a specified service point shall be assessed in accordance with the approved fee schedule.

(e) *Schedule of fees to be established.* (1) Each agency shall establish a schedule of fees for official services which the agency is delegated or designated the authority to perform. The schedule

501

shall be in a standard format in accordance with the instructions. Such schedules may include fees for nonofficial services provided by the agency, but they shall be clearly identified and will not be subject to approval by the Service.

(2) The schedule shall be published and made available by the agency to all users of its services.

(f) *Request for approval of fees*—(1) *Time requirement.* A request for approval of a new or revised fee shall be submitted to the Service not less than 60 days in advance of the proposed effective date for the fee. Failure to submit a request within the prescribed time period may be considered grounds for postponment or denial of the request.

(2) *Contents of request.* Each request shall show (i) the present fee, if any, and the proposed fee, together with data showing in detail how the fee was developed, and (ii) the proposed effective date.

(g) *Review of request*—(1) *Approval action.* If upon review the Service finds that the request and supporting data justify the new or revised fee, the request will be marked "approved" and returned to the agency.

(2) *Denial action.* If the Service finds that the request and supporting data do not justify the new or revised fee, approval of the request will be withheld pending receipt of any additional supporting data which the agency has to offer. If the data are not submitted within a reasonable period, the request shall be denied. In the case of a denial of a request, the agency shall be notified of the reason for denial.

(Approved by the Office of Management and Budget under control numbers 0580–0003 and 0580–0012)

[45 FR 15810, Mar. 11, 1980; 45 FR 55119, Aug. 18, 1980, as amended at 48 FR 44453, Sept. 29, 1983; 50 FR 30131, July 24, 1985]

§ 800.71 **Fees assessed by the Service.**

(a) *Official inspection and weighing services.* The fees shown in Schedule A of paragraph (a)(1) of this section apply to official inspection and weighing services performed by FGIS in the U.S. and Canada. The fees shown in Schedule B of paragraph (a)(2) of this section apply to official domestic inspection and weighing services performed by delegated States and designated agencies, including land carrier shipments to Canada and Mexico. The fees charged to delegated States by the Service are set forth in the State's Delegation of Authority document. Failure of a delegated State or designated agency to pay the appropriate fees to the Service within 30 days after becoming due will result in an automatic termination of the delegation or designation. The delegation or designation may be reinstated by the Service if fees that are due, plus interest and any further expenses incurred by the Service because of the termination, are paid within 60 days of the termination.

(1) *Schedule A—Fees for official inspection and weighing services performed in the United States and Canada.* For each calendar year, FGIS will calculate *Schedule A* fees as defined in paragraph (b) of this section. FGIS will publish a notice in the FEDERAL REGISTER and post *Schedule A* fees on the Agency's public website.

(2) *Schedule B—Fees for Supervision of Official Inspection and Weighing Services Performed by Delegated States and Designated Agencies in the United States.* The Service will assess a supervision fee per metric ton of domestic U.S. grain shipments inspected or weighed, or both, including land carrier shipments to Canada and Mexico. For each calendar year, the Service will calculate Schedule B fees as defined in paragraph (b) of this section. The Service will publish a notice in the FEDERAL REGISTER and post Schedule B fees on the Agency's public website.

(b) *Annual review of fees.* For each calendar year, starting with 2021, the Service will review fees included in this section and publish fees each year according to the following:

(1) *Tonnage fees.* Tonnage fees in Schedule A in paragraph (a)(1) of this section will consist of the national tonnage fee and local tonnage fees and the Service will calculate and round the fee to the nearest $0.001 per metric ton. All outbound grain officially inspected and/or weighed by the Field Offices in New Orleans, League City, Portland, and Toledo will be assessed the national tonnage fee plus the appropriate

local tonnage fee. Export grain officially inspected and/or weighed by delegated States and official agencies, excluding land carrier shipments to Canada and Mexico, will be assessed the national tonnage fee only. The fees will be set according to the following:

(i) *National tonnage fee.* The national tonnage fee is the national program administrative costs for the previous fiscal year divided by the average yearly tons of export grain officially inspected and/or weighed by delegated States and designated agencies, excluding land carrier shipments to Canada and Mexico, and outbound grain officially inspected and/or weighed by the Service during the previous 5 fiscal years.

(ii) *Local tonnage fee.* The local tonnage fee is the Field Office administrative costs for the previous fiscal year divided by the average yearly tons of outbound grain officially inspected and/or weighed by the Field Office during the previous 5 fiscal years. The local tonnage fee is calculated individually for each Field Office.

(2) *Supervision fee.* Supervision fee in Schedule B in paragraph (a)(2) of this section will be set according to the following:

(i) *Operating reserve adjustment.* The operating reserve adjustment is the supervision program costs for the previous fiscal year divided by 2 less the end of previous fiscal year operating reserve balance.

(ii) *Supervision tonnage fee.* The supervision tonnage fee is the sum of the prior fiscal year program costs plus operating reserve adjustment divided by the average yearly tons of domestic U.S. grain shipments inspected or weighed, or both, including land carrier shipments to Canada and Mexico during the previous 5 fiscal years. If the calculated value is zero or a negative value, the Service will suspend the collection of supervision tonnage fees for one calendar year.

(3) *Operating reserve.* In order to maintain an operating reserve not less than 3 and not more than 6 months, the Service will review the value of the operating reserve at the end of each fiscal year and adjust fees according to the following:

(i) *Less than 4.5 months.* If the operating reserve is less than 4.5 times the monthly operating expenses, the Service will increase all fees in Schedule A in paragraph (a)(1) of this section by 2 percent for each $1,000,000, rounded down, that the operating reserve is less than 4.5 times the monthly operating expense, with a maximum increase of 5 percent annually. Except for fees based on tonnage or hundredweight, all fees will be rounded to the nearest $0.10.

(ii) *Greater than 4.5 months.* If the operating reserve is greater than 4.5 times the monthly operating expenses, the Service will decrease all fees in Schedule A in paragraph (a)(1) of this section by 2 percent for each $1,000,000, rounded down, that the operating reserve is greater than 4.5 times the monthly operating expense, with a maximum decrease of 5 percent annually. Except for fees based on tonnage or hundredweight, all fees will be rounded to the nearest $0.10.

(c) *Periodic review.* The Service will periodically review and adjust all fees in Schedules A and B in paragraphs (a)(1) and (2) of this section, respectively, as necessary to ensure they reflect the true cost of providing and supervising official service. This process will incorporate any fee adjustments from paragraph (b) of this section.

(d) *Miscellaneous fees for other services.* For each calendar year, the Service will review fees included in this section and publish fees in the FEDERAL REGISTER and on the Agency's public website.

(1) *Registration certificates and renewals.* The fee for registration certificates and renewals will be published annually in the FEDERAL REGISTER and on the Agency's public website, and the Service will calculate the fee using the noncontract hourly rate published pursuant to 7 CFR 800.71(a)(1) multiplied by five. If you operate a business that buys, handles, weighs, or transports grain for sale in foreign commerce, or you are also in a control relationship with respect to a business that buys, handles, weighs, or transports grain for sale in interstate commerce, you must complete an application and pay the published fee.

(2) *Designation amendments.* The fee for amending designations will be published annually in the FEDERAL REGISTER and on the Agency's public

website. The Service will calculate the fee using the cost of publication plus one hour at the noncontract hourly rate. If you submit an application to amend a designation, you must pay the published fee.

[81 FR 49860, July 29, 2016, as amended at 81 FR 96340, Dec. 30, 2016; 83 FR 6453, Feb. 14, 2018; 83 FR 11633, Mar. 16, 2018; 83 FR 66585, Dec. 27, 2018; 86 FR 49469, Sept. 3, 2021]

§ 800.72 Explanation of additional service fees for services performed in the United States only.

(a) When transportation of the service representative to the service location (at other than a specified duty point) is more than 25 miles from an FGIS office, the actual transportation cost in addition to the applicable hourly rate for each service representative will be assessed from the FGIS office to the service point and return. When commercial modes of transportation (e.g., airplanes) are required, the actual expense incurred for the round-trip travel will be assessed. When services are provided to more than one applicant, the travel and other related charges will be prorated between applicants.

(b) In addition to a 2-hour minimum charge for service on Saturdays, Sundays, and holidays, an additional charge will be assessed when the revenue from the services in § 800.71(a)(1), schedule A, table 2, does not equal or exceed what would have been collected at the applicable hourly rate. The additional charge will be the difference between the actual unit fee revenue and the hourly fee revenue. Hours accrued for travel and standby time shall apply in determining the hours for the minimum fee.

[61 FR 43305, Aug. 22, 1996, as amended at 81 FR 49862, July 29, 2016]

§ 800.73 Computation and payment of service fees; general fee information.

(a) *Computing hourly rates.* The applicable hourly rate will be assessed in quarter hour increments for:

(1) Travel from the FGIS field office or assigned duty station to the service point and return;

(2) The performance of the requested service, less mealtime.

(b) *Application of fees when service is delayed or dismissed by the applicant.* The applicable hourly rate will be assessed for the entire period of scheduled service when:

(1) Service has been requested at a specified location;

(2) A service representative is on duty and ready to provide service but is unable to do so because of a delay not caused by the Service; and

(3 FGIS officials determine that the service representative cannot be utilized to provide service elsewhere without cost to the Service.

(c) *Application of fees when an application for service is withdrawn or dismissed.* The applicable hourly rate will be assessed to the applicant for the entire period of scheduled service if the request is withdrawn or dismissed after the service representative departs for the service point, or if the service request is not canceled by 2 p.m., local time, the business day preceding the date of scheduled service. However, the applicable hourly rate will not be assessed to the applicant if FGIS officials determine that the service representative can be utilized elsewhere or released without cost to the Service.

(d) *To whom fees are assessed.* Fees for inspection, weighing, and related services performed by service representatives, including additional fees as provided in § 800.72, shall be assessed to and paid by the applicant for the service.

(e) *Advance payment.* As necessary, the Administrator may require that fees shall be paid in advance of the performance of the requested service. Any fees paid in excess of the amount due shall be used to offset future billings, unless a request for a refund is made by the applicant.

(f) *Form of payment.* Bills for fees assessed under the regulations in this part for official services performed by FGIS shall be paid by check, draft, or money order, payable to the U.S. Department of Agriculture, Agricultural Marketing Service.

[61 FR 43305, Aug. 22, 1996, as amended at 69 FR 26490, May 13, 2004]

KINDS OF OFFICIAL SERVICES

§ 800.75 Kinds of official inspection and weighing services.

(a) *General.* Paragraphs (b) through (m) of this section describe the kinds of official service available. Each kind of service has several levels. §§ 800.115, 800.116, 800.117, and 800.118 explain Original Services, §§ 800.125, 800.126, 800.127, 800.128, and 800.129 explain Reinspection Services and Review of Weighing Services, and §§ 800.135, 800.136, 800.137, 800.138, and 800.139 explain Appeal Inspection Services. The results of each official service listed in paragraphs (b) through (j) will be certificated according to § 800.160.

(b) *Official sample-lot inspection service.* This service consists of official personnel (1) sampling an identified lot of grain and (2) analyzing the grain sample for grade, official factors, or official criteria, or any combination thereof, according to the regulations, Official U.S. Standards for Grain, instructions, and the request for inspection.

(c) *Warehouseman's sample-lot inspection service.* This service consists of (1) a licensed warehouseman sampler (i) sampling an identified lot of grain using an approved diverter-type mechanical sampler and (ii) sending the sample to official personnel and (2) official personnel analyzing the grain sample for grade, official factors, official criteria, or any combination thereof, according to the regulations, Official U.S. Standards for Grain, instructions, and the request for inspection.

(d) *Submitted sample inspection service.* This service consists of an applicant or an applicant's agent submitting a grain sample to official personnel, and official personnel analyzing the grain sample for grade, official factors, official criteria, or any combination thereof, according to the regulations, Official U.S. Standards for Grain, instructions, and the request for inspection.

(e) *Official sampling service.* This service consists of official personnel (1) sampling an identified lot of grain and (2) forwarding a representative portion(s) of the sample along with a copy of the certificate, as requested by the applicant.

(f) *Official stowage examination service.* (1) This service consists of official personnel visually determining if an identified carrier or container is clean; dry; free of infestation, rodents, toxic substances, and foreign odor; and is suitable to store or carry grain.

(2) A stowage examination may be obtained as a separate service or with one or more other services. Approval of the stowage space is required for official sample-lot inspection services on all export lots of grain and all official sample-lot inspection services performed on outbound domestic lots of grain which are sampled and inspected at the time of loading. Also, approval of the stowage space is required for any weighing services performed on all outbound land carriers.

(g) *Class X weighing service.* This service consists of official personnel (1) completely supervising the loading or unloading of an identified lot of grain and (2) physically weighing or completely supervising approved weighers weighing the grain.

(h) *Class Y weighing service.* This service consists of (1) approved weighers physically weighing the grain and (2) official personnel partially or completely supervising the loading or unloading of an identified lot of grain.

(i) *Checkweighing service (sacked grain).* This service consists of official personnel or approved weighers under the supervision of official personnel (1) physically weighing a selected number of sacks from a grain lot and (2) determining the estimated total gross, tare, and new weights, or the estimated average gross or net weight per filled sack according to the regulations, instructions, and request by the applicant.

(j) *Checkloading service.* This service consists of official personnel (1) performing a stowage examination; (2) computing the number of filled grain containers loaded aboard a carrier; and (3) if practicable, sealing the carrier for security.

(k) *Test weight reverification service.* This service consists of official personnel (1) comparing the weight of elevator test weights with known weights; (2) correcting the elevator test weights, when necessary; and (3) issuing a Report of Test.

(1) *Railroad track scale testing service.* This service consists of official personnel (1) testing railroad track scales with Service-controlled test cars and (2) issuing a Report of Test.

(m) *Hopper and truck scale testing service.* This service consists of official personnel (1) testing hopper and truck scales and (2) issuing a Report of Test.

(The information collection requirements contained in this section were approved by the Office of Management and Budget under control number 0580–0011)

[50 FR 45392, Oct. 31, 1985]

§ 800.76   Prohibited services; restricted services.

(a) *Prohibited services.* No agency shall perform any inspection function or provide any inspection service on the basis of unofficial standards, procedures, factors, or criteria if the agency is designated or authorized to perform the service or provide the service on an official basis under the Act. No agency shall perform official and unofficial weighing on the same mode of conveyance at the same facility.

(b) *Restricted services*—(1) *Not standardized grain.* When an inspection or weighing service is requested on a sample or a lot of grain which does not meet the requirements for grain as set forth in the Official U.S. Standards for Grain, a certificate showing the words "Not Standardized Grain" shall be issued according to the instructions.

(2) *Grain screening.* The inspection or weighing of grain screenings may be obtained from an agency or field office according to the instructions.

[50 FR 45393, Oct. 31, 1985, as amended at 60 FR 65235, Dec. 19, 1995; 63 FR 45677, Aug. 27, 1998]

INSPECTION METHODS AND PROCEDURES

§ 800.80   Methods and order of performing official inspection services.

(a) *Methods*—(1) *General.* All official inspection services shall be performed in accordance with methods and procedures prescribed in the regulations and the instructions.

(2) *Lot inspection services.* A lot inspection service shall be based on a representative sampling and examination of the grain in the entire lot, except as provided in § 800.85, and an ac-

curate analysis of the grain in the sample.

(3) *Stowage examination service.* A stowage examination service shall be based on a thorough and accurate examination of the carrier or container into which grain will be loaded.

(4) *Submitted sample inspection service.* A submitted sample inspection service shall be based on a submitted sample of sufficient size to enable official personnel to perform a complete analysis for grade. If a complete analysis for grade cannot be performed because of an inadequate sample size or other conditions, the request for service shall be dismissed or a factor only inspection may be performed upon request.

(5) *Reinspection and appeal inspection service.* A reinspection, appeal inspection, or Board appeal inspection service shall be based on an independent review of official grade information, official factor information, or other information consistent with the scope of the original inspection.

(b) *Order of service.* Official inspection services shall be performed, to the extent practicable, in the order in which they are received. Priority shall be given to inspections required for export grain. Priority may be given to other kinds of inspection services under the Act with the specific approval of the Service.

(c) *Recording receipt of documents.* Each document submitted by or on behalf of an applicant for inspection services shall be promptly stamped or similarly marked by official personnel to show the date of receipt.

(d) *Conflicts of interest.* No official personnel shall perform or participate in performing an official inspection service on grain or on a carrier or container in which they have a direct or indirect financial interest.

[50 FR 49669, Dec. 4, 1985]

§ 800.81   Sample requirements; general.

(a) *Samples for official sample-lot inspection service*—(1) *Original official sample-lot inspection service.* For original sample-lot inspection purposes, an official sample shall be obtained by official personnel; representative of the grain in the lot; and protected from manipulation, substitution, and improper or careless handling.

(2) *Official sample-lot reinspection and appeal inspection service.* For an official sample-lot reinspection service or an official appeal sample-lot inspection service, the sample(s) on which the reinspection or appeal is determined shall (i) be obtained by official personnel and (ii) otherwise meet the requirements of paragraph (a)(1) of this section. If the reinspection or appeal inspection is determined on the basis of official file sample(s), the samples shall meet the requirements of § 800.82(d).

(3) *New sample.* Upon request and if practicable, a new sample shall be obtained and examined as a part of a reinspection or appeal inspection. The provision for a new sample shall not apply if obtaining the new sample involves a change in method of sampling.

(b) *Representative sample.* A sample shall not be considered representative unless it (1) has been obtained by official personnel, (2) is of the size prescribed in the instructions, and (3) has been obtained, handled, and submitted in accordance with the instructions. A sample which fails to meet the requirements of this paragraph may, upon request of the applicant, be inspected as a submitted sample.

(c) *Protecting samples.* Official personnel shall protect official samples, warehouseman's samples, and submitted samples from manipulation, substitution, or improper and careless handling which may deprive the samples of their representativeness or which may change the physical or chemical properties of the grain, as appropriate, from the time of sampling or receipt until the inspection services are completed and the file samples have been discarded.

(d) *Restriction on sampling.* Official personnel shall not perform an original inspection or a reinspection service on an official sample or a warehouseman's sample unless the grain from which the sample was obtained was located within the area of responsibility assigned to the agency or field office at the time of sampling, except as provided for in § 800.117, or on a case-by-case basis as determined by the Administrator.

(e) *Disposition of samples—(1) Excess grain.* Any grain in excess of the quantity specified in the instructions for the requested service, the file samples,

and samples requested by interested persons shall be returned to the lot from which the grain was obtained or to the owner of the lot or the owner's order.

(2) *Inspection samples.* Inspection samples, after they have served their intended purpose, shall be disposed of as follows:

(i) Samples which contain toxic substances or materials shall be kept out of food and feed channels, and

(ii) Official personnel shall dispose of samples obtained or submitted to them according to procedures established by the Service. Complete and accurate records of disposition shall be maintained.

(Approved by the Office of Management and Budget under control number 0580–0013)

[50 FR 49669, Dec. 4, 1985, as amended at 68 FR 19138, Apr. 18, 2003]

§ 800.82 **Sampling provisions by level of service.**

(a) *Original inspection service—(1) Official sample-lot inspection service.* Each original inspection service shall be performed on the basis of one or more official samples obtained by official personnel from grain in the lot and forwarded to the appropriate agency or field office.

(2) *Warehouseman's sample-lot inspection.* Each original warehouseman's sample-lot inspection service shall be performed on the basis of samples obtained by a licensed warehouseman and sent to the appropriate agency or field office in whose circuit the warehouse is located.

(3) *Submitted sample service.* Each original submitted sample inspection service shall be performed on the basis of the sample as submitted.

(b) *Reinspection, and appeal inspection services—(1) Official sample-lot inspection service.* Each of these inspection services shall be performed on the basis of official samples as available, including file samples, at the time the service is requested. In performing these services, a sample obtained with an approved diverter-type mechanical sampler or with a pelican sampler generally shall be used with respect to quality factors and official criteria, and a sample obtained with a probe at the time of the reinspection or appeal, generally, shall be used with respect to heating, musty, sour, insect infestation, and other condition and odor factors. In instances where original inspection results are based on samples obtained by probe, the decision as to whether file samples or new samples obtained by probe are to be used shall be made by the official personnel performing the service.

(2) *Warehouseman's sample-lot inspection service.* Each reinspection service and appeal inspection service on a warehouseman's sample shall be performed on an analysis of the official file sample.

(3) *Submitted sample service.* Each reinspection service and appeal inspection service on a submitted sample shall be performed on an analysis of the official file sample.

(c) *Board appeal inspection services.* Board appeal inspection services shall be performed on an analysis of the official file sample.

(d) *Use of file samples*—(1) *Requirements for use.* A file sample that is retained by official personnel in accordance with the procedures prescribed in the instructions may be considered representative for a reinspection service, appeal inspection service, and a Board appeal inspection service if (i) the file sample has remained at all times in the custody and control of the official personnel that performed the inspection service in question; and (ii) the official personnel who performed the original inspection service and those who are to perform the reinspection, the appeal inspection, or the Board appeal inspection service determine that the samples were representative at the time the original inspection service was performed and that the quality or condition of the grain in the samples has not changed.

(2) *Certificate statement.* When the results of a reinspection, appeal inspection, or Board appeal inspection service are based on an official file sample, the certificate for the reinspection service, the appeal inspection service, and the Board appeal inspection service shall show a statement, as specified in the instructions, indicating that the results are based on the official file sample.

[50 FR 49670, Dec. 4, 1985]

§ 800.83 Sampling provisions by kind of movement.

(a) *Export cargo movements*—(1) *Bulk grain.* Except as may be approved by the Administrator on a shipment-by-shipment basis in an emergency, each inspection for official grade, official factor, or official criteria on an export cargo shipment of bulk grain shall be performed on official samples obtained from the grain (i) as the grain is being loaded aboard the final carrier; (ii) after the final elevation of the grain prior to loading and as near to the final loading spout as is physically practicable (except as approved by the Administrator when representative samples can be obtained before the grain reaches the final loading spout); and (iii) by means of a diverter-type mechanical sampler approved by the Service and operated in accordance with instructions. If an approved diverter-type mechanical sampler is not properly installed at an elevator or facility as required, each certificate issued at that elevator or facility for an export cargo shipment of bulk grain shall show a statement indicating the type of approved sampling method used, as prescribed in the instructions.

(2) *Sacked grain.* Each inspection for official grade, official factor, or official criteria on an export cargo shipment of sacked grain shall be performed on official samples obtained from the grain by any sampling method approved by the Service and operated in accordance with instructions.

(b) *Other movements.* Each inspection for official grade, official factor, or official criteria on a domestic cargo movement ("In," "Out," or en route barge movement), a movement in a land carrier (any movement in a railcar, truck trailer, truck/trailer combination, or container), or a "LOCAL" movement of bulk or sacked grain shall be performed on official samples obtained from the grain by any sampling method approved by the Service and operated in accordance with the instructions.

[50 FR 49670, Dec. 4, 1985]

§ 800.84 Inspection of grain in land carriers, containers, and barges in single lots.

(a) *General.* The inspection of bulk or sacked grain loaded or unloaded from any carrier or container, except shiplot grain, must be conducted in accordance with the provision in this section and procedures prescribed in the instructions. Applicant must provide written instructions to official personnel, reflecting contract requirements for quality and quantity for the inspection of multiple carriers graded on a composite grade or average grade basis.

(b) *Single and multiple grade procedure*—(1) *Single grade.* When grain in a carrier(s) is/are offered for inspection as one lot and the grain is found to be uniform in condition, the grain must be sampled, inspected, graded, and certified as one lot. For the purpose of this paragraph, condition only includes the factors heating and odor.

(i) *Composite grade.* Grain loaded in multiple carriers offered for inspection may be combined into a single sample for grade analysis and certified as a single lot, *provided that* the grain in each individual carrier is inspected and found uniform in respect to odor, condition, and insect infestation, and sampling is performed at the individual loading location in a reasonably continuous operation. The maximum number of individual units that may be combined to form a composite grade analysis is 20 containers, 5 railcars, or 15 trucks. Composite analysis must be restricted to carriers inspected within the official service provider's area of responsibility.

(ii) *Average grade.* Grain loaded in multiple carriers offered for inspection may be graded individually, then averaged for certification as a single lot, *provided that:* the grain in each individual carrier is inspected and graded as an individual unit; the grain is found to be uniform in respect to odor, condition, and insect infestation; and sampling is performed at the individual loading location in a reasonably continuous operation. The maximum number of individual units that may be combined to form an average grade analysis is 20 containers, 5 railcars, or 15 trucks. Average grade analysis is restricted to carriers inspected within the official service provider's area of responsibility.

(2) *Multiple grade.* When grain in a carrier is offered for inspection as one lot and the grain is found to be not uniform in condition because portions of the grain are heating or have an odor, the grain in each portion will be sampled, inspected, and graded separately; but the results must be shown on one certificate. The certificate must show the approximate quantity or weight of each portion, the location of each portion in the carrier or container, and the grade of the grain in each portion. The requirements of this section are not applicable when an applicant requests that the grade of the entire carrier be based on a determination of heating or odor when only a portion of the carrier is found to be heating or have an odor.

(3) *Infested.* If any portion of grain in a lot is found to be infested, according to applicable provisions of the Official U.S. Standards for Grain, the entire lot shall be considered infested. When grain in railcars or trucks with permanently enclosed tops is considered infested, the applicant for inspection shall be promptly notified and given the option of (i) receiving a grade certificate with a special grade designation indicating that the entire lot is infested or (ii) fumigating the grain in the lot in accordance with instructions and receiving a grade certificate without the special grade designation.

(c) *One certificate per carrier: exceptions.* Except as provided in this paragraph, one official certificate must be issued for the inspection of the grain in each truck, trailer, truck/trailer(s) combination, container, railcar, barge, or similarly-sized carrier, or composite/average grade analysis on multiple carrier units. The requirements of this paragraph are not applicable:

(1) When grain is inspected in a combined lot under § 800.85;

(2) When grain is inspected under paragraph (d) of this section; or

(3) When certification is at the option of the applicant in accordance with instructions.

(d) *Bulkhead lots.* If grain in a carrier is offered for official inspection as two or more lots and the lots are separated by bulkheads or other partitions, the grain in each lot shall be sampled, inspected, and graded separately in accordance with paragraphs (a) and (b) of this section. An official certificate shall be issued for each lot inspected. Each certificate shall show the term "Bulkhead Lot," the approximate quantity or weight of the grain in the lot, the location of the lot in the carrier, and the grade of the grain in the lot.

(e) *Bottom not sampled.* If bulk grain offered for official inspection is at rest in a carrier or container and is fully accessible for sampling in an approved manner, except that the bottom of the carrier or container cannot be reached with each probe, the grain shall be sampled as thoroughly as possible with an approved probe. The grain in the resulting samples shall be inspected, graded, and certificated, except that each certificate shall show a statement, as specified in the instructions, indicating the depth probed. Any inspection which is based on a sample that does not represent the entire carrier or container does not meet the mandatory inspection requirements of section 5(a)(1) of the Act.

(f) *Partial inspection—heavily loaded—* (1) *General.* When an "In" movement of bulk grain is offered for inspection at rest in a carrier or container and is loaded in such a manner that it is possible to secure only door-probe or shallow-probe samples, the container shall be considered to be "heavily loaded," and the request for inspection either shall be dismissed or a partial inspection shall be made. If the request is for the inspection of an "Out" movement of grain, the request shall be dismissed on the grounds that the grain is not accessible for a correct "Out" inspection.

(2) *Certification procedure.* If a partial inspection is made, the grain will be sampled as thoroughly as possible with an approved probe and inspected, graded, and a "partial inspection—heavily loaded" certificate issued. The certificate shall show the words "Partial inspection—heavily loaded" in the space provided for remarks. The type of samples that were obtained shall be described in terms of "door probe" or "shallow probe."

(3) *Reinspection and appeal inspection procedure.* A request for a reinspection or an appeal inspection service on grain in a carrier or container that is certificated as "partial inspection—heavily loaded" shall be dismissed in accordance with § 800.48(a)(4).

(4) *Restriction.* No "partial inspection—heavily loaded" certificate shall be issued for sacked grain or any inspection other than the inspections described in paragraphs (f)(1) through (4) of this section and § 800.85(h)(2).

(g) *Part lots*—(1) *General.* If a portion of the grain in a carrier or container is removed, the grain that is removed and the grain remaining shall be considered separate lots. When an official inspection service is requested on either portion, the grain shall be sampled, inspected, graded, and a "part-lot" inspection certificate issued.

(2) *Grain remaining in carrier or container.* The certificate for grain remaining in a carrier or container shall show (i) the following completed statement: "Partly unloaded; results based on portion remaining in (show carrier or container identification)," (ii) the term "Part lot" following the quantity information, (iii) the identification of the carrier or container, and (iv) the estimated amount and location of the part lot.

(3) *Grain unloaded from carrier or container.* If grain is sampled by official personnel during unloading, the certificate for the grain that is unloaded shall show (i) the completed statement: "Part lot; results based on portion removed from (show carrier identification)" and (ii) the term "Part lot" following the quantity information. If the grain is not sampled by official personnel during unloading, the certificate may, upon request of the applicant, show a completed statement such as "Applicant states grain is ex-car " or "Applicant states grain is ex-barge ," but the certificate shall not otherwise show a carrier or container identification or the term "Part lot."

(h) *Identification for compartmented cars.* The identification for compartments in a compartmented railcar shall, in the absence of readily visible markings, be stated in terms of the location of the grain in a compartment, with the first compartment at the brake end of the car being identified as B-1, and the remaining compartments being numbered consecutively towards the other end of the car.

[50 FR 49671, Dec. 4, 1985, as amended at 57 FR 11428, Apr. 3, 1992; 78 FR 43756, July 22, 2013]

§ 800.85 Inspection of grain in combined lots.

(a) *General.* The official inspection for grade of bulk or sacked grain loaded aboard, or being loaded aboard, or discharged from two or more carriers or containers (including barges designed for loading aboard a ship) as a combined lot shall be performed according to the provisions of this section and procedures prescribed in the instructions.

(b) *Application procedure*—(1) For inspection during loading, unloading, or at rest. Applications for official inspection of grain as a combined lot must:

(i) Be filed in accordance with § 800.116;

(ii) Show the estimated quantity of grain that is to be certified as one lot;

(iii) Show the contract grade, and if applicable; other inspection criteria required by the contract; and

(iv) Identify each carrier into which grain is being loaded or from which grain is being unloaded.

(2) *Recertification.* An application for recertification as a combined lot of grain that has been officially inspected and certificated as two or more single lots shall (i) be filed not later than 2 business days after the latest inspection date of the single lots and (ii) show information specified in paragraph (b)(1) of this section.

(c) *Inspection procedure; general—land carriers and barges*—(1) *Inspection during loading, or unloading, or at rest.* Grain in two or more land carriers or barges that are to be officially inspected as a combined lot, must be sampled in a reasonably continuous operation. Representative samples must be obtained from the grain in each individual carrier and inspected in accordance with procedures as prescribed in the instructions.

(2) *Recertification.* Grain that has been officially inspected and certified as two or more single, composite, or average quality lots may be recertified as a combined lot provided that:

(i) The grain in each lot was sampled in a reasonably continuous operation;

(ii) The original inspection certificates issued for the single, composite, or average quality lots have been surrendered to official personnel;

(iii) Representative file samples of the single, composite, or average quality lots are available;

(iv) The grain in the single, composite, or average quality lots is of the same grade or better grade and quality than as specified in the written instructions provided by the shipper;

(v) Official personnel who performed the inspection service for the single, composite, or average quality lots and the official personnel who are to recertify the grain as a combined lot must determine that the samples used as a basis for the inspection of the grain in the single, composite, or average quality lots were representative at the time of sampling and have not changed in quality or condition; and

(vi) The quality or condition of the grain meets uniformity requirements established by the Service for official inspection of grain in combined lots.

(d) *Weighted or mathematical average.* Official factor and official criteria information shown on a certificate for grain in a combined lot shall, subject to the provisions of paragraphs (e) through (g) of this section, be based on the weighted or mathematical averages of the analysis of the sublots in the lot and shall be determined in accordance with the instructions.

(e) *Infested grain.* If the grain in a combined lot is offered for official inspection as it is being loaded aboard a carrier and the grain, or a portion of the grain, in a lot is found to be infested, according to applicable provisions of the Official U.S. Standards for Grain, the applicant shall be notified and may exercise options specified in the instructions. When grain in railcars or trucks with permanently enclosed tops is considered infested, the applicant shall be given the option of (1) receiving a grade certificate with a special grade designation indicating that the entire lot is infested or (2) fumigating the grain in the lot in accordance with instructions and receiving a grade certificate without the special grade designation.

(f) *Grain uniform in quality.* Samples obtained from grain officially inspected as a combined lot shall be examined for uniformity of quality. If the grain in the samples is found to be uniform in quality and the grain is loaded aboard or is unloaded from the carriers in a reasonably continuous operation, the grain in the combined lot shall be officially inspected and certificated as one lot. The requirements of this paragraph (f) and paragraph (c) of this section with respect to reasonably continuous loading or unloading do not apply to grain which is at rest in carriers when the grain is offered for inspection.

(g) *Grain not uniform in quality.* When grain officially inspected as a combined lot is found to be not uniform in quality or if the grain is not loaded or unloaded in a reasonably continuous operation, the grain in each portion, and any grain which is loaded or unloaded at different times, shall be officially sampled, inspected, graded, and certificated as single lots.

(h) *Special certification procedures*—(1) *Grain not uniform in quality.* When grain in a combined lot is found to be not uniform in quality under paragraph (g) of this section, the official inspection certificate for each portion of different quality shall show (i) the grade, identification, and approximate quantity of the grain and (ii) other information required by the instructions.

(2) *Partial inspection.* When an inbound movement of bulk grain is offered for official inspection at rest as a combined lot and all carriers are not fully accessible for sampling, the request for official inspection either shall be dismissed or a combined lot inspection shall be made on those carriers that are accessible. Those lots that are not accessible shall be handled in accordance with § 800.84. If the request is for an official inspection service on an outbound movement of grain at rest in a combined lot, the request shall be dismissed on the ground that the grain is not accessible for a correct "Out" inspection.

(3) *Official mark.* If grain in a combined lot is inspected for grade as it is being loaded aboard two or more carriers, upon request of the applicant, the following mark shall be shown on the inspection certificate: "Loaded under continuous official inspection" or "Loaded under continuous official inspection and weighing."

(4) *Combined-lot certification; general.* Each official certificate for a combined-lot inspection service must show the identification for the "combined lot" or, at the request of the applicant, the identification of each carrier in the combined lot. If the identification of each carrier is not shown, the statement "Carrier identification available on the official work record" must be shown on the inspection certificate in the space provided for remarks. The identification and any seal information for the carriers may be shown in the Remarks section on the reverse side of the inspection certificate, provided that the statement "See reverse side" is shown on the face of the certificate in the space provided for remarks, or on an additional page.

(5) *Recertification.* If a request for a combined-lot inspection service is filed after the grain has been officially inspected and certified as single, composite, or average quality lots, the combined-lot inspection certificate must show, in addition to the requirements of paragraph (h)(4) of this section the following:

(i) The date of inspection of the grain in the combined lot (if the single, composite, or average quality lots were inspected on different dates, the latest of the dates must be shown);

(ii) A serial number other than the serial numbers of the official inspection certificates that are to be superseded;

(iii) The location of the grain, if at rest, or the name(s) of the elevator(s) from which or into which the grain in the combined lot was loaded or unloaded;

(iv) A statement showing the approximate quantity of grain in the combined lot;

(v) A completed statement showing the identification of any superseded certificates; and

(vi) If at the time of issuing the combined-lot inspection certificate the superseded certificates are not in the custody of the official personnel, a statement indicating that the superseded certificates have not been surrendered must be clearly shown in the space provided for remarks. If the superseded certificates are in the custody of official personnel, the superseded certificates must be clearly marked "Void."

(i) *Further combining.* After a combined-lot inspection certificate has been issued, there shall be no further combining and no dividing of the certificate.

(j) *Limitation.* No combined-lot inspection certificate shall be issued (1) for any official inspection service other than as described in this section or (2) which shows a quantity of grain in excess of the quantity in the single lots.

[50 FR 49672, Dec. 4, 1985, as amended at 78 FR 43756, July 22, 2013]

§ 800.86 Inspection of shiplot, unit train, and lash barge grain in single lots.

(a) *General.* Official inspection for grade of bulk or sacked grain aboard, or being loaded aboard, or being unloaded from a ship, unit train, or lash barges as a single lot shall be performed according to the provisions of this section and procedures prescribed in the instructions.

(b) *Application procedure.* Applications for the official inspection of shiplot, unit train, and lash barges as a single lot shall:

(1) Be filed in advance of loading or unloading;

(2) Show the estimated quantity of grain to be certificated;

(3) Show the contract grade and official criteria if applicable; and

(4) Identify the carrier and stowage area into which the grain is being loaded, or from which the grain is being unloaded, or in which the grain is at rest.

(c) *Inspection procedures*—(1) *General information.* Shiplot, unit train, and lash barge grain officially inspected as a single lot shall be sampled in a reasonably continuous operation. Representative samples shall be obtained from the grain offered for inspection and inspected and graded in accordance with a statistical acceptance sampling and inspection plan according to the provisions of this section and procedures prescribed in the instructions.

(2) *Tolerances.* The probability of accepting or rejecting portions of the lot during loading or unloading is dependent on inspection results obtained from preceding portions and the applied breakpoints and procedures. Breakpoints shall be periodically reviewed and revised based on new estimates of inspection variability. Tables 1 through 24 list the breakpoints for all grains.

TABLE 1—GRADE LIMITS (GL) AND BREAKPOINTS (BP) FOR SIX-ROWED MALTING BARLEY

Grade	Minimum limits of—						Maximum limits of—											
	Test weight per bushel (pounds)		Suitable malting types (percent)		Sound barley [1] (percent)		Damaged kernels [1] (percent)		Wild oats (percent)		Foreign material (percent)		Other grains (percent)		Skinned and broken kernels (percent)		Thin barley (percent)	
	GL	BP	GL	BP	GL	BP	GL	BP	GL	BP	GL	BP	GL	BP	GL	BP	GL	BP
U.S. No. 1	47.0	−0.5	97.0	−1.0	98.0	−0.8	2.0	0.8	1.0	0.6	0.5	0.1	2.0	0.8	4.0	1.1	7.0	0.6
U.S. No. 2	45.0	−0.5	97.0	−1.0	98.0	−0.8	3.0	0.9	1.0	0.6	1.0	0.4	3.0	0.9	6.0	1.4	10.0	0.9
U.S. No. 3	43.0	−0.5	95.0	−1.3	96.0	−1.1	4.0	1.1	2.0	0.8	2.0	0.5	5.0	1.3	8.0	1.5	15.0	0.9
U.S. No. 4	43.0	−0.5	95.0	−1.3	93.0	−1.1	5.0	1.3	3.0	0.9	3.0	0.6	5.0	1.3	10.0	1.6	15.0	0.9

[1] Injured-by-frost kernels and injured-by-mold kernels are not considered damaged kernels or considered against sound barley.

TABLE 2—GRADE LIMITS (GL) AND BREAKPOINTS (BP) FOR TWO-ROWED MALTING BARLEY

Grade	Minimum limits of—						Maximum limits of—											
	Test weight per bushel (pounds)		Suitable malting types (percent)		Sound barley [1] (percent)		Damaged kernels [1] (percent)		Wild oats (percent)		Foreign material (percent)		Other grains (percent)		Skinned and broken kernels (percent)		Thin barley (percent)	
	GL	BP	GL	BP	GL	BP	GL	BP	GL	BP	GL	BP	GL	BP	GL	BP	GL	BP
U.S. No. 1	50.0	−0.5	97.0	−1.0	98.0	−0.8	2.0	0.8	1.0	0.6	0.5	0.1	2.0	0.8	4.0	1.1	5.0	0.4
U.S. No. 2	48.0	−0.5	97.0	−1.0	98.0	−0.8	3.0	0.9	1.0	0.6	1.0	0.4	3.0	0.9	6.0	1.4	7.0	0.5
U.S. No. 3	48.0	−0.5	95.0	−1.3	96.0	−1.1	4.0	1.1	2.0	0.8	2.0	0.5	5.0	1.3	8.0	1.5	10.0	0.9
U.S. No. 4	48.0	−0.5	95.0	−1.3	93.0	−1.1	5.0	1.3	3.0	0.9	3.0	0.6	5.0	1.3	10.0	1.6	10.0	0.9

[1] Injured-by-frost kernels and injured-by-mold kernels are not considered damaged kernels or considered against sound barley.

NOTE: Malting barley must not be infested in accordance with §810.107(b) and must not contain any special grades as defined in §810.206. Six- and two-rowed barley varieties not meeting the above requirements must be graded in accordance with standards established for the class Barley.

TABLE 3—GRADE LIMITS (GL) AND BREAKPOINTS (BP) FOR BARLEY

Grade	Minimum limits of—				Maximum limits of—								
	Test weight per bushel (pounds)		Sound barley (percent)		Damaged kernels [1] (percent)		Heat damaged kernels (percent)		Foreign material (percent)		Broken kernels (percent)		Thin barley (percent)
	GL	BP	GL	BP	GL	BP	GL	BP	GL	BP	GL	BP	GL	BP
U.S. No. 1	47.0	−0.5	97.0	−1.1	2.0	0.8	0.2	0.1	1.0	0.4	4.0	1.0	10.0	0.9
U.S. No. 2	45.0	−0.5	94.0	−1.4	4.0	1.0	0.3	0.1	2.0	0.4	8.0	1.5	15.0	0.9
U.S. No. 3	43.0	−0.5	90.0	−1.6	6.0	1.4	0.5	0.2	3.0	0.5	12.0	1.8	25.0	1.3
U.S. No. 4	40.0	−0.5	85.0	−2.2	8.0	1.5	1.0	0.5	4.0	0.5	18.0	1.8	35.0	1.9
U.S. No. 5	36.0	−0.5	75.0	−2.2	10.0	1.8	3.0	0.6	5.0	0.6	28.0	2.4	75.0	2.3

[1] Includes heat-damaged kernels. Injured-by-frost kernels and injured-by-mold kernels are not considered damaged kernels.

TABLE 4—BREAKPOINTS FOR BARLEY SPECIAL GRADES AND FACTORS

Special grade or factor	Grade or range limit	Breakpoint
Dockage	As specified by contract or load order	0.23
Two-rowed Barley	Not more than 10.0% of Six-rowed in Two-rowed	1.8
Six-rowed Barley	Not more than 10.0% of Two-rowed in Six-rowed	1.8
Malting (Blue Aleurone Layers)	Not less than 90.0%	−1.3
Malting (White Aleurone Layers)	Not less than 90.0%	−1.3
Smutty	More than 0.20%	0.06
Garlicky	3 or more in 500 grams	2⅓
Ergoty	More than 0.10%	0.13
Infested	Same as in §810.107	0
Blighted	More than 4.0%	1.1
Injured-by-Frost Kernels	Not more than 1.9%	0.1
Injured-by-Heat Kernels	Not more than 0.2%	0.04
Frost-damaged Kernels	Not more than 0.4%	0.05
Heat-damaged Kernels	Not more than 0.1%	0.1
Other Grains	Not more than 25.0%	2.4
Moisture	As specified by contract or load order grade	0.5

TABLE 5—GRADE LIMITS (GL) AND BREAKPOINTS (BP) FOR CORN

Grade	Minimum test weight per bushel (pounds)		Maximum limits of—					
			Damaged kernels					
			Heat-damaged kernels (percent)		Total (percent)		Broken corn and foreign material (percent)	
	GL	BP	GL	BP	GL	BP	GL	BP
U.S. No. 1	56.0	−0.4	0.1	0.1	3.0	1.0	2.0	0.2
U.S. No. 2	54.0	−0.4	0.2	0.2	5.0	1.3	3.0	0.3
U.S. No. 3	52.0	−0.4	0.5	0.3	7.0	1.5	4.0	0.3
U.S. No. 4	49.0	−0.4	1.0	0.5	10.0	1.8	5.0	0.4
U.S. No. 5	46.0	−0.4	3.0	0.9	15.0	2.1	7.0	0.4

TABLE 6—BREAKPOINTS FOR CORN SPECIAL GRADES AND FACTORS

Special grade or factor	Grade limit	Breakpoint
Flint	95 percent or more of flint corn	−1.0
Flint and Dent	More than 5 percent, but less than 95 percent of flint corn	1.0 or −1.0
Infested	Same as in §810.107	0
Corn of other colors:		
White	Not more than 2.0 percent	0.8
Yellow	Not more than 5.0 percent	1.0
Waxy	95 percent or more	−3.0
High BCFM	As specified by contract or load order grade	10 percent of the load order grade
Moisture	As specified by contract or load order grade	0.4

TABLE 7—GRADE LIMITS (GL) AND BREAKPOINTS (BP) FOR FLAXSEED

| Grade | Minimum test weight per bushel (pounds) | | Maximum limits of-damaged kernels | | | |
| | | | Heat-damaged kernels (percent) | | Total (percent) | |
	GL	BP	GL	BP	GL	BP
U.S. No. 1	49.0	−0.1	0.2	0.1	10.0	0.9
U.S. No. 2	47.0	−0.1	0.5	0.1	15.0	1.1

TABLE 8—BREAKPOINTS FOR FLAXSEED SPECIAL GRADES AND FACTORS

Special grade or factor	Grade limit	Breakpoint
Moisture	As specified by load order or contract grade	0.4
Dockage	0.99 percent or above	0.32

TABLE 9—GRADE LIMITS (GL) AND BREAKPOINTS (BP) FOR MIXED GRAIN

Grade	Maximum Limits of—				
	Moisture (percent)	Damaged kernels			
		Total (percent)		Heat-damaged kernels (percent)	
		GL	BP	GL	BP
U.S. Mixed Grain	16.0	15.0	0.6	3.0	0.4

NOTE: There is no tolerance for U.S. Sample grade Mixed Grain.

TABLE 10—BREAKPOINTS FOR MIXED GRAIN SPECIAL GRADES AND FACTORS

Special grade or factor	Grade limit	Breakpoint
Smutty	15 or more in 250 grams (wheat, rye, or triticale predominates)	6
	More than 0.2% (all other mixtures)	0.05
Ergoty	More than 0.30% (rye wheat predominates)	0.13
	More than 0.10% (all other mixtures)	0
Garlicky	2 or more per 1,000 grams (wheat, rye, or triticale predominates)	1
	4 or more per 500 grams (all other mixtures)	2
Infested	Same as in § 810.107	0
Blighted	More than 4.0% (barley predominates)	1.1
Treated	Same as in § 810.805	0
Moisture	As specified by contract or load order grade	0.5

TABLE 11—GRADE LIMITS (GL) AND BREAKPOINTS (BP) FOR OATS

| Grade | Minimum limits of— | | | | Maximum limits of— | | | | | |
| | Test weight per bushel (pounds) | | Sound Oats (percent) | | Heat-damaged kernels (percent) | | Foreign material (percent) | | Wild Oats (percent) | |
	GL	BP	GL	BP	GL	BP	GL	BP	GL	BP
U.S. No. 1	36.0	−0.5	97.0	−0.8	0.1	0.1	2.0	0.4	2.0	0.6
U.S. No. 2	33.0	−0.5	94.0	−1.2	0.3	0.4	3.0	0.4	3.0	0.8
U.S. No. 3[1]	30.0	−0.5	90.0	−1.4	1.0	0.5	4.0	0.5	5.0	1.1
U.S. No. 4[2]	27.0	−0.5	80.0	−1.9	3.0	0.8	5.0	0.5	10.0	1.4

[1] Oats that are Slightly Weathered shall be graded not higher than U.S. No. 3.
[2] Oats that are Badly Stained or Materially Weathered shall be graded not higher than U.S. No. 4.

TABLE 12—BREAKPOINTS FOR OATS SPECIAL GRADES AND FACTORS

Special grade or factors	Grade limit	Breakpoint
Heavy	38 pounds or more	−0.5
Extra Heavy	40 pounds or more	−0.5
Moisture	As specified by contract or load order grade	0.5
Thin	More than 20.0%	0.5
Smutty	More than 0.2%	0.05

TABLE 12—BREAKPOINTS FOR OATS SPECIAL GRADES AND FACTORS—Continued

Special grade or factors	Grade limit	Breakpoint
Ergoty	More than 0.10%	0.10
Garlicky	4 or more in 500 grams	2⅓
Infested	Same as in §810.107	0
Bleached	Same as in §810.1005	0

TABLE 13—GRADE LIMITS (GL) AND BREAKPOINTS (BP) FOR RYE

Grade	Minimum test weight per bushel (pounds)		Foreign Material				Damaged kernels(percent)				Thin rye (percent)	
			Foreign matter other than wheat (percent)		Total (percent)		Heat-damaged (percent)		Total (percent)			
	GL	BP	GL	BP	GL	BP	GL	BP	GL	BP	GL	BP
U.S. No. 1	56.0	−0.5	1.0	0.4	3.0	0.8	0.2	0.1	2.0	0.8	10.0	0.6
U.S. No. 2	54.0	−0.5	2.0	0.5	6.0	1.1	0.2	0.1	4.0	1.1	15.0	0.8
U.S. No. 3	52.0	−0.5	4.0	0.8	10.0	1.4	0.5	0.4	7.0	1.4	25.0	0.9
U.S. No. 4	49.0	−0.5	6.0	0.8	10.0	1.4	3.0	0.8	15.0	2.0		

TABLE 14—BREAKPOINTS FOR RYE SPECIAL GRADES AND FACTORS

Special grade or factor	Grade limit	Breakpoint
Moisture	As specified by contract or load order grade	0.3
Light Garlicky	2 or more per 1,000 grams	1⅓
Garlicky	More than 6 per 1,000 grams	7⅓
Ergoty	More than 0.30%	0.10
Plump	Not more than 5.0% through 0.064 × 3/8 sieve	0.5
Light Smutty	More than 14 per 250 grams	6
Smutty	More than 30 per 250 grams	10
Infested	Same as in §810.107	0
Dockage	As specified by contract or load order grade	0.2

TABLE 15—GRADE LIMITS (GL) AND BREAKPOINTS (BP) FOR SORGHUM

Grade	Minimum test weight per bushel (pounds)		Damaged kernels				Broken kernels and foreign material			
			Heat-damaged (percent)		Total (percent)		Total (percent)		Foreign material (percent)	
	GL	BP	GL	BP	GL	BP	GL	BP	GL	BP
U.S. No. 1	57.0	−0.4	0.2	0.1	2.0	1.1	3.0	0.5	1.0	0.4
U.S. No. 2	55.0	−0.4	0.5	−0.4	5.0	1.8	6.0	0.6	2.0	0.5
U.S. No. 3 [1]	53.0	−0.4	1.0	0.5	10.0	2.3	8.0	0.7	3.0	0.6
U.S. No. 4	51.0	−0.4	3.0	0.8	15.0	2.8	10.0	0.8	4.0	0.7

[1] Sorghum that is distinctly discolored shall be graded not higher than U.S. No. 3.

TABLE 16—BREAKPOINTS FOR SORGHUM SPECIAL GRADES AND FACTORS

Special grade or factors	Grade limit	Breakpoint
Class Tannin	Not less than 90.0%	−1.9
Sorghum	Not less than 97.0%	−1.0
White	Not less than 98.0%	−0.9
Smutty	20 or more in 100 grams	8
Infested	Same as in §810.107	0
Dockage	0.99% and above	0.32
Moisture	As specified by contract or load order grade	0.5

TABLE 17—GRADE LIMITS (GL) AND BREAKPOINTS (BP) FOR SOYBEANS

Grade	Maximum limits of—									
	Damaged kernels				Foreign material (percent)		Splits (percent)		Soybeans of other colors (percent)	
	Heat-damaged (percent)		Total (percent)							
	GL	BP	GL	BP	GL	BP	GL	BP	GL	BP
U.S. No. 1	0.2	0.2	2.0	0.8	1.0	0.2	10.0	1.6	1.0	0.7
U.S. No. 2	0.5	0.3	3.0	0.9	2.0	0.3	20.0	2.2	2.0	1.0
U.S. No. 3[1]	·1.0	0.5	5.0	1.2	3.0	0.4	30.0	2.5	5.0	1.6
U.S. No. 4[2]	3.0	0.9	8.0	1.5	5.0	0.5	40.0	2.7	10.0	2.3

[1] Soybeans that are purple mottled or stained which will not be graded higher than U.S. No. 3.
[2] Soybeans that are materially weathered which will not be graded not higher than U.S. No. 4.

TABLE 18—BREAKPOINTS FOR SOYBEAN SPECIAL GRADES AND FACTORS

Special grade or factor	Grade limit	Breakpoint
Garlicky	5 or more per 1,000 grams	2
Infested	Same as in § 810.107	0
Soybeans of other colors	Not more than 10.0%	2.3
Moisture	As specified by contract or load order grade	0.3
Test Weight	As specified by contract or load order	−0.4

TABLE 20—BREAKPOINTS FOR SUNFLOWER SEED SPECIAL GRADES AND FACTORS

Special grade or factor	Grade limit	Breakpoint
Moisture	As specified by contract or load order grade	0.5
Foreign Material	1.25% and less	0.27
	1.26% and above	0.39
Admixture	As specified by contract or load order grade	0.6

TABLE 21—GRADE LIMITS (GL) AND BREAKPOINTS (BP) FOR TRITICALE

Grade	Minimum test weight per bushel (percent)		Maximum limits of—											
			Damaged kernels				Foreign material				Shrunken and broken kernels (percent)		Defects[3] (percent)	
			Heat-damaged (percent)		Total[1] (percent)		Material other than wheat or rye (percent)		Total[2] (percent)					
	GL	BP	GL	BP	GL	BP	GL	BP	GL	BP	GL	BP	GL	BP
U.S. No. 1	48.0	−0.5	0.2	0.1	2.0	0.8	1.0	0.4	2.0	0.6	5.0	0.8	5.0	1.3
U.S. No. 2	45.0	−0.5	0.2	0.1	4.0	1.1	2.0	0.5	4.0	0.9	8.0	0.8	8.0	1.3
U.S. No. 3	43.0	−0.5	0.5	0.4	8.0	1.5	3.0	0.6	7.0	1.2	12.0	1.6	12.0	2.3
U.S. No. 4	41.0	−0.5	3.0	0.8	15.0	2.0	4.0	0.8	10.0	1.4	20.0	2.3	20.0	2.3

[1] Includes heat-damaged kernels.
[2] Includes material other than wheat or rye.
[3] Defects includes damaged kernels (total), foreign material (total), and shrunken and broken kernels. The sum of these three factors may not exceed the limit for defects for each numerical grade.

TABLE 22—BREAKPOINTS FOR TRITICALE SPECIAL GRADES AND FACTORS

Special grade or factor	Grade limit	Breakpoint
Garlicky	2 or more per 1,000 grams	1⅓
Ergoty	More than 0.10%	0.1
Smutty	More than 14 per 250 grams	6
Infested ....,	Same as in § 810.107	0
Dockage	0.99% or above	0.32
Moisture	As specified by contract or load order grade	0.5

518

TABLE 23—GRADE LIMITS (GL) AND BREAKPOINTS (BP) FOR WHEAT

	Minimum limits of—		Maximum limits of—							
	Test weight per bushel		Damaged kernels			Shrunken and broken kernels (percent)	Defects [3] (percent)	Wheat of other classes [4]		
Grade	Hard red spring wheat or white club wheat [1] (pounds)	All other classes and sub-classes (pounds)	Heat-damaged kernels (percent)	Total [2] (percent)	Foreign material (percent)				Contrasting classes (percent)	Total [5] (percent)
	GL BP	GL BP	GL BP	GL BP	GL BP	GL BP	GL BP	GL BP	GL BP	GL BP
U.S. No. 1	58.0 −0.3	60.0 −0.3	0.2 0.2	2.0 1.0	0.4 0.2	3.0 0.3	3.0 0.7	1.0 0.7	3.0 1.6	
U.S. No. 2	57.0 −0.3	58.0 −0.3	0.2 0.2	4.0 1.5	0.7 0.3	5.0 0.4	5.0 0.9	2.0 1.0	5.0 2.1	
U.S. No. 3	55.0 −0.3	56.0 −0.3	0.5 0.3	7.0 1.9	1.3 0.4	8.0 0.5	8.0 1.2	3.0 1.3	10.0 2.9	
U.S. No. 4	53.0 −0.3	54.0 −0.3	1.0 0.4	10.0 2.3	3.0 0.6	12.0 0.6	12.0 1.4	10.0 2.3	10.0 2.9	
U.S. No. 5	50.0 −0.3	51.0 −0.3	3.0 0.7	15.0 2.7	5.0 0.7	20.0 0.7	20.0 1.5	10.0 2.3	10.0 2.9	

[1] These requirements also apply when Hard Red Spring or White Club wheat predominate in a sample of Mixed wheat.
[2] Includes heat-damaged kernels.
[3] Defects include damaged kernels (total), foreign material, and shrunken and broken kernels. The sum of these factors may not exceed the limit for defects for each numerical grade.
[4] Unclassed wheat of any grade may contain not more than 10.0 percent of wheat of other classes.
[5] Includes contrasting classes.

TABLE 24—BREAKPOINTS FOR WHEAT SPECIAL GRADES AND FACTORS

Special grade or factor	Grade limit	Break-point
Moisture	As specified by contract or load order grade	0.3
Garlicky	More than 2 bulblets per 1,000 grams	1⅓
Light smutty	More than 5 smut balls per 250 grams	3
Smutty	More than 30 smut balls per 250 grams	10
Infested	Same as in §810.107	0
Ergoty	More than 0.05%	0.03
Treated	Same as In §810.2204	0
Dockage	As specified by contract or load order grade	0.2
Protein	As specified by contract or load order grade	0.5
Class and Subclass Hard red spring:		
DNS	75% or more DHV	−5.0
NS	25% or more DHV but less than 75% DHV	−5.0
Durum:		
HADU	75% or more HVAC	−5.0
ADU	60% or more HVAC but less than 75% of HVAC	−5.0
Soft white:		
SWH	Not more than 10% white club wheat	2.0
WHCB	Not more than 10% of other soft white wheat	2.0
WWH	More than 10% WHCB and more than 10% of other soft white wheat	−3.0

(3) *Grain accepted by the inspection plan.* Grain which is offered for inspection as part of a single lot and accepted by a statistical acceptance sampling and inspection plan according to the provisions of this section and procedures prescribed in the instructions shall be certificated as a single lot provided it was sampled in a reasonably continuous operation. Official factor and official criteria information shown on the certificate shall be based on the weighted or mathematical averages of the analysis of sublots.

(4) *Grain rejected by the inspection plan.* When grain which is offered for inspection as part of a single lot is re-jected by the plan or is not sampled in a reasonably continuous operation, the grain in each portion shall be certificated separately. If any portion of grain is not accepted by the plan and designated a material portion, the applicant shall be promptly notified and have the option of:

(i) Removing the material portion from the carrier; or

(ii) Requesting the material portion be separately certificated; or

(iii) Requesting either a reinspection or an appeal inspection of the material portion; or

519

(iv) Requesting a reinspection service and/or an appeal inspection service on the entire lot.

(5) *Reinspection service and appeal inspection service.* A reinspection or an appeal inspection may be requested on a material portion. A Board appeal inspection may also be requested on a material portion after the reinspection or appeal inspection. A reinspection, an appeal inspection, and a Board appeal inspection may be requested on the total sublots in the lot.

(i) *Material portions.* A material portion designated by the plan may be reinspected or appeal inspected once in the field, but not both, and once at the Board of Appeals and Review. The reinspection or appeal inspection result shall, unless a material error is found, be averaged with the original inspection determination. The Board appeal inspection result shall, unless a material error is found, be averaged with the previous inspection result. The inspection plan tolerances shall be reapplied to the material portion grain to determine acceptance or rejection. If a material error is found, the reinspection or appeal inspection result shall replace the original inspection result or the Board appeal result shall replace the previous inspection result. For purposes of this section, a material error is defined as results differing by more than two standard deviations. Acceptance or rejection of that portion of grain shall be based on the reinspection or appeal inspection and on the Board appeal inspection result alone when a material error is found.

(ii) *Entire lot.* The applicant may request a reinspection service, an appeal inspection service, and a Board appeal inspection service on the entire lot. Inspection results for these services shall replace the previous inspection results. The tolerances shall be reapplied to all portions of the entire lot to determine acceptance or rejection.

(d) *Infested grain*—(1) *Available options.* If gain or any portion of grain in a single shiplot, unit train, or lash barge lot is found to be infested, according to the provisions of the Official U.S. Standards for Grain, the applicant shall be promptly notified and have the option of:

(i) Unloading the portion of infested grain from the lot and an additional amount of other grain in common stowage with the infested grain; or

(ii) When applicable, completing the loading and treating all infested grain in the lot; or

(iii) When applicable, treating the infested grain for the purpose of destroying the insects, subject to subsequent examination by official personnel; or

(iv) Continue loading without treating the infested grain, in which case all of the infested grain in the lot and all grain in common stowage areas with the infested grain will be officially certificated as infested according to the provisions of the Official U.S. Standards for Grain.

(2) *Exception.* If infested grain in loaded into common stowage with a lot, or a portion of a lot, which has not been officially certificated as being infested, the applicant loading the infested grain may not use the option in paragraph (d)(1)(i) of this section.

(3) *With treatment.* If infested grain is treated with a fumigant in accordance with the instructions and the treatment is witnessed by official personnel, the official sampling, inspection, grading, and certification of the lot shall continue as though the infested condition did not exist.

(e) *Special certification procedures*—(1) *Rejected grain.* When grain is rejected by the inspection plan under paragraph (c)(4) of this section, the official inspection certificate for each different portion of different quality shall show:

(i) A statement that the grain has been loaded aboard with grain of other quality;

(ii) The grade, location, or other identification and approximate quanity of grain in the portions; and

(iii) Other information required by the regulations and the instructions.

The requirement of paragraph (e)(1)(i) of this section does not apply to grain that is inspected as it is unloaded from the carrier or to portions loaded in separate carriers or stowage space.

(2) *Common stowage.* (i) *Without separation.* When bulk grain is offered for official inspection as it is loaded aboard a ship and is loaded without separation in a stowage area with other grain or another commodity, the

official inspection certificate for the grain in each lot shall show the kind, the grade, if known, and the location of the other grain, or the kind and location of the other commodity in the adjacent lots.

(ii) *With separation.* When separations are laid between lots, the official inspection certificates shall show the kind of material used in the separations and the locations of the separations in relation to each lot.

(iii) *Exception.* The common stowage requirements of this paragraph are not applicable to the first lot in a stowage area unless a second lot is loaded, in whole or in part, in the stowage area prior to issuing the official inspection certificate for the first lot.

(3) *Protein.* A special statement indicating the actual protein range of a lot shall be shown on the official inspection certificate if the difference between the lowest and highest protein determinations for the lot exceeds 1.0 percent when protein is officially determined and a specific range limit is not established by the contract grade.

(4) *Part lot.* If part of a lot of grain in an inbound carrier is unloaded and part is left in the carrier, the unloaded grain shall be officially inspected and certificated in accordance with the provisions of §800.84(g).

(5) *Official mark.* If the grain in a single lot is officially inspected for grade as it is being loaded, upon request, the following official mark shall be shown on the inspection certificate: "Loaded under continuous official inspection."

[55 FR 24042, June 13, 1990; 55 FR 46131, Nov. 1, 1990, as amended at 56 FR 4675, Feb. 5, 1991; 57 FR 58965, 58970, Dec. 14, 1992; 61 FR 18490, Apr. 26, 1996; 63 FR 20056, Apr. 23, 1998; 64 FR 6783, Feb. 11, 1999; 71 FR 52405, Sept. 6, 2006; 71 FR 77853, Dec. 27, 2006; 73 FR 39732, July 20, 2007; 82 FR 20543, May 3, 2017]

§800.87 New inspections.

(a) *Identity lost.* An applicant may request official personnel to perform a new original inspection service on an identified lot of grain, or on an identified carrier or container, if the identity of the lot or the carrier or container has been lost.

(b) *Identity not lost.* If the identity of the grain or the carrier or container is not lost, a new original inspection shall not be performed on the same identified lot of grain or carrier or container in the same assigned area of responsiblity within 5 business days after the last official inspection.

[50 FR 49674, Dec. 4, 1985]

§800.88 Loss of identity.

(a) *Lots.* Except as noted in paragraph (d) of this section, the indentity of a lot of grain shall be considered lost if (1) a portion of the grain is unloaded, transferred, or otherwise removed from the carrier or container in which the grain was located at the time of the original inspection; or (2) a portion of grain or other material, including additives, is added to the lot after the orginal inspection was performed, unless the addition of the additive was performed in accordance with the regulations and the instructions. At the option of official personnel performing a reinspection, appeal inspection, or Board appeal inspection service, the identity of grain in a closed carrier or container shall be considered lost if the carrier or container is not sealed or if the seal record is incomplete.

(b) *Carriers and containers.* The indentity of a carrier or container shall be considered lost when (1) the stowage area is cleaned, painted, treated, fumigated, or fitted after the original inspection was performed; or (2) the identification of the carrier or container has been changed since the original inspection was performed.

(c) *Submitted samples.* The identity of a submitted sample of grain shall be considered lost when (1) the identifying number, mark, or symbol for the sample is lost or destroyed or (2) the samples have not been retained and protected by official personnel as prescribed in the instructions.

(d) *Additives.*[1] If additives are applied during loading to outbound, including export, grain after sampling or during unloading to inbound grain before sampling for the purpose of insect or fungi

---

[1] Elevators, other handlers of grain, and their agents are responsible for the additive's proper usage and application. Compliance with this section does not excuse compliance with applicable Federal, State, and local laws.

control, dust suppression, or identification, the inspection certificate shall show a statement showing the type and purpose of the additive application, except that no statement is required to be shown when the additive is a fumigant applied for the purpose of insect control.

[52 FR 6495, Mar. 4, 1987, as amended at 58 FR 3212, Jan. 8, 1993; 59 FR 52077, Oct. 14, 1994]

WEIGHING PROVISIONS AND PROCEDURES

§ 800.95 Methods and order of performing weighing services.

(a) *Methods.* All Class X or Class Y weighing, checkweighing, checkloading, stowage examination, and other weighing services shall be performed by official personnel or approved weighers using approved weighing equipment and according to procedures prescribed in the regulations and the instructions.

(b) *Order of service.* Weighing services shall be performed, to the extent practicable, in the order in which requests are received. Official personnel must mark or stamp the date received on each written request for service. Precedence will be given to requests for weighing required by sections 5(a)(1) or 5(a)(2) of the Act.

[52 FR 6495, Mar. 4, 1987]

§ 800.96 Weighing procedures.

(a) *Inbound.* Inbound grain that is to be weighed must be routed directly from the carrier and cannot be cleaned, dried, or otherwise processed to remove or add other grain or material en route. Except as noted in paragraph (c) of this section, the identity of an inbound lot shall be considered lost when a portion of the lot is transferred or otherwise removed prior to weighing or a portion of grain or other material is added to the lot prior to weighing. When loss of identity occurs, no amount shall be shown in the "Net Weight" portion of the weight certificate for the lot.

(b) *Outbound.* Outbound grain that has been weighed must be routed directly from the scale to the carrier and cannot be cleaned, dried, or otherwise processed to remove or add other grain or material en route. Except as noted in paragraph (c) of this section, the

identity of an outbound lot will be considered lost if a portion of the lot is transferred or otherwise removed from the lot after weighing or a portion of grain or other material is added to the lot after weighing. When loss of identity occurs, no amount shall be shown in the "Net Weight" portion of the weight certificate for the lot.

(c) *Exceptions*—(1) *Spills.* (i) *Outbound.* (A) *Replaced.* If a spill occurs in handling and loading of outbound grain and the spilled grain is retrieved, or is replaced in kind, and is loaded on board during the loading operations, the weight certificate shall show the weight of the grain that was physically loaded on board. Upon request of the applicant, an additional certificate may be issued by the agency or the field office to show the weight of the additional grain that was used to replace a spill.

(B) *Not replaced.* If a spill occurs in the handling and loading of outbound grain and the spilled grain is not retrieved or is not replaced during the loading operation, the weight certificate shall show the weight of the grain that was actually weighed, minus the estimated amount of the grain that was spilled. Upon request of the applicant, an additional certificate may be issued showing the estimated amount of grain that was spilled. The applicant may, upon request, have the total amount that was weighed shown on the weight certificate with the estimated amount of the spilled grain noted.

(ii) *Inbound.* If a spill occurs in the handling of inbound grain and the grain is not retrieved and weighed, the weight certificate shall show the weight of the grain that was actually unloaded from the carrier and a statement regarding the spill as prescribed in the instructions.

(2) *Additives.*[1] If additives are applied during loading to outbound, including export, grain after weighing or during unloading to inbound grain before weighing for the purpose of insect or fungi control, dust suppression, or

---

[1] Elevators, other handlers of grain, and their agents are responsible for the additive's proper usage and application. Compliance with this section does not excuse compliance with applicable Federal, State, and local laws.

identification, the weight certificate shall show the actual weight of the grain after the application of the additive for inbound grain or the actual weight of the grain prior to the application of the additive for outbound or export grain and a statement showing the type and purpose of the additive application, except that no statement is required to be shown when the additive is a fumigant applied for the purpose of insect control.

(3) *Dust.* If dust is removed during the handling of grain, the weight certificate shall not be adjusted to reflect the weight of the removed dust.

(4) *Commingled carriers.* If grain from two or more identified carriers becomes mixed, (i) the combined weight of the grain shall be shown in the "Net Weight" block of one certificate with all carrier identification shown in the identification of carrier section of the certificate, or (ii) upon request of the applicant, a certificate shall be issued for each carrier with the "Net Weight" block crossed out, and with the total combined weight unloaded and the identification of the other carrier(s) shown in the "Remarks" section.

(5) *Unremoved grain.* If, after unloading an inbound carrier, there is sound grain remaining in the carrier that could have been removed with reasonable effort, the weight certificate shall show the weight of the grain that was actually unloaded from the carrier and a statement regarding the grain remaining in the carrier.

[52 FR 6495, Mar. 4, 1987, as amended at 58 FR 3212, Jan. 8, 1993; 59 FR 52077, Oct. 14, 1994]

§800.97 Weighing grain in containers, land carriers, barges, and shiplots.

(a) *General.* The weighing of grain loaded or unloaded from any carrier shall be conducted according to this section and the instructions.

(b) *Procedure*—(1) *General.* If grain in a carrier is offered for inspection or weighing service as one lot, the grain must be weighed at the individual weighing location in a reasonably continuous operation and certified as one lot. The identification of the carrier(s) must be recorded on the scale tape or ticket and the weight certificate.

(2) *Sacked grain.* If sacked grain is offered for weighing and the grain is not fully accessible, the request for weighing service shall be dismissed.

(3) *Part lots.* If a portion of an inbound lot of grain is unloaded and a portion is left in the carrier because it is not uniform in quality or condition, or the lot is unloaded in other than a reasonably continuous operation, the portion that is removed and the portion remaining in the carrier shall be considered as part lots and shall be weighed and certificated as part lots.

(c) *Certification of trucklots, carlots, and bargelots*—(1) *Basic requirement.* One official certificate must be issued for the weighing of the grain in each container, truck, trailer, truck/trailer(s) combination, railroad car, barge, or similarly sized carrier. This requirement is not applicable to multiple grain carriers weighed as a single lot or combined lot under §800.98.

(2) *Part-lot weight certificates.* A part-lot weight certificate shall show (i) the weight of the portion that is unloaded and (ii) the following statement: "Part-lot: The net weight stated herein reflects a partial unload."

(d) *Certification of shiplot grain*—(1) *Basic requirement.* The certificate shall show (i) if applicable, a statement that the grain has been loaded aboard with other grain, (ii) the official weight, (iii) the stowage or other identification of the grain, and (iv) other information required by the regulations and the instructions.

(2) *Common stowage*—(i) *Without separation.* If bulk grain is offered for weighing as it is being loaded aboard a ship and is loaded without separation in a stowage area with other grain or another commodity, the weight certificate for the grain in each lot shall show that the lot was loaded aboard with other grain or another commodity without separation and the relative location of the grain.

(ii) *With separation.* If separations are laid between adjacent lots, the weight certificates shall show the kind of material used in the separations and the location of the separations in relation to each lot.

(iii) *Exception.* The common stowage requirements of this paragraph shall not be applicable to the first lot in a stowage area unless a second lot has been loaded, in whole or in part, in the stowage area before issuing the official weight certificate for the first lot.

(3) *Official mark.* If the grain is officially weighed in a reasonably continuous operation, upon request by the applicant, the following statement may be shown on the weight certificate: "Loaded under continuous official weighing."

[52 FR 6496, Mar. 4, 1987, as amended at 78 FR 43757, July 22, 2013]

§ 800.98   **Weighing grain in combined lots.**

(a) *General.* The weighing of bulk or sacked grain loaded aboard, or being loaded aboard, or unloaded from two or more carriers as a combined lot shall be conducted according to this section and the instructions.

(b) *Weighing procedure*—(1) *Single lot weighing.* (i) Single lots of grain that are to be weighed as a combined lot may be weighed at multiple locations, provided that:

(A) The lots are contained in the same type of carrier; and

(B) Weighing is performed at each individual location in a reasonably continuous operation.

(ii) The grain loaded into or unloaded from each carrier must be weighed in accordance with procedures prescribed in the instructions. In the case of sacked grain, a representative weight sample must be obtained from the grain in each carrier unless otherwise specified in the instructions.

(2) *Recertification.* Grain that has been weighed and certified as two or more single lots may be recertified as a combined lot, provided that the original weight certificates issued for the single lots have been or will be surrendered to the appropriate agency or field office, and the official personnel who performed the weighing service for the single lots and the official personnel who are to recertify the grain as a combined lot determine that the weight of the grain in the lots has not since changed, and in the case of sacked grain, that the weight samples used as a basis for weighing the single lots were representative at the time of the weighing.

(3) *Grain uniform in quality.* An applicant may request that grain be weighed and certificated as a combined lot whether or not the grain is uniform in quality for the purpose of inspection under the Act.

(c) *Certification procedures*—(1) *General.* Each certificate for a combined-lot Class X or Class Y weighing service shall show the identification for the "Combined lot" or, at the request of the applicant, the identification of each carrier in the combined lot. The identification and any seal information for the carriers may be shown on the reverse side of the weight certificate, provided the statement "See reverse side" is shown on the face of the certificate in the space provided for remarks.

(2) *Recertification.* If a request for a combined-lot Class X or Class Y weighing service is filed after the grain in the single lots has been weighed and certified, the combined-lot weighing certificate must show the following:

(i) The date of weighing the grain in the combined lot (if the single lots were weighed on different dates, the latest dates must be shown);

(ii) A serial number, other than the serial numbers of the weight certificates that are to be superseded;

(iii) The name of the elevator(s) from which or into which the grain in the combined lot was loaded or unloaded;

(iv) A statement showing the weight of the grain in the combined lot;

(v) A completed statement showing the identification of any superseded certificate as follows: "This combined-lot certificate supersedes certificate Nos. __, dated ____; and

(vi) If at any time of issuing the combined-lot weight certificate, the superseded certificates are not in the custody of the agency or field office, the statement "The superseded certificates identified herein have not been surrendered" must be shown clearly in the space provided for remarks beneath the statement identifying the superseded certificates. If the superseded certificates are in the custody of the agency or field office, the superseded certificates must be clearly marked "Void."

(3) *Part lot.* If a part of a combined lot of grain in inbound carriers is unloaded and a part is left in the carriers, the grain that is unloaded shall be certificated in accordance with the provisions in § 800.97(c)(2).

(4) *Official mark.* When grain is weighed as a combined lot in one continuous operation, upon request by the applicant, the following statement shall be shown on the weight certificate: "Loaded under continuous official weighing," or "Loaded under continuous official inspection and weighing."

(5) *Further combining.* After a combined-lot weight certificate has been issued, there shall be no further combining and no dividing of the certificate.

(6) *Limitations.* No combined-lot weight certificate shall be issued (i) for any weighing service other than as described in this section or (ii) which shows a weight of grain different from the total of the combined single lot.

[52 FR 6496, Mar. 4, 1987, as amended at 78 FR 43757, July 22, 2013]

§ 800.99 Checkweighing sacked grain.

(a) *General.* Each checkweighing service performed on a lot of sacked grain to determine the weight of the grain shall be made on the basis of one or more official weight samples obtained from the grain by official personnel according to this section and procedures prescribed in the instructions.

(b) *Representative sample.* No official weight sample shall be considered to be representative of a lot of sacked grain unless the sample is of the size prescribed in the instructions and has been obtained and weighed according to the procedures prescribed in the instructions.

(c) *Protecting samples and data.* Official personnel and other employees of an agency or the Service shall protect official weight samples and data from manipulation, substitution, and improper and careless handling which might deprive the samples and sample data of their representativeness.

(d) *Restriction on weighing.* No agency shall weigh any lot of sacked grain unless at the time of obtaining the official weight sample the grain from which the sample was obtained was located within the area of responsibility assigned to the agency, except as otherwise provided for in § 800.117, or on a case-by-case basis as determined by the Administrator.

(e) *Equipment and labor.* Each applicant for weighing services shall provide necessary labor for obtaining official weight samples and place the samples in a position for weighing and shall supply suitable weighing equipment approved by the Service, pursuant to the regulations and the instructions.

(f) *Disposition of official weight samples.* In weighing sacked grain in lots, the grain in the official weight samples shall be returned to the lots from which the samples were obtained.

(g) *Provisions by kinds of service*—(1) *"IN" movements.* Each checkweighing on an "IN" movement of sacked grain shall be based on an official weight sample obtained while the grain is at rest in the carrier or during unloading, in accordance with procedures prescribed in the instructions.

(2) *"OUT" movements (export).* Each checkweighing of sacked export grain shall be based on an official weight sample obtained as the grain is being loaded aboard the final carrier, as the grain is being sacked, or while the grain is at rest in a warehouse or holding facility, in accordance with procedures prescribed in the instructions.

(3) *"OUT" movements (other than export).* Each checkweighing of an "OUT" movement of nonexport sacked grain shall be based on an official weight sample obtained from the grain as the grain is being loaded in the carrier, or while the grain is at rest in the carrier, or while the grain is at rest in a warehouse or holding facility, or while the grain is being sacked, in accordance with procedures prescribed in the instructions.

(4) *"LOCAL" weighing.* Each checkweighing of a "LOCAL" movement of sacked grain shall be based on an official weight sample obtained while the grain is at rest or while the grain is being transferred, in accordance with procedures prescribed in the instructions.

[52 FR 6497, Mar. 4, 1987, as amended at 68 FR 19138, Apr. 18, 2003]

ORIGINAL SERVICES

§ 800.115 Who may request original services.

(a) *General.* Any interested person may request original inspection and weighing services. The kinds of inspection and weighing services are described in § 800.75.

(b) *Class Y weighing services.* A request for Class Y weighing services at an export elevator at an export port location shall cover all lots shipped or received in a specific type of carrier. At all other elevators, the request shall cover all lots shipped from or to a specific location in a specific type of carrier. Each request shall be for a contract period of at least 3 months, but a facility may, upon satisfactory notification, exempt specific unit trains from the request.

(c) *Contract services.* Any interested person may enter into a contract with an agency or the Service whereby the agency or Service will provide original services for a specified period and the applicant will pay a specified fee.

(Approved by the Office of Management and Budget under control number 0580-0012)

[50 FR 45393, Oct. 31, 1985]

§ 800.116   How to request original services.

(a) *General.* Except as otherwise provided for in § 800.117, requests for original services shall be filed with an agency or field office authorized to operate in the area in which the original service is to be performed. All requests shall include the information specified in § 800.46. Verbal requests shall be confirmed in writing when requested by official personnel, as specified in § 800.46. Copies of request forms may be obtained from the agency or field office upon request. If the information specified by § 800.46 is not available at the time the request is filed, official personnel may, at their discretion, withhold service pending receipt of the required information. An official certificate shall not be issued unless the information as required by § 800.46 has been submitted, or official personnel determine that sufficient information has been made available so as to perform the requested service. A record that sufficient information was made available must be included in the record of the official service.

(b) *Request requirements.* Except as provided for in § 800.117, requests for original services, other than submitted sample inspections, must be made to the agency or field office responsible for the area in which the service will be provided. Requests for submitted sample inspections may be made with any agency, or any field office that provides original inspection service. Requests for inspection or Class X weighing of grain during loading, unloading, or handling must be received in advance of loading so official personnel can be present. All requests will be considered filed when official personnel receive the request. A record shall be maintained for all requests. All requests for service that is to be performed outside normal business hours must be received by 2 p.m. the preceding day.

(Approved by Office of Management and Budget under control number 0580-0013)

[68 FR 19139, Apr. 18, 2003]

§ 800.117   Who shall perform original services.

(a) *General.* Original services shall be performed by the agency or field office assigned the area in which the service will be provided, except as provided in paragraph (b) of this section.

(b) *Exceptions for official agencies to provide service*—(1) *Timely service.* If the assigned official agency cannot provide service within 6 hours of a request, the service may be provided by another official agency upon approval from the Service.

(2) *Barge probe service.* Any official agency may provide probe sampling and inspection service for barge-lots of grain with no restrictions due to geographical locations.

(3) *Written agreement.* If the assigned official agency agrees in writing with the adjacent official agency to waive the current geographic area restriction at the request of the applicant for service, the adjacent official agency may provide service at a particular location upon providing written notice to the Service, and the Service determines that the written agreement conforms to the provisions in the Act.

(c) *Interim service at other than export port locations.* If the assigned official agency is not available on a regular basis to provide original services, and no official agency within a reasonable proximity is willing to provide such services on an interim basis, the services shall be provided by authorized employees of the Secretary, or other persons licensed by the Secretary, until the services can be provided on a regular basis by an official agency, as provided in § 800.196.

[68 FR 19139, Apr. 18, 2003, as amended at 81 FR 49862, July 29, 2016]

§ 800.118 Certification.

Official certificates shall be issued according to § 800.160. Upon request, a combination inspection and Class X weighing certificate may be issued when both services are performed in a reasonably continuous operation at the same location by the same agency or field office. An official certificate shall not be issued unless the information as required by § 800.46 has been submitted, or official personnel determine that sufficient information has been made available so as to perform the requested service. A record that sufficient information was made available must be included in the record of the official service.

(Approved by Office of Management and Budget under Control Number 0580–0013)

[68 FR 19139, Apr. 18, 2003]

OFFICIAL REINSPECTION SERVICES AND REVIEW OF WEIGHING SERVICES

§ 800.125 Who may request reinspection services or review of weighing services.

(a) *General.* Any interested person may request a reinspection or review of weighing service, except as provided for in § 800.86(c)(5). Only one reinspection service or review of weighing service may be performed on any original service. When more than one interested person requests a reinspection or review of weighing service, the first person to file is the applicant of record.

(b) *Kind and scope of request.* A reinspection or review of weighing service is limited to the kind and scope of the original service. If the request specifies a different kind or scope, the request shall be dismissed but may be resubmitted as a request for original services: Provided, however, that an applicant for service may request a reinspection of a specific factor(s), official

grade and factors, or official criteria. In addition, reinspections for grade may include a review of any pertinent factor(s), as deemed necessary by official personnel. Official criteria are considered separately from official grade or official factors when determining the kind and scope. When requested, a reinspection for official grade or official factors and official criteria may be handled separately even though both sets of results are reported on the same certificate. Moreover, a reinspection or review of weighing may be requested on either the inspection or Class X weighing results when both results are reported on a combination inspection and Class X weight certificate.

(Approved by the Office of Management and Budget under control number 0580–0013)

[50 FR 45393, Oct. 31, 1985, as amended at 54 FR 5924, Feb. 7, 1989; 55 FR 24048, June 13, 1990; 68 FR 61328, Oct. 28, 2003]

§ 800.126 How to request reinspection or review of weighing services.

(a) *General.* Requests shall be made with the agency or field office that performed the original service. All requests shall include the information specified in § 800.46. Verbal requests shall be confirmed in writing when requested by official personnel. Copies of request forms may be obtained from the agency or field office. If at the time the request is filed the documentation required by § 800.46 is not available, official personnel may, at their discretion, withhold services pending the receipt of the required documentation. A reinspection certificate or the results of a review of weighing service shall not be issued unless (1) the documentation requested under § 800.46 has been submitted or (2) official personnel determine sufficient information has been made available so as to perform the requested service. A record that sufficient information was made available shall be included in the record of the official service.

(b) *Request requirements.* Requests will be considered filed on the date they are received by official personnel. A record shall be maintained for all requests.

(1) *Reinspection services.* Requests shall be received (i) before the grain has left the specified service point where the grain was located when the

527

original inspection was performed; (ii) no later than the close of business on the second business day following the date of the original inspection; and (iii) before the identity of the grain has been lost. If a representative file sample, as prescribed in § 800.82, is available, official personnel may waive the requirements pursuant to this subparagraph. The requirements of paragraph (b)(1)(i) of this section may be waived only upon written consent of the applicant and all interested persons. The requirements of paragraph (b)(1)(ii) and (iii) of this section may be waived at the request of the applicant or other interested persons. The requirement of paragraph (b)(1)(ii) of this section may also be waived upon satisfactory showing by an interested person of evidence of fraud or that because of distance or other good cause, the time allowed for filing was not sufficient. A record of each waiver shall be included in the record of the reinspection service.

(2) *Review of weighing services.* Requests shall be received no later than 90 calendar days after the date of the original Class X or Class Y weighing service.

(Approved by the Office of Management and Budget under control number 0580-0012)

[50 FR 45394, Oct. 31, 1985]

§ 800.127 Who shall perform reinspection or review of weighing services.

Reinspection or review of weighing services shall be performed by the agency or field office that performed the original service.

[50 FR 45394, Oct. 31, 1985]

§ 800.128 Conflicts of interest..

Official personnel cannot perform or participate in performing or issue an official certificate for a reinspection or a review of weighing service if they participated in the original service unless there is only one qualified person available at the time and place of the reinspection or review of weighing.

[50 FR 45394, Oct. 31, 1985]

§ 800.129 Certificating reinspection and review of weighing results.

(a) *General.* Except as provided in paragraph (a)(1) of this paragraph, official certificates shall be issued according to § 800.160 and the instructions. Except as provided in paragraph (b)(2) of this section, only the result of the reinspection service shall be reported.

(1) *Results of material portion sublots.* When results of a reinspection on a material portion do not detect a material error, they shall be averaged with the original inspection results. For purposes of this section, a material error is defined as results differing by more than two standard deviations. The averaged inspection results shall replace the original inspection results recorded on the official inspection log. Reinspection results shall replace the original inspection results recorded on the official inspection log if a material error is detected. No certificates will be issued unless requested by the applicant or deemed necessary by official personnel.

(2) *Reporting review of weighing results.* When the review of weighing service results indicate that the original weighing results were correct, the applicant will be notified in writing. When the original weighing service results are incorrect, a corrected weight certificate or, if applicable, a corrected combination inspection and Class X weight certificate will be issued according to the provisions of § 800.165.

(b) *Required statements on reinspection certificates.* Each reinspection certificate shall show the statements required by this section, § 800.161, and applicable instructions.

(1) Each reinspection certificate must clearly show (i) the term "Reinspection" and (ii) a statement identifying the superseded certificate. The superseded certificate will be considered null and void as of the date of the reinspection certificate.

(2) When official grade or official factors, Class X weighing results, and official criteria are reported on the same certificate, the reinspection certificate shall show a statement indicating that the reinspection results are based on official grade, or official factors, or official criteria and that all other results are those of the original service.

(3) If the superseded certificate is in the custody of the agency or field office, the superseded certificate shall be marked "Void." If the superseded certificate is not in the custody of the

agency or field office at the time the reinspection certificate is issued, a statement indicating that the superseded certificate has not been surrendered shall be shown on the reinspection certificate.

(4) As of the date of issuance of the official certificate, the superseded certificate for the original service will be void and shall not be used to represent the grain.

(5) When certificates are issued under paragraph (a)(1) of this section, the reinspection certificate shall show a statement indicating that the results replaced the original results and that the reinspection certificate is not valid for trading purposes.

[50 FR 45394, Oct. 31, 1985, as amended at 55 FR 24048, June 13, 1990]

APPEAL INSPECTION SERVICES

§ 800.135 Who may request appeal inspection services.

(a) *General.* Any interested person may request appeal inspection or Board appeal inspection services, except as provided for in § 800.86(c)(5). When more than one interested person requests an appeal inspection or Board appeal inspection service, the first person to file is the applicant of record. Only one appeal inspection may be obtained from any original inspection or reinspection service. Only one Board appeal inspection may be obtained from an appeal inspection. Board appeal inspections will be performed on the basis of the official file sample. Board appeal inspections are not available on stowage examination services.

(b) *Kind and scope of request.* An appeal inspection service is limited to the kind and scope of the original or reinspection service; or, in the case of a Board Appeal inspection service, the kind and scope of the appeal inspection service. If the request specifies a different kind or scope, the request shall be dismissed but may be resubmitted as a request for original services: Provided, however, that an applicant for service may request an appeal or Board Appeal inspection of a specific factor(s), official grade and factors, or official criteria. In addition, appeal and Board Appeal inspections for grade may include a review of any pertinent

factor(s), as deemed necessary by official personnel. Official criteria are considered separately from official grade or official factors when determining kind and scope. When requested, an appeal inspection for grade, or official factors, and official criteria may be handled separately even though both results are reported on the same certificate. Moreover, an appeal inspection may be requested on the inspection results when both inspection and Class X weighing results are reported on a combination inspection and Class X weight certificate.

(Approved by the Office of Management and Budget under control number 0580–0013)

[50 FR 45395, Oct. 31, 1985, as amended at 55 FR 24048, June 13, 1990; 68 FR 61328, Oct. 28, 2003]

§ 800.136 How to request appeal inspection services.

(a) *General.* Requests shall be filed with the field office responsible for the area in which the original service was performed. Requests for Board appeal inspections may be filed with the Board of Appeals and Review or the field office that performed the appeal inspection. All requests shall include the information specified in § 800.46. Verbal requests shall be confirmed in writing when requested by official personnel as specified in § 800.46. Copies of request forms may be obtained from the field office upon request. If at the time the request is filed the documentation required by § 800.46 is not available, official personnel may, at their discretion, withhold service pending the receipt of the required documentation. An appeal inspection certificate will not be issued unless (1) documentation requested under § 800.46 has been submitted or (2) office personnel determine that sufficient information has been made available so as to perform the request. A record that sufficient information has been made available must be included in the record of the official service.

(b) *Filing requirements.* Requests will be considered filed on the date they are received by official personnel. A record shall be maintained for all requests. Requests must be filed (1) before the grain has left the specified service point where the grain was located when the original inspection was performed, (2) no later than the close of business on the second business day following the date of the last inspection, and (3) before the identity of the grain has been lost. If a representative file sample as prescribed in § 800.82 is available, official personnel may waive the requirements pursuant to this paragraph. The requirements of paragraph (b)(1) of this section may be waived only upon written consent of the applicant and all interested persons. The requirements of paragraphs (b)(2) and (b)(3) of this section may be waived at the request of the applicant or other interested persons. The requirement of paragraph (b)(2) of this section may also be waived upon satisfactory showing by an interested person of evidence of fraud or that because of distance or other good cause, the time allowed for filing was not sufficient. A record of each waiver shall be included in the record of the appeal inspection service.

(Approved by the Office of Management and Budget under control number 0580–0012)

[50 FR 45395, Oct. 31, 1985, as amended at 54 FR 5924, Feb. 7, 1989]

§ 800.137   Who shall perform appeal inspection services.

(a) *Appeal.* Appeal inspection services shall be performed by the field office responsible for the area in which the original inspection was performed.

(b) *Board appeal.* Board appeal inspection services shall be performed only by the Board of Appeals and Review. The field office that performed the appeal inspection service will act as a liaison between the Board of Appeals and Review and the applicant.

[50 FR 45395, Oct. 31, 1985]

§ 800.138   Conflict of interest.

Official personnel cannot perform or participate in performing or issue an official certificate for an appeal inspection if they participated in the original inspection, reinspection, or, in the case of a Board appeal inspection, the appeal inspection service unless there is only one qualified person available at the time and place of the appeal inspection.

[50 FR 45395, Oct. 31, 1985]

§ 800.139   Certificating appeal inspections.

(a) *General.* Except as provided in paragraphs (b) of this section, official certificate shall be issued according to § 800.160 and the instructions. Except as provided in paragraph (c)(2) of this section, only the results of the appeal inspection service shall be reported.

(b) *Results of material portion sublots.* When results of an appeal inspection performed by a field office or the Board of Appeals and Review on a material portion do not detect a material error, they shall be averaged with the previous inspection results recorded on the official inspection log for the identified sample. For purposes of this section, a material error is defined as results differing by more than two standard deviations. The appeal or Board appeal inspection result shall replace the previous inspection results recorded on the official inspection log for the identified sample if a material error is detected. No certificate will be issued unless requested by the applicant or deemed necessary by inspection personnel.

(c) *Required statements.* Each appeal certificate shall show the statements required by this section, § 800.161, and applicable instructions.

(1) Each appeal inspection certificate shall clearly show (i) the term "Appeal" or "Board appeal" and (ii) a statement identifying the superseded certificate. The superseded certificate will be considered null and void as of the date of the appeal inspection certificate.

(2) When official grade or official factors, Class X weighing results, and official criteria are reported on the same certificate, the appeal inspection certificate shall show a statement indicating that appeal or Board appeal inspection results are based on official grade, official factors, or official criteria and that all other results are those of the original, reinspection, or, in the case of a Board appeal, the appeal inspection results.

(3) Superseded certificates held by the Service shall be marked "Void." If the superseded certificate is not in the custody of the Service at the time the appeal certificate is issued, a statement indicating that the superseded

certificate has not been surrendered shall be shown on the appeal certificate.

(4) As of the date of issuance of the appeal or Board appeal certificate, the superseded certificate for the original, reinspection, or appeal inspection service will be void and shall not be used to represent the grain.

(5) When certificates are issued under paragraph (b) of this section, the appeal inspection certificate shall show a statement indicating that the results replace the original inspection, reinspection, or, in the case of a Board appeal, the appeal inspection results and that the appeal inspection certificate is not valid for trading purposes.

(d) *Finality of Board appeal inspections.* A Board appeal inspection will be the final appeal inspection service.

[50 FR 45395, Oct. 31, 1985, as amended at 55 FR 24048, June 13, 1990]

OFFICIAL RECORDS AND FORMS
(GENERAL)

SOURCE: Sections 800.145 through 800.159 appear at 50 FR 18986, May 6, 1985, unless otherwise noted.

§ 800.145 Maintenance and retention of records—general requirements.

(a) *Preparing and maintaining records.* The records specified in §§ 800.146–800.159 shall be prepared and maintained in a manner that will facilitate (1) the daily use of records and (2) the review and audit of the records to determine compliance with the Act, the regulations, the standards, and the instructions.

(b) *Retaining records.* Records shall be retained for a period not less than that specified in §§ 800.146–800.159. In specific instances, the Administrator may require that records be retained for a period of not more than 3 years in addition to the specified retention period. In addition, records may be kept for a longer time than the specified retention period at the option of the agency, the contractor, the approved scale testing organization, or the individual maintaining the records.

(Approved by the Office of Management and Budget under control number 0580–0011)

§ 800.146 Maintenance and retention of records issued by the Service under the Act.

Agencies, contractors, and approved scale testing organizations shall maintain complete records of the Act, regulations, the standards, any instructions issued by the Service, and all amendments and revisions thereto. These records shall be maintained until superseded or revoked.

(Approved by the Office of Management and Budget under control number 0580–0011)

§ 800.147 Maintenance and retention of records on delegations, designations, contracts, and approval of scale testing organizations.

Agencies, contractors, and approved scale testing organizations shall maintain complete records of their delegation, designation, contract, or approval. These records consist of a copy of the delegation or designation documents, a copy of the current contract, or a copy of the notice of approval, respectively, and all amendments and revisions thereto. These records shall be maintained until superseded, terminated, revoked, or cancelled.

(Approved by the Office of Management and Budget under control number 0580–0011)

§ 800.148 Maintenance and retention of records on organization, staffing, and budget.

(a) *Organization.* Agencies, contractors, and approved scale testing organizations shall maintain complete records of their organization. These records shall consist of the following documents: (1) If it is a business organization, the location of its principal office; (2) if it is a corporation, a copy of the articles of incorporation, the names and addresses of officers and directors, and the names and addresses of shareholders; (3) if it is a partnership or an unincorporated association, the names and addresses of officers and members, and a copy of the partnership agreement or charter; and (4) if it is an individual, the individual's place of residence. These records shall be maintained for 5 years.

(b) *Staffing.* Agencies, contractors, and approved scale testing organizations shall maintain complete records of their employees. These records consist of (1) the name of each current employee, (2) each employee's principal duty, (3) each employee's principal duty station, (4) information about the training that each employee has received, and (5) related information required by the Service. These records shall be maintained for 5 years.

(c) *Budget.* Agencies, contractors, and approved scale testing organizations shall maintain complete records of their budget. These records consist of actual income generated and actual expenses incurred during the current year. Complete accounts for receipts from (1) official inspection, weighing, equipment testing, and related services; (2) the sale of grain samples; and (3) disbursements from receipts shall be available for use in establishing or revising fees for services under the Act. Budget records shall also include detailed information on the disposition of grain samples obtained under the Act. These records shall be maintained for 5 years.

(Approved by the Office of Management and Budget under control number 0580-0011)

§ 800.149   Maintenance and retention of records on licenses and approvals.

(a) *Licenses.* Agencies, contractors, and approved scale testing organizations shall maintain complete records of licenses. These records consist of current information showing (1) the name of each licensee, (2) the scope of each license, (3) the termination date of each license, and (4) related information required by the Service. These records shall be maintained for the tenure of the licensee.

(b) *Approvals.* Agencies shall maintain complete records of approvals of weighers. These records consist of current information showing the name of each approved weigher employed by or at each approved weighing facility in the area of responsibility assigned to an agency or field office. These records shall be maintained for the tenure of the weigher's employment as an approved weigher.

(Approved by the Office of Management and Budget under control number 0580-0011)

§ 800.150   Maintenance and retention of records on fee schedules.

Agencies, contractors, and approved scale testing organizations shall maintain complete records on fee schedules. These records consist of (a) a copy of the current fee schedule; (b) in the case of an agency, data showing how the fees in the schedule were developed; (c) superseded fee schedules; and (d) related information required by the Service. These records shall be maintained for 5 years.

(Approved by the Office of Management and Budget under control number 0580-0011)

§ 800.151   Maintenance and retention of records on space and equipment.

(a) *Space.* Agencies shall maintain complete records on space. These records consist of (1) a description of space that is occupied or used at each location, (2) the name and address of the owner of the space, (3) financial arrangements for the space, and (4) related information required by the Service. These records shall be maintained for 5 years.

(b) *Equipment.* Agencies shall maintain complete records on equipment. These records consist of (1) the description of each piece of equipment used in performing official inspection or Class X or Class Y weighing services under the Act, (2) the location of the equipment, (3) the name and address of owner of the equipment, (4) the schedules for equipment testing and the results of the testing, and (5) related information required by the Service. These records shall be maintained for 5 years.

(Approved by the Office of Management and Budget under control number 0580-0011)

§ 800.152   Maintenance and retention of file samples.

(a) *General.* The Service and agencies shall maintain complete file samples for their minimum retention period (calendar days) after the official function was completed or the results otherwise reported.

(b) *Minimum retention period.* Upon request by an agency and with the approval of the Service, specified file samples or classes of file samples may be retained for shorter periods of time.

Carrier	In	Out	Export	Other
(1) Trucks	3	5	30	
(2) Railcars	5	10	30	
(3) Ships & Barges	5	25	90	
(4) Ships and Barges (short voyage—5 days or less)	5	25	60	
(5) Containers	5	60	60	
(6) Bins & Tanks				3
(7) Submitted Samples				3

(c) *Special retention periods.* In specific instances, the Administrator may require that file samples be retained for a period of not more than 90 calendar days. File samples may be kept for a longer time than the regular retention period at the option of the Service, the agency, or the individual maintaining the records.

[50 FR 18986, May 6, 1985, as amended at 78 FR 43757, July 22, 2013]

(Approved by the Office of Management and Budget under control number 0580–0011)

§ 800.153 Maintenance and retention of records on official inspection, Class X or Class Y weighing, and equipment testing service.

Agencies and approved scale testing organizations shall maintain complete detailed official inspection work records, copies of official certificates, and equipment testing work records for 5 years.

(Approved by the Office of Management and Budget under control number 0580–0011)

§ 800.154 Availability of official records.

(a) *Availability to officials.* Each agency, contractor, and approved scale testing organization shall permit authorized representatives of the Comptroller General, the Secretary, or the Administrator to have access to and to copy, without charge, during customary business hours any records maintained under §§ 800.146–800.159.

(b) *Availability to the public*—(1) *Agency, contractor, and approved scale testing organization records.* The following official records will be available, upon request by any person, for public inspection during customary business hours:

(i) Copies of the Act, the regulations, the standards, and the instructions;

(ii) The delegation, designation, contract, or approval issued by the Service;

(iii) Organization and staffing records;

(iv) A list of licenses and approvals; and

(v) The approved fee schedule of the agency, if applicable.

(2) *Service. records*—Records of the Service are available in accordance with the Freedom of Information Act (5 U.S.C. 552(a)(3)) and the regulations of the Secretary of Agriculture (7 CFR, part 1, subpart A).

(c) *Locations where records may be examined or copied*—(1) *Agency, contractor, and approved scale testing organization records.* Records of agencies, contractors, and approved scale testing organizations available for public inspection shall be retained at the principal place of business of the agency, contractor, or approved scale testing and certification organization.

(2) *Service records.* Records of the Service available for public inspection shall be retained at each field office and at the headquarters of the Service in Washington, DC.

§ 800.155 Detailed work records—general requirements.

(a) *Preparation.* Detailed work records shall be prepared for each official inspection, Class X or Class Y weighing, and equipment testing service performed or provided under the Act. The records shall (1) be on standard forms prescribed in the instructions; (2) be typed or legibly written in English; (3) be concise, complete, and accurate; (4) show all information and data that are needed to prepare the corresponding official certificates or official report; (5) show the name or initials of the individual who made each determination; and (6) show other information required by the Service to monitor or supervise the service provided.

(b) *Use.* Detailed work records shall be used as a basis for (1) issuing official certificates or official forms, (2) approving inspection and weighing equipment for the performance of official inspection or Class X or Class Y weighing services, (3) monitoring and supervising activities under the Act, (4) answering inquiries from interested persons, (5) processing complaints, and (6) billing and accounting. These records may be used to report results of official inspection or Class X or Class Y weighing services in advance of issuing an official certificate.

(c) *Standard forms.* The following standard forms shall be furnished by the Service to an agency: Official Export Grain Inspection and Weight Certificates (singly or combined), official inspection logs, official weight loading logs, official scale testing reports, and official volume of work reports. Other forms used by an agency in the performance of official services, including certificates, shall be furnished by the agency.

(Approved by the Office of Management and Budget under control number 0580-0011)

§ 800.156 Official inspection records.

(a) *Pan tickets.* The record for each kind of official inspection service identified in § 800.76 shall, in addition to the official certificate, consist of one or more pan tickets as prescribed in the instructions. Activities that are performed as a series during the course of an inspection service may be recorded on one pan ticket or on separate pan tickets. The original copy of each pan ticket shall be retained by the agency or field office that performed the inspection.

(b) *Inspection logs.* The record of an official inspection service for grain in a combined lot and shiplot shall include the official inspection log as prescribed in the instructions. The original copy of each inspection log shall be retained by the agency or field office that performed the inspection. If the inspection is performed by an agency, one copy of the inspection log shall be promptly sent to the appropriate field office.

(c) *Other forms.* Any detailed test that cannot be completely recorded on a pan ticket or an inspection log shall be recorded on other forms prescribed in the instructions. If the space on a pan ticket or an inspection log does not permit

showing the full name for an official factor or an official criteria, an approved abbreviation may be used.

(d) *File samples*—(1) *General.* The record for an official inspection service based, in whole or in part, on an examination of a grain in a sample shall include one or more file samples as prescribed in the instructions.

(2) *Size.* Each file sample shall consist of an unworked portion of the official sample or warehouseman's sample obtained from the lot of grain and shall be large enough to permit a reinspection, appeal inspection, or Board appeal inspection for the kind and scope of inspection for which the sample was obtained. In the case of a submitted sample inspection, if an undersized sample is received, the entire sample shall be retained.

(3) *Method.* Each file sample shall be retained in a manner that will preserve the representativeness of the sample from the time it is obtained or received by the agency or field office until it is discarded. High moisture samples, infested samples, and other problem samples shall be retained according to the instructions.

(4) *Uniform system.* To facilitate the use of file samples, agencies shall establish and maintain a uniform file sample system according to the instructions.

(5) *Forwarding samples.* Upon request by the supervising field office or the Board of Appeals and Review, each agency shall furnish file samples (i) for field appeal or Board appeal inspection service, or (ii) for monitoring or supervision. If, at the request of the Service, an agency locates and forwards a file sample for an appeal inspection, the agency may, upon request, be reimbursed at the rate prescribed in § 800.71 by the Service.

(Approved by the Office of Management and Budget under control number 0580-0011)

§ 800.157 Official weighing records.

(a) *Scale ticket, scale tape, or other weight records.* In addition to the official certificate, the record for each Class X or Class Y weighing service shall consist of a scale ticket, a scale tape, or any other weight record prescribed in the instructions.

(b) *Weighing logs.* The record of a Class X or Class Y weighing service performed on bulk grain in a combined lot or bulk shiplot grain shall include the official weighing log as prescribed in the instructions. The original copy of each weighing log shall be retained by the field office or agency that performed the weighing.

(Approved by the Office of Management and Budget under control number 0580–0011)

§ 800.158 Equipment testing work records.

The record for each official equipment testing service or activity consists of an official equipment testing report as prescribed in the instructions. Upon completion of each official equipment test, one or more copies of the completed testing report may, upon request, be issued to the owner or operator of the equipment. The testing report shall show the (a) date the test was performed, (b) name of the organization and personnel that performed the test, (c) names of the Service employees who monitored the testing, (d) identification of equipment that was tested, (e) results of the test, (f) names of any interested persons who were informed of the test results, (g) number or other identification of the approval tag or label affixed to the equipment, and (h) other information required by the instructions.

(Approved by the Office of Management and Budget under control number 0580–0011)

§ 800.159 Related official records.

(a) *Volume of work report.* Field offices and agencies shall prepare periodic reports showing the kind and the volume of inspection and weighing services that they performed. The report shall be prepared and copies shall be submitted to the Service according to the instructions.

(b) *Record of withdrawals and dismissals.* Field offices and agencies shall maintain a complete record of requests for official inspection or weighing services that are withdrawn by the applicant or that are conditionally withheld or dismissed. The record shall be prepared and maintained according to the instructions.

(c) *Licensee record.* Licensees, including licensed warehouse samplers, shall (1) keep the license issued to them by the Service and (2) keep or have reasonable access to a complete record of the Act, the standards, the regulations, and the instructions.

(Approved by the Office of Management and Budget under control number 0580–0011)

OFFICIAL CERTIFICATES

SOURCE: Sections 800.160 through 800.166 appear at 50 FR 45396, Oct. 31, 1985, unless otherwise noted.

§ 800.160 Official certificates; issuance and distribution.

(a) *Required issuance.* An official certificate shall be issued for each inspection service and each weighing service except as provided §§ 800.84, 800.129, and 800.139 and paragraph (b) of this section.

(b) *Distribution*—(1) *General*—(i) *Export.* The original and at least three copies of each certificate will be distributed to the applicant or applicant's order. One copy of each certificate shall be retained by the agency, field office, or Board of Appeals and Review.

(ii) *Nonexport.* The original and at least one copy of each certificate will be distributed to the applicant or to the applicant's order. In the case of inbound trucklot grain, one copy shall be delivered by the applicant to the person who owned the grain at the time of delivery. One copy of each certificate shall be retained by the agency, field office, or Board of Appeals and Review.

(iii) *Local movements of shiplot grain.* When shiplot grain is offered for inspection as a single lot and a portion of the lot is returned to the elevator, certificates representing the inspection service shall not be issued unless (A) requested by the applicant or (B) deemed necessary by official personnel.

(2) *Reinspection and appeal inspection services.* In addition to the distribution requirements of paragraph (b) of this section, one copy of each reinspection or appeal inspection certificate shall be distributed to each interested person of record or the interested person's order and to the agency or field office that issued the superseded certificate.

(3) *Additional copies.* Additional copies of certificates will be furnished to the applicant or interested person upon request. Fees for extra copies may be assessed according to the fee schedules established by the agency or the Service.

(c) *Prompt issuance.* The results of the inspection or weighing service shall be reported to the applicant on the date the inspection or weighing service is completed. Certificates shall be issued as soon as possible, but no later than the close of business on the next business day. Upon request of an agency or a field office, the requirements of this paragraph may be waived by the Service when results have been reported before issuing the certificate.

(d) *Who may issue certificates*—(1) *Authority.* Certificates for inspection or Class X weighing services may be issued only be official personnel who are specifically licensed or authorized to perform and certify the results reported on the certificate. Certificates for Class Y weighing services may be issued only by individuals who are licensed or authorized or are approved to perform and certify the results.

(2) *Exception.* The person in the best position to know whether the service was performed in an approved manner and that the determinations are accurate and true should issue the certificate. If the service is performed by one person, the certificate should be issued by that person. If the service is performed by two or more persons, the certificate should be issued by the person who made the majority of the determinations or the person who makes the final determination. Supervisory personnel may issue a certificate when the individual is licensed or authorized to perform the service being certificated.

(e) *Name requirement.* On export certificates, the typewritten name and signature of the individual issuing the certificate shall appear on the original and all copies. On all other certificates, the name or signature of the individual issuing the certificate shall appear on the original and all copies. Upon request by the applicant, the name and signature may be shown on all other certificates.

(f) *Authorization to affix names*—(1) *Requirements.* The name or signature of official personnel may be affixed to official certificates which are prepared from work records signed or initialed by the person whose name will be shown. An agent affixing the name and signature shall (i) be employed by the agency or Service; (ii) have been designated to affix names and signatures; and (iii) hold a power of attorney from the person whose name and signature will be affixed. The power of attorney shall be on file with the agency or Service.

(2) *Initialing.* When a name or signature is affixed by an authorized agent, the initials of the agent shall appear directly below or following the signature of the person.

(g) *Advance information.* Upon request, the contents of an official certificate may be furnished in advance to the applicant and any other interested party, or to their order, and any additional expense shall be borne by the requesting party.

(h) *Certification after dismissal.* An official certificate cannot be issued for a service after the request has been withdrawn or dismissed.

(Approved by the Office of Management and Budget under control number 0580–0011)

[50 FR 45396, Oct. 31, 1985, as amended at 57 FR 11428, Apr. 3, 1992]

§ 800.161   Official certificate requirements.

(a) *General.* Official certificates shall show the information and statements required by § 800.161 through § 800.165 and the instructions. The Administrator shall approve any other information and statements reported. Information shall be reported in a uniform, accurate, and concise manner, be in English, be typewritten or handwritten in ink, and be clearly legible.

(b) *Required format.* Official certificates shall be uniform in size, shape, color, and format and conform to requirements prescribed in the instructions. Upon request and for good cause, the Service may approve special design certificates. All information and statements shall be shown on the front of the certificate, except that on domestic grain certificates, (1) approved abbreviations for official factors and official criteria, with their meanings, may be shown on the back and (2) the identification of carriers or containers in a combined-lot inspection may be shown on the back if ample space is not available on the front. When information is recorded on the back of the certificate, the statement "See reverse side" must be shown on the front.

(c) *Required information.* Each official certificate shall show the following information in accordance with the instructions:

(1) For an agency issuing export certificates or the Federal Grain Inspection Service, "United States Department of Agriculture—Federal Grain Inspection Service;"

(2) For a designated agency, the name of the agency, as applicable;

(3) Captions identifying the kind of service;

(4) A preprinted serial number and lettered prefix;

(5) "Original" or "copy," as applicable;

(6) "Divided lot," "duplicate," or "corrected," as applicable;

(7) The identification of the carrier or container;

(8) The date the service was performed;

(9) The date and method of sampling;

(10) The kind of movement and the level of service performed;

(11) The grade and kind or "Not Standardized Grain," as applicable;

(12) The results of the service performed;

(13) The location of the issuing office;

(14) The location of the grain when the service was performed;

(15) A space for remarks;

(16) Whether a reinspection or appeal inspection service was based in whole or in part on file samples when file samples are used;

(17) A statement reflecting the results of a stowage examination, when applicable;

(18) Seal records, when applicable; and

(19) The name of the person issuing the certificate.

(d) *Required statements.* Each official certificate shall include the following statements according to the instructions: (1) A statement that the certificate is issued under the authority of the United States Grain Standards Act; (2) a nonnegotiability statement; (3) a warning statement; and (4) a statement referencing the certificate number and date. Each official certificate for an official sample-lot inspection service shall include a caption "U.S. Grain Standards Act" and a USDA-FGIS shield ghosted across the front. Each official certificate for a warehouseman's sample-lot inspection, a submitted sample inspection, or Class Y weighing service shall include a statement that the certificate does not meet the requirements of section 5 of the Act of warehouseman's sample-lot inspection, the word "QUALIFIED;" for submitted sample inspections, the words "Not Officially Sampled;" for Class Y weighing, the words "Class Y Weighing" screened across the front.

(e) *Permissive information and statements*—(1) *Certificates.* Information and statements requested by the applicant but not required by the regulations or instructions may be shown on the certificate if the information or statements have been approved in the instructions or on a case-by-case basis by the Administrator.

(2) *Letterhead.* Information and statements requested by the applicant but not required by the regulations or instructions may be shown on letterhead stationary of the Service or an agency when (i) ample space is not available for reporting the information or statements on the certificate, (ii) letterhead stationary is determined to be more suitable than the official certificate, and (iii) the certificate is referenced on the letterhead stationary and distributed according to § 800.160. Letterhead stationary of the Service shall be used for all export grain.

(Approved by the Office of Management and Budget under control number 0580–0011)

§ 800.162 **Certification of grade; special requirements.**

(a) *General.* Except as provided in paragraph (c) of this section, each official certificate for grade shall show:

(1) The grade and factor information required by the Official U.S. Standards for Grain;

(2) The test weight of the grain, if applicable;

(3) The moisture content of the grain;

(4) The results for each official factor for which a determination was made;

(5) The results for each official factor that determined the grade when the grain is graded other that U.S. No. 1;

(6) Any other factor information considered necessary to describe the grain; and

(7) Any additional factor results requested by the applicant for official factors defined in the Official U.S. Standards for Grain.

(b) *Cargo shipments.* Each official certificate for grade representing a cargo shipment shall show, in addition to the requirements of paragraph (a) of this section, the results of all official grade factors defined in the Official United States Standards for Grain for the type of grain being inspected.

(c) Test weight for canola and soybeans. Official canola inspection certificates will show, in addition to the requirements of paragraphs (a) and (b) of this section, the official test weight per bushel only upon request by the applicant. Official soybean inspection certificates will show, in addition to the requirements of paragraphs (a) and (b) of this section, the official test weight per bushel unless the applicant requests that test weight not be determined. Upon request, soybean test weight results will not be determined and/or reported on the official certificate.

(d) *Aflatoxin test for corn.* Official corn export certificates shall show, in addition to the requirements of paragraphs (a), (b), and (c) of this section, the official aflatoxin test results if required under § 800.15(b).

(Approved by the Office of Management and Budget under control number 0580-0011)

[50 FR 45396, Oct. 31, 1985, as amended at 52 FR 24437, June 30, 1987; 57 FR 2439, Jan. 22, 1992; 57 FR 3273, Jan. 29, 1992; 57 FR 56439, Nov. 30, 1992; 71 FR 52405, Sept. 6, 2006]

§ 800.163 Divided-lot certificates.

(a) *General.* When shiplot grain is offered for inspection or Class X weighing as a single lot and is certificated as a single lot, the applicant may exchange the official certificate for two or more divided-lot certificates. This applies to original inspection, reinspection, appeal inspection, Board appeal inspection, and Class X weighing services.

(b) *Application.* Requests for divided-lot certificates shall be made (1) in writing; (2) by the applicant who filed the inital request; (3) to the office that issued the outstanding certificate; (4) within 5 business days of the outstanding certificate date; and (5) before the identity of the grain has been lost.

(c) *Quantity restrictions.* Divided-lot certificates shall not show an aggregate quantity different than the total quantity shown on the superseded certificate.

(d) *Surrender of certificate.* The certificate that will be superseded shall (1) be in the custody of the agency or the Service; (2) be marked "Void;" and (3) show the identification of the divided-lot certificates.

(e) *Certification requirements.* The same information and statements, including permissive statements, that were shown on the superseded certificate shall be shown on each divided-lot certificate. Divided-lot certificates shall show (1) a statement indicating the grain was inspected or weighed as an undivided lot; (2) the terms "Divided Lot-Original," and the copies shall show "Divided Lot-Copies;" (3) the same serial number with numbered suffix (for example, 1764-1, 1764-2, 1764-3, and the like); and (4) the quantity specified by the request.

(f) *Issuance and distribution.* Divided-lot certificates shall be issued no later than the close of business on the next business day after the request and be distributed according to § 800.160.

(g) *Limitations.* No divided-lot certificate can be issued (1) for grain in any shipment other than shiplot grain inspected or weighed as a single lot or (2) for an export certificate which has been superseded by another export certificate. After divided-lot certificates have been issued, further dividing or combining is prohibited except with the approval of the Service.

(h) *Use of superseded certificate prohibited.* As of the date of the divided-lot certificate, the superseded certificate will be void and shall not be used or represent the grain.

(Approved by the Office of Management and Budget under control number 0580–0011)

§ 800.164 Duplicate certificates.

Upon request, a duplicate certificate may be issued for a lost or destroyed official certificate.

(a) *Application*. Requests for duplicate certificates shall be filed: (1) in writing; (2) by the applicant who requested the service covered by the lost or destroyed certificate; and (3) with the office that issued the initial certificate.

(b) *Certification requirements*. The same information and statements, including permissive statements, that were shown on the lost or destroyed certificate shall be shown on the duplicate certificate. Duplicate certificates shall show (1) the terms "Duplicate-Original" and the copies shall show "Duplicate-Copies" and (2) a statement that the certificate was issued in lieu of a lost or destroyed certificate.

(c) *Issuance*. Duplicate certificates shall be issued as promptly as possible and distributed according to § 800.160.

(d) *Limitations*. Duplicate certificates will not be issued for certificates that have been superseded.

(Approved by the Office of Management and Budget under control number 0580–0011)

§ 800.165 Corrected certificates.

(a) *General*. The accuracy of the statements and information shown on official certificates shall be verified by the individual whose name or signature is shown on the certificate, or by the authorized agent who affixed the name or signature. Errors found during this process shall be corrected according to this section.

(b) *Who may correct*. Only official personnel or their authorized agents may make corrections, erasures, additions, or other changes to official certificates.

(c) *Corrections prior to issuance*—(1) *Export certificates*. No corrections, erasures, additions, or other changes can be made to an export certificate. If any error is found prior to issuance, a new certificate shall be prepared and issued and the incorrect certificate marked "Void."

(2) *Other than export certificates*. No corrections, erasures, additions, or other changes shall be made to other than export certificates which involve identification, grade, gross, tare, or net weight. If errors are found, a new certificate shall be prepared and issued and the incorrect certificate marked "Void." Otherwise, errors may be corrected provided that (i) the corrections are neat and legible, (ii) each correction is initialed by the individual who corrects the certificate, and (iii) the corrections and initials are shown on the original and all copies.

(d) *Corrections after issuance*—(1) *General*. If errors are found on a certificate at any time up to a maximum of 1 year after issuance, the errors shall be corrected by obtaining the incorrect certificate and replacing it with a corrected certificate. When the incorrect certificate cannot be obtained, a corrected certificate can be issued superseding the incorrect one.

(2) *Certification requirements*. The same statements and information, including permissive statements, that were shown on the incorrect certificate, along with the correct statement or information, shall be shown on the corrected certificate. According to this section and the instructions, corrected certificates shall show (i) the terms "Corrected-Original" and "Corrected-Copy;" (ii) a statement identifying the superseded certificate and the corrections; (iii) a statement indicating the superseded certificate was not surrended if the incorrect certificate was not surrendered; and (iv) a new serial number. In addition, the incorrect certificate shall be marked "Void" when submitted.

(e) *Limitations*. Corrected certificates cannot be issued for a certificate that has been superseded by another certificate or on the basis of a subsequent analysis for quality.

(f) *Use of superseded certificate prohibited*. As of the date of issuance of the corrected certificate, the superseded certificate will be void and shall not be used to represent the grain.

(Approved by the Office of Management and Budget under control number 0580–0011)

§ 800.166   Reproducing certificates.

Official certificates may be photo copied or similarly reproduced.

(Approved by the Office of Management and Budget under control number 0580-0011.)

LICENSES AND AUTHORIZATIONS (FOR INDIVIDUALS ONLY)

§ 800.170   When a license or authorization or approval is required.

(a) *Requirement.* (1) Any individual who performs or represents that he or she is licensed or authorized to perform any or all inspection or Class X weighing services under the Act must be licensed or authorized by the Service to perform each service.

(2) Any individual who performs or represents that he or she is licensed or authorized, or an approved weigher, to perform Class Y weighing services under the Act must be licensed or authorized, or approved, by the Service to perform this service.

(b) *Excepted activities.* A license or authorization, or approval for weighing, under the Act and regulations is not required for (1) opening or closing a carrier or container of grain, or transporting or filing official samples, or similar laboring functions; (2) typing or filing official inspection and weighing certificates or other official forms or performing similar clerical functions; (3) performing official equipment testing functions with respect to official inspection equipment; (4) performing inspection, weighing, or scale testing functions that are not conducted for the purposes of the Act; or (5) performing scale testing functions by a State or municipal agency or by the employees of such agencies.

(c) *30-day waiver.* A prospective applicant for a license as a sampler, inspection technician, or weighing technician may, for a period of time not to exceed 30 calendar days, help perform those official sampling, inspection, or Class X or Class Y weighing services for which the applicant desires to be licensed, under the direct physical supervision of an individual who is licensed to perform the services. The supervising individual shall be fully responsible for each function performed by the prospective applicant and shall initial any

work form prepared by the prospective applicant.

(d) *No fee by Service.* No fee will be assessed by the Service for licensing an individual employed by an agency or contractor.

(e) *Fee by agency.* At the request of the Service, an agency may help examine an applicant for a warehouse sampler's license for competency and may assess a fee in accordance with the provisions of § 800.70. The fee shall be paid by the applicant or by the elevator that employs the applicant.

(Secs. 9, 18, Pub. L. 94-582, 90 Stat. 2875 and 2884 (7 U.S.C. 79a and 87e))

[45 FR 15810, Mar. 11, 1980, as amended at 46 FR 30325, June 5, 1981]

§ 800.171   Who may be licensed or authorized.

(a) *Prohibitions.* No person may be licensed or authorized who has a conflict of interest as defined in section 11 of the Act or specified in § 800.187.

(b) *Exceptions to prohibitions*—(1) *Conflict by agency.* An employee of an agency that has a conflict of interest that is waived by the Administrator under section 11(b)(5) of the Act may be licensed: *Provided,* That the employee has no conflict of interest other than the agency conflict of interest.

(2) *Warehouse samplers.* A qualified employee of an elevator may be licensed to perform specified sampling services under the Act in accordance with the provisions of § 800.174(a)(2).

(c) *General qualifications*—(1) *Inspection and weighing.* To obtain a license to perform inspection or weighing services under the Act, an individual must be employed by an agency to perform the services and must otherwise be found competent in accordance with this section and § 800.173.

(2) *Specified technical services.* To obtain a license to perform specified sampling, inspection testing, weighing, and similar services under the Act, an individual must (i) be employed by an agency to perform the services, or (ii) enter into or be employed under a contract with the Service to perform the services, and (iii) otherwise be found competent in accordance with this section and § 800.173.

(3) *Warehouse sampler.* To obtain a warehouse sampler's license, an applicant must be employed by an elevator to perform sampling services and otherwise be found competent in accordance with this section and § 800.173.

(4) *Requirements.* To be considered competent, an individual must (i) meet the qualifications specified in § 800.173; and (ii) have available the equipment and facilities necessary to perform the services for which the individual is to be licensed.

(d) *Competency determinations*—(1) *Agency samplers and technicians.* The competency of an applicant for a license as a sampler, inspection technician, or weighing technician shall be determined by (i) the chief inspector or the chief weighmaster, as applicable, of the agency that employs the applicant or, in the case of a warehouse sampler, the agency that is assigned the area in which the elevator that employs the sampler is located, and (ii) the field office supervisor.

(2) *Inspectors, weighers, contract samplers, and technicians.* The competency of an applicant for a license as an inspector or weigher or any license issued under the terms of a contract with the Service shall be determined by the Service.

(3) *Examinations.* A determination of competency of an applicant for a license shall include an evaluation of the results of examinations or reexaminations under § 800.173.

[45 FR 15810, Mar. 11, 1980, as amended at 49 FR 36072, Sept. 14, 1984]

§ 800.172   Applications for licenses.

(a) *General.* An application for a license, the renewal of a license, or the return of a suspended license shall be made to the Service on forms furnished by the Service. Each application shall (1) be in English, (2) be typewritten or legibly written in ink, (3) show all information prescribed by the application form, and (4) be signed by the applicant.

(b) *Additional information.* An applicant shall furnish any additional information considered necessary by the Service for consideration of an application.

(c) *Withdrawal.* An application for a license may be withdrawn by an applicant at any time.

(d) *Review of applications*—(1) *General procedure.* Each application shall be reviewed to determine whether the applicant and the application comply with the Act and the regulations.

(2) *Application and applicant in compliance.* If it is determined that the applicant and the application comply with the Act and the regulations, the requested license shall be granted.

(3) *Application not in compliance.* If an application does not comply with this section and the noncompliance prevents a satisfactory review by the Service, the applicant shall be provided an opportunity to submit any needed information. If the needed information is not submitted by the applicant within a reasonable time, the application may be dismissed.

(4) *Applicant not in compliance.* If it is determined that an applicant does not comply with the provisions of the Act and §§ 800.171, 800.173, and 800.187 at the time the application is submitted, the applicant shall be provided an opportunity to comply. If the applicant cannot comply within a reasonable period of time, the application shall be dismissed.

(e) *Procedure for dismissal.* If a dismissal involves an application for a renewal of a license or for the return of a suspended license, the dismissal shall be performed in accordance with the provisions of § 800.179. All other dismissals shall be performed by promptly notifying the applicant and the employer of the applicant of the reasons for the dismissal.

(Approved by the Office of Management and Budget under control number 0580–0012)

[45 FR 15810, Mar. 11, 1980, as amended at 48 FR 44453, Sept. 29, 1983; 54 FR 5924, Feb. 7, 1989]

§ 800.173   Examinations and reexaminations.

(a) *General.* Applicants for a license and individuals who are licensed to perform any or all official inspection or Class X or Class Y weighing services shall, at the discretion of the Service, submit to examinations or reexaminations to determine their competency to perform the official inspection or weighing functions for which they desire to be, or are, licensed.

(b) *Time and place of examinations and reexaminations.* Examinations or reexaminations under this section shall be conducted by official personnel designated by the Service and shall be given at a reasonable time and place in accordance with the instructions.

(c) *Scope of examinations and reexaminations.* Examinations or reexaminations may include oral or written tests on the applicable provisions of the Act, the regulations, the Official U.S. Standards for Grain, the procedures for the inspection and weighing of grain under the Act, the instructions, on-site performance evaluations, and vision or olfactory examinations.

(d) *Competency standards*—(1) *Inspection.* An individual may be found to be incompetent to perform official inspection services if the individual (i) has a color-vision deficiency; (ii) cannot meet the physical requirements necessary to perform the functions; (iii) cannot readily distinguish between the different kinds and classes of grain, or the different conditions in grain, including heating, musty, sour, insect infestation, and smut; (iv) cannot demonstrate a technical ability to operate grain sampling, testing, and grading equipment; (v) does not have a working knowledge of applicable provisions of the Act, the regulations, the Official U.S. Standards for Grain, and the instructions; (vi) cannot determine work-related mathematical computations; or (vii) cannot prepare legible records in English.

(2) *Weighing.* An individual may be found to be incompetent to perform Class X or Class Y weighing services under the Act if the individual (i) does not meet the requirements of paragraphs (d)(1)(ii), (v), (vi), and (vii) of this section or (ii) cannot demonstrate a technical ability to operate grain weighing equipment.

§ 800.174 Issuance and possession of licenses and authorizations.

(a) *Scope of licenses and authorizations.* Subject to the provisions of § 800.171, eligible individuals may be licensed or authorized by the Service to perform one or more services specified in this paragraph.

(1) *Official samplers.* Individuals employed by an agency or the Service or employed under the terms of a contract with the Service may be licensed or authorized, as applicable, to perform or supervise the performance of stowage examinations, grain sampling, and related technical services and to issue official certificates for the services performed by them.

(2) *Licensed warehouse samplers.* Elevator or warehouse employees may be licensed to sample grain and perform stowage examinations. No elevator employee shall be licensed to (i) sample export grain for inspection under the Act, (ii) test or grade grain, or (iii) certify the results of any inspection service under the Act.

(3) *Official inspection technicians.* Individuals employed by an agency or the Service or employed under the terms of a contract with the Service may be licensed or authorized to perform or supervise the performance of stowage examinations, grain sampling, or all or specified noninterpretive laboratory-testing services and to issue official certificates for the services performed by them.

(4) *Official inspectors.* Individuals employed by an agency or the Service may be licensed or authorized to perform and supervise the performance of stowage examinations, sampling, laboratory-testing, grading, and related services and to issue official certificates for the services performed by them.

(5) *Official weighing technicians.* Individuals who are employed by an agency or the Service to observe the loading, unloading, and handling of grain that has been or is to be weighed under the Act may be licensed or authorized to perform and supervise the performance of grain handling and stowage examination services and to issue official certificates for the services performed by them.

(6) *Official weighers.* Individuals employed by an agency or the Service may be licensed or authorized to perform and supervise the performance of grain handling, stowage examination,

official weighing (Class X), and supervision of weighing (Class Y), and related services and to issue official certificates for the services performed by them.

(7) *Authorized scale tester.* Individuals employed by the Service may be authorized to test and supervise the testing of scales used for Class X and Class Y weighing services and to approve and certify scales based on the results of these tests.

(b) *Condition for issuance—*(1) *Compliance with the Act.* Each license is issued on the condition that the licensee will, during the term of the license, comply with the Act, the regulations, and the instructions.

(2) *Possession of license.* Each license shall be the property of the Service, but each licensee shall have the right to possess the license subject to the provisions of §§ 800.173, 800.186, and 800.187.

(c) *Duplicate license.* Upon satisfactory proof of the loss or destruction of a license, a duplicate will be issued by the Service.

(d) *Retention of licenses.* Each license shall be retained by the holder of the license in a manner that the license can be examined upon request by service personnel.

§ 800.175 Termination of licenses.

(a) *Term of license.* Each license shall terminate in accordance with the termination date shown on the license and as specified in paragraph (b) of this section. The termination date for a license shall be no less than 5 years or more than 6 years after the issuance date for the initial license; thereafter, every 5 years. Upon request of a licensee and for good cause shown, the termination date may be advanced or delayed by the Administrator for a period not to exceed 60 days.

(b) *Termination schedule for licenses.* Subject to the provisions of paragraph (a) of this section, licenses shall terminate on the last day of the month shown in the following schedule:

Last names beginning with	Termination date
A	January.
B	February.
C, D	March.
E, F, G	April.

Last names beginning with	Termination date
H, I, J	May.
K, L	June.
M	July.
N, O, P, Q	August.
S	September.
R, T, U, V	October.
W	November.
X, Y, Z	December.

(c) *Termination notices.* The Service shall issue notice of termination to licensees and to their employers at least 60 days before the termination date. The notice shall (1) provide detailed instructions for requesting renewal of licenses; (2) state whether a reexamination will be required; and (3) if a reexamination will be required, show the nature and scope of the reexamination. Failure to receive a notice from the Service shall not exempt a licensee from the responsibility of having the license renewed on or before the termination date.

(d) *Renewal of licenses.* Licenses that are renewed shall show the permanent license number, the date of renewal, and the word "Renewed."

(e) *Termination of suspended licenses.* Any suspension of a license, including voluntary suspension or suspension by change in employment, shall not affect the termination date of the license. If a licensee applies for renewal of the license prior to the termination date, the license will not terminate during the period of suspension.

(f) *Surrender of license.* Each license that is terminated, suspended, or canceled under the provisions of §§ 800.175 through 800.178 or is suspended, revoked, or not renewed for cause under the provisions of § 800.179 shall be promptly surrendered to the field office.

(g) *Marking terminated, canceled, or revoked licenses.* Each terminated, canceled, or revoked license surrendered to the Service shall be marked "Canceled."

[45 FR 15810, Mar. 11, 1980, as amended at 81 FR 49863, July 29, 2016]

§ 800.176 Voluntary cancellation or suspension of licenses.

Upon request by a licensee, the Service may cancel a license or suspend a license for a period of time not to exceed 1 year. A license that has been voluntarily suspended shall be returned by the Service upon request by the licensee within 1 year, subject to the provisions of § 800.172; a license that has been cancelled shall be considered void and shall not be subject to return or renewal.

§ 800.177 Automatic suspension of license by change in employment.

A license issued to an individual who is employed by an agency shall be automatically suspended when the individual ceases to be employed by the agency. If the individual is reemployed by the agency or employed by another agency within 1 year of the suspension date and the license has not terminated in the interim, upon request of the licensee, the license will be reinstated subject to the provisions of §§ 800.172 and 800.173.

§ 800.178 Summary revocation of licenses.

Licenses may be summarily revoked upon a finding that the licensee has been convicted of any offense either prohibited by section 13 of the Act or prohibited by Title 18 of the United States Code, with respect to the performance of services under the Act.

§ 800.179 Refusal of renewal, suspension, or revocation of licenses for cause.

(a) *General.* A license may be suspended or revoked or may be refused renewal or return (if suspended) for causes prescribed in section 9 of the Act.

(b) *Procedure for summary action.* Under section 9 of the Act, any license may, without first affording the licensee (hereafter in this section "respondent") an opportunity for a hearing, be summarily suspended pending final determination, whenever the action is considered to be in the best interest of the official inspection system. Such action shall be effective upon receipt of notice from the Service by the respondent. Within 30 calendar days after issuing a notice of summary action, the Service shall afford the respondent an opportunity for a hearing as provided under paragraph (c) of this section. Pending final determination, the Service may terminate the action if alternative employment arrangements satisfactory to the Service can be and are made for the respondent by the employer of the respondent.

(c) *Procedure for other than summary action.* Except as provided for in paragraph (a) of this section, before the Service refuses to renew, or suspends or revokes a license, or refuses to return a suspended license, the respondent shall be (1) notified of the proposed action and the reasons therefor, and (2) afforded (i) an opportunity to express his/her views on the proposed action in an informal manner, or (ii) at the request of the respondent, a hearing in accordance with the provisions of the Rules of Practice Governing Formal Adjudicatory Proceedings Instituted by the Secretary under Various Statutes (7 CFR, part 1, subpart H).

§ 800.180 Summary cancellation of licenses.

A license may be summarily canceled when (a) the license has been under voluntary or automatic suspension for a period of 1 year and there has been no request for return of the license or a request for return of the license has been dismissed in accordance with § 800.172; or (b) the licensee has died or fails to surrender the license in accordance with § 800.175(f).

DUTIES AND CONDUCT OF LICENSED AND AUTHORIZED PERSONNEL

§ 800.185 Duties of official personnel and warehouse samplers.

(a) *General.* Official personnel and warehouse samplers shall, when performing official services or duties under the Act, comply with the Act, the regulations, and the instructions.

(b) *Inspection and weighing services.* Official personnel shall perform requested official inspection and Class X and Class Y weighing services (1) without discrimination, (2) as soon as practicable, and (3) in accordance with methods and procedures prescribed in the instructions.

(c) *Sealing carriers or containers.* Upon request, or in accordance with the instructions, official personnel shall (1) when feasible, affix security seals to doors, hatch covers, and similar openings on carriers or containers that contain grain that has been officially inspected or Class X or Class Y weighed under the Act and (2) show seal records on certificates and other official forms in accordance with the provisions of §800.161.

(d) *Scope of operations.* Official personnel and warehouse samplers shall operate only within the scope of their license or authorization and except as otherwise provided in §800.117, operate only within the area of responsibility assigned to the official agency, field office, or contractor which employs them. Official personnel and warehouse samplers may perform official inspection or weighing services in a different area of responsibility with the specific consent of the Service.

(e) *Working materials.* Official personnel and warehouse samplers shall be responsible for maintaining a working knowledge of the applicable provisions of the Act, the regulations, the Official U.S. Standards for Grain, the instructions, and all amendments and revisions thereto.

(f) *Observation of services.* Official personnel and warehouse samplers shall permit any person (or the person's agent) who has a financial interest in grain that is being inspected or weighed under the Act, or in equipment that is being tested under the Act, to observe the performance of any or all official inspection, or Class X or Class Y weighing. Appropriate areas in the elevator may be specified by the Service in conjunction with the elevator management for observing each service. The areas shall be safe, shall afford a clear and unobstructed view of the performance of the services, but shall not permit a close over-the-shoulder type of observation by the interested person or the person's agent.

(g) *Reporting violations.* Official personnel and warehouse samplers shall in accordance with the instructions promptly report (1) information which shows or tends to show a violation of any provision of the Act, the regulations, or the instructions, and (2) information on any instructions which have been issued to them by any official personnel or other persons which are contrary to the Act, the regulations, or the instructions.

(h) *Related duties.* Official personnel and warehouse samplers shall, when practicable, assist in training other employees who desire to become licensed.

(i) *Instructions by Service.* Official personnel and warehouse samplers shall carry out all written instructions or oral directives issued to them by the Service and, upon request, inform the Service regarding inspection, weighing, or equipment testing services performed by them. Oral directives from the Service not found in written instructions shall be confirmed in writing, upon request.

(Approved by the Office of Management and Budget under control number 0580–0013)

[45 FR 15810, Mar. 11, 1980; 45 FR 55119, Aug. 18, 1980, as amended at 48 FR 44453, 44454, Sept. 29, 1983; 54 FR 5924, Feb. 7, 1989; 68 FR 19139, Apr. 18, 2003]

§800.186   Standards of conduct.

(a) *General.* Official personnel and warehouse samplers must maintain high standards of honesty, integrity, and impartiality to assure proper performance of their duties and responsibilities and to maintain public confidence in the services provided by them.

(b) *Prohibited conduct; official personnel and warehouse samplers.* No official personnel or warehouse sampler shall:

(1) Perform any official inspection, Class X or Class Y weighing, or equipment testing service unless licensed or authorized to do so;

(2) Engage in criminal, dishonest, or notoriously disgraceful conduct, or other conduct prejudicial to the Department or the Service;

(3) Report for duty in an intoxicated or drugged condition, or consume intoxicating beverages or incapacitating drugs while on duty;

(4) Smoke in prohibited areas in elevators or perform official services in an unsafe manner that could endanger official personnel working on or about the premises;

(5) Make unwarranted criticisms or accusations against other official personnel, warehouse samplers, or employees of the Department; and

(6) Refuse to testify or respond to questions in connection with official inquiries or investigations.

(7) Coerce or attempt to coerce any person into providing any special or undue benefit to official personnel, approved weighers, or warehouse samplers.

(c) *Prohibited conduct; official personnel.* In addition to the conduct prohibited by paragraph (b) of this section, no official personnel shall:

(1) Solicit contributions from other official personnel or warehouse samplers for an employee of the Service, or make such a contribution. Nothing in this paragraph shall preclude the occasional voluntary giving or acceptance of gifts of a nominal value on special occasions;

(2) Take any action that might (i) create the appearance of a loss of impartiality or (ii) adversely affect the confidence of the public in the integrity of the inspection, weighing, or equipment testing services performed under the Act;

(3) Except as provided in § 800.76(a), engage in any outside (unofficial) work or activity that:

(i) May impair their efficiency in performing official functions; or

(ii) Consists in whole or in part of unofficial acts of sampling, stowage examination, inspection testing, equipment testing, inspection, or weighing services similar to the official services for which the employing agency is designated; or

(iii) May result in the acquisition of property interests that could create a conflict of interest as defined in section 11 of the Act; or

(iv) May tend to bring criticism on or otherwise embarrass the Department or the Service;

(4) Issue to other official personnel, warehouse samplers, or approved weighers any instructions or directives inconsistent with the Act, the regulations, the Official U.S. Standards for Grain, or the instructions;

(5) Organize or help establish a general or specialized farm organization, or act as an officer or business agency in, recruit members for, or accept office space or contributions from such an organization;

(6) Advocate that any general or specialized farm organization better represents the interest of farmers than any other organization or individual, or recommend that the responsibilities of any government agency be carried out through a general or specialized farm organization. Nothing in paragraph (c)(5) of this section shall prevent official personnel from holding membership in a general or specialized farm organization or prohibit official personnel from participating in the operation of local groups or organizations that conduct government-authorized programs.

[45 FR 15810, Mar. 11, 1980, as amended at 48 FR 44454, Sept. 29, 1983; 60 FR 65235, Dec. 19, 1995; 63 FR 45677, Aug. 27, 1998]

§ 800.187   **Conflicts of interest**

(a) *General.* Warehouse samplers are exempt from the conflict-of-interest provisions of this section.

(b) *What constitutes a gratuity.* For the purposes of these regulations, the term "gratuity" shall include any favor, entertainment, gift, tip, loan, payment for unauthorized or fictitious work, unusual discount, or anything of monetary value. The term shall not include (1) the occasional exchange of a cup of coffee or similar social courtesies of nominal value in a business or work relationship if the exchange is wholly free of any embarrassing or improper implications; (2) the acceptance of unsolicited advertising material such as pencils, pens, and note pads of nominal value if the material is wholly free of any embarrassing or improper implications; and (3) the exchange of the usual courtesies in an obvious family or personal relationship (including those between official personnel and their parents, spouses, children, or close personal friends) when the circumstances make it clear that the exchange is the result of the family or personal relationship, rather than a business or work relationship.

(c) *Conflicts.* In addition to the conflicts of interest prohibited by section 11 of the Act, the activities specified in this paragraph shall also be considered to be a conflict of interest. Accordingly, no official personnel shall, during the term of their license or authorization (including any period of suspension):

(1) Accept any gratuity.

(2) Accept any fee or charge or other thing of monetary value, in addition to the published fee or charge, for the performance of official inspection or weighing services under circumstances in which the acceptance could result, or create the appearance of resulting, in (i) the use of their office or position for undue private gain, (ii) an undertaking to give undue preferential treatment to any group or any person, or (iii) any other loss of independence or impartiality in the performance of official inspection or Class X or Class Y weighing services.

(3) Knowingly perform, or participate in performing, an inspection or weighing service on grain in which they have a direct or indirect financial interest.

(4) Engage in the business by buying, selling, transporting, cleaning, elevating, storing, binning, mixing, blending, drying, treating, fumigating, or other preparation of grain (other than a grower of grain, or in the disposition of inspection samples); or in the business of cleaning, treating, or fitting carriers or containers for transporting or storing grain; the merchandising for nonfarm use of equipment for cleaning, drying, treating, fumigating, or otherwise processing, handling, or storing grain; or the merchandising of grain inspection or weighing equipment (other than buying or selling by official personnel of the equipment for use in the performance of their official services).

(5) Seek or hold any appointive or elective office in a grain industry organization or association. This provision does not apply to organizations of official inspectors or official weighers.

(6) Participate in any transaction involving the purchase or sale of corporate stocks or bonds, grain or grain-related commodities, or other property for speculative or income purposes if the transaction could reasonably be construed to interfere with the proper and impartial performance of official inspection for Class X or Class Y weighing services. Official personnel are not prohibited from (i) producing grain as a grower and selling the grain; (ii) making bona fide investments in governmental obligations, banking institutions, savings and loan associations, and other tangibles and intangibles that are clearly not involved in the production, transportation, storage, marketing, or processing of grain; or (iii) borrowing money from banks or other financial institutions on customary terms.

(d) *Reports of interests.* Official personnel shall report information regarding their employment or other business or financial interests which may be required by the Service.

(e) *Avoiding conflicts of interest.* Official personnel shall not acquire any financial interest or engage in any activity that would result in a violation of this § 800.187, or § 800.186, or section 11 of the Act and shall not permit their spouses, minor children, or blood relatives who reside in their immediate households to acquire any such interest or engage in any such activity. For the purpose of this section, the interest of a spouse, minor child, or blood relative who is a resident of the immediate household of official personnel shall be considered to be an interest of the official personnel.

(f) *Disposing of a conflict of interest—* (1) *Remedial action.* Upon being informed that a conflict of interest exists and that remedial action is required, an applicant for a license and official personnel shall take immediate action to end the conflict of interest and inform the Service of the action taken.

(2) *Hardship cases.* Applicants and official personnel who believe that remedial action will cause undue personal hardship may request an exception by forwarding to the Service a written statement setting forth the facts, circumstances, and reasons for requesting an exception.

(3) *Failure to terminate.* If a final determination is made by the Service that a conflict of interest does exist and should not be excepted, failure to terminate the conflict of interest shall subject: (i) An applicant for a license to a dismissal of the application; (ii) An employee of the Service to disciplinary

action; and (iii) A licensee to license revocation.

(Approved by the Office of Management and Budget under control number 0580–0012)

[45 FR 15810, Mar. 11, 1980, as amended at 48 FR 44453 and 44454, Sept. 29, 1983; 54 FR 5924, Feb. 7, 1989]

§ 800.188 Crop year, variety, and origin statements.

No official personnel shall certify or otherwise state in writing (a) the year of production of grain, including use of terms such as "new crop" or "old crop"; (b) the place or geographical area where the grain was grown; or (c) the variety of the grain.

§ 800.189 Corrective actions for violations.

(a) *Criminal prosecution.* Official personnel and warehouse samplers who commit an offense prohibited by section 13 of the Act are subject to criminal prosecution in accordance with section 14 of the Act.

(b) *Administrative action*—(1) *Other than Service employees.* In addition to possible criminal prosecution, licensees and warehouse samplers are subject to administrative action in accordance with sections 9 and 14 of the Act.

(2) *Service employees.* In addition to possible criminal prosecution, employees of the Service are subject to disciplinary action by the Service.

DELEGATIONS, DESIGNATIONS, APPROVALS, CONTRACTS, AND CONFLICTS OF INTEREST

AUTHORITY: Sections 800.195 through 800.199 were issued under secs. 8, 9, 10, 13, and 18, Pub. L. 94–582, 90 Stat. 2870, 2875, 2877, 2880, and 2884, 7 U.S.C. 79, 79a, 79b, 84, 87, and 87e.

§ 800.195 Delegations.

(a) *General.* Eligible States may be delegated authority to perform official services (excluding appeal inspection) at export port locations within their respective States.

(b) *Restrictions.* Only the Service or the delegated State may perform official inspection, Class X, and Class Y weighing services at an export port location within the State. If official inspection services, at export port locations within the State, are performed

by the Service, only the Service may perform Class X and Class Y weighing services at the locations. If official inspection services are performed by a delegated State, either the State or the Service may perform Class X and Class Y weighing services at the export port locations within the State.

(c) *Who can apply.* States which: (1) Were performing official inspection at an export port location under the Act on July 1, 1976, or; (2)(i) performed official inspection at an export port location at any time prior to July 1, 1976; (ii) were designated under section 7(f) of the Act on December 22, 1981, to perform official inspections; and (iii) operate in a State from which total annual exports of grain do not exceed, as determined by the Administrator, 5 per centum of the total amount of grain exported from the United States annually may apply to the Service for a delegation.

(d) *When and how to apply.* A request for authority to operate as a delegated State should be filed with the Service not less than 90 calendar days before the State proposes to perform the official service. A request for authority to operate as a delegated State shall show: (1) The export port location(s) where the State proposes to perform official inspection, Class X, and Class Y weighing services; (2) the estimated annual volume of inspection and weighing services for each location; and (3) the schedule of fees the State proposes to assess. A request for a revision to a delegation shall (i) be filed with the Service not less than 90 calendar days before the desired effective date, and (ii) specify the change desired.

(e) *Review of eligibility and criteria for delegation.* Each applicant for authority to operate as a delegated State shall be reviewed to determine whether the applicant meets the eligibility conditions contained in paragraph (c) of this section and the criteria contained in section 7(f)(1)(A) of the Act. The requested delegation may be granted if the Service determines that the applicant meets the eligibility conditions and criteria. If an application is dismissed, the Service shall notify the applicant promptly, in writing, of the reason(s) for the dismissal.

(f) *Responsibilities*—(1) *Providing official services.* Each delegated State shall be responsible for providing each official service authorized by the delegation at all export elevators at export port locations in the State. The State shall perform each official service according to the Act, regulations, and instructions.

(2) *Staffing, licensing, and training.* Delegated States shall employ official personnel on the basis of job qualifications rather than political affiliations. The State shall employ sufficient personnel to provide the services normally requested in an accurate and timely manner. The State shall only use personnel licensed by the Service for the performance of official services and shall train and assist its personnel in acquiring and maintaining the necessary skills. The State shall keep the Service informed of the employment status of each of its licensees and any substantial change in a licensee's duties.

(3) *Rotation of personnel.* Where feasible, each delegated State shall rotate licensees among elevators and other facilities as is necessary to preserve the integrity of the official inspection and weighting systems.

(4) *Supervision.* The State and its officials shall be responsible for the actions of the official personnel employed by the State, for direct supervision of the daily activities of such personnel, and for the conduct of official services and related activities in the State. The State shall supervise official activities according to the Act, regulations, and instructions and shall take action necessary to ensure that its employees are not performing prohibited functions and are not involved in any action prohibited by the Act, regulations, or instructions. Each State shall report to the Service information which shows or may show a violation of any provision of the Act, regulations, or instructions and information on any instructions which have been issued to State personnel by Service personnel or by any other person which are contrary to or inconsistent with the Act, regulations, or instructions.

(5) *Conflict of interest.* (i) *General.* The delegated State and any commissioner, director, employee, or other related person or entity shall not have a conflict of interest, as defined in section 11 of the Act and § 800.199 of the regulations. A conflict of interest may be waived pursuant to § 800.199(d).

(ii) *Unofficial activities.* The delegated State or personnel employed by the State shall not perform any unofficial service that is the same as any of the official services covered by the delegation.

(6) *Fees.* The delegated State shall charge fees according to § 800.70.

(7) *Facilities and equipment.* (i) *General.* The laboratory and office facilities of each delegated State shall be: Located; equipped; and large enough so that requested services are provided in an orderly and timely manner.

(ii) *Equipment testing.* Each delegated State shall test the equipment that it uses for official services according to the instructions.

(8) *Security.* Each delegated State shall provide sufficient security to assure that official samples, records, equipment, and forms are reasonably secure from theft, alteration, or misuse.

(9) *Certificate control system.* Each delegated State shall establish a certificate control system for all official certificates it receives, issues, voids, or otherwise renders useless. The system shall provide for: (i) Recording the numbers of the official certificates printed or received; (ii) protecting unused certificates from fraudulent or unauthorized use; and (iii) maintaining a file copy of each certificate issued, voided, or otherwise rendered useless in a manner that would permit retrieval.

(10) *Records.* Each delegated State shall maintain the records specified in §§ 800.145 through 800.159.

(11) *Notification to Secretary.* A delegated State shall notify the Secretary of its intention to temporarily discontinue official inspection and/or weighing services for any reason, except in the case of a major disaster. The delegated State must provide written notification to the Service no less than 72 hours in advance of the discontinuation date.

(g) *Termination*—(1) *Automatic termination.* Failure to pay the user fees prescribed by the Service for supervisory costs related to official inspection and weighing services within 30 days after due shall result in the automatic termination of the delegation. The delegation shall be reinstated if fees currently due, plus interest and any further expenses incurred by the Service because of the termination, are paid within 60 days after the termination.

(2) *Voluntary cancellation.* A State may request that its delegation be canceled by giving 90 days written notice to the Service.

(3) *Revocation.* (i) *Without hearing.* The Administrator may revoke the delegation of a State without first affording the State opportunity for a hearing. Unless otherwise provided, the revocation shall be effective when the State receives a notice from the Service regarding the revocation and the reason(s) therefor.

(ii) *Informal conference.* At the discretion of the Administrator, before the delegation of a State is revoked under paragraph (g)(3)(i) of this section, the Service may (A) notify the State of the proposed action and the reason(s) therefor, and (B) afford the State an opportunity to express its views in an informal conference before the Administrator.

(4) *Review.* At least once every 5 years, a delegated State shall submit to a review of its delegation by the Service in accordance with the criteria and procedures for delegation prescribed in section 7(e) of the Act, this section of the regulations, and the instructions. The Administrator may revoke the delegation of a State according to this subsection if the State fails to meet or comply with any of the criteria for delegation set forth in the Act, regulations, and instructions.

(h) *Provision of services following termination.* If a State's delegation is terminated, official services at the export port locations in the State shall be provided by the Service.

(The information collection requirements contained in paragraph (d) were approved by the Office of Management and Budget under control number 0580–0012; paragraphs (f)(2) and (f)(4) were approved under control number 0580–0011)

[49 FR 30915, Aug. 2, 1984, as amended at 50 FR 18988, May 6, 1985; 54 FR 5924, Feb. 7, 1989; 60 FR 65236, Dec. 19, 1995; 81 FR 49863, July 29, 2016]

§ 800.196 Designations.

(a) *General.* Eligible persons or governmental agencies may be designated to perform official services (excluding appeal inspection) within a specified area (other than export port locations).

(b) *Restrictions*—(1) *General.* If official inspection services are performed in an area by a designated agency, Class X and Class Y weighing services in that area may be performed only by the designated agency if the agency applies for designation to provide weighing services and is found qualified by the Service. If the agency designated to provide official inspection services is found not qualified or does not apply, the Class X and Class Y weighing services may be performed by another available agency that is found qualified and is designated by the Service, or the official services may be performed by the Service.

(2) *Interim authority.* (i) *By agency.* A designated agency may perform official services outside its assigned area on an interim basis when authorized by the Service.

(ii) *By Service.* Official inspection services and/or Class X and Class Y weighing services may be performed by the Service in an area (other than export port locations) on an interim basis in accordance with sections 7(h) and 7A(c) of the Act.

(c) *Who can apply.* Any State or local governmental agency or any person may apply, subject to sections 7 and 7A of the Act, to the Service for designation as an official agency to perform official inspection services (excluding appeal inspection) and/or Class X and Class Y weighing services in a given area (other than export port locations) in the United States.

(d) *When and how to apply.* An application for designation should be filed with the Service, according to the provisions of the FEDERAL REGISTER notice which requests applicants for designation to perform official services in existing or new geographic areas. The application for designation:

(1) Shall be submitted on a form furnished by the Service;

(2) Shall be typewritten or legibly written in English;

(3) Shall show or be accompanied by documents which show all information

requested on the form, or otherwise required by the Service; and

(4) Shall be signed by the applicant or its chief operating officer.

(e) *Review of conditions and criteria for designation*—(1) *Application.* Each application for a designation shall be reviewed to determine whether it complies with paragraph (d) of this section. If an application is not in compliance, the applicant shall be provided an opportunity to submit the needed information. If the needed information is not submitted within a reasonable time, as determined by the Service, the application may be dismissed. When an application is dismissed, the Service shall notify the applicant, in writing, of the reason(s) for the dismissal.

(2) *Applicant.* Each applicant for authority to operate as as designated agency shall be reviewed to determine whether the applicant meets the conditions and criteria contained in sections 7(f)(1)(A) and (B) of the Act, §800.199 of the regulations, and paragraph (g) of this section. The requested designation may be granted if the Service determines that:

(i) The requested action is consistent with the need for official services;

(ii) The applicant meets the conditions and criteria specified in the Act and regulations;

(iii) The applicant is better able than any other applicant to provide official services; and

(iv) The applicant addresses concerns identified during consultations that the Service conducts with applicants for service to the satisfaction of the Service.

(f) *Area of responsibility*—(1) *General.* Each agency shall be assigned an area of responsibility by the Service. Each area shall be identified by geographical boundaries and, in the case of a State or local government, shall not exceed the jurisdictional boundaries of the State or the local government, unless otherwise approved by the Service. The area of responsibility may not include any export elevators at export port locations or any portion of an area of responsibility assigned to another agency that is performing the same functions, except as otherwise provided in §800.117. A designated agency may perform official services at locations outside its assigned area of responsibility only after obtaining approval from the Service, or in accordance with provisions set forth in §800.117.

(2) *Amending.* A request for an amendment to an assigned area of responsibility shall (i) be submitted to the Service in writing; (iii) specify the change desired; (iii) be signed by the applicant or its chief operating officer; and (iv) be accompanied by the fee prescribed by the Service. The assigned area may be amended if the Service determines that the amendment is consistent with the provisions and objectives of the Act, regulations, and instructions. Upon a finding of need, the Service may initiate action to change an assigned area of responsibility.

(3) *Specified service points.* An agency may change its specified service points by notifying the Service in advance. Interested persons may obtain a list of specified service points within an agency's area of responsibility by contacting the agency. The list shall include all specified service points and shall identify each specified service point which operates on an intermittent or seasonal basis.

(g) *Responsibilities*—(1) *Providing official services.* Insofar as practicable, each agency shall be responsible for providing at all locations in its assigned area each service authorized by the designation. An agency may, subject to Service approval, make arrangements with a neighboring agency to provide official services requested infrequently. The agency shall perform all official services according to the Act, regulations, and instructions in effect at the time of designation or which may be promulgated subsequently.

(2) *Fees.* The agency shall charge fees according to §800.70.

(3) *Staffing, licensing, and training*—(i) *General.* The agency shall employ sufficient personnel to provide the official services normally requested in an accurate and timely manner. Each agency shall only use personnel licensed by the Service for the performance of official services and shall train and assist its personnel in acquiring and maintaining the necessary skills. Each agency shall keep the Service informed of the employment status of each of its licensees

and any substantial change in a licensee's duties.

(ii) *State agencies.* State agencies shall employ official personnel on the basis of job qualifications rather than political affiliations.

(4) *Rotation of personnel.* Where feasible, each agency shall rotate licensees among elevators and other facilities as is necessary to preserve the integrity of the official inspection and weighing systems.

(5) *Supervision.* The agency and its officials shall be responsible for the actions of the official personnel employed by the agency, for direct supervision of the daily activities of such personnel, and for the conduct of official services and related activities at the agency. The agency shall supervise official activities, in accordance with the Act, regulations, and instructions, and shall take action necessary to ensure that its employees are not performing prohibited functions and are not involved in any action prohibited by the Act, regulations, or instructions. Each agency shall report to the responsible field office information which shows or may show a violation of any provision of the Act, regulations, or instructions and information on any instructions which have been issued to agency personnel by Service personnel or by any other person which are inconsistent with the Act, regulations, or instructions.

(6) *Conflict of interest*—(i) *General.* Each agency and any officer, director, stockholder, employee, or other related entity shall not have a conflict of interest, as defined in Section 11 of the Act and § 800.199 of the regulations. A conflict of interest may be waived pursuant to § 800.199(d). The agency shall advise the Service immediately of any proposed change in name, ownership, officers or directors, or control of the agency and, if a trust, any change affecting the trust agreement.

(ii) *Unofficial activities.* Except as provided in § 800.76(a), the agency or personnel employed by the agency shall not perform any unofficial service that is the same as the official services covered by the designation.

(7) *Facilities and equipment*—(i) *General.* The laboratory and office facilities of each agency shall be: Located;

equipped; and large enough so that requested services are provided in an orderly and timely manner.

(ii) *Equipment testing.* Each agency shall test the equipment it uses for official services according to the instructions.

(8) *Security.* Each agency shall provide sufficient security to ensure that official samples, records, equipment, and forms are reasonably secure from theft, alteration, or misuse.

(9) *Certificate control system.* Each agency shall establish a certificate control system for all official certificates it receives, issues, voids, or otherwise renders useless. The system shall provide for (i) recording the numbers of the official certificates printed or received; (ii) protecting unused certificates from fraudulent or unauthorized use; and (iii) maintaining a file copy of each certificate issued, voided, or otherwise rendered useless in a manner that would permit retrieval.

(10) *Records.* Each agency shall maintain the records specified in §§ 800.145 through 800.159.

(h) *Termination and renewal*—(1) *Triennial*—(i) *Termination.* A designation shall terminate at a time specified by the Administrator, but not later than 5 years after the effective date of the designation. A notice of termination shall be issued by the Service to a designated agency at least 120 calendar days in advance of the termination date. The notice shall provide instructions for requesting renewal of the designation. Failure to receive a notice from the Service shall not exempt a designated agency from the responsibility of having its designation renewed on or before the specified termination date.

(ii) *Renewal.* Designations may be renewed, upon application, in accordance with criteria and procedures for designation prescribed in section 7(f) of the Act and this section of the regulations. The Administrator may decline to renew a designation if:

(A) The requesting agency fails to meet or comply with any of the criteria for designation set forth in the Act, regulations, and instructions, of

(B) The Administrator determines that another qualified applicant is better able to provide official services in the assigned area.

(2) *Automatic termination.* Failure to pay the user fees prescribed by the Service for supervisory costs related to official inspection and weighing services within 30 days after due shall result in the automatic termination of the designation. The designation shall be reinstated if fees currently due, plus interest and any further expenses incurred by the Service because of the termination, are paid within 60 days after the termination.

(3) *Voluntary cancellation.* An agency may request that its designation be canceled by giving 90 days written notice to the Service.

(4) *Suspension or revocation of designation.* (i) *General.* A designation is subject to suspension or revocation, under section 7(g)(3) of the Act, by the Service, whenever the Administrator determines that:

(A) The agency has failed to meet one or more of the criteria specified in section 7(f) of the Act or the regulations for the performance of official functions, or otherwise has not complied with any provision of the Act, regulations, or instructions, or

(B) Has been convicted of any violation of other Federal law involving the handling or official inspection of grain.

(ii) *Summary suspension.* The Service may, without first affording the agency (hereafter referred to in this paragraph as the "respondent") an opportunity for a hearing, suspend a designation or refuse to reinstate a designation when the suspension period has expired, pending final determination of the proceeding whenever the Service has reason to believe there is cause for revocation of the designation and considers such action to be in the best interest of the official inspection and weighing system. A suspension or refusal to reinstate a suspended designation shall be effective upon the respondent's receipt of a notice from the Service. Within 30 calendar days following the issuance of a notice of such action, the Service shall afford the respondent an opportunity for a hearing under paragraph (h)(4)(iii) of this section. The Service may terminate the action if it finds

that alternative managerial, staffing, financial, or operational arrangements satisfactory to the Service can be and are made by the respondent.

(iii) *Other than summary suspension.* Except as provided in paragraph (h)(4)(ii) of the section, before the Service revokes or suspends a designation, the respondent shall be: (A) Notified by the Service of the proposed action and the reason(s) therefor, and (B) afforded an opportunity for a hearing in accordance with the Rules of Practice Governing Formal Adjudicatory Proceedings Instituted by the Secretary Under Various Statutes (7 CFR part 1, subpart H). Before initiating formal adjudicatory proceedings, the Service may, at its discretion, afford the respondent an opportunity to present its views on the proposed action and the reason(s) therefor in an informal conference. If, as a result of the informal conference, a consent agreement is reached, no formal adjudicatory proceedings shall be initiated.

(i) *Provision of services following suspension or termination.* If the designation of an agency is suspended, terminated, or the renewal of a designation is not granted, the Service shall attempt, upon a finding of need, to arrange for a replacement agency. If a qualified replacement agency cannot be designated on a timely basis, a qualified agency, if available, shall be designated on an interim basis. If a qualified agency is not available on an interim basis, the Service shall provide needed services on an interim basis.

(Approved by the Office of Management and Budget under control number 0580–0013)

[49 FR 30915, Aug. 2, 1984, as amended at 50 FR 18989, May 6, 1985; 54 FR 5924, Feb. 7, 1989; 60 FR 65236, Dec. 19, 1995; 63 FR 45677, Aug. 27, 1998; 68 FR 19139, Apr. 18, 2003; 81 FR 49863, July 29, 2016]

§800.197 Approval as a scale testing and certification organization.

(a) *Who may apply.* Any State, local government, or person may request approval to perform scale testing and certification under the Act.

(b) *When and how to apply.* A request for approval to perform scale testing and certification under the Act should be filed with the Service not less than 90 calendar days before the requested action's effective date. A request for approval to perform scale testing and certification shall:

(1) Show or be accompanied by documents which show all information required by the Service;

(2) Certify that each employee scheduled to perform official scale testing and certification services is competent to test weighing equipment and has a working knowledge of the regulations and instructions applicable to such services;

(3) Be accompanied by the fee prescribed in § 800.71; and

(4) Be signed by the applicant or its chief operating officer.

(c) *Review of applicant.* The review of an applicant for authority to perform scale testing and certification shall include an evaluation of the applicant's policies and procedures for testing and certifying scales for Class X and Class Y weighing.

(d) *Termination*—(1) *Voluntary.* A scale testing and certification organization may request cancellation of its approval by notifying the Service.

(2) *Suspension or revocation of approval*—(i) *General.* An approval is subject to suspension or revocation whenever the Administrator determines that the approved organization has violated any provision of the Act or regulations, or has been convicted of any violation involving the handling, weighing, or inspection of grain under Title 18 of the United States Code.

(ii) *Summary suspension.* The Service may, without first affording the organization an opportunity for a hearing, suspend an approval or refuse to reinstate an approval when the suspension period has expired, pending final determination of the proceeding whenever the Service has reason to believe there is cause for revocation of the approval and considers such action to be in the best interest of the official weighing system. A suspension or refusal to reinstate a suspended approval shall be effective when the organization receives a notice from the Service. Within 30 calendar days following the issuance of a notice of such action, the Service shall give the organization an opportunity for a hearing under paragraph (d)(2)(iii) of this section. The Service may terminate its action if it finds

that alternative managerial, staffing, or operational arrangements satisfactory to the Service can be and are made by the organization.

(iii) *Other than summary suspension.* Except as provided in paragraph (d)(2)(ii) of this section, before the Service revokes or suspends an approval, the organization shall be notified by the Service of the proposed action and the reason(s) therefor and shall be given an opportunity for a hearing. Before the Service initiates a hearing, it may, at its discretion, give the organization an opportunity to present its views on the proposed action and the reason(s) therefor in an informal conference. If a consent agreement is reached during the informal conference, no formal adjudicatory proceedings shall be initiated.

(The information collection requirements contained in paragraph (b) were approved by the Office of Management and Budget under control number 0580–0012)

[49 FR 30915, Aug. 2, 1984, as amended at 54 FR 5924, Feb. 7, 1989]

§ 800.198 Contracts.

(a) *Services contracted and who may apply.* The Service may enter into a contract with any person, State, or governmental agency to perform on an occasional basis:

(1) Specified official sampling, laboratory testing, or other similar objective technical activities involved in the testing of grain for official factors or official criteria, and

(2) Monitoring activities in foreign ports with respect to export grain that has been inspected and weighed under the Act.

(b) *Restrictions*—(1) *Conflict of interest.* A person, State or governmental agency with a conflict of interest prohibited by section 11 of the Act or § 800.199 shall not be eligible to enter into a contract with the Service.

(2) *Appeal service.* An agency or employees of agencies shall not be eligible to enter into a contract with the Service to obtain samples for, or to perform other services involved in appeal inspection or Board appeal inspection services. However, agencies may forward file samples to the Service in accordance with § 800.156(d).

(3) *Monitoring services.* Agencies, employees of agencies, organizations, employees of organizations, and other persons that regularly provide official services to persons who export grain from the United States are eligible to enter into a contract with the Service to perform monitoring services on export grain in foreign ports only if they are under Service employees' direct supervision during monitoring activities.

(c) *When and how to apply.* An application for a contractual arrangement shall: (1) Be typewritten or legibly written in English; (2) conform to the invitation to bid or other instructions issued by the Service or be filed on a form furnished by the Service; (3) show or be accompanied by documents which show any information requested by the Service; and (4) be signed by the applicant or its chief operating officer. All contracts shall be issued by the Department and shall follow Departmental procedures.

(d) *Termination and renewal.* A contract with the Service shall terminate annually unless othewise provided in the contract. A contract may be renewed in accordance with Departmental procedures.

(e) *Cancellation.* A contract may, upon request of the governmental agency or person that entered into the contract with the Service, be canceled by the Department in accordance with the terms of the contract or Departmental procedures and regulations.

(The information collection requirements contained in paragraph (c) were approved by the Office of Management and Budget under control number 0580–0012)

[49 FR 30915, Aug. 2, 1984, as amended at 50 FR 18989, May 6, 1985; 54 FR 5924, Feb. 7, 1989]

§800.199 Conflict-of-interest provisions.

(a) *Meaning of terms.* For the purpose of this section, the following terms shall have the meaning given for them below:

(1) *Grain business.* The term "grain business" shall include (i) any entity that is engaged in the commercial transportation, storage, merchandising or other commercial handling of grain, which includes: The commercial buying, selling, transporting, cleaning, elevating, storing, binning, mixing, blending, drying, treating, fumigating, or other preparation of grain (other than as a grower of grain or the disposition of inspection samples); the cleaning, treating, or fitting of carriers or containers for transporting or storing of grain; the merchandising of equipment for cleaning, drying, treating, fumigating, or other processing, handling, or storing of grain; the merchandising of grain inspection and weighing equipment (other than the buying or selling by an agency or official personnel of the equipment for their exclusive use in the performance of their official inspection or Class X or Class Y weighing services); and the commercial use of official inspection and Class X or Class Y weighing services and (ii) any board of trade, chamber of commerce, grain exchange, or other trade group composed, in whole or in part, of one or more such entities.

(2) *Interest.* The term "interest" when used with respect to an individual, shall include the interest of a spouse, minor child, or blood relative who resides in the immediate household of the individual.

(3) *Related.* The term "related" when used in reference to a business or governmental entity means an entity that owns or controls another entity, or is owned or controlled by another entity, or both entities are owned or controlled by another entity.

(4) *Substantial stockholder.* The term "substantial stockholder" means any person holding 2 per centum or more, or 100 shares or more of the voting stock of the corporation, whichever is the lesser interest.

(b) *Prohibited conflicts of interest.* Unless waived on a case-by-case basis by the Administrator under section 11(b)(5) or the Act, the following conflicts of interest for a business or association are prohibited:

(1) *Agency and contractor.* No agency or contractor, or any member, director, officer, or employee thereof, and no business or governmental entity related to any such agency or contractor, shall be employed in or otherwise engaged in, or directly or indirectly have any stock or other financial interest in, any grain business or otherwise have any conflict of interest specified in §800.187(b).

(2) *Grain business.* No grain business or governmental entity conducting any such business, or any member, director, officer, or employee thereof, and no other business or governmental entity related to any such entity, shall operate or be employed by, or directly or indirectly have any stock or other financial interest in, any agency or contractor.

(3) *Stockholder in any agency or contractor.* No substantial stockholder in any agency or contractor shall be employed in or otherwise engaged in, or be a substantial stockholder in, any grain business, or directly or indirectly have any other kind of financial interest in any such business or otherwise have any conflict of interest specified in § 800.187(b).

(4) *Stockholder of a grain business.* No substantial stockholder in any grain business shall operate or be employed by or be a substantial stockholder in, or directly or indirectly have any other kind of financial interest in an incorporated agency or contractor.

(5) *Gratuity.* No person described in paragraph (b)(1) of this section shall give to or accept from a person described in paragraph (b)(2) of this section any gratuity, and no person described in paragraph (b)(2) of this section shall give to or accept from a person described in paragraph (b)(1) of this section any gratuity. A "gratuity" is defined in § 800.187(a).

(c) *Exempt conflicts of interest—*(1) *Agency and contractor.* An agency or contractor may use laboratory or office space or inspection, weighing, transportation, or office equipment that is owned or controlled, in whole or in part, by a grain business or related entity when the use of the space or equipment is approved by the Service for the performance of onsite official services under the Act.

(2) *Financial institution.* A bona fide financial institution that has a financial relationship with one or more grain businesses or related entities may have a financial relationship with an agency, contractor, or related agency.

(3) *Grain business.* A grain business or related entity may furnish laboratory or office space or inspection, weighing, transportation, or office equipment for use by an agency, contractor, or field office when use of the space or equipment is approved by the Service for the performance of onsite official inspection or weighing services.

(d) *Disposition of a conflict of interest.* Upon being informed that a prohibited conflict of interest exists in the ownership, management, or operation of an agency and that remedial action is required, the agency shall take immediate action to resolve that conflict of interest and inform the Service of the action taken. An agency which believes that remedial action will cause undue economic hardship or other irreparable harm may request a waiver by forwarding to the Service a written statement setting forth the facts, the circumstances, and the reasons for requesting a waiver.

[49 FR 30915, Aug. 2, 1984]

SUPERVISION, MONITORING, AND EQUIPMENT TESTING

§ 800.215 Activities that shall be supervised.

(a) *General.* Supervision of the activities described in this section shall be performed in accordance with the instructions.

(b) *Administrative activities.* Administrative activities subject to supervision include but are not limited to (1) providing staffing, equipment, and facilities for performing authorized services; (2) dismissing requests for services and withholding requested services; (3) maintaining official records; (4) assessing and collecting fees; (5) rotating official personnel; (6) implementing instructions for (i) recruiting official personnel, (ii) training and supervising official and approved personnel, (iii) work performance and work production standards; and (7) supervising and monitoring.

(c) *Technical activities—*(1) *Equipment testing activities.* Equipment testing activities subject to supervision include but are not limited to (i) implementing (A) the equipment performance requirements in parts 801 and 802 of this chapter and (B) the instructions for the operation of equipment used under the

Act and for performing equipment-testing activities and (ii) performing equipment-testing activities by official personnel or by approved scale testing organizations.

(2) *Inspection activities.* Inspection activities subject to supervision include but are not limited to (i) implementing (A) the Official U.S. Standards for Grain, (B) official criteria, and (C) instructions for the performance of inspection activities and (ii) performing stowage examination, sampling, laboratory testing, grading, and certification activities by official personnel.

(3) *Weighing activities.* Weighing activities subject to supervision include but are not limited to (i) implementing (A) uniform weighing procedures and (B) instructions for the performance of weighing activities and (ii) performing (A) stowage examination, sampling (sacked grain), weighing, and certification activities by official personnel and (B) by approved weighers of weighing activities.

(4) *Testing of prototype equipment activities.* Prototype or proposed equipment is tested to determine whether the equipment will improve the performance of activities under the Act. Prototype equipment-testing activities subject to supervision include but are not limited to (i) implementing instructions for the testing of prototype equipment, (ii) testing prototype equipment by official personnel, and (iii) approving or denying the use of prototype equipment for use under the Act.

§ 800.216 Activities that shall be monitored.

(a) *General.* Each of the administrative and technical activities identified in § 800.215 and the elevator and merchandising activities identified in this section shall be monitored in accordance with the instructions.

(b) *Grain merchandising activities.* Grain merchandising activities subject to monitoring for compliance with the Act include but are not limited to (1) failing to promptly forward an export certificate; (2) describing grain by other than official grades; (3) falsely describing export grain; (4) falsely making or using official certificates, forms, or marks; (5) making false qual-

ity or quantity representations about grain; and (6) selling export grain without a certificate of registration.

(c) *Grain handling activities.* Grain handling activities subject to monitoring for compliance with the Act include, but are not limited to:

(1) Shipping export grain without inspection or weighing;

(2) Violating any Federal law with respect to the handling, weighing, or inspection of grain;

(3) Deceptively loading, handling, weighing, or sampling grain; and

(4) Exporting grain without a certificate of registration.

(d) *Recordkeeping activities.* Elevator and merchandising recordkeeping activities subject to monitoring for compliance with the Act include those that are identified in section 12(d) of the Act and § 800.25 of the regulations.

(e) *Other activities.* Other activities subject to monitoring for compliance with the Act include but are not limited to (1) resolving conflicts of interest by official agencies or their employees; (2) providing access to elevator facilities and records; (3) improperly influencing or interfering with official personnel; (4) falsely representing that a person is official personnel; (5) using false means in filing an application for services under the Act; and (6) preventing interested persons from observing the loading, Class X or Class Y weighing, or official sampling of grain.

[45 FR 15810, Mar. 11, 1980; 45 FR 55119, Aug. 18, 1980, as amended at 50 FR 2273, Jan. 16, 1985; 81 FR 49863, July 29, 2016]

§ 800.217 Equipment that shall be tested.

(a) *General.* Testing of equipment and prototype equipment described in this section shall be performed in accordance with the instructions.

(b) *Inspection equipment.* Each unit of equipment used in the official sampling, testing, or grading of grain, or in monitoring the official inspection of grain, shall be examined to determine whether the equipment is functioning in an approved manner. In addition, each unit of equipment for which official performance requirements have been established shall be tested for accuracy. For the purpose of this paragraph, diverter-type mechanical samplers used in obtaining warehouseman's samples shall be considered to be official inspection equipment used under the Act.

(c) *Weighing equipment.* Each unit of equipment used in the Class X or Class Y weighing of grain or in monitoring the Class X or Class Y weighing of grain, each related grain handling system, and each related computer system shall be examined to determine whether it is functioning in an approved manner. In addition, each unit of equipment for which official performance requirements have been established shall be tested for accuracy.

(d) *Prototype equipment*—(1) *At request of interested party.* Upon request of a financially interested party and with the concurrence of the Administrator, prototype grain inspection or weighing equipment may be tested by the Service for official use.

(2) *Determination by Service.* Upon a determination of need, the Service may develop, contract for, or purchase and test prototype grain inspection or weighing equipment for official use.

§ 800.218 Review of rejection or disapproval of equipment.

Any person desiring to complain of a rejection or disapproval of equipment by official personnel or of any alleged discrepancy in the testing of equipment under the Act by official personnel or by approved scale testing organizations may file a complaint with the Service.

§ 800.219 Conditional approval on use of equipment.

(a) *Approval.* Equipment that is in use under the Act on the effective date of this section shall be considered conditionally to have been adopted and approved by the Service.

(b) *Limitation on approval.* This conditional approval shall not bar a later rejection or disapproval of the equipment by the Service upon a determination that the equipment (1) should be rejected for official use, or (2) is not functioning in an approved manner, or (3) is not producing results that are accurate within prescribed tolerances, or (4) is

producing results that are otherwise not consistent with the objectives of the Act.

PART 801—OFFICIAL PERFORMANCE REQUIREMENTS FOR GRAIN INSPECTION EQUIPMENT

AUTHORITY: 7 U.S.C. 71–87k

SOURCE: 51 FR 7050, Feb. 28, 1986, unless otherwise noted.

§ 801.1 Applicability.

The requirements set forth in this part 801 describe certain specifications, tolerances, and other technical requirements for official grain inspection equipment and related sample handling systems used in performing inspection services under the Act.

§ 801.2 Meaning of terms.

(a) *Construction.* Words used in the singular form in this part shall be considered to imply the plural and vice versa, as appropriate.

(b) *Definitions.* The definitions of terms listed in the part 800 shall have the same meaning when the terms are used in this part 801. For the purposes of this part, the following terms shall have the meanings given for them below.

(1) *Avoirdupois weight.* A unit of weight based on a pound of 16 ounces.

(2) *Barley pearler.* An approved laboratory device used to mechanically dehull kernels of barley or other grain.

(3) *Deviation from standard.* In testing inspection equipment for accuracy, the variation between (i) the individual test result from the equipment that is

being tested and (ii) the reference standard or the individual test result from the standard (or National standard) equipment, as applicable.

(4) *Direct comparison method.* An equipment testing procedure wherein transfer standards are tested at the same time and place to compare the performance of two or more units of the same inpsection equipment. One unit of the equipment used in the test shall be standard inspection equipment. (See also sample exchange method).

(5) *Diverter-type mechanical sampler (primary).* An approved device used to obtain representative portions from a flowing stream of grain.

(6) *Diverter-type mechanical sampler (secondary).* An approved device used to subdivide the portions of grain obtained with a diverter-type mechanical sampler (primary).

(7) *Divider.* An approved laboratory device used to mechanically divide a sample of grain into two or more representative portions.

(8) *Dockage tester.* An approved laboratory device used to mechanically separate dockage and/or foreign material from grain.

(9) *Maintenance tolerance.* An allowance established for use in determining whether inspection equipment should be approved for use in performing official inspection services.

(10) *Mean deviation from standard.* In testing inspection equipment for accuracy, the variation between (i) the average for the test results from the equipment that is being tested and (ii) the reference standard or the average of the test results from the standard (or National standard) equipment, as applicable.

(11) *Metric weight.* A unit of weight based on the kilogram of 1,000 grams.

(12) *Moisture meter.* An approved laboratory device used to indicate directly or through conversion and/or correction tables the moisture content of grain including cereal grains and oil seeds.

(13) *National standard inspection equipment.* A designated approved unit of inspection equipment used as the reference in determining the accuracy of standard inspection equipment.

(14) *Official inspection equipment.* Equipment approved by the Service and used in performing official inspection services.

(15) *Sample exchange method.* An equipment testing procedure wherein transfer standards are tested to compare the performance of two or more units of the same inspection equipment installed at different locations. One unit of the equipment used in the test shall be standard inspection equipment. (See also direct comparison method.)

(16) *Sieves.* Approved laboratory devices with perforations for use in separating particles of various sizes.

(17) *Standard inspection equipment.* An approved unit of inspection equipment that is designated by the Service for use in determining the accuracy of official inspection equipment.

(18) *Test weight.* The avoirdupois weight of the grain or other material in a level-full Winchester bushel.

(19) *Test weight apparatus.* An approved laboratory device used to measure the test weight (density) of a sample of grain.

(20) *Transfer standard.* The medium (device or material) by which traceability is transferred from one inspection equipment standard unit to another unit.

(21) *Winchester bushel.* A container that has a capacity of 2,150.42 cubic inches (32 dry quarts).

§801.3 Tolerances for barley pearlers.

The maintenance tolerances for barley pearlers used in performing official inspection services shall be:

Item	Tolerance
Timer switch:	
0 to 60 seconds	±5 seconds, deviation from standard clock
61 to 90 seconds ...	±7 seconds, deviation from standard clock
Over 90 seconds	±10 seconds, deviation from standard clock
Pearled portion	±1.0 gram, mean deviation from standard barley pearler using barley

§801.4 Tolerances for dockage testers.

The maintenance tolerances for dockage testers used in performing official inspection services shall be:

Item	Tolerance
Air separation	±0.10 percent, mean deviation from standard dockage tester using Hard Red Winter wheat
Riddle separation	±0.10 percent, mean deviation from standard dockage tester using Hard Red Winter wheat
Sieve separation	±0.10 percent, mean deviation from standard dockage tester using Hard Red Winter wheat
Total dockage separation.	±0.15 percent, mean deviation from standard dockage tester using Hard Red Winter wheat

§ 801.5 Tolerance for diverter-type mechanical samplers.

The maintenance tolerance for diverter-type mechanical samplers (primary, or primary and secondary in combination) used in performing official inspection services shall be ±10 percent, mean deviation from standard sampling device using corn or the same type of grain that the system will be used to sample.

§ 801.6 Tolerances for moisture meters.

(a) The maintenance tolerances for Motomco 919 moisture meters used in performing official inspection services shall be:

(1) Headquarters standard meters:

Moisture range	Tolerance	
	Direct comparison	Sample exchange
Low	±0.05 percent moisture, mean deviation from National standard moisture meter using Hard Red Winter wheat	
Mid	±0.05 percent moisture, mean deviation from National standard moisture meter using Hard Red Winter wheat	
High	±0.05 percent moisture, mean deviation from National standard moisture meter using Hard Red Winter wheat	

(2) All other than Headquarters standard meters:

Moisture range	Tolerance	
	Direct comparison	Sample exchange
Low	±0.15 percent moisture, mean deviation from standard moisture meter using Hard Red Winter wheat	±0.20 percent moisture, mean deviation from standard moisture meter using Hard Red Winter wheat
Mid	±0.10 percent moisture, mean deviation from standard moisture meter using Hard Red Winter wheat	±0.15 percent moisture, mean deviation from standard moisture meter using Hard Red Winter wheat
High	±0.15 percent moisture, mean deviation from standard moisture meter using Hard Red Winter wheat	±0.20 percent moisture, mean deviation from standard moisture meter using Hard Red Winter wheat

(b) The maintenance tolerances for GAC 2100 moisture meters used in performing official inspection services shall be:

(1) Headquarters standard meters. By direct comparison using mid-range Hard Red Winter wheat, ±0.05% mean deviation for the average of the Headquarters standard moisture meters.

(2) All other than Headquarters standard meters. By sample exchange using mid-range Hard Red Winter wheat, ±0.15% mean deviation from the standard meter.

[63 FR 34554, June 25, 1998]

§ 801.7 Reference methods and tolerances for near-infrared spectroscopy (NIRS) analyzers.

(a) *Reference methods.* (1) The chemical reference protein determinations used to reference and calibrate official NIRS instruments shall be performed in accordance with "Comparison of Kjeldahl Method for Determination of

Crude Protein in Cereal Grains and Oil-seeds with Generic Combustion Method: Collaborative Study," July/August 1993, Ronald Bicsak, Journal of AOAC International Vol. 76, No. 4, 1993, and subsequently approved by the AOAC International as the Combustion method, AOAC International Method 992.23. This incorporation by reference was approved by the Director of the Federal Register in accordance with 5 U.S.C. 552(a) and 1 CFR part 51. Copies may be obtained from Director, Technical Services Division, Federal Grain Inspection Service, 10383 North Executive Hills Blvd., Kansas City, MO 64153–1394. Copies may be inspected at the above address or at the National Archives and Records Administration (NARA). For information on the availability of this material at NARA, call 202–741–6030, or go to: *http://www.archives.gov/ federal_register/ code_of_federal_regulations/ ibr_locations.html.*

(2) The chemical reference starch determination used to reference and calibrate official NIRS instruments shall be performed in accordance with the Corn Refiners Association Method A–20, Analysis for Starch in Corn, Second revision, April 15, 1986, Standard Analytical Methods of the Member Companies of the Corn Refiners Association, Inc. This incorporation by reference was approved by the Director of the Federal Register in accordance with 5 U.S.C. 552(a) and 1 CFR part 51. Copies may be obtained from Director, Technical Services Division, Federal Grain Inspection Service, 10383 North Executive Hills Blvd., Kansas City, MO 64153–1394. Copies may be inspected at the above address or at the National Archives and Records Administration (NARA). For information on the availability of this material at NARA, call 202–741–6030, or go to: *http:// www.archives.gov/federal_register/ code_of_federal_regulations/ ibr_locations.html.*

(b) *Tolerances*—(1) *NIRS wheat protein analyzers.* The maintenance tolerances for the NIRS analyzers used in performing official inspections for determination of wheat protein content shall be ±0.15 percent mean deviation from the national standard NIRS instruments, which are referenced and

calibrated to the Combustion method, AOAC International Method 992.23.

(2) *NIRS soybean oil and protein analyzers.* The maintenance tolerances for the NIRS analyzers used in performing official inspections for determination of soybean oil shall be ±0.20 percent mean deviation from the national standard NIRS instruments, which are referenced and calibrated to the FGIS solvent oil extraction method; and for determination of protein content shall be ±0.20 percent mean deviation from the national standard NIRS instruments, which are referenced and calibrated to the Combustion method, AOAC International Method 992.23.

(3) *NIRS corn oil, protein, and starch analyzers.* The maintenance tolerances for the NIRS analyzers used in performing official inspections for determination of corn oil shall be ±0.20 percent mean deviation from the national standard NIRS instruments, which are referenced and calibrated to the FGIS solvent oil extraction method; for determination of protein content shall be ±0.30 percent mean deviation from the national standard NIRS instruments, which are referenced and calibrated to the Combustion method, AOAC International Method 992.23; and for determination of starch content shall be ±0.35 percent mean deviation from the national standard NIRS instruments, which are referenced and calibrated to the Starch method, Corn Refiners Association Method A–20.

(4) NIRS barley protein analyzers. The maintenance tolerances for the NIRS analyzers used in performing official inspections for determination of barley protein content are 0.20 percent mean deviation from the national standard NIRS instruments, which are referenced and calibrated to the Combustion method, AOAC International Method 992.23.

[63 FR 35505, June 30, 1998, as amended at 69 FR 18803, Apr. 9, 2004; 71 FR 65373, Nov. 8, 2006]

§801.8 Tolerances for sieves.

The maintenance tolerances for sieves used in performing official inspection services shall be:

(a) Thickness of metal: ±0.0015 inch.

(b) Accuracy of perforation: ±0.001 inch from design specification.

(c) Sieving accuracy:

Sieve description	Tolerance	
	Direct comparison	Sample exchange
.064 × ⅜ inch oblong	±0.2 percent, mean deviation from standard sieve using wheat.	±0.3 percent, mean deviation from standard sieve using wheat
⁵⁄₆₄ × ¾ inch slotted	±0.3 percent, mean deviation from standard sieve using barley.	±0.5 percent, mean deviation from standard sieve using barley
5⁵⁄₆₄ × ¾ inch slotted	±0.5 percent, mean deviation from standard sieve using barley.	±0.7 percent, mean deviation from standard sieve using barley
⁶⁄₆₄ × ¾ inch slotted	±0.7 percent, mean deviation from standard sieve using barley.	±1.0 percent, mean deviation from standard sieve using barley

§ 801.9 Tolerances for test weight apparatuses.

The maintenance tolerances for test weight per bushel apparatuses used in performing official inspection services shall be:

Item	Tolerance
Beam/scale accuracy	±0.10 pound per bushel deviation at any reading, using test weights
Overall accuracy	±0.15 pound per bushel, mean deviation from standard test weight apparatus using wheat

§ 801.10 [Reserved]

§ 801.11 Related design requirements.

(a) *Suitability.* The design, construction, and location of official sampling and inspection equipment and related sample handling systems shall be suitable for the official sampling and inspection activities for which the equipment is to be used.

(b) *Durability.* The design, construction, and material used in official sampling and inspection equipment and related sample handling systems shall assure that, under normal operating conditions, operating parts will remain fully operable, adjustments will remain reasonably constant, and accuracy will be maintained between equipment test periods.

(c) *Marking and identification.* Official sampling and inspection equipment for which tolerances have been established shall be permanently marked to show the manufacturer's name, initials, or trademark; the serial number of the equipment; and the model, the type, and the design or pattern of the equipment. Operational controls for mechanical samplers and related sample handling systems, including but not limited to pushbuttons and switches, shall be conspicuously identified as to the equipment or activity controlled by the pushbutton or switch.

(d) *Repeatability.* Official inspection equipment when tested in accordance with §§ 800.217 and 800.219 shall, within the tolerances prescribed in §§ 801.3 through 801.10, be capable of repeating its results when the equipment is operated in its normal manner.

(e) *Security.* Mechanical samplers and related sample handling systems shall provide a ready means of sealing to deter unauthorized adjustments, removal, or changing of component parts or timing sequence without removing or breaking the seals; and otherwise be designed, constructed, and installed in a manner to prevent deception by any person.

(f) *Installation requirements.* Official sampling and inspection equipment and related sample handling systems shall be installed (1) at a site approved by the Service, (2) according to the manufacturer's instructions, and (3) in such a manner that neither the operation nor the performance of the equipment or system will be adversely affected by the foundation, supports, or any other characteristic of the installation.

§ 801.12 Design requirements incorporated by reference.

(a) *Moisture meters.* All moisture meters approved for use in official grain moisture determination and certification shall meet applicable requirements contained in the FGIS Moisture Handbook and the General Code and Grain Moisture Meters Code of the 1991 edition of the National Institute of Standards and Technology's (NIST)

Handbook 44, "Specifications, Tolerances, and Other Technical Requirements for Weighing and Measuring Devices." Pursuant to the provisions of 5 U.S.C. 552(a), the materials in Handbook 44 are incorporated by reference as they exist on the date of approval and a notice of any change in these materials will be published in the FEDERAL REGISTER.

The NIST Handbook is for sale by the Superintendent of Documents, U.S. Government Printing Office, Washington, DC 20403. It is also available for inspection at the National Archives and Records Administration (NARA). For information on the availability of this material at NARA, call 202–741–6030, or go to: *http://www.archives.gov/federal_register/code_of_federal_regulations/ibr_locations.html.*

The following Handbook 44 requirements are not incorporated by reference:

General Code (1.10.)
  G-S.5.5. Money Values, Mathematical Agreement
  G-T.1. Acceptance Tolerances
  G-UR.3.3. Position of Equipment
  G-UR.3.4. Responsibility, Money-Operated Devices
Grain Moisture Meters (5.56.)
  N.1.1. Transfer Standards
  N.1.2. Minimum Test
  N.1.3. Temperature Measuring Equipment
  T.2. Tolerance Values
  T.3. For Test Weight Per Bushel Indications or Recorded Representations
  UR.3.2. Other Devices not used for Commercial Measurement
  UR.3.7. Location
  UR.3.11. Posting of Meter Operating Range

  (b) [Reserved]

[57 FR 2673, Jan. 23, 1992, as amended at 69 FR 18803, Apr. 9, 2004]

PART 802—OFFICIAL PERFORMANCE AND PROCEDURAL REQUIREMENTS FOR GRAIN WEIGHING EQUIPMENT AND RELATED GRAIN HANDLING SYSTEMS

Sec.
802.0 Applicability.
802.1 Qualified laboratories.

  AUTHORITY: Pub. L. 94–582, 90 Stat. 2867, as amended (7 U.S.C. 71 *et seq.*).

§ 802.0 Applicability.

(a) The requirements set forth in this part 802 describe certain specifications, tolerances, and other technical requirements for grain weighing equipment and related grain handling systems used in performing Class X and Class Y weighing services, official inspection services, and commercial services under the Act. All scales used for official grain weight and inspection certification services provided by FGIS must meet applicable requirements contained in the FGIS Weighing Handbook, the General Code, the Scales Code, the Automatic Bulk Weighing Systems Code, and the Weights Code of the 2008 edition of National Institute of Standards and Technology (NIST) Handbook 44, "Specifications, Tolerances, and Other Technical Requirements for Weighing and Measuring Devices" (Handbook 44); and NIST Handbook 105–1 (1990 Edition), "Specifications and Tolerances for Reference Standards and Field Standard Weights and Measures," (Handbook 105–1). These requirements are confirmed to be met by having National Type Evaluation Program type approval. Scales used for commercial purposes will be required to meet only the applicable requirements of the 2008 edition of the NIST Handbook-44. Pursuant to the provisions of 5 U.S.C. 552(a), with the exception of the Handbook 44 requirements listed in paragraph (b), the materials in Handbooks 44 and 105–1 are incorporated by reference as they exist on the date of approval and a notice of any change in these materials will be published in the FEDERAL REGISTER. This incorporation by reference was approved by the Director of the Federal Register on March 8, 2011, in accordance with 5 U.S.C. 552(a) and 1 CFR part 51. The NIST Handbooks are for sale by the National Conference of Weights and Measures (NCWM), 1135 M Street, Suite 110, Lincoln, Nebraska 68508. Information on these materials may be obtained from NCWM by calling 402–434–4880, by E-mailing *nfo@ncwm.net,* or on the Internet at *http://www.nist.gov/owm.*

(b) The following Handbook 44 requirements are not incorporated by reference:

Scales (2.20)

S.1.8. Computing Scales
S.1.8.2. Money-Value Computation
S.1.8.3. Customer's Indications
S.1.8.4. Recorded Representations, Point of Sale
S.2.5.2. Jeweler's, Prescription, & Class I & II Scales
S.3.3. Scoop Counterbalance
N.1.3.2. Dairy-Product Test Scales
N.1.5. Discrimination Test (Not adopted for Grain Test Scales only)
N.1.8. Material Tests
N.3.1.2. Interim Approval
N.3.1.3. Enforcement Action For Inaccuracy
N.4. Coupled-in-Motion Railroad Weighing Systems
N.6. Nominal Capacity of Prescription Scales
T.1.2. Postal and Parcel Post Scales
T.2.3. Prescription Scales
T.2.4. Jewelers' Scales (all sections)
T.2.5. Dairy—Product—Test Scales (all sections)
T.N.3.9. Materials Test on Customer—Operated Bulk—Weighing Systems for Recycled Materials
UR.1.4. Grain Test Scales: Value of Scale Divisions
UR.3.1. Recommended Minimum Load
UR.3.1.1. Minimum Load, Grain Dockage

Automatic Bulk Weighing Systems (2.22)

N.1.3. Decreasing-Load Test

[75 FR 76255, Dec. 8, 2010]

§ 802.1 Qualified laboratories.

(a) *Metrology laboratories.* (1) Any State metrology laboratory currently approved by the NBS ongoing certification program having auditing capability is automatically approved by the Service.

(2) Any county or city weights and measures jurisdiction approved by NBS or by their respective NBS-Certified State laboratory as being equipped with appropriate traceable standards and trained staff to provide valid calibration is approved by the Service. The State approval may be documented by a certificate or letter. The jurisdiction must be equipped to provide suitable certification documentation.

(3) Any commercial industrial laboratory primarily involved in the business of sealing and calibrating test weights (standards) will be approved by the Service provided:

(i) It requests written authority to perform tolerance testing of weights used within the Service's program(s) through their approved State jurisdic-tion. Copies of its request and written reference regarding the State decision shall be provided to the Service. A positive decision by the State will be required as a prerequisite to the Service's granting approval to any commercial laboratory to tolerance test the weights used in testing scales under the jurisdiction of the Service;

(ii) It has NBS traceable standards (through the State) and trained staff to perform calibrations in a manner prescribed by NBS and/or the State;

(iii) It is equipped to provide suitable certification documentation;

(iv) It permits the Service to make onsite visits to laboratory testing space.

(4) Approval of the commercial industrial laboratory will be at the Service's discretion. Once it has obtained approval, the commercial industrial laboratory maintains its site in a manner prescribed by the State and the Service.

(b) *Type evaluation laboratories.* Any State measurement laboratory currently certified by NBS in accordance with its program for the Certification of Capability of State Measurement Laboratories to conduct evaluations under the National Type Evaluation Program is approved by the Service.

(Approved by the Office of Management and Budget under control number 0580–0011)

[51 FR 7052, Feb. 28, 1986, as amended at 54 FR 5925, Feb. 7, 1989]

PART 810—OFFICIAL UNITED STATES STANDARDS FOR GRAIN

Subpart A—General Provisions

TERMS DEFINED

Sec.
810.101 Grains for which standards are established.
810.102 Definition of other terms.

PRINCIPLES GOVERNING THE APPLICATION OF STANDARDS

810.103 Basis of determination.
810.104 Percentages.

GRADES, GRADE REQUIREMENTS, AND GRADE DESIGNATIONS

810.105 Grades and grade requirements.
810.106 Grade designations.

AUTHORITY: 7 U.S.C. 71-87k.

SOURCE: 52 FR 24418, June 30, 1987, unless otherwise noted.

Subpart A—General Provisions

NOTE: Compliance with the provisions of these standards does not excuse failure to comply with the provisions of the Federal Food, Drug, and Cosmetic Act, or other Federal laws.

TERMS DEFINED

§810.101 Grains for which standards are established.

Grain refers to barley, canola, corn, flaxseed, mixed grain, oats, rye, sorghum, soybeans, sunflower seed, triticale, and wheat. Standards for these food grains, feed grains, and oilseeds are established under the United States Grain Standards Act.

[57 FR 3274, Jan. 29, 1992]

§810.102 Definition of other terms.

Unless otherwise stated, the definitions in this section apply to all grains. All other definitions unique to a particular grain are contained in the appropriate subpart for that grain.

(a) *Distinctly low quality.* Grain that is obviously of inferior quality because it is in an unusual state or condition, and that cannot be graded properly by use of other grading factors provided in the standards. Distinctly low quality includes the presence of any objects too large to enter the sampling device; i.e., large stones, wreckage, or similar objects.

(b) *Moisture.* Water content in grain as determined by an approved device according to procedures prescribed in FGIS instructions.

(c) *Stones.* Concreted earthy or mineral matter and other substances of similar hardness that do not disintegrate in water.

(d) *Test Weight per bushel.* The weight per Winchester bushel (2,150.42 cubic inches) as determined using an approved device according to procedures prescribed in FGIS instructions. Test weight per bushel in the standards for corn, mixed grain, oats, sorghum, and soybeans is determined on the original sample. Test weight per bushel in the standards for barley, flaxseed, rye, sunflower seed, triticale, and wheat is determined after mechanically cleaning the original sample. Test weight per bushel is recorded to the nearest tenth pound for corn, rye, sorghum, soybeans, triticale, and wheat. Test weight per bushel for all other grains, if applicable, is recorded in whole and half pounds with a fraction of a half pound disregarded. Test weight per bushel is not an official factor for canola.

(e) *Whole kernels.* Grain with ¼ or less of the kernel removed.

[52 FR 24418, June 30, 1987, as amended at 60 FR 61196, Nov. 29, 1995; 71 FR 52406, Sept. 6, 2006; 72 FR 39732, July 20, 2007]

PRINCIPLES GOVERNING THE APPLICATION OF STANDARDS

§810.103 Basis of determination.

(a) *Distinctly low quality.* The determination of distinctly low quality is made on the basis of the lot as a whole at the time of sampling when a condition exists that may or may not appear in the representative sample and/or the sample as a whole.

(b) *Certain quality determinations.* Each determination of rodent pellets, bird droppings, other animal filth, broken glass, castor beans, cockleburs, crotalaria seeds, dockage, garlic, live insect infestation, large stones, moisture, temperature, an unknown foreign substance(s), and a commonly recognized harmful or toxic substance(s) is made on the basis of the sample as a whole. When a condition exists that may not appear in the representative sample, the determination may be made on the basis of the lot as a whole at the time of sampling according to procedures prescribed in FGIS instructions.

(c) *All other determinations.* The basis of determination for all other factors is contained in the individual standards.

§810.104 Percentages.

(a) *Rounding.* Percentages are determined on the basis of weight and are rounded as follows:

(1) When the figure to be rounded is followed by a figure greater than or equal to 5, round to the next higher figure; e.g., report 6.36 as 6.4, 0.35 as 0.4, and 2.45 as 2.5.

(2) When the figure to be rounded is followed by a figure less than 5, retain the figure; e.g., report 8.34 as 8.3, and 1.22 as 1.2.

(b) *Recording.* The percentage of dockage in flaxseed and sorghum is reported in whole percent with fractions

567

of a percent being disregarded. Dockage in barley and triticale is reported in whole and half percent with a fraction less than one-half percent being disregarded. Dockage in wheat and rye is reported in whole and tenth percents to the nearest tenth percent. Foreign material in sunflower seed is reported to the nearest one-half percent. Ranges of sunflower seed foreign material are reported as follows: 0.0 to 0.24 is reported as 0.0 percent, 0.25 to 0.74 as 0.5 percent, 0.75 to 1.24 as 1.0 percent, and the like. Foreign material and fines in mixed grain is reported in whole percent. The percentage of smut in barley, sclerotinia and stones in canola, and ergot in all grains is reported to the nearest hundredth percent. The percentage when determining the identity of all grains is reported to the nearest whole percent. Also reported to the nearest whole percent are the classes and subclasses in wheat; flint corn; flint and dent corn; waxy corn; classes in barley; and the percentage of each kind of grain in mixed grain. Plump barley shall be expressed in terms of the range in which it falls. Ranges shall be: Below 50 percent, 50 to 55 percent, 56 to 60 percent, 61 to 65 percent, and the like. All other percentages are reported in tenths percent.

[52 FR 24418, June 30, 1987; 52 FR 28534, July 31, 1987, as amended at 54 FR 24157, June 6, 1989; 57 FR 3274, Jan. 29, 1992; 59 FR 10573, Mar. 7, 1994; 61 FR 18491, Apr. 26, 1996; 63 FR 20056, Apr. 23, 1998]

GRADES, GRADE REQUIREMENTS, AND GRADE DESIGNATIONS

§ 810.105 Grades and grade requirements.

The grades and grade requirements for each grain (except mixed grain) and shown in the grade table(s) of the respective standards. Mixed grain grade requirements are not presented in tabular form.

§ 810.106 Grade designations.

(a) *Grade designations for grain.* The grade designations include in the following order:

(1) The letters "U.S.";

(2) The abbreviation "No." and the number of the grade or the words "Sample grade";

(3) When applicable, the subclass;

(4) The class or kind of grain;

(5) When applicable, the special grade(s) except in the case of bright, extra heavy, and heavy oats or plump rye, the special grades, "bright", "extra heavy", "heavy" and "plump" will precede the word "oats" or "rye" as applicable; and

(6) When applicable, the word "dockage" together with the percentage thereof.

When applicable, the remarks section of the certificate will include in the order of predominance; in the case of a mixed class, the name and approximate percentage of the classes; in the case of sunflower seed, the percentage of admixture; in the case of mixed grain, the grains present in excess of 10.0 percent of the mixture and when applicable, the words *Other grains* followed by a statement of the percentage of the combined quantity of those kinds of grains, each of which is present in a quantity less than 10.0 percent; in the case of barley, if requested, the word "plump" with the percentage range thereof; in the case of wheat, if requested, the percentage of protein content.

(b) *Optional grade designations.* In addition to paragraph (a) of this Section, grain may be certificated under certain conditions as described in FGIS instructions when supported by official analysis, as "U.S. No. 2 or better (*type of grain*)", "U.S. No. 3 or better (*type of grain*)", and the like.

[52 FR 24418, June 30, 1987, as amended at 53 FR 15017, Apr. 27, 1988]

SPECIAL GRADES, SPECIAL GRADE REQUIREMENTS, AND SPECIAL GRADE DESIGNATIONS

§ 810.107 Special grades and special grade requirements.

A special grade serves to draw attention to a special factor or condition present in the grain and, when applicable, is supplemental to the grade assigned under § 810.106. Except for the special grade "infested," the special grades are identified and requirements are established in each respective grain standards.

(a) *Infested wheat, rye, and triticale.* Tolerances for live insects responsible for infested wheat, rye, and triticale are defined according to sampling designations as follows:

(1) *Representative sample*. The representative sample consists of the work portion, and the file sample if needed and when available. These grains will be considered infested if the representative sample (other than shiplots) contains two or more live weevils, or one live weevil and one or more other live insects injurious to stored grain, or two or more live insects injurious to stored grain.

(2) *Lot as a whole (stationary)*. The lot as a whole is considered infested when two or more live weevils, or one live weevil and one or more other live insects injurious to stored grain, or two or more other live insects injurious to stored grain are found in, on, or about the lot (excluding submitted samples and shiplots).

(3) *Sample as a whole (continuous loading/unloading of shiplots and bargelots)*. The minimum sample size for bargelots and shiplots is 500 grams per each 2,000 bushels of grain. The sample as a whole is considered infested when a component (as defined in FGIS instructions) contains two or more live weevils, or one live weevil and one or more other live insects injurious to stored grain, or two or more other live insects injurious to stored grain.

(b) *Infested barley, canola, corn, oats, sorghum, soybeans, sunflower seed, and mixed grain*. Tolerances for live insects responsible for infested barley, canola, corn, oats, sorghum, soybeans, sunflower seed, and mixed grain are defined according to sampling designations as follows:

(1) *Representative sample*. The representative sample consists of the work portion, and the file sample if needed and when available. These grains will be considered infested if the representative sample (other than shiplots) contains two or more live weevils, or one live weevil and five or more other live insects injurious to stored grain, or ten or more other live insects injurious to stored grain.

(2) *Lot as a whole (stationary)*. The lot as a whole is considered infested when two or more live weevils, or one live weevil and five or more other live insects injurious to stored grain, or ten

or more other live insects injurious to stored grain are found in, on, or about the lot (excluding submitted samples and shiplots).

(3) *Sample as a whole (continuous loading/unloading of shiplots and bargelots)*. The minimum sample for shiplots and bargelots is 500 grams per each 2,000 bushels of grain. The sample as a whole is considered infested when a component (as defined in FGIS instructions) contains two or more live weevils, or one live weevil and five or more other live insects injurious to stored grain, or ten or more other live insects injurious to stored grain.

[52 FR 24441, June 30, 1987, as amended at 57 FR 3274, Jan. 29, 1992]

§ 810.108 Special grade designations.

Special grade designations are shown as prescribed in § 810.106. Multiple special grade designations will be listed in alphabetical order. In the case of treated wheat, the official certificate shall show whether the wheat has been scoured, limed, washed, sulfured, or otherwise treated.

Subpart B—United States Standards for Barley

TERMS DEFINED

§ 810.201 Definition of barley.

Grain that, before the removal of dockage, consists of 50 percent or more of whole kernels of cultivated barley (*Hordeum vulgare* L.) and not more than 25 percent of other grains for which standards have been established under the United States Grain Standards Act. The term "barley" as used in these standards does not include hull-less barley or black barley.

§ 810.202 Definition of other terms.

(a) *Black barley*. Barley with black hulls.

(b) *Broken kernels*. Barley with more than ¼ of the kernel removed.

(c) *Classes*. There are two classes of barley: Malting barley and Barley.

(1) *Malting barley* is divided into the following two subclasses:

(i) *Six-rowed Malting barley* has a minimum of 95.0 percent of a six-rowed suitable malting type that contains not more than 1.9 percent injured-by-frost kernels, 0.4 percent frost-damaged kernels, 0.2 percent injured-by-heat kernels, 0.1 percent heat-damaged kernels, 1.9 percent injured-by-mold kernels, and 0.4 percent mold-damaged kernels. Six-rowed Malting barley must not be infested, blighted, ergoty, garlicky, or smutty as defined in § 810.107(b) and § 810.206.

(ii) *Two-rowed Malting barley* has a minimum of 95.0 percent of a two-rowed suitable malting type that contains not more than 1.9 percent injured-by-frost kernels, 0.4 percent frost-damaged kernels, 0.2 percent injured-by-heat kernels, 0.1 percent heat-damaged kernels, 1.9 percent injured-by-mold kernels, and 0.4 percent mold-damaged kernels. Two-rowed Malting barley must not be infested, blighted, ergoty, garlicky, or smutty as defined in § 810.107(b) and § 810.206.

(2) *Barley.* Any barley of a six-rowed or two-rowed type. The class Barley is divided into the following three sub-classes:

(i) *Six-rowed barley.* Any Six-rowed barley that contains not more than 10.0 percent of two-rowed varieties.

(ii) *Two-rowed barley.* Any Two-rowed barley with white hulls that contains not more than 10.0 percent of six-rowed varieties.

(iii) *Barley.* Any barley that does not meet the requirements for the sub-classes Six-rowed barley or Two-rowed barley.

(d) *Damaged kernels.* Kernels, pieces of barley kernels, other grains, and wild oats that are badly ground-damaged, badly weather-damaged, diseased, frost-damaged, germ-damaged, heat-damaged, injured-by-heat, insect-bored, mold-damaged, sprout-damaged, or otherwise materially damaged.

(e) *Dockage.* All matter other than barley that can be removed from the original sample by use of an approved device according to procedures prescribed in FGIS instructions. Also, underdeveloped, shriveled, and small pieces of barley kernels removed in properly separating the material other than barley and that cannot be recovered by properly rescreening or recleaning.

(f) *Foreign material.* All matter other than barley, other grains, and wild oats that remains in the sample after removal of dockage.

(g) *Frost-damaged kernels.* Kernels, pieces of barley kernels, other grains, and wild oats that are badly shrunken and distinctly discolored black or brown by frost.

(h) *Germ-damaged kernels.* Kernels, pieces of barley kernels, other grains, and wild oats that have dead or discolored germ ends.

(i) *Heat-damaged kernels.* Kernels, pieces of barley kernels, other grains, and wild oats that are materially discolored and damaged by heat.

(j) *Injured-by-frost kernels.* Kernels and pieces of barley kernels that are distinctly indented, immature or shrunken in appearance or that are light green in color as a result of frost before maturity.

(k) *Injured-by-heat kernels.* Kernels, pieces of barley kernels, other grains, and wild oats that are slightly discolored as a result of heat.

(l) *Injured-by-mold kernels.* Kernels, pieces of barley kernels containing slight evidence of mold.

(m) *Mold-damaged kernels.* Kernels, pieces of barley kernels, other grains, and wild oats that are weathered and contain considerable evidence of mold.

(n) *Other grains.* Black barley, corn, cultivated buckwheat, einkorn, emmer, flaxseed, guar, hull-less barley, nongrain sorghum, oats, Polish wheat, popcorn, poulard wheat, rice, rye, safflower, sorghum, soybeans, spelt, sunflower seed, sweet corn, triticale, and wheat.

(o) *Plump barley.* Barley that remains on top of a $\frac{6}{64} \times \frac{3}{4}$ slotted-hole sieve after sieving according to procedures prescribed in FGIS instructions.

(p) *Sieves.* (1) $\frac{5}{64} \times \frac{3}{4}$ slotted-hole sieve. A metal sieve 0.032 inch thick with slotted perforations 0.0781 ($\frac{5}{64}$) inch by 0.750 ($\frac{3}{4}$) inch.

(2) 5-½ $\frac{6}{64} \times \frac{3}{4}$ slotted-hole sieve. A metal sieve 0.032 inch thick with slotted perforations 0.0895 (5-½/64) inch by 0.750 ($\frac{3}{4}$) inch.

(3) $\frac{6}{64} \times \frac{3}{4}$ slotted-hole sieve. A metal sieve 0.032 inch thick with slotted perforations 0.0937 ($\frac{6}{64}$) inch by 0.750 ($\frac{3}{4}$) inch.

(q) *Skinned and broken kernels.* Barley kernels that have one-third or more of the hull removed, or that the hull is loose or missing over the germ, or broken kernels, or whole kernels that have a part or all of the germ missing.

(r) *Sound barley.* Kernels and pieces of barley kernels that are not damaged, as defined under (d) of this section.

(s) *Suitable malting type.* Varieties of malting barley that are recommended by the American Malting Barley Association and other malting type(s) used by the malting and brewing industry. The varieties are listed in AMSs instructions.

(t) *Thin barley.* Thin barley shall be defined for the appropriate class as follows:

(1) *Malting barley.* Six-rowed Malting barley that passes through a ⁵⁄₆₄ × ¾ slotted-hole sieve and Two-rowed Malting barley which passes through a ⁵·⁵⁄₆₄ × ¾ slotted-hole sieve in accordance with procedures prescribed in AMSs instructions.

(2) *Barley.* Six-rowed barley, Two-rowed barley, or Barley that passes

through a ⁵⁄₆₄ × ¾ slotted-hole sieve in accordance with procedures prescribed in AMSs instructions.

(u) *Wild oats.* Seeds of *Avena fatua* L. and *A. sterilis* L.

[52 FR 24418, June 30, 1987; 52 FR 28534, July 31, 1987, as amended at 61 FR 18491, Apr. 26, 1996; 82 FR 20543, May 3, 2017]

PRINCIPLES GOVERNING THE APPLICATION OF STANDARDS

§ 810.203   Basis of determination.

All other determinations. Each determination of heat-damaged kernels, injured-by-heat kernels, and white or blue aleurone layers in Six-rowed barley is made on pearled, dockage-free barley. Other determinations not specifically provided for under the *General Provisions* are made on the basis of the grain when free from dockage, except the determination of odor is made on either the basis of the grain as a whole or the grain when free from dockage.

GRADES AND GRADE REQUIREMENTS

§ 810.204   Grades and grade requirements for Six-rowed Malting barley.

Grade	Minimum limits of—			Maximum limits of—					
	Test weight per bushel (pounds)	Suitable malting types (percent)	Sound barley¹ (percent)	Damaged kernels¹ (percent)	Wild oats (percent)	Foreign material (percent)	Other grains (percent)	Skinned and broken kernels (percent)	Thin barley (percent)
U.S. No. 1	47.0	97.0	98.0	2.0	1.0	0.5	2.0	4.0	7.0
U.S. No. 2	45.0	97.0	98.0	3.0	1.0	1.0	3.0	6.0	10.0
U.S. No. 3	43.0	95.0	96.0	4.0	2.0	2.0	5.0	8.0	15.0
U.S. No. 4	43.0	95.0	93.0	5.0	3.0	3.0	5.0	10.0	15.0

¹ Injured-by-frost kernels and injured-by-mold kernels are not considered damaged kernels or considered against sound barley.

NOTE: Malting barley must not be infested in accordance with § 810.107(b) and must not contain any special grades as defined in § 810.206. Six-rowed Malting barley varieties not meeting the requirements of this section must be graded in accordance with standards established for the class Barley.

[82 FR 20543, May 3, 2017]

§ 810.205   Grades and grade requirements for Two-rowed Malting barley.

Grade	Minimum limits of—			Maximum limits of—					
	Test weight per bushel (pounds)	Suitable malting types (percent)	Sound barley¹ (percent)	Damaged kernels¹ (percent)	Wild oats (percent)	Foreign material (percent)	Other grains (percent)	Skinned and broken kernels (percent)	Thin barley (percent)
U.S. No. 1	50.0	97.0	98.0	2.0	1.0	0.5	2.0	4.0	5.0

	Minimum limits of—			Maximum limits of—					
Grade	Test weight per bushel (pounds)	Suitable malting types (percent)	Sound barley [1] (percent)	Damaged kernels [1] (percent)	Wild oats (percent)	Foreign material (percent)	Other grains (percent)	Skinned and broken kernels (percent)	Thin barley (percent)
U.S. No. 2	48.0	97.0	98.0	3.0	1.0	1.0	3.0	6.0	7.0
U.S. No. 3	48.0	95.0	96.0	4.0	2.0	2.0	5.0	8.0	10.0
U.S. No. 4	48.0	95.0	93.0	5.0	3.0	3.0	5.0	10.0	10.0

[1] Injured-by-frost kernels and injured-by-mold kernels are not considered damaged kernels or considered against sound barley.

NOTE: Malting barley must not be infested in accordance with § 810.107(b) and must not contain any special grades as defined in § 810.206. Six-rowed Malting barley and Six-rowed Blue Malting barley varieties not meeting the requirements of this section must be graded in accordance with standards established for the class Barley.

[82 FR 20544, May 3, 2017]

§ 810.206　Grades and grade requirements for barley.

	Minimum limits of—		Maximum Limits of—				
Grade	Test weight per bushel (pounds)	Sound barley (percent)	Damaged kernels [1] (percent)	Heat damaged kernels (percent)	Foreign material (percent)	Broken kernels (percent)	Thin barley (percent)
U.S. No. 1	47.0	97.0	2.0	0.2	1.0	4.0	10.0
U.S. No. 2	45.0	94.0	4.0	0.3	2.0	8.0	15.0
U.S. No. 3	43.0	90.0	6.0	0.5	3.0	12.0	25.0
U.S. No. 4	40.0	85.0	8.0	1.0	4.0	18.0	35.0
U.S. No. 5	36.0	75.0	10.0	3.0	5.0	28.0	75.0

U.S. Sample Grade:
U.S. Sample grade shall be barley that:
(a) Does not meet the requirements for the grades 1, 2, 3, 4, or 5; or
(b) Contains 8 or more stones or any number of stones which have an aggregate weight in excess of 0.2 percent of the sample weight, 2 or more pieces of glass, 3 or more crotalaria seeds (*Crotalaria* spp.), 2 or more caster beans (*Ricinus communis* L.), 4 or more particles of unknown foreign substance(s) or commonly recognized harmful or toxic substance(s), 8 or more cocklebur (*Xanthium* spp.) or similar seeds singly or in combination, 10 or more rodent pellets, bird droppings, or equivalent quantity of other animal filth per 1⅛ to 1¼ quarts of barley; or
(c) Has a musty, sour, or commercially objectionable foreign odor (except smut or garlic odor); or
(d) Is heating or otherwise of distinctly low quality.
[1] Includes heat-damaged kernels. Injured-by-frost kernels and injured-by-mold kernels are not considered damaged kernels.

[61 FR 18492, Apr. 26, 1996]

SPECIAL GRADES AND SPECIAL GRADE
REQUIREMENTS

§ 810.207　Special grades and special grade requirements.

(a) *Blighted barley.* Barley that contains more than 4.0 percent of fungus-damaged and/or mold-damaged kernels.

(b) *Ergoty barley.* Barley that contains more than 0.10 percent ergot.

(c) *Garlicky barley.* Barley that contains three or more green garlic bulblets, or an equivalent quantity of dry or partly dry bulblets in 500 grams of barley.

(d) *Smutty barley.* Barley that has kernels covered with smut spores to give a smutty appearance in mass, or which contains more than 0.20 percent smut balls.

[52 FR 24418, June 30, 1987, as amended at 52 FR 24441, June 30, 1987]

Subpart C—United States Standards for Canola—Terms Defined

SOURCE: 57 FR 3274, Jan. 29, 1992, unless otherwise noted.

§ 810.301　Definition of canola.

Seeds of the genus *Brassica* from which the oil shall contain less than 2 percent erucic acid in its fatty acid profile and the solid component shall contain less than 30.0 micromoles of any one or any mixture of 3-butenyl glucosinolate, 4-pentenyl glucosinolate, 2-hydroxy-3-butenyl, or 2-hydroxy-4-pentenyl glucosinolate, per gram of air-dried, oil free solid. Before the removal of dockage, the seed shall contain not more than 10.0% of other grains for which standards have been established under the United States Grain Standards Act.

§810.302 **Definitions of other terms.**

(a) *Conspicuous Admixture.* All matter other than canola, including but not limited to ergot, sclerotinia, and stones, which is conspicuous and readily distinguishable from canola and which remains in the sample after the removal of machine separated dockage. Conspicuous admixture is added to machine separated dockage in the computation of total dockage.

(b) *Damaged kernels.* Canola and pieces of canola that are heat-damaged, sprout-damaged, mold-damaged, distinctly green damaged, frost damaged, rimed damaged, or otherwise materially damaged.

(c) *Distinctly green kernels.* Canola and pieces of canola which, after being crushed, exhibit a distinctly green color.

(d) *Dockage.* All matter other than canola that can be removed from the original sample by use of an approved device according to procedures prescribed in FGIS instructions. Also, underdeveloped, shriveled, and small pieces of canola kernels that cannot be recovered by properly rescreening or recleaning. Machine separated dockage is added to conspicuous admixture in the computation of total dockage.

(e) *Ergot.* Sclerotia (sclerotium, sing.) of the fungus, *Claviceps* species, which are associated with some seeds other than canola where the fungal organism has replaced the seed.

(f) *Heat-damaged kernels.* Canola and pieces of canola which, after being crushed, exhibit that they are discolored and damaged by heat.

(g) *Inconspicuous admixture.* Any seed which is difficult to distinguish from canola. This includes, but is not limited to, common wild mustard (*Brassica kaber* and *B. juncea*), domestic brown mustard (*Brassica juncea*), yellow mustard (*B. hirta*), and seed other than the mustard group.

(h) *Sclerotia (Sclerotium, sing.).* Dark colored or black resting bodies of the fungi *Sclerotinia* and *Claviceps*.

(i) *Sclerotinia.* Genus name which includes the fungus *Sclerotinia sclerotiorum* which produces sclerotia. Canola is only infrequently infected, and the sclerotia, unlike sclerotia of ergot, are usually associated within the stem of the plants.

PRINCIPLES GOVERNING THE APPLICATION OF STANDARDS

§810.303 **Basis of determination.**

Each determination of conspicuous admixture, ergot, sclerotinia, stones, damaged kernels, heat-damaged kernels, distinctly green kernels, and inconspicuous admixture is made on the basis of the sample when free from dockage. Other determinations not specifically provided for under the general provisions are made on the basis of the sample as a whole, except the determination of odor is made on either the basis of the sample as a whole or the sample when free from dockage. The content of glucosinolates and erucic acid is determined on the basis of the sample according to procedures prescribed in FGIS instructions.

GRADES AND GRADE REQUIREMENTS

§810.304 **Grades and grade requirements for canola.**

Grading factors	Grades, U.S. Nos.		
	1	2	3
	Maximum percent limits of:		
Damaged kernels:			
Heat damaged	0.1	0.5	2.0
Distinctly green	2.0	6.0	20.0
Total	3.0	10.0	20.0
Conspicuous admixture:			
Ergot	0.05	0.05	0.05
Sclerotinia	0.05	0.10	0.15
Stones	0.05	0.05	0.05
Total	1.0	1.5	2.0
Inconspicuous admixture	5.0	5.0	5.0
	Maximum count limits of:		
Other material:			
Animal filth	3	3	3
Glass	0	0	0
Unknown foreign substance	1	1	1

573

Grading factors	Grades, U.S. Nos.		
	1	2	3
U.S. Sample grade Canola that: (a) Does not meet the requirements for U.S. Nos. 1, 2, 3; or (b) Has a musty, sour, or commercially objectionable foreign odor; or (c) Is heating or otherwise of distinctly low quality.			

SPECIAL GRADES AND SPECIAL GRADE
REQUIREMENTS

§ 810.305 Special grades and special grade requirements.

Garlicky canola. Canola that contains more than two green garlic bulblets or an equivalent quantity of dry or partly dry bulblets in approximately a 500 gram portion.

NONGRADE REQUIREMENTS

§ 810.306 Nongrade requirements.

Glucosinolates. Content of glucosinolates in canola is determined according to procedures prescribed in FGIS instructions.

Subpart D—United States Standards for Corn

TERMS DEFINED

§ 810.401 Definition of corn.

Grain that consists of 50 percent or more of whole kernels of shelled dent corn and/or shelled flint corn (*Zea mays* L.) and not more than 10.0 percent of other grains for which standards have been established under the United States Grain Standards Act.

§ 810.402 Definition of other terms.

(a) *Broken corn.* All matter that passes readily through a 12/64 round-hole sieve and over a 6/64 round-hole sieve sample according to procedures prescribed in FGIS instructions.

(b) *Broken corn and foreign material.* All matter that passes readily through a 12/64 round-hole sieve and all matter other than corn that remains in the sieved after sieving according to procedures prescribed in FGIS instructions.

(c) *Classes.* There are three classes for corn: Yellow corn, White corn, and Mixed corn.

(1) *Yellow corn.* Corn that is yellow-kerneled and contains not more than 5.0 percent of corn of other colors. Yellow kernels of corn with a slight tinge of red are considered yellow corn.

(2) *White corn.* Corn that is white-kerneled and contains not more than 2.0 percent of corn of other colors. White kernels of corn with a slight tinge of light straw or pink color are considered white corn.

(3) *Mixed corn.* Corn that does not meet the color requirements for either of the classes Yellow corn or White corn and includes white-capped Yellow corn.

(d) *Damaged kernels.* Kernels and pieces of corn kernels that are badly ground-damaged, badly weather-damaged, diseased, frost-damaged, germ-damaged, heat-damaged, insect-bored, mold-damaged, sprout-damaged, or otherwise materially damaged.

(e) *Foreign material.* All matter that passes readily through a 6/64 round-hole sieve and all matter other than corn that remains on top of the 12/64 round-hole sieve according to procedures prescribed in FGIS instructions.

(f) *Heat-damaged kernels.* Kernels and pieces of corn kernels that are materially discolored and damaged by heat.

(g) *Sieves*—(1) *12/64 round-hole sieve.* A metal sieve 0.032 inch thick with round perforations 0.1875 (12/64) inch in diameter which are ¼ inch from center to center. The perforations of each row shall be staggered in relation to the adjacent row.

(2) *6/64 round-hole sieve.* A metal sieve 0.032 inch thick with round perforations 0.0937 (6/64) inch in diameter which are 5/32 inch from center to center. The perforations of each row shall be staggered in relation to the adjacent row.

[52 FR 24418, June 30, 1987, as amended at 52 FR 24437, June 30, 1987; 52 FR 28534, July 31, 1987]

PRINCIPLES GOVERNING THE
APPLICATION OF STANDARDS

§ 810.403 Basis of determination.

Each determination of class, damaged kernels, heat-damaged kernels, waxy corn, flint corn, and flint and dent corn is made on the basis of the grain after the removal of the broken corn and foreign material. Other determinations not specifically provided for under the general provisions are made on the basis of the grain as a whole, except the determination of odor is made on either the basis of the grain as a whole or the grain when free from broken corn and foreign material.

[52 FR 24418, June 30, 1987; 52 FR 28534, July 31, 1987]

GRADES AND GRADE REQUIREMENTS

§810.404 Grades and grade requirements for corn.

Grade	Minimum test weight per bushel (pounds)	Maximum limits of		
		Damaged kernels		Broken corn and foreign material (percent)
		Heat damaged kernels (percent)	Total (percent)	
U.S. No. 1	56.0	0.1	3.0	2.0
U.S. No. 2	54.0	0.2	5.0	3.0
U.S. No. 3	52.0	0.5	7.0	4.0
U.S. No. 4	49.0	1.0	10.0	5.0
U.S. No. 5	46.0	3.0	15.0	7.0

U.S. Sample Grade
  U.S. Sample grade is corn that:
    (a) Does not meet the requirements for the grades U.S. Nos. 1, 2, 3, 4, or 5; or
    (b) Contains stones with an aggregate weight in excess of 0.1 percent of the sample weight, 2 or more pieces of glass, 3 or more crotalaria seeds (Crotalaria spp.), 2 or more castor beans (Ricinus communis L.), 4 or more particles of an unknown foreign substance(s) or a commonly recognized harmful or toxic substance(s), 8 or more cockleburs (Xanthium spp.), or similar seeds singly or in combination, or animal filth in excess of 0.20 percent in 1,000 grams; or
    (c) Has a musty, sour, or commercially objectionable foreign odor; or
    (d) Is heating or otherwise of distinctly low quality.

[60 FR 61196, Nov. 29, 1995]

SPECIAL GRADES AND SPECIAL GRADE REQUIREMENTS

§810.405 Special grades and special grade requirements.

(a) *Flint corn.* Corn that consists of 95 percent or more of flint corn.

(b) *Flint and dent corn.* Corn that consists of a mixture of flint and dent corn containing more than 5.0 percent but less than 95 percent of flint corn.

(c) *Waxy corn.* Corn that consists of 95 percent or more waxy corn, according to procedures prescribed in FGIS instructions.

[52 FR 24418, June 30, 1987, as amended at 52 FR 24441, June 30, 1987; 52 FR 28534, July 31, 1987]

Subpart E—United States Standards for Flaxseed

TERMS DEFINED

§810.601 Definition of flaxseed.

Grain that, before the removal of dockage, consists of 50 percent or more of common flaxseed (*Linum usitatissimum* L.) and not more than 20 percent of other grains for which standards have been established under the United States Grain Standards Act and which, after the removal of dockage, contains 50 percent or more of whole flaxseed.

§810.602 Definition of other terms.

(a) *Damaged kernels.* Kernels and pieces of flaxseed kernels that are badly ground-damaged, badly weather-damaged, diseased, frost-damaged, germ-damaged, heat-damaged, insect-bored, mold-damaged, sprout-damaged, or otherwise materially damaged.

(b) *Dockage.* All matter other than flaxseed that can be removed from the original sample by use of an approved device according to procedures prescribed in FGIS instructions. Also, underdeveloped, shriveled, and small pieces of flaxseed kernels removed in properly separating the material other than flaxseed and that cannot be recovered by properly rescreening or recleaning.

(c) *Heat-damaged kernels.* Kernels and pieces of flaxseed kernels that are materially discolored and damaged by heat.

(d) *Other grains.* Barley, corn, cultivated buckwheat, einkorn, emmer, guar, hull-less barley, nongrain sorghum, oats, Polish wheat, popcorn, poulard wheat, rice, rye, safflower, sorghum, soybeans, spelt, sunflower seed, sweet corn, triticale, wheat, and wild oats.

PRINCIPLES GOVERNING THE
APPLICATION OF STANDARDS

§ 810.603   Basis of determination.

Other determinations not specifically provided for under the general provisions are made on the basis of the grain when free from dockage, except the determination of odor is made on either the basis of the grain as a whole or the grain when free from dockage.

GRADES AND GRADE REQUIREMENTS

§ 810.604   Grades and grade requirements for flaxseed.

Grade	Minimum test weight per bushel (pounds)	Maximum limits of damaged kernels—	
		Heat damaged kernels (percent)	Total (percent)
U.S. No. 1	49.0	0.2	10.0
U.S. No. 2	47.0	0.5	15.0

Grade	Minimum test weight per bushel (pounds)	Maximum limits of damaged kernels—	
		Heat damaged kernels (percent)	Total (percent)

U.S. Sample grade—

U.S. Sample grade is flaxseed that:

(a) Does not meet the requirements for the grades U.S. Nos. 1 or 2; or

(b) Contains 8 or more stones which have an aggregate weight in excess of 0.2 percent of the sample weight, 2 or more pieces of glass, 3 or more crotalaria seeds (*Crotalaria* spp.), 2 or more castor beans (*Ricinus communis* L.), 4 or more particles of an unknown foreign substance(s) or a commonly recognized harmful or toxic substance(s), 10 or more rodent pellets, bird dropping, or equivalent quantity of other animal filth per 1⅛ to 1¼ quarts of flaxseed; or

(c) Has musty, sour, or commercially objectionable foreign odor (except smut or garlic odor), or

(d) Is heating or otherwise of distinctly low quality.

Subpart F—United States Standards for Mixed Grain

TERMS DEFINED

§ 810.801   Definition of mixed grain.

Any mixture of grains for which standards have been established under the United States Grain Standards Act, provided that such mixture does not come within the requirements of any of the standards for such grains; and that such mixture consists of 50 percent or more of whole kernels of grain and/or whole or broken soybeans which will not pass through a 5/64 triangular-hole sieve and/or whole flaxseed that passes through such a sieve after sieving according to procedures prescribed in FGIS instructions.

§ 810.802   Definition of other terms.

(a) *Damaged kernels.* Kernels and pieces of grain kernels for which standards have been established under the Act, that are badly ground-damaged, badly weather-damaged, diseased, frost-damaged, germ-damaged, heat-damaged, insect-bored, mold-damaged, sprout-damaged, or otherwise materially damaged.

(b) *Foreign material and fines.* All matter other than whole flaxseed that passes through a 5/64 triangular-hole sieve, and all matter other than grains for which standards have been established under the Act, that remains in the sieved sample.

(c) *Grades.* U.S. Mixed Grain, or U.S. Sample grade Mixed Grain, and special grades.

(d) *Heat-damaged kernels.* Kernels and pieces of grain kernels for which standards have been established under the Act, that are materially discolored and damaged by heat.

(e) *Sieve—⁵⁄₆₄ triangular-hole sieve.* A metal sieve 0.032 inch thick with equilateral triangular perforations the inscribed circles of which are 0.0781 (⁵⁄₆₄) inch in diameter.

PRINCIPLES GOVERNING THE APPLICATION OF STANDARDS

§ 810.803 Basis of determination.

Each determination of damaged and heat-damaged kernels, and the percentage of each kind of grain in the mixture is made on the basis of the sample after removal of foreign material and fines. Other determinations not specifically provided for under the general provisions are made on the basis of the grain as a whole, except the determination of odor is made on either the basis of the grain as a whole or the grain when free from foreign material and fines.

GRADES AND GRADE REQUIREMENTS

§ 810.804 Grades and grade requirements for mixed grain.

(a) *U.S. Mixed Grain (grade).* Mixed grain with not more than 15.0 percent of damaged kernels, and not more than 3.0 percent of heat-damaged kernels, and that otherwise does not meet the requirements for the grade U.S. Sample grade Mixed Grain.

(b) *U.S. Sample grade Mixed Grain.* Mixed grain that:

(1) Does not meet the requirements for the grade U.S. Mixed Grain; or

(2) Contains more than 16.0 percent moisture; or

(3) Contains 8 or more stones that have an aggregate weight in excess of 0.2 percent of the sample weight, 2 or more pieces of glass, 3 or more Crotalaria seeds (*Crotalaria* spp.), 2 or more castor beans (*Ricinus communis* L.), 8 more cockleburs (*Xanthium* spp.) or similar seeds singly or in combination, 4 or more pieces of an unknown foreign substance(s) or a recognized harmful or toxic substance(s), 10 or more rodent pellets, bird droppings, or an equivalent quantity of other animal filth per 1,000 grams of mixed grain; or

(4) Is musty, sour, or heating; or

(5) Has any commercially objectionable foreign odor except smut or garlic; or

(6) Is otherwise of distinctly low quality.

SPECIAL GRADES AND SPECIAL GRADE REQUIREMENTS

§ 810.805 Special grades and special grade requirements.

(a) *Blighted mixed grain.* Mixed grain in which barley predominates and that contains more than 4.0 percent of fungus-damaged and/or mold-damaged barley kernels.

(b) *Ergoty mixed grain.* (1) Mixed grain in which rye or wheat predominates and that contains more than 0.30 percent ergot, or

(2) Any other mixed grain that contains more than 0.10 percent ergot.

(c) *Garlicky mixed grain.* (1) Mixed grain in which wheat, rye, or triticale predominates, and that contains 2 or more green garlic bulblets, or an equivalent quantity of dry or partly dry bulblets in 1,000 grams of mixed grain; or

(2) Any other mixed grain that contains 4 or more green garlic bulblets, or an equivalent quantity of dry or partly dry bulblets, in 500 grams of mixed grain.

(d) *Smutty mixed grain.* (1) Mixed grain in which rye, triticale, or wheat predominates, and that contains 15 or more average size smut balls, or an equivalent quantity of smut spores in 250 grams of mixed grain, or

(2) Any other mixed grain that has the kernels covered with smut spores to give a smutty appearance in mass, or that contains more than 0.2 percent smut balls.

(e) *Treated mixed grain.* Mixed grain that has been scoured, limed, washed, sulfured, or treated in such a manner that its true quality is not reflected by the grade designation U.S. Mixed Grain or U.S. Sample grade Mixed Grain.

[52 FR 24418, June 30, 1987, as amended at 52 FR 24441, June 30, 1987]

Subpart G—United States Standards for Oats

TERMS DEFINED

§ 810.1001 Definition of oats.

Grain that consists of 50 percent or more of oats (*Avena sativa* L. and *A. byzantina* C. Koch) and may contain, singly or in combination, not more than 25 percent of wild oats and other grains for which standards have been established under the United States Grain Standards Act.

§ 810.1002 Definition of other terms.

(a) *Fine seeds.* All matter that passes through a ⁵⁄₆₄ triangular-hole sieve after sieving according to procedures prescribed in FGIS instructions.

(b) *Foreign material.* All matter other than oats, wild oats, and other grains.

(c) *Heat-damaged kernels.* Kernels and pieces of oat kernels, other grains, and wild oats that are materially discolored and damaged by heat.

(d) *Other grains.* Barley, corn, cultivated buckwheat, einkorn, emmer, flaxseed, guar, hull-less barley, nongrain sorghum, Polish wheat, popcorn, poulard wheat, rice, rye, safflower, sorghum, soybeans, spelt, sunflower seed, sweet corn, triticale, and wheat.

(e) *Sieves*—(1) ⁵⁄₆₄ *triangular-hole sieve.* A metal sieve 0.032 inch thick with equilateral triangular perforations the inscribed circles of which are 0.0781 (⁵⁄₆₄) inch in diameter.

(2) *0.064 × ³⁄₈ oblong-hole sieve.* A metal sieve 0.032 inch thick with oblong perforations 0.064 inch by 0.375 (³⁄₈) inch.

(f) *Sound oats.* Kernels and pieces of oat kernels (except wild oats) that are not badly ground-damaged, badly weather-damaged, diseased, frost-damaged, germ-damaged, heat-damaged, insect-bored, mold-damaged, sprout-damaged, or otherwise materially damaged.

(g) *Wild oats.* Seeds of *Avena fatua* L. and *A. sterillis* L.

PRINCIPLES GOVERNING THE APPLICATION OF STANDARDS

§ 810.1003 Basis of determination.

Other determinations not specifically provided for under the general provisions are made on the basis of the grain as a whole.

GRADES AND GRADE REQUIREMENTS

§ 810.1004 Grades and grade requirements for oats.

Grade	Minimum limits—		Maximum limits—		
	Test weight per bushel (pounds)	Sound oats (percent)	Heat-damaged kernels (percent)	Foreign material (percent)	Wild oats (percent)
U.S. No. 1	36.0	97.0	0.1	2.0	2.0
U.S. No. 2	33.0	94.0	0.3	3.0	3.0
U.S. No. 3 [1]	30.0	90.0	1.0	4.0	5.0
U.S. No. 4 [2]	27.0	80.0	3.0	5.0	10.0

U.S. Sample grade—
  U.S. Sample grade are oats which:
    (a) Do not meet the requirements for the grades U.S. Nos. 1, 2, 3, or 4; or
    (b) Contain 8 or more stones which have an aggregate weight in excess of 0.2 percent of the sample weight, 2 or more pieces of glass, 3 or more crotalaria seeds (*Crotalaria* spp.), 2 or more castor beans (*Ricinus communis* L.), 4 or more particles of an unknown foreign substance(s) or a commonly recognized harmful or toxic substance(s), 8 or more cocklebur (*Xanthium* spp.) or similar seeds singly or in combination, 10 or more rodent pellets, bird droppings, or equivalent quantity of other animal filth per 1⅛ to 1¼ quarts of oats; or
    (c) Have a musty, sour, or commercially objectionable foreign odor (except smut or garlic odor); or
    (d) Are heating or otherwise of distinctly low quality.

[1] Oats that are slightly weathered shall be graded not higher than U.S. No. 3.
[2] Oats that are badly stained or materially weathered shall be graded not higher than U.S. No. 4.

SPECIAL GRADES AND SPECIAL GRADE
REQUIREMENTS

§810.1005 Special grades and special grade requirements.

(a) *Bleached oats.* Oats that in whole or in part, have been treated with sulfurous acid or any other bleaching agent.

(b) *Bright oats.* Oats, except bleached oats, that are of good natural color.

(c) *Ergoty oats.* Oats that contain more than 0.10 percent ergot.

(d) *Extra-heavy oats.* Oats that have a test weight per bushel of 40 pounds or more.

(e) *Garlicky oats.* Oats that contain 4 or more green garlic bulblets or an equivalent quantity of dry or partly dry bulblets in 500 grams of oats.

(f) *Heavy oats.* Oats that have a test weight per bushel of 38 pounds or more but less than 40 pounds.

(g) *Smutty oats.* Oats that have kernels covered with smut spores to give a smutty appearance in mass, or that contain more than 0.2 percent of smut balls.

(h) *Thin oats.* Oats that contain more than 20.0 percent of oats and other matter, except fine seeds, that pass through a 0.064 × ⅜ oblong-hole sieve but remain on top of a ⁵⁄₆₄ triangular-hole sieve after sieving according to procedures prescribed in FGIS instructions.

[52 FR 24418, June 30, 1987, as amended at 52 FR 24441, June 30, 1987]

Subpart H—United States Standards for Rye

TERMS DEFINED

§810.1201 Definition of rye.

Grain that, before the removal of dockage, consists of 50 percent or more of common rye (*Secale cereale* L.) and not more than 10 percent of other grains for which standards have been established under the United States Grain Standards Act and that, after the removal of dockage, contains 50 percent or more of whole rye.

§810.1202 Definition of other terms.

(a) *Damaged kernels.* Kernels, pieces of rye kernels, and other grains that are badly ground-damaged, badly weather-damaged, diseased, frost-damaged, germ-damaged, heat-damaged, insect-bored, mold-damaged, sprout-damaged, or otherwise materially damaged.

(b) *Dockage.* All matter other than rye that can be removed from the original sample by use of an approved device in accordance with procedures prescribed in FGIS instructions. Also, underdeveloped, shriveled, and small pieces of rye kernels removed in properly separating the material other than rye and that cannot be recovered by properly rescreening and recleaning.

(c) *Foreign material.* All matter other than rye that remains in the sample after the removal of dockage.

(d) *Heat-damaged kernels.* Kernels, pieces of rye kernels, and other grains that are materially discolored and damaged by heat.

(e) *Other grains.* Barley, corn, cultivated buckwheat, einkorn, emmer, flaxseed, guar, hull-less barley, nongrain sorghum, oats, Polish wheat, popcorn, poulard wheat, rice, safflower, sorghum, soybeans, spelt, sunflower seed, sweet corn, triticale, wheat, and wild oats.

(f) *Sieve—0.064 × ⅜ oblong-hole sieve.* A metal sieve 0.032 inch thick with oblong perforations 0.064 by 0.375 (⅜) inch.

(g) *Thin rye.* Rye and other matter that passes through a 0.064 × ⅜ oblong-hole sieve after sieving according to procedures prescribed in FGIS instructions.

PRINCIPLES GOVERNING THE
APPLICATION OF STANDARDS

§810.1203 Basis of determination.

Other determinations not specifically provided for under the general provisions are made on the basis of the grain when free from dockage, except the determination of odor is made on either the basis of the grain as a whole or the grain when free from dockage.

GRADES AND GRADE REQUIREMENTS

§ 810.1204 Grades and grade requirements for rye.

| Grade | Minimum test weight per bushel (pounds) | Maximum limits of— | | | | |
| | | Foreign material | | Damaged kernels | | Thin Rye (percent) |
		Foreign matter other than wheat (percent)	Total (percent)	Heat damaged (percent)	Total (percent)	
U.S. No. 1	56.0	1.0	3.0	0.2	2.0	10.0
U.S. No. 2	54.0	2.0	6.0	0.2	4.0	15.0
U.S. No. 3	52.0	4.0	10.0	0.5	7.0	25.0
U.S. No. 4	49.0	6.0	10.0	3.0	15.0	

U.S. Sample grade—
   U.S. Sample grade is rye that:
     (a) Does not meet the requirements for the grades U.S. Nos. 1, 2, 3, or 4; or
     (b) Contains 8 or more stones or any numbers of stones which have an aggregate weight in excess of 0.2 percent of the sample weight, 2 or more pieces of glass, 3 or more crotalaria seeds (*Crotalaria* spp.), 2 or more castor beans (*Ricinus communis* L.), 4 or more particles of an unknown foreign substance(s) or a commonly recognized harmful or toxic substance(s), 2 or more rodent pellets, bird droppings, or equivalent quantity of other animal filth per 1⅛ or 1¼ quarts of rye; or
     (c) Has a musty, sour, or commercially objectionable foreign odor (except smut or garlic odor); or
     (d) Is heating or otherwise of distinctly low quality.

SPECIAL GRADES AND SPECIAL GRADE REQUIREMENTS

§ 810.1205 Special grades and special grade requirements.

(a) *Ergoty rye.* Rye that contains more than 0.30 percent of ergot.

(b) *Garlicky rye.* Rye that contains in a 1,000-gram portion more than six green garlic bulblets or an equivalent quantity of dry or partly dry bulblets.

(c) *Light garlicky rye.* Rye that contains in a 1,000-gram portion two or more, but not more than six, green garlic bulblets or an equivalent quantity of dry or partly dry bulblets.

(d) *Light smutty rye.* Rye that has an unmistakable odor of smut, or that contains in a 250-gram portion smut balls, portions of smut balls, or spores of smut in excess of a quantity equal to 14 smut balls but not in excess of a quantity equal to 30 smut balls of average size.

(e) *Plump rye.* Rye that contains not more than 5.0 percent of rye and other matter that passes through a 0.064 × ⅜ oblong-hole sieve.

(f) *Smutty rye.* Rye that contains in a 250-gram portion smut balls, portions of smut balls, or spores of smut in excess of a quantity equal to 30 smut balls of average size.

[52 FR 24418, June 30, 1987, as amended at 52 FR 24441, June 30, 1987]

Subpart I—United States Standards for Sorghum

TERMS DEFINED

§ 810.1401 Definition of sorghum.

Grain that, before the removal of dockage, consists of 50 percent or more of whole kernels of sorghum (*Sorghum bicolor* (L.) Moench) excluding nongrain sorghum and not more than 10.0 percent of other grains for which standards have been established under the United States Grain Standards Act.

§ 810.1402 Definition of other terms.

(a) *Broken kernels.* All matter which passes through a 5/64 triangular-hole sieve and over a 2-1/2/64 round-hole sieve according to procedures prescribed in FGIS instructions.

(b) *Broken kernels and foreign material.* The combination of broken kernels and foreign material as defined in paragraph (a) and (f) of this section.

(c) *Classes.* There are four classes of sorghum: Sorghum, Tannin sorghum, White sorghum, and Mixed sorghum.

(1) *Sorghum.* Sorghum which lacks a pigmented testa (subcoat) and contains less than 98.0 percent White sorghum and not more than 3.0 percent Tannin sorghum. The pericarp color of this class may appear white, yellow, red, pink, orange or bronze.

(2) *Tannin sorghum.* Sorghum which has a pigmented testa (subcoat) and contains not more than 10 percent of kernels without a pigmented testa.

(3) *White sorghum.* Sorghum which lacks a pigmented testa (subcoat) and contains not less than 98.0 percent kernels with a white pericarp, and contains not more than 2.0 percent of sorghum of other classes. This class includes sorghum containing spots that, singly or in combination, cover 25.0 percent or less of the kernel.

(4) *Mixed sorghum.* Sorghum which does not meet the requirements for any of the classes Sorghum, Tannin sorghum, or White sorghum.

(d) *Damaged kernels.* Kernels, pieces of sorghum kernels and other grains that are badly ground damaged, badly weather damaged, diseased, frost-damaged, germ-damaged, heat-damaged, insect-bored, mold-damaged, sprout-damaged, or otherwise materially damaged.

(e) *Dockage.* All matter other than sorghum that can be removed from the original sample by use of an approved device according to procedures prescribed in FGIS instructions. Also, underdeveloped, shriveled, and small pieces of sorghum kernels removed in properly separating the material other than sorghum.

(f) *Foreign material.* All matter, except sorghum, which passes over the number 6 riddle and all matter other than sorghum that remains on top of the 5/64 triangular-hole sieve according to procedures prescribed in FGIS instructions.

(g) *Heat-damaged kernels.* Kernels, pieces of sorghum kernels, and other grains that are materially discolored and damaged by heat.

(h) *Nongrain sorghum.* Seeds of broomcorn, Johnson-grass, *Sorghum almum* Parodi, and sudangrass; and seeds of Sorghum bicolor (L.) Moench that appear atypical of grain sorghum.

(i) *Pericarp.* The pericarp is the outer layers of the sorghum grain and is fused to the seedcoat.

(j) *Sieves*—(1) *1.98 mm (5/64 (0.0781) inches) triangular-hole sieve.* A metal sieve 0.81 mm (0.032 inches) thick with equilateral triangular perforations the inscribed circles of which are 1.98 mm (0.0781 inches) in diameter.

(2) *0.99 mm (2 1/2 /64 (0.0391) inches) round-hole sieve.* A metal sieve 0.81 mm (0.032 inch) thick with round holes 0.99 mm (0.0391 inches) in diameter.

[52 FR 24418, June 30, 1987, as amended at 52 FR 24437, June 30, 1987; 52 FR 28534, July 31, 1987; 57 FR 58971, Dec. 14, 1992; 72 FR 39732, July 20, 2007]

PRINCIPLES GOVERNING THE APPLICATION OF STANDARDS

§810.1403 Basis of determination.

Each determination of broken kernels and foreign material is made on the basis of the grain when free from dockage. Each determination of class, damaged kernels, heat-damaged kernels, and stones is made on the basis of the grain when free from dockage and that portion of the broken kernels, and foreign material that will pass through a 1.98 mm (5/64 inches) triangular-hole sieve. Other determinations not specifically provided for in the general provisions are made on the basis of the grain as a whole except the determination of odor is made on either the basis of the grain as a whole or the grain when free from dockage, broken kernels, and foreign material removed by the 1.98 mm (5/64 inches) triangular-hole sieve.

[57 FR 58971, Dec. 14, 1992]

GRADES AND GRADE REQUIREMENTS

§810.1404 Grades and grade requirements for sorghum.

Grading factors	Grades U.S. Nos.[1]			
	1	2	3	4
Minimum pound limits of				
Test weight per bushel	57.0	55.0	53.0	51.0
Maximum percent limits of				
Damaged kernels:				
Heat (part of total)	0.2	0.5	1.0	3.0
Total	2.0	5.0	10.0	15.0
Broken kernels and foreign material:				
Foreign material (part of total)	1.0	2.0	3.0	4.0
Total	3.0	6.0	8.0	10.0
Maximum count limits of				
Other material:				
Animal filth	9	9	9	9
Castor beans	1	1	1	1
Crotalaria seeds	2	2	2	2
Glass	1	1	1	1

Grading factors	Grades U.S. Nos. [1]			
	1	2	3	4
Stones [2]	7	7	7	7
Unknown foreign substance	3	3	3	3
Cockleburs	7	7	7	7
Total [3]	10	10	10	10

U.S. Sample grade is sorghum that:
(a) Does not meet the requirements for U.S. Nos. 1, 2, 3, or 4; or
(b) Has a musty, sour, or commercially objectionable foreign odor (except smut odor); or
(c) Is badly weathered, heating, or distinctly low quality.

[1] Sorghum which is distinctly discolored shall not grade higher than U.S. No. 3.
[2] Aggregate weight of stones must also exceed 0.2 percent of the sample weight.
[3] Includes any combination of animal filth, castor beans, crotalaria seeds, glass, stones, unknown foreign substance or cockleburs.

[72 FR 39733, July 20, 2007]

SPECIAL GRADES AND SPECIAL GRADE
REQUIREMENTS

§ 810.1405 Special grades and special grade requirements.

Smutty sorghum. Sorghum that has kernels covered with smut spores to give a smutty appearance in mass, or that contains 20 or more smut balls in 100 grams of sorghum.

[52 FR 24418, June 30, 1987, as amended at 52 FR 24441, June 30, 1987]

Subpart J—United States Standards for Soybeans

TERMS DEFINED

§ 810.1601 Definition of soybeans.

Grain that consists of 50 percent or more of whole or broken soybeans (*Glycine max* (L.) Merr.) that will not pass through an 8/64 round-hole sieve and not more than 10.0 percent of other grains for which standards have been established under the United States Grain Standards Act.

§ 810.1602 Definition of other terms.

(a) *Classes.* There are two classes for soybeans: Yellow soybeans and Mixed soybeans.

(1) *Yellow soybeans.* Soybeans that have yellow or green seed coats and which in cross section, are yellow or have a yellow tinge, and may include not more than 10.0 percent of soybeans of other colors.

(2) *Mixed soybeans.* Soybeans that do not meet the requirements of the class Yellow soybeans.

(b) *Damaged kernels.* Soybeans and pieces of soybeans that are badly ground-damaged, badly weather-damaged, diseased, frost-damaged, germ-damaged, heat-damaged, insect-bored, mold-damaged, sprout-damaged, stinkbug-stung, or otherwise materially damaged. Stinkbug-stung kernels are considered damaged kernels at the rate of one-fourth of the actual percentage of the stung kernels.

(c) *Foreign material.* All matter that passes through an 8/64 round-hole sieve and all matter other than soybeans remaining in the sieved sample after sieving according to procedures prescribed in FGIS instructions.

(d) *Heat-damaged kernels.* Soybeans and pieces of soybeans that are materially discolored and damaged by heat.

(e) *Purple mottled or stained.* Soybeans that are discolored by the growth of a fungus; or by dirt; or by a dirt-like substance(s) including nontoxic inoculants; or by other nontoxic substances.

(f) *Sieve—8/64 round-hole sieve.* A metal sieve 0.032 inch thick perforated with round holes 0.125 (8/64) inch in diameter.

(g) *Soybeans of other colors.* Soybeans that have green, black, brown, or bicolored seed coats. Soybeans that have green seed coats will also be green in cross section. Bicolored soybeans will have seed coats of two colors, one of which is brown or black, and the brown or black color covers 50 percent of the seed coats. The hilum of a soybean is not considered a part of the seed coat for this determination.

(h) *Splits.* Soybeans with more than ¼ of the bean removed and that are not damaged.

PRINCIPLES GOVERNING THE
APPLICATION OF STANDARDS

§ 810.1603 Basis of determination.

Each determination of class, heat-damaged kernels, damaged kernels, splits, and soybeans of other colors is made on the basis of the grain when free from foreign material. Other determinations not specifically provided for under the general provisions are made on the basis of the grain as a whole.

GRADES AND GRADE REQUIREMENTS

§810.1604 Grades and grade requirements for soybeans.

Grading factors	Grades U.S. Nos.			
	1	2	3	4
	Maximum percent limits of:			
Damaged kernels:				
Heat (part of total)	0.2	0.5	1.0	3.0
Total	2.0	3.0	5.0	8.0
Foreign material	1.0	2.0	3.0	5.0
Splits	10.0	20.0	30.0	40.0
Soybeans of other colors: [1]	1.0	2.0	5.0	10.0
	Maximum count limits of:			
Other material:				
Animal filth	9	9	9	9
Caster beans	1	1	1	1
Crotalaria seeds	2	2	2	2
Glass	0	0	0	0
Stones [2]	3	3	3	3
Unknown foreign substance	3	3	3	3
Total [3]	10	10	10	10

U.S. Sample grade are Soybeans that:
(a) Do not meet the requirements for U.S. Nos. 1, 2, 3, or 4; or
(b) Have a musty, sour, or commercially objectionable foreign odor (except smut or garlic odor); or
(c) Are heating or of distinctly low quality.

[1] Disregard for Mixed soybeans.
[2] In addition to the maximum count limit, stones must exceed 0.1 percent of the sample weight.
[3] Includes any combination of animal filth, castor beans, crotalaria seeds, glass, stones, and unknown substances. The weight of stones is not applicable for total other material.

[71 FR 52406, Sept. 6, 2006]

SPECIAL GRADES AND SPECIAL GRADE
REQUIREMENTS

§810.1605 Special grades and special grade requirements.

(a) *Garlicky soybeans.* Soybeans that contain 5 or more green garlic bulblets or an equivalent quantity of dry or partly dry bulblets in a 1,000 gram portion.

(b) *Purple mottled or stained soybeans.* Soybeans with pink or purple seed coats as determined on a portion of approximately 400 grams with the use of an FGIS Interpretive Line Photograph.

[52 FR 24418, June 30, 1987, as amended at 52 FR 24441, June 30, 1987; 59 FR 10573, Mar. 7, 1994]

Subpart K—United States Standards for Sunflower Seed

TERMS DEFINED

§810.1801 Definition of sunflower seed.

Grain that, before the removal of foreign material, consists of 50.0 percent or more of cultivated sunflower seed (*Helianthus annuus* L.) and not more than 10.0 percent of other grains for which standards have been established under the United States Grain Standards Act.

§810.1802 Definition of other terms.

(a) *Cultivated sunflower seed.* Sunflower seed grown for oil content. The term seed in this and other definitions related to sunflower seed refers to both the kernel and hull which is a fruit or achene.

(b) *Damaged sunflower seed.* Seed and pieces of sunflower seed that are badly ground-damaged, badly weather-damaged, diseased, frost-damaged, heat-

583

damaged, mold-damaged, sprout-damaged, or otherwise materially damaged.

(c) *Dehulled seed.* Sunflower seed that has the hull completely removed from the sunflower kernel.

(d) *Foreign material.* All matter other than whole sunflower seeds containing kernels that can be removed from the original sample by use of an approved device and by handpicking a portion of the sample according to procedures prescribed in FGIS instructions.

(e) *Heat-damaged sunflower seed.* Seed and pieces of sunflower seed that are materially discolored and damaged by heat.

(f) *Hull (Husk).* The ovary wall of the sunflower seed.

(g) *Kernel.* The interior contents of the sunflower seed that are surrounded by the hull.

PRINCIPLES GOVERNING THE
APPLICATION OF STANDARDS

§ 810.1803 Basis of determination.

Each determination of heat-damaged kernels, damaged kernels, test weight per bushel, and dehulled seed is made on the basis of the grain when free from foreign material. Other determinations not specifically provided for in the general provisions are made on the basis of the grain as a whole, except the determination of odor is made on either the basis of the grain as a whole or the grain when free from foreign material.

GRADES AND GRADE REQUIREMENTS

§ 810.1804 Grades and grade requirements for sunflower seed.

Grade	Minimum test weight per bushel (pounds)	Maximum limits of—		
		Damaged Sunflower Seed		Dehulled seed (percent)
		Heat Damaged (percent)	Total (Percent)	
U.S. No. 1	25.0	0.5	5.0	5.0
U.S. No. 2	25.0	1.0	10.0	5.0

U.S. Sample grade—
    U.S. Sample grade is sunflower seed that:
      (a) Does not meet the requirements for the grades U.S. Nos. 1 or 2; or
      (b) Contains 8 or more stones which have an aggregate weight in excess of 0.20 percent of the sample weight, 2 or more pieces of glass, 3 or more crotalaria seeds (*Crotalaria* spp.), 2 or more castor beans (*Ricinus communis* L.), 4 or more particles of an unknown foreign substance(s), or a commonly recognized harmful or toxic substance(s), 10 or more rodent pellets, bird droppings, or equivalent quantity of other animal filth per 600 grams of sunflower seed; or
      (c) Has a musty, sour, or commercially objectionable foreign odor; or
      (d) Is heating or otherwise of distinctly low quality.

Subpart L—United States Standards for Triticale

TERMS DEFINED

§ 810.2001 Definition of triticale.

Grain that, before the removal of dockage, consists of 50 percent or more of triticale (*X Triticosecale* Wittmack) and not more than 10 percent of other grains for which standards have been established under the United States Grain Standards Act and that, after the removal of dockage, contains 50 percent or more of whole triticale.

§ 810.2002 Definition of other terms.

(a) *Damaged kernels.* Kernels, pieces of triticale kernels, and other grains that are badly ground-damaged, badly weather-damaged, diseased, frost-damaged, germ-damaged, heat-damaged, insect-bored, mold-damaged, sprout-damaged, or otherwise materially damaged.

(b) *Defects.* Damaged kernels, foreign material, and shrunken and broken kernels. The sum of these three factors may not exceed the limit for the factor defects for each numerical grade.

(c) *Dockage.* All matter other than triticale that can be removed from the original sample by use of an approved device according to procedures prescribed in FGIS instructions. Also, underdeveloped, shriveled, and small pieces of triticale kernels removed in properly separating the material other than triticale and that cannot be recovered by properly rescreening or recleaning.

(d) *Foreign material.* All matter other than triticale.

(e) *Heat-damaged kernels.* Kernels, pieces of triticale kernels, and other grains that are materially discolored and damaged by heat.

(f) *Other grains.* Barley, corn, cultivated buckwheat, einkorn, emmer, flaxseed, guar, hull-less barley, nongrain sorghum, oats, Polish wheat, popcorn, poulard wheat, rice, rye, safflower, sorghum, soybeans, spelt, sunflower seed, sweet corn, wheat, and wild oats.

(g) *Shrunken and broken kernels.* All matter that passes through a 0.064 × 3/8 oblong-hole sieve after sieving according to procedures prescribed in FGIS instructions.

(h) *Sieve—0.064 × ⅜ oblong-hole sieve.* A metal sieve 0.032 inch thick with oblong perforations 0.064 inch by 0.375 (3/8) inch.

[52 FR 24418, June 30, 1987; 52 FR 28534, July 31, 1987]

PRINCIPLES GOVERNING THE
APPLICATION OF STANDARDS

§ 810.2003 Basis of determination.

Each determination of heat-damaged kernels, damaged kernels, material other than wheat or rye, and foreign material (total) is made on the basis of the grain when free from dockage and shrunken and broken kernels. Other determinations not specifically provided for under the general provisions are made on the basis of the grain when free from dockage except the determination of odor is made on either the basis of the grain as a whole or the grain when free from dockage.

GRADES AND GRADE REQUIREMENTS

§ 810.2004 Grades and grade requirements for triticale.

Grade	Minimum test weight per bushel (pounds)	Maximum limits of—						
		Damaged Kernels		Foreign material			Shrunken and broken kernels (percent)	Defects [3] (percent)
		Heat damaged (percent)	Total [1] (percent)	Material other than wheat or rye (percent)	Total [2] (percent)			
U.S. No. 1	48.0	0.2	2.0	1.0	2.0	5.0	5.0
U.S. No. 2	45.0	0.2	4.0	2.0	4.0	8.0	8.0
U.S. No. 3	43.0	0.5	8.0	3.0	7.0	12.0	12.0
U.S. No. 4	41.0	3.0	15.0	4.0	10.0	20.0	20.0

U.S. Sample grade—
  U.S. Sample grade is triticale that:
    (a) Does not meet the requirements for the grades U.S. Nos. 1, 2, 3, or 4; or
    (b) Contains 8 or more stones or any number of stones which have an aggregate weight in excess of 0.2 percent of the sample weight, 2 or more pieces of glass, 3 or more crotalaria seeds (*Crotalaria* spp.), 2 or more castor beans (*Ricinus communis* L.), 4 or more particles of an unknown foreign substance(s) or a commonly recognized harmful or toxic substance(s), 2 or more rodent pellets, bird droppings, or equivalent quantity of other animal filth per 1⅛ to 1¼ quarts of triticale; or
    (c) Has a musty, sour, or commercially objectionable foreign odor (except smut or garlic odor); or
    (d) Is heating or otherwise of distinctly low quality.

[1] Includes heat-damaged kernels.
[2] Includes material other than wheat or rye.
[3] Defects include damaged kernels (total), foreign material (total) and shrunken and broken kernels. The sum of these three factors may not exceed the limit for defects for each numerical grade.

[52 FR 24418, June 30, 1987; 52 FR 28534, July 31, 1987]

SPECIAL GRADES AND SPECIAL GRADE
REQUIREMENTS

§ 810.2005 Special grades and special grade requirements.

(a) *Ergoty triticale.* Triticale that contains more than 0.10 percent of ergot.

(b) *Garlicky triticale.* Triticale that contains in a 1,000 gram portion more than six green garlic bulblets or an equivalent quantity of dry or partly dry bulblets.

(c) *Light garlicky triticale.* Triticale that contains in a 1,000 gram portion two or more, but not more than six, green garlic bulblets or an equivalent quantity of dry or partly dry bulblets.

(d) *Light smutty triticale.* Triticale that has an unmistakable odor of smut, or that contains in a 250 gram portion smut balls, portions of smut balls, or spores of smut in excess of a quantity equal to 14 smut balls, but not in excess of a quantity equal to 30 smut balls of average size.

(e) *Smutty triticale.* Triticale that contains in a 250 gram portion smut balls, portions of smut balls, or spores of smut in excess of a quantity equal to 30 smut balls of average size.

[52 FR 24418, June 30, 1987, as amended at 52 FR 24441, June 30, 1987]

Subpart M—United States Standards for Wheat

TERMS DEFINED

§ 810.2201 Definition of wheat.

Grain that, before the removal of dockage, consists of 50 percent or more common wheat (*Triticum aestivum* L.), club wheat (*T. compactum* Host.), and durum wheat (*T. durum* Desf.) and not more than 10 percent of other grains for which standards have been established under the United States Grain Standards Act and that, after the removal of the dockage, contains 50 percent or more of whole kernels of one or more of these wheats.

§ 810.2202 Definition of other terms.

(a) *Classes.* There are eight classes for wheat: Durum wheat, Hard Red Spring wheat, Hard Red Winter wheat, Soft Red Winter wheat, Hard White wheat, Soft White wheat, Unclassed wheat, and Mixed wheat.

(1) *Durum wheat.* All varieties of white (amber) durum wheat. This class is divided into the following three subclasses:

(i) *Hard Amber Durum wheat.* Durum wheat with 75 percent or more of hard and vitreous kernels of amber color.

(ii) *Amber Durum wheat.* Durum wheat with 60 percent or more but less than 75 percent of hard and vitreous kernels of amber color.

(iii) *Durum wheat.* Durum wheat with less than 60 percent of hard vitreous kernels of amber color.

(2) *Hard Red Spring wheat.* All varieties of Hard Red Spring wheat. This class shall be divided into the following three subclasses.

(i) *Dark Northern Spring wheat.* Hard Red Spring wheat with 75 percent or more of dark, hard, and vitreous kernels.

(ii) *Northern Spring wheat.* Hard Red Spring wheat with 25 percent or more but less than 75 percent of dark, hard, and vitreous kernels.

(iii) *Red Spring wheat.* Hard Red Spring wheat with less than 25 percent of dark, hard, and vitreous kernels.

(3) *Hard Red Winter wheat.* All varieties of Hard Red Winter wheat. There are no subclasses in this class.

(4) *Soft Red Winter wheat.* All varieties of Soft Red Winter wheat. There are no subclasses in this class.

(5) *Hard White wheat.* All hard endosperm white wheat varieties. There are no subclasses in this class.

(6) *Soft White wheat.* All soft endosperm white wheat varieties. This class is divided into the following three subclasses:

(i) *Soft White wheat.* Soft endosperm white wheat varieties which contain not more than 10 percent of white club wheat.

(ii) *White Club wheat.* Soft endosperm white club wheat varieties containing not more than 10 percent of other soft white wheats.

(iii) *Western White wheat.* Soft White wheat containing more than 10 percent of white club wheat and more than 10 percent of other soft white wheats.

(7) *Unclassed wheat.* Any variety of wheat that is not classifiable under other criteria provided in the wheat standards. There are no subclasses in this class. This class includes any

wheat which is other than red or white in color.

(8) *Mixed wheat.* Any mixture of wheat that consists of less than 90 percent of one class and more than 10 percent of one other class, or a combination of classes that meet the definition of wheat.

(b) *Contrasting Classes.* Contrasting classes are:

(1) Durum wheat, Soft White wheat, and Unclassed wheat in the classes Hard Red Spring wheat and Hard Red Winter wheat.

(2) Hard Red Spring wheat, Hard Red Winter wheat, Hard White wheat, Soft Red Winter wheat, Soft White wheat, and Unclassed wheat in the class Durum wheat.

(3) Durum wheat and Unclassed wheat in the class Soft Red Winter wheat.

(4) Durum wheat, Hard Red Spring wheat, Hard Red Winter wheat, Soft Red Winter wheat, and Unclassed wheat in the class Soft White wheat.

(5) Durum wheat, Soft Red Winter wheat, and Unclassed wheat in the class Hard White wheat.

(c) *Damaged kernels.* Kernels, pieces of wheat kernels, and other grains that are badly ground-damaged, badly weather-damaged, diseased, frost-damaged, germ-damaged, heat-damaged, insect-bored, mold-damaged, sprout-damaged, or otherwise materially damaged.

(d) *Defects.* Damaged kernels, foreign material, and shrunken and broken kernels. The sum of these three factors may not exceed the limit for the factor defects for each numerical grade.

(e) *Dockage.* All matter other than wheat that can be removed from the original sample by use of an approved device according to procedures prescribed in FGIS instructions. Also, underdeveloped, shriveled, and small pieces of wheat kernels removed in properly separating the material other than wheat and that cannot be recovered by properly rescreening or recleaning.

(f) *Foreign material.* All matter other than wheat that remains in the sample after the removal of dockage and shrunken and broken kernels.

(g) *Heat-damaged kernels.* Kernels, pieces of wheat kernels, and other grains that are materially discolored and damaged by heat which remain in the sample after the removal of dockage and shrunken and broken kernels.

(h) *Other grains.* Barley, corn, cultivated buckwheat, einkorn, emmer, flaxseed, guar, hull-less barley, nongrain sorghum, oats, Polish wheat, popcorn, poulard wheat, rice, rye, safflower, sorghum, soybeans, spelt, sunflower seed, sweet corn, triticale, and wild oats.

(i) *Shrunken and broken kernels.* All matter that passes through a 0.064 × ⅜ oblong-hole sieve after sieving according to procedures prescribed in the FGIS instructions.

(j) *Sieve—0.064 × ⅜ oblong-hole sieve.* A metal sieve 0.032 inch thick with oblong perforations 0.064 inch by 0.375 (⅜) inch.

[52 FR 24418, June 30, 1987, as amended at 54 FR 48736, Nov. 27, 1989; 57 FR 58966, Dec. 14, 1992; 71 FR 8235, Feb. 18, 2006; 78 FR 27858, May 13, 2013]

PRINCIPLES GOVERNING THE APPLICATION OF STANDARDS

§ 810.2203   Basis of determination.

Each determination of heat-damaged kernels, damaged kernels, foreign material, wheat of other classes, contrasting classes, and subclasses is made on the basis of the grain when free from dockage and shrunken and broken kernels. Other determinations not specifically provided for under the general provisions are made on the basis of the grain when free from dockage, except the determination of odor is made on either the basis of the grain as a whole or the grain when free from dockage.

[52 FR 24418, June 30, 1987; 52 FR 28534, July 31, 1987]

GRADES AND GRADE REQUIREMENTS

§ 810.2204   Grades and grade requirements for wheat.

(a) Grades and grade requirements for all classes of wheat, except Mixed wheat.

GRADES AND GRADE REQUIREMENTS

Grading factors	Grades U.S. Nos.				
	1	2	3	4	5
Minimum pound limits of:					
Test weight per bushel:					
Hard Red Spring wheat or White Club wheat	58.0	57.0	55.0	53.0	50.0
All other classes and subclasses	60.0	58.0	56.0	54.0	51.0
Maximum percent limits of:					
Defects:					
Damaged kernels.					
Heat (part of total)	0.2	0.2	0.5	1.0	3.0
Total	2.0	4.0	7.0	10.0	15.0
Foreign material	0.4	0.7	1.3	3.0	5.0
Shrunken and broken kernels	3.0	5.0	8.0	12.0	20.0
Total [1]	3.0	5.0	8.0	12.0	20.0
Wheat of other classes: [2].					
Contrasting classes	1.0	2.0	3.0	10.0	10.0
Total [3]	3.0	5.0	10.0	10.0	10.0
Stones	0.1	0.1	0.1	0.1	0.1
Maximum count limits of:					
Other material in one kilogram:					
Animal filth	1	1	1	1	1
Castor beans	1	1	1	1	1
Crotalaria seeds	2	2	2	2	2
Glass	0	0	0	0	0
Stones	3	3	3	3	3
Unknown foreign substances	3	3	3	3	3
Total [4]	4	4	4	4	4
Insect-damaged kernels in 100 grams	31	31	31	31	31

U.S. Sample grade is Wheat that:
(a) Does not meet the requirements for U.S. Nos. 1, 2, 3, 4, or 5; or
(b) Has a musty, sour, or commercially objectionable foreign odor (except smut or garlic odor); or
(c) Is heating or of distinctly low quality.

[1] Includes damaged kernels (total), foreign material, shrunken and broken kernels.
[2] Unclassed wheat of any grade may contain not more than 10.0 percent of wheat of other classes.
[3] Includes contrasting classes.
[4] Includes any combination of animal filth, castor beans, crotalaria seeds, glass, stones, or unknown foreign substance.

(b) *Grades and grade requirements for Mixed wheat.* Mixed wheat is graded according to the U.S. numerical and U.S. Sample grade requirements of the class of wheat that predominates in the mixture, except that the factor wheat of other classes is disregarded.

[52 FR 24418, June 30, 1987, as amended at 52 FR 24442, June 30, 1987; 57 FR 58966, Dec. 14, 1992; 71 FR 8235, Feb. 18, 2006]

SPECIAL GRADES AND SPECIAL GRADE REQUIREMENTS

§ 810.2205 Special grades and special grade requirements.

(a) *Ergoty wheat.* Wheat that contains more than 0.05 percent of ergot.

(b) *Garlicky wheat.* Wheat that contains in a 1,000 gram portion more than two green garlic bulblets or an equivalent quantity of dry or partly dry bulblets.

(c) *Light smutty wheat.* Wheat that has an unmistakable odor of smut, or which contains, in a 250-gram portion, smut balls, portions of smut balls, or spores of smut in excess of a quantity equal to 5 smut balls, but not in excess of a quantity equal to 30 smut balls of average size.

(d) *Smutty wheat.* Wheat that contains, in a 250 gram portion, smut balls, portions of smut balls, or spores of smut in excess of a quantity equal to 30 smut balls of average size.

(e) *Treated wheat.* Wheat that has been scoured, limed, washed, sulfured, or treated in such a manner that the true quality is not reflected by either the numerical grades or the U.S. Sample grade designation alone.

[52 FR 24418, June 30, 1987, as amended at 52 FR 24442, June 30, 1987; 57 FR 58967, Dec. 14, 1992]

PART 868—GENERAL REGULATIONS AND STANDARDS FOR CERTAIN AGRICULTURAL COMMODITIES

Subpart A—Regulations

DEFINITIONS

589

AUTHORITY: 7 U.S.C. 1621–1627.

Subpart A—Regulations

SOURCE: 53 FR 3722, Feb. 9, 1988, unless otherwise noted. Redesignated at 60 FR 16364, Mar. 30, 1995.

DEFINITIONS

§ 868.1 Meaning of terms.

(a) *Construction.* Words used in the singular form are considered to imply the plural and vice versa, as appropriate.

(b) *Definitions.* For the purpose of these regulations, unless the context requires otherwise, the following terms have the meanings given for them in this paragraph.

(1) *Act.* The Agricultural Marketing Act of 1946, as amended (secs. 202–208, 60 Stat. 1087, as amended, 7 U.S.C. 1621 *et seq.*).

(2) *Administrator.* The Administrator of the Agricultural Marketing Service, or any person to whom the Administrator's authority has been delegated.

(3) *Appeal inspection service.* A review by the Service of the result(s) of an original inspection or retest inspection service.

(4) *Applicant.* An interested person who requests any inspection service with respect to a commodity.

(5) *Authorized inspector.* A Department employee authorized by the Administrator to inspect a commodity in accordance with the Act, regulations, standards, and instructions.

(6) *Board appeal inspection service.* A review by the Board of Appeals and Review of the result(s) of an original inspection or appeal inspection service on graded commodities.

(7) *Board of Appeals and Review or Board.* The Board of Appeals and Review of the Service that performs Board appeal inspection services.

(8) *Business day.* The established field office working hours, any Monday through Friday that is not a holiday, or the working hours and days established by a cooperator.

(9) *Carrier.* A truck, trailer, truck/trailer(s) combination, railroad car, barge, ship, or other container used to transport bulk, sacked, or packaged commodity.

(10) *Commodity.* Agricultural commodities and products thereof that the Secretary has assigned to the Service for inspection under the Act, including but not limited to dry beans, grain, hops, lentils, oilseeds, dry peas, split peas, and rice.

(11) *Continuous inspection.* The conduct of inspection services in an approved plant where one or more official inspection personnel are present during the processing of a commodity to make in-process examinations of the preparation, processing, packing, and warehousing of the commodity and to determine compliance with applicable sanitation requirements.

(12) *Contract service.* Any service performed under a contract between an applicant and the Service.

(13) *Contractor.* Any person who enters into a contract with the Service or with a cooperator to perform specified inspection services.

(14) *Cooperator.* An agency or department of the Federal Government which has an interagency agreement or State agency which has a reimbursable agreement with the Service.

(15) *Cooperator inspection service.* The inspection service provided by a cooperator under the regulations. Under this service, inspection certificates are issued by the cooperator and all fees and charges are collected by the cooperator, except as provided in the agreement.

(16) *Department.* The United States Department of Agriculture.

(17) *Factor.* A quantified physical or chemical property identified in official standards, specifications, abstracts, contracts, or other documents whose measurement describes a specific quality of a commodity.

(18) *Field office.* An office of the Service designated to perform, monitor, or supervise inspection services.

(19) *Grade.* A grade designating a level of quality as defined in the commodity standards promulgated pursuant to the Act.

(20) *Graded commodity.* Commodities for which the Service has promulgated Standards under the Act and commodities which are tested by the Service at a field office or by a cooperator for specific physical factors using approved equipment and an inspector's interpretation of visual conditions.

(21) *Holiday.* The legal public holidays specified in paragraph (a) of section 6103, title 5, of the United States Code (5 U.S.C. 6103(a)) and any other day declared to be a holiday by Federal Statute or Executive Order. Under section 6103 and Executive Order 10357, as amended, if the specified legal public holiday falls on a Saturday, the preceding Friday shall be considered to be the holiday, or if the specified legal public holiday falls on a Sunday, the following Monday shall be considered to be the holiday.

(22) *Inspection certificate.* A written or printed official document which is approved by the Service and which shows the results of an inspection service performed under the Act.

(23) *Inspection service.* (i) Applying such tests and making examinations of a commodity and records by official personnel as may be necessary to determine the kind, class, grade, other quality designation, the quantity, or condition of commodity; performing condition of container, carrier stowage examinations; and any other services as related to commodities, as necessary; and (ii) issuing an inspection certificate.

(24) *Instructions.* The Notices, Instructions, Handbooks, and other directives issued by the Service.

(25) *Interagency agreement.* An agreement between the Service and other agencies or departments of the Federal Government to conduct commodity inspection services as authorized in the Act.

(26) *Interested person.* Any person having a contract or other financial interest in a commodity as the owner, seller, purchaser, warehouseman, carrier, or otherwise.

(27) *Licensee.* Any person licensed by the Service.

(28) *Nongraded commodity.* Nonprocessed commodities which are chemically tested for factors not included in the Standards under the Act or the U.S. Grain Standards Act (7 U.S.C. 71 *et seq.*) and processed commodities.

(29) *Nonregular workday.* Any Sunday or holiday.

(30) *Official inspector.* Any official personnel who performs, monitors, or supervises the performance of inspection service and certifies the results of inspection of the commodity.

(31) *Official personnel.* Any authorized Department employee or person licensed by the Administrator to perform all or specified functions under the Act.

(32) *Official sampler.* Any official personnel who performs, monitors, or supervises the performance of sampling of a commodity.

(33) *Official technician.* Any official personnel who performs, monitors, or supervises the performance of specified inspection services and certifies the results thereof, other than certifying the grade of a commodity.

(34) *Origin.* The geographical area or place where the commodity is grown.

(35) *Original inspection service.* An initial inspection of a community.

(36) *Person.* Any individual, partnership, association, corporation, or other business entity.

(37) *Plant.* The premises, buildings, structure, and equipment (including but not limited to machines, utensils, vehicles, and fixtures located in or about the premises) used or employed in the preparation, processing, handling, transporting, and storage of commodities.

(38) *Regular workday.* Any Monday through Saturday that is not a holiday.

(39) *Regulations.* The regulations in this part.

(40) *Reimbursable agreement.* An agreement between the Service and State agencies to conduct commodity inspection services authorized pursuant to the Act.

(41) *Retest inspection service.* To test, using the same laboratory procedures, a factor(s) of nongraded commodities previously tested.

(42) *Secretary.* The Secretary of Agriculture of the United States or any person to whom the Secretary's authority has been delegated.

(43) *Service.* The Federal Grain Inspection Service of the Agricultural Marketing Service, of the United States Department of Agriculture.

(44) *Service representative.* An employee authorized by the Service or a person licensed by the Administrator.

(45) *Specification.* A document which clearly and accurately describes the essential and technical requirements for items, materials, or services including requested inspection procedures.

(46) *Standards.* The commodity standards in this part that describe the physical and biological condition of a commodity at the time of inspection.

(47) *Submitted sample.* A sample submitted by or for an applicant for inspection.

(48) *Test.* A procedure to measure a factor using specialized laboratory equipment involving the application of established scientific principles and laboratory procedures.

[53 FR 3722, Feb. 9, 1988, as amended at 60 FR 5835, Jan. 31, 1995. Redesignated at 60 FR 16364, Mar. 30, 1995, as amended at 63 FR 29531, June 1, 1998; 70 FR 69250, Nov. 15, 2005]

ADMINISTRATION

§868.5 Administrator.

The Administrator, under the authority delegated by the Secretary, is charged with administering the programs and functions authorized under the Act and the regulations concerning those commodities assigned by the Secretary to the Service.

§868.6 Nondiscrimination—policy and provisions.

In implementing, administering, and enforcing the Act and the regulations, standards, and instructions, it is the policy of the Service to promote adherence to the provisions of the Civil Rights Act of 1964 (42 U.S.C. 2000a *et seq.*).

§868.7 Procedures for establishing regulations and standards.

Notice of proposals to prescribe, amend, or revoke regulations and standards shall be published in accordance with applicable provisions of the Administrative Procedures Act (5 U.S.C. 551 *et seq.*). Any interested person desiring to file a petition for the issuance, amendment, or revocation of regulations or standards may do so in accordance with 7 CFR 1.28 of the regulations of the Office of the Secretary of Agriculture.

§868.8 Complaints and reports of alleged violations.

(a) *General.* Except as provided in paragraph (b) of this section, complaints and reports of violations involving the Act or the regulations, standards, and instructions issued under the Act should be filed with the Service in accordance with 7 CFR 1.133 of the regulations of the Office of the Secretary of Agriculture and these regulations and the instructions.

(b) *Retest inspection and appeal inspection service.* Complaints involving the results of inspection services shall, to the extent practicable, be submitted as requests for retest inspection, appeal inspection, or Board appeal inspection services as set forth in these regulations.

(Approved by the Office of Management and Budget under control number 0580–0011)

§868.9 Provisions for hearings.

Opportunities shall be provided for hearings either in accordance with the Rules of Practice Governing Formal Adjudicatory Proceedings Instituted by the Secretary under Various Statutes (7 CFR part 1, subpart H) or in accordance with FGIS procedures as appropriate.

§868.10 Information about the Service, Act, and regulations.

Information about the Service, Act, regulations, standards, rules of practice, instructions, and other matters related to the inspection of commodities may be obtained by telephoning or writing the U.S. Department of Agriculture, Federal Grain Inspection Service, P.O. Box 96454, Washington, DC 20090–6454, or any field office or cooperator.

§868.11 Public information.

(a) *General.* This section is issued in accordance with §§1.1 through 1.23 of the regulations of the Secretary in part 1, subpart A, of subtitle A of title 7 (7 CFR 1.1 through 1.23), and appendix A thereto, implementing the Freedom of Information Act (5 U.S.C. 552). The Secretary's regulations, as implemented by this section, govern the availability of records of the Service to the public.

(b) *Public inspection and copying.* Materials maintained by the Service, including those described in 7 CFR 1.5, will be made available, upon a request which has not been denied, for public inspection and copying at the U.S. Department of Agriculture, Federal Grain Inspection Service, 1400 Independence Avenue SW., Washington, DC 20250. The public may request access to these materials 8:00 a.m.–4:30 p.m. Monday through Friday except for holidays.

(c) *Indexes.* The Service shall maintain an index of all material required to be made available in 7 CFR 1.5. Copies of these indexes will be maintained at the location given in paragraph (b) of this section. Notice is hereby given that quarterly publication of these indexes is unnecessary and impracticable because the material is voluminous and does not change often enough to justify the expense of quarterly publication. However, upon specific request, copies of any index will be provided at a cost not to exceed the direct cost of duplication.

(d) *Requests for records.* Requests for records under 5 U.S.C. 552(a)(3) shall be made in accordance with 7 CFR 1.6 and shall be addressed as follows: Office of the Administrator, Federal Grain Inspection Service, FOIA Request, U.S. Department of Agriculture, P.O. Box 96454, Washington, DC 20090–6454.

(e) *FOIA Appeals.* Any person whose request, under paragraph (d) of this section, is denied shall have the right to appeal such denial in accordance with 7 CFR 1.13. Appeals shall be addressed to the Administrator, Federal Grain Inspection Service, FOIA Appeal, U.S. Department of Agriculture, P.O. Box 96454, Washington, DC 20090–6454.

(f) *Disclosure of information.* FGIS employees or persons acting for FGIS under the Act shall not, without the consent of the applicant, divulge or make known in any manner any facts or information acquired pursuant to the Act, regulations, or instructions except as authorized by the Administrator, by a court of competent jurisdiction, or otherwise by law.

[53 FR 3722, Feb. 9, 1988, as amended 54 FR 5923, Feb. 7, 1989. Redesignated at 60 FR 16364, Mar. 30, 1995]

§ 868.12   Identification.

All official personnel shall have in their possession and present upon request, while on duty, the means of identification furnished to them by the Department.

§ 868.13   Regulations not applicable for certain purposes.

These regulations do not apply to the inspection of grain under the United States Grain Standards Act, as amended (7 U.S.C. 71 *et seq.*) or the inspection of commodities under the United States Warehouse Act, as amended (7 U.S.C. 241 *et seq.*).

CONDITIONS FOR OBTAINING OR WITHHOLDING SERVICE

§ 868.20   Availability of services.

(a) *Original inspection service.* Original inspection services are available according to this section and §§ 868.40 through 868.44.

(b) *Retest inspection and appeal inspection services.* Retest inspection, appeal inspection, and Board appeal inspection services are available according to §§ 868.50 through 868.52 and §§ 868.60 through 868.63.

(c) *Proof of authorization.* A cooperator or the Service may request satisfactory proof that an applicant is an interested person or their authorized agent.

[53 FR 3722, Feb. 9, 1988. Redesignated and amended at 60 FR 16364, Mar. 30, 1995]

§ 868.21   Requirements for obtaining service.

(a) *Consent and agreement by applicant.* In submitting a request for inspection service, the applicant and the owner of the commodity consent to the requirements specified in paragraphs (b) through (j) of this section.

(b) *Written confirmation.* Verbal requests for inspection service shall be confirmed in writing upon request. Each written request shall be made in English and shall include:

(1) The date filed;

(2) The identification, quantity, and location of the commodity;

(3) The type of service(s) requested;

(4) The name and mailing address of the applicant and, if made by an authorized agent, the agent's name and mailing address; and

(5) Any other relevant information that the official with whom the application is filed may request.

A written request or a written confirmation of a verbal request shall be signed by the applicant or a duly authorized agent.

(c) *Names and addresses of interested persons.* When requested, each applicant for inspection service shall show on the application form the name and mailing address of each known interested person.

(d) *Surrender of superseded certificates.* Superseded certificates must be promptly surrendered.

(e) *Accessibility*—(1) *Commodities.* Each commodity lot inspected shall be arranged so the entire lot may be examined or, if necessary, a representative sample, as appropriate, can be obtained. If the entire lot is not accessible for examination or a representative sample cannot be obtained, the inspection shall be restricted to an examination or sampling of the accessible portion and the results certified as stated in § 868.34.

(2) *Origin records.* When an applicant requests origin inspection, the records indicating the origin of the commodity to be inspected shall be made accessible for examination and verification by official personnel.

(f) *Plant examination.* Plant surveys shall be performed upon request. Survey results shall be reported in writing to a designated plant official. If the plant is approved as a result of the survey, inspection service may begin or continue at a time agreed upon by the plant management and the cooperator or Service. If the plant is not approved as a result of the survey, inspection service shall be conditionally withheld pursuant to the procedures in § 868.24.

(g) *Working space.* An applicant must provide adequate and separate space when inspection service is performed at a plant.

(h) *Loading and unloading conditions.* Each applicant for inspection service shall provide or arrange for suitable conditions in the—

(1) Loading and unloading areas and the truck and railroad holding areas;

(2) Pier or dock areas;

(3) Deck and stowage areas of a carrier;

(4) Other service areas; and

(5) Equipment used in loading or unloading, processing, and handling the commodity.

Suitable conditions are those which will facilitate accurate inspection, maintain the quantity and the quality of the commodity that is to be inspected, and not be hazardous to the health and safety of official personnel as prescribed in the instructions.

(i) *Timely arrangements.* Requests for inspection service shall be made in a timely manner; otherwise, official personnel may not be available to provide the requested service. "Timely manner" shall mean not later than 2 p.m., local time, of the preceding business day.

(j) *Payment of bills.* Each applicant for inspection service shall pay bills for the service pursuant to §§ 868.90–868.92.

(Approved by the Office of Management and Budget under control number 0580–0012)

[53 FR 3722, Feb. 9, 1988. Redesignated and amended at 60 FR 16364, Mar. 30, 1995]

§ 868.22 Withdrawal of request for inspection service by applicant.

An applicant may withdraw a request for inspection service any time before official personnel release results, either verbally or in writing. Reimbursement of expenses, if any, shall be made pursuant to § 868.26.

[53 FR 3722, Feb. 9, 1988. Redesignated and amended at 60 FR 16364, Mar. 30, 1995]

§ 868.23 Dismissal of request for inspection service.

(a) *Conditions for dismissal*—(1) *General.* A cooperator or the Service shall dismiss requests for inspection service when:

(i) Performing the requested service is not practicable or possible.

(ii) The cooperator or the Service lacks authority under the Act or regulations to provide the inspection service requested or is unable to comply with the Act, regulations, standards, or instructions.

(iii) Sufficient information is not available to make an accurate determination.

(2) *Original inspection service.* A request for original inspection service shall be dismissed if an original inspection has already been performed and circumstances do not prevent a retest inspection, appeal inspection, or Board appeal inspection from being performed on the same lot.

(3) *Retest inspection service.* A request for a retest inspection service shall be dismissed by official personnel when:

(i) The factor requested was not tested during the original inspection;

(ii) The condition of the commodity has undergone a material change;

(iii) A representative file sample is not available;

(iv) The applicant requests that a new sample be obtained;

(v) The request is for a graded commodity; or

(vi) The reasons for the retest inspection are frivolous.

(4) *Appeal inspection service.* A request for an appeal inspection service shall be dismissed by official personnel when:

(i) The scope is different from the scope of the original inspection service;

(ii) The condition of the commodity has undergone a material change;

(iii) The request specifies a file sample and a representative file sample is not available;

(iv) The applicant requests that a new sample be obtained and a new sample cannot be obtained; or

(v) The reasons for the appeal inspection are frivolous.

(5) *Board appeal inspection service.* A request for a Board appeal inspection service shall be dismissed by official personnel when:

(i) The scope is different from the scope of the original inspection service;

(ii) The condition of the commodity has undergone a material change;

(iii) A representative file sample is not available;

(iv) The applicant requests that a new sample be obtained; or

(v) The reasons for the Board appeal inspection are frivolous.

(b) *Procedure for dismissal.* The cooperator or the Service shall notify the applicant of the proposed dismissal of service. If correctable, the applicant will be afforded reasonable time to take. corrective action or to demonstrate there is no basis for the dismissal. If corrective action has not been adequate, the applicant will be notified of the decision to dismiss the request for service, and any results of service shall not be released.

§ 868.24 Conditional withholding of service.

(a) *Conditional withholding.* A cooperator or the Service shall conditionally withhold service when an applicant fails to meet any requirement prescribed in § 868.21.

(b) *Procedure for withholding.* The cooperator or the Service shall notify the applicant of the reason for the proposal to conditionally withhold service. The applicant will then be afforded reasonable time to take corrective action or to demonstrate that there is no basis

for withholding service. If corrective action has not been adequate, the applicant will be notified of the decision to withhold service; and any results of service shall not be released.

[53 FR 3722, Feb. 9, 1988. Redesignated and amended at 60 FR 16364, Mar. 30, 1995]

§ 868.25 Denial or withdrawal of service.

(a) *General.* Service may be denied or withdrawn because of (1) any willful violation of the Act, regulations, standards, or instructions or (2) any interference with or obstruction of any official personnel in the performance of their duties by intimidation, threat, assault, or any other improper means.

(b) The Rules of Practice Governing Formal Adjudicatory Proceedings Instituted by the Secretary under Various Statutes (7 CFR part 1, subpart H) shall be followed in the denial or withdrawal of service.

§ 868.26 Expenses of the cooperator or the Service.

For any request that has been withdrawn, dismissed, or withheld under §§ 868.22, 868.23, or 868.24, respectively, each applicant shall pay expenses incurred by the cooperator or the Service.

[53 FR 3722, Feb. 9, 1988. Redesignated and amended at 60 FR 16364, Mar. 30, 1995]

INSPECTION METHODS AND PROCEDURES

§ 868.30 Methods and order of performing inspection service.

(a) *Methods*—(1) *General.* All sampling and inspection services performed by official personnel shall be made in accordance with the regulations, standards, and the instructions.

(2) *Lot inspection service.* A lot inspection service shall be based on official personnel obtaining representative samples, examining the commodity in the entire lot, and making an accurate analysis of the commodity on the basis of the samples.

(3) *Submitted sample inspection service.* A submitted sample inspection service shall be based on a submitted sample of sufficient size to enable official personnel to perform an accurate, complete analysis. The sample size will be prescribed in the instructions. If a

complete analysis cannot be performed because of an inadequate sample size or other conditions, the request shall be dismissed or a factor only inspection may be performed upon request.

(b) *Order of service.* Inspection services shall be performed, to the extent practicable, in the order in which requests for service are received.

(c) *Recording receipt of documents.* Each document submitted by or on behalf of an applicant for inspection service shall be promptly stamped or similarly marked by official personnel to show the date of receipt.

(d) *Conflicts of interest.* (1) Official personnel shall not perform or participate in performing an inspection service on a commodity or a carrier or container in which the official personnel have a direct or indirect financial interest.

(2) Official personnel shall not perform, participate in performing, or issue a certificate if the official personnel participated in a previous inspection or certification of the lot unless there is only one authorized person available at the time and place of the requested inspection service.

§868.31 Kinds of inspection services.

(a) *General.* The inspection of commodities shall be according to the—

(1) Standards of class, grade, other quality designation, quantity, or condition for such commodities promulgated by the Administrator; or

(2) Specifications prescribed by Federal agencies; or

(3) Specifications of trade associations or organizations; or

(4) Other specifications as requested by applicant; or

(5) The instructions.

The kinds of services provided and the basis for performing the services include those specified in paragraphs (b) through (m) of this section. Some or all of these services are provided when performing a complete inspection service.

(b) *Quality inspection service.* This service consists of official personnel—

(1) Obtaining representative sample(s) of an identified commodity lot;

(2) Examining, grading, or testing the sample(s);

(3) Examining relevant records for the lot; and

(4) Certifying the results.

(c) *Submitted sample inspection service.* This service consists of official personnel grading or testing a sample submitted by the applicant and certifying the results.

(d) *Examination service.* This service consists of official personnel examining supplies without the use of special laboratory equipment or procedures to determine conformance to requirements requested by the applicant and certifying the results.

(e) *Checkweighing service (container).* This service consists of official personnel—

(1) Weighing a selected number of containers from a commodity lot;

(2) Determining the estimated total gross, tare, and net weights or the estimated average gross or net weight per filled container; and

(3) Certifying the results.

(f) *Bulk weighing service.* This service consists of official personnel—

(1) Completely supervising the loading or the unloading of an identified lot of bulk or containerized commodity,

(2) Physically weighing or completely supervising the weighing of the commodity; and

(3) Certifying the results.

(g) *Checkloading service.* This service consists of official personnel—

(1) Performing a stowage examination;

(2) Computing the number of filled commodity containers loaded aboard the carrier;

(3) Observing the condition of commodity containers loaded aboard the carrier;

(4) If practicable, sealing the carrier; and

(5) Certifying the results.

(h) *Checkcounting service.* This service consists of official personnel determining the total number of filled outer containers in a lot to determine that the number of containers shown by the applicant is correct and certifying the results.

(i) *Condition inspection service.* This service consists of official personnel determining the physical condition of the commodity by determining whether an identifiable commodity lot is water damaged, fire damaged, or has rodent or bird contamination, insect infestation, or any other deteriorating condition and certifying the results.

(j) *Condition of food containers service.* This service consists of official personnel determining the degree of acceptability of the containers with respect to absence of defects which affect the serviceability, including appearance as well as usability, of the container for its intended purpose and certifying the results.

(k) *Observation of loading service.* This service consists of official personnel determining that an identified lot has been moved from a warehouse or carrier and loaded into another warehouse or carrier and certifying the results.

(l) *Plant approval service.*[1] This service consists of official personnel performing a plant survey to determine if the plant premises, facilities, sanitary conditions, and operating methods are suitable to begin or continue inspection service.

(m) *Stowage examination service.* This service consists of official personnel visually determining if an identified carrier or container is clean; dry; free of infestation, rodents, toxic substances and foreign odor; and suitable to store or carry commodities and certifying the results.

§ 868.32 Who shall inspect commodities.

Official commodity inspections shall be performed only by official personnel.

§ 868.33 Sample requirements; general.

(a) *Samples for lot inspection service—* (1) *Original lot inspection service.* The sample(s) on which the original inspection is determined shall be—

(i) Obtained by official personnel;

(ii) Representative of the commodity in the lot;

(iii) Protected by official personnel from manipulation, substitution, and improper or careless handling; and

(iv) Obtained within the prescribed area of responsibility of the cooperator

---

[1] Compliance with the requirements in this paragraph does not excuse failure to comply with all applicable sanitation rules and regulations of city, county, State, Federal, or other agencies having jurisdiction over such plants and operations.

or field office performing the inspection service.

(2) *Retest lot inspection service.* The sample(s) on which the retest is determined shall meet the requirements of paragraph (a)(1) of this section. The retest inspection shall be performed on the basis of a file sample(s), and the samples shall meet the requirements prescribed in § 868.35(e).

(3) *Appeal lot inspection service.* For an appeal lot inspection service, the sample(s) on which the appeal is determined shall meet the requirements of paragraph (a)(1) of this section. If the appeal inspection is performed on the basis of a file sample(s), the samples shall meet the requirements prescribed in § 868.35(e). In accordance with § 868.61(b), an applicant may request that a new sample be obtained and examined as part of the appeal inspection service.

(4) *Board appeal lot inspection service.* A Board appeal lot inspection service shall be performed on the basis of file sample.

(b) *Sampler requirement.* An official sampler shall sample commodities and forward the samples to the appropriate cooperator or field office or other location as specified. A sampling report signed by the sampler shall accompany each sample. The report shall include the identity, quantity, and location of the commodity sampled; the name and mailing address of the applicant; and all other information regarding the lot as may be required.

(c) *Representative sample.* A sample shall not be considered representative of a commodity lot unless the sample—

(1) Has been obtained by official personnel;

(2) Is of the size prescribed in the instructions; and

(3) Has been obtained, handled, and submitted in accordance with the instructions.

(d) *Protecting samples.* Official personnel shall protect samples from manipulation, substitution, and improper and careless handling which would deprive the samples of their representativeness or which would change the physical and chemical properties of the commodity from the time of sampling until inspection services are completed and file samples have been discarded.

[53 FR 3722, Feb. 9, 1988. Redesignated and amended at 60 FR 16364, Mar. 30, 1995]

§ 868.34 Partial inspection.

When the entire lot is not accessible for examination or a representative sample cannot be obtained from the entire lot, the certificate shall state the estimated quantity of the commodity in the accessible portion and the quantity of the entire lot. The inspection shall be limited to the accessible portion. In addition, the words "Partial Inspection" shall be printed or stamped on the certificate.

§ 868.35 Sampling provisions by level of service.

(a) *Original inspection service*—(1) *Lot inspection service.* Each original lot inspection service shall be made on the basis of one or more representative samples obtained by official personnel from the commodity in the lot and forwarded to the appropriate location.

(2) *Submitted sample service.* Each original submitted sample inspection service shall be performed on the basis of the sample as submitted.

(b) *Retest inspection service.* Each retest inspection service performed on a commodity lot or a submitted sample shall be based on an analysis of the file sample.

(c) *Appeal inspection service*—(1) *Lot inspection service.* Each appeal inspection service on a commodity lot shall be made on the basis of a file sample or, upon request, a new sample.

(2) *Submitted sample service.* Each appeal inspection service on the commodity in a submitted sample shall be based on an analysis of the file sample.

(d) *Board appeal inspection service.* Each Board appeal inspection service performed on a commodity lot or submitted sample shall be based on an analysis of the file sample.

(e) *Use of file samples*—(1) *Requirements for use.* A file sample that is retained by official personnel in accordance with the procedures prescribed in the instructions shall be considered representative for retest inspection, appeal inspection, and Board appeal inspection service if: (i) The file samples have remained at all times in the custody and control of the official personnel that performed the inspection service and (ii) the official personnel who performed the inspection service in question and those who are to perform the retest inspection, the appeal inspection, or the Board appeal inspection service determines that the samples were representative of the commodity at the time the inspection service was performed and that the quality or condition of the commodity in the samples has not since changed.

(2) *Certificate statement.* The certificate for a retest inspection, appeal inspection, or Board appeal inspection service which is based on a file sample shall show the statement "Results based on file sample."

§ 868.36 Loss of identity.

(a) *Lots.* The identity of a packaged lot, bulk lot, or sublot of a commodity shall be considered lost if:

(1) A portion of the commodity is unloaded, transferred, or otherwise removed from the carrier or location after the time of original inspection, unless the identity is preserved; or

(2) More commodity or other material, including a fumigant or insecticide, is added to the lot after the original inspection was performed, unless the addition of the fumigant or insecticide was performed in accordance with the instructions; or

(3) At the option of official personnel performing an appeal inspection or Board appeal inspection service, the identity of a commodity in a closed carrier or container may be considered lost if the carrier or container is not sealed or the seal record is incomplete.

(b) *Carriers and containers.* The identity of a carrier or container shall be considered lost if (1) the stowage area is cleaned, treated, fumigated, or fitted after the original inspection was performed or (2) the identification has been changed since the original inspection.

(c) *Submitted sample.* The identity of a submitted sample of a commodity shall be considered lost if:

(1) The identifying number, mark, or symbol for the sample is lost or destroyed; or

(2) The sample has not been retained and protected by official personnel as

prescribed in the regulations and the instructions.

ORIGINAL INSPECTION SERVICE

§ 868.40 Who may request original inspection service.

Any interested person may apply for inspection service.

§ 868.41 Contract service.

Any interested person may enter into a contract with a cooperator or the Service whereby the cooperator or Service will provide original inspection services for a specified period, and the applicant will pay a specific fee.

§ 868.42 How to request original inspection service.

(a) *General.* Requests may be made verbally or in writing. Verbal requests shall be confirmed in writing when requested by official personnel. All written requests shall include the information specified in § 868.21. Copies of request forms may be requested from the cooperator or the Service. If all required documentation is not available when the request is made, it shall be provided as soon as it is available. At their discretion, official personnel may withhold inspection service pending receipt of the required documentation.

(b) *Request requirements.* Requests for original inspection service, other than submitted sample inspections, must be made with the cooperator or the Service responsible for the area in which the service will be provided. Requests for submitted sample inspections may be made with any cooperator or any field office that provides original inspection service. Requests for inspection of commodities during loading, unloading, handling, or processing shall be received far enough in advance so official personnel can be present.

(Approved by the Office of Management and Budget under control number 0580-0012)

[53 FR 3722, Feb. 9, 1988. Redesignated and amended at 60 FR 16364, Mar. 30, 1995]

§ 868.44 New original inspection.

When circumstances prevent a retest inspection, appeal inspection, or Board appeal inspection, an applicant may request a new original inspection on any previously inspected lot; except that a new original inspection may not be performed on an identifiable commodity lot which, as a result of a previous inspection, was found to be contaminated with filth, other than insect fragments in nongraded processed products, or to contain a deleterious substance. A new original inspection shall be based on a new sample and shall not be restricted to the scope of any previous inspection. A new original inspection certificate shall not supersede any previously issued certificate.

RETEST INSPECTION SERVICE

§ 868.50 Who may request retest inspection service.

(a) *General.* Any interested person may request a retest inspection service on nongraded commodities. When more than one interested person requests a retest inspection service, the first interested person to file is the applicant of record. Only one retest inspection service may be performed on any original inspection service.

(b) *Scope of request.* A retest inspection service may be requested for any or all quality factors tested but shall be limited to analysis of the file sample.

(Approved by the Office of Management and Budget under control number 0580-0012)

§ 868.51 How to request retest inspection service.

(a) *General.* Requests shall be made with the field office responsible for the area in which the original inspection service was performed. Verbal requests shall be confirmed in writing, upon request, as specified in § 868.21. Copies of request forms may be obtained from the field office upon request. If at the time the request is filed and the documentation required by § 868.21 is not available, official personnel may, at their discretion, withhold service pending the receipt of the required documentation.

(b) *Request requirements.* Requests will be considered filed on the date they are received by official personnel.

(Approved by the Office of Management and Budget under control number 0580-0012)

[53 FR 3722, Feb. 9, 1988. Redesignated and amended at 60 FR 16364, Mar. 30, 1995]

§868.52 Certificating retest inspection results.

(a) *General.* Retest inspection certificates shall be issued according to §868.70 and instructions. The certificate shall show the results of the factor(s) retested and the original results not included in the retest service.

(b) *Required statements on retest certificates.* Each retest inspection certificate shall show the statements required by this section, §868.71, and the instructions.

(1) Each retest inspection certificate shall clearly show the term "Retest" and a statement identifying the superseded original certificate. The superseded certificate shall be considered null and void as of the date of the retest certificate. When applicable, the certificate shall also show a statement as to which factor(s) result is based on the retest inspection service and that all other results are those of the original inspection service.

(2) If the superseded certificate is in the custody of the Service, the superseded certificate shall be marked "Void." If the superseded certificate is not in the custody of the Service at the time the retest certificate is issued, a statement indicating that the superseded certificate has not been surrendered shall be shown on the retest certificate.

[53 FR 3722, Feb. 9, 1988. Redesignated and amended at 60 FR 16364, Mar. 30, 1995]

APPEAL INSPECTION SERVICE

§868.60 Who may request appeal inspection service.

(a) *General.* Any interested person may request appeal inspection or Board appeal inspection service. When more than one interested person requests an appeal inspection or Board appeal inspection service, the first interested person to file is the applicant of record. Only one appeal inspection may be obtained from any original inspection or retest inspection service for nongraded commodities. Only one Board appeal inspection may be obtained from any original or appeal inspection service for graded commodities. Board appeal inspection shall be performed on the basis of the file sample.

(b) *Kind and scope of request.* When the results for more than one kind of service are reported on a certificate, an appeal inspection or Board appeal inspection service, as applicable, may be requested on any or all kinds of services reported on the certificate. The scope of an appeal inspection service will be limited to the scope of the original inspection or, in the case of a Board appeal inspection service, the original or appeal inspection service. A request for appeal inspection of a retest inspection will be based upon the scope of the original inspection. If the request specifies a different scope, the request shall be dismissed. Provided, however, that an applicant for service may request an appeal or Board appeal inspection of specific factor(s) or official grade and factors. In addition, appeal and Board appeal inspection for grade may include a review of any pertinent factor(s), as deemed necessary by official personnel.

(Approved by the Office of Management and Budget under control number 0580–0013)

[53 FR 3722, Feb. 9, 1988. Redesignated at 60 FR 16364, Mar. 30, 1995 and amended at 70 FR 69250, Nov. 15, 2005]

§868.61 How to request appeal inspection service.

(a) *General.* Requests shall be made with the field office responsible for the area in which the original service was performed. Requests for Board appeal inspections may be made with the Board of Appeals and Review or the field office that performed the appeal inspection. Verbal requests must be confirmed in writing, upon request, as specified in §868.21. Copies of request forms may be obtained from the field office upon request. If at the time the request is made the documentation required by §868.21 is not available, official personnel may, at their discretion, withhold service pending the receipt of the required documentation.

(b) *Request requirements.* (1) This subparagraph is applicable to rice inspection only. Except as may be agreed upon by the interested persons, the application shall be made: (i) Before the rice has left the place where the inspection being appealed was performed and (ii) no later than the close of business on the second business day following the date of the inspection being appealed. However, the Administrator may extend the time requirement as deemed necessary.

(2) Subject to the limitations of paragraph (b)(3) of this section, the applicant may request that an appeal inspection be based on: (i) The file sample or (ii) a new sample. However, an appeal inspection shall be based on a new sample only if the lot can positively be identified by official personnel as the one that was previously inspected and the entire lot is available and accessible for sampling and inspection. Board appeals shall be on the basis of the file sample.

(3) An appeal inspection shall be limited to a review of the sampling procedure and an analysis of the file sample when, as a result of a previous inspection, the commodity was found to be contaminated with filth (other than insect fragments in nongraded processed products) or to contain a deleterious substance. If it is determined that the sampling procedures were improper, a new sample shall be obtained if the lot can be positively identified as the lot which was previously inspected and the entire lot is available and accessible for sampling and inspection.

(Approved by the Office of Management and Budget under control number 0580-0012)

[53 FR 3722, Feb. 9, 1988. Redesignated and amended at 60 FR 16364, Mar. 30, 1995]

§ 868.62 Who shall perform appeal inspection service.

(a) *Appeal.* For graded commodities, the appeal inspection service shall be performed by the field office responsible for the area in which the original inspection was performed. For nongraded commodities, the appeal inspection service shall be performed by the Service's Commodity Testing Laboratory.

(b) *Board appeal.* Board appeal inspection service shall be performed only by the Board of Appeals and Review. The field office will act as a liaison between the Board of Appeals and Review and the applicant.

§ 868.63 Certificating appeal inspection results.

(a) *General.* An appeal inspection certificate shall be issued according to

§ 868.70 and instructions. Except as provided in paragraph (b)(2) of this section, only the results of the appeal inspection or Board appeal inspection service shall be shown on the appeal inspection certificate.

(b) *Required statements.* Each appeal inspection certificate shall show the statements required by this section, § 868.71, and instructions.

(1) Each appeal inspection certificate shall clearly show: (i) The term "Appeal" or "Board Appeal" and (ii) a statement identifying the superseded certificate. The superseded certificate shall be considered null and void as of the date of the appeal inspection or Board appeal inspection certificate.

(2) When the results for more than one kind of service are reported on a certificate, the appeal or Board appeal inspection certificate shall show a statement of which kind of service(s) results are based on the appeal or Board appeal inspection service and that all other results are those of the original inspection, retest inspection, or appeal inspection service.

(3) If the superseded certificate is in the custody of the Service, the superseded certificate shall be marked "Void." If the superseded original inspection, retest inspection, or appeal inspection certificate is not in the custody of the Service at the time the appeal certificate is issued, a statement indicating that the superseded certificate has not been surrendered shall be shown on the appeal certificate.

(c) *Finality of Board appeal inspection.* A Board appeal inspection shall be the final appeal inspection service except that for nongraded commodities an appeal shall be the final appeal inspection.

[53 FR 3722, Feb. 9, 1988. Redesignated and amended at 60 FR 16364, Mar. 30, 1995]

OFFICIAL CERTIFICATES

§ 868.70 Official certificates; issuance and distribution.

(a) *Required issuance.* An inspection certificate shall be issued to show the results of each kind and each level of inspection service.

(b) *Distribution*—(1) *Original.* The original and one copy of each inspection certificate shall be distributed to the applicant or the applicant's order. In addition, one copy of each inspection certificate shall be filed with the office providing the inspection; and, if the inspection is performed by a cooperator, one copy shall be forwarded to the appropriate field office. If requested by the applicant prior to issuance of the inspection certificate, additional copies not to exceed a total of three copies will be furnished at no extra charge.

(2) *Retest and appeal inspection service.* In addition to the distribution requirements in paragraph (b)(1) of this section, one copy of each retest or appeal inspection certificate will be distributed to each interested person of record or the interested person's order and to the cooperator or field office that issued the superseded certificate.

(3) *Additional copies.* Additional copies of certificates will be furnished to the applicant or interested person upon request. Fees for extra copies in excess of three may be assessed according to the fee schedules established by the cooperator or the Service.

(c) *Prompt issuance.* An inspection certificate shall be issued before the close of business on the business day following the date the inspection is completed.

(d) *Who may issue a certificate*—(1) *Authority.* Certificates for inspection services may be issued only by official personnel who are specifically authorized or licensed to perform and certify the results reported on the certificate.

(2) *Exception.* The person in the best position to know whether the service was performed in an approved manner and that the determinations are accurate and true should issue the certificate. If the inspection is performed by one person, the certificate should be issued by that person. If an inspection is performed by two or more persons, the certificate should be issued by the person who makes the majority of the determinations or the person who makes the final determination. Supervisory personnel may issue a certificate when the individual is licensed or authorized to perform the inspection being certificated.

(e) *Name requirement.* The name or the signature, or both, of the person who issued the inspection certificate shall be shown on the original and all copies of the certificate.

(f) *Authorization to affix names*—(1) *Requirements.* The names or the signatures, or both, of official personnel may be affixed to official certificates which are prepared from work records signed or initialed by the person whose name will be shown. The agent affixing the name or signature, or both, shall: (i) Be employed by a cooperating agency or the Service, (ii) have been designated to affix names or signatures, or both, and (iii) hold a power of attorney from the person whose name or signature, or both, will be affixed. The power of attorney shall be on file with the employing cooperating agency or the Service as appropriate.

(2) *Initialing.* When a name or signature, or both, is affixed by an authorized agent, the initials of the agent shall appear directly below or following the name or signature of the person.

(g) *Advance information.* Upon request, the contents of an official certificate may be furnished in advance to the applicant and any other interested person, or to their order, and any additional expense shall be borne by the requesting party.

(h) *Certification; when prohibited.* An official certificate shall not be issued for service after the request for an inspection service has been withdrawn or dismissed.

§868.71 Official certificate requirements.

Official certificates shall—

(a) Be on standard printed forms prescribed in the instructions;

(b) Be in English;

(c) Be typewritten or handwritten in ink and be clearly legible;

(d) Show the results of inspection services in a uniform, accurate, and concise manner;

(e) Show the information required by §§868.70–868.75; and

(f) Show only such other information and statements of fact as are provided in the instructions authorized by the Administrator.

[53 FR 3722, Feb. 9, 1988. Redesignated and amended at 60 FR 16364, Mar. 30, 1995]

§ 868.72 Certification of results.

(a) *General*. Each official certificate shall show the results of the inspection service.

(b) *Graded commodities*. Each official certificate for graded commodities shall show—

(1) The class, grade, or any other quality designation according to the official grade standards;

(2) All factor information requested by the applicant; and

(3) All grade determining factors for commodities graded below the highest quality grade.

§ 868.73 Corrected certificates.

(a) *General*. The accuracy of the statements and information shown on official certificates must be verified by the individual whose name or signature, or both, is shown on the official certificate or by the authorized agent who affixed the name or signature, or both. Errors found during this process shall be corrected according to this section.

(b) *Who may correct*. Only official personnel or their authorized agents may make corrections, erasures, additions, or other changes to official certificates.

(c) *Corrections prior to issuance*. No corrections, erasures, additions, or other changes shall be made which involve identification, quality, or quantity. If such errors are found, a new official certificate shall be prepared and issued and the incorrect certificate marked "Void." Otherwise, errors may be corrected provided that—

(1) The corrections are neat and legible;

(2) Each correction is initialed by the individual who corrects the certificate; and

(3) The corrections and initials are shown on the original and all copies.

(d) *Corrections after issuance*—(1) *General*. If errors are found on an official certificate at any time up to a maximum of 1 year after issuance, the errors shall be corrected by obtaining the incorrect certificate and replacing it with a corrected certificate. When the incorrect certificate cannot be obtained, a corrected certificate can be issued superseding the incorrect one.

(2) *Certification requirements*. The same statements and information, including permissive statements, that were shown on the incorrect certificate, along with the correct statement or information, shall be shown on the corrected certificate. According to this section and the instructions, corrected certificates shall show—

(i) The terms "Corrected Original" and "Corrected Copy,"

(ii) A statement identifying the superseded certificate and the corrections,

(iii) A statement indicating the superseded certificate was not surrendered when the incorrect certificate was not submitted; and

(iv) A new serial number.

In addition, the incorrect certificate shall be marked "Void" when submitted.

(e) *Limitations*. Corrected certificates cannot be issued for a certificate that has been superseded by another certificate or on the basis of a subsequent analysis for quality.

§ 868.74 Divided-lot certificates.

(a) *General*. When commodities are offered for inspection and are certificated as a single lot, the applicant may exchange the inspection certificate for two or more divided-lot certificates.

(b) *Application*. Requests for divided-lot certificates shall be made—

(1) In writing;

(2) By the applicant who made the initial request;

(3) To the office that issued the outstanding certificate;

(4) Within 5 business days of the outstanding certificate date; and

(5) Before the identity of the commodity has been lost.

(c) *Quantity restrictions*. Divided-lot certificates shall not show an aggregate quantity different than the total quantity shown on the superseded certificate.

(d) *Surrender of certificate*. The certificate that will be superseded shall—

(1) Be in the custody of the cooperator or the Service;

(2) Be marked "Void," and

(3) Show the identification of the divided-lot certificates.

(e) *Certification requirements.* The same information and statements, including permissive statements, that were shown on the superseded certificate shall be shown on each divided-lot certificate. Divided-lot certificates shall show—

(1) A statement indicating the commodity was inspected as an undivided lot;

(2) The terms "Divided-Lot Original," and the copies shall show "Divided-Lot Copy;"

(3) The same serial number with numbered suffix (for example, 1764–1, 1764–2, 1764–3, and so forth); and

(4) The quantity specified by the request.

(f) *Issuance and distribution.* Divided-lot certificates shall be issued no later than the close of business on the next business day after the request and be distributed according to § 868.70(b).

(g) *Limitations.* After divided-lot certificates have been issued, further dividing or combining is prohibited except with the approval of the Service.

(Approved by the Office of Management and Budget under control number 0580–0012)

[53 FR 3722, Feb. 9, 1988. Redesignated and amended at 60 FR 16364, Mar. 30, 1995]

§ 868.75 Duplicate certificates.

Upon request, a duplicate certificate may be issued for a lost or destroyed official certificate.

(a) *Application.* Requests for duplicate certificates shall be filed—

(1) In writing;

(2) By the applicant who requested the service covered by the lost or destroyed certificate; and

(3) With the office that issued the initial certificate.

(b) *Certification requirements.* The same information and statements, including permissive statements, that were shown on the lost or destroyed certificate shall be shown on the duplicate certificate. Duplicate certificates shall show: (1) The terms "Duplicate Original," and the copies shall show "Duplicate Copy" and (2) a statement that the certificate was issued in lieu of a lost or destroyed certificate.

(c) *Issuance.* Duplicate certificates shall be issued as promptly as possible and distributed according to § 868.70(b).

(d) *Limitations.* Duplicate certificates shall not be issued for certificates that have been superseded.

(Approved by the Office of Management and Budget under control number 0580–0012)

[53 FR 3722, Feb. 9, 1988. Redesignated and amended at 60 FR 16364, Mar. 30, 1995]

LICENSED INSPECTORS, TECHNICIANS, AND SAMPLERS

§ 868.80 Who may be licensed.

(a) *Inspectors.* The Administrator may license any person to inspect commodities and to perform related services if the individual—

(1) Is employed by a cooperator, is a contractor, or is employed by a contractor;

(2) Possesses the qualifications prescribed in the instructions; and

(3) Has no interest, financial or otherwise, direct or indirect in merchandising, handling, storing, or processing the kind of commodities or related products to be inspected.

The Administrator may require applicants to be examined for competency at a specific time and place and in a prescribed manner.

(b) *Technicians or samplers.* The Administrator may license any person as a technician to perform official specified laboratory functions, including sampling duties and related services, or as a sampler to draw samples of commodities and perform related services if the individual: (1) Possesses proper qualifications as prescribed in the instructions and (2) has no interest, financial or otherwise direct or indirect in merchandising, handling, storing, or processing the kind of commodities or related products to be chemically analyzed, mechanically tested, sampled, and so forth. The Administrator may require applicants to be examined for competency at a specific time and place and in a prescribed manner.

(Approved by the Office of Management and Budget under control number 0580–0012)

[53 FR 3722, Feb. 9, 1988. Redesignated and amended at 60 FR 16364, Mar. 30, 1995; 63 FR 29531, June 1, 1998]

§ 868.81 Licensing procedures.

(a) *Application.* An application for a license, the renewal of a license, or the return of a suspended license shall be submitted to the Service on forms furnished by the Service. Each application shall be in English, be typewritten or legibly written in ink, show all information prescribed by the application form, and be signed by the applicant.

(b) *Examinations and reexaminations.* Applicants for a license and individuals who are licensed to perform any or all inspection services shall, at the discretion of the Service, submit to examinations or reexaminations to determine their competency to perform the inspection functions for which they desire to be or are licensed.

(c) *Termination*—(1) *Procedure.* Each license shall terminate according to the termination date shown on the license and as specified by the schedule in this paragraph. The termination date for a license shall be no less than 3 years or more than 4 years after the issuance date for the initial license; thereafter, every 3 years. Upon request of a licensee and for good cause shown, the termination date may be advanced or delayed by the Administrator for a period not to exceed 60 days.

TERMINATION SCHEDULE

Last name beginning with	Termination date
A	January.
B	February.
C, D	March.
E, F, G	April.
H, I, J	May.
K, L	June.
M	July.
N, O, P, Q	August.
R	September.
S, T, U, V	October.
W	November.
X, Y, Z	December.

The Service shall issue a termination notice 60 days before the termination date. The notice shall give detailed instructions for requesting renewal of license, state whether a reexamination is required, and, if a reexamination is required, give the scope of the examination. Failure to receive a notice from the Service shall not exempt a licensee from the responsibility of having the license renewed by the termination date.

(2) *Exception.* The license of an individual under contract with the Service shall terminate upon termination of the contract.

(d) *Surrender of license.* Each license that is terminated or which is suspended or revoked under § 868.84 shall be promptly surrendered to the Administrator or other official of the Service designated by the Administrator.

(Approved by the Office of Management and Budget under control number 0580–0012)

[53 FR 3722, Feb. 9, 1988. Redesignated and amended at 60 FR 16364, Mar. 30, 1995]

§ 868.82 **Voluntary cancellation or suspension of license.**

Upon request by a licensee, the Service may cancel a license or suspend a license for a period of time not to exceed 1 year. A license that has been voluntarily suspended shall be returned by the Service upon request by the licensee within 1 year, subject to the provisions of § 868.81(a) and (b); a license that has been cancelled shall be considered void and shall not be subject to return or renewal.

[53 FR 3722, Feb. 9, 1988. Redesignated and amended at 60 FR 16364, 16365, Mar. 30, 1995]

§ 868.83 **Automatic suspension of license by change in employment.**

A license issued to an individual shall be automatically suspended when the individual ceases to be employed by the cooperator. If the individual is reemployed by the cooperator or employed by another cooperator within 1 year of the suspension date and the license has not terminated in the interim, upon request of the licensee, the license will be reinstated subject to the provisions of § 868.81(a) and (b).

[53 FR 3722, Feb. 9, 1988. Redesignated and amended at 60 FR 16364, 16365, Mar. 30, 1995]

§ 868.84 **Suspension or revocation of license.**

(a) *General.* (1) An inspector's, technician's, or sampler's license may be suspended or revoked if the licensee:

(i) Willfully, carelessly, or through incompetence fails to perform the duties specified in the Act, regulations, standards, or the instructions or

(ii) Becomes incapable of performing required duties.

(2) A license may not be suspended or revoked until the individual:

(i) Has been served notice, in person or by registered mail, that suspension or revocation of the license is under consideration for reasons set out in the notice and

(ii) Has been given an opportunity for a hearing.

(b) *Procedure for summary action.* In cases where the public health, interest, or safety require, the Administrator may summarily suspend an inspector's, technician's, or sampler's license without prior hearing. In such cases, the licensee shall be advised of the factors which appear to warrant suspension or revocation of the license. The licensee shall be accorded an opportunity for a hearing before the license is finally suspended or revoked.

(c) *Procedures for other than summary action.* Except in cases of willfulness or those described in paragraph (b) of this section, the Administrator, before instituting proceedings for the suspen-

sion or revocation of a license, shall provide the licensee an opportunity to demonstrate or achieve compliance with the Act, regulations, standards, and instructions. If the licensee does not demonstrate or achieve compliance, the Administrator may institute proceedings to suspend or revoke the license.

(The information collection requirements contained in paragraph (c) have been approved by the Office of Management and Budget under control number 0580–0012)

FEES

§868.90 Fees for certain Federal inspection services.

(a) The fees shown in Table 1 apply to Federal Commodity Inspection Services specified below.

TABLE 1—HOURLY RATES [1] [3]

[Fees for inspection of commodities other than rice]

Hourly Rates (per service representative):	
Monday to Friday	$34.20
Saturday, Sunday, and Holidays	44.40
Miscellaneous Processed Commodities: [2]	
(1) Additional Tests (cost per test, assessed in addition to the hourly rate):	
(i) Aflatoxin Test (Thin Layer Chromatography)	51.40
(ii) Falling Number	12.50
(iii) Aflatoxin Test Kit	7.50
Graded Commodities (Beans, Peas, Lentils, Hops, and Pulses):	
(1) Additional Tests—Unit Rates (Beans, Peas, Lentils):	
(i) Field run (per lot or sample)	23.00
(ii) Other than field run (per lot or sample)	13.75
(iii) Factor analysis (per factor)	5.65
(2) Additional Tests—Unit Rates (Hops): (i) Lot or sample (per lot or sample)	29.30
(3) Additional Tests—Unit Rates (Nongraded Nonprocessed Commodities): (i) Factor analysis (per factor)	5.65
(4) Stowage Examination (service-on-request) [4] (i) Ship (per stowage space) (minimum $252.50 per ship)	50.50
(ii) Subsequent ship examinations (same as original) (minimum $151.50 per ship)	
(iii) Barge (per examination)	40.50
(iv) All other carriers (per examination)	15.50

[1] Fees for original commodity inspection and appeal inspection services include, but are not limited to, sampling, grading, weighing, stowage examinations, pre-inspection conferences, sanitation inspections, and other services requested by the applicant and that are performed within 25 miles of the field office. Travel and related expenses (commercial transportation costs, mileage, and per diem) will be assessed in addition to the hourly rate for service beyond the 25-mile limit. Refer to §868.92. Explanation of service fees and additional fees, for all other service fees except travel and per diem.

[2] When performed at a location other than the Commodity Testing Laboratory.

[3] Faxed and extra copies of certificates will be charged at $1.50 per copy.

[4] If performed outside of normal business hours, 1½ times the applicable unit fee will be charged.

(b) In addition to the fees, if any, for sampling or other requested service, a fee will be assessed for each laboratory test (original, retest, or appeal) listed in table 2 of this section.

(c) If a requested test is to be reported on a specified moisture basis, a fee for a moisture test will also be assessed.

(d) Laboratory tests referenced in table 2 of this section will be charged at the applicable laboratory fee.

TABLE 2—FEES FOR LABORATORY TEST SERVICES [1]

Laboratory tests	Fees
(1) Aflatoxin (Quantitative—HPLC)	$182.00
(2) Aflatoxin (Quantitative—Test Kit)	87.00
(3) Aflatoxin (Qualitative—Test Kit)	47.00
(4) Appearance and odor	7.00
(5) Ash	17.00
(6) Brix	16.00
(7) Calcium	27.00
(8) Carotenoid Color	27.00
(9) Cold test (oil)	20.00
(10) Color test (syrups)	13.00
(11) Cooking tests (pasta)	13.00
(12) Crude fat	20.00
(13) Crude fiber	27.00
(14) Falling number	24.00
(15) Free fatty acid	24.00
(16) Insoluble impurities (oils and shortenings)	9.00
(17) Iron enrichment	30.00
(18) Lovibond color	20.00
(19) Moisture	13.00
(20) Moisture and volatile matter	17.00
(21) Oxidative stability index (OSI)	54.00
(22) Peroxide Value	27.00
(23) Popping ratio	38.00
(24) Protein	16.00
(25) Sanitation (light filth)	47.00
(26) Sieve test	11.00
(27) Smoke Point	43.00
(28) Solid fat index	168.00
(29) Visual exam	22.00
(30) Vomitoxin (Qualitative—Test Kit)	61.00
(31) Vomitoxin (Quantitative—Test Kit)	81.00
(32) Other laboratory analytical services (per hour per service representative)	67.00

[1] When laboratory tests/services are provided for AMS by a private laboratory, the applicant will be assessed a fee, which, as nearly as practicable, covers the costs to AMS for the service provided.

[61 FR 66535, Dec. 18, 1996, as amended 66 FR 17777, Apr. 4, 2001; 69 FR 1894, Jan. 13, 2004]

§ 868.91 Fees for certain Federal rice inspection services.

The fees for services in paragraph (a) apply to Federal inspection services. Starting with fiscal year 2022, calculations provided in paragraph (b) will be used to determine annual fee rates.

(a) Fees for services are published on the Service's website.

(b) For each fiscal year, starting with 2022, the Administrator will calculate the rates for services, issue a public notice, and publish fees on the Service's website with an effective date of October 1 of each year.

(1) For each year, the Administrator will calculate the rates for services, per hour per inspection program employee using the following formulas:

(i) *Regular rate.* The Service's total inspection program personnel direct pay divided by direct hours, which is then multiplied by the next year's percentage of cost of living increase, plus the benefits rate, plus the operating rate, plus the allowance for bad debt rate. If applicable, actual travel expenses may also be added to the cost of providing the service.

(ii) *Overtime rate.* The Service's total inspection program personnel direct pay divided by direct hours, which is then multiplied by the next year's percentage of cost of living increase and then multiplied by 1.5, plus the benefits rate, plus the operating rate, plus an allowance for bad debt. If applicable, actual travel expenses may also be added to the cost of providing the service.

(iii) *Holiday rate.* The Service's total inspection program personnel direct pay divided by direct hours, which is then multiplied by the next year's percentage of cost of living increase and then multiplied by 2, plus the benefits rate, plus the operating rate, plus an allowance for bad debt. If applicable, actual travel expenses may also be added to the cost of providing the service.

(2) For each year, based on previous year/historical actual costs, the Administrator will calculate the benefits, operating, and allowance for bad debt components of the regular, overtime, and holiday rates as follows:

(i) *Benefits rate.* The Service's total inspection program direct benefits costs divided by the total hours (regular, overtime, holiday) worked, which is then multiplied by the next year's percentage of cost of living increase. Some examples of direct benefits are health insurance, retirement, life insurance, and Thrift Savings Plan (TSP) retirement basic and matching contributions.

(ii) *Operating rate.* The Service's total inspection program operating costs divided by total hours (regular, overtime, and holiday) worked, which is then multiplied by the percentage of inflation.

(iii) *Allowance for bad debt rate.* Total allowance for bad debt, divided by total hours (regular, overtime, holiday) worked.

(3) The Administrator will use the most recent economic factors released by the Office of Management and Budget for budget development purposes to derive the cost of living expenses and percentage of inflation factors used in the formulas in this section.

[85 FR 5302, Jan. 30, 2020]

§ 868.92 Explanation of service fees and additional fees.

(a) *Costs included in the fees.* Fees for official services in §§ 868.90 and 868.91 include—

(1) The cost of performing the service and related supervision and administrative costs;

(2) The cost of first-class mail service;

(3) The cost of overtime and premium pay; and

(4) The cost of certification except as provided in § 868.92(c).

(b) *Computing hourly rates.* Hourly fees will be assessed in quarter hour increments for—

(1) Travel from the FGIS field office or assigned duty location to the service point and return; and

(2) The performance of the requested service, less mealtime.

(c) *Additional fees.* Fees in addition to the applicable hourly or unit fee will be assessed when—

(1) An applicant requests more than the original and three copies of a certificate;

(2) An applicant requests onsite typing of certificates or typing of certificates at the FGIS field office during other than normal working hours; and

(3) An applicant requests the use of express-type mail or courier service.

(d) *Application of fees when service is delayed by the applicant.* Hourly fees will be assessed when—

(1) Service has been requested at a specified location;

(2) A Service representative is on duty and ready to provide service but is unable to do so because of a delay not caused by the Service; and

(3) FGIS officials determine that the Service representative(s) cannot be utilized elsewhere or cannot be released without cost to the Service.

(e) *Application of fees when an application for service is withdrawn or dismissed.* Hourly fees will be assessed to the applicant for the scheduled service if the request is withdrawn or dismissed after the Service representative departs for the service point or if the request for service is not withdrawn or dismissed by 2 p.m. of the business day preceding the date of scheduled service. However, hourly fees will not be assessed to the applicant if FGIS officials determine that the Service representative can be utilized elsewhere or if the Service representative can be released without cost to the Service.

(f) *To whom fees are assessed.* Fees for official services including additional fees as provided in § 868.92(c) shall be assessed to and paid by the applicant for the Service.

(g) *Advance payment.* As necessary, the Administrator may require that fees shall be paid in advance of the performance of the requested service. Any fees paid in excess of the amount due shall be used to offset future billings, unless a request for a refund is made by the applicant.

(h) *Time and form of payment*—(1) *Fees for Federal inspection service.* Bills for fees assessed under the regulations for official services performed by FGIS shall be paid by check, draft, or money order, payable to U.S. Department of Agriculture, Federal Grain Inspection Service.

(2) *Fees for cooperator inspection service.* Fees for inspection services provided by a cooperator shall be paid by the applicant to the cooperator in accordance with the cooperator's fee schedule.

[53 FR 3722, Feb. 9, 1988. Redesignated and amended at 60 FR 16364, 16365, Mar. 30, 1995; 61 FR 66536, Dec. 18, 1996; 85 FR 5302, Jan. 30, 2020]

Subpart B—Marketing Standards

SOURCE: 62 FR 6706, Feb. 13, 1997, unless otherwise noted.

§ 868.101 General information.

The Agricultural Marketing Service, (AMS) of the U.S. Department of Agriculture (USDA) facilitates the fair and efficient marketing of agricultural products by maintaining voluntary grade standards for Beans, Whole Dry Peas, Split Peas, and Lentils, which provide a uniform language for describing the quality of these commodities in the marketplace. These standards may cover (but are not limited to) terms, classes, quality levels, performance criteria, and inspection requirements. Procedures contained in this part set forth the process which AMS will follow in developing, issuing, revising, suspending, or terminating the U.S. standards for Beans, Whole Dry Peas, Split Peas, and Lentils. Communications about AMS standards in general should be addressed to the Administrator, AMS, USDA, 1400 Independence Avenue, SW., Washington, DC 20250–3601.

§ 868.102 Procedures for establishing and revising grade standards.

(a) AMS will develop, revise, suspend, or terminate grade standards if it determines that such action is in the public interest. AMS encourages inter-ested parties to participate in the review, development, and revision of grade standards. Interested parties include growers, producers, processors, shippers, distributors, consumers, trade associations, companies, and State or Federal agencies. Such persons may at any time recommend that AMS develop, revise, suspend, or terminate a grade standard. Requests for action should be in writing, and should be accompanied by a draft of the suggested change, as appropriate.

(b) AMS will:

(1) Determine the need for new or revised standards;

(2) Collect technical, marketing, or other appropriate data;

(3) Conduct research regarding new or revised standards, as appropriate; and

(4) Draft the proposed standards.

(c) If AMS determines that new standards are needed, existing standards need to be revised, or the suspension or termination of existing standards is justified, AMS will undertake the action with input from interested parties.

§ 868.103 Public notification of grade standards action.

(a) After developing a standardization proposal, AMS will publish a notice in the FEDERAL REGISTER proposing new or revised standards or suspending or terminating existing standards. The notice will provide a sufficient comment period for interested parties to submit comments.

(b) AMS will simultaneously issue a news release about these actions, notifying the affected industry and general public. AMS will also distribute copies of proposals to anyone requesting a copy or to anyone it believes may be interested, including other Federal, State, or local government agencies.

(c) All comments received within the comment period will be made part of the public record maintained by AMS, will be available to the public for review, and will be considered by AMS before final action is taken on the proposal.

(d) Based on the comments received, AMS's knowledge of standards, grading, marketing, and other technical factors, and any other relevant information, AMS will decide whether the proposed actions should be implemented.

(e) If AMS concludes that the changes as proposed or with appropriate modifications should be adopted, AMS will publish the final changes in the FEDERAL REGISTER as a final notice. AMS will make the grade standards and related information available in printed form and electronic media.

(f) If AMS determines that proposed changes are not warranted, or otherwise are not in the public interest, AMS will either publish in the FEDERAL REGISTER a notice withdrawing the proposal, or will revise the proposal and again seek public input.

Subpart C—United States Standards for Rough Rice

NOTE TO THE SUBPART: Compliance with the provisions of these standards does not excuse failure to comply with the provisions of the Federal Food, Drug, and Cosmetic Act, or other Federal laws.

SOURCE: 42 FR 40869, Aug. 12, 1977; 42 FR 64356, Dec. 23, 1977, unless otherwise noted.

TERMS DEFINED

§ 868.201   Definition of rough rice.

Rice (*Oryza sativa* L.) which consists of 50 percent or more of paddy kernels (see § 868.202(i)) of rice.

[34 FR 7863, May 17, 1969. Redesignated and amended at 60 FR 16364, 16365, Mar. 30, 1995]

§ 868.202   Definition of other terms.

For the purposes of these standards, the following terms shall have the meanings stated below:

(a) *Broken kernels.* Kernels of rice which are less than three-fourths of whole kernels.

(b) *Chalky kernels.* Whole or large broken kernels of rice which are one-half or more chalky.

(c) *Classes.* The following four classes:

Long Grain Rough Rice
Medium Grain Rough Rice
Short Grain Rough Rice
Mixed Rough Rice

Classes shall be based on the percentage of whole kernels, large broken kernels, and types of rice.

(1) "Long grain rough rice" shall consist of rough rice which contains more than 25 percent of whole kernels and which after milling to a well-milled degree, contains not more than 10 percent of whole or broken kernels of medium or short grain rice.

(2) "Medium grain rough rice" shall consist of rough rice which contains more than 25 percent of whole kernels and which after milling to a well-milled degree, contains not more than 10 percent of whole or large broken kernels of long grain rice or whole kernels of short grain rice.

(3) "Short grain rough rice" shall consist of rough rice which contains more than 25 percent of whole kernels and which, after milling to a well-milled degree, contains not more than 10 percent of whole or large broken kernels of long grain rice or whole kernels of medium grain rice.

(4) "Mixed rough rice" shall consist of rough rice which contains more than 25 percent of whole kernels and which, after milling to a well-milled degree, contains more than 10 percent of "other types" as defined in paragraph (h) of this section.

(d) *Damaged kernels.* Whole or broken kernels of rice which are distinctly discolored or damaged by water, insects, heat, or any other means, and whole or large broken kernels of parboiled rice in non-parboiled rice. "Heat-damaged kernels" (see paragraph (e) of this section) shall not function as damaged kernels.

(e) *Heat-damaged kernels.* Whole or large broken kernels of rice which are materially discolored and damaged as a result of heating, and whole or large broken kernels of parboiled rice in non-parboiled rice which are as dark as, or darker in color than, the interpretive line for heat-damaged kernels.

(f) *Milling yield.* An estimate of the quantity of whole kernels and total milled rice (whole and broken kernels combined) that are produced in the milling of rough rice to a well-milled degree.

(g) *Objectionable seeds.* Seeds other than rice, except seeds of *Echinochloa crusgalli* (commonly known as barnyard grass, watergrass, and Japanese millet).

(h) *Other types.* (1) Whole kernels of:

(i) Long grain rice in medium or short grain rice,

(ii) Medium grain rice in long or short grain rice,

(iii) Short grain rice in long or medium grain rice, and

(2) Large broken kernels of long grain rice in medium or short grain rice and large broken kernels of medium or short grain rice in long grain rice.

NOTE: Broken kernels of medium grain rice in short grain rice and large broken kernels of short grain rice in medium grain rice shall not be considered other types.

(i) *Paddy kernels.* Whole or broken unhulled kernels of rice.

(j) *Red rice.* Whole or large broken kernels of rice on which there is an appreciable amount of red bran.

(k) *Seeds.* Whole or broken seeds of any plant other than rice.

(l) *Smutty kernels.* Whole or broken kernels of rice which are distinctly infected by smut.

(m) *Types of rice.* The following three types:

Long grain
Medium grain
Short grain

Types shall be based on the length-width ratio of kernels of rice that are unbroken and the width, thickness, and shape of kernels of rice that are broken as prescribed in FGIS instructions.

(n) *Ungelatinized kernels.* Whole or large broken kernels of parboiled rice with distinct white or chalky areas due to incomplete gelatinization of the starch.

(o) *Whole and large broken kernels.* Rice (including seeds) that (1) passes over a 6 plate (for southern production), or (2) remains on top of a 6 sieve (for western production).

(p) *Whole kernels.* Unbroken kernels of rice and broken kernels of rice which are at least three-fourths of an unbroken kernel.

(q) *6 sieve.* A metal sieve 0.032-inch thick, perforated with rows of round holes 0.0938 (⁶⁄₆₄) inch in diameter.

(r) *6 plate.* A laminated metal plate 0.142-inch thick, with a top lamina 0.051-inch thick, perforated with rows of round holes 0.0938 (⁶⁄₆₄) inch in diameter, and a bottom lamina 0.091-inch thick, without perforations.

[42 FR 40869, Aug. 12, 1977; 42 FR 64356, Dec. 23, 1977, as amended at 47 FR 34516, Aug. 10, 1982; 54 FR 21403, May 18, 1989; 54 FR 51344, Dec. 14, 1989. Redesignated at 60 FR 16364, Mar. 30, 1995]

PRINCIPLES GOVERNING APPLICATION OF STANDARDS

§ 868.203   Basis of determination.

The determination of seeds, objectionable seeds, heat-damaged kernels, red rice and damaged kernels, chalky kernels, other types, color, and the special grade Parboiled rough rice shall be on the basis of the whole and large broken kernels of milled rice that are produced in the milling of rough rice to a well-milled degree. When determining class, the percentage of (a) whole kernels of rough rice shall be determined on the basis of the original sample, and (b) types of rice shall be determined on the basis of the whole and large broken kernels of milled rice that are produced in the milling of rough rice to a well-milled degree. Smutty kernels shall be determined on the basis of the rough rice after it has been cleaned and shelled as prescribed in FGIS instructions, or by any method that is approved by the Administrator as giving equivalent results. All other determinations shall be on the basis of the original sample. Mechanical sizing of kernels shall be adjusted by handpicking as prescribed in FGIS instructions, or by any method that is approved by the Administrator as giving equivalent results.

[42 FR 40869, Aug. 12, 1977; 42 FR 64356, Dec. 23, 1977, as amended at 47 FR 34516, Aug. 10, 1982; 54 FR 21403, May 18, 1989. Redesignated at 60 FR 16364, Mar. 30, 1995]

§ 868.204   Interpretive line samples.

Interpretive line samples showing the official scoring line for factors that are determined by visual examinations shall be maintained by the Federal

Grain Inspection Service, U.S. Department of Agriculture, and shall be available for reference in all inspection offices that inspect and grade rice.

[42 FR 40869, Aug. 12, 1977; 42 FR 64356, Dec. 23, 1977, as amended at 47 FR 34516, Aug. 10, 1982. Redesignated at 54 FR 21403, May 18, 1989, and 60 FR 16364, Mar. 30, 1995]

§ 868.205   Milling requirements.

In determining milling yield (see § 868.202(f)) in rough rice, the degree of milling shall be equal to, or better than, that of the interpretive line sample for "well-milled" rice.

[42 FR 40869, Aug. 12, 1977. Redesignated at 54 FR 21413, May 18, 1989, and further redesignated and amended at 60 FR 16364, 16365, Mar. 30, 1995]

§ 868.206   Milling yield determination.

Milling yield shall be determined by the use of an approved device in accordance with procedures prescribed in FGIS instructions. For the purpose of this paragraph, "approved device" shall include the McGill Miller No. 3 and any other equipment that is approved by the Administrator as giving equivalent results.

NOTE: Milling yield shall not be determined when the moisture content of the rough rice exceeds 18.0 percent.

[42 FR 40869, Aug. 12, 1977; 42 FR 64356, Dec. 23, 1977, as amended at 47 FR 34516, Aug. 10, 1982; Redesignated and amended at 54 FR 21403, May 18, 1989, and further redesignated at 60 FR 16364, Mar. 30, 1995]

§ 868.207   Moisture.

Water content in rough rice as determined by an approved device in accordance with procedures prescribed in the FGIS instructions. For the purpose of this paragraph, "approved device" shall include the Motomco Moisture Meter and any other equipment that is approved by the Administrator as giving equivalent results.

[42 FR 40869, Aug. 12, 1977; 42 FR 64356, Dec. 23, 1977, as amended at 47 FR 34516, Aug. 10, 1982. Redesignated at 54 FR 21403, May 18, 1989, as amended at 54 FR 51344, Dec. 14, 1989. Redesignated at 60 FR 16364, Mar. 30, 1995]

§ 868.208   Percentages.

(a) *Rounding.* Percentages are determined on the basis of weight and are rounded as follows:

(1) When the figure to be rounded is followed by a figure greater than or equal to 5, round to the next higher figure; e.g., report 6.36 as 6.4, 0.35 as 0.4, and 2.45 as 2.5.

(2) When the figure to be rounded is followed by a figure less than 5, retain the figure; e.g., report 8.34 as 8.3 and 1.22 as 1.2.

(b) *Recording.* All percentages, except for milling yield, are stated in whole and tenth percent to the nearest tenth percent. Milling yield is stated to the nearest whole percent.

[54 FR 21403, May 18, 1989. Redesignated at 60 FR 16364, Mar. 30, 1995]

§ 868.209   Information.

Requests for the Rice Inspection Handbook, Equipment Handbook, or for information concerning approved devices and procedures, criteria for approved devices, and requests for approval of devices should be directed to the U.S. Department of Agriculture, Federal Grain Inspection Service, P.O. Box 96454, Washington, DC 20090–6454, or any field office or cooperator.

[54 FR 21404, May 18, 1989. Redesignated at 60 FR 16364, Mar. 30, 1995]

GRADES, GRADE REQUIREMENTS, AND GRADE DESIGNATIONS

§ 868.210   Grades and grade requirements for the classes of Rough Rice. (See also § 868.212.)

Grade	Maximum limits of—							
	Seeds and heat-damaged kernels		Heat-damaged kernels (singly) (Number in 500 grams)	Red rice and damaged kernels (singly or combined) (Percent)	Chalky kernels [1] [2]		Other types [3] (Percent)	Color requirements [1] (minimum)
	Total (singly or combined) (Number in 500 grams)	Heat-damaged kernels and objectionable seeds (singly or combined) (Number in 500 grams)			In long grain rice (Percent)	In medium or short grain rice (Percent)		
U.S. No. 1	4	3	1	0.5	1.0	2.0	1.0	Shall be white or creamy.
U.S. No. 2	7	5	2	1.5	2.0	4.0	2.0	May be slightly gray.
U.S. No. 3	10	8	5	2.5	4.0	6.0	3.0	May be light gray.
U.S. No. 4	27	22	15	4.0	6.0	8.0	5.0	May be gray or slight rosy.
U.S. No. 5	37	32	25	6.0	10.0	10.0	10.0	May be dark gray or rosy.
U.S. No. 6	75	75	75	[4] 15.0	15.0	15.0	10.0	May be dark gray or rosy.

U.S. Sample grade

    U.S. Sample grade shall be rough rice which: (a) does not meet the requirements for any of the grades from U.S. No. 1 to U.S. No. 6, inclusive; (b) contains more than 14.0 percent of moisture; (c) is musty, or sour, or heating; (d) has any commercially objectionable foreign odor; or (e) is otherwise of distinctly low quality.

[1] For the special grade Parboiled rough rice, see § 868.212(b).
[2] For the special grade Glutinous rough rice, see § 868.212(d).
[3] These limits do not apply to the class Mixed Rough Rice.
[4] Rice in grade U.S. No. 6 shall contain not more than 6.0 percent of damaged kernels.

[56 FR 55978, Oct. 31, 1991. Redesignated and amended at 60 FR 16364, Mar. 30, 1995]

§ 868.211   Grade designation and other certificate information.

(a) *Rough rice.* The grade designation for all classes of Rough rice shall be included on the certificate grade-line in the following order:

(1) The letters "U.S.;"

(2) The number of the grade or the words "Sample grade," as warranted;

(3) The words "or better," when applicable and requested by the applicant prior to inspection;

(Approved by the Office of Management and Budget under control number 0580-0013)

(4) The class;

(5) Each applicable special grade (see § 868.213); and

(6) A statement of the milling yield.

(b) *Mixed rough rice information.* For the class Mixed Rough rice, the following information shall be included in the Results section of the certificate in the following order:

(1) The percentage of whole kernels of each type in the order of predominance;

(2) The percentage of large broken kernels of each type in the order of predominance;

(3) The percentage of material removed by the No. 6 sieve or the No. 6 sizing plate; and

(4) The percentage of seeds, when applicable.

(c) *Large broken kernels.* Large broken kernels, other than long grain, in Mixed Rough rice shall be certified as "medium or short grain."

[74 FR 55442, Oct. 28, 2009]

SPECIAL GRADES, SPECIAL GRADE RE-
QUIREMENTS, AND SPECIAL GRADE DES-
IGNATIONS

§868.212 Special grades and require-ments.

A special grade, when applicable, is supplemental to the grade assigned under §868.210. Such special grades for rough rice are established and determined as follows:

(a) *Infested rough rice.* Tolerances for live insects for infested rough rice are defined according to sampling designations as follows:

(1) *Representative sample.* The representative sample consists of the work portion, and the file sample if needed and when available. The rough rice (except when examined according to paragraph (a)(3) of this section will be considered infested if the representative sample contains two or more live weevils, or one live weevil and one or more other live insects injurious to stored rice or five or more other live insects injurious to stored rice.

(2) *Lot as a whole (stationary).* The lot as a whole is considered infested when two or more live weevils, or one live weevil and one or more other live insects injurious to stored rice, or five or more other live insects injurious to stored rice, or 15 or more live Angoumois moths or other live moths injurious to stored rice are found in, on, or about the lot.

(3) *Sample as a whole during continuous loading/unloading.* The minimum sample size for rice being sampled during continuous loading/unloading is 500 grams per each 100,000 pounds of rice. The sample as a whole is considered infested when a component (as defined in FGIS instructions) contains two or more live weevils, or one live weevil and one or more other live insects injurious to stored rice, or five or more other live insects injurious to stored rice.

(b) *Parboiled rough rice.* Parboiled rough rice shall be rough rice in which the starch has been gelatinized by soaking, steaming, and drying. Grades U.S. No. 1 to U.S. No. 6 inclusive, shall contain not more than 10.0 percent of ungelatinized kernels. Grades U.S. No. 1 and U.S. No. 2 shall contain not more than 0.1 percent, grades U.S. No. 3 and

U.S. No. 4 not more than 0.2 percent, and grades U.S. No. 5 and U.S. No. 6 not more than 0.5 percent of nonparboiled rice. If the rice is: (1) Not distinctly colored by the parboiling process, it shall be considered "Parboiled Light"; (2) distinctly but not materially colored by the parboiling process, it shall be considered "Parboiled"; (3) materially colored by the parboiling process, it shall be considered "Parboiled Dark." The color levels for "Parboiled Light," "Parboiled," and "Parboiled Dark" rice shall be in accordance with the interpretive line samples for parboiled rice.

NOTE: The maximum limits for "Chalky kernels," "Heat-damaged kernels," "Kernels damaged by heat," and the "Color require-ments" shown in §868.210 are not applicable to the special grade "Parboiled rough rice."

(c) *Smutty rough rice.* Smutty rough rice shall be rough rice which contains more than 3.0 percent of smutty kernels.

(d) *Glutinous rough rice.* Glutinous rough rice shall be special varieties of rice (Oryza sativa L. glutinosa) which contain more than 50 percent chalky kernels. Grade U.S. No. 1 shall contain not more than 1.0 percent of nonchalky kernels, grade U.S. No. 2 not more than 2.0 percent of nonchalky kernels, grade U.S. No. 3 not more than 4.0 percent of nonchalky kernels, grade U.S. No. 4 not more than 6.0 percent of nonchalky kernels, grade U.S. No. 5 not more than 10.0 percent of nonchalky kernels, and grade U.S. No. 6 not more than 15.0 percent of nonchalky kernels.

NOTE: The maximum limits for "Chalky kernels" in §868.210 are not applicable to the special grade "Glutinous rough rice."

(e) *Aromatic rough rice.* Aromatic rough rice shall be special varieties of rice (Oryza sativa L. scented) that have a distinctive and characteristic aroma; e.g., basmati and jasmine rice.

[42 FR 40869, Aug. 12, 1977, as amended at 54 FR 21406, May 18, 1989; 56 FR 55978, Oct. 31, 1991; 58 FR 68016, Dec. 23, 1993. Redesignated and amended at 60 FR 16364, 16365, Mar. 30, 1995]

§ 868.213 Special grade designation.

The grade designation for infested, parboiled, smutty, glutinous, or aromatic rough rice shall include, following the class, the word(s) "Infested," "Parboiled Light," "Parboiled," "Parboiled Dark," "Smutty," "Glutinous," or "Aromatic," as warranted, and all other information prescribed in § 868.211.

[58 FR 68016, Dec. 23, 1993. Redesignated and amended at 60 FR 16364, Mar. 30, 1995]

Subpart D—United States Standards for Brown Rice for Processing

NOTE TO THE SUBPART: Compliance with the provisions of these standards does not excuse failure to comply with the provisions of the Federal Food, Drug, and Cosmetic Act, or other Federal laws.

SOURCE: 42 FR 40869, Aug. 12, 1977; 42 FR 64356, Dec. 23, 1977, unless otherwise noted. Redesignated at 60 FR 16364, Mar. 30, 1995.

TERMS DEFINED

§ 868.251 Definition of brown rice for processing.

Rice (*Oryza sativa* L.) which consists of more than 50.0 percent of kernels of brown rice, and which is intended for processing to milled rice.

§ 868.252 Definition of other terms.

For the purposes of these standards, the following terms shall have the meanings stated below:

(a) *Broken kernels.* Kernels of rice which are less than three-fourths of whole kernels.

(b) *Brown rice.* Whole or broken kernels of rice from which the hulls have been removed.

(c) *Chalky kernels.* Whole or broken kernels of rice which are one-half or more chalky.

(d) *Classes.* There are four classes of brown rice for processing.

Long Grain Brown Rice for Processing.
Medium Grain Brown Rice for Processing.
Short Grain Brown Rice for Processing.
Mixed Brown Rice for Processing.

Classes shall be based on the percentage of whole kernels, broken kernels, and types of rice.

(1) "Long-grain brown rice for processing" shall consist of brown rice for processing which contains more than 25.0 percent of whole kernels of brown rice and not more than 10.0 percent of whole or broken kernels of medium- or short-grain rice.

(2) "Medium-grain brown rice for processing" shall consist of brown rice for processing which contains more than 25.0 percent of whole kernels of brown rice and not more than 10.0 percent of whole or broken kernels of long-grain rice or whole kernels of short-grain rice.

(3) "Short-grain brown rice for processing" shall consist of brown rice for processing which contains more than 25.0 percent of whole kernels of brown rice and not more than 10.0 percent of whole or broken kernels of long-grain rice or whole kernels of medium-grain rice.

(4) "Mixed brown rice for processing" shall be brown rice for processing which contains more than 25.0 percent of whole kernels of brown rice and more than 10.0 percent of "other types" as defined in paragraph (i) of this section.

(e) *Damaged kernels.* Whole or broken kernels of rice which are distinctly discolored or damaged by water, insects, heat, or any other means (including parboiled kernels in nonparboiled rice and smutty kernels). "Heat-damaged kernels" (see paragraph (f) of this section) shall not function as damaged kernels.

(f) *Heat-damaged kernels.* Whole or broken kernels of rice which are materially discolored and damaged as a result of heating and parboiled kernels in nonparboiled rice which are as dark as, or darker in color than, the interpretive line for heat-damaged kernels.

(g) *Milling yield.* An estimate of the quantity of whole kernels and total milled rice (whole and broken kernels combined) that is produced in the milling of brown rice for processing to a well-milled degree.

(h) *Objectionable seeds.* Whole or broken seeds other than rice, except seeds of *Echinochloa crusgalli* (commonly known as barnyard grass, watergrass, and Japanese millet).

(i) *Other types.* (1) Whole kernels of:

(i) Long grain rice in medium or short grain rice and medium or short grain rice in long grain rice,

(ii) Medium grain rice in long or short grain rice,

(iii) Short grain rice in long or medium grain rice,

(2) Broken kernels of long grain rice in medium or short grain rice and broken kernels of medium or short grain rice in long grain rice.

NOTE: Broken kernels of medium grain rice in short grain rice and broken kernels of short grain rice in medium grain rice shall not be considered other types.

(j) *Paddy kernels.* Whole or broken unhulled kernels and whole or broken kernels of rise having a portion or portions of the hull remaining which cover one-half (½) or more of the whole or broken kernel.

(k) *Red rice.* Whole or broken kernels of rice on which the bran is distinctly red in color.

(l) *Related material.* All by-products of a paddy kernel, such as the outer glumes, lemma, palea, awn, embryo, and bran layers.

(m) *Seeds.* Whole or broken seeds of any plant other than rice.

(n) *Smutty kernels.* Whole or broken kernels of rice which are distinctly infected by smut.

(o) *Types of rice.* There are three types of brown rice for processing:

Long grain
Medium grain
Short grain

Types shall be based on the length/width ratio of kernels of rice that are unbroken and the width, thickness, and shape of kernels of rice that are broken as prescribed in FGIS instructions.

(p) *Ungelantinized kernels.* Whole or broken kernels of parboiled rice with distinct white or chalky areas due to incomplete gelatization of the starch.

(q) *Unrelated material.* All matter other than rice, related material, and seeds.

(r) *Well-milled kernels.* Whole or broken kernels of rice from which the hulls and practically all of the embryos and the bran layers have been removed.

(s) *Whole kernels.* Unbroken kernels of rice and broken kernels of rice which are at least three-fourths of an unbroken kernel.

(t) *6 plate.* A laminated metal plate 0.142-inch thick, with a top lamina 0.051-inch thick, perforated with rows of round holes 0.0938 (⁶⁄₆₄) inch in diameter, and a bottom lamina 0.091-inch thick, without perforations.

(u) *6½ sieve.* A metal sieve 0.032-inch thick, perforated with rows of round holes 0.1016 (6½/64) inch in diameter.

[13 FR 9479, Dec. 31, 1948, as amended at 44 FR 73008, Dec. 17, 1979; 47 FR 34516, Aug. 10, 1982; 54 FR 21403, 21406, May 18, 1989; 54 FR 51344, Dec. 14, 1989. Redesignated at 60 FR 16364, Mar. 30, 1995]

PRINCIPLES GOVERNING APPLICATION OF STANDARDS

§ 868.253 Basis of determination.

The determination of kernels damaged by heat, heat-damaged kernels, parboiled kernels in nonparboiled rice, and the special grade Parboiled brown rice for processing shall be on the basis of the brown rice for processing after it has been milled to a well-milled degree. All other determinations shall be on the basis of the original sample. Mechanical sizing of kernels shall be adjusted by handpicking as prescribed in FGIS instructions, or by any method which gives equivalent results.

[42 FR 40869, Aug. 12, 1977; 42 FR 64356, Dec. 23, 1977, as amended at 47 FR 34516, Aug. 10, 1982; 54 FR 21403, 21406, May 18, 1989, and further redesignated at 60 FR 16364, Mar. 30, 1995]

§ 868.254 Broken kernels determination.

Broken kernels shall be determined by the use of equipment and procedures prescribed in FGIS instructions, or by any method which gives equivalent results.

[42 FR 40869, Aug. 12, 1977; 42 FR 64356, Dec. 23, 1977, as amended at 47 FR 34516, Aug. 10, 1982; 54 FR 21403, May 18, 1989. Redesignated at 54 FR 21406, May 18, 1989, and further redesignated at 60 FR 16364, Mar. 30, 1995]

§ 868.255 Interpretive line samples.

Interpretive line samples showing the official scoring line for factors that are determined by visual observation shall be maintained by the Federal Grain Inspection Service, U.S. Department of Agriculture, and shall be available for reference in all inspection offices that inspect and grade rice.

[42 FR 40869, Aug. 12, 1977; 42 FR 64356, Dec. 23, 1977, as amended at 47 FR 34516, Aug. 10, 1982; 54 FR 21403, May 18, 1989. Redesignated at 54 FR 21406, May 18, 1989, and further redesignated at 60 FR 16364, Mar. 30, 1995]

§ 868.256 Milling requirements.

In determining milling yield (see § 868.252(g)) in brown rice for processing, the degree of milling shall be equal to, or better than, that of the interpretive line sample for "well-milled" rice.

[42 FR 40869, Aug. 12, 1977. Redesignated at 21406, May 18, 1989, and further redesignated and amended at 60 FR 16364, 16365, Mar. 30, 1995]

§ 868.257 Milling yield determination.

Milling yield shall be determined by the use of an approved device in accordance with procedures prescribed in FGIS instructions. For the purpose of this paragraph, "approved device" shall include the McGill Miller No. 3 and any other equipment that is approved by the Administrator as giving equivalent results.

NOTE: Milling yield shall not be determined when the moisture content of the brown rice for processing exceeds 18.0 percent.

[42 FR 40869, Aug. 12, 1977; 42 FR 64356, Dec. 23, 1977, as amended at 47 FR 34516, Aug. 10, 1982; 54 FR 21403, May 18, 1989. Redesignated at 54 FR 21406, May 18, 1989, and further redesignated at 60 FR 16364, Mar. 30, 1995]

§ 868.258 Moisture.

Water content in brown rice for processing as determined by an approved device in accordance with procedures prescribed in FGIS instructions. For the purpose of this paragraph, "ap-proved device" shall include the Motomco Moisture Meter and any other equipment that is approved by the Administrator as giving equivalent results.

[42 FR 40869, Aug. 12, 1977; 42 FR 64356, Dec. 23, 1977, as amended at 47 FR 34516, Aug. 10, 1982; 54 FR 21403, May 18, 1989. Redesignated at 54 FR 21406, May 18, 1989, and further redesignated at 60 FR 16364, Mar. 30, 1995]

§ 868.259 Percentages.

(a) *Rounding.* Percentages are determined on the basis of weight and are rounded as follows:

(1) When the figure to be rounded is followed by a figure greater than or equal to 5, round to the next higher figure; e.g., report 6.36 as 6.4, 0.35 as 0.4, and 2.45 as 2.5.

(2) When the figure to be rounded is followed by a figure less than 5, retain the figure, e.g., report 8.34 as 8.3 and 1.22 and 1.2.

(b) *Recording.* All percentages, except for milling yield, are stated in whole and tenth percent to the nearest whole percent. Milling yield is stated to the nearest whole percent.

[54 FR 21406, May 18, 1989. Redesignated at 60 FR 16364, Mar. 30, 1995]

§ 868.260 Information.

Requests for the Rice Inspection Handbook, Equipment Handbook, or for information concerning approved devices and procedures, criteria for approved devices, and requests for approval of devices should be directed to the U.S. Department of Agriculture, Federal Grain Inspection Service, P.O. Box 96454, Washington, DC 20090–6454, or any field office or cooperator.

[54 FR 21406, May 18, 1989. Redesignated at 60 FR 16364, Mar. 30, 1995]

GRADES, GRADE REQUIREMENTS, AND GRADE DESIGNATIONS

§ 868.261 Grade and grade requirements for the classes of brown rice for processing. (See also § 868.263.)

Grade	Paddy kernels		Seeds and heat-damaged kernels			Red rice and damaged kernels (singly or combined) (percent)	Chalky kernels[1][2] (percent)	Broken kernels removed by a 6 plate or a 6½ sieve[3] (percent)	Other types[4]	Wellmilled kernels (percent)
	Percent	Number in 500 grams	Total (singly or combined) (number in 500 grams)	Heat-damaged kernels (number in 500 grams)	Objectionable seeds (number in 500 grams)					
U.S. No. 1	—	20	10	1	2	1.0	2.0	1.0	1.0	1.0
U.S. No. 2	2.0	—	40	2	10	2.0	4.0	2.0	2.0	3.0
U.S. No. 3	2.0	—	70	4	20	4.0	6.0	3.0	5.0	10.0
U.S. No. 4	2.0	—	100	8	35	8.0	8.0	4.0	10.0	10.0
U.S. No. 5	2.0	—	150	15	50	15.0	15.0	6.0	10.0	10.0
U.S. Sample grade	U.S. Sample grade shall be brown rice for processing which (a) does not meet the requirements for any of the grades from U.S. No. 1 to U.S. No. 5, inclusive; (b) contains more than 14.5 percent of moisture; (c) is musty, or sour, or heating; (d) has any commercially objectionable foreign odor; (e) contains more than 0.2 percent of related material or more than 0.1 percent of unrelated material; (f) contains two or more live weevils or other live insects; or (g) is otherwise of distinctly low quality.									

[1] For the special grade Parboiled brown rice for processing, see § 868.263(a).
[2] For the special grade Glutinous brown rice for processing, see § 868.263(c).
[3] Plates should be used for southern production rice and sieves should be used for western production rice, but any device or method which gives equivalent results may be used.
[4] These limits do not apply to the class Mixed Brown Rice for Processing.

[56 FR 55979, Oct. 31, 1991. Redesignated and amended at 60 FR 16364, Mar. 30, 1995]

§ 868.262 Grade designation and other certificate information.

(a) *Brown rice for processing.* The grade designation for all classes of Brown rice for processing shall be included on the certificate grade-line in the following order:

(1) The letters "U.S.;"

(2) The number of the grade or the words "Sample grade," as warranted;

(3) The words "or better," when applicable and requested by the applicant prior to inspection;

(Approved by the Office of Management and Budget under control number 0580–0013)

(4) The class; and

(5) Each applicable special grade (see § 868.264).

(b) *Mixed Brown rice for Processing information.* For the class Mixed Brown rice for processing, the following information shall be included in the Results section of the certificate in the following order:

(1) The percentage of whole kernels of each type in the order of predominance;

(2) The percentage of broken kernels of each type in the order of predominance, when applicable; and

(3) The percentage of seeds, related material, and unrelated material.

(c) *Broken kernels.* Broken kernels, other than long grain in Mixed Brown rice for processing shall be certified as "medium or short grain."

[74 FR 55442, Oct. 28, 2009]

SPECIAL GRADES, SPECIAL GRADE REQUIREMENTS, AND SPECIAL GRADE DESIGNATIONS

§ 868.263 Special grades and special grade requirements.

A special grade, when applicable, is supplemental to the grade assigned under § 868.262. Such special grades for brown rice for processing are established and determined as follows:

(a) *Parboiled brown rice for processing.* Parboiled brown rice for processing shall be rice in which the starch has been gelatinized by soaking, steaming, and drying. Grades U.S. Nos. 1 to 5, inclusive, shall contain not more than 10.0 percent of ungelatinized kernels.

Grades U.S. No. 1 and U.S. No. 2 shall contain not more than 0.1 percent, grades U.S. No. 3 and U.S. No. 4 not more than 0.2 percent, and grade U.S. No. 5 not more than 0.5 percent of non-parboiled rice.

NOTE: The maximum limits for "chalky kernels," "Heat-damaged kernels," and "Kernels damaged by heat" shown in § 868.261 are not applicable to the special grade "Parboiled brown rice for processing."

(b) *Smutty brown rice for processing.* Smutty brown rice for processing shall be rice which contains more than 3.0 percent of smutty kernels.

(c) *Glutinous brown rice for processing.* Glutinous brown rice for processing shall be special varieties of rice (Oryza sativa L. glutinosa) which contain more than 50 percent chalky kernels. Grade U.S. No. 1 shall contain not more than 1.0 percent of nonchalky kernels, grade U.S. No. 2 not more than 2.0 percent of nonchalky kernels, grade U.S. No. 3 not more than 4.0 percent of nonchalky kernels, grade U.S. No. 4 not more than 6.0 percent of nonchalky kernels, and grade U.S. No. 5 not more than 10.0 percent of nonchalky kernels.

NOTE: The maximum limits for "Chalky kernels" in § 868.261 are not applicable to the special grade "Glutinous brown rice for processing."

(d) *Aromatic brown rice for processing.* Aromatic brown rice for processing shall be special varieties of rice (Oryza sativa L. scented) that have a distinctive and characteristic aroma; e.g., basmati and jasmine rice.

[42 FR 40869, Aug. 12, 1977; 42 FR 64356, Dec. 23, 1977, as amended at 56 FR 55979, Oct. 31, 1991; 58 FR 68016, Dec. 23, 1993. Redesignated and amended at 60 FR 16364, 16365, Mar. 30, 1995]

§ 868.264 Special grade designation.

The grade designation for parboiled, smutty, glutinous, or aromatic brown rice for processing shall include, following the class, the word(s) "Parboiled," "Smutty," "Glutinous," or "Aromatic," as warranted, and all other information prescribed in § 868.262.

[58 FR 68016, Dec. 23, 1993. Redesignated and amended at 60 FR 16364, Mar. 30, 1995]

Subpart E—United States Standards for Milled Rice

NOTE TO THE SUBPART: Compliance with the provisions of these standards does not excuse failure to comply with the provisions of the Federal Food, Drug, and Cosmetic Act, or other Federal laws.

SOURCE: 42 FR 40869, Aug. 12, 1977; 42 FR 64356, Dec. 23, 1977, unless otherwise noted. Redesignated at 60 FR 16364, Mar. 30, 1995.

TERMS DEFINED

§ 868.301 Definition of milled rice.

Whole or broken kernels of rice (*Oryza sativa* L.) from which the hulls and at least the outer bran layers have been removed and which contain not more than 10.0 percent of seeds, paddy kernels, or foreign material, either singly or combined.

[48 FR 24859, June 3, 1983. Redesignated at 60 FR 16364, Mar. 30, 1995]

§ 868.302 Definition of other terms.

For the purposes of these standards, the following terms shall have the meanings stated below:

(a) *Broken kernels.* Kernels of rice which are less than three-fourths of whole kernels.

(b) *Brown rice.* Whole or broken kernels of rice from which the hulls have been removed.

(c) *Chalky kernels.* Whole or broken kernels of rice which are one-half or more chalky.

(d) *Classes.* There are seven classes of milled rice. The following four classes shall be based on the percentage of whole kernels, and types of rice:

Long Grain Milled Rice.
Medium Grain Milled Rice.
Short Grain Milled Rice.
Mixed Milled Rice.

The following three classes shall be based on the percentage of whole kernels and of broken kernels of different size:

Second Head Milled Rice.
Screenings Milled Rice.
Brewers Milled Rice.

(1) "Long grain milled rice" shall consist of milled rice which contains more than 25.0 percent of whole kernels of milled rice and in U.S. Nos. 1 through 4 not more than 10.0 percent of whole or broken kernels of medium or short grain rice. U.S. No. 5 and U.S. No.

6 long grain milled rice shall contain not more than 10.0 percent of whole kernels of medium or short grain milled rice (broken kernels do not apply).

(2) "Medium grain milled rice" shall consist of milled rice which contains more than 25.0 percent of whole kernels of milled rice and in U.S. Nos. 1 through 4 not more than 10.0 percent of whole or broken kernels of long grain rice or whole kernels of short grain rice. U.S. No. 5 and U.S. No. 6 medium grain milled rice shall contain not more than 10.0 percent of whole kernels of long or short grain milled rice (broken kernels do not apply).

(3) "Short grain milled rice" shall consist of milled rice which contains more than 25.0 percent of whole kernels of milled rice and in U.S. Nos. 1 through 4 not more than 10.0 percent of whole or broken kernels of long grain rice or whole kernels of medium grain rice. U.S. No. 5 and U.S. No. 6 short grain milled rice shall contain not more than 10.0 percent of whole kernels of long or medium grain milled rice (broken kernels do not apply).

(4) "Mixed milled rice" shall consist of milled rice which contains more than 25.0 percent of whole kernels of milled rice and more than 10.0 percent of "other types" as defined in paragraph (i) of this section. U.S. No. 5 and U.S. No. 6 mixed milled rice shall contain more than 10.0 percent of whole kernels of "other types" (broken kernels do not apply).

(5) "Second head milled rice" shall consist of milled rice which, when determined in accordance with §868.303, contains:

(i) Not more than (a) 25.0 percent of whole kernels, (b) 7.0 percent of broken kernels removed by a 6 plate, (c) 0.4 percent of broken kernels removed by a 5 plate, and (d) 0.05 percent of broken kernels passing through a 4 sieve (southern production); or

(ii) Not more than (a) 25.0 percent of whole kernels, (b) 50.0 percent of broken kernels passing through a 6½ sieve, and (c) 10.0 percent of broken kernels passing through a 6 sieve (western production).

(6) "Screenings milled rice" shall consist of milled rice which, when de-

termined in accordance with §868.303, contains:

(i) Not more than (a) 25.0 percent of whole kernels, (b) 10.0 percent of broken kernels removed by a 5 plate, and (c) 0.2 percent of broken kernels passing through a 4 sieve (southern production); or

(ii) Not more than (a) 25.0 percent of whole kernels and (b) 15.0 percent of broken kernels passing through a 5½ sieve; and more than (c) 50.0 percent of broken kernels passing through a 6½ sieve and (d) 10.0 percent of broken kernels passing through a 6 sieve (western production).

(7) "Brewers milled rice" shall consist of milled rice which, when determined in accordance with §868.303, contains not more than 25.0 percent of whole kernels and which does not meet the kernel-size requirements for the class Second Head Milled Rice or Screenings Milled Rice.

(e) *Damaged kernels.* Whole or broken kernels of rice which are distinctly discolored or damaged by water, insects, heat, or any other means, and parboiled kernels in nonparboiled rice. "Heat-damaged kernels" (see paragraph (g) of this section) shall not function as damaged kernels.

(f) *Foreign material.* All matter other than rice and seeds. Hulls, germs, and bran which have separated from the kernels of rice shall be considered foreign material.

(g) *Heat-damaged kernels.* Whole or broken kernels of rice which are materially discolored and damaged as a result of heating and parboiled kernels in nonparboiled rice which are as dark as, or darker in color than, the interpretive line for heat-damaged kernels.

(h) *Objectionable seeds.* Seeds other than rice, except seeds of *Echinochloa crusgalli* (commonly known as barnyard grass, watergrass, and Japanese millet).

(i) *Other types.* (1) Whole kernels of:
(i) Long grain rice in medium or short grain rice, (ii) medium grain rice in long or short grain rice, (iii) Short grain rice in long or medium grain rice, and (2) broken kernels of long grain rice in medium or short grain rice and broken kernels of medium or short grain rice in long grain rice, except in U.S. No. 5 and U.S. No. 6 milled rice. In

U.S. No. 5 and U.S. No. 6 milled rice, only whole kernels will apply.

NOTE: Broken kernels of medium grain rice in short grain rice and broken kernels of short grain rice in medium grain rice shall not be considered other types.

(j) *Paddy Kernels.* Whole or broken unhulled kernels of rice; whole or broken kernels of brown rice, and whole or broken kernels of milled rice having a portion or portions of the hull remaining which cover one-eighth (⅛) or more of the whole or broken kernel.

(k) *Red rice.* Whole or broken kernels of rice on which there is an appreciable amount of red bran.

(l) *Seeds.* Whole or broken seeds of any plant other than rice.

(m) *Types of rice.* There are three types of milled rice as follows:

Long grain.
Medium grain.
Short grain.

Types shall be based on the length-width ratio of kernels of rice that are unbroken and the width, thickness, and shape of kernels that are broken, prescribed in FGIS instructions.

(n) *Ungelatinized kernels.* Whole or broken kernels of parboiled rice with distinct white or chalky areas due to incomplete gelatinization of the starch.

(o) *Well-milled kernels.* Whole or broken kernels of rice from which the hulls and practically all of the germs and the bran layers have been removed.

NOTE: This factor is determined on an individual kernel basis and applies to the special grade Undermilled milled rice only.

(p) *Whole kernels.* Unbroken kernels of rice and broken kernels of rice which are at least three-fourths of an unbroken kernel.

(q) *5 plate.* A laminated metal plate 0.142-inch thick, with a top lamina, 0.051-inch thick, perforated with rows of round holes 0.0781 (⁵⁄₆₄) inch in diameter, ⁵⁄₃₂ inch from center to center, with each row staggered in relation to the adjacent rows, and a bottom lamina 0.091-inch thick, without perforations.

(r) *6 plate.* A laminated metal plate 0.142-inch thick, with a top lamina 0.051-inch thick, perforated with rows of round holes 0.0938 (⁶⁄₆₄) inch in diameter, ⁵⁄₃₂ inch from center to center, with each row staggered in relation to the adjacent rows, and a bottom lam-

ina 0.091-inch thick, without perforations.

(s) *2½ sieve.* A metal sieve 0.032-inch thick, perforated with rows of round holes 0.0391 (2½/64) inch in diameter, 0.075-inch from center to center, with each row staggered in relation to the adjacent rows.

(t) *4 sieve.* A metal sieve 0.032-inch thick, perforated with rows of round holes 0.0625 (⁴⁄₆₄) inch in diameter, ⅛ inch from center to center, with each row staggered in relation to the adjacent rows.

(u) *5 sieve.* A metal sieve 0.032-inch thick, perforated with rows of round holes 0.0781 (⁵⁄₆₄) inch in diameter, ⁵⁄₃₂ inch from center to center, with each row staggered in relation to the adjacent rows.

(v) *5½ sieve.* A metal sieve 0.032-inch thick, perforated with rows of round holes 0.0859 (5½/64) inch in diameter, ⁹⁄₆₄ inch from center to center, with each row staggered in relation to the adjacent rows.

(w) *6 sieve.* A metal sieve 0.032-inch thick, perforated with rows of round holes 0.0938 (⁶⁄₆₄) inch in diameter, ⁵⁄₃₂ inch from center to center, with each row staggered in relation to the adjacent rows.

(x) *6½ sieve.* A metal sieve 0.032-inch thick, perforated with rows of round holes 0.1016 (6½/64) inch in diameter, ⁵⁄₃₂ inch from center to center, with each row staggered in relation to the adjacent rows.

(y) *30 sieve.* A woven wire cloth sieve having 0.0234-inch openings, with a wire diameter of 0.0153 inch, and meeting the specifications of American Society for Testing and Materials Designation E-11-61, prescribed in FGIS instructions.

[13 FR 9479, Dec. 31, 1948, as amended at 44 FR 73008, Dec. 17, 1979; 47 FR 34516, Aug. 10, 1982; 54 FR 21403, 21406, May 18, 1989; 54 FR 51345, Dec. 14, 1989. Redesignated and amended at 60 FR 16364, Mar. 30, 1995]

PRINCIPLES GOVERNING APPLICATION OF STANDARDS

§ 868.303 Basis of determination.

All determinations shall be on the basis of the original sample. Mechanical sizing of kernels shall be adjusted by handpicking, as prescribed in FGIS instructions, or by any method which gives equivalent results.

[42 FR 40869, Aug. 12, 1977; 42 FR 64356, Dec. 23, 1977, as amended at 47 FR 34516, Aug. 10, 1982; 54 FR 21403, 21406, May 18, 1989. Redesignated at 60 FR 16364, Mar. 30, 1995]

§ 868.304 Broken kernels determination.

Broken kernels shall be determined by the use of equipment and procedures prescribed in FGIS instructions or by any method which gives equivalent results.

[42 FR 40869, Aug. 12, 1977; 42 FR 64356, Dec. 23, 1977, as amended at 47 FR 34516, Aug. 10, 1982; 54 FR 21403, May 18, 1989. Redesignated at 54 FR 21406, May 18, 1989 and 60 FR 16364, Mar. 30, 1995]

§ 868.305 Interpretive line samples.

Interpretive line samples showing the official scoring line for factors that are determined by visual observation shall be maintained by the Federal Grain Inspection Service, U.S. Department of Agriculture, and shall be available for reference in all inspection offices that inspect and grade rice.

[42 FR 40869, Aug. 12, 1977; 42 FR 64356, Dec. 23, 1977, as amended at 47 FR 34516, Aug. 10, 1982. Redesignated at 54 FR 21406, May 18, 1989 and 60 FR 16364, Mar. 30, 1995]

§ 868.306 Milling requirements.

The degree of milling for milled rice; *i.e.*, "hard milled," "well-milled," and "reasonably well-milled," shall be equal to, or better than, that of the interpretive line samples for such rice.

[67 FR 61250, Sept. 30, 2002]

§ 868.307 Moisture.

Water content in milled rice as determined by an FGIS approved device in accordance with procedures prescribed in FGIS instructions.

[67 FR 61250, Sept. 30, 2002]

§ 868.308 Percentages.

(a) *Rounding.* Percentages are determined on the basis of weight and are rounded as follows:

(1) When the figure to be rounded is followed by a figure greater than or equal to 5, round to the next higher figure; e.g., report 6.36 as 6.4, 0.35 as 0.4, and 2.45 as 2.5.

(2) When the figure to be rounded is followed by a figure less than 5, retain the figure, e.g., report 8.34 as 8.3 and 1.22 and 1.2.

(b) *Recording.* The percentage of broken kernels removed by a 5 plate in U.S. Nos. 1 and 2 Milled Rice and the percentage of objectionable seeds in U.S. No. 1 Brewers Milled Rice is reported to the nearest hundredth percent. The percentages of all other factors are recorded to the nearest tenth of a percent.

[54 FR 21406, May 18, 1989. Redesignated at 60 FR 16364, Mar. 30, 1995]

§ 868.309 Information.

Requests for the Rice Inspection Handbook, Equipment Handbook, or for information concerning approved devices and procedures, criteria for approved devices, and requests for approval of devices should be directed to the U.S. Department of Agriculture, Federal Grain Inspection Service, P.O. Box 96454, Washington, DC 20090–6454, or any field office or cooperator.

[54 FR 21407, May 18, 1989. Redesignated at 60 FR 16364, Mar. 30, 1995]

§ 868.310 Grades and grade requirements for the classes Long Grain Milled Rice, Medium Grain Milled Rice, Short Grain Milled Rice, and Mixed Milled Rice. (See also § 868.315.)

GRADES, GRADE REQUIREMENTS, AND GRADE DESIGNATIONS

Maximum limits of—

Grade	Seeds, heat damaged, and paddy kernels (singly or combined) Total (number in 500 grams)	Heat damaged kernels and objectionable seeds (number in 500 grams)	Red rice and damaged kernels (singly or combined) (percent)	Chalky kernels [1][2] In long grain rice (percent)	In medium or short grain rice (percent)	Broken kernels Total (percent)	Removed by a 5 plate [3] (percent)	Removed by a 6 plate [3] (percent)	Through a 6 sieve [3] (percent)	Other types [4] Whole kernels (percent)	Whole and broken kernels (percent)	Color requirements [1]	Minimum milling requirements [5]
U.S. No. 1	2	1	0.5	1.0	2.0	4.0	0.04	0.1	0.1	....	1.0	White or creamy.	Well Milled.
U.S. No. 2	4	2	1.5	2.0	4.0	7.0	0.06	0.2	0.2	....	2.0	Slightly gray.	Well Milled.
U.S. No. 3	7	5	2.5	4.0	6.0	15.0	0.1	0.8	0.5	....	3.0	Light gray	Reasonably well milled.
U.S. No. 4	20	15	4.0	6.0	8.0	25.0	0.4	1.0	0.7	....	5.0	Gray or slightly rosy.	Reasonably well milled.
U.S. No. 5	30	25	6.0[5]	10.0	10.0	35.0	0.7	3.0	1.0	10.0		Dark gray or rosy.	Reasonably well milled.
U.S. No. 6	75	75	15.0[6]	15.0	15.0	50.0	1.0	4.0	2.0	10.0		Dark gray or rosy.	Reasonably well milled.
U.S. Sample grade:													

U.S. Sample grade shall be milled rice of any of these classes which: (a) Does not meet the requirements for any of the grades from U.S. No. 1 to U.S. No. 6, inclusive; (b) contains more than 15.0 percent of moisture; (c) is musty or sour, or heating; (d) has any commercially objectionable foreign odor; (e) contains more than 0.1 percent of foreign material; (f) Contains two or more live or dead weevils or other insects, insect webbing, or insect refuse; (g) is otherwise of distinctly low quality.

¹ For the special grade Parboiled milled rice, see §868.315(c).
² For the special grade Glutinous milled rice, see §868.315(e).
³ Plates should be used for southern production rice; and sieves should be used for western production rice, but any device or method which gives equivalent results may be used.
⁴ These limits do not apply to the class Mixed Milled Rice.
⁵ For the special grade Undermilled milled rice, see §868.315(d).
⁶ Grade U.S. No. 6 shall contain not more than 6.0 percent of damaged kernels.

[67 FR 61250, Sept. 30, 2002, as amended at 70 FR 37255, June 29, 2005]

§ 868.311 Grades and grade requirements for the class Second Head Milled Rice. (See also § 868.315.)

GRADES, GRADE REQUIREMENTS, AND GRADE DESIGNATIONS

Grade	Seeds, heat-damaged, and paddy kernels (singly or combined)		Red rice and damaged kernels (singly or combined) (percent)	Chalky kernels [1] [3] (percent)	Color require- ments [1]	Minimum milling requirements [2]
	Total (number in 500 grams)	Heat-dam- aged ker- nels and objection- able seeds (number in 500 grams)				
U.S. No. 1	15	5	1.0	4.0	White or Creamy.	Well milled.
U.S. No. 2	20	10	2.0	6.0	Slightly gray	Well milled.
U.S. No. 3	35	15	3.0	10.0	Light gray	Reasonably well milled.
U.S. No. 4	50	25	5.0	15.0	Gray or slightly gray.	Reasonably well milled.
U.S. No. 5	75	40	10.0	20.0	Dark gray or rosy.	Reasonably well milled.

U.S. Sample grade:

 U.S. Sample grade shall be milled rice of this class which: (a) Does not meet the requirements for any of the grades from U.S. No. 1 to U.S. No. 5, inclusive; (b) contains more than 15.0 percent of moisture; (c) is musty or sour, or heating; (d) has any commercially objectionable foreign odor; (e) contains more than 0.1 percent of foreign material; (f) contains two or more live or dead weevils or other insects, insect webbing, or insect refuse; or (g) is otherwise of distinctly low quality.

[1] For the special grade Parboiled milled rice, see § 868.315(c).
[2] For the special grade Undermilled milled rice, see § 868.315(d).
[3] For the special grade Glutinous milled rice, see § 868.315(e).

[67 FR 61251, Sept. 30, 2002]

§ 868.312 Grade and grade requirements for the class Screenings Milled Rice. (See also § 868.315.)

GRADES, GRADE REQUIREMENTS, AND GRADE DESIGNATIONS

Grade	Paddy kernels and seeds		Chalky kernels [1] [3] (percent)	Color requirements [1]	Minimum milling requirements [2]
	Total (number in 500 grams)	Objectionable seeds (number in 500 grams)			
U.S. No. 1 [4] [5]	30	20	5.0	White or Creamy	Well milled.
U.S. No. 2 [4] [5]	75	50	8.0	Slightly gray	Well milled.
U.S. No. 3 [4] [5]	125	90	12.0	Light gray or slightly rosy	Reasonably well milled.
U.S. No. 4 [4] [5]	175	140	20.0	Gray or rosy	Reasonably well milled.
U.S. No. 5	250	200	30.0	Dark gray or very rosy ...	Reasonably well milled.

U.S. Sample grade:

 U.S. Sample grade shall be milled rice of this class which: (a) Does not meet the requirements for any of the grades from U.S. No. 1 to U.S. No. 5, inclusive; (b) contains more than 15.0 percent of moisture; (c) is musty or sour, or heating; (d) has any commercially objectionable foreign odor; (e) has a badly damaged or extremely red appearance (f) contains more than 0.1 percent of foreign material; (g) contains two or more live or dead weevils or other insects, insect webbing, or insect refuse; or (h) is otherwise of distinctly low quality.

[1] For the special grade Parboiled milled rice, see § 868.315(c).
[2] For the special grade Undermilled milled rice, see § 868.315(d).
[3] For the special grade Glutinous milled rice, see § 868.315(e).
[4] Grades U.S. No. 1 to U.S. No. 4, inclusive, shall contain not more than 3.0 percent of heat-damaged kernels, kernels damaged by heat and/or parboiled kernels in nonparboiled rice.
[5] Grades U.S. No. 1 to U.S. No. 4, inclusive, shall contain not more than 1.0 percent of material passing through a 30 sieve.

[67 FR 61251, Sept. 30, 2002]

§ 868.313   Grades and grade requirements for the class Brewers Milled Rice. (See also § 868.315.)

GRADES, GRADE REQUIREMENTS, AND GRADE DESIGNATIONS

| Grade | Maximum limits of— paddy kernels and seeds | | Color requirements [1] | Minimum milling requirements [2] |
	Total (singly or combined) (percent)	Objectionable seeds (percent)		
U.S. No. 1 [3] [4]	0.5	0.05	White or Creamy	Well milled.
U.S. No. 2 [3] [4]	1.0	0.1	Slightly gray	Well milled.
U.S. No. 3 [3] [4]	1.5	0.2	Light gray or slightly rosy	Reasonably well milled.
U.S. No. 4 [3] [4]	3.0	0.4	Gray or rosy	Reasonably well milled.
U.S. No. 5	5.0	1.5	Dark gray or very rosy	Reasonably well milled.

U.S. Sample grade:

     U.S. Sample grade shall be milled rice of this class which: (a) Does not meet the requirements for any of the grades from U.S. No. 1 to U.S. No. 5, inclusive; (b) contains more than 15.0 percent of moisture; (c) is musty or sour, or heating; (d) has any commercially objectionable foreign odor; (e) has a badly damaged or extremely red appearance; (f) contains more than 0.1 percent of foreign material; (g) contains more than 15.0 percent of broken kernels that will pass through a 2½ sieve; (h) contains two or more live or dead weevils or other insects, insect webbing, or insect refuse; or (h) is otherwise of distinctly low quality.

    [1] For the special grade Parboiled milled rice, see § 868.315(c).
    [2] For the special grade Undermilled milled rice, see § 868.315(d).
    [3] Grades U.S. No. 1 to U.S. No. 4, inclusive, shall contain not more than 3.0 percent of heat-damaged kernels, kernels damaged by heat and/or parboiled kernels in nonparboiled rice.
    [4] Grades U.S. No. 1 to U.S. No. 4, inclusive, shall contain not more than 1.0 percent of material passing through a 30 sieve. This limit does not apply to the special grade Granulated brewers milled rice.

[67 FR 61252, Sept. 30, 2002]

§ 868.314   Grade designation and other certificate information.

(a) *Milled rice.* The grade designation for all classes of Milled rice shall be included on the certificate grade-line in the following order:

(1) The letters "U.S.;"

(2) The number of the grade or the words "Sample grade," as warranted;

(3) The words "or better," when applicable and requested by the applicant prior to inspection;

(Approved by the Office of Management and Budget under control number 0580–0013)

(4) The class; and

(5) Each applicable special grade (see § 868.316).

(b) *Mixed Milled rice information.* For the class Mixed Milled rice, the following information shall be included in the Results section of the certificate in the following order:

(1) The percentage of whole kernels of each type in the order of predominance;

(2) The percentage of broken kernels of each type in the order of predominance, when applicable; and

(3) The percentage of seeds and foreign material.

(c) *Broken kernels.* Broken kernels, other than long grain in Mixed Milled rice shall be certified as "medium or short grain."

[74 FR 55442, Oct. 28, 2009]

SPECIAL GRADES, SPECIAL GRADE REQUIREMENTS, AND SPECIAL GRADE DESIGNATIONS

§ 868.315   Special grades and special grade requirements.

A special grade, when applicable, is supplemental to the grade assigned under § 868.314. Such special grades for milled rice are established and determined as follows:

(a) *Coated milled rice.* Coated milled rice shall be rice which is coated, in whole or in part, with substances that are safe and suitable as defined in the regulation issued pursuant to the Federal Food, Drug, and Cosmetic Act at 21 CFR 130.3(d).

(b) *Granulated brewers milled rice.* Granulated brewers milled rice shall be milled rice which has been crushed or granulated so that 95.0 percent or more will pass through a 5 sieve, 70.0 percent or more will pass through a 4 sieve, and not more than 15.0 percent will pass through a 2½ sieve.

627

(c) *Parboiled milled rice.* Parboiled milled rice shall be milled rice in which the starch has been gelatinized by soaking, steaming, and drying. Grades U.S. No. 1 to U.S. No. 6, inclusive, shall contain not more than 10.0 percent of ungelatinized kernels. Grades U.S. No. 1 and U.S. No. 2 shall contain not more than 0.1 percent, grades U.S. No. 3 and U.S. No. 4 not more than 0.2 percent, and grades U.S. No. 5 and U.S. No. 6 not more than 0.5 percent of nonparboiled rice. If the rice is: (1) Not distinctly colored by the parboiling process, it shall be considered "Parboiled Light"; (2) distinctly but not materially colored by the parboiling process, it shall be considered "Parboiled"; (3) materially colored by the parboiling process, it shall be considered "Parboiled Dark." The color levels for "Parboiled Light," "Parboiled," and "Parboiled Dark" shall be in accordance with the interpretive line samples for parboiled rice.

NOTE: The maximum limits for "Chalky kernels," "Heat-damaged kernels," "Kernels damaged by heat," and the "Color requirements" in §§ 868.310, 868.311, 868.312, and 868.313 are not applicable to the special grade "Parboiled milled rice."

(d) *Undermilled milled rice.* Undermilled milled rice shall be milled rice which is not equal to the milling requirements for "hard milled," "well milled," and "reasonably well milled" rice (see § 868.306). Grades U.S. No. 1 and U.S. No. 2 shall contain not more than 2.0 percent, grades U.S. No. 3 and U.S. No. 4 not more than 5.0 percent, grade U.S. No. 5 not more than 10.0 percent, and grade U.S. No. 6 not more than 15.0 percent of well-milled kernels. Grade U.S. No. 5 shall contain not more than 10.0 percent of red rice and damaged kernels (singly or combined) and in no case more than 6.0 percent of damaged kernels.

(e) *Glutinous milled rice.* Glutinous milled rice shall be special varieties of rice (Oryza sativa L. glutinosa) which contain more than 50 percent chalky kernels. For long grain, medium grain, and short grain milled rice, grade U.S. No. 1 shall contain not more than 1.0 percent of nonchalky kernels, grade

U.S. No. 2 not more than 2.0 percent of nonchalky kernels, grade U.S. No. 3 not more than 4.0 percent of nonchalky kernels, grade U.S. No. 4 not more than 6.0 percent of nonchalky kernels, grade U.S. No. 5 not more than 10.0 percent of nonchalky kernels, and grade U.S. No. 6 not more than 15.0 percent of nonchalky kernels. For second head milled rice, grade U.S. No. 1 shall contain not more than 4.0 percent of nonchalky kernels, grade U.S. No. 2 not more than 6.0 percent of nonchalky kernels, grade U.S. No. 3 not more than 10.0 percent of nonchalky kernels, grade U.S. No. 4 not more than 15.0 percent of nonchalky kernels, and grade U.S. No. 5 not more than 20.0 percent of nonchalky kernels. For screenings milled rice, there are no grade limits for percent of nonchalky kernels. For brewers milled rice, the special grade "Glutinous milled rice" is not applicable.

NOTE: The maximum limits for "Chalky kernels," shown in §§ 868.310, 868.311, and 868.312 are not applicable to the special grade "Glutinous milled rice."

(f) *Aromatic milled rice.* Aromatic milled rice shall be special varieties of rice (Oryza sativa L. scented) that have a distinctive and characteristic aroma; e.g., basmati and jasmine rice.

(Secs. 203, 205, 60 Stat. 1087, 1090 as amended; 7 U.S.C. 1622, 1624)

[42 FR 40869, Aug. 12, 1977; 42 FR 64356, Dec. 23, 1977, as amended at 48 FR 24859, June 3, 1983; 54 FR 21403, 21407, May 18, 1989; 56 FR 55981, Oct. 31, 1991; 58 FR 68016, Dec. 23, 1993. Redesignated and amended at 60 FR 16364, 16365, Mar. 30, 1995; 67 FR 61252, Sept. 30, 2002]

§ 868.316 Special grade designation.

The grade designation for coated, granulated brewers, parboiled, undermilled, glutinous, or aromatic milled rice shall include, following the class, the word(s) "Coated," "Granulated," "Parboiled Light," "Parboiled," "Parboiled Dark," "Undermilled," "Glutinous," or "Aromatic," as warranted, and all other information prescribed in § 868.314.

[58 FR 68016, Dec. 23, 1993. Redesignated and amended at 60 FR 16364, 16365, Mar. 30, 1995]

SUBCHAPTER B—FAIR TRADE PRACTICES

PART 869—REGULATIONS FOR THE UNITED STATES WAREHOUSE ACT

AUTHORITY: 7 U.S.C. 241 *et seq.*

SOURCE: 67 FR 50783, Aug. 5, 2002, unless otherwise noted. Redesignated at 84 FR 45645, Aug. 30, 2019.

EDITORIAL NOTE: Nomenclature changes to part 869 appear at 84 FR 45646, Aug. 30, 2019.

Subpart A—General Provisions

§ 869.1 Applicability.

(a) The regulations of this part set forth the terms and conditions under which the Secretary of Agriculture through the Agricultural Marketing Service (AMS) will administer the United States Warehouse Act (USWA or the Act) and sets forth the standards and the terms and conditions a participant must meet for eligibility to act under the USWA. The extent the provisions of this part are more restrictive, or more lenient, with respect to the same activities governed by State law, the provisions of this part shall prevail.

(b) Additional terms and conditions may be set forth in applicable licensing agreements, provider agreements and other documents.

(c) Compliance with State laws relating to the warehousing, grading, weighing, storing, merchandising or other similar activities is not required with respect to activities engaged in by a warehouse operator in a warehouse subject to a license issued in accordance with this part.

§ 869.2 Administration.

(a) AMS will administer all provisions and activities regulated under the Act under the general direction and supervision of AMS's Director, Warehouse and Commodity Management Division, or a designee.

(b) AMS may waive or modify the licensing or authorization requirements or deadlines in cases where lateness or failure to meet such requirements does not adversely affect the licensing or authorizations operated under the Act.

(c) AMS will provide affected licensees or authorized providers with changes to their licensing or provider agreements before the effective date.

(d) Licensing and authorization agreement updates will be available at:

(1) AMS's USWA website, and

(2) The following address: Director, Warehouse and Commodity Management Division, Fair Trade Practices Program, AMS, USDA, Stop 3601, 1400 Independence Avenue SW, Washington, DC 20250-3601.

[67 FR 50783, Aug. 5, 2002. Redesignated and amended at 84 FR 45645, 45646 Aug. 30, 2019]

§ 869.3   Definitions.

Words used in this part will be applicable to the activities authorized by this part and will be used in all aspects of administering the Act.

Access means the ability, when authorized, to read, change, and transfer warehouse receipts or other applicable document information retained in a central filing system.

Agricultural product means an agriculturally-produced product stored or handled for the purposes of interstate or foreign commerce, including a processed product of such agricultural product, as determined by AMS.

Central filing system (CFS) means an electronic system operated and maintained by a provider, as a disinterested third party, authorized by AMS where information relating to warehouse receipts, USWA documents and other electronic documents is recorded and maintained in a confidential and secure fashion independent of any outside influence or bias in action or appearance.

Certificate means a USWA document that bears specific assurances under the Act or warrants a person to operate or perform in a certain manner and sets forth specific responsibilities, rights, and privileges granted to the person under the Act.

Control of the facility means ultimate responsibility for the operation and integrity of a facility by ownership, lease, or operating agreement.

Department means the Department of Agriculture.

Electronic document means any document that is generated, sent, received, or stored by electronic, optical, or similar means, including, but not limited to, electronic data interchange, advanced communication methods, electronic mail, telegram, telex, or telecopy.

Electronic warehouse receipt (EWR) means a warehouse receipt that is authorized by AMS to be issued or transmitted under the Act in the form of an electronic document.

Examiner means an individual designated by AMS for the purpose of examining warehouses or for any other activities authorized under the Act.

Financial assurance means the surety or other financial obligation authorized by AMS that is a condition of receiving a license or authorization under the Act.

Force majeure means severe weather conditions, fire, explosion, flood, earthquake, insurrection, riot, strike, labor dispute, act of civil or military, nonavailability of transportation facilities, or any other cause beyond the control of the warehouse operator or provider that renders performance impossible.

Holder means a person that has possession in fact or by operation of law of a warehouse receipt, USWA electronic document, or any electronic document.

License means a license issued under the Act by AMS.

Licensing agreement means the document and any amendment or addenda to such agreement executed by the warehouse operator and AMS specifying licensing terms and conditions specific to the warehouse operator and the agricultural product licensed to be stored.

Non-storage agricultural product means an agricultural product received temporarily into a warehouse for conditioning, transferring or assembling for shipment, or lots of an agricultural product moving through a warehouse for current merchandising or milling use, against which no warehouse receipts are issued and no storage charges assessed.

Official Standards of the United States means the standards of the quality or condition for an agricultural product, fixed and established under (7 U.S.C. 51) the United States Cotton Standards Act, (7 U.S.C. 71) the United States Grain Standards Act, (7 U.S.C. 1622) the Agricultural Marketing Act of 1946, or other applicable official United States Standards.

Other electronic documents (OED) means those electronic documents, other than an EWR or USWA electronic document, that may be issued or transferred, related to the shipment, payment or financing of agricultural products that AMS has authorized for inclusion in a provider's CFS.

Person means a person as set forth in 1 U.S.C. 1, a State; or a political subdivision of a State.

Provider means a person authorized by AMS, as a disinterested third party, which maintains one or more confidential and secure electronic systems independent of any outside influence or bias in action or appearance.

Provider agreement means the document and any amendment or addenda to such agreement executed by the provider and AMS that sets forth the provider's responsibilities concerning the provider's operation or maintenance of a CFS.

Receipt means a warehouse receipt issued in accordance with the Act, including an electronic warehouse receipt.

Schedule of charges means the tariff or uniform rate or amount charged by an authorized person for specific services offered or rendered under the Act.

Schedule of fees means the fees charged and assessed by AMS for licensing, provider agreements or services furnished under the Act to help defray the costs of administering the Act, and as such are shown in a schedule of fees attached to the licensing or provider agreement.

Service license means the document and any amendment to such document, issued under the Act by AMS to individuals certified competent by the licensed warehouse operator to perform inspection, sampling, grading

classifying, or weighing services according to established standards and procedures, set forth in §868.202, at the specific warehouse license.

Stored agricultural products means all agricultural products received into, stored within, or delivered out of the warehouse that are not classified as a non-storage agricultural product under this part.

User means a person that uses a provider's CFS.

USWA electronic document means a USWA electronic document initiated by AMS to be issued, transferred or transmitted that is not identified as an EWR or OED in the appropriate licensing or provider agreement or as determined by AMS.

Warehouse means a structure or other authorized storage facility, as determined by AMS, in which any agricultural product may be stored or handled for the purpose of interstate or foreign commerce.

Warehouse capacity means the maximum quantity of an agricultural product that the warehouse will accommodate when stored in a manner customary to the warehouse as determined by AMS.

Warehouse operator means a person lawfully engaged in the business of storing or handling agricultural products.

Warehousing activities and practices means any legal, operational, managerial or financial duty that a warehouse operator has regarding an agricultural product.

[67 FR 50783, Aug. 5, 2002. Redesignated and amended at 84 FR 45645, 45646 Aug. 30, 2019]

§869.4 Fees.

(a) AMS will assess persons covered by the Act fees to cover the costs of administering the Act.

(b) Warehouse operators, licensees, applicants, or providers must pay:

(1) An annual fee as provided in the applicable licensing or provider agreement; and

(2) Fees that AMS assesses for specific services, examinations and audits, or as provided in the applicable licensing or provider agreement.

(c) The schedule of fees showing the current fees or any annual fee changes will be provided as an addendum to the applicable licensing or provider agreement or/and:

(1) Will be available at AMS's website, or

(2) May be requested at the following address: Director, Warehouse and Commodity Management Division, Fair Trade Practices Program, AMS, USDA, Stop 3601, 1400 Independence Avenue SW, Washington, DC 20250-3601.

(d) At the sole discretion of AMS, these fees may be waived.

[67 FR 50783, Aug. 5, 2002. Redesignated and amended at 84 FR 45645, 45646 Aug. 30, 2019]

§ 869.5 Penalties.

If a person fails to comply with any requirement of the Act, the regulations set forth in this part or any applicable licensing or provider agreement, AMS may assess, after an opportunity for a hearing as provided in § 869.8, a civil penalty:

(a) Of not more than the amount specified in § 3.91(b)(10)(i) of this title per violation, if an agricultural product is not involved in the violation; or

(b) Of not more than 100 percent of the value of the agricultural product, if an agricultural product is involved in the violation.

[67 FR 50783, Aug. 5, 2002, as amended at 75 FR 17560, Apr. 7, 2010. Redesignated and amended at 84 FR 45645, 45646 Aug. 30, 2019]

§ 869.6 Suspension, revocation and liquidation.

(a) AMS may, after an opportunity for a hearing as provided in § 869.8, suspend, revoke or liquidate any license or agreement issued under the Act, for any violation of or failure to comply with any provision of the Act, regulations or any applicable licensing or provider agreement.

(b) The reasons for a suspension, revocation or liquidation under this part include, but are not limited to:

(1) Failure to perform licensed or authorized services as provided in this part or in the applicable licensing or provider agreement;

(2) Failure to maintain minimum financial requirements as provided in the applicable licensing or provider agreement;

(3) Failure to submit a proper annual financial statement within the established time period as provided in the

applicable licensing or provider agreement.

(4) Failure to maintain control of the warehouse or provider system.

(5) The warehouse operator or provider requests closure, cancellation or liquidation. and

(6) Commission of fraud against AMS, any depositor, EWR or OED holder or user, or any other function or operation under this part.

(c) AMS retains USWA's full authority over a warehouse operator or provider for one year after such license revocation or provider agreement termination or until satisfaction of any claims filed against such warehouse operator or provider are resolved, whichever is later.

(d) Upon AMS's determination that continued operation of a warehouse by a warehouse operator or an electronic provider system by a provider is likely to result in probable loss of assets to storage depositors, or loss of data integrity to EWR or OED holders and users. AMS may immediately suspend, close, or take control and begin an orderly liquidation of such warehouse inventory or provider system data as provided in this part or in the applicable licensing or provider agreement.

(e) Any disputes involving probable loss of assets to storage depositors, or loss of data integrity to EWR or OED holders and users will be determined by AMS for the benefit of the depositors, or EWR or OED holders and users and such determinations shall be final.

[67 FR 50783, Aug. 5, 2002. Redesignated and amended at 84 FR 45645, 45646 Aug. 30, 2019]

§ 869.7 Return of suspended or revoked certificates of licensing or certificates of authorization.

(a) When a license issued to a warehouse operator or service license ends or is suspended or revoked by AMS, such certificates of licensing and applicable licensing agreement and certificates of authorization must be immediately surrendered and returned to AMS.

(b) When an agreement with a provider ends or is suspended or revoked by AMS, such certificates of authorization and applicable provider agreement must be immediately surrendered to AMS

§869.8 Appeals.

(a) Any person who is subject to an adverse determination made under the Act may appeal the determination by filing a written request with AMS at the following address: Director, Warehouse and Commodity Management Division, Fair Trade Practices Program, AMS, USDA, Stop 3601, 1400 Independence Avenue SW, Washington, DC 20250–3601.

(b) Any person who believes that they have been adversely affected by a determination under this part must seek review by AMS within twenty-eight calendar days of such determination, unless provided with notice by AMS of a different deadline.

(c) The appeal process set forth in this part is applicable to all licensees and providers under any provision of the Act, regulations or any applicable licensing agreement as follows:

(1) AMS will notify the person in writing of the nature of the suspension, revocation or liquidation action;

(2) The person must notify AMS of any appeal of its action within twenty-eight calendar days;

(3) The appeal and request must state whether:

(i) A hearing is requested,

(ii) The person will appear in person at such hearing, or

(iii) Such hearing will be held by telephone;

(4) AMS will provide the person a written acknowledgment of their request to pursue an appeal;

(5) When a person requests an appeal and does not request a hearing AMS will allow that person:

(i) To submit in writing the reasons why they believe AMS's determination to be in error,

(ii) Twenty-eight calendar days from the receipt of the acknowledgment to file any statements and documents in support of their appeal, unless provided with notice by AMS of a different deadline, and

(iii) An additional fourteen calendar days to respond to any new issues raised by AMS in response to the person's initial submission, unless provided with notice by AMS of a different deadline;

(6) If the person requests to pursue an appeal and requests a hearing, AMS will:

(i) Notify the person of the date of the hearing,

(ii) Determine the location of the hearing, when the person asks to appear in person,

(iii) Notify the person of the location of the hearing,

(iv) Afford the person twenty-eight calendar days from the receipt of the notification of the scheduling of the hearing to submit any statements and documents in support of the appeal, unless provided with notice by AMS of a different deadline, and

(v) Allow the person an additional fourteen calendar days from the date of the hearing to submit any additional material, unless provided with notice by AMS of a different deadline;

(7) Determinations of AMS will be final and no further appeal within USDA will be available except as may be specified in the final determination of AMS; and

(8) A person may not initiate an action in any court of competent jurisdiction concerning a determination made under the Act prior to the exhaustion of the appeal process set forth in this section.

[67 FR 50783, Aug. 5, 2002. Redesignated and amended at 84 FR 45645, 45646 Aug. 30, 2019]

§869.9 Dispute resolution and arbitration of private parties.

(a) A person may initiate legal action in any court of competent jurisdiction concerning a claim for noncompliance or an unresolved dispute with respect to activities authorized under the Act.

(b) Any claim for noncompliance or an unresolved dispute between a warehouse operator or provider and another party with respect to activities authorized under the Act may be resolved by the parties through mutually agreed-upon arbitration procedures or as may be prescribed in the applicable licensing or provider agreement. No arbitration determination or award will affect AMS's authority under the Act.

(c) In no case will USDA provide assistance or representation to parties involved in an arbitration proceeding arising with respect to activities authorized under the Act.

§ 869.10 Posting of certificates of licensing, certificates of authorization or other USWA documents.

(a) The warehouse operator must post, in a conspicuous place in the principal place where warehouse receipts are issued, any applicable certificate furnished by AMS that the warehouse operator is an authorized licensee under the Act.

(b) Immediately upon receipt of their certificate of service licensing or any modification or extension thereof under the Act, the licensee and warehouse operator must jointly post the same, and thereafter, except as otherwise provided in the regulations in this part or as prescribed in the applicable licensing agreement, keep such certificate of licensing conspicuously posted in the office where all or most of the services are done, or in such place as may be designated by AMS.

(c) The provider must post, in a conspicuous place in the principal place of business, any applicable certificate of authorization furnished by AMS that the provider is authorized to offer and provide specific services under the Act.

§ 869.11 Lost or destroyed certificates of licensing, authorization or agreements.

AMS will replace lost or destroyed certificates of licensing, certificate of authorization or applicable agreement upon satisfactory proof of loss or destruction. AMS will mark such certificates or agreements as duplicates.

§ 869.12 Safe keeping of records.

Each warehouse operator or provider must take necessary precautions to safeguard all records, either paper or electronic format, from destruction.

§ 869.13 Information of violations.

Every person licensed or authorized under the Act must immediately furnish AMS any information they may have indicating that any provision of the Act or the regulations in this part has been violated.

§ 869.14 Bonding and other financial assurance requirements.

(a) As a condition of receiving a license or authorization under the Act, the person applying for the license or authorization must execute and file with AMS a bond or provide such other financial assurance as AMS determines appropriate to secure the person's compliance with the Act.

(b) Such bond or assurance must be for a period of not less than one year and in such amount as required by AMS.

(c) Failure to provide for, or renew, a bond or a financial assurance instrument will result in the immediate and automatic revocation of the warehouse operator's license or provider's agreement.

(d) If AMS determines that a previously accepted bond or other financial assurance is insufficient, AMS may immediately suspend or revoke the license or authorization covered by the bond or other financial assurance if the person that filed the bond or other financial assurance does not provide such additional bond or other financial assurance as AMS determines appropriate.

(e) To qualify as a suitable bond or other financial assurance, the entity issuing the bond or other financial assurance must be subject to service of process in lawsuits or legal actions on the bond or other financial assurance in the State in which the warehouse is located.

Subpart B—Warehouse Licensing

§ 869.100 Application.

(a) An applicant for a license must submit to AMS information and documents determined by AMS to be sufficient to conclude that the applicant can comply with the provisions of the Act. Such documents must include a current review or an audit-level financial statement prepared according to generally accepted accounting standards as defined by the American Institute of Certified Public Accountants. For any entity that is not an individual, a document that establishes proof of the existence of the entity, such as:

(1) For a partnership, an executed partnership agreement; and

(2) For a corporation:

(i) Articles of incorporation certified by the Secretary of State of the applicable State of incorporation;

(ii) Bylaws; and

(iii) Permits to do business; and

(3) For a limited partnership, an executed limited partnership agreement; and

(4) For a limited liability company:

(i) Articles of organization or similar documents; and

(ii) Operating agreement or similar agreement.

(b) The warehouse facilities of an operator licensed under the Act must, as determined by AMS, be:

(1) Physically and operationally suitable for proper storage of the applicable agricultural product or agricultural products specified in the license;

(2) Operated according to generally accepted warehousing activities and practices in the industry for the applicable agricultural product or agricultural products stored in the facility; and

(3) Subject to the warehouse operator's control of the facility including all contiguous storage space with respect to such facilities.

(c) As specified in individual licensing agreements, a warehouse operator must:

(1) Meet the basic financial requirements determined by AMS; and

(2) Meet the net worth requirements determined by AMS;

(d) In order to obtain a license, the warehouse operator must correct any exceptions made by the warehouse examiner at the time of the original warehouse examination.

(e) AMS may issue a license for the storage of two or more agricultural products in a single warehouse as provided in the applicable licensing agreements. The amount of the bond or financial assurance, net worth, and inspection and license fees will be determined by AMS in accordance with the licensing agreements applicable to the specific agricultural product, based upon the warehouses' total capacity for storing such product, that would require:

(1) The largest bond or financial assurance;

(2) The greatest amount of net worth; and

(3) The greatest amount of fees.

§ 869.101 Financial records and reporting requirements.

(a) Warehouse operators must maintain complete, accurate, and current financial records that must be available to AMS for review or audit at AMS's request as may be prescribed in the applicable licensing agreement.

(b) Warehouse operators must, annually, present a financial statement as may be prescribed in the applicable licensing agreement to AMS.

§ 869.102 Financial assurance requirements.

(a) Warehouse operators must file with AMS financial assurances approved by AMS consisting of:

(1) A warehouse operator's bond; or

(2) Obligations that are unconditionally guaranteed as to both interest and principal by the United States, in a sum equal at their par value to the amount of the bond otherwise required to be furnished, together with an irrevocable power of attorney authorizing AMS to collect, sell, assign and transfer such obligations in case of any default in the performance of any of the conditions required in the licensing agreement; or

(3) An irrevocable letter of credit issued in the favor of AMS with a term of not less than two years; or

(4) A certificate of participation in, and coverage by, an indemnity or insurance fund as approved by AMS, established and maintained by a State, backed by the full faith and credit of the applicable State, which guarantees depositors of the licensed warehouse full indemnification for the breach of any obligation of the licensed warehouse operator under the terms of the Act. If a warehouse operator files a bond or financial assurance in the form of a certification of participation in an indemnity or insurance fund, the certification may only be used to satisfy any deficiencies in assets above the minimum net worth requirement as prescribed in the applicable licensing agreement. A certificate of participation and coverage in this fund must be furnished to AMS annually; or

(5) Other alternative instruments and forms of financial assurance approved by AMS as may be prescribed in the applicable licensing agreement.

(b) The warehouse operator may not withdraw obligations required under this section until one year after license termination or until satisfaction of any claims against the obligations, whichever is later.

§ 869.103 Amendments to license.

AMS will issue an amended license upon:

(a) Receipt of forms prescribed and furnished by AMS outlining the requested changes to the license;

(b) Payment of applicable licensing and examination fees;

(c) Receipt of bonding or other financial assurance if required in the applicable licensing agreement; and

(d) Receipt of a report on the examination of the proposed facilities pending inclusion or exclusion, if determined necessary by AMS.

§ 869.104 Insurance requirements.

Each warehouse operator must comply fully with the terms of insurance policies or contracts covering their licensed warehouse and all products stored therein, and must not commit any acts, nor permit others to do anything, that might impair or invalidate such insurance.

§ 869.105 Care of agricultural products.

Each warehouse operator must at all times, including during any period of suspension of their license, exercise such care in regard to stored and non-storage agricultural products in their custody as required in the applicable licensing agreement.

§ 869.106 Excess storage and transferring of agricultural products.

(a) If at any time a warehouse operator stores an agricultural product in a warehouse subject to a license issued under the Act in excess of the warehouse capacity for which it is licensed, such warehouse operator must immediately notify AMS of such excess storage and the reason for the storage.

(b) A warehouse operator who desires to transfer stored agricultural products to another warehouse may do so either by physical movement, by other methods as may be provided in the applicable licensing agreement, or as authorized by AMS.

§ 869.107 Warehouse charges and tariffs.

(a) A warehouse operator must not make any unreasonable or exorbitant charge for services rendered.

(b) A warehouse operator must follow the terms and conditions for each new or revised warehouse tariff or schedule of charges and rates as prescribed in the applicable licensing agreement.

§ 869.108 Inspections and examinations of warehouses.

(a) Warehouse operators must permit any agent of the Department to enter and inspect or examine, on any business day during the usual hours of business, any licensed warehouse, the offices of the warehouse operator, the books, records, papers, and accounts.

(b) Routine and special inspections and examinations will be unannounced.

(c) Warehouse operators must provide safe access to all storage facilities.

(d) Warehouse operators must inform any agent of the Department, upon arrival, of any hazard.

(e) Agents of the Department must accomplish inspections and examinations of warehouses in a manner that is efficient and cost-effective without jeopardizing any inspection and examination integrity.

§ 869.109 Disaster loss to be reported.

If at any time a disaster or loss occurs at or within any licensed warehouse, the warehouse operator must report immediately the occurrence of the disaster or loss and the extent of damage, to AMS.

§ 869.110 Conditions for delivery of agricultural products.

(a) In the absence of a lawful excuse, a warehouse operator will, without unnecessary delay, deliver the agricultural product stored or handled in the warehouse on a demand made by:

(1) The holder of the warehouse receipt for the agricultural product; or

(2) The person that deposited the agricultural product, if no warehouse receipt has been issued.

(b) Prior to delivery of the agricultural product, payment of the accrued charges associated with the storage or handling of the agricultural product, including satisfaction of the warehouse operator's lien, must be made if requested by the warehouse operator.

(c) When the holder of a warehouse receipt requests delivery of an agricultural product covered by the warehouse receipt, the holder must surrender the warehouse receipt to the warehouse operator before obtaining the agricultural product.

(d) A warehouse operator must cancel each warehouse receipt surrendered to the warehouse operator upon the delivery of the agricultural product for which the warehouse receipt was issued and in accordance with the applicable licensing agreement.

(e) For the purpose of this part, unless prevented from doing so by force majeure, a warehouse operator will deliver or ship such agricultural products stored or handled in their warehouse as prescribed in the applicable licensing agreement.

§ 869.111 Fair treatment.

(a) Contingent upon the capacity of a warehouse, a warehouse operator will deal in a fair and reasonable manner with persons storing, or seeking to store, an agricultural product in the warehouse if the agricultural product is:

(1) Of the kind, type, and quality customarily stored or handled in the area in which the warehouse is located;

(2) Tendered to the warehouse operator in a suitable condition for warehousing; and

(3) Tendered in a manner that is consistent with the ordinary and usual course of business.

(b) Nothing in this section will prohibit a warehouse operator from entering into an agreement with a depositor of an agricultural product to allocate available storage space.

§ 869.112 Terminal and futures contract markets.

(a) AMS may issue service licenses to weigh-masters or their deputies to perform services relating to warehouse receipts that are deliverable in satisfaction of futures contracts in such contract markets or as may be prescribed in any applicable licensing agreement.

(b) AMS may authorize a registrar of warehouse receipts issued for an agricultural product in a warehouse licensed under the Act that operates in any terminal market or in any futures contract market the official designated by officials of the State in which such market is located if such individual is not:

(1) An owner or employee of the licensed warehouse;

(2) The owner of, or an employee of the owner of, such agricultural product deposited in any such licensed warehouse; or

(3) As may be prescribed in any applicable licensing or provider agreement.

Subpart C—Inspectors, Samplers, Classifiers, and Weighers

§ 869.200 Service licenses.

(a) AMS may issue to a person a license for:

(1) Inspection of any agricultural product stored or handled in a warehouse subject to the Act;

(2) Sampling of such an agricultural product;

(3) Classification of such an agricultural product according to condition, grade, or other class and certify the condition, grade, or other class of the agricultural product;

(4) Weighing of such an agricultural product and certify the weight of the agricultural product; or

(5) Performing two or more services specified in paragraphs (a)(1), (a)(2), (a)(3) or (a)(4) of this section.

(b) Each person seeking a license to perform activities described in this section must submit an application on forms furnished by AMS that contain, at a minimum, the following information:

(1) The name, location and license number of the warehouses where the applicant would perform such activities;

(2) A statement from the warehouse operator that the applicant is competent and authorized to perform such activities at specific locations; and

(3) Evidence that the applicant is competent to inspect, sample, classify, according to grade or weigh the agricultural product.

(c) The warehouse operator will promptly notify AMS in writing of any changes with respect to persons authorized to perform such activities at the licensed warehouse.

§ 869.201 Agricultural product certificates; format.

Each inspection, grade, class, weight or combination certificate issued under the Act by a licensee to perform such services must be:

(a) In a format prescribed by AMS;

(b) Issued and maintained in a consecutive order; and

(c) As prescribed in the applicable licensing or provider agreement and authorized by AMS.

§ 869.202 Standards of grades for other agricultural products.

Official Standards of the United States for any kind, class or grade of an agricultural product to be inspected must be used if such standards exist. Until Official Standards of the United States are fixed and established for the kind of agricultural product to be inspected, the kind, class and grade of the agricultural product must be stated, subject to the approval of AMS. If such standards do not exist for such an agricultural product, the following will be used:

(a) State standards established in the State in which the warehouse is located,

(b) In the absence of any State standards, in accordance with the standards, if any, adopted by the local board of trade, chamber of commerce, or by the agricultural product trade generally in the locality in which the warehouse is located, or

(c) In the absence of the standards set forth in paragraphs (a) and (b) of this section, in accordance with any standards approved for the purpose by AMS.

Subpart D—Warehouse Receipts

§ 869.300 Warehouse receipt requirements.

(a) Warehouse receipts may be:

(1) Negotiable or non-negotiable;

(2) For a single unit, multiple units, identity preserved or commingled lot; and

(3) In a paper or electronic format that, besides complying with the requirements of the Act, must be in a format as prescribed in the applicable licensing or provider agreement and authorized by AMS.

(b) The warehouse operator must:

(1) At the request of a depositor of an agricultural product stored or handled in a warehouse licensed under the Act, issue a warehouse receipt to the depositor;

(2) Not issue a warehouse receipt for an agricultural product unless the agricultural product is actually stored in their warehouse at the time of issuance;

(3) Not issue a warehouse receipt until the quality, condition and weight of such an agricultural product is ascertained by a licensed inspector and weigher;

(4) Not directly or indirectly compel or attempt to compel the depositor to request the issuance of a warehouse receipt omitting the statement of quality or condition;

(5) Not issue an additional warehouse receipt under the Act for a specific identity-preserved or commingled agricultural product lot (or any portion thereof) if another warehouse receipt representing the same specific identity-preserved or commingled lot of the agricultural product is outstanding. No two warehouse receipts issued by a warehouse operator may have the same warehouse receipt number or represent the same agricultural product lot;

(6) When issuing a warehouse receipt and purposefully omitting any information, notate the blank to show such intent;

(7) Not deliver any portion of an agricultural product for which they have issued a negotiable warehouse receipt until the warehouse receipt has been surrendered to them and canceled as prescribed in the applicable licensing agreement;

(8) Not deliver more than 90% of the receipted quantity of an agricultural product for which they have issued a non-negotiable warehouse receipt until such warehouse receipt has been surrendered or the depositor or the depositor's agent has provided a written order for the agricultural product and

the warehouse receipt surrendered upon final delivery; and

(9) Deliver, upon proper presentation of a warehouse receipt for any agricultural product, and payment or tender of all advances and charges, to the depositor or lawful holder of such warehouse receipt the agricultural product of such identity, quantity, grade and condition as set forth in such warehouse receipt.

(c) In the case of a lost or destroyed warehouse receipt, a new warehouse receipt upon the same terms, subject to the same conditions, and bearing on its face the number and the date of the original warehouse receipt may be issued.

§869.301 Notification requirements.

Warehouse operators must file with AMS the name and genuine signature of each person authorized to sign warehouse receipts for the licensed warehouse operator, and will promptly notify AMS of any changes with respect to persons authorized to sign.

§869.302 Paper warehouse receipts.

Paper warehouse receipts must be issued as follows:

(a) On distinctive paper specified by AMS;

(b) Printed by a printer authorized by AMS; and

(c) Issued, identified and maintained in a consecutive order.

§869.303 Electronic warehouse receipts.

(a) Warehouse operators issuing EWR under the Act may issue EWR's for the agricultural product stored in their warehouse. Warehouse operators issuing EWR's under the Act must:

(1) Only issue EWR's through one AMS-authorized provider annually;

(2) Inform AMS of the identity of their provider, when they are a first time user of EWR's, 60 calendar days in advance of issuing an EWR through that provider. AMS may waive or modify this 60-day requirement as set forth in §869.2(b);

(3) Before issuing an EWR, request and receive from AMS a range of consecutive warehouse receipt numbers that the warehouse will use consecutively for issuing their EWR's;

(4) When using an authorized provider, issue and cancel all warehouse receipts as EWR's;

(5) Cancel an EWR only when they are the holder of the warehouse receipt;

(6) Be the holder of an EWR to correct information contained within any required data field;

(7) Receive written authorization from AMS at least 30 calendar days before changing providers. Upon authorization, they may request their current provider to transfer their EWR data from its Central Filing System (CFS) to the CFS of the authorized provider whom they select; and

(8) Notify all holders of EWR's by inclusion in the CFS at least 30 calendar days before changing providers, unless otherwise required or allowed by AMS.

(b) An EWR establishes the same rights and obligations with respect to an agricultural product as a paper warehouse receipt and possesses the following attributes:

(1) The holder of an EWR will be entitled to the same rights and privileges as the holder of a paper warehouse receipt.

(2) Only the current holder of the EWR may transfer the EWR to a new holder.

(3) The identity of the holder must be confidential and included as information for every EWR.

(4) Only one person may be designated as the holder of an EWR at any one time.

(5) A warehouse operator may not issue an EWR on a specific identity-preserved or commingled lot of agricultural product or any portion thereof while another valid warehouse receipt representing the same specific identity-preserved or commingled lot of agricultural product remains not canceled. No two warehouse receipts issued by a warehouse operator may have the same warehouse receipt number or represent the same agricultural product lot.

(6) An EWR may only be issued to replace a paper warehouse receipt if requested by the current holder of the paper warehouse receipt.

(7) Holders and warehouse operators may authorize any other user of their provider or the provider itself to act on

their behalf with respect to their activities with this provider. This authorization must be in writing, and acknowledged and retained by the warehouse operator and provider.

(c) A warehouse operator not licensed under the Act may, at the option of the warehouse operator, issue EWRs in accordance with this subpart, except this option does not apply to a warehouse operator that is licensed under State law to store agricultural products in a warehouse if the warehouse operator elects to issue an EWR under State law.

[67 FR 50783, Aug. 5, 2002. Redesignated and amended at 84 FR 45645, 45646 Aug. 30, 2019]

Subpart E—Electronic Providers

§ 869.400 Administration.

This subpart sets forth the regulations under which AMS may authorize one or more electronic systems under which:

(a) Electronic documents relating to the shipment, payment, and financing of the sale of agricultural products may be issued or transferred; or

(b) Electronic receipts may be issued and transferred.

§ 869.401 Electronic warehouse receipt and USWA electronic document providers.

(a) To establish a USWA-authorized system to issue and transfer EWR's and USWA electronic documents, each applicant must submit to AMS information and documents determined by AMS to be sufficient to determine that the applicant can comply with the provisions of the Act. Each provider operating pursuant to this section must meet the following requirements:

(1) Have and maintain a net worth as specified in the applicable provider agreement;

(2) Maintain two insurance policies; one for "errors and omissions" and another for "fraud and dishonesty." Each policy's minimum coverage and maximum deductible amounts and applicability of other forms of financial assurances as set forth in § 869.14 will be prescribed in the applicable provider agreement. Each policy must contain a clause requiring written notification to

AMS 30 days prior to cancellation or as prescribed by AMS;

(3) Submit a current review or an audit level financial statement prepared according to generally accepted accounting standards as defined by the American Institute of Certified Public Accountants;

(4) For any entity that is not an individual, a document that establishes proof of the existence, such as:

(i) For a partnership, an executed partnership agreement; and

(ii) For a corporation:

(A) Articles of incorporation certified by the Secretary of State of the applicable State of incorporation;

(B) Bylaws; and

(C) Permits to do business; and

(iii) For a limited partnership, an executed limited partnership agreement; and

(iv) For a limited liability company:

(A) Articles of organization or similar documents; and

(B) Operating agreement or similar agreement.

(5) Meet any additional financial requirements as set forth in the applicable provider agreement;

(6) Pay user fees annually to AMS, as set and announced annually by AMS prior to April 1 of each calendar year; and

(7) Operate a CFS as a neutral third party in a confidential and secure fashion independent of any outside influence or bias in action or appearance.

(b) The provider agreement will contain, but not be limited to, these basic elements:

(1) Scope of authority;

(2) Minimum document and warehouse receipt requirements;

(3) Liability;

(4) Transfer of records protocol;

(5) Records;

(6) Conflict of interest requirements;

(7) USDA common electronic information requirements;

(8) Financial requirements

(9) Terms of insurance policies or assurances;

(10) Provider's integrity statement;

(11) Security audits; and

(12) Submission, authorization, approval, use and retention of documents.

(c) AMS may suspend or terminate a provider's agreement for cause at any time.

(1) Hearings and appeals will be conducted in accordance with procedures as set forth in §§ 869.6 and 869.8.

(2) Suspended or terminated providers may not execute any function pertaining to USDA, USWA documents, or USWA or State EWR's during the pendency of any appeal or subsequent to this appeal if the appeal is denied, except as authorized by AMS.

(3) The provider or AMS may terminate the provider agreement without cause solely by giving the other party written notice 60 calendar days prior to termination.

(d) Each provider agreement will be automatically renewed annually on April 30th as long as the provider complies with the terms contained in the provider agreement, the regulations in this subpart, and the Act.

[67 FR 50783, Aug. 5, 2002. Redesignated and amended at 84 FR 45645, 45646 Aug. 30, 2019]

§ 869.402 Providers of other electronic documents.

(a) To establish a USWA-authorized system to issue and transfer OED, each applicant must submit to AMS information and documents determined by AMS to be sufficient to determine that the applicant can comply with the provisions of the Act. Each provider operating pursuant to this section must meet the following requirements:

(1) Have and maintain a net worth as specified in the applicable provider agreement;

(2) Maintain two insurance policies; one for 'errors and omissions' and another for 'fraud and dishonesty'. Each policy's minimum coverage and maximum deductible amounts and applicability of other forms of financial assurances as set forth in § 869.14 will be prescribed in the applicable provider agreement. Each policy must contain a clause requiring written notification to AMS 30 days prior to cancellation or as prescribed by AMS;

(3) Submit a current review or an audit level financial statement prepared according to generally accepted accounting standards as defined by the American Institute of Certified Public Accountants;

(4) For any entity that is not an individual, a document that establishes proof of the existence, such as:

(i) For a partnership, an executed partnership agreement; and

(ii) For a corporation:

(A) Articles of incorporation certified by the Secretary of State of the applicable State of incorporation;

(B) Bylaws; and

(C) Permits to do business; and

(iii) For a limited partnership, an executed limited partnership agreement; and

(iv) For a limited liability company:

(A) Articles of organization or similar documents; and

(B) Operating agreement or similar agreement.

(5) Meet any additional financial requirements as set forth in the applicable provider agreement;

(6) Pay user fees annually to AMS, as set and announced annually by AMS prior to April 1 of each calendar year; and

(7) Operate a CFS as a neutral third party in a confidential and secure fashion independent of any outside influence or bias in action or appearance.

(b) The provider agreement will contain, but not be limited to, these basic elements:

(1) Scope of authority;

(2) Minimum document and warehouse receipt requirements;

(3) Liability;

(4) Transfer of records protocol;

(5) Records;

(6) Conflict of interest requirements;

(7) USDA common electronic information requirements;

(8) Financial requirements;

(9) Terms of insurance policies or assurances;

(10) Provider's integrity statement;

(11) Security audits; and

(12) Submission, authorization, approval, use and retention of documents.

(c) AMS may suspend or terminate a provider's agreement for cause at any time.

(1) Hearings and appeals will be conducted in accordance with procedures as set forth in §§ 869.6 and 869.8.

(2) Suspended or terminated providers may not execute any function pertaining to USDA, USWA documents, USWA or State EWR's or OED's during

the pendency of any appeal or subsequent to this appeal if the appeal is denied, except as authorized by AMS.

(d) Each provider agreement will be automatically renewed annually on April 30th as long as the provider complies with the terms contained in the provider agreement, the regulations in this subpart, and the Act.

(e) In addition to audits prescribed in this section the provider must submit a copy of any audit, examination or investigative report prepared by any Federal regulatory agency with respect to the provider including agencies such as, but not limited to, the Comptroller of the Currency, Department of the Treasury, the Federal Trade Commission, and the Commodity Futures Trading Commission.

[67 FR 50783, Aug. 5, 2002. Redesignated and amended at 84 FR 45645, 45647 Aug. 30, 2019]

§ 869.403 Audits.

(a) No later than 120 calendar days following the end of the provider's fiscal year, the provider authorized under §§ 869.401 and 869.402 must submit to AMS an annual audit level financial statement and an electronic data processing audit that meets the minimum requirements as provided in the applicable provider agreement. The electronic data processing audit will be used by AMS to evaluate current computer operations, security, disaster recovery capabilities of the system, and compatibility with other systems authorized by AMS.

(b) Each provider will grant the Department unlimited, free access at any time to all records under the provider's control relating to activities conducted under this part and as specified in the applicable provider agreement.

[67 FR 50783, Aug. 5, 2002. Redesignated and amended at 84 FR 45645, 45647 Aug. 30, 2019]

§ 869.404 Schedule of charges and rates.

(a) A provider authorized under § 869.401 or § 869.402 must furnish AMS with copies of its current schedule of charges and rates for all services as they become effective.

(b) Charges and rates assessed any user by the provider must be in effect for a minimum period of one year.

(c) Providers must furnish AMS and all users a 60-calendar day advance notice of their intent to change any charges and rates.

[67 FR 50783, Aug. 5, 2002. Redesignated and amended at 84 FR 45645, 45647 Aug. 30, 2019]

PART 870—ECONOMIC ASSISTANCE ADJUSTMENT FOR TEXTILE MILLS

Sec.
870.1   Applicability.
870.3   Eligible upland cotton.
870.5   Eligible domestic users.
870.7   Upland cotton Domestic User Agreement.
870.9   Payment.

AUTHORITY: 7 U.S.C. 9037(c).

SOURCE: 86 FR 54530, Oct. 1, 2021, unless otherwise noted.

§ 870.1 Applicability.

(a) These regulations specify the terms and conditions under which the Commodity Credit Corporation (CCC) will make payments to eligible domestic users who have entered into an Upland Cotton Domestic User Agreement with CCC to participate in the upland cotton domestic user program.

(b) CCC will specify the forms to be used in administering the Economic Adjustment Assistance for Textile Mills program.

§ 870.3 Eligible upland cotton.

(a) For purposes of this subpart, eligible upland cotton is baled upland cotton, regardless of origin, that is opened by an eligible domestic user, and is either:

(1) Baled lint, including baled lint classified by USDA's Agricultural Marketing Service as Below Grade;

(2) Loose samples removed from upland cotton bales for classification purposes that have been rebaled;

(3) Semi-processed motes that are of a quality suitable, without further processing, for spinning, papermaking, or production of non-woven fabric; or

(4) Re-ginned (processed) motes.

(b) Eligible upland cotton must not be:

(1) Cotton for which a payment, under the provisions of this subpart, has been made available;

(2) Raw (unprocessed) motes, pills, linters, or other derivatives of the lint cleaning process; or

(3) Textile mill wastes.

§870.5 Eligible domestic users.

(a) For purposes of this subpart, a person regularly engaged in the business of opening bales of eligible upland cotton for the purpose of spinning, papermaking, or processing of non-woven cotton fabric in the United States, who has entered into an agreement with CCC to participate in the upland cotton user program, will be considered an eligible domestic user.

(b) Applications for payment under this subpart must contain documentation required by the provisions of the Upland Cotton Domestic User Agreement and other instructions that CCC issues.

§870.7 Upland cotton Domestic User Agreement.

(a) Payments specified in this subpart will be made available to eligible domestic users who have entered into an Upland Cotton Domestic User Agreement with CCC and who have complied with the terms and conditions in this subpart, the Upland Cotton Domestic User Agreement, and instructions issued by CCC.

(b) Upland Cotton Domestic User Agreements may be obtained from the Warehouse and Commodity Management Division, P.O. Box 419205, Stop 9148, Kansas City, MO 64141–6205. In order to participate in the program authorized by this subpart, domestic users must execute the Upland Cotton Domestic User Agreement and forward the original and one copy to KCCO.

§870.9 Payment.

(a) The payment rate for purposes of calculating payments as specified in this subpart is 3 cents per pound.

(b) Payments specified in this subpart will be determined by multiplying the payment rate, of 3 cents per pound, by

(1) In the case of baled upland cotton, whether lint, loose samples or reginned motes, but not semi-processed motes, the net weight of the cotton used (gross weight minus the weight of bagging and ties);

(2) In the case of unbaled reginned motes consumed, without rebaling, for an end use in a continuous manufacturing process, the weight of the reginned motes after final cleaning; and

(3) In the case of semi-processed motes which are of a quality suitable, without further processing, for spinning, papermaking, or manufacture of non-woven cotton fabric, 25 percent of the weight (gross weight minus the weight of bagging and ties, if baled) of the semi-processed motes; provided further, that with respect to semi-processed motes that are used prior to August 18, 2010, payment may be allowed by CCC in its sole discretion at 100 percent of the weight as determined appropriate for a transition of the program to the 25 percent factor.

(c) In all cases, the payment will be determined based on the amount of eligible upland cotton that an eligible domestic user consumed during the immediately preceding calendar month. For the purposes of this subpart, eligible upland cotton will be considered consumed by the domestic user on the date the bale is opened for consumption, or if not baled, the date consumed, without further processing, in a continuous manufacturing process.

(d) Payments specified in this subpart will be made available upon application for payment and submission of supporting documentation, as required by the CCC-issued provisions of the Upland Cotton Domestic User Agreement.

(e) All payments received by the eligible domestic user of upland cotton must be used for purposes specified in 7 U.S.C. 9037(c)(3), which include but are not limited to, acquisition, construction, installation, modernization, development, conversion, or expansion of land, plant, buildings, equipment, facilities, or machinery. Such capital expenditures must be directly attributable and certified as such by the user for the purpose of manufacturing upland cotton into eligible cotton products in the United States.

PARTS 871–899 [RESERVED]

FINDING AIDS

A list of CFR titles, subtitles, chapters, subchapters and parts and an alphabetical list of agencies publishing in the CFR are included in the CFR Index and Finding Aids volume to the Code of Federal Regulations which is published separately and revised annually.

Table of CFR Titles and Chapters
Alphabetical List of Agencies Appearing in the CFR
List of CFR Sections Affected

Table of CFR Titles and Chapters
(Revised as of January 1, 2022)

Title 1—General Provisions

Title 2—Grants and Agreements

648

649

Title 5—Administrative Personnel—Continued

Title 6—Domestic Security

Title 7—Agriculture

652

Title 15—Commerce and Foreign Trade—Continued

Title 20—Employees' Benefits

Title 21—Food and Drugs

Title 22—Foreign Relations

Title 23—Highways

Title 24—Housing and Urban Development

Title 24—Housing and Urban Development—Continued

Title 25—Indians

Title 26—Internal Revenue

Title 27—Alcohol, Tobacco Products and Firearms

Title 28—Judicial Administration

Title 28—Judicial Administration—Continued

Title 29—Labor

Title 30—Mineral Resources

Title 31—Money and Finance: Treasury

Title 31—Money and Finance: Treasury—Continued

Title 32—National Defense

Title 33—Navigation and Navigable Waters

Title 34—Education

Title 35 [Reserved]

Title 36—Parks, Forests, and Public Property

Title 37—Patents, Trademarks, and Copyrights

661

Title 41—Public Contracts and Property Management—Continued

Title 42—Public Health

Title 43—Public Lands: Interior

Title 44—Emergency Management and Assistance

Title 45—Public Welfare

Title 46—Shipping

Title 49—Transportation

Title 50—Wildlife and Fisheries

665

Title 50—Wildlife and Fisheries—Continued

Alphabetical List of Agencies Appearing in the CFR

(Revised as of January 1, 2022)

Agency	CFR Title, Subtitle or Chapter
Administrative Conference of the United States	1, III
Advisory Council on Historic Preservation	36, VIII
Advocacy and Outreach, Office of	7, XXV
Afghanistan Reconstruction, Special Inspector General for	5, LXXXIII
African Development Foundation	22, XV
Federal Acquisition Regulation	48, 57
Agency for International Development	2, VII; 22, II
Federal Acquisition Regulation	48, 7
Agricultural Marketing Service	7, I, VIII, IX, X, XI; 9, II
Agricultural Research Service	7, V
Agriculture, Department of	2, IV; 5, LXXIII
Advocacy and Outreach, Office of	7, XXV
Agricultural Marketing Service	7, I, VIII, IX, X, XI; 9, II
Agricultural Research Service	7, V
Animal and Plant Health Inspection Service	7, III; 9, I
Chief Financial Officer, Office of	7, XXX
Commodity Credit Corporation	7, XIV
Economic Research Service	7, XXXVII
Energy Policy and New Uses, Office of	2, IX; 7, XXIX
Environmental Quality, Office of	7, XXXI
Farm Service Agency	7, VII, XVIII
Federal Acquisition Regulation	48, 4
Federal Crop Insurance Corporation	7, IV
Food and Nutrition Service	7, II
Food Safety and Inspection Service	9, III
Foreign Agricultural Service	7, XV
Forest Service	36, II
Information Resources Management, Office of	7, XXVII
Inspector General, Office of	7, XXVI
National Agricultural Library	7, XLI
National Agricultural Statistics Service	7, XXXVI
National Institute of Food and Agriculture	7, XXXIV
Natural Resources Conservation Service	7, VI
Operations, Office of	7, XXVIII
Procurement and Property Management, Office of	7, XXXII
Rural Business-Cooperative Service	7, XVIII, XLII
Rural Development Administration	7, XLII
Rural Housing Service	7, XVIII, XXXV
Rural Utilities Service	7, XVII, XVIII, XLII
Secretary of Agriculture, Office of	7, Subtitle A
Transportation, Office of	7, XXXIII
World Agricultural Outlook Board	7, XXXVIII
Air Force, Department of	32, VII
Federal Acquisition Regulation Supplement	48, 53
Air Transportation Stabilization Board	14, VI
Alcohol and Tobacco Tax and Trade Bureau	27, I
Alcohol, Tobacco, Firearms, and Explosives, Bureau of	27, II
AMTRAK	49, VII
American Battle Monuments Commission	36, IV
American Indians, Office of the Special Trustee	25, VII
Animal and Plant Health Inspection Service	7, III; 9, I
Appalachian Regional Commission	5, IX
Architectural and Transportation Barriers Compliance Board	36, XI

Agency	CFR Title, Subtitle or Chapter
Defense Logistics Agency	32, I, XII; 48, 54
Engineers, Corps of	33, II; 36, III
National Imagery and Mapping Agency	32, I
Navy, Department of	32, VI; 48, 52
Secretary of Defense, Office of	2, XI; 32, I
Defense Contract Audit Agency	32, I
Defense Intelligence Agency	32, I
Defense Logistics Agency	32, XII; 48, 54
Defense Nuclear Facilities Safety Board	10, XVII
Delaware River Basin Commission	18, III
Denali Commission	45, IX
Disability, National Council on	5, C; 34, XII
District of Columbia, Court Services and Offender Supervision Agency for the	5, LXX; 28, VIII
Drug Enforcement Administration	21, II
East-West Foreign Trade Board	15, XIII
Economic Affairs, Office of the Under-Secretary for	15, XV
Economic Analysis, Bureau of	15, VIII
Economic Development Administration	13, III
Economic Research Service	7, XXXVII
Education, Department of	2, XXXIV; 5, LIII
Bilingual Education and Minority Languages Affairs, Office of	34, V
Career, Technical, and Adult Education, Office of	34, IV
Civil Rights, Office for	34, I
Educational Research and Improvement, Office of	34, VII
Elementary and Secondary Education, Office of	34, II
Federal Acquisition Regulation	48, 34
Postsecondary Education, Office of	34, VI
Secretary of Education, Office of	34, Subtitle A
Special Education and Rehabilitative Services, Office of	34, III
Educational Research and Improvement, Office of	34, VII
Election Assistance Commission	2, LVIII; 11, II
Elementary and Secondary Education, Office of	34, II
Emergency Oil and Gas Guaranteed Loan Board	13, V
Emergency Steel Guarantee Loan Board	13, IV
Employee Benefits Security Administration	29, XXV
Employees' Compensation Appeals Board	20, IV
Employees Loyalty Board	5, V
Employment and Training Administration	20, V
Employment Policy, National Commission for	1, IV
Employment Standards Administration	20, VI
Endangered Species Committee	50, IV
Energy, Department of	2, IX; 5, XXIII; 10, II, III, X
Federal Acquisition Regulation	48, 9
Federal Energy Regulatory Commission	5, XXIV; 18, I
Property Management Regulations	41, 109
Energy, Office of	7, XXIX
Engineers, Corps of	33, II; 36, III
Engraving and Printing, Bureau of	31, VI
Environmental Protection Agency	2, XV; 5, LIV; 40, I, IV, VII
Federal Acquisition Regulation	48, 15
Property Management Regulations	41, 115
Environmental Quality, Office of	7, XXXI
Equal Employment Opportunity Commission	5, LXII; 29, XIV
Equal Opportunity, Office of Assistant Secretary for	24, I
Executive Office of the President	3, I
Environmental Quality, Council on	40, V
Management and Budget, Office of	2, Subtitle A; 5, III, LXXVII; 14, VI; 48, 99
National Drug Control Policy, Office of	2, XXXVI; 21, III
National Security Council	32, XXI; 47, II
Presidential Documents	3
Science and Technology Policy, Office of	32, XXIV; 47, II
Trade Representative, Office of the United States	15, XX

669

674

List of CFR Sections Affected

All changes in this volume of the Code of Federal Regulations (CFR) that were made by documents published in the FEDERAL REGISTER since January 1, 2017 are enumerated in the following list. Entries indicate the nature of the changes effected. Page numbers refer to FEDERAL REGISTER pages. The user should consult the entries for chapters, parts and subparts as well as sections for revisions.

For changes to this volume of the CFR prior to this listing, consult the annual edition of the monthly List of CFR Sections Affected (LSA). The LSA is available at *www.govinfo.gov*. For changes to this volume of the CFR prior to 2001, see the "List of CFR Sections Affected, 1949–1963, 1964–1972, 1973–1985, and 1986–2000" published in 11 separate volumes. The "List of CFR Sections Affected 1986–2000" is available at *www.govinfo.gov*.

○